DATE DUE

DEC 1 8 1999			
GAYLORD			PRINTED IN U.S.A.

Gene Manipulations in Fungi

Contributors

Herbert N. Arst, Jr.
D. J. Ballance
Debra A. Barnes
J. W. Bennett
Ramunas Bigelis
George Boguslawski
Marian Carlson
Mary Case
W. LaJean Chaffin
Ronald L. Cihlar
Jan Cybis
Karl Esser
Douglas E. Eveleigh
John R. S. Fincham
P. M. Green
Lawrence T. Grossman
Michael E. S. Hudspeth

Layne Huiet
John A. Kinsey
C. P. Kurtzman
Linda L. Lasure
K. D. Macdonald
Bland S. Montenecourt
C. H. O'Donnell
Diane Puetz
Jasper Rine
Claudio Scazzocchio
Jeremy Thorner
Paul Tudzynski
B. Gillian Turgeon
G. Turner
A. Upshall
Richard L. Weiss
Barbara E. Wright

O. C. Yoder

Gene Manipulations in Fungi

Edited by

J. W. Bennett
Department of Biology
Tulane University
New Orleans, Louisiana

Linda L. Lasure
Bioproduct Research
Miles Laboratories, Inc.
Elkhart, Indiana

1985

ACADEMIC PRESS, INC.
HARCOURT BRACE JOVANOVICH, PUBLISHERS
Orlando   San Diego   New York
Austin   London   Montreal   Sydney
Tokyo   Toronto

COPYRIGHT © 1985 BY ACADEMIC PRESS, INC.
ALL RIGHTS RESERVED.
NO PART OF THIS PUBLICATION MAY BE REPRODUCED OR
TRANSMITTED IN ANY FORM OR BY ANY MEANS, ELECTRONIC
OR MECHANICAL, INCLUDING PHOTOCOPY, RECORDING, OR
ANY INFORMATION STORAGE AND RETRIEVAL SYSTEM, WITHOUT
PERMISSION IN WRITING FROM THE PUBLISHER.

ACADEMIC PRESS, INC.
Orlando, Florida 32887

United Kingdom Edition published by
ACADEMIC PRESS INC. (LONDON) LTD.
24–28 Oval Road, London NW1 7DX

LIBRARY OF CONGRESS CATALOG CARD NUMBER: 85-5979

ISBN 0–12–088640–5 (alk. paper)
ISBN 0–12–088641–3 (paperback: alk. paper)

PRINTED IN THE UNITED STATES OF AMERICA

85 86 87 88   9 8 7 6 5 4 3 2 1

Contents

Contributors xiii

Preface xvii

Acknowledgments xix

I Historical Perspective: Mutants to Models

1 From Auxotropic Mutants to DNA Sequences
John R. S. Fincham

I.	The Establishment of the Paradigm: One Gene–One Polypeptide	3
II.	Cluster Genes and Gene Clusters	10
III.	Systems of Regulation of Gene Activity	13
IV.	Cloning and Sequencing	22
	References	28

2 Molecular Taxonomy of the Fungi
C. P. Kurtzman

I.	Introduction	35
II.	Nucleic Acid Isolation and Purification	37

III.	DNA Base Composition	38
IV.	DNA Relatedness	41
V.	Mitochondrial DNA Relatedness	49
VI.	Ribosomal RNA Relatedness	51
VII.	Comparison of Relatedness from Nucleic Acid Studies with That Determined by Other Methodologies	53
	References	56

3 Fungal Mitochondrial Genomes

Lawrence T. Grossman and Michael E. S. Hudspeth

I.	Introduction	66
II.	Physical Organization	66
III.	Genes and the Genetic Code	70
IV.	Gene Order	74
V.	Transcription	76
VI.	Pseudogenes	81
VII.	Mitochondrial Plasmids	83
VIII.	Effects of Alteration of mtDNA	84
IX.	Nuclear–Cytoplasmic Interactions	88
	References	91

4 Modeling the Environment for Gene Expression

Barbara E. Wright

I.	Introduction	105
II.	Modeling Metabolism under Steady-State Conditions	106
III.	Modeling Metabolism during Product Accumulation	115
	References	121

II Yeasts

5 *Saccharomyces cerevisiae* as a Paradigm for Modern Molecular Genetics of Fungi

Jasper Rine and Marian Carlson

I.	Introduction	126
II.	Neoclassical Genetics	127
III.	Transformation	133
IV.	Cloning a Gene	141
V.	Manipulation of a Cloned Gene	141

	VI. Other Uses of Cloned Genes	152
	VII. An Agenda for Progress	153
	References	155

6 Yeast Transformation
George Boguslawski

I.	Introduction	161
II.	Specific Transformation Systems in Yeast	167
III.	Applications	178
IV.	Conclusions	187
	References	188

7 Use of the *LYS2* Gene for Gene Disruption, Gene Replacement, and Promoter Analysis in *Saccharomyces cerevisiae*
Debra A. Barnes and Jeremy Thorner

I.	Introduction	197
II.	Characterization of the *LYS2* Gene and Its Product	203
III.	Genetic Manipulations Utilizing the *LYS2* Gene	210
IV.	Conclusions and Prospectus	222
	References	223

III Molds

8 Molecular Biology of the *qa* Gene Cluster of *Neurospora*
Layne Huiet and Mary Case

I.	Introduction	229
II.	Molecular Analysis of the *qa* Cluster	230
III.	Transformation of the *qa* Gene Cluster in *Neurospora*	236
IV.	Summary	242
	References	242

9 *Neurospora* Plasmids
John A. Kinsey

I.	Introduction	245
II.	Naturally Occurring Plasmids	245

	III.	Replicating Plasmids Constructed in the Laboratory 248
	IV.	Overview and Prospects for the Future 253
		References 256

10 Cloning and Transformation in *Aspergillus*

G. Turner and D. J. Ballance

	I.	Introduction 259
	II.	Transformation Methodology 260
	III.	Fate of Transforming DNA after Entry 265
	IV.	The Quest for Replicating Vectors 270
	V.	Development of Vectors 273
	VI.	Future Prospects 275
		References 275

11 Expression of *Aspergillus* Genes in *Neurospora*

Richard L. Weiss, Diane Puetz, and Jan Cybis

	I.	Transformation of *Neurospora crassa* 280
	II.	Isolation of the *Aspergillus nidulans* Gene for Ornithine Carbamoyltransferase 281
	III.	Transformation of the *Neurospora crassa arg-12* Mutant .. 282
	IV.	Control of Gene Expression 288
	V.	Subcellular Localization of the Active Enzyme 289
	VI.	Conclusions 290
		References 291

12 Gene Dosage Effects and Antibiotic Synthesis in Fungi

C. H. O'Donnell, A. Upshall, and K. D. Macdonald

	I.	Introduction 293
	II.	Methods for the Amplification of Genetic Material in Microorganisms 295
	III.	Studies with Disomic Strains of *Aspergillus nidulans* 297
		References 305

13 Formal Genetics and Molecular Biology of the Control of Gene Expression in *Aspergillus nidulans*

Herbert N. Arst, Jr., and Claudio Scazzocchio

	I.	Formal Genetic Methodology of *Aspergillus nidulans* as Applied to the Study of Control Systems 310

	II.	The Metabolic Versatility of *A. nidulans* and Its Exploitation. 311
	III.	Regulatory Genes 313
	IV.	Putative Receptor Sites 327
	V.	The Spatial Organization of Functionally Related Genes ... 330
	VI.	At What Level Does Regulation of Gene Expression Occur? . 335
		References 337

14 A Cloning Strategy in Filamentous Fungi

P. M. Green and Claudio Scazzocchio

Text 345
References 352

IV Applications

15 Primary Metabolism and Industrial Fermentations

Ramunas Bigelis

	I.	Primary Metabolites 358
	II.	Genetic Approaches to the Production of Primary Metabolites 359
	III.	Organic Acids 360
	IV.	Amino Acids 368
	V.	Polysaccharides 372
	VI.	Lipids 374
	VII.	Nucleotides and Nucleic Acid–Related Compounds 376
	VIII.	Vitamins 378
	IX.	Polyols 380
	X.	Ethanol 382
	XI.	The Promise of Biotechnology 383
		References 385

16 Mitochondrial DNA for Gene Cloning in Eukaryotes

Paul Tudzynski and Karl Esser

	I.	Introduction 403
	II.	Fungal Plasmids 404
	III.	Cloning Vectors of Mitochondrial Origin 409
	IV.	Biotechnological Implications 412
		References 413

17 Molecular Bases of Fungal Pathogenicity to Plants

O. C. Yoder and B. Gillian Turgeon

I.	Introduction	417
II.	Infection Structures	418
III.	Cutinase	421
IV.	Pisatin Demethylase	422
V.	Toxins	424
VI.	Cloning and Analysis of Pathologically Important Genes from Fungi	427
	References	441

18 Morphogenesis and Dimorphism of *Mucor*

Ronald L. Cihlar

I.	Introduction	449
II.	Dimorphism of *Mucor*	450
III.	Molecular Analysis of *Mucor*	457
IV.	Perspectives—Molecular Genetics	462
	References	464

19 Toward Gene Manipulations with Selected Human Fungal Pathogens

W. LaJean Chaffin

I.	Introduction	469
II.	*Cryptococcus neoformans*	471
III.	*Histoplasma capsulatum*	473
IV.	*Wangiella dermatitidis*	474
V.	*Candida albicans*	476
VI.	Concluding Remarks	486
	References	487

20 Fungal Carbohydrases: Amylases and Cellulases

Bland S. Montenecourt and Douglas E. Eveleigh

I.	Introduction	491
II.	Fungal Amylases	494
III.	Fungal Cellulases	501
IV.	Summary and Outlook	505
	References	508

V Postscript

21 Prospects for a Molecular Mycology
J. W. Bennett

- I. Introduction ... 515
- II. A Primer in Mycology ... 516
- III. Molecular Mycology ... 519
- References ... 526

Appendixes

I Fungal Taxonomy
Linda L. Lasure and J. W. Bennett

- I. Introduction ... 531
- II. Outline of Fungal Taxonomy ... 534
- References ... 534

II Conventions for Gene Symbols
J. W. Bennett and Linda L. Lasure

- I. *Aspergillus nidulans* ... 538
- II. *Neurospora crassa* ... 539
- III. *Saccharomyces cerevisiae* ... 540
- IV. Other Fungi ... 542
- References ... 542

Index ... 545

Contributors

Numbers in parentheses indicate the pages on which the authors' contributions begin.

Herbert N. Arst, Jr.[1] (309), Department of Genetics, University of Newcastle upon Tyne, Claremont Place, Newcastle upon Tyne NE1 7RU, United Kingdom

D. J. Ballance (259), Department of Microbiology, University of Bristol, Bristol BS8 1TD, United Kingdom

Debra A. Barnes (197), Department of Microbiology and Immunology, University of California, Berkeley, California 94720

J. W. Bennett (515, 531, 537), Department of Biology, Tulane University, New Orleans, Louisiana 70118

Ramunas Bigelis (257), Biotechnology Group, Miles Laboratories, Inc., Elkhart, Indiana 46515

George Boguslawski (161), Biosynthesis Research Laboratory, Biotechnology Group, Miles Laboratories, Inc., Elkhart, Indiana 46515

Marian Carlson (125), Department of Human Genetics and Development, College of Physicians and Surgeons, Columbia University, New York, New York 10032

Mary Case (229), Department of Genetics, University of Georgia, Athens, Georgia 30602

W. LaJean Chaffin (469), Department of Microbiology, Texas Tech University Health Sciences Center, Lubbock, Texas 79430

Ronald L. Cihlar (449), Department of Microbiology, Schools of Medicine and Dentistry, Georgetown University, Washington, D.C. 20007

Jan Cybis (279), Institute of Biochemistry and Biophysics, Polish Academy of Sciences, 02-532 Warsaw, Poland

Karl Esser (403), Lehrstuhl für Allegemeine Botanik, Ruhr-Universität, D-4630 Bochum 1, Federal Republic of Germany

---

[1]Present address: Department of Bacteriology, Royal Postgraduate Medical School, Hammersmith Hospital, Ducane Road, London W.12, United Kingdom.

Douglas E. Eveleigh (491), Department of Biochemistry and Microbiology, Cook College, Rutgers University, New Brunswick, New Jersey 08903

John R. S. Fincham[2] (3), Department of Genetics, University of Edinburgh, Edinburgh, United Kingdom

P. M. Green[3] (345), Department of Biology, University of Essex, Colchester CO4 3SQ, United Kingdom

Lawrence T. Grossman (65), Department of Cellular and Molecular Biology, Division of Biological Sciences, The University of Michigan, Ann Arbor, Michigan 48109

Michael E. S. Hudspeth[4] (65), Department of Cellular and Molecular Biology, Division of Biological Sciences, The University of Michigan, Ann Arbor, Michigan 48109

Layne Huiet[5] (229), Department of Genetics, University of Georgia, Athens, Georgia 30602

John A. Kinsey (245), Department of Microbiology, Kansas University School of Medicine, Kansas City, Kansas 66103

C. P. Kurtzman (35), Northern Regional Research Center, Agricultural Research Service, U.S. Department of Agriculture, Peoria, Illinois 61604

Linda L. Lasure (531, 537), Bioproduct Research, Miles Laboratories, Inc., Elkhart, Indiana 46515

K. D. Macdonald (293), Chemical Defence Establishment, Salisbury, Wiltshire SP4 0JQ, United Kingdom

Bland S. Montenecourt (491), Department of Biology and Biotechnology Research Center, Lehigh University, Bethelehem, Pennsylvania 18015

C. H. O'Donnell[6] (293), Department of Biological Sciences, University of Lancaster, Lancaster LA1 4YQ, United Kingdom

Diane Puetz (279), Molecular Biology Institute, University of California at Los Angeles, Los Angeles, California 90024

Jasper Rine (125), Department of Biochemistry, University of California, Berkeley, California 94720

Claudio Scazzocchio[7] (309, 345), Department of Biology, University of Essex, Colchester CO4 3SQ, United Kingdom

[2]Present address: Department of Genetics, University of Cambridge, Cambridge CB2 3EH, United Kingdom.

[3]Present address: Paediatric Research Unit, The Prince Philip Research Laboratories, Guys Tower, London Bridge, London SE1 9RT, United Kingdom.

[4]Present address: Department of Biological Sciences, Northern Illinois University, De Kalb, Illinois 60115-2861.

[5]Present address: CSIRO Division of Plant Industry, Canberra City, A.C.T. 2601, Australia.

[6]Present address: Biosoft, Elsevier Publications, 68 Hills Road, Cambridge CB2 1LA, United Kingdom.

[7]Present address: Bâtiment 409, Institut de Microbiologie, Université Paris XI, Centre d'Orsay, 91405 Orsay, France.

Jeremy Thorner (197), Department of Microbiology and Immunology, University of California, Berkeley, California 94720

Paul Tudzynski (403), Lehrstuhl für Allgemeine Botanik, Ruhr-Universität, D-4630 Bochum 1, Federal Republic of Germany

B. Gillian Turgeon (417), Department of Plant Pathology, Cornell University, Ithaca, New York 14853

G. Turner (259), Department of Microbiology, University of Bristol, Bristol BS8 1TD, United Kingdom

A. Upshall[8] (293), Department of Biological Sciences, University of Lancaster, Lancaster LA1 4YQ, United Kingdom

Richard L. Weiss (279), Department of Chemistry and Biochemistry, University of California at Los Angeles, Los Angeles, California 90024

Barbara E. Wright (105), Department of Microbiology, University of Montana, Missoula, Montana 59812

O. C. Yoder (417), Department of Plant Pathology, Cornell University, Ithaca, New York 14853

---

[8]Present address: Zymo Genetics, Inc., 2121 N. 35th Street, Seattle, Washington 98103.

Preface

The fungi constitute a morphologically and physiologically diverse group of eukaryotes, now usually classified in a separate kingdom. Among the attributes that distinguish the fungi are a hyphal or yeastlike form; a rigid, frequently chitinous cell wall; apical growth; absorptive, heterotrophic metabolism; reproduction by spores; and small genome size.

The economic importance of the fungi is a reflection of their diverse, and often unique, metabolic capabilities. In nature, they are major agents of disease and decay. The ability to secrete enzymes into the environment makes filamentous fungi attractive to exploiters of modern biotechnology. In industry, they are the source of many important natural products such as citric acid and penicillin. In the laboratory, they are model systems for basic research.

Scientists who are otherwise separated by discipline, organism, geography, and objectives nevertheless share common goals in applying modern genetic techniques to fungi. With the important exception of *Saccharomyces cerevisae*, the molecular biology of fungi is just beginning. Chapter after chapter in this volume emphasizes the need for improved transformation systems, appropriate vectors, and broadly applicable selectable markers.

"Gene Manipulations in Fungi" combines a review of classical fungal genetics, contemporary research, and responsible speculation about the future. Yeast is, without question, the primary model system for eukaryotes. It is rapidly replacing *Escherichia coli* as the organism of choice for advanced studies in molecular genetics. Elegant research is also being conducted with the molds *Aspergillus nidulans* and *Neurospora crassa*.

We have included a number of topics of economic importance. Although the ability to manipulate genomes from plant and animal pathogens, or from most industrial fungi, is currently very limited, intense research is under way. It is useful for applied scientists to understand the state of the art. All of these economically important fungi offer special challenges to the virtuosity of the molecular biologist. Novel adaptations of recombinant DNA techniques are required, and we hope this book will stimulate innovative approaches. Where appropriate, useful methodologies are described in the text; elsewhere, literature citations are provided.

Molecular biology generates optimism, even euphoria. Many of our authors project their enthusiasm. We share their optimism and are pleased to have brought together such a diverse group of mycophiles. We predict that "Gene Manipulations in Fungi" will be only the first of many books celebrating the marriage between molecular biology and mycology.

J. W. Bennett
Linda L. Lasure

Acknowledgments

Our thanks to the American Society for Microbiology and Miles Laboratories for having sponsored a conference on Gene Manipulations in the Exploitation and Study of Fungi, held in South Bend, Indiana, during May, 1983. It was this conference and the enthusiasm of the participants that were the inspiration for this book. Special acknowledgments go to Arny Demain, then Chairman of the ASM Conference Committee, who first urged us to organize the Conference on Gene Manipulations in Fungi, and to Jim Lovett, who first put us in touch with Academic Press.

A number of scientists have reviewed chapters for this book. We especially thank Ross Kiester, Alan Lambowitz, Bob Metzenberg, Roger Milkman, Pete Magee, Claudio Scazzocchio, Michael Schectman, and Paul Szaniszlo. In writing "Prospects for a Molecular Mycology" (Chapter 21), conversations with Bob Metzenberg, Ron Morris, Claudio Scazzocchio, Bill Timberlake, and Olen Yoder were particularly useful. Acknowledgments also go to A. J. Clutterbuck, David Perkins, and Fred Sherman for their prompt and careful reviews of Appendix II, "Conventions for Gene Symbols."

Patti Ailes, Terri Collins, Pat Crickenberger, Pam Milone, and Phyllis Vandegrift provided valuable secretarial assistance. The editors at Academic Press were always available with technical and moral support.

Finally, J. W. B. thanks her sons, John, Dan, and Mark, for their patience.

I
Historical Perspective: Mutants to Models

1

From Auxotrophic Mutants to DNA Sequences

JOHN R. S. FINCHAM[1]
Department of Genetics
University of Edinburgh
Edinburgh, United Kingdom

    I. The Establishment of the Paradigm: One Gene–One Polypeptide 3
       A. The Identification of Metabolic Lesions in Auxotrophs 3
       B. One Gene–One Enzyme 5
       C. Suppressor Mutations 7
       D. The Allelic Complementation Paradox and Its Resolution 8
       E. Beyond Auxotrophs 9
   II. Cluster Genes and Gene Clusters 10
       A. Multifunctional Gene Products 10
       B. True Gene Clusters 12
  III. Systems of Regulation of Gene Activity 13
  IV. Cloning and Sequencing 22
       A. Gaining Access to the Gene 22
       B. New Information from DNA Sequences 26
       C. Chromatin Structure and Control of Transcription 27
       References 28

I. THE ESTABLISHMENT OF THE PARADIGM: ONE GENE–ONE POLYPEPTIDE

A. The Identification of Metabolic Lesions in Auxotrophs

    The idea of congenital defects in metabolism is due to Garrod (1909), who was concerned with human disease. In the 1930s it resurfaced in the work at the John

---

[1] Present address: Department of Genetics, University of Cambridge, Cambridge CB2 3EH, United Kingdom.

Innes Horticultural Institution (initiated by J. B. S. Haldane) on the genetics of flower color, and in B. Ephrussi and G. W. Beadle's studies on *Drosophila* eye color mutants. These pioneering contributions prepared the ground for the generalization that each metabolic reaction was the responsibility of a single gene. However, it was not until Beadle and E. L. Tatum, in the early 1940s, hit on the idea of using the fungus *Neurospora crassa* as an experimental organism for biochemical genetics that the idea became seriously testable. Tatum and Beadle's (1942) paper, reporting the isolation and genetic testing of the first *Neurospora* auxotrophic mutant (requiring *p*-aminobenzoic acid for growth), marked the first step in the modern phase of biochemical genetics, which has continued to grow exponentially up to the present day.

That the new era started when and where it did was the consequence of a happy coincidence of men and circumstances. *Neurospora* spp. had been discovered by C. L. Shear and B. O. Dodge, a mycologist of unusual genetic understanding (Shear and Dodge, 1927). Dodge demonstrated the suitability of *N. crassa* for experimental genetics, working out the mating-type system and demonstrating Mendelian segregation in asci. He succeeded in interesting T. H. Morgan, then the chairman of the Biology Division at CalTech, and Morgan encouraged his junior associate, C. C. Lindegren, to take up work on the organism. Lindegren (1936) worked out the first linkage maps, established procedures for tetrad analysis and centromere mapping, and generally established the formal genetics of *N. crassa* on a sound footing. Beadle, then at Stanford, was looking for new ways to follow up the ideas provoked by his *Drosophila* eye color studies, and was alerted to the possibility of making auxotrophic mutants by E. L. Tatum, who was knowledgeable about the variation in nutritional requirements found within certain groups of bacteria. The attraction of *Neurospora*, with its excellent genetics and rapid life cycle, was enhanced by the demonstration that it had no requirements for organic nutrients other than biotin and sugar (or other carbon source). Showing that single additional requirements could result from single radiation-induced mutations was straightforward in principle though, by the total-isolation methods available at that time, hard work in practice. But as the number of tested isolates mounted into the tens of thousands, the yield of nutritionally exacting, or auxotrophic, mutants reached the hundreds, and the biochemical genetics of *Neurospora* was established with a wealth of experimental material for investigation (Beadle, 1945; Horowitz *et al.*, 1945).

The concept of one gene—one metabolic step soon received clear support. Auxotrophs nearly all responded to single growth supplements, and each therefore appeared to be blocked in a single biosynthetic pathway. Within a pathway, it was often possible to obtain evidence as to the position of the mutational "block," by showing either the accumulation of the last intermediate preceding the block or a growth response to all intermediates beyond the block, or in both ways. The testing of metabolic intermediates as nutritional supplements for arginine auxotrophs by Adrian Srb and Norman Horowitz (1944) was the first example of the use of auxotrophs to elucidate or confirm a metabolic pathway. In the years that followed, comparable studies on other pathways, both in *Neurospora* and, even more, in *Escherichia coli,* were a great stimulus and aid to metabolic biochemistry. Indeed, for more than a decade after Beadle and Tatum's original breakthrough, its benefit to biochemistry was, to most, more obvious than its central importance to genetics. Advance on the genetic front remained for awhile somewhat in abeyance.

B. One Gene–One Enzyme

Beadle and Tatum themselves showed some caution in attributing functions to genes, and it was their younger and more boldly speculative colleague David Bonner who probably did most to propagate the one gene–one enzyme concept in the early days. In the 1946 Cold Spring Harbor Symposium, Bonner had a somewhat critical reception, especially from Max Delbruck (1946), who pointed out that the study of auxotrophs, which by definition respond to externally supplied nutrients, gave no information about any genes whose mutational losses of function could not be so repaired. This criticism stimulated the earliest work on temperature-conditional mutants. Norman Horowitz and Urs Leupold (1951), who used *E. coli* to supplement the *Neurospora* data, were able to show that a high proportion (about 50% in *Neurospora* and 75% in *E. coli*) of mutants that would grow on minimal medium at 25 but not 35°C, *were* able to grow at the higher temperature on "complete" medium. The screen for temperature-conditionals was not selecting for genes with especially simple functions but only for genes with at least one function capable of being made temperature-sensitive by mutation. It was thus possible to argue that auxotrophic mutations could identify at least a substantial proportion of all genes essential for growth, and certainly not only a small atypical class. This went some way toward meeting Delbruck's objection to the methodology even if it left the door open to the possibility of there being some genes of multiple primary function. Since that time, of course, studies of conditional mutants of various kinds in bacteria, yeasts, and filamentous fungi have vastly extended the range of gene functions that are accessible to study.

In 1951 the way in which genes determined enzymes was left rather vague— necessarily so, since the determination of protein properties by precisely specified amino acid sequence was not yet established. It was clearly envisaged, however, that genes were not merely switches that turned the synthesis of specific proteins on or off, but were somehow determinants of protein structure, especially, perhaps, three-dimensional structure. To quote Bonner's (1946) paper: "These various considerations suggest the view that the gene controls biochemical reactions by imposing, directly or indirectly, a specific configuration on the enzyme. . . ." On this view, it should be possible to show not only that certain mutants had lost enzyme activity, but also that some mutants had *altered* enzymes. Up to 1948, however, there were no examples of even the simple loss of activity of specific enzymes in auxotrophic mutants. In that year Herschel Mitchell and Joseph Lein (1948) reported that a *trp-3* (then called *td*) mutant was lacking tryptophan synthetase (then called tryptophan desmolase). This was latter confirmed with a more extensive series of *trp-3* mutants by Charles Yanofsky (1952). Meanwhile, I (Fincham, 1951) had found a deficiency of NADP-specific glutamate dehydrogenase in *N. crassa am* mutants, which were partially blocked in ammonia assimilation into α-amino groups. The first evidence of qualitative changes of enzyme properties was obtained by Bonner and Yanofsky with Sig Suskind (Suskind *et al.*, 1955), who showed that many *trp-3* mutants produced altered forms of tryptophan synthetase that had lost enzymic activity while retaining the antigenic properties of the normal enzyme. Later (Fincham, 1957) I was able to report that a partial revertant obtained in an *am* mutant strain produced a glutamate dehydrogenase with altered kinetic properties, apparently as a result of a second mutation in the *am* locus. Subsequent studies, of which my own (Fincham, 1962) was only one of several, established that a series of allelic forms of a gene could (and, plausibly, generally did) specify a series of structurally different forms of the same enzyme.

By the 1960s the biochemists had prepared the ground for the next stage of our understanding of gene–enzyme relationships. The determination of all the properties of an enzyme by the amino acid sequence of its polypeptide chain(s) became firmly established. Qualitative changes in enzyme properties could be most simply explained as the result of amino acid replacements. The principles of fine-structure gene mapping, with sites of mutation within a gene ordered in linear array, had already been established in the fungus *Aspergillus nidulans* (Pritchard, 1955) and in bacteriophage T4 (Benzer, 1958). The idea of the linear sequence of amino acid residues in the protein being specified by a corresponding sequence of mutable elements in the gene (the principle of colinearity) was irresistible even before there was any direct evidence for it. The direct evidence was first provided by two prokaryotic systems: *E. coli* tryptophan synthetase α-polypeptide (Yanofsky *et al.*, 1964) and bacteriophage T4 head protein (Sarabhai *et al.*, 1964). The running had indeed been taken over by workers on bacteria and bacteriophages, and the molecular genetics of fungi was relegated to what was, at best, a supporting role.

To the extent that fungal genetics was able to retain some place in the limelight, this was largely due to a spectacular development, or resurgence, of studies on *Saccharomyces cerevisiae*, particularly in the laboratory of Fred Sherman. The work of Sherman, mainly with John Stewart, on the effects of mutation on the cytochrome *c* protein of yeast, shares with the numerous studies of human haemoglobin defects the distinction of providing most of the textbook examples of the effects of mutation on primary protein structure in eukaryotic organisms. Iso-1 cytochrome c of yeast and its protein-coding gene *CYC1* provided the first fungal verification of the colinearity principle (reviewed in Sherman and Stewart, 1971); the *Neurospora* glutamate dehydrogenase-*am* system followed some years later (Brett *et al.*, 1976).

It often happens in science that discoveries, if they are to bring clarification, need to be made in the right sequence. A discovery made out of turn may well sow confusion. The history of the theory of the gene–enzyme relationship provides at least two examples. We now know of the existence of genes which can indeed mutate to affect, even to eliminate, the activities of two or several enzymes simultaneously. These genes surfaced sufficiently late in the day to be comfortably accommodated within the one gene–one polypeptide paradigm, and it seems clear that they have regulatory rather than structure-determining functions (see Section III). Their premature discovery would certainly have led to some wrong conclusions. A second example is the phenomenon of suppressor mutation, which came to light some years before it could be properly interpreted.

C. Suppressor Mutations

At the same time as he was demonstrating that *trp-3* mutants of *Neurospora* were deficient in tryptophan synthetase, Yanofsky discovered that mutations at several other genetic loci could restore enzyme activity to some of the *trp-3* mutants. In at least one case (Yanofsky, 1952) the restored enzyme seemed to have the same properties as the tryptophan synthetase of the wild type, replacing, as was later shown (Suskind and Jordan, 1959), the defective serologically cross-reacting protein made by the original mutant. Evidence of this sort, seeming to show that structural information for a particular enzyme protein, supposedly a single polypeptide chain, could be contributed by two or more different genes, was rather baffling at the time and the problem was, in effect, put on one side for a number of years. In the early 1960's, Seymour Benzer and Alan Garen, working on bacteriophage T4 and *E. coli* respectively, proposed and substantiated the concepts of "nonsense" and "missense" suppression mediated by alterations in the coding specificity of transfer RNA. The study of nonsense (i.e., chain termination) suppression in the prokaryote systems and later, under the name of supersuppression, in *Saccharomyces cerevisiae* (Hawthorne and Mortimer, 1963) and *Schizosaccharomyces pombe* (Hawthorne and Leupold, 1974) led to a complete and satisfying explanation in terms of changes in the anticodon sequences of specific tRNA species. The responsibility for the determination of the amino acid sequence of each polypeptide chain still rested with a single gene. An incidental result of this piece of clarification was the identification of genes for tRNA molecules with the consequential refining of the one gene–one polypeptide concept to "one gene–one RNA transcript" (a formula itself subject to refinement as we discover more about the possibilities of processing of single transcripts to form multiple functional products). The molecular nature of some of the anticodon changes was first defined in *Saccharomyces* by Piper *et al.* (1976), several years after the comparable work on *E. coli.*

D. The Allelic Complementation Paradox and Its Resolution

Another phenomenon that was initially difficult to reconcile with single genes for single polypeptide chains was allelic complementation—that is to say, the ability of certain pairs of mutants, mapping close together and individually deficient in the same enzyme, to cooperate in diploids or heterokaryons to produce significant amounts of enzyme activity. Dow Woodward (Woodward *et al.*, 1958), working on *ad-4* (adenylosuccinase) mutants of *N. crassa,* and John Pateman and myself (Fincham and Pateman, 1957), with *am* (glutamate dehydrogenase) mutants of the same species, discovered the effect independently. In both gene–enzyme systems the difficulty of interpretation arose from the sporadic nature of complementary interactions among the series of mutants. Had the series been clearly divisible into two mutually complementary groups (like the $r_{II}A$ and $r_{II}B$ bacteriophage T4 mutants of Seymour Benzer) the situation could have been interpreted in terms of two adjacent genes–two polypeptide chains–one enzyme. But the fact that most pairwise combinations of *ad-4* or *am* mutants were noncomplementary, together with the significant finding that the allelic complementation generally yielded demonstrably abnormal (often thermolabile) enzyme, demanded a different explanation. This was suggested by the physical biochemical studies, several of which were published at about that time, showing that many, probably most, enzymic proteins were dimers or higher oligomers of identical polypeptide chains. The active enzyme molecules due to complementation could be plausibly explained as hybrids, in which polypeptide chains from different mutants somehow corrected, albeit somewhat precariously, each other's defects—perhaps, as first suggested by Catcheside and Overton (1959), through effects on polypeptide conformation. The conformation–correction hypothesis came to seem especially likely in the case of the *am* mutant series in that in several of the *am* mutants the enzymic defect was shown to be a shift in balance between inactive and active conformational alternatives (Fincham, 1962).

That allelic complementation did indeed work through the formation of hybrid enzyme was first shown in the case of *E. coli* alkaline phosphate (Schlesinger and Levinthal, 1963), where the enzyme was a dimer that could be dissociated and reassociated *in vitro. In vitro* complementation has been demonstrated in several fungal systems as well, the report by Woodward (1959) taking priority. The *Neurospora am* system has proved complicated and interesting, glutamate dehydrogenase being a hexamer. Hybrids of different subunit compositions have been physically separated from *in vitro*-hybridized mixtures of purified mutant proteins and their allosteric properties investigated. The result has been a general confirmation of the conformation–correction hypothesis (Coddington *et al.*, 1966; Watson and Wootton, 1978), but the exact nature of the conformational constraints operating between the enzyme monomers, and the effects of the complementable mutations, will be understood only after the three-dimensional structure of *Neurospora* glutamate dehydrogenase has been determined by crystallography.

Although the "hybrid oligomer with conformational correction" model of allelic complementation certainly applies to some cases, it was perhaps accepted too easily as *the* general mechanism. The model predicted that allelic complementation would be confined to gene–enzyme systems in which the enzyme was oligomeric, with more than one monomer of the same kind, and that only those mutants able to form complete, or nearly complete, polypeptide chains would be able to complement. But we know of many genes in which mutants with relatively early chain termination can show complementation with certain point mutants. Complementation maps in such cases show a characteristically polarized pattern, with segments of different lengths (representing noncomplementation with different but overlapping sets of point mutants) extending from one end of the map. The *aro* (Giles *et al.*, 1967) and *pyr-3* (Radford, 1969) series in *Neurospora* are good examples. It is now clear that these polarized segments are formed by chain-termination mutants (identified as such by their susceptibility to "nonsense" suppression). This kind of situation is highly suggestive of the construction of hybrid enzyme monomers by piecing together overlapping polypeptide fragments. Examples of this mode of complementation have been demonstrated physicochemically in the cases of ribonuclease (Richards, 1958), *E. coli* β-galactosidase (Goldberg, 1969), and *E. coli* tryptophan synthetase α-subunit (Jackson and Yanofsky, 1969). None of the fungal examples is as well documented as yet.

E. Beyond Auxotrophs

There is, of course, a sense in which Max Delbruck's early criticism of generalization on the basis of auxotrophic mutants was correct. There is certainly a host of cellular functions which, if damaged by mutation, cannot be repaired by any nutritional supplement. Mutants other than auxotrophs have to be used if more than a limited range of genes is to be made available for study.

A considerable number of gene functions have been revealed through analysis of mutations that confer resistance to drugs of various kinds, mostly metabolite analogs or antibiotics. In *Neurospora* (Pall, 1970, 1971) and *Saccharomyces* (Grenson *et al.*, 1970) several amino acid uptake systems were distinguished and shown to be under separate gene control through the analysis of mutants resistant to different amino acid analogues. The genes identified presumably code for membrane components. Resistance to cycloheximide has been attributed to mutation in a gene coding for a ribosomal protein—L21 in *Podospora anserina* (Crouzet and Begueret, 1980) and L29 in *Saccharomyces cerevisiae* (Stocklein and Pipersberg, 1980). The *S. cerevisiae* gene for ribosomal protein L3 has been identified and isolated through the trichodermin resistance conferred by one of its mutations (Fried and Warner, 1981). Another route to the identification of genes for yeast ribosomal proteins is via the analysis of certain suppressor mutants (Ishiguro *et al.*, 1981). In *Aspergillus nidulans,* resistance to benomyl or thiabendazole follows from an alteration in tubulin structure and has led to the identification of the gene coding for β-tubulin (Morris *et al.*, 1979).

Useful as drug resistance has been, it can only serve to identify as many gene functions as there are specific kinds of drug. Temperature-conditional mutants, on the other hand, can lead the investigator to any gene with an essential function. The extensive collection of heat-sensitive mutants of *S. cerevisiae* made by L. H. Hartwell and colleagues is impressive testimony to the effectiveness of this approach. These mutants, unable to go through mitosis at the restrictive temperature, were subdivided, on the basis of relatively simple isotopic labeling procedures, into sets in which the synthesis of different classes of macromolecules—proteins, DNA, RNA—was blocked by the upward temperature shift (Hartwell, 1967). Some of the most interesting were those that, at the restrictive temperature, were arrested at specific stages of the cell budding cycle (Hartwell, 1976; for *cold*-sensitive cell-cycle mutants, see Moir and Botstein, 1982). Many of the cell-cycle mutants were affected at specific stages of DNA replication and some of these have now been associated with single specific enzymes; *cdc9* mutants, for example, are affected in DNA ligase (McCready and Cox, 1982). A similar approach can be applied to filamentous fungi such as *A. nidulans* (see, e.g., Oakley and Morris, 1983).

Through the exercise of ingenuity and insight screens can be devised for the isolation of mutants, other than auxotrophs, corresponding to a wide range of different genes. The most powerful method of all is undoubtedly the use of cloned DNA fragments for complementation of temperature-conditional lethals, but this depends on the development of "shuttle vectors" appropriate to the species under study (see Section IV,A).

II. CLUSTER GENES AND GENE CLUSTERS

A. Multifunctional Gene Products

After the discovery of the bacterial operon in the late 1950's by Jacob and Monod, fungal geneticists rather expected to find analogous multigene transcription units in their own organisms. Possible examples were not long in coming to light. The problem posed in each case was whether one was dealing with a cluster of genes coding for different enzymes operating in the same biosynthetic pathway or with a single gene coding for a single enzyme of complex function. Three series of *N. crassa* mutants will serve as examples: *pyr-3* (Suyama et al., 1959; Radford, 1969), *his-3* (Webber, 1960; Catcheside, 1965), and *aro-1/9/5/4/2* (Giles et al., 1967). Each of these three cases showed the following features:

1. A number of enzyme activities could be lost individually by mutation, the numbers being three for *pyr-3*, three for *his-3*, and five for the *aro* complex. Mutants with single enzyme deficiencies fell into discrete groups, one for each activity, that complemented each other in all combinations. The complementation groups corresponded to adjacent but nonoverlapping segments of the fine-structure genetic map. These findings, taken by themselves, were suggestive of gene clusters, with one gene for each enzyme activity.

2. Within some of the putative genes there was sporadic allelic complementation, an observation that could be accommodated within the theory of allelic complementation reviewed above. More significantly, a number of mutants were each deficient in two or more enzyme activities since they failed to complement two or more groups of single-deficiency mutants. Genetically, the multiple-deficiency mutants mapped within the same segments as single-deficiency mutants; in general their complementation relationships implied that they represented mutational lesions that knocked out not only the activity of the "gene" within which they fell but also the activities of all the genes to one side; the secondarily affected "genes" had therefore to be viewed as in some sense downstream of the lesion. It was later shown that the pleiotropic mutants with polarized complementation patterns were often suppressible by "nonsense" suppressors and must therefore have chain-termination mutations. All of this was certainly suggestive of an operon in the bacterial mode. The main problem was that of understanding how the polar effects (which, incidentally, were absolute rather than merely quantitative as in the bacterial examples) could work. The close coupling of transcription into mRNA and translation into polypeptide, on which the polar effects in bacterial operons depend, could hardly occur in a eukaryote, where the two processes are separated by the nuclear membrane.

The difficulty in explaining the "downstream" effects of chain-terminating mutations does not arise if one supposes that the "stream" is the stream of translation, not of transcription—in other words that the multiple enzyme activities are all due to one polypeptide chain of complex function. This hypothesis has been confirmed in the cases of *his-3* (Minson and Creaser, 1969) and *aro* (Lumsden and Coggins, 1977) and seems highly probable for *pyr-3* also (Mackoff, 1977).

"Multiheaded" or "multiple-domain" proteins are now very well known to biochemists, and the idea that each separately folded domain of a long polypeptide chain can enjoy a degree of functional autonomy is no longer surprising. Such proteins are, however, often deceptive in the earlier stages of their characterization through their tendency to undergo protease-catalyzed cleavage in the relatively exposed regions connecting domains (see, e.g., Mackoff, 1977). This propensity caused a good deal of confusion in the *Neurospora* examples under review.

A considerable number of genes, now sometimes called "cluster genes," coding for polypeptide chains of multiple enzymic function are now known, not only in fungi but in *Drosophila* and mammals also. Table I lists some of the fungal examples. Further studies combining genetic and protein structural analysis should be very fruitful in terms of understanding the functions and possible evolution of complex proteins.

B. True Gene Clusters

Some of the apparent examples of functionally related gene clusters in fungi turned out to be real (Table II). They can be distinguished from "cluster genes" on the basis of three criteria.

First, the products of different genes in the same cluster are different polypeptide chains. This, in itself, may be a deceptive criterion since, as noted in the preceding section, multifunctional enzymes are sometimes rather easily degraded to monofunctional fragments.

The second criterion is the failure to find any mutation mapping within one single-enzyme segment and exerting a pleiotropic effect on another. Pleiotropic mutations eliminating some or all of the enzyme activities encoded in the cluster may occur, but these can be shown to be deletions or, in some cases, mutations in trans-acting positive regulatory genes either linked to or distant from the cluster. Deletions, and trans as opposed to cis effects of mutations, can be identified by standard genetic tests—crossing to point mutants and complementation analysis, respectively.

Finally, the presence of separate units of translation within a gene cluster can be shown unequivocally by DNA sequence determination, provided that the DNA of the gene cluster can be cloned. This final stage of analysis has been reached in the case of the *Neurospora qa* (quinate degradation) cluster (Stroman *et al.*, 1978). It has been shown in this case that the genes are separate units of transcription as well as of translation (Patel *et al.*, 1981; see also St. John and Davis, 1981, on the *S. cerevisiae gal* cluster).

What is the significance of clustering of functionally related but separately transcribed genes? There are two possibilities, not mutually exclusive. First, it may reflect a common mechanism of regulation exercised, perhaps, through conformational changes in a chromatin domain within which the cluster is embedded. Some of the known clusters are, in fact, subject to common regulation by trans-acting controlling genes (Tables II and IV). Second, the clustering of a set of complementary genes, collectively performing some specialized function, may reflect their evolution. Linkage may be selected for since it stabilizes the association of mutually adapted alleles at the different loci.

III. SYSTEMS OF REGULATION OF GENE ACTIVITY

While the genetic basis of enzyme structure continues to excite interest in particular cases, it is now essentially a solved problem. It has been recognized now for over 20 years that the next problem is the regulation of gene activity.

The initial stimulus to work in this area was provided by the highly successful regulator–operator–promoter model, devised for the *E. coli lac* operon by Jacob and Monod. Some of the earliest attempts to explore the relevance of this model to fungi were those of Douglas and Hawthorne (1964, 1966) on galactose utilization in *Saccharomyces* and Pateman and Cove (1967) on nitrate utilization in *A. nidulans*. *Saccharomyces* and *Aspergillus* have continued to hold the lead in the field of fungal gene regulation. *Neurospora crassa,* in spite of some notable studies, especially in the area of phosphate nutrition (see, e.g., Metzenberg and Chia, 1979), has taken a secondary role. *Aspergillus nidulans* has a special advantage in its omnivorous tendencies; it can seemingly use almost any organic source of carbon or nitrogen, with induction (and/or derepression) of the appropriate enzymes in each case.

Mutants affected in regulation have in general been recognized either by their failure to adapt to certain special nutrients or by their resistance to inhibition by metabolite analogues; such phenotypes correspond respectively to failure to induce or derepress the relevant enzyme system and constitutive formation of an enzyme (or permease) that is normally repressed by the analogue. Classical genetic analysis of such mutants, combined with relatively simple measurements of enzyme activities (or sometimes of nutrient uptake), have identified three kinds of components of regulatory systems.

First, there are the genes subject to regulation; in general, these have been genes whose protein products (enzymes or permeases) are needed, and normally evoked, when the fungus is called upon to use some special source of carbon, nitrogen, phosphate, or sulphate.

Second (Table III), there are regulatory genes, capable of acting from a distance (i.e., in trans) on the regulated genes. The products of the regulators may be either negatively acting repressors or (more often in the fungal systems so far described) positively acting activators of transcription. Their mutation can lead to either enzyme-negative or enzyme-constitutive phenotypes; in the case of a positively acting regulator gene, null mutations will generally be recessive, and constitutive mutations dominant in heterokaryons or diploids.

Third, there are the sites or tracts of chromatin, adjacent to the regulated genes but outside their protein-coding sequences, that have critical cis-limited effects

TABLE I.

"Cluster Genes" in Fungi

Species	Gene	No. of enzyme activities	Length of polypeptide chain	Enzyme activities[a]
Saccharomyces cerevisiae	HIS4	3	799 residues	Phosphoribosyl-AMP cyclohydrolase, phosphoribosyl-ATP pyrophosphorylase, histidinol dehydrogenase (1, 2)
Neurospora crassa	his3+	3		
Neurospora crassa	trp-1+	3	762 residues	(a) Glutamine amidotransferase (for anthranilate synthesis), (b) phosphoribosylanthranilate isomerase, (c) indoleglycerolphosphate synthetase (3, 4, 5, 6)
Aspergillus nidulans	trpC+	3		
Schizosaccharomyces pombe	trp1+	3		
Saccharomyces cerevisiae	TRP3	2 (a,c only)	?	

(*continued*)

TABLE I (*Continued*)

Species	Gene	No. of enzyme activities	Length of polypeptide chain	Enzyme activities[a]
Neurospora crassa	*trp-3*$^+$	2	76 kilodaltons	Indoleglycerolphosphate → indole + serine → tryptophan (two component reactions of tryptophan synthetase) (7, 8)
Saccharomyces cerevisiae	*TRP5*	2	76 kilodaltons	
Saccharomyces cerevisiae	*ARGB/C*	2	?	Acetylglutamate kinase / acetylglutamyl-phosphate reductase (9)
Neurospora crassa	*aro-1,2,4,5,9*	5	150 kilodaltons	Five sequential steps from deoxyheptulosonic acid phosphate to *enol*pyruvyl shikimic acid phosphate (10, 11)
Schizosaccharomyces pombe	*aro3*	5	?	

(*continued*)

TABLE I (*Continued*)

Species	Gene	No. of enzyme activities	Length of polypeptide chain	Enzyme activities[a]
Saccharomyces cerevisiae	*DUR1,2*	2	?	Urea carboxylase / allophanate hydrolase (12)
Saccharomyces cerevisiae	*ACC1,2*	2	?	Acetyl-CoA carboxylase/biotin:apocarboxylase ligase (13)
Saccharomyces cerevisiae	*FAS1*	3	185 kilodaltons[b]	Acyl carrier protein/condensing enzyme/β-ketoacyl reductase
Saccharomyces cerevisiae	*FAS2*	3	180 kilodaltons[b]	β-Hydroxyacyl dehydratase / reductase/acyltransferase (14, 15)

(*continued*)

TABLE I (*Continued*)

Species	Gene	No. of enzyme activities	Length of polypeptide chain	Enzyme activities[a]
Saccharomyces cerevisiae	ADE3	3	110–120 kilodaltons	Tetrahydrofolate synthetase / cyclohydrolase methenyltetrahydrofolate dehydrogenase (16)
Saccharomyces cerevisiae	URA2	2	?	Carbamoylphosphate synthetase / aspartate transcarbamoyl transferase (17)
Neurospora crassa	pyr3[+]	2	380 kilodaltons	Carbamoylphosphate synthetase / aspartate transcarbamoyl transferase (18)
Schizosaccharomyces pombe	ade10[+]	2	?	AICAR-formyltransferase / IMP cyclohydrolase (19)

[a] References: (1) Donahue *et al.* (1982), (2) Minson and Creaser (1969), (3) Schechtman and Yanofsky (1983), (4) Käfer (1978), (5) Thuriaux *et al.* (1982), (6) Paluh and Zalkin (1983), (7) Matchett and De Moss (1975), (8) Dethuler and Kirschner (1979), (9) Minet *et al.* (1980), (10) Lumsden and Coggins (1977), (11) Strauss (1979), (12) Cooper *et al.* (1980), (13) Mishima *et al.* (1980), (14) Kuhn *et al.* (1972), (15) Schweitzer *et al.* (1973), (16) De Mata and Rabinowitz (1980), (17) Denis-Duphil *et al.* (1981), (18) Mackoff *et al.* (1978), (19) Richter and Heslot (1982).

[b] The FAS1 and FAS2 products form a 2.3×10^6 dalton fatty acid synthetase complex.

TABLE II.
Gene Clusters in Fungi

Species	Clustered genes	Overall reaction and enzyme encoded	Pleiotropic mutations[a]
Neurospora crassa	qa-1F–qa-1S–qa-3–qa-4–qa-2	Quinate → procatechuate	All three enzymes deficient in certain qa-1 mutants. (qa-1S, qa-1F are regulator genes) (1, 2, 3)
	qa-3	Dehydroquinase (catabolic)	
	qa-4	Quinate dehydrogenase	
	qa-2	Dihydroshikimate dehydrase	
Saccharomyces cerevisiae	GAL7–GAL10–GAL1	Galactose → UDPglucose	All three enzymes deficient in certain gal4 mutants (gal4 is an unlinked regulator gene) (4)
	GAL7	Galactose-1-phosphate-UDP transferase	
	GAL10	Galactokinase	
	GAL1	UDPgalactose epimerase	
Aspergillus nidulans	prnA–prnD–(prn[b])–prnB–prnC	L-Proline → glutamate	Both enzymes and (to a lesser extent) prnB product deficient in certain prnA mutants (prnA is regulator gene) (5)
	prnD	Proline oxidase	
	prnB	L-Proline uptake	
	prnC	Δ^1-Pyrroline-5-carboxylate dehydrogenase	
Aspergillus nidulans	crnA–niaD–niiA	$NO_3^- \to NH_4^+$	Both enzymes (but not uptake system) deficient in certain nirA mutants (nirA[b] is an unlinked regulator gene) (6)
	niaD	Nitrate reductase	
	niiA	Nitrite reductase	
	crnA	Nitrate uptake	

[a] References: (1) Chaleff (1974) and Valone et al. (1971), (2) Valone et al. (1971), (3) Huiet (1984), (4) St. John and Davis (1981), (5) Jones et al. (1981), (6) Brownlee and Arst (1983).
[b] See Table III.

TABLE III.
Examples of trans-Acting Regulatory Genes

Species	Gene	Genes regulated	Effects of mutation[a]
Saccharomyces cerevisiae	AAS1 AAS2 AAS3	Numerous genes for enzymes of histidine, tryptophan, arginine, lysine etc. biosynthesis	Enzymes nonderepressible on amino acid starvation; mutants hypersensitive to inhibition by amino acid analogues (1)
	TRA3 TRA5	Numerous genes for enzymes of histidine, tryptophan, arginine, lysine etc. biosynthesis	Enzymes permanently derepressed; mutants resistant to analogues, e.g., triazole alanine (1)
	GAL4	GAL7, GAL10, GAL1[b]	Recessive mutations lead to failure of induction by galactose of all three enzymes; one dominant mutation ("Gal81ʳ"), apparently within GAL4, causes constitutive expression of all three (2)
	GAL80	GAL7, GAL10, GAL1[b]	Recessive mutations give constitutive expression of all three enzymes
	ARGRI ARGRII ARGRIII	ARG2, 3, 5, 6, 1/10, coding for six enzymes of arginine synthesis	Recessive mutations in any of three genes render all six enzymes nonrepressible by arginine; also make arginase and ornithine δ-transaminase noninducible (3)
	CPA81	CPA1 (arginine-specific carbamoylphosphate synthetase)	Recessive mutations render the enzyme much less repressible by arginine (4)

(continued)

TABLE III (*Continued*)

Species	Gene	Genes regulated	Effects of mutation[a]
Aspergillus nidulans	*areA*	Many genes concerned in utilization of N sources other than NH$_3$; e.g., *amdS*, *prnB,C*, *niiA*, *uapA*, *gabA*	Recessive *areA* mutants unable to use N sources other than NH$_3$; dominant mutations render some or all of the controlled genes nonrepressible by NH$_3$ and/or glucose (5)
	nirA	*niaD*, *niiA* (nitrate and nitrite reductase)	Incompletely recessive *nirA* mutants do not form nitrite, nitrate reductases in response to nitrate; semidominant (*nirA*^{c}) mutants produce both enzymes in absence of nitrate (still subject to *areA*-mediated NH$_3$ repression); semidominant *nirA*$^-$ mutants are NH$_3$-derepressed (but still require nitrate induction) (6, 7)

(*continued*)

TABLE III (*Continued*)

Species	Gene	Genes regulated	Effects of mutation[a]
	amdR (*intA*)[c]	*amdS* (acetamidase), *gabA* (γ-aminobutyrate uptake), *gatA* (γ-amino transaminase), *lamA* (lactamase for 2-pyrrolidone and 2-piperidone)	Recessive *amdR* mutants unable to induce acetamidase, GABA permease or transaminase; *amdR*[c] semidominant mutants produce them constitutively and can override the effect of *areA* mutations with respect to these gene functions (8, 9)
	uaY	Several genes involved in uptake and degradation of purines e.g., *uapA*, *hxA* (purine hydroxylase I) *uaZ* (urate oxidase)	*uaY* mutants unable to utilize purines in haploids; incompletely recessive in diploids (10)

[a] References: (1) Wolfner *et al.* (1975), (2) Oshima (1982), (3) Bechet *et al.* (1970), (4) Piérard *et al.* (1979), (5) Arst and Cove (1973), (6) Pateman and Cove (1967) and Cove (1969), (7) Tollervey and Arst (1981), (8) Arst (1976), (9) Arst *et al.* (1978), (10) Scazzocchio *et al.* (1982).
[b] See Table II.
[c] Arst's symbol.

on the requirements of these genes for expression and, in particular, on their response to the trans-acting regulators (Table IV).

The effects of mutations on fungal systems of gene regulation (some of which are summarized in Table III) suggest two important generalizations. First, as already noted, *positively* acting regulation seems to predominate. Second, many genes are subject to a multiplicity of controls. The superimposition of the ammonium-mediated nitrogen-catabolite repression on the more specific inductive controls of utilization of particular nitrogen sources in *Aspergillus* is a good example; indeed, utilization of a compound such as acetamide, which can act as source of both nitrogen and carbon, is subject to carbon-catabolite repression as well.

In no case is the mechanism of regulation understood in molecular detail. However, it is probable that all workers in the field have the same general hypothesis in mind. According to the obvious model, a regulatory gene encodes a protein that binds to a cis-acting site adjacent to each controlled gene so as to activate (or in some cases to inhibit) the initiation of transcription by RNA polymerase. The properties of the regulatory protein are pictured as subject to allosteric regulation by one or more small molecules the concentrations of which signal the nutritional needs of the organism.

The validation of this picture, which obviously leans heavily on much more fully substantiated bacterial models, is still at a very early stage. In principle it should be possible to isolate the trans-acting regulatory proteins through their binding to DNA, especially if the relevant DNA can be cloned. The report of Grove and Marzluf (1981) of the identification of *Neurospora nit-2* product (apparently corresponding to *areA* in *Aspergillus*) may be the forerunner of many similar studies (see Arst and Scazzocchio, Chapter 13).

It is already apparent that some of the systems of gene regulation in fungi are of great complexity. In several cases—galactose utilization in *Saccharomyces* (for a recent overview, see Guarente *et al.*, 1982) and phosphate nutrition in both *Neurospora* (Littlewood *et al.*, 1975) and Saccharomyces (Toh-e *et al.*, 1981)—"cascades" of regulatory genes appear to be involved, the regulator gene being itself regulated by a regulated gene. Another complication is the finding of regulating genes mediating the induction of two or more pathways simultaneously—integrator genes as Arst (1976) has called them.

IV. CLONING AND SEQUENCING

A. Gaining Access to the Gene

Over the last decade fungal genetics, like genetics in general, has been turned almost literally upside down by the development of methods for cloning and

TABLE IV.

Some Examples of Mutations in cis-Acting Control Regions

Species	Gene controlled	cis-Acting mutation	Effect on gene activity[a]
Saccharomyces cerevisiae	CARGA (arginase)	CARGA$^+$ O$^-$	No longer repressed by arginine[b] (1)
	CARGB (ornithine δ-transaminase)	CARGB$^+$ O$^-$	No longer repressed by arginine[b] (2)
	CPAI (carbamoyl phosphate synthetase)	CPAI O$^-$	No longer repressed by arginine[b] (3)
	ARGB/C (acetylglutamate kinase/reductase)	ARG B/C-O^c	Overproduction of the enzyme(s)[c] (4)
	DUR1/2 (urea carboxylase/allophanate hydrolase)	durO^h	Constitutive production of enzymes[b] (5)
	ADR2 (glucose-repressible alcohol dehydrogenase)		Overproduction, repressible by glucose[b] (6,7)

(continued)

TABLE IV (*Continued*)

Species	Gene controlled	cis-Acting mutation	Effect on gene activity[a]
	CYC7 (iso-2 cytochrome c)	CYC7-H2	Constitutive overproduction[b] (8)
	HIS4 (see Table I)	his4-912	Loss of all 3 enzyme activities (9)
Neurospora crassa	qa-2 (catabolic dehydroquinase)	qa-2[ai]	No longer dependent on qa-1F[d] (10)
Aspergillus nidulans	prnB (proline uptake)	prn[d] (see Table I)	Resistance to repressor effect of areA[r] (see Table III) (11)
	uapA (uric acid uptake)	uap-100	Resistance to repressor effect of areA[r] (12)
	amdS (acetamidase)	amd+ amdI9	Resistance to repressor effect of areA[r] (13)
	amdS (acetamidase)	amdI66	Greatly increased positive response to mutation in trans-acting regulatory gene amdA (14)
	gabA (γ-aminobutyric acid uptake)	gabI	Resistance to repressor effect of areA[r] (15)

[a] References: (1) Wiame and Dubois (1975), (2) Deschamps and Wiame (1979), (3) Thuriaux *et al.* (1972), (4) Jacobs *et al.* (1980), (5) Lemoine *et al.* (1978), (6) Russell *et al.* (1983), (7) Ciriacy and Williamson (1981), (8) Errede *et al.* (1980), (9) Chaleff and Fink (1980), (10) Tyler *et al.* (1984), (11) Arst *et al.* (1980), (12) Arst and Scazzocchio (1975), (13) Hynes (1975), (14) Hynes (1982), (15) Bailey *et al.* (1979).
[b] Due to insertion of Ty1 element 161 bp upstream of coding sequence.
[c] Probably one enzyme catalyzing two sequential steps—see Table I.
[d] Several different mutations upstream of qa-2—some deletions, some base-pair substitutions.

sequencing of specific genes. The new methods are to some degree independent of the previous genetic methodology based on mutation and genetic mapping but, in *Saccharomyces* especially, it has been the combination of the old and new approaches that has been most rewarding.

There is no need here to review the now-standard procedures for making "libraries" of genomic fragments in plasmid of bacteriophage vectors. It may be useful, however, to consider briefly the different ways of identifying particular items in a library, and the extent to which each has proved useful in studies on fungi.

The first fungal genes to be cloned and sequenced were obtained by reverse-transcription of abundant species of mRNA. M. J. Holland and J. P. Holland (1979) and Holland *et al.* (1981) were in this way able to obtain the sequences of several genes coding for enzymes of the major glycolytic pathway. The limitation of this approach is that it can be applied only to genes that are abundantly transcribed.

A second possibility arises if one is fortunate enough to be able to deduce a part of the mRNA coding sequence from an analysis of the altered amino acid sequence generated by two mutually compensating frameshift mutations. Montgomery *et al.* (1978) were able to screen successfully for the *S. cerevisiae CYC1* (iso-1 cytochrome *c*) gene with a synthetic DNA, only 13 nucleotides in length, deduced from such an analysis. More recently, Kinnaird *et al.* (1982) applied the same principle to the cloning of the *N. crassa am* (glutamate dehydrogenase) gene, the synthetic probe in this case being a 17-mer. This method depends on a large amount of preliminary genetic analysis and protein chemistry and, again, it is applicable only to genes that are at least moderately well transcribed, for only these will have protein products obtainable in sufficient quantity for amino acid sequence analysis.

Undoubtedly the most powerful general method of gene capture is that based on complementation of a defective mutant by a cloned DNA fragment. In principle any gene can be obtained in this way, provided only that its function can be selected for in an appropriate mutant, and that it can be transcribed and the transcript translated in the mutant cell into which it is introduced. The problem of getting fungal as well as bacterial cells to take up DNA is now solved, following the discovery of calcium precipitation as a general recipe for rendering the DNA assimilable.

The pioneering work showing the feasibility of the complementation method was carried out in John Carbon's laboratory at Santa Barbara (Clarke and Carbon, 1978). Yeast DNA fragments were cloned in an *E. coli* plasmid vector and screened for ability to complement various *E. coli* auxotrophs. This method worked for several yeast genes but failed with a number of others. There are two possible reasons for failure, one being "lack of fit" between yeast promoter sequences and bacterial RNA polymerases and the second (an absolute barrier where it exists) the inability of bacterial cells to splice-out intron sequences. Neither of these difficulties arises if the cloned DNA sequences can be tested for complementing ability in mutant cells of the species from which the DNA was derived in the first place. In relation to fungi this requires vector molecules able to replicate in fungal cells, and the *Saccharomyces* "two-micron" plasmid, first reported by Sinclair *et al.* (1967), came to have the same importance for gene cloning in yeast as ColE1 had earlier in *E. coli*. In one respect the 2-μm plasmid was inferior to its *E. coli* counterpart—it could not be grown in yeast cells to such high copy number as ColE1 could achieve in *E. coli* and thus, in its original form, it was not very suitable for the preparation of large amounts of cloned DNA. Jean Beggs (1978) obtained the best of both worlds by constructing a hybrid vector that combined the replication sequences of both the 2-μm plasmid and ColE1 and consequently was able to multiply both in yeast and (to high copy number) in *E. coli*. The current generation of "shuttle" vectors incorporate also gene sequences that can be positively selected for in each species—e.g., ampicillin and tetracycline resistance in *E. coli* and ability to synthesize leucine (*LEU2*, one of Carbon's clones) in *leu2* mutant yeast (Beggs, 1978).

Perhaps the most notable gene capture so far, inasmuch as it involved a gene whose product was entirely unknown and whose function had to be selected for in a somewhat roundabout way, is the cloning of an *S. cerevisiae* DNA sequence carrying information for α mating type (Nasmyth and Tatchell, 1980). The clone initially isolated, which turned out to come from the normally nontranscribed *HML*α "cassette" locus, was then used as a probe in the cloning of the second cassette, *HMRa*, and the expressed mating type locus, *MATa/*α, with both of which it had considerable flanking sequence in common. It is interesting to note that these hitherto obscure genes are now providing some of the most promising material for study of the role of chromatin structure in the regulation of gene expression (Nasmyth, 1982) (see Section IV,C).

Shuttle vectors suitable for selection of *Aspergillus* genes in *Aspergillus* and *Neurospora* genes in *Neurospora* are still at the development stage and their advent is eagerly awaited (see Part III, this volume). Happily, the *Saccharomyces* 2-μm plasmid will replicate well in *Schizosaccharomyces pombe*, and so the whole functional genome of the latter species is also wide open to cloning (Beach *et al.*, 1982).

B. New Information from DNA Sequences

To some, the merely exploratory sequence determination of a cloned gene may seem a somewhat unimaginative exercise. Yet the sequencing of fungal genes, usually undertaken with few special expectations, has yielded much highly interesting information, some of which I will briefly note.

First, there is the expected confirmation of the use of the universal genetic code in the nuclear genes of fungi, in contrast to the totally unexpected departures from it in fungal mitochondrial genomes. (The genetic and molecular mapping of mitochondrial DNA in yeast opened up a whole new world of genetics, but is largely neglected in this review.) There are large differences between *Saccharomyces* on the one hand and *Schizosaccharomyces* and *Neurospora* on the other in codon usage (Bennetzen and Hall, 1982), lending weight to the view that *Saccharomyces* is a stranger in the Ascomycetes. The abundant use in one species of a codon that is rarely used in another (e.g., AGA is a common arginine codon in yeast but is hardly used at all in the *Neurospora* genes examined so far) may well be one source of difficulty in getting genes from filamentous fungi expressed in *Saccharomyces*.

Second, there is the not unexpected finding that fungal genes sometimes contain introns (though less abundantly and on a smaller scale than in vertebrate animals), and that these conform to the higher eukaryote consensus in the sequences just internal to their splicing junctions. Yeast nuclear gene introns differ from those of vertebrates in having a (so far) invariant internal sequence TACTAACA towards their 3' ends (Langford and Gallwitz, 1983). In *Neurospora*, the genes sequenced so far show what appears to be a variable but significant agreement with a consensus GCT^G_AAC between 10 and 20 bases upstream of the 3' splice site (Kinnaird and Fincham, 1983). Langford and Gallwitz showed that the yeast internal consensus was needed for splicing, and they suggest that it performs the same function as is ascribed to the higher eukaryote U_1 small nuclear RNA (Lerner *et al.*, 1980)—namely that of helping to stabilize, by complementary base-pairing, an association between the 5' consensus sequence and the 3' end of the intron prior to splicing. The *Neurospora* internal consensus, relatively unimpressive though it is, at least has the invariant CT-AC supposedly necessary for bonding in reverse orientation to the invariant GT-AG at the 5' end of the intron.

Third, there are preliminary indications of sequences which may determine transcription start points and termination points (for a brief review, see Kinnaird and Fincham, 1983). More information and in particular more experiments on the effects of deleting particular sequences are clearly needed.

C. Chromatin Structure and Control of Transcription

Many different studies on gene transcription in higher eukaryotes have made it clear that it is highly correlated with certain features of chromatin structure. These include a general sensitivity to nuclease attack and a special hypersensitivity of certain sites, particularly sites in the neighborhood of transcription start points. Transcriptionally active chromatin has also been reported to be associated with certain nonhistone proteins, of which some (the "high-mobility group" or HMG proteins 14 and 17) appear to be universally present while others, as yet uncharacterized, may be specific for certain genes or classes of genes. The cause–effect relationships among these various features of active chromatin have yet to be determined. Much current speculation is based on the idea that switches in DNA conformation, for instance between straight DNA duplex and cruciform structure (in regions with inverted repeats) or between B- and Z-DNA, may be instrumental in turning transcription on or off, and that the role of some of the active-chromatin proteins may be to bind to and stabilize particular conformations of specific DNA sequences.

This is arguably the most exciting new frontier of molecular genetics and, although most of the advances so far have been made with animal systems, the fungi offer what appear to be extremely good experimental systems for its further extension. The transcription of many fungal genes is, as we saw earlier, subject to stringent regulation by easily manipulated environmental and/or genetic factors, and most of them may fairly easily be cloned so as to provide DNA probes for the relevant chromatin structures. At least in *S. cerevisiae* (and one hopes in other fungi as well before long) chosen genes may be inserted into cloning plasmids and replicated in cells of defined genotype grown under defined conditions, and the characteristics of the chromatin in which they find themselves can then be investigated (Lohr, 1983). Again, given the availability of the DNA, there are obvious possibilities of identifying some of the products of trans-acting regulatory genes through their binding to specific cis-acting sequences.

Veterans such as myself may well look back with nostalgia to the time when it was possible to draw novel and important conclusions from simple growth tests. The world of fungal biochemical genetics today is much more complicated and far more demanding technically. But for those who have the new molecular technology at their fingertips the excitement is as great as it has ever been.

REFERENCES

Arst, H. N., Jr. (1976). Integrator gene in *Aspergillus nidulans*. *Nature (London)* **262**, 231–234.
Arst, H. N., Jr., and Cove, D. J. (1973). Nitrogen metabolite repression in *Aspergillus nidulans*. *Mol. Gen. Genet.* **126**, 111–141.
Arst, H. N., Jr., and Scazzocchio, C. (1975). Initiator constitutive mutants: an 'up promoter' effect in *Aspergillus nidulans*. *Nature (London)* **254**, 31–34.
Arst, H. N., Jr., Penfold, H. A., and Bailey, C. R. (1978). Lactam utilization in *Aspergillus nidulans:* evidence for a fourth gene under the control of the integrator gene *int A*. *Mol. Gen. Genet.* **166**, 321–327.
Arst, H. N., Jr., MacDonald, D. W., and Jones, S. A. (1980). Regulation of proline transport in *Aspergillus nidulans*. *J. Gen. Microbiol.* **116**, 285–294.
Bailey, C. R., Penfold, H. A., and Arst, H. N., Jr. (1979). *Cis*-dominant regulatory mutations affecting the expression of GABA permease of *Aspergillus nidulans*. *Mol. Gen. Genet.* **169**, 79–83.
Beach, D., Piper, M., and Nurse, P. (1982). Construction of a *Schizosaccharomyces pombe* gene

bank in a yeast bacterial shuttle vector and its use to isolate genes by complementation. *Mol. Gen. Genet.* **187**, 326–329.

Beadle, G. W. (1945). Genetics and metabolism in *Neurospora*. *Physiol. Rev.* **25**, 643–663.

Bechet, J., Grenson, M., and Wiame, J.-M. (1970). Mutations affecting the repressibility of arginine biosynthetic enzymes of *Saccharomyces*. *Eur. J. Biochem.* **12**, 31–39.

Beggs, J. (1978). Transformation of yeast by a replicating hybrid plasmid. *Nature (London)* **275**, 104–109.

Bennetzen, J., and Hall, B. D. (1982). Codon selection in yeast. *J. Biol. Chem.* **237**, 3026–3031.

Benzer, S. (1958). The elementary units of heredity. *In* "The Chemical Basis of Heredity" (W. D. McElroy and B. Glass, eds.), pp. 70–93. John Hopkins Press, Baltimore, Maryland.

Bonner, D. (1946). Biochemical mutations in *Neurospora*. *Cold Spring Harbor Symp. Quant. Biol.* **11**, 14–24.

Brett, M., Chambers, G. K., Holder, A. A., Fincham, J. R. S., and Wootton, J. C. (1976). Mutational amino acid replacements in *Neurospora crassa* NADP-specific glutamate dehydrogenase. *J. Mol. Biol.* **106**, 1–22.

Brownlee, A. G., and Arst, H. N., Jr. (1983). Nitrate uptake in *Aspergillus nidulans* and the involvement of the third gene of the nitrate assimilation gene cluster. *J. Bacteriol.* **155**, 1138–1146.

Catcheside, D. G. (1965). Multiple enzymic functions of a gene in *Neurospora crassa*. *Biophys. Biochem. Res. Commun.* **18**, 648–651.

Catcheside, D. G., and Overton, A. (1959). Complementation between alleles in heterokaryons. *Cold Spring Harbor Symp. Quant. Biol.* **23**, 137–140.

Chaleff, D. T., and Fink, G. R. (1980). Genetic events associated with an insertion mutation in yeast. *Cell (Cambridge, Mass.)* **21**, 227–237.

Chaleff, R. S. (1974). The inducible quinate-shikimate catabolic pathway in *Neurospora crassa*: genetic organization. *J. Gen. Microbiol.* **81**, 337–353.

Ciriacy, M., and Williamson, V. M. (1981). Analysis of mutations affecting Ty-mediated gene expression in *Saccharomyces cervisiae*. *Mol. Gen. Genet.* **182**, 159–163.

Clarke, L., and Carbon, J. (1978). Functional expression of cloned yeast DNA in *Escherichia coli*: specific complementation of argininosuccinate lyase (*argH*) mutations. *J. Mol. Biol.* **120**, 517–532.

Coddington, A., Sundaram, T. K., and Fincham, J. R. S. (1966). Multiple active varieties of *Neurospora* glutamate dehydrogenase formed by hybridization between two inactive mutant proteins *in vivo* and *in vitro*. *J. Mol. Biol.* **17**, 503–512.

Cooper, T. G., Lam, C., and Turoscy, V. (1980). Structural analysis of the *dur* loci in *S. cerevisiae*: two domains of a single multifunctional gene. *Genetics* **94**, 555–580.

Cove, D. J. (1969). Evidence for a near limiting concentration of a regulator. *Nature (London)* **224**, 272–273.

Crouzet, M., and Beguerct, J. (1980). A new mutant form of the ribosomal protein L21 in the fungus *Podospora anserina*: identification of the structural gene for this protein. *Mol. Gen. Genet.* **180**, 177–183.

Delbruck, M. (1946). Comment appended to Bonner (1946).

De Mata, Z. S., and Rabinowitz, J. C. (1980). Formyl-methenyl-methylenetetrahydrofolate synthetase (combined) from yeast. Biochemical characterisation of the protein from an *ade3* mutant lacking the formyltetrahydrofolate synthetase function. *J. Biol. Chem.* **255**, 2569–2577.

Denis-Duphil, M., Mathieu-Shire, Y., and Herve, G. (1981). Proteolytically induced changes in the molecular form of the carbamoylphosphate synthetase–uracil-transcarbamylase complex coded for by the *URA2* locus in *Saccharomyces cerevisiae*. *J. Bacteriol.* **148**, 659–669.

Deschamps, J., and Wiame, J.-M. (1979). Mating type effect on *cis*-mutations leading to constitutivity of ornithine transaminase in diploid cells of *Saccharomyces cerevisiae*. *Genetics* **92**, 749–758.

Dethuler, M., and Kirschner, K. (1979). Tryptophan synthase from *Saccharomyces cerevisiae* is a dimer of two polypeptide chains of Mr 76,000 each. *Eur. J. Biochem.* **102**, 159–165.

Donahue, T. F., Farabaugh, P. J., and Fink, G. R. (1982). The nucleotide sequence of the *HIS4* region of yeast. *Gene* **18**, 47–57.

Douglas, H. C., and Hawthorne, D. C. (1964). Enzymatic expression and genetic linkage of genes controlling galactose utilization in *Saccharomyces*. *Genetics* **49**, 837–844.

Douglas, H. C., and Hawthorne, D. C. (1966). Regulation of genes controlling synthesis of the galactose pathway enzymes in yeast. *Genetics* **54**, 911–916.

Errede, B., Cardillo, T. S., Sherman, F., Dubois, E., Deschamps, T., and Wiame, J.-M. (1980). Mating signals control expression of mutations resulting from insertion of a transposable repetitive element adjacent to diverse yeast genes. *Cell (Cambridge, Mass.)* **22**, 427.

Fincham, J. R. S. (1951). The occurrence of glutamic dehydrogenase in *Neurospora* and its apparent absence in certain mutant strains. *J. Gen. Microbiol.* **5**, 793–806.

Fincham, J. R. S. (1957). A modified glutamic dehydrogenase as a result of gene mutation in *Neurospora crassa*. *Biochem. J.* **15**, 721–728.

Fincham, J. R. S. (1962). Genetically determined multiple forms of glutamic dehydrogenase in *Neurospora crassa*. *J. Mol. Biol.* **4**, 257–274.

Fincham, J. R. S., and Pateman, J. A. (1957). Formation of an enzyme through complementary action of mutant 'alleles' in different nuclei of a heterocaryon. *Nature (London)* **179**, 741–742.

Fried, H. M., and Warner, J. R. (1981). Cloning of yeast gene for trichodermin resistance and ribosomal protein L3. *Proc. Natl. Acad. Sci. U.S.A.* **78**, 238–242.

Garrod, A. E. (1909). "Inborn Errors of Metabolism." Oxford Univ. Press, Oxford.

Giles, N. H., Case, M. E., Partridge, C. W. H., and Ahmed, S. I. (1967). A gene cluster in *Neurospora crassa* coding for an aggregate of five aromatic synthetic enzymes. *Proc. Natl. Acad. Sci. U.S.A.* **58**, 1453–1460.

Goldberg, M. (1969). Tertiary structure of *Escherichia coli* β-galactosidase. *J. Mol. Biol.* **46**, 441–446.

Grenson, M., Hou, C., and Crabeel, M. (1970). Multiplicity of the amino acid permeases in *Saccharomyces cerevisiae*. IV. Evidence for a general amino acid permease. *J. Bacteriol.* **103**, 770–777.

Grove, G., and Marzluf, G. A. (1981). Identification of the product of the major regulatory gene of the nitrogen control circuit of *Neurospora crassa* as a nuclear DNA binding protein. *J. Biol. Chem.* **256**, 463–470.

Guarente, L., Yokum, R. R., and Gifford, P. (1982). A *GAL10–CYC1* hybrid yeast promoter identifies the *GAL4* regulatory region as an upstream site. *Proc. Natl. Acad. Sci. U.S.A.* **79**, 7410–7414.

Hartwell, L. H. (1967). Macromolecule synthesis in temperature-sensitive mutants of yeast. *J. Bacteriol.* **93**, 1662–1670.

Hartwell, L. H. (1976). Sequential function of gene products relative to DNA synthesis in the cell division cycle in yeast. *J. Mol. Biol.* **104**, 803–817.

Hawthorne, D. C., and Leupold, V. (1974). Suppressor mutations in yeast. *Curr. Top. Microbiol. Immunol.* **64**, 1–47.

Hawthorne, D. C., and Mortimer, R. K. (1963). Super-suppressors in yeast. *Genetics* **48**, 617–620.

Holland, M. J., and Holland, J. P. (1979). The primary structure of a gene coding for glyceraldehyde-3-phosphate dehydrogenase from *Saccharomyces cerevisiae*. *J. Biol. Chem.* **254**, 5466–5474.

Holland, M. J., Holland, J. P., Thill, G. P., and Jackson, K. A. (1981). The primary structure of yeast enolase and glyceraldehyde-3-phosphate dehydrogenase genes. *J. Biol. Chem.* **256,** 1385–1395.

Horowitz, N. H., and Leupold, U. (1951). Some recent studies bearing on the one gene–one enzyme hypothesis. *Cold Spring Harbor Symp. Quant. Biol.* **16,** 65–72.

Horowitz, N. H., Bonner, D., Mitchell, H. K., Tatum, E. L., and Beadle, G. W. (1945). Genic control of biochemical reactions in *Neurospora. Am. Nat.* **79,** 304–317.

Huiet, L. (1984). Molecular analysis of the *Neurospora qa-1* regulatory region indicates that two interacting genes control gene expression. *Proc. Natl. Acad. Sci. U.S.A.* **81,** 1174–1178.

Hynes, M. J. (1975). A *cis*-dominant regulatory mutation affecting enzyme induction in the eukaryote *Aspergillus nidulans. Nature (London)* **253,** 210–212.

Hynes, M. J. (1982). A *cis*-dominant mutation in *Aspergillus nidulans* affecting the expression of the *amdS* gene in the presence of mutations in the unlinked gene *amdA. Genetics* **102,** 139–147.

Ishiguro, J., Ono, B.-I., Masurekar, M., McLaughlin, C. S., and Sherman, F. (1981). Altered ribosomal S11 from the *SUP46* suppressor of yeast. *J. Mol. Biol.* **147,** 391–398.

Jackson, D. A., and Yanofsky, C. (1969). Restoration of enzymic activity by complementation *in vitro* between mutant α subunits of tryptophan synthetase and between mutant subunits and fragments of the α-subunit. *J. Biol. Chem.* **244,** 4539–4546.

Jacobs, P., Jauniaux, J.-C., and Grenson, M. (1980). A *cis*-dominant regulatory mutation linked to the *argB–argC* gene cluster in *Saccharomyces cerevisiae. J. Mol. Biol.* **139,** 691–704.

Jones, S. A., Arst, H. N., Jr., and MacDonald, D. W. (1981). Gene roles in the *prn* cluster of *Aspergillus nidulans. Curr. Genet.* **3,** 49–56.

Käfer, E. (1978). The anthranilate synthetase complex and the trifunctional *trpC* gene of *Aspergillus. Can. J. Genet. Cytol.* **19,** 723–728.

Kinnaird, J. H., and Fincham, J. R. S. (1983). The complete nucleotide sequence of the *Neurospora crassa am* (NADP-specific glutamate dehydrogenase) gene. *Gene* **26,** 253–260.

Kinnaird, J. H., Keighren, M. A., Kinsey, J. A., Eaton, M., and Fincham, J. R. S. (1982). Cloning of the *am* (glutamate dehydrogenase) gene of *Neurospora crassa* through the use of a synthetic DNA probe. *Gene* **20,** 387–396.

Kuhn, J., Castorph, H., and Schweitzer, E. (1972). Gene linkage and gene–enzyme relations in the fatty acid synthetase system of *Saccharomyces cerevisiae. Eur. J. Biochem.* **24,** 492–497.

Langford, C. J., and Gallwitz, D. (1983). Evidence for an intron-contained sequence required for the splicing of yeast RNA polymerase II transcripts. *Cell (Cambridge, Mass.)* **33,** 519–527.

Lemoine, Y., Dubois, E. L., and Wiame, J.-M. (1978). The regulation of urea amidolyase of *Saccharomyces cerevisiae*. Mating type influence on a constitutivity mutation acting in *cis*. *Mol. Gen. Genet.* **166,** 251–258.

Lerner, M. R., Boyle, J. A., Mount, S. M., Wolin, S. L., and Steitz, J. A. (1980). Are snRNPs involved in splicing? *Nature (London)* **283,** 220–224.

Leupold, U., and Horowitz, N. H. (1952). Über temperatursuntanten bei *Escherichia coli* und ihre bedentung für die "eingen-ein enzym" Hypothese. *Z. Indukt. Abstamm. Vererbungsl.* **84,** 306–319.

Lindegren, C. C. (1936). A six-point map of the sex-chromosome of *Neurospora crassa. J. Genet.* **32,** 243–256.

Littlewood, B. S., Chia, W., and Metzenberg, R. L. (1975). Genetic control of phosphate metabolizing enzymes in *Neurospora crassa:* relationships among regulatory mutations. *Genetics* **79,** 419–434.

Lohr, D. (1983). The chromatin structure of an actively expressed single copy yeast gene. *Nucleic Acids Res.* **11,** 6755–6773.

Lumsden, J., and Coggins, J. R. (1977). The subunit structure of the *arom* multienzyme complex of *Neurospora crassa*. A possible pentafunctional polypeptide chain. *Biochem. J.* **161,** 599–607.

McCready, S. J., and Cox, B. S. (1982). The role of the *cdc9* ligase in replication and excision repair in *Saccharomyces cerevisiae. Curr. Genet.* **6,** 29–30.

Mackoff, A. J. (1977). Characterization of the aspartate carbamoyl transferase fragment generated by protease action on the *pyrimidine*-3 gene product of *Neurospora crassa. Biochim. Biophys. Acta* **485,** 314–329.

Mackoff, A. J., Buxton, F. P., and Radford, A. (1978). A possible model for the structure of the *Neurospora* carbamoyl phosphate synthase–aspartate carbamoyltransferase complex enzyme. *Mol. Gen. Genet.* **161,** 297–304.

Matchett, W. H., and De Moss, J. A. (1975). The subunit structure of tryptophan synthetase from *Neurospora crassa. J. Biol. Chem.* **250,** 2941–2946.

Metzenberg, R. L., and Chia, W. (1979). Genetic control of phosphorus assimulation in *Neurospora crassa:* dose-dependent dominance and recessiveness in constitutive mutants. *Genetics* **93,** 625–643.

Minet, M., Jauniaux, J.-C., and Grenson, M. (1980). Organization and expression of a two-gene cluster in arginine biosynthesis of *Saccharomyces cerevisiae. J. Mol. Biol.* **139,** 691–704.

Minson, A. C., and Creaser, E. H. (1969). Purification of a trifunctional enzyme catalysing three steps of the histidine pathway from *Neurospora crassa. Biochem. J.* **114,** 49–56.

Mishima, M., Roggenkamp, R., and Schweitzer, E. (1980). Yeast mutants defective in acetyl-CoA carboxylase and biotin: apocarboxylase ligase. *Eur. J. Biochem.* **111,** 79–87.

Mitchell, H. K., and Lein, J. (1948). A *Neurospora* mutant deficient in the synthesis of tryptophan. *J. Biol. Chem.* **175,** 481–482.

Moir, D., and Botstein, D. (1982). Determination of the order of gene function in the yeast nuclear division pathway using *cs* and *ts* mutants. *Genetics* **100,** 565–577.

Montgomery, D. L., Hall, B. D., Gillam, S., and Smith, M. (1978). Identification and isolation of the yeast cytochrome *c* gene. *Cell (Cambridge, Mass.)* **14,** 673–680.

Morris, N. R., Lai, M. H., and Oakley, C. E. (1979). Identification of a gene for β-tubulin in *Aspergillus nidulans. Cell (Cambridge, Mass.)* **16,** 437–442.

Nasmyth, K. A. (1982). The regulation of yeast mating type by *SIR:* an action at a distance affecting both transcription and transposition. *Cell (Cambridge, Mass.)* **30,** 567–578.

Nasmyth, K. A., and Tatchell, K. (1980). The structure of transposable yeast mating type loci. *Cell (Cambridge, Mass.)* **19,** 753–764.

Oakley, B. R., and Morris, N. R. (1983). A mutation in *Aspergillus nidulans* that blocks the transition from interphase to prophase. *J. Cell Biol.* **96,** 1155–1158.

Oshima, Y. (1982). Regulatory circuits for gene expression: the metabolism of galactose and phosphate. *In* "Molecular Biology of the Yeast Saccharomyces: Metabolism and Gene Expression" (J. N. Strathern, E. W. Jones, and J. R. Broach, eds.), pp. 159–180. Cold Spring Harbor Lab. Press, Cold Spring Harbor, New York.

Pall, M. L. (1970). Amino acid transport in *Neurospora crassa.* II. Properties of a basic amino acid transport system. III. Acidic amino acid transport. *Biochim. Biophys. Acta* **203,** 138–149, 513–520.

Pall, M. L. (1971). Amino acid transport in *Neurospora crassa.* IV. Properties and regulation of a methionine transport system. *Biochim. Biophys. Acta* **233,** 201–214.

Paluh, J. L., and Zalkin, A. (1983). Isolation of *Saccharomyces cerevisiae TRP3. J. Bacteriol.* **153,** 345–349.

Patel, V. B., Schweitzer, M., Dykstra, C. C., Kushner, S. R., and Giles, N. H. (1981). Genetic organization and transcriptional regulation in the *qa* gene cluster of *Neurospora crassa. Proc. Natl. Acad. Sci. U.S.A.* **78,** 5783–5787.

Pateman, J. A., and Cove, D. J. (1967). Regulation of nitrate reduction in *Aspergillus nidulans. Nature (London)* **215,** 1234–1237.

Piérard, A., Messenguy, F., Feller, A., and Hilger, F. (1979). Dual regulation of the synthesis of the arginine pathway carbamoyl phosphate synthetase of *Saccharomyces cerevisiae* by specific and general controls of amino acid biosynthesis. *Mol. Gen. Genet.* **174**, 163–171.

Piper, P. W., Wasserstein, M., Engback, F., Katloft, K., Celis, J. E., Zeuthen, J., Liebman, S., and Sherman, F. (1976). Nonsense suppressors of *Saccharomyces cerevisiae* can be generated by mutation of the tyrosine tRNA anticodon. *Nature (London)* **262**, 757–761.

Pritchard, R. H. (1955). The linear arrangement of a series of alleles in *Aspergillus nidulans*. *Heredity* **9**, 343–371.

Radford, A. (1969). Polarized complementation at the *pyrimidine-3* locus of *Neurospora*. *Mol. Gen. Genet.* **104**, 288–294.

Richards, F. M. (1958). On the enzymic activity of subtilisin-modified ribonuclease. *Proc. Natl. Acad. Sci. U.S.A.* **44**, 162–166.

Richter, R., and Heslot, H. (1982). Genetic and functional analysis of the complex locus *ade10* in *Schizosaccharomyces pombe*. *Curr. Genet.* **5**, 233–244.

Russell, D. W., Smith, M., Cox, D., Williamson, V. M., and Young, E. T. (1983). DNA sequences of two yeast promoter-up mutants. *Nature (London)* **304**, 652–654.

St. John, T. P., and Davis, R. W. (1981). The organization and transcription of the galactose gene cluster of *Saccharomyces*. *J. Mol. Biol.* **152**, 285–315.

Sarabhai, A. S., Stretton, A. O. W., Brenner, S., and Bolle, A. (1964). Co-linearity of the gene with the polypeptide chain. *Nature (London)* **201**, 13–17.

Scazzocchio, C., Sdrin, N., and Ong, G. (1982). Positive regulation in a eukaryote; a study of the *uaY* gene of *Aspergillus nidulans*. *Genetics* **100**, 185–208.

Schechtman, M. G., and Yanofsky, C. (1983). Structure of the trifunctional *trp-1* gene from *Neurospora crassa* and its aberrant expression in *Escherichia coli*. *J. Mol. Appl. Genet.* **2**, 83–99.

Schlesinger, M. T., and Levinthal, C. (1963). Hybrid protein formation in *E. coli* alkaline phosphatase leading to *in vitro* complementation. *J. Mol. Biol.* **7**, 1–12.

Schweitzer, E., Kniep, B., Castorph, H., and Holzner, V. (1973). Pantetheine-free mutants of yeast fatty acid synthesis complex. *Eur. J. Biochem.* **39**, 353–362.

Shear, C. L., and Dodge, B. O. (1927). Life histories and heterothallism of the red breadmoulds of the *Monilia sitophila* group. *J. Agric. Res. (Washington, D.C.)* **34**, 1019–1042.

Sherman, F., and Stewart, J. W. (1971). Genetics and biosynthesis of cytochrome *c*. *Annu. Rev. Genet.* **5**, 257–296.

Sinclair, J. H., Stephens, B. J., Sanghavi, P., and Rabinowitz, M. (1967). Mitochondrial satellite and circular DNA filaments in yeast. *Science (Washington, D.C.)* **156**, 1234–1237.

Srb, A. M., and Horowitz, N. H. (1944). The ornithine cycle in *Neurospora* and its genetic control. *J. Biol. Chem.* **154**, 129–139.

Stocklein, W., and Pipersberg, W. (1980). Altered ribosomal protein L29 in cycloheximide-resistant strains of *Saccharomyces cerevisiae*. *Curr. Genet.* **1**, 177–184.

Strauss, A. (1979). The genetic fine structure of the complex locus *aro3* involved in early aromatic amino acid biosynthesis in *Schizosaccharomyces pombe*. *Mol. Gen. Genet.* **172**, 233–241.

Stroman, P., Reinart, W., Case, M. E., and Giles, N. H. (1978). Organization of the *qa* gene cluster in *Neurospora crassa*: direction of transcription of the *qa-3* gene. *Genetics* **92**, 67–74.

Suskind, S. R., and Jordan, E. (1959). Enzymatic activity of a genetically altered tryptophan synthetase in *Neurospora crassa*. *Science (Washington, D.C.)* **129**, 1614–1615.

Suskind, S. R., Yanofsky, C., and Bonner, D. M. (1955). Allelic strains of *Neurospora* lacking tryptophan synthetase: a preliminary immunochemical characterization. *Proc. Natl. Acad. Sci. U.S.A.* **41**, 577–582.

Suyuma, Y., Munkres, K. D., and Woodward, V. W. (1959). Genetic analysis of the *pyr-3* locus of *Neurospora crassa*: the bearing of recombination and gene conversion on intra-allelic linearity. *Genetica* **30**, 293–311.

Tatum, E. L., and Beadle, G. W. (1942). Genetic control of biochemical reactions in *Neurospora:* an "aminobenzoicless" mutant. *Proc. Natl. Acad. Sci. U.S.A.* **28,** 234–243.

Thuriaux, P., Ramos, R., Piérard, A., Grenson, M., and Wiame, J.-M. (1972). Regulation of the carbamoyl phosphate synthetase belonging to the arginine biosynthetic pathway of *Saccharomyces cerevisiae*. *J. Mol. Biol.* **67,** 277–287.

Thuriaux, P., Heyer, W.-D., and Strauss, A. (1982). Organization of the complex locus *trp1* in the fission yeast *Schizosaccharomyces pombe*. *Curr. Genet.* **6,** 13–18.

Toh-e, A., Inouye, S., and Oshima, Y. (1981). Structure and function of the *PHO82-pho4* locus controlling the synthesis of repressible acid phosphatases of *Saccharomyces cerevisiae*. *J. Bacteriol.* **145,** 221–232.

Tollervey, D. W., and Arst, H. N., Jr. (1981). Mutations to constitutivity and derepression are separate and separable in a regulatory gene of *Aspergillus nidulans*. *Curr. Genet.* **4,** 63–68.

Tyler, B. M., Geever, R. F., Case, M. E., and Giles, N. H. (1984). *Cis*-acting and *trans*-acting regulatory mutations define two types of promoters controlled by the *qa-1F* gene of *Neurospora*. *Cell (Cambridge, Mass.)* **36,** 493–502.

Valone, J. A., Jr., Case, M. E., and Giles, N. H. (1971). Constitutive mutants in a regulatory gene exercising positive control of quinic acid catabolism in *Neurospora crassa*. *Proc. Natl. Acad. Sci. U.S.A.* **74,** 3508–3512.

Watson, D. H., and Wootton, J. C. (1978). Subunit ratios of separated hybrid hexamers of *Neurospora* NADP-specific glutamate dehydrogenase containing mutually complementing mutationally modified monomers. *Biochem. J.* **175,** 1125–1133.

Webber, B. B. (1960). Genetical and biochemical studies of histidine-requiring mutants of *Neurospora crassa*. II. Evidence concerning heterogeneity among *his-3* mutants. *Genetics* **45,** 1617–1626.

Wiame, J.-M., and Dubois, E. L. (1976). The regulation of enzyme synthesis in arginine metabolism of *Saccharomyces cerevisiae*. *Proc.—Int. Symp. Genet. Ind. Microorg.*, 2nd, Sheffield, Engl., 1974 pp. 391–406.

Wolfner, M., Yep, D., Messenguy, F., and Fink, G. R. (1975). Integration of amino acid biosynthesis into the cell cycles of *Saccharomyces cerevisiae*. *J. Mol. Biol.* **96,** 273–290.

Woodward, D. O. (1959). Complementation *in vitro* between adenylosuccinase mutants of *Neurospora crassa*. *Proc. Natl. Acad. Sci. U.S.A.* **45,** 846–850.

Woodward, D. O., Partridge, C. W. H., and Giles, N. H. (1958). Complementation at the *ad-4* locus in *Neurospora crassa*. *Proc. Natl. Acad. Sci. U.S.A.* **44,** 1237–1244.

Yanofsky, C. (1952). The effects of gene change on tryptophan desmolase formation. *Proc. Natl. Acad. Sci. U.S.A.* **38,** 215–226.

Yanofsky, C., Carlton, B. C., Guest, J. R., Helinski, D. R., and Henning, U. (1964). On the colinearity of gene structure and protein structure. *Proc. Natl. Acad. Sci. U.S.A.* **51,** 266–272.

2

Molecular Taxonomy of the Fungi

C. P. KURTZMAN
Northern Regional Research Center
Agricultural Research Service
U.S. Department of Agriculture
Peoria, Illinois

I. Introduction	35
II. Nucleic Acid Isolation and Purification	37
III. DNA Base Composition	38
A. Methodology	38
B. Taxonomic Uses of G + C Values	39
IV. DNA Relatedness	41
A. Methodology	41
B. Extent of Resolution Correlated with Genetic Hybridization	44
C. Comparisons of Species	47
V. Mitochondrial DNA Relatedness	49
VI. Ribosomal RNA Relatedness	51
A. Methodology	51
B. Comparisons of Taxa	52
VII. Comparison of Relatedness from Nucleic Acid Studies with That Determined by Other Methodologies	53
References	56

I. INTRODUCTION

In a book concerned with the manipulation of genetic material in fungi, one may reasonably ask why there is a chapter on taxonomy. One might argue that the name or classification means little and that it is the potential of the organism that is our primary interest. While this may be true, a classification system founded on phylogeny and combined with criteria allowing reliable identification will permit much more efficient exploitation of useful organisms than is possible at present. The following examples serve to illustrate this point of view. I recently happened onto three papers by the same group of authors, each paper describing the same fermentation but by a "different yeast." Had these authors checked on the taxonomy of the species, they would have learned that DNA relatedness studies published 4 years earlier showed all three to be conspecific.

Laboratories around the world are increasingly seeking to improve industrial fermentations as well as other biological processes through the combination of characters not found in organisms in nature. One means to do this has been through the fusion of different organisms, each of which has some unique character. Here, a good classification system can be of real value because protoplasts fused from closely related species are generally more stable than fusion products from distantly related taxa (Ferenczy, 1981). In the case of properties added through recombinant DNA technology, one might suspect that expression of the newly inserted gene sequence would be more efficient in organisms more closely related to the donor species. It seems likely that as molecular techniques become increasingly commonplace in taxonomic studies a much better understanding of species relationships will result with benefits extending well beyond the realm of the systematist.

Now, let us direct our attention to the classification of the fungi. These heterotrophic organisms represent a phylogenetically diverse grouping that has been separated into five taxonomic subdivisions: Mastigomycotina, Zygomycotina, Ascomycotina, Basidiomycotina, and Deuteromycotina (Ainsworth, 1973). Further reduction to lesser taxonomic groups, e.g., classes, orders, families, genera, and species, has been made primarily based on differences in the morphology of sexual and/or vegetative structures. Among the yeasts, assignment to genera and especially to species is based not only on morphology but also on the ability to ferment or assimilate various carbon compounds and on the ability to utilize nitrate as the sole source of nitrogen. In still other taxa, such as the rusts, smuts, and certain imperfect fungi, speciation has been based in part on the type of host plant parasitized.

The species, the primary unit of taxonomy, represents, when defined in terms of genetics, a product of nature rather than the creation of the taxonomist. Dobzhansky (1976) noted that among sexually reproducing and outbreeding organisms, species can be defined as Mendelian populations or arrays of populations that are reproductively isolated from other population arrays. Owing to tradition as well as to technical difficulties, relatively few taxa have been separated from one another because of demonstrated infertility. However, the problems stemming from unknown sexual stages, lack of complementary mating types, and loss of fertility largely can be circumvented because genetic divergence also may be measured from the extent of nucleic acid complementarity between taxa. Consequently, it is now practical to define species as well as to assess present taxonomic criteria through measurements of DNA relatedness, RNA relatedness, and changes in nucleic acid fragment patterns generated with restriction endonucleases. The aim of this chapter will be to discuss methods now in use and to assess the extent of taxonomic resolution provided.

II. NUCLEIC ACID ISOLATION AND PURIFICATION

The methodology for isolation and purification of fungal DNA has been dealt with in detail elsewhere, and the following account will serve only as a brief review to direct attention to some of the more pertinent references. For many fungi, DNA isolation is relatively straightforward. Cell wall breakage may be effected by mechanical or lytic means. Mechanical methods have the advantage of being relatively rapid, but unless cell breakage is carefully monitored, considerable shearing of the DNA may result. The Braun cell homogenizer (Bronwill Scientific, Inc.), one of the more frequently used mechanical devices, causes breakage by its shaking of a buffered cell suspension in the presence of 0.45-mm glass beads (Bak and Stenderup, 1969; Price *et al.*, 1978; Kurtzman *et al.*, 1980a). An inexpensive variation of this technique involves placing the buffered cell and bead mixture into a test tube and "vortexing" the mixture (Lipinski *et al.*, 1976; Van Etten and Freer, 1978). Less frequently reported methods include grinding acetone-dried cells in a ball mill (Martini and Phaff, 1973) and grinding cells in liquid nitrogen with a mortar and pestle (Storck and Alexopoulos, 1970; Mendonça-Hagler *et al.*, 1974). Sonication is seldom used for cell breakage because of the danger of extreme DNA shearing.

Lytic methods offer the possibility of isolating DNA with a higher molecular weight than might be obtained by mechanical cell disruption. Meyer and Phaff (1969) and Bicknell and Douglas (1970) used a modification of the Smith and Halvorson (1967) technique in which yeast cells undergo autolysis in the presence of saline EDTA, 2-mercaptoethanol, and sodium lauryl sulfate. The occasional failure of this method has prompted the widespread use of mechanical disruption. The enzyme preparations Glusulase and Zymolyase may be used to prepare osmotically fragile spheroplasts by cell wall digestion. Following this step, high-molecular-weight DNA is isolated from the lysed spheroplasts. Cryer *et al.* (1975) have given details of these procedures. Zymolyase generally promotes more rapid cell wall dissolution than Glusulase, but neither enzyme preparation is particularly effective with basidiomycetes. Brady (1981), however, reported the relatively rapid dissolution of *Rhodotorula* cell walls by an enzyme preparation from culture filtrates of *Penicillium lilacinum*. Another success in this area was achieved by Rhodes and Kwon-Chung (1982), who produced protoplasts from blastospores of *Cryptococcus neoformans* with the multienzyme preparation mutanase.

Following rupture or lysis of the cells, DNA isolation and purification may be carried out by the methods of Marmur (1961) and Bernardi *et al.* (1970) as detailed by Price *et al.* (1978). Basically, this consists of making the cell suspension 1 M in sodium perchlorate and 1% in sarcosate, and emulsifying with an equal volume of chloroform–isoamyl alcohol (24 : 1, vol/vol). After incubation and phase separation by centrifugation, DNA is precipitated from the aqueous phase with cold ethanol. The DNA is treated with α-amylase, pancreatic RNase, and pronase. Following enzyme removal, the DNA is again ethanol-precipitated, spooled, redissolved, and further treated with α-amylase, pancreatic RNase, and T_1 RNase. Additional purification can include hydroxylapatite fractionation and banding in a cesium chloride gradient.

Some fungi have active DNases that escape denaturation during the usual extraction procedures and cause sufficient degradation of the DNA to prevent spooling. Several methods have produced varying degrees of success in the deactivation of nucleases, and they merit testing if problems are encountered. Diethyl pyrocarbonate used at 0.5–1.0% at the time of cell breakage gave encouraging results in studies of *Lipomyces* spp. (D. L. Holzschu and H. J. Phaff, personal communication) and *Sporidiobolus* (Holzschu et al., 1981). Ehrenberg *et al.* (1976) discussed the interaction of diethyl pyrocarbonate with nucleic acids and emphasized some of the precautions to be taken. C. P. Kurtzman and co-workers (unpublished observations) found that the spermine–spermidine–sucrose buffer of Morris, as cited by Timberlake (1978), inactivated nucleases much more effectively in *Aspergillus* spp. and *Neurospora* spp. than the sucrose buffer frequently used in yeast studies (Price et al., 1978). Heating the cells to 60°C for 20 min was also reasonably effective in deactivating nucleases in *Aspergillus* and *Neurospora* (C. P. Kurtzman et al., unpublished observations). Higher temperatures, if they do not result in denaturation of the DNA, may be even more effective.

Unless care is taken to isolate intact nuclei and mitochondria, the preceding methods will give a mixture of DNA from these two organelles. Because of its lower guanine + cytosine (G + C) content, mitochondrial DNA may be separated from nuclear DNA through preparative cesium chloride density centrifugation. Methods for assessing DNA purity include chemical analyses as well as spectrophotometric measurements and have been summarized by Price *et al.* (1978) and Johnson (1981).

Techniques for isolation of RNA have been presented by Gillespie and Spiegelman (1965), Bicknell and Douglas (1970), and Johnson (1981). These references also include procedures for RNA–DNA reassociation experiments. Although the cellular RNA concentration is manyfold higher than that of DNA, the nearly ubiquitous occurrence of RNases and their marked stability make isolation of undegraded RNA more difficult than the isolation of high-molecular-weight DNA.

III. DNA BASE COMPOSITION

A. Methodology

Nuclear DNA base composition, expressed as molar percentages of G + C, may be determined from thermal denaturation profiles (melts), buoyant density in cesium salt gradients generated by ultracentrifugation, chemical analysis, absorbance ratios, or high-pressure liquid chromatography of nucleotides or free bases (Johnson, 1981). Most investigators use thermal denaturation (Marmur and Doty, 1962), but Bak (1973) showed that results obtained by this method can be greatly affected by sample impurities and/or minor DNA species, and must therefore be interpreted with caution. Cesium chloride buoyant density determinations (Schildkraut *et al.*, 1962) generally show the greatest accuracy, since they are unbiased by the presence of contaminating RNA, mitochondrial DNA, and other impurities, such as carbohydrates and proteins.

B. Taxonomic Uses of G + C Values

Nuclear DNA base composition is the simplest of the molecular techniques that allow resolution of taxa. However, the taxonomic uses of G + C values are mainly exclusionary because fungus species range in G + C content from approximately 30 to 70 mol%, and overlap between unrelated species is inevitable. For example, *Debaryomyces hansenii* and *D. marama* have G + C contents of 39.0 and 39.1 mol%, respectively, yet they show only 8.4% base sequence complementarity (Price *et al.*, 1978). Consequently, since similar G + C content does not necessarily indicate conspecificity, we must ask how dissimilar G + C values need to be in order to state with relative certainty that two strains represent different species.

The answer to this question depends to some extent on the method used for determination of base composition. When buoyant density is used, the span of G + C values among strains of a species is about 1% (Price *et al.*, 1978; Kurtzman *et al.*, 1980a,b), which is approximately coincident with the accuracy of the method, i.e., a standard deviation of 0.00–0.50 mol%. Thermal denaturation shows greater variation, and strains of a species may have a G + C range of as much as 2% (Meyer *et al.*, 1978). Thus, strains showing a difference in G + C contents in excess of 1.0–1.5% by buoyant density or 2.0–2.5% by thermal denaturation may be expected to be different species.

An examination of values in Table I shows considerable overlap in G + C content among the subdivisions of fungi. Only among the yeasts is it possible to predict the subdivision from G + C content. Ascomycetous species range from 28 to 52%, whereas basidiomycetous species range from 49 to 68%. Thus, only a small region overlaps the yeasts, whereas the overlap among the higher fungi is much greater. Species within a genus frequently have G + C values that lie within a range of 10%. Clustering may also occur for higher orders of classification. For example, the Oomycetes have somewhat greater ratios than the Zygomycetes (Storck and Alexopoulos, 1970).

Despite the limited taxonomic resolution provided by comparisons of G + C content, some insight into relationships within the genus *Humicola* has been demonstrated with this technique by the studies of Lepidi *et al.* (1972), De

TABLE I.

Nuclear DNA Base Composition of Selected Fungus Taxa

Species	G + C (mol%)
Mastigomycotina	
Phytophthora infestans	54[a]
Saprolegnia parasitica	59[b]
Allomyces arbuscula	62.2[c]
Zygomycotina	
Cunninghamella echinata	34[d]
Mucor rouxii	37[d]
Syncephalastrum racemosum	48[e]
Achlya racemosa	52[b]
Ascomycotina	
Pichia kluyveri	30[f]
Debaryomyces vanriji	33.2[g]
Debaryomyces hansenii	38.6[g]
Metschnikowia reukaufii	42.2[h]
Metschnikowia pulcherrima	48.3[h]
Pichia rhodanensis	51.9[i]
Neurospora crassa	54[d]
Gelasinospora calospora	55[a]
Ceratocystis ulmi	56[d]
Chaetomium globosum	58[d]
Basidiomycotina	
Filobasidiella neoformans	51.5[j]
Coprinus lagopus	53.0[a]
Lenzites saepiaria	54.0[k]
Daedalea confragosa	57[d]
Polyporus brumalis	59[k]
Rhodosporidium toruloides	60[l]
Schizophyllum commune	61[d]
Sporidiobolus salmonicolor	63[e]

[a] Storck and Alexopoulos (1970).
[b] Green and Dick (1972).
[c] Ojha et al. (1977).
[d] Villa and Storck (1968).
[e] Storck (1966).
[f] Nakase and Komagata (1968).
[g] Price et al. (1978).
[h] Meyer and Phaff (1969).
[i] Kurtzman et al. (1980a).
[j] Storck et al. (1969).
[k] Storck et al. (1971).
[l] Nakase and Komagata (1971).

Bertoldi et al. (1973), and De Bertoldi (1976). Species of *Humicola* range in G + C content from 28.5 to 56.9 mol%. On the basis of morphological differences and distinctive G + C values, De Bertoldi (1976) described several new species. The wide range of G + C content, however, suggests the genus to be polyphyletic. Furthermore, strains of several species differ by 5 to 7 mol% in G + C

(De Bertoldi et al., 1973), suggesting that the strains actually represent separate species.

Guého (1979) examined species assigned to *Geotrichum* and demonstrated from G + C values, as well as through other tests, that the genus is comprised of both ascomycetous and basidiomycetous anamorphs. Definition of species in these two groups must await comparisons of DNA complementarity because many of the presently assigned species have nearly identical G + C values.

In contrast to the preceding studies, Davison and co-workers (1980) reported 34 medically important species comprising the genera *Trichophyton, Microsporum*, and *Epidermophyton* to have G + C contents of 48.7–50.3 mol%. Such a narrow range would almost certainly indicate that some of these taxa are conspecific, a possibility that can be resolved by comparisons of DNA complementarity. Neish and Green (1976) faced a similar dilemma in their study of *Saprolegnia;* all species had the same G + C values and were indistinguishable by this methodology.

IV. DNA RELATEDNESS

A. Methodology

The methods for assessing DNA relatedness vary, but short of actual sequencing rely on measuring the extent and stability of renatured DNA strands from the test pair, i.e., the fidelity of complementary base pairing. Depending on the method, the DNA may or may not need to be labeled with radioisotopes. Because of technical difficulties, heavy-isotope labels are seldom used. An interesting variation of the DNA–DNA hybridization technique was employed by Bak and Stenderup (1969). Radioactive complementary RNA was synthesized *in vitro* with each test DNA serving as a template, and the relatedness between taxa was then assessed from hybridization of single-stranded DNA with the synthesized RNA.

In vivo labeling of DNA is frequently done with ^{14}C, ^{3}H, or ^{32}P. Each isotope has its particular advantages and disadvantages. The DNA may also be labeled *in vitro* with ^{125}I (Commerford, 1971; Mendonça-Hagler and Phaff, 1975; Holzschu et al., 1979) or by nick translation (Kelley et al., 1970; Maniatis et al., 1975). DNA labeled with ^{125}I *in vitro* must be highly purified, which requires the extra purification step of banding in CsCl in a preparative ultracentrifuge. Following labeling, DNA reassociation reactions may be carried out by the membrane method, which involves immobilizing single strands of one of the DNA species onto nitrocellulose filters and allowing sheared single strands of the second DNA to react with the immobilized DNA (Denhardt, 1966). Alternatively, both DNA species may be allowed to react in free solution, and the extent of reassociation may be assessed by percent binding of the resulting duplexes to hydroxylapatite (Brenner, 1973) or resistance to hydrolysis by S_1 nuclease (Crosa et al., 1973). Price et al. (1978) and Johnson (1981) provide references and detailed procedures for these methods.

In addition to following DNA reassociation with isotopically labeled preparations, the reactions may also be monitored spectrophotometrically by measuring the kinetics of duplex formation. This technique has been refined to the point where results are comparable to data obtained with isotopes (Subirana and Doty, 1966; De Ley et al., 1970; Seidler and Mandel, 1971; Kurtzman et al., 1980a). Relatedness may also be estimated from the thermal stability of the renatured heterologous DNA as compared with renatured homologous DNA. This may be done spectrophotometrically by monitoring thermal melts or with labeled DNA by thermal elution from hydroxylapatite columns (Ullman and McCarthy, 1973). Brenner et al. (1972) equated one degree of difference in thermal stability with 1% mismatch in base pairs. Lower thermal stability usually correlates well with lower DNA homology.

A number of factors affect DNA renaturation reactions and so may have a marked effect on the results obtained. Highly purified DNA preparations are essential. Contaminating protein or RNA can result in falsely high or low relatedness values, depending on whether the contaminants act as single-stranded or double-stranded DNA in the assay system (McCarthy and Church, 1970; Brenner, 1973).

Several classes of DNA occur in eukaryotes and their presence affects estimates of relatedness between taxa, since it is the intent of most studies to measure genomic similarity from complementarity of single-copy nuclear DNA sequences. Christiansen et al. (1971) reported up to 15% rapidly renaturing sequences in the nuclear DNA (nDNA) of yeasts, which may be interpreted as analogous to the repeated sequences found in higher eukaryotes (Britten and Kohne, 1968). However, the presence of rapidly renaturing, plasmidlike 2-µm circular DNA in *Saccharomyces cerevisiae* (Bak et al., 1972), with a buoyant density similar to that of nDNA, suggests that the proportion of repeated sequences in nDNA is smaller than first thought. Similar circular DNAs have been reported from *Hansenula wingei* and *Torulopsis glabrata* (O'Connor et al., 1975, 1976). Linear DNA plasmids have been discovered in *Kluyveromyces lactis* (Gunge et al., 1981). These plasmids, which are lighter than the densities of both nuclear and mitochondrial DNA, appear to be associated with the killer character of this yeast. Among the filamentous fungi, kinetic studies also indicate the nuclear genome to be comprised primarily of unique sequences. Dusenbery (1975) reported 64% of the total cellular DNA of *Phycomyces blakesleeanus* to represent a slowly reassociating fraction that is probably made up of unique sequences. A similar figure, 60%, was reported for *Allomyces arbuscula* (Ojha et al., 1977). Hudspeth et al. (1977) determined 82% of the

2. Molecular Taxonomy of the Fungi

genome of *Achlya bisexualis* to be single-copy DNA, whereas this class represented 97–98% for *Aspergillus nidulans* (Timberlake, 1978) and 90% for *Neurospora crassa* (Krumlauf and Marzluf, 1980).

Nuclear DNA represents the preponderance of cellular DNA in yeasts and has a kinetic complexity of 6.5–14 × 10^9 daltons or about two to five times that found in *Escherichia coli* (Christiansen *et al.*, 1971; Petes *et al.*, 1973). Genome sizes of the filamentous fungi cited above are estimated to range from 1.7–2.7 × 10^{10} daltons (Dusenbery, 1975; Ojha *et al.*, 1977; Hudspeth *et al.*, 1977; Timberland, 1978; Krumlauf and Marzluf, 1980). Between 5 and 20% of the total DNA in *S. cerevisiae* is mitochondrial in origin. It is arranged in closed circles with a G + C content noticeably lower than the nDNA (Hartwell, 1974). Sedimentation studies (Blamire *et al.*, 1972) and measurements of contour length (Hollenberg *et al.*, 1970; Petes *et al.*, 1973) suggest a molecular mass of 46 × 10^6 daltons. The mitochondrial DNA (mtDNA) contour length of several other yeasts has been reported to be 25–50% that of *S. cerevisiae* (O'Connor *et al.*, 1975, 1976).

If comparisons of taxa are to be made from single-copy nDNA, mtDNA, plasmidlike circular DNA, and repeat sequence nDNA must be removed. Circular plasmidlike DNA as well as circular mtDNA may be relatively easily separated from nDNA by centrifugation in CsCl in the presence of ethidium bromide. Williamson and Fennell (1975) emphasized that resolution is poor if mtDNA should break into linear fragments during isolation, a not uncommon happening. However, they point out that linear mtDNA may be effectively resolved in CsCl gradients when reacted with 4',6-diamidino-2-phenylindole (DAPI), a fluorescent dye that preferentially binds with AT-rich DNA. Repeated sequences may be removed from nDNA by partial reassociation of sheared DNA fragments followed by fractionation with hydroxylapatite. Since the rate of reassociation is affected by concentration, multiple copies duplex first and can be removed because of their ability to bind to hydroxylapatite (Britten and Kohne, 1968). Price *et al.* (1978) used this method to remove repeated sequences from yeast DNA for taxonomic studies. The fractionation undoubtedly also removes mtDNA and plasmidlike DNA because both occur as multiple copies, and their small molecular weight would allow rapid reassociation when in significant concentration. The same effect may be obtained in spectrophotometric determinations of relatedness if the time between melting and cooling to the reassociation temperature (T_{m-25}) is not too rapid. Under their conditions, Kurtzman *et al.* (1980a) found a cooling period of 12 min to allow duplexing of most repetitive sequences as well as of mitochondrial DNA. Relatedness values of strains common to the work of Price *et al.* (1978) and of Kurtzman *et al.* (1980a) were essentially identical, showing that similar results can be obtained with rather different methodologies.

For most studies, DNA fragment size and the ionic strength of the incubation medium may be maintained at any constant value within a fairly wide range without significantly affecting the results (Brenner, 1973). However, increasing ionic strength increases the reaction rate. Seidler *et al.* (1975) and Kurtzman *et al.* (1980a) used relatively high concentrations of SSC (SSC = standard saline citrate, 150 mM NaCl and 15 mM sodium citrate, pH 7.0) when assessing DNA relatedness spectrophotometrically to complete the determination within a few hours. Thus, when C_0t [C_0t = initial nucleotide concentration in moles per liter × time in seconds (Britten and Kohne, 1968)] is used as a measure of the extent of reassociation, the ionic strength of the reaction mixture must be known if values obtained in different studies are to be compared. For example, the incubation mixture used by Kurtzman *et al.* (1980a) allowed reassociation to proceed nearly five times faster than occurred in the buffer system employed by Price *et al.* (1978). As noted in the definition of C_0t, initial nucleotide concentration also affects the reaction rate; consequently, a solution containing 100 μg of DNA per milliliter would reassociate twice as fast as one containing 50 μg of DNA per milliliter.

Optimal DNA reassociation occurs at 25–30°C below the midpoint of the melting curve (T_m), and this is dependent on base composition, since GC base pairs have greater thermal stability than do AT pairs (Marmur and Doty, 1962; Marmur *et al.*, 1963). If reassociation occurs at a temperature higher than optimum, any mismatched sequences are less likely to pair, whereas temperatures under the optimum may allow the mismatch to appear complementary. Consequently, we must view DNA relatedness values obtained from whole-genome comparisons as an estimate of overall base sequence similarity under defined conditions, rather than as a certain percentage of the DNA that has precisely the same base sequence.

B. Extent of Resolution Correlated with Genetic Hybridization

Guidelines establishing the extent of DNA relatedness exhibited between strains of microbial species have resulted primarily from studies of bacteria. For example, on the basis of extensive comparisons between species of enteric bacteria, Brenner (1973) proposed that strains showing 70% or greater DNA relatedness should be considered members of the same species. Johnson (1973) compared numerous anaerobic bacteria and arrived at essentially the same guidelines. Among the fungi, Price *et al.* (1978) examined species in four yeast genera and proposed that strains having 80% or greater DNA relatedness were conspecific, an opinion expressed earlier by Martini and Phaff (1973) from their studies in *Kluyveromyces*. However, these guidelines were based on strains that had been characterized by traditional morphological and physiological tests. Many fungi, on the other hand, have known sexual stages; with such species, hybridization

2. Molecular Taxonomy of the Fungi

offers a means for correlating DNA complementarity with actual biological phenomena.

Johannsen (1980) and van der Walt and Johannsen (1979) approached this problem through hybridization studies in the essentially homothallic genus *Kluyveromyces,* where they were able to compare their results with DNA relatedness data obtained earlier by Martini (1973) and Martini and Phaff (1973); van der Walt and Johannsen mated auxotrophic yeast strains, attained by N-methyl-N'-nitro-N-nitrosoguanidine mutagenesis, on media that supported growth of prototrophic recombinant clones. Utilizing this method, they observed various combinations of prototrophic colonies at a frequency of $10^{-2}-10^{-8}$. Taxa with high DNA relatedness showed a higher frequency of recombination, but in a few cases taxa having low DNA complementarity also gave a considerable number of recombinants. Lack of crosses through successive generations of progeny has made interpretation of these data difficult. For example, mixes of *K. thermotolerans* and *K. marxianus,* which have significantly different nuclear DNA base composition values, 46.2 mol% G + C and 41.2 mol% G + C, respectively, and essentially no DNA homology (H. L. Presley and H. J. Phaff, personal communication; Martini and Phaff, 1973; Martini *et al.,* 1972), yielded recombinant clones. The F_1 generation from a cross of *K. marxianus* and *K. thermotolerans* appears to be most similar to *K. marxianus* on the basis of DNA composition, i.e., F_1 40.6 mol% G + C; *K. marxianus* 41.2 mol% G + C; *K. thermotolerans* 46.2 mol% G + C (Martini *et al.,* 1972; H. L. Presley, personal communication). These data suggest that the F_1 generation isolated from mixes of *K. marxianus* and *K. thermotolerans* by Johannsen and van der Walt (1978) represents an aneuploid yeast strain. In addition, other laboratories have not been able to confirm the fertility of the F_1 generation isolated from this cross (S. R. Snow, personal communication) or a similar cross between *K. fragilis* and *K. dobzhanskii* (Douglas *et al.,* 1969). Consequently, interspecific fusion seems to have occurred, but there appears to be insufficient genetic complementarity to permit analysis of meiotic products.

Kurtzman and co-workers (1980a,b) were able to correlate genetic relatedness and DNA complementarity by working with heterothallic strains, thereby circumventing problems inherent in the *Kluyveromyces* study, where auxotrophic mutants of homothallic strains were employed. In addition, Kurtzman *et al.* (1980a,b) isolated progeny from crosses by micromanipulation. In the first of these studies, *Pichia amylophila* and *P. mississippiensis* were shown to mate well, but ascospores were not viable. The two species showed only about 25% DNA relatedness, thus correlating lack of fertility with low DNA complementarity. However, the relationship between *Issatchenkia scutulata* var. *scutulata* and *I. scutulata* var. *exigua* was somewhat different. These two taxa also showed only about 25% DNA relatedness, but genetic crosses gave 3–6% viable ascospores. Furthermore, crosses between these F_1 progeny gave 17% viability among F_2 progeny. Reciprocal crosses among the F_1 progeny were fertile, as were backcrosses to the parents. From this, it appears that all stocks had essentially homologous chromosomes and that progeny were neither amphidiploids nor aneuploids. Clearly then, the lower limit of DNA–DNA homology values suggesting species delimitation is not yet well defined, but base sequence divergence, as estimated from whole-genome comparisons, may be as great as 75% before genetic exchange can no longer occur.

Data in Table II summarize the preceding findings and provide further comparisons of other taxa. In all of these studies, progressively less DNA relatedness parallels decreasing mating competence and fertility. Exceptions to this trend include gene changes that prevent mating between highly related strains and amphidiploidy and aneuploidy where moderate to high relatedness between strains might occur, but the disparate chromosome complements could cause infertility.

In addition to providing information on DNA divergence and speciation, the data in Table II demonstrate the limits of taxonomic resolution provided by

TABLE II.

Correlation of Mating Reaction and DNA Complementarity among Closely Related Heterothallic Ascomycetous and Basidiomycetous Yeasts

Species	Mating reaction	DNA relatedness (%)
Filobasidiella neoformans × *Filobasidiella bacillispora*	Fair conjugation 0–30% basidiospore viability (F_1 progeny; F_2 not determined)	55–63[a]
Issatchenkia scutulata var. *scutulata* × *Issatchenkia scutulata* var. *exigua*	Good conjugation Ascospores viable: $F_1 = 5\%$; $F_2 = 17\%$	21–26[b]
Pichia amylophila × *Pichia mississippiensis*	Good conjugation Ascospores not viable	20–27[c]
Hansenula bimundalis × *Hansenula americana*	Poor conjugation Ascospores not produced	21[d]
Hansenula alni × *Hansenula wingei*	Poor conjugation Ascospores not produced	6[e]
Issatchenkia orientalis × *Issatchenkia occidentalis*	Infrequent conjugation Ascospores not produced	3–8[b]

[a] Aulakh *et al.* (1981).
[b] Kurtzman *et al.* (1980b).
[c] Kurtzman *et al.* (1980a).
[d] Kurtzman (1984a).
[e] Phaff *et al.* (1979); Fuson *et al.* (1979).

DNA–DNA reassociation. Resolution appears to go no further than the detection of sibling species. Beyond this, all species, regardless of extent of kinship, show essentially less than 10% complementarity. Perhaps this should not be surprising since it has already been pointed out that detectable DNA homology presupposes a relatively high degree of nucleotide complementarity (Wilson *et al.*, 1977; Baumann and Baumann, 1978).

C. Comparisons of Species

Bak and Stenderup (1969) appear to have been the first to make extensive use of DNA reassociation as a means for determining relatedness among fungi. Their study included several phenotypically similar *Candida* species of medical interest. Specifically, their data showed sufficiently high base sequence complementarity (about 65–81%) to regard *C. albicans*, *C. stellatoidea*, and *C. claussenii* as conspecific.

Additional advances in the clarification of imperfect taxa through DNA studies were made by Meyer and Phaff (1972), who found high degrees of homology between *C. lusitaniae* and *C obtusa* and between *C. salmonicola* and *C. sake;* by Meyer *et al.* (1975), who demonstrated *C. cloacae* and *C. subtropicalis* to be synonyms of *C. maltosa;* and by Kurtzman *et al.* (1980b), who showed *C. krusei* and *C. sorbosa* to represent the imperfect states of *Issatchenkia orientalis* and *I. occidentalis,* respectively. Kurtzman *et al.* (1979) and Manachini (1979) demonstrated *C. utilis* to be the imperfect form of *Hansenula jadinii*. The work of Mendonça-Hagler and Phaff (1975) provided clear evidence that *Saccharomyces telluris, Torulopsis bovina, T. pintolopesii,* and *C. sloofii* are conspecific. This study also convincingly demonstrated that pseudomycelium formation provides no distinction between *Torulopsis* and *Candida*.

Speciation in a number of ascosporogenous yeast genera has been clarified as a result of DNA relatedness studies. Price *et al.* (1978) demonstrated that the four species described in the genus *Schwanniomyces* were, in fact, conspecific. These same authors also showed that nine of the rough-spored *Saccharomyces* (*Torulaspora*) species exhibited 98% or greater DNA relatedness and had been erroneously separated because of minor differences in the assimilation and fermentation of various carbon compounds.

DNA relatedness studies have had an enormous impact on our assessment of the criteria used to define species and genera among the yeasts (Table III). Yarrow and Meyer (1978) combined the genera *Candida* and *Torulopsis* when it was found that presence or absence of pseudohyphae might occur in a single species. Similarly, the 75% relatedness detected between *Pichia lindneri* and *Hansenula minuta* prompted Kurtzman (1984b) to propose that the two genera, which are separated on ability to assimilate nitrate, be combined.

Rather few DNA comparisons have been made among fungi other than yeasts.

TABLE III.

DNA Relatedness between Yeast Species Differing in Traditional Taxonomic Characteristics

Species	Characteristic (+ or −)	DNA relatedness (%)
	Pseudohyphae	
Candida slooffii	+	80[a]
Torulopsis pintolopesii	−	
	True hyphae	
Hansenula wingei	+	78[b]
Hansenula canadensis	−	
	Glucose fermentation	
Debaryomyces formicarius	+	96[c]
Debaryomyces vanriji	−	
	Lactose assimilation	
Schwanniomyces castellii	+	97[c]
Schwanniomyces occidentalis	−	
	Nitrate assimilation	
Hansenula minuta	+	75[d]
Pichia lindneri	−	
	Nitrate assimilation	
Sterigmatomyces halophilus	+	100[e]
Sterigmatomyces indicus	−	

[a] Mendonça-Hagler and Phaff (1975).
[b] Fuson *et al.* (1979).
[c] Price *et al.* (1978).
[d] Kurtzman (1984b).
[e] Kurtzman *et al.* (1980c).

Of these, Mendonça-Hagler *et al.* (1974) examined species of *Ceratocystis* as possible teleomorphs of *Sporothrix schenckii*. The pairing of *C. minor* and *S. schenckii* gave 75% complementarity, thus indicating that these two species represent the perfect–imperfect forms of the same taxon.

Dutta and colleagues (Dutta, 1976; Dutta *et al.*, 1976; Williams *et al.*, 1981) examined speciation in the genus *Neurospora* and reported that least-related pair, *N. crassa* and *N. dodgei*, to show 61% complementarity. This figure is really quite high when viewed against our experience with the yeasts, where relatedness this great is indicative of conspecificity. Surprisingly, an apparent control pairing between *N. crassa* and *Gelasinospora austosteira* resulted in 56% complementarity. Using the same methodology as Dutta *et al.* (1976), Ojha *et al.* (1975) found very high relatedness among some of the zoosporic fungi. For example, *Allomyces arbuscula* showed 74% homology with *A. microgynus* and

58% with *Blastocladiella emersonii*. In further comparisons, *Saprolegnia ferax* was reported to have 44% complementarity with *Neurospora crassa*.

In an effort to understand relationships among the agronomically and industrially important yellow-green aspergilli, Kurtzman et al. (1984) compared *Aspergillus flavus, A. parasiticus, A. oryzae*, and *A. sojae*. The first two species commonly infest cereal grains and peanuts, in which they may produce the potent carcinogen aflatoxin. The latter two species, which are not known to produce aflatoxin (Wang and Hesseltine, 1982), are widely used throughout the Orient as koji molds for fermentation of sake, miso, and soy sauce. Although mycologists have suggested that the four species are closely related, they usually have been retained as separate taxa (Blochwitz, 1929; Saito, 1943; Thom and Raper, 1945; Raper and Fennell, 1965).

The DNA comparisons showed the extent of base sequence complementarity between *A. flavus* and *A. oryzae* to be 100% (Fig. 1). Similarly, *A. parasiticus* and *A. sojae* showed 91% relatedness. Complementarity between these two groups, demonstrated in the *A. flavus–A. parasiticus* pairing, was 70%. By contrast, *A. tamarii* showed 40–55% relatedness with the foregoing species, while *A. leporis* exhibited less than 10% similarity with them. Of further interest was the demonstration that certain strains initially identified as *A. flavus* represent a new species (C. P. Kurtzman, M. J. Smiley, C. J. Robnett, and D. T. Wicklow, in preparation). This new species, which also produces aflatoxin (Hesseltine *et al.*, 1970), was isolated predominantly from insects. The interpretation of these data is that *A. flavus, A. oryzae, A. parasiticus,* and *A. sojae* represent variants of the same species. Owing to a few morphological differences, as well as to the need to keep toxic and nontoxic strains separate, the "species" will be maintained as varieties of *A. flavus*.

The data are also consistent with the proposal (Blochwitz, 1929; Saito, 1943; Wicklow, 1984) that *A. oryzae* and *A. sojae* developed from naturally occurring *A. flavus* and *A. parasiticus* after initial colonization of koji starters. The domesticated strains show reduced sporulation (e.g., few and smaller, uniseriate conidial heads), produce conidiophores and conidia that are often less conspicuously roughened and more variable in size than those of the naturally occurring species, and do not produce sclerotia or aflatoxin. This is in keeping with the proposed transition from survival in nature to survival in koji fermentations, where improved fitness in the new environment would depend more on rapid conidium germination and ability to efficiently hydrolyze starches or proteins.

V. MITOCHONDRIAL DNA RELATEDNESS

In a few studies, relatedness among the fungi has been examined through comparisons of fragment patterns of mtDNA generated with restriction endo-

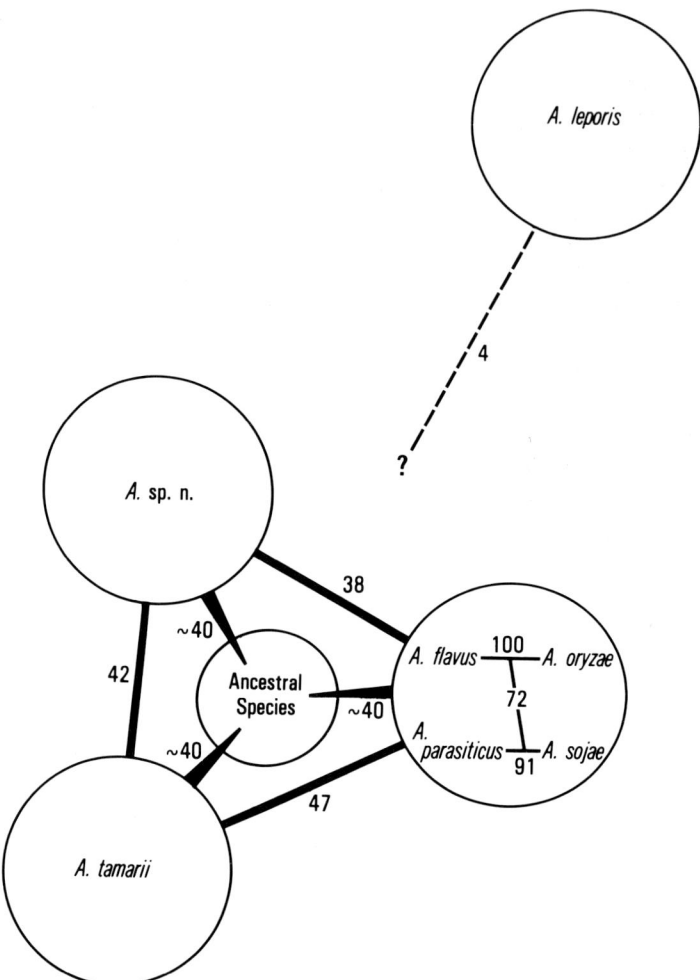

Fig. 1. Diagrammatic representation of the extent of nuclear DNA complementarity between *Aspergillus flavus, A. oryzae, A. parasiticus, A. sojae,* and other phenotypically similar aspergilli. Numbers along connecting lines represent percentage of relatedness (standard deviation ≤5%). Relatedness between these taxa and *Neurospora crassa* Shear *et* Dodge, included for comparative purposes, averaged 3% (Kurtzman *et al.*, 1984).

nucleases. The small size of the mitochondrial genome, as discussed earlier, makes such comparisons practical, whereas the larger nuclear genome would present more fragments than could be reasonably handled. McArthur and Clark-Walker (1983) used mtDNA restriction patterns to correlate perfect–imperfect relationships between the yeast genera *Dekkera* and *Brettanomyces*. They found identical restriction patterns for the pair *D. bruxellensis/B. lambicus*, as well as for the imperfects *B. abstinens/B. custersii* and *B. anomalus/B. clausenii*. This strongly suggests that the pairs are conspecific. Size differences in mtDNA among the other species assigned to these genera prevented an unambiguous assessment of their relationships.

An analysis of mtDNA from seven species of the genus *Aspergillus* was undertaken by Kozlowski and Stepień (1982). Of particular interest was the inclusion of *A. oryzae* and *A. tamarii*, species also compared by Kurtzman *et al.* (1984) through reassociation of nuclear DNA. Restriction patterns of mtDNA from these two species showed considerable similarity, as would be expected from the nuclear DNA study. One surprise in this study was the great similarity of mtDNAs from *A. tamarii* and *A. wentii*.

In both of the preceding studies, it was pointed out that one potential difficulty with restriction pattern analysis is that mtDNA polymorphisms, which arise from insertions or deletions, will give the erroneous appearance of greater sequence divergence than really exists. Because the rate of change in mtDNA in some organisms may be up to tenfold more rapid than that of nuclear DNA, the resolution afforded by mtDNA patterns may not be sufficient to recognize the more divergent strains of a species (Groot *et al.*, 1975; Brown *et al.*, 1979).

VI. RIBOSOMAL RNA RELATEDNESS

A. Methodology

The rather narrow resolution afforded by whole-genome DNA reassociation and mtDNA restriction analysis does not allow verification of species assignments within genera or an understanding of intergeneric relationships. The DNA coding for ribosomal RNA (rRNA) appears to be among the most highly conserved sequences known, and it offers a means for assessing affinities above the species level. One of the most common methods for measuring rRNA complementarity consists of immobilizing single-stranded DNA on nitrocellulose filters and incubating the filters in a buffer solution with radiolabeled rRNA (Bicknell and Douglas, 1970; Kennell, 1971; Johnson, 1981). A modification of this method was developed by Baharaeen *et al.* (1983) in which complementary DNA was synthesized on 25 S rRNA fragments and then allowed to hybridize with rRNA in solution.

Woese and collaborators (Fox *et al.*, 1980) have effectively assessed phylogenetic relationships among the prokaryotes by cataloging 16 S rRNA oligonucleotides generated with T_1 ribonuclease. Johnson and Harich (1983) reported good correlation of values obtained by this method with those from experiments in which rRNAs are hybridized to DNA bound to nitrocellulose. Application of the oligonucleotide cataloging method to fungi is unlikely in the near future because of the greater number of fragments generated from the 18 S ribosomal subunit of eukaryotes (G. E. Fox, personal communication).

Another method that holds considerable promise for phylogenetic studies is the comparison of sequences from 5 S rRNA subunits. Results from this technique will be discussed in the next section.

B. Comparisons of Taxa

Bicknell and Douglas (1970) were among the first to employ rRNA comparisons to assess phylogenetic relatedness among fungi. Their study focused on the genus *Saccharomyces,* some species of which now have been assigned to the genera *Zygosaccharomyces, Kluyveromyces,* and *Torulaspora.* In general, the more highly related species clustered into groups corresponding to present generic assignments. Similar results were obtained for this same group of species by Adoutte-Panvier *et al.* (1980) through electrophoretic and immunochemical comparisons of ribosomal proteins. Bicknell and Douglas also showed that 25 S rRNA comparisons cannot resolve the phylogeny of closely related species because of the highly conserved nature of the sequences. In studies of other yeast species, Segal and Eylan (1974a,b, 1975) reached similar conclusions.

The preceding studies have estimated relatedness from the extent of complementarity of 25 S rRNA. Another approach has been to compare the actual sequence of 5 S rRNA nucleotides. Hori (1976) sequenced the 5 S rRNA from a number of prokaryotes and eukaryotes including several yeasts. Using nucleotide substitution as an evolutionary clock, Hori suggested that prokaryotes and eukaryotes diverged approximately 2.5×10^9 years ago. Andersen *et al.* (1982) compared the 5 S rRNA sequence of *Phycomyces blakesleeanus* with that of other eukaryotes and observed this fungus to show greater similarity with *Tetrahymena thermophila* than with certain ascomycetes.

Walker and Doolittle (1982, 1983) also used 5 S rRNA sequences to investigate relatedness among the basidiomycetes. Their work identified two distinct clusters that correlated with the presence or absence of septal dolipores rather than with the traditional separation of these species into the classes Heterobasidiomycetae and Homobasidiomycetae. Further, the data suggest that capped dolipores evolved from capless dolipores, which may have evolved from single septal pores.

Estimates of relatedness also have been made from the molecular weight of rRNA. Lovett and Haselby (1971) found the molecular weight of the 18 S component from all major classes of fungi to be essentially the same. However, the 25 S component from the Myxomycetes and Oomycetes had a significantly higher molecular weight than did the 25 S component from the Chytridiomycetes, Zygomycetes, Ascomycetes, and Basidiomycetes. Similarly, Fraser and Buczacki (1983) reported the molecular weight of rRNA from the Plasmodiophorales to be sufficiently distinct to separate them from other fungus groups.

VII. COMPARISON OF RELATEDNESS FROM NUCLEIC ACID STUDIES WITH THAT DETERMINED BY OTHER METHODOLOGIES

Earlier in this chapter, essentially in Sections IV,C and VI,B, examples were given in which species defined by traditional methods did not coincide with those determined from molecular comparisons. It might be well to consider these examples in more detail. Because most of the DNA relatedness studies have been done with yeasts, most of this discussion will, of necessity, concern that group of fungi. The ability to ferment or assimilate specific carbon compounds is central to present schemes for yeast identification. As is clear from genetic studies (Winge and Roberts, 1949; Lindegren and Lindegren, 1949; Kurtzman and Smiley, 1976; Starmer *et al.*, 1978), the utilization of carbon compounds is frequently a single gene function. Ability to assimilate nitrate has been of cardinal importance in the definition of many species and genera since Stelling-Dekker (1931) first proposed its use over 50 years ago for the separation of *Hansenula* (NO_3^+) and *Pichia* (NO_3^-), yet genetic studies have shown that nitrate assimilation is under the control of only a few genes (Marzluf, 1981). As a result, we see that the genes controlling utilization of compounds in these standard tests represent only a minute portion of the whole genome. This also appears true for formation of hyphae and pseudohyphae (Table III). Consequently, physiological tests and presence of hyphae should be regarded as convenient laboratory means for separating taxa rather than as the foundation upon which taxa are defined.

Nucleic acid studies have not proved entirely supportive of some conclusions drawn from numerical analysis, or from proton magnetic resonance (PMR) spectra and serology of cell wall mannans. For example, using numerical analysis, Campbell (1973) proposed reduction of *Hansenula wingei, H. canadensis,* and both varieties of *H. bimundalis* to a single species. Fuson and co-workers (1979) showed from DNA sequence studies that *H. canadensis* and *H. wingei* are conspecific, whereas this latter species and *H. bimundalis* show little relatedness. Kurtzman (1984a) demonstrated only 21% complementarity between the two varieties of *H. bimundalis*. This disparity of the two methods probably results from the relatively limited amount of phenotypic data available for numerical analysis, whereas DNA studies have the whole genome as a data base. This also appears to be the case for separations made using proton magnetic resonance, since Spencer and Gorin (1969) found the spectra of cell wall mannans from *H. beckii* and *H. canadensis* to be quite similar, but different from the pattern shared by *H. wingei* and the varieties of *H. bimundalis*. In further comparisons of this group, Tsuchiya *et al.* (1974) reported the cell surface antigens of *H. wingei, H. canadensis,* and *H. beckii* to be indistinguishable. The reason for this discrepancy is not at all clear, but Ballou (1974) has shown for *Saccharomyces* mannans that single gene changes can impart significant differences in immunological reaction. The difference between the two main PMR spectral types of mannan in *S. cerevisiae* was also found to be controlled by a single gene (Spencer *et al.,* 1971).

Differentiation of many ascosporogenous genera is based on ascospore shape and surface structure. An opportunity to test the soundness of this convention in the genera *Schwanniomyces, Saccharomyces, Debaryomyces,* and *Pichia* arose through a comparison of the DNA data of Price *et al.* (1978) with the scanning electron microscopy studies by Kurtzman and Kreger-van Rij (1976), Kurtzman and Smiley (1974, 1979), and Kurtzman *et al.* (1972, 1975). All four species of *Schwanniomyces* had high DNA relatedness (Price *et al.*, 1978) and similar spore architecture (Kurtzman *et al.,* 1972; Kreger-van Rij, 1977). The least related, *S. persoonii* at 80% complementarity, showed fewer and less pronounced spore protuberances than the other three species, all four of which are now regarded as conspecific. Ascospore topography for many species in the *Torulaspora* section of *Saccharomyces* [van der Walt's (1970) group III *Saccharomyces* spp.] was essentially identical and also frequently indistinguishable from the majority of species in *Debaryomyces*. Although many species within their respective genera were found to be conspecific by DNA homology, others proved distinct despite spore similarity. The few species with recognizably different spores showed little DNA relatedness. Thus, species with unlike spores can be expected to show little DNA complementarity, but no prediction can be made concerning species with similar spores. A few exceptions exist among other groups. Ascospores of *Pichia ohmeri* may be either spheroidal or hat-shaped, depending on mating type (Wickerham and Burton, 1954; Fuson *et al.,* 1980), and spore shape in *Saccharomycopsis lipolytica* is mating type–dependent (Wicherham *et al.,* 1969). Spore shapes among *Kluyveromyces* spp. with high degrees of DNA complementarity may be either spheroidal or kidney-shaped (H. L. Presley and H. J. Phaff, personal communication).

Kreger-van Rij and Veenhuis (1973) demonstrated by transmission electron microscopy a basic difference in hyphal septa among ascomycetous yeasts. Three classes of septum ultrastructure were observed: a single central pore or closure line, plasmodesmata, and a dolipore configuration. All mycelial species of *Hansenula* and *Pichia* have the central pore, whereas plasmodesmata are common to all *Saccharomycopsis* species except *S. lipolytica*. Dolipores were found in species of *Ambrosiozyma*. Future comparisons of these species by molecular techniques that detect distant relatedness might show ultrastructural differences in septa to be of evolutionary significance, as now appears to be the case among the lower basidiomycetes (Walker and Doolittle, 1982, 1983).

One issue that needs attention concerns the means for detecting phylogenetic relationships of species within genera. As we have seen, whole-genome DNA complementarity is too specific, whereas rRNA comparisons are probably too broad in resolution to detect closely related species (Bicknell and Douglas, 1970). However, the DNA coding for certain enzymes such as glutamine synthetase and superoxide dismutase, although less conserved than that coding for rRNA, is still less divergent than is apparent from comparisons of whole-genome DNA (Baumann and Baumann, 1978; Baumann et al., 1980). By determining changes in the amino acid sequence of these enzymes through such techniques as quantitative microcomplement fixation, an estimate of intermediate relatedness may be found. Immunological studies of protein similarity are very sparse for yeasts, and the resultant information is highly dependent on the kinds of protein studied. Lachance and Phaff (1979) used exo-β-glucanases from species of *Kluyveromyces* but found that the enzyme was poorly conserved and that immunological distances between most species were too great to be reliable for determining evolutionary relationships in the genus.

Electrophoresis of allozymes presents another means for estimating molecular diversity, and, as with immunological studies, resolution depends on the extent of sequence conservation. The proportion of point mutations that are electrophoretically detectable is estimated at approximately 0.27, because of the redundancy in the genetic code and the large proportion of amino acids that are electrically neutral (Baptist et al., 1971; Selander, 1976; Holzschu, 1981). The nearly universal occurrence of extensive protein polymorphism in natural populations has led some workers to believe that variation would prove physiologically important and, therefore, under selective control; others regard it as without phenotypic effect and thus selectively neutral (Selander, 1976). Regardless, data derived by this technique do provide insight into evolutionary processes and taxonomy. Baptist and Kurtzman (1976) utilized comparative enzyme patterns to separate sexually active strains of *Cryptococcus laurentii* var. *laurentii* from nonreactive strains and from the varieties *magnus* and *flavescens*. Holzschu (1981) studied the evolutionary relationships among some 400 strains of various cactophilic species of *Pichia*. His study of the banding patterns in starch gels of 14 metabolic enzymes allowed a determination of the genetic distances among the various yeast populations.

As we have seen from numerous examples, application of molecular techniques has provided insight into evolutionary relationships among the fungi that were, in many cases, previously unsuspected. These new findings have changed many of our older concepts, sometimes painfully, and this has resulted in a considerable number of taxonomic revisions. While these name changes are inconvenient and may even seem capricious, they are necessary if our taxonomy is to reflect phylogeny, a supposition implicit in the assignment of species to genera and of genera to higher orders of classification.

REFERENCES

Adoutte-Panvier, A., Davies, J. E., Gritz, L. R., and Littlewood, B. S. (1980). Studies of ribosomal proteins of yeast species and their hybrids: gel electrophoresis and immunochemical cross-reactions. *Mol. Gen. Genet.* **179**, 273–282.
Ainsworth, G. C. (1973). Introduction and keys to higher taxa. *In* "The Fungi: An Advanced Treatise" (G. C. Ainsworth, F. K. Sparrow, and A. S. Sussman, eds.), Vol. 4A, pp. 1–7. Academic Press, New York.
Andersen, J., Andresini, W., and Delihas, N. (1982). On the phylogeny of *Phycomyces blakesleeanus* nucleotide sequence of 5 S ribosomal RNA. *J. Biol. Chem.* **257**, 9114–9118.
Aulakh, H. S., Straus, S. E., and Kwon-Chung, K. J. (1981). Genetic relatedness of *Filobasidiella neoformans (Cryptococcus neoformans)* and *Filobasidiella bacillispora (Cryptococcus bacillisporus)* as determined by deoxyribonucleic acid base composition and sequence homology studies. *Int. J. Syst. Bacteriol.* **31**, 97–103.
Baharaeen, S., Melcher, U., and Vishniac, H. S. (1983). Complementary DNA-25S ribosomal RNA hybridization: an improved method for phylogenetic studies. *Can. J. Microbiol.* **29**, 546–551.
Bak, A. L. (1973). DNA base composition in mycoplasma, bacteria, and yeast. *Curr. Top. Microbiol. Immunol.* **61**, 89–149.
Bak, A. L., and Stenderup, A. (1969). Deoxyribonucleic acid homology in yeasts. Genetic relatedness within the genus *Candida*. *J. Gen. Microbiol.* **59**, 21–30.
Bak, A. L., Christiansen, C., and Christiansen, G. (1972). Circular, repetitive DNA in yeast. *Biochim. Biophys. Acta* **269**, 527–530.
Ballou, C. E. (1974). Some aspects of the structure, immunochemistry, and genetic control of yeast mannans. *Adv. Enzymol. Relat. Areas Mol. Biol.* **40**, 239–270.
Baptist, J. N., and Kurtzman, C. P. (1976). Comparative enzyme patterns in *Cryptococcus laurentii* and its taxonomic varieties. *Mycologia* **68**, 1195–1203.
Baptist, J. N., Shaw, C. R., and Mandel, M. (1971). Comparative zone electrophoresis of enzymes of *Pseudomonas solanacearum* and *Pseudomonas cepacia*. *J. Bacteriol.* **108**, 799–803.
Baumann, L., and Baumann, P. (1978). Studies of relationship among terrestrial *Pseudomonas, Alcaligenes,* and enterobacteria by an immunological comparison of glutamine synthetase. *Arch. Microbiol.* **119**, 25–30.
Baumann, L., Bang, S. S., and Baumann, P. (1980). Study of relationship among species of *Vibrio, Photobacterium,* and terrestrial enterobacteria by an immunological comparison of glutamine synthetase and superoxide dismutase. *Curr. Microbiol.* **4**, 133–138.
Bernardi, G., Faures, M., Piperno, G., and Slonimski, P. P. (1970). Mitochondrial DNAs from respiratory-sufficient and cytoplasmic respiratory-deficient mutants of yeast. *J. Mol. Biol.* **48**, 23–43.
Bicknell, J. N., and Douglas, H. C. (1970). Nucleic acid homologies among species of *Saccharomyces*. *J. Bacteriol.* **101**, 505–512.
Blamire, J., Cryer, D. R., Finkelstein, D. B., and Marmur, J. (1972). Sedimentation properties of yeast nuclear and mitochondrial DNA. *J. Mol. Biol.* **67**, 11–24.
Blochwitz, A. (1929). Die Gattung *Aspergillus*. Neue Spezies. *Ann. Mycol.* **2**, 205–240.

2. Molecular Taxonomy of the Fungi

Brady, R. J. (1981). A technique for preparation of spheroplasts of *Rhodotorula*. *Abstr., Annu. Meet., Am. Soc. Microbiol.* K65, p. 148.

Brenner, D. J. (1973). Deoxyribonucleic acid reassociation in the taxonomy of enteric bacteria. *Int. J. Syst. Bacteriol.* **23,** 298–307.

Brenner, D. J., Steigerwalt, A. G., and Fanning, G. R. (1972). Differentiation of *Enterobacter aerogenes* from klebsiellae by deoxyribonucleic acid reassociation. *Int. J. Syst. Bacteriol.* **22,** 193–200.

Britten, R. J., and Kohne, D. E. (1968). Repeated sequences in DNA. *Science (Washington, D.C.)* **161,** 529–540.

Brown, W. M., George, M., Jr., and Wilson, A. C. (1979). Rapid evolution of animal mitochondrial DNA. *Proc. Natl. Acad. Sci. U.S.A.* **76,** 1967–1971.

Campbell, I. (1973). Numerical analysis of *Hansenula, Pichia,* and related yeast genera. *J. Gen. Microbiol.* **77,** 427–441.

Christiansen, C., Bak, A. L., Stenderup, A., and Christiansen, G. (1971). Repetitive DNA in yeast. *Nature (London), New Biol.* **231,** 176–177.

Commerford, S. L. (1971). Iodination of nucleic acids *in vitro*. *Biochemistry* **10,** 1993–1999.

Crosa, J. H., Brenner, D. J., and Falkow, S. (1973). Use of a single-strand-specific nuclease for analysis of bacterial and plasmid deoxyribonucleic acid homo- and heteroduplexes. *J. Bacteriol.* **115,** 904–911.

Cryer, D. R., Eccleshall, R., and Marmur, J. (1975). Isolation of yeast DNA. *Methods Cell Biol.* **12,** 39–44.

Davison, F. D., Mackenzie, D. W. R., and Owens, R. J. (1980). Deoxyribonucleic acid base compositions of dermatophytes. *J. Gen. Microbiol.* **118,** 465–470.

De Bertoldi, M. (1976). New species of *Humicola:* an approach to genetic and biochemical classification. *Can. J. Bot.* **54,** 2755–2768.

De Bertoldi, M., Lepidi, A. A., and Nuti, M. P. (1973). Significance of DNA base composition in classification of *Humicola* and related genera. *Trans. Br. Mycol. Soc.* **60,** 77–85.

De Ley, J., Cattoir, H., and Reynaerts, A. (1970). The quantitative measurement of DNA hybridization from renaturation rates. *Eur. J. Biochem.* **12,** 133–142.

Denhardt, D. T. (1966). A membrane filter technique for the detection of complementary DNA. *Biochem. Biophys. Res. Commun.* **23,** 641–646.

Dobzhansky, T. (1976). Organismic and molecular aspects of species formation. *In* "Molecular Evolution" (F. J. Ayala, ed.), pp. 95–105. Sinauer, Sunderland, Massachusetts.

Douglas, H. C., Grindall, D. P., and Talbott, H. (1969). Electrophoretic variants of phosphoglucomutase in *Saccharomyces* species. *J. Bacteriol.* **99,** 287–290.

Dusenbery, R. L. (1975). Characterization of the genome of *Phycomyces blakesleeanus*. *Biochim. Biophys. Acta* **378,** 363–377.

Dutta, S. K. (1976). DNA homologies among heterothallic species of *Neurospora*. *Mycologia* **68,** 388–401.

Dutta, S. K., Sheikh, I., Choppala, J., Aulakh, G. S., and Nelson, W. H. (1976). DNA homologies among homothallic, pseudo-homothallic and heterothallic species of *Neurospora*. *Mol. Gen. Genet.* **147,** 325–330.

Ehrenberg, L., Fedorcsak, I., and Solymosy, F. (1976). Diethyl pyrocarbonate in nucleic acid research. *Prog. Nucleic Acid Res.* **16,** 189–262.

Ferenczy, L. (1981). Microbial protoplast fusion. *In* "Genetics as a Tool in Microbiology" (S. W. Glover and D. A. Hopwood, eds.), Symposium of the Society of General Microbiology, Vol. 31, pp. 1–34. Cambridge Univ. Press, London and New York.

Fox, G. E., Stackebrandt, E., Hespell, R. B., Gibson, J., Maniloff, J., Dyer, T. A., Wolfe, R. S., Balch, W. E., Tanner, R. S., Magrum, L. J., Zablen, L. B., Balkemore, R., Gupta, R., Bonen, L., Lewis, B. J., Stahl, D. A., Leuhrsen, K. R., Chen, K. N., and Woese, C. R. (1980). The phylogeny of prokaryotes. *Science (Washington, D.C.)* **209,** 457–463.

Fraser, R. S. S., and Buczacki, S. T. (1983). Ribosomal RNA molecular weights and the affinities of the Plasmodiophorales. *Trans. Br. Mycol. Soc.* **80,** 107–112.
Fuson, G. B., Price, C. W., and Phaff, H. J. (1979). Deoxyribonucleic acid sequence relatedness among some members of the yeast genera *Hansenula*. *Int. J. Syst. Bacteriol.* **29,** 64–69.
Fuson, G. B., Price, C. W., and Phaff, H. J. (1980). Deoxyribonucleic acid base sequence relatedness among strains of *Pichia ohmeri* that produce dimorphic ascospores. *Int. J. Syst. Bacteriol.* **30,** 217–219.
Gillespie, D., and Spiegelman, S. (1965). A quantitative assay for DNA–RNA hybrids with DNA immobilized on a membrane. *J. Mol. Biol.* **12,** 829–842.
Green, B. R., and Dick, M. W. (1972). DNA base composition and the taxonomy of the Oomycetes. *Can. J. Microbiol.* **18,** 963–968.
Groot, G. S. P., Flavell, R. A., and Sanders, J. P. M. (1975). Sequence homology of nuclear and mitochondrial DNAs of different yeasts. *Biochim. Biophys. Acta* **378,** 186–194.
Guého, E. (1979). Deoxyribonucleic acid base composition and taxonomy in the genus *Geotrichum* Link. *Antonie van Leeuwenhoek* **45,** 199–210.
Gunge, N., Tamuru, A., Ozawa, F., and Sakaguchi, K. (1981). Isolation and characterization of linear deoxyribonucleic acid plasmids from *Kluyveromyces lactis* and the plasmid-associated killer character. *J. Bacteriol.* **145,** 382–390.
Hartwell, L. H. (1974). *Saccharomyces cerevisiae* cell cycle. *Bacteriol. Rev.* **38,** 164–198.
Hesseltine, C. W., Shotwell, O. L., Smith, M., Ellis, J. J., Vandegraft, E., and Shannon, G. (1970). Production of various aflatoxins by strains of the *Aspergillus flavus* series. In "Toxic Micro-organisms" (M. Herzberg, ed.), pp. 202–210. U.S. Gov. Print. Off., Washington, D.C.
Hollenberg, C. P., Borst, P., and van Brugger, E. F. J. (1970). Mitochondrial DNA. V. A 25-μ closed circular duplex molecule in wild-type yeast mitochondria. Structure and kinetic complexity. *Biochim. Biophys. Acta* **209,** 1–15.
Holzschu, D. L. (1981). Molecular taxonomy and evolutionary relationships among cactophilic yeasts. Ph.D. Thesis, Univ. of California, Davis.
Holzschu, D. L., Presley, H. L., Miranda, M., and Phaff, H. J. (1979). Identification of *Candida lusitaniae* as an opportunistic yeast in humans. *J. Clin. Microbiol.* **10,** 202–205.
Holzschu, D. L., Tredick, J., and Phaff, H. J. (1981). Validation of the yeast *Sporidiobolus ruinenii* based on its deoxyribonucleic acid relatedness to other species of the genus *Sporidiobolus*. *Curr. Microbiol.* **5,** 73–76.
Hori, H. (1976). Molecular evolution of 5S RNA. *Mol. Gen. Genet.* **145,** 119–123.
Hudspeth, M. E. S., Timberlake, W. E., and Goldberg, R. B. (1977). DNA sequence organization in the water mold *Achlya*. *Proc. Natl. Acad. Sci. U.S.A.* **74,** 4332–4336.
Johannsen, E. (1980). Hybridization studies within the genus *Kluyveromyces* van der Walt emend. van der Walt. *Antonie van Leeuwenhoek* **46,** 177–189.
Johannsen, E., and van der Walt, J. P. (1978). Interfertility as basis for the delimination of *Kluyveromyces marxianus*. *Arch. Microbiol.* **118,** 45–48.
Johnson, J. L. (1973). Use of nucleic acid homologies in the taxonomy of anaerobic bacteria. *Int. J. Syst. Bacteriol.* **23,** 308–315.
Johnson, J. L. (1981). Genetic characterization. In "Manual of Methods for General Bacteriology" (P. Gerhardt, ed.), pp. 450–472. Am. Soc. Microbiol., Washington, D.C.
Johnson, J. L., and Harich, B. (1983). Comparisons of procedures for determining ribosomal ribonucleic acid similarities. *Curr. Microbiol.* **9,** 111–120.
Kelly, R. B., Cozzrelli, N. R., Deutscher, M. P., Lehman, I. R., and Kornberg, A. (1970). Enzymatic synthesis of deoxyribonucleic acids. *J. Biol. Chem.* **245,** 39–45.
Kennell, D. E. (1971). Principles and practices of nucleic acid hybridization. *Prog. Nucleic Acid Res. Mol. Biol.* **11,** 259–301.

Kozlowski, M., and Stepień, P. P. (1982). Restriction enzyme analysis of mitochondrial DNA of members of the genus *Aspergillus* as an aid in taxonomy. *J. Gen. Microbiol.* **128,** 471–476.

Kreger-van Rij, N. J. W. (1977). Electron microscopy of sporulation in *Schwanniomyces alluvius*. *Antonie van Leeuwenhoek* **43,** 55–64.

Kreger-van Rij, N. J. W., and Veenhuis, M. (1973). Electron microscopy of septa in ascomycetous yeasts. *Antonie van Leeuwenhoek* **39,** 481–490.

Krumlauf, R., and Marzluf, G. A. (1980). Genome organization and characterization of the repetitive and inverted repeat DNA sequences in *Neurospora crassa*. *J. Biol. Chem.* **255,** 1138–1145.

Kurtzman, C. P. (1984a). Resolution of varietal relationships within the species *Hansenula anomala, Hansenula bimundalis,* and *Pichia nakazawae* through comparisons of DNA relatedness. *Mycotaxon* **19,** 271–279.

Kurtzman, C. P. (1984b). Synonomy of the yeast genera *Hansenula* and *Pichia* demonstrated through comparisons of deoxyribonucleic acid relatedness. *Antonie van Leeuwenhoek* **50,** 209–217.

Kurtzman, C. P., and Kreger-van Rij, N. J. W. (1976). Ultrastructure of ascospores from *Debaryomyces melissophilus,* a new taxonomic combination. *Mycologia* **68,** 422–425.

Kurtzman, C. P., and Smiley, M. J. (1974). A taxonomic re-evaluation of the round-spored species of *Pichia*. *Proc. Int. Symp. Yeasts, 4th, Vienna* **I,** 231–232.

Kurtzman, C. P., and Smiley, M. J. (1976). Heterothallism in *Pichia kudriavzevii* and *Pichia terricola*. *Antonie van Leeuwenhoek* **42,** 355–363.

Kurtzman, C. P., and Smiley, M. J. (1979). Taxonomy of *Pichia carsonii* and its synonyms *Pichia vini* and *P. vini* var. *melibiosi:* comparison by DNA reassociation. *Mycologia* **71,** 658–662.

Kurtzman, C. P., Smiley, M. J., and Baker, F. L. (1972). Scanning electron microscopy of ascospores of *Schwanniomyces*. *J. Bacteriol.* **112,** 1380–1382.

Kurtzman, C. P., Smiley, M. J., and Baker, F. L. (1975). Scanning electron microscopy of ascospores of *Debaryomyces* and *Saccharomyces*. *Mycopathol. Mycol. Appl.* **55,** 29–34.

Kurtzman, C. P., Johnson, C. J., and Smiley, M. J. (1979). Determination of conspecificity of *Candida utilis* and *Hansenula jadinii* through DNA reassociation. *Mycologia* **71,** 844–847.

Kurtzman, C. P., Smiley, M. J., Johnson, C. J., Wickerham, L. J., and Fuson, G. B. (1980a). Two new and closely related heterothallic species. *Pichia amylophilia* and *Pichia mississippiensis:* Characterization by hybridization and deoxyribonucleic acid reassociation. *Int. J. Syst. Bacteriol.* **30,** 208–216.

Kurtzman, C. P., Smiley, M. J., and Johnson, C. J. (1980b). Emendation of the genus *Issatchenkia* Kudriavzev and comparison of species by deoxyribonucleic acid reassociation, mating reaction, and ascospore ultrastructure. *Int. J. Syst. Bacteriol.* **30,** 503–513.

Kurtzman, C. P., Smiley, M. J., Johnson, C. J., and Hoffman, M. J. (1980c). Deoxyribonucleic acid relatedness among species of *Sterigmatomyces*. *Abstr. Int. Symp. Yeasts, 5th* Y-5.2.5(L), p. 246.

Kurtzman, C. P., Smiley, M. J., Robnett, C. J., Axt, A., and Wicklow, D. T. (1984). DNA relatedness among the agronomically and industrially important fungi *Aspergillus flavus, A. oryzae, A. parasiticus,* and *A. sojae*. *Abstr., Annu. Meet., Am. Soc. Microbiol.* 011. p. 190.

Lachance, M. A., and Phaff, H. J. (1979). Comparative study of molecular size and structure of exo-β-glucanases from *Kluyveromyces* and other yeast genera: evolutionary and taxonomic implications. *Int. J. Syst. Bacteriol.* **29,** 70–78.

Lepidi, A. A., Nuti, M. P., De Bertoldi, M., and Santulli, M. (1972). Classification of the genus *Humicola* Traaen: II. The DNA base composition of some strains within the genus. *Mycopathol. Mycol. Appl.* **47,** 153–159.

Lindegren, C. C., and Lindegren, G. (1949). Unusual gene-controlled combinations of carbohydrate fermentations in yeast hybrids. *Proc. Natl. Acad. Sci. U.S.A.* **35,** 23–27.

Lipinski, C., Ferro, A. J., and Mills, D. (1976). Macromolecule synthesis in a mutant of *Saccharomyces cerevisiae* inhibited by S-adenosylmethionine. *Mol. Gen. Genet.* **144,** 301–306.
Lovett, J. S., and Haselby, J. A. (1971). Molecular weights of the ribosomal ribonucleic acid of fungi. *Arch. Microbiol.* **80,** 191–204.
McArthur, C. R., and Clark-Walker, G. D. (1983). Mitochondrial DNA size diversity in the *Dekkera/Brettanomyces* yeasts. *Curr. Genet.* **7,** 29–35.
McCarthy, B. J., and Church, R. B. (1970). The specificity of molecular hybridization reactions. *Annu. Rev. Biochem.* **39,** 131–150.
Manachini, P. L. (1979). DNA sequence similarity, cell wall mannans, and physiological characteristics in some strains of *Candida utilis, Hansenula jadinii,* and *Hansenula petersonii. Antonie van Leeuwenhoek* **45,** 451–463.
Maniatis, T., Jeffrey, A., and Kleid, D. G. (1975). Nucleotide sequence of the rightward operator of phage. *Proc. Natl. Acad. Sci. U.S.A.* **72,** 1184–1188.
Marmur, J. (1961). A procedure for the isolation of DNA from microorganisms. *J. Mol. Biol.* **3,** 208–218.
Marmur, J., and Doty, P. (1962). Determination of the base composition of DNA from its thermal denaturation temperature. *J. Mol. Biol.* **5,** 109–118.
Marmur, J., Rownd, R., and Schildkraut, C. L. (1963). Denaturation and renaturation of DNA. *Prog. Nucleic Acid Res. Mol. Biol.* **1,** 231–300.
Martini, A. (1973). Ibridazioni DNA/DNA tra specie di lieviti del genere *Kluyveromyces. Ann. Fac. Agrar., Univ. Studi Perugia* **28,** 1–15.
Martini, A., and Phaff, H. J. (1973). The optical determination of DNA–DNA homologies in yeasts. *Ann. Micr.* **23,** 59–68.
Martini, A., Phaff, H. J., and Douglass, S. A. (1972). Deoxyribonucleic acid base composition of species in the yeast genus *Kluyveromyces* van der Walt emend. v. d. Walt. *J. Bacteriol.* **111,** 481–487.
Marzluf, G. A. (1981). Regulation of nitrogen metabolism and gene expression in fungi. *Microbiol. Rev.* **45,** 437–461.
Mendonça-Hagler, L. C., and Phaff, H. J. (1975). Deoxyribonucleic acid base composition and DNA/DNA hybrid formation in psychrophobic and related yeasts. *Int. J. Syst. Bacteriol.* **25,** 222–229.
Mendonça-Hagler, L. C., Travassos, L. R., Lloyd, K. O., and Phaff, H. J. (1974). Deoxyribonucleic acid base composition and hybridization studies on the human pathogen *Sporothrix schenckii* and *Ceratocystis* species. *Infect. Immuno.* **8,** 674–680.
Meyer, S. A., and Phaff, H. J. (1969). Deoxyribonucleic acid base composition in yeasts. *J. Bacteriol.* **97,** 52–56.
Meyer, S. A., and Phaff, H. J. (1972). DNA base composition and DNA–DNA homology studies as tools in yeast systematics. *In* "Yeasts, Models in Science and Technics" (A. Kochová-Kratochvilová and E. Minarik, eds.), pp. 375–386. Publ. House Slovak Acad. Sci., Bratislava, Czechoslovakia.
Meyer, S. A., Anderson, K., Brown, R. E., Smith, M. T., Yarrow, D., Mitchell, G., and Ahearn, D. G. (1975). Physiological and DNA characterization of *Candida maltosa,* a hydrocarbon utilizing yeast. *Arch. Microbiol.* **104,** 225–231.
Meyer, S. A., Smith, M. T., and Simione, F. P., Jr. (1978). Systematics of *Hanseniaspora* Zikes and *Kloeckera* Janke. *Antonie van Leeuwenhoek* **44,** 79–96.
Nakase, T., and Komagata, K. (1968). Taxonomic significance of base composition of yeast DNA. *J. Gen. Appl. Microbiol.* **14,** 345–357.
Nakase, T., and Komagata, K. (1971). Significance of DNA base composition in the classification of yeast genera *Cryptococcus* and *Rhodotorula. J. Gen. Appl. Microbiol.* **17,** 121–130.

Neish, G. A., and Green, B. R. (1976). Nuclear and satellite DNA base composition and the taxonomy of *Saprolegnia* (Oomycetes). *J. Gen. Microbiol.* **96**, 215–219.

O'Connor, R. M., McArthur, C. R., and Clark-Walker, G. D. (1975). Closed-circular DNA from mitochondrial-enriched fractions of four *petite-* negative yeasts. *Eur. J. Biochem.* **53**, 137–144.

O'Connor, R. M., McArthur, C. R., and Clark-Walker, G. D. (1976). Respiratory-deficient mutants of *Torulopsis glabrata*, a yeast with circular mitochondrial deoxyribonucleic acid of 6 μm. *J. Bacteriol.* **126**, 959–968.

Ojha, M., Dutta, S. K., and Turian, G. (1975). DNA nucleotide sequence homologies between some zoosporic fungi. *Mol. Gen. Genet.* **136**, 151–165.

Ojha, M., Turler, H., and Turian, G. (1977). Characterization of *Allomyces* genome. *Biochim. Biophys. Acta* **478**, 377–391.

Petes, T. D., Beyers, B., and Fangman, W. L. (1973). Size and structure of yeast chromosomal DNA *Proc. Natl. Acad. Sci. U.S.A.* **70**, 3072–3076.

Phaff, H. J., Miller, M. W., and Miranda, M. (1979). *Hansenula alni*, a new heterothallic species of yeast from exudates of alder trees. *Int. J. Syst. Bacteriol.* **29**, 60–63.

Price, C. W., Fuson, G. B., and Phaff, H. J. (1978). Genome comparison in yeast systematics: delimitation of species within the genera *Schwanniomyces, Saccharomyces, Debaryomyces*, and *Pichia. Microbiol. Rev.* **42**, 161–193.

Raper, K. B., and Fennell, D. I. (1965). "The Genus *Aspergillus*." Williams & Wilkins, Baltimore, Maryland.

Rhodes, J. C., and Kwon-Chung, K. J. (1982). A new efficient method for protoplast formation in *Cryptococcus neoformans. Abstr., Annu. Meet., Am. Soc. Microbiol.* F61, p. 336.

Saito, K. (1943). On the scientific name of aspergilli isolated in Japan. *Nippon Jozo Kyokai Zasshi* **38**, 412–414.

Schildkraut, C. L., Marmur, J., and Doty, P. (1962). Determination of the base composition of deoxyribonucleic acid from its buoyant density in CsCl. *J. Mol. Biol.* **4**, 430–433.

Segal, E., and Eylan, E. (1974a). Genetic relatedness of *Candida albicans* to asporogenous and ascosporogenous yeasts as reflected by nucleic acid homologies. *Microbios* **9**, 25–33.

Segal, E., and Eylan, E. (1974b). Nucleic acid homologies between *Candida albicans* and *Hansenula* species. *Microbios* **10**, 133–138.

Segal, E., and Eylan, E. (1975). Nucleic acid homology studies among *Candida albicans, Syringospora albicans*, and *Leucosporidium* species. *Microbios* **12**, 111–117.

Seidler, R. J., and Mandel, M. (1971). Quantitative aspects of DNA renaturation: DNA base composition, state of chromosome replication, and polynucleotide homologies. *J. Bacteriol.* **106**, 608–614.

Seidler, R. J., Knittel, M. D., and Brown, C. (1975). Potential pathogens in the environment. Cultural reactions and nucleic acid studies on *Klebsiella pneumoniae* from chemical and environmental sources. *Appl. Microbiol.* **29**, 819–825.

Selander, R. K. (1976). Genetic variation in natural populations. *In* "Molecular Evolution" (F. J. Ayala, ed.), pp. 21–46. Sinauer, Sunderland, Massachusetts.

Smith, D., and Halvorson, H. O. (1967). The isolation of DNA from yeast. *In* "Nucleic Acids," Part A (L. Grossman, and K. Moldave, eds.), Methods in Enzymology, Vol. 12, pp. 538–541. Academic Press, New York.

Spencer, J. F. T., and Gorin, P. A. J. (1969). Systematics of the genera *Hansenula* and *Pichia:* proton magnetic resonance spectra of their mannans as an aid in classification. *Can. J. Microbiol.* **15**, 375–382.

Spencer, J. F. T., Gorin, P. A. J., and Rank, G. H. (1971). The genetic control of the two types of mannan produced by *Saccharomyces cerevisiae. Can. J. Microbiol.* **17**, 1451–1454.

Starmer, W. T., Phaff, H. J., Miranda, M., and Miller, M. W. (1978). *Pichia amethionina,* a new heterothallic yeast associated with the decaying stems of cereoid cacti. *Int. J. Syst. Bacteriol.* **28,** 433–441.

Stelling-Dekker, N. M. (1931). Die sporogenen Hefen. *Verh. K. Akad. Wet. Amsterdam, Afd. Natuurk., Reeks 2* **28,** 1.

Storck, R. (1966). Nucleotide composition of nucleic acids of fungi. II. Deoxyribonucleic acids. *J. Bacteriol.* **91,** 277–230.

Storck, R., and Alexopoulos, C. J. (1970). Deoxyribonucleic acid of fungi. *Bacteriol. Rev.* **34,** 126–154.

Storck, R., Alexopoulos, A., and Phaff, H. J. (1969). Nucleotide composition of deoxyribonucleic acid of some species of *Cryptococcus, Rhodotorula,* and *Sporobolomyces. J. Bacteriol.* **98,** 1069–1072.

Storck, R., Nobles, M. K., and Alexopoulos, C. J. (1971). The nucleotide composition of deoxyribonucleic acid of some species of *Hymenochaetaceae* and *Polyporaceae. Mycologia* **63,** 38–49.

Subirana, J. A., and Doty, P. (1966). Kinetics of renaturation of renatured DNA. I. Spectrophotometric results. *Biopolymers* **4,** 171–187.

Thom, C., and Raper, K. B. (1945). "A Manual of the Aspergilli." Williams & Wilkins, Baltimore, Maryland.

Timberlake, W. E. (1978). Low repetitive DNA content in *Aspergillus nidulans. Science (Washington, D.C.)* **202,** 973–975.

Tsuchiya, T., Fukazawa, Y., Taguchi, M., Nakase, T., and Shinoda, T. (1974). Serological aspects of yeast classification. *Mycopathol. Mycol. Appl.* **53,** 77–91.

Ullman, J. S., and McCarthy, B. J. (1973). The relationships between mismatched base pairs and the thermal stability of DNA duplexes. I. Effects of depurination and chain scission. *Biochim. Biophys. Acta* **294,** 405–415.

van der Walt, J. P. (1970). *Saccharomyces* Meyen emend. Reess. *In* "The Yeasts: A Taxonomic Study" (J. Lodder, ed.), pp. 555–718. North-Holland Publ., Amsterdam.

van der Walt, J. P., and Johannsen, E. (1979). A comparison of interfertility and in vitro DNA–DNA reassociation as criteria for speciation in the genus *Kluyveromyces. Antonie van Leeuwenhoek* **45,** 281–291.

Van Etten, J. L., and Freer, S. N. (1978). Simple procedure for disruption of fungal spores. *Appl. Environ. Microbiol.* **35,** 622–623.

Villa, V. D., and Storck, R. (1968). Nucleotide composition of nuclear and mitochondrial deoxyribonucleic acid of fungi. *J. Bacteriol.* **96,** 184–190.

Walker, W. F., and Doolittle, W. F. (1982). Redividing the basidiomycetes on the basis of 5S rRNA sequences. *Nature (London)* **299,** 723–724.

Walker, W. F., and Doolittle, W. F. (1983). 5S rRNA sequences from eight basidiomycetes and fungi imperfecti. *Nucleic Acids Res.* **11,** 7625–7630.

Wang, H. L., and Hesseltine, C. W. (1982). Oriental fermented foods. *In* "Prescott & Dunn Industrial Microbiology" (G. Reed, ed.), 4th Ed., pp. 492–538. AVI, Westport, Connecticut.

Wickerham, L. J., and Burton, K. A. (1954). A clarification of the relationship of *Candida guilliermondii* to other yeasts by a study of their mating types. *J. Bacteriol.* **68,** 594–597.

Wickerham, L. J., Kurtzman, C. P., and Herman, A. I. (1969). Sexuality in *Candida lipolytica. In* "Recent Trends in Yeast Research" (D. G. Ahearn, ed.), Vol. 1, pp. 81–92. Georgia State Univ. Press, Atlanta.

Wicklow, D. T. (1984). Adaptation in wild and domesticated yellow-green aspergilli. *In* "Toxigenic Fungi—Their Toxins and Health Hazard" (Y. Ueno, ed.), pp. 78–86. Elsevier, Amsterdam.

Williams, N. P., Mukhopadhyay, D., and Dutta, S. K. (1981). Homologies of *Neurospora* homothallic species using repeated and nonrepeated DNA sequences. *Experientia* **37,** 1157–1158.

Williamson, D. H., and Fennell, D. J. (1975). The use of fluorescent DNA-binding agent for detecting and separating yeast mitochondria DNA. *Methods Cell Biol.* **12,** 335–351.
Wilson, A. C., Carlson, S. S., and White, T. J. (1977). Biochemical evolution. *Annu. Rev. Biochem.* **46,** 573–639.
Winge, O., and Roberts, C. (1949). Inheritance of enzymatic characters in yeast, and the phenomenon of long-term adaptation. *C. R. Trav. Lab. Carlsberg* **24,** 263–315.
Yarrow, D., and Meyer, S. A. (1978). Proposal for amendment of the diagnosis of the genus *Candida* Berkhout nom. cons. *Int. J. Syst. Bacteriol.* **28,** 611–615.

3

Fungal Mitochondrial Genomes[1]

LAWRENCE I. GROSSMAN AND MICHAEL E. S. HUDSPETH[2]
Department of Cellular and Molecular Biology
Division of Biological Sciences
The University of Michigan
Ann Arbor, Michigan

I. Introduction	66
II. Physical Organization	66
A. Size and Shape of Mitochondrial Genomes	66
B. Organizational Features	66
C. A Note on DNA Isolation	70
III. Genes and the Genetic Code	70
A. Protein Structural Genes	70
B. Structural RNA Genes	71
C. Introns and Unidentified Reading Frames	72
D. Genetic Code Variation	74
IV. Gene Order	74
V. Transcription	76
A. Promoters	76
B. Message Processing and Splicing	78
VI. Pseudogenes	81
VII. Mitochondrial Plasmids	83
VIII. Effects of Alteration of mtDNA	84
A. Petite Mutations	84
B. *poky* Mutations	85
C. Stopper Mutations	85
D. Ragged Mutations	86
E. Senescence in *Podospora*	87
IX. Nuclear–Cytoplasmic Interactions	88
References	91

[1]Gene symbols: We have referred to mitochondrial genes according to their encoded products rather than their genetic loci. This description directly equates genic regions of one organism with another, even though the genetic terminology may vary, and avoids nonintuitive designations used for historical reasons, such as COII and *oxi3*.

[2]Present address: Department of Biological Sciences, Northern Illinois University, De Kalb, Illinois 60115-2861.

I. INTRODUCTION

Until quite recently, work on fungal mitochondrial genomes was limited largely to the ascomycetes *Saccharomyces cerevisiae, Neurospora crassa,* and *Aspergillus nidulans.* In the past few years, however, there has been a significant broadening of the base of organisms to include other yeasts and other filamentous fungi.

This chapter reviews recent progress in the characterization of mitochondrial genomes. In keeping with the overall thrust of this volume, we will limit ourselves to fungi, except insofar as comparisons with other organisms are germane. Reviews covering both mitochondrial and chloroplast literature have appeared within the past several years (Dujon, 1981; Gray, 1982; Gray and Doolittle, 1982; Wallace, 1982; Clayton, 1984).

II. PHYSICAL ORGANIZATION

A. Size and Shape of Mitochondrial Genomes

Despite the similar content of coding information in all metazoan and fungal mitochondrial genomes examined (see below), genome size in fungi varies about sixfold. The smallest fungal genomes observed are in *Torulopsis* [18.9 kilobase pairs (kb)] (Clark-Walker *et al.*, 1980) and *Schizosaccharomyces pombe* (19 kb) (Del Giudice *et al.*, 1981), barely larger than the minimal 16.5-kb size found in animals, and the largest is in the basidiomycete *Suillus grisellus* (121 kb) (T. Bruns, unpublished observations). Larger mitochondrial genomes are found only in higher plants (Ward *et al.*, 1981; Lonsdale, 1985). The sizes of many fungal mitochondrial genomes are listed in Table I. With the exception of ciliated protozoans (Goddard and Cummings, 1975; Goldbach *et al.*, 1979), the slime mold *Physarum polycephalum* (Kawano *et al.*, 1982), and a single yeast, *Hansenula mrakii* (Weslowski and Fukuhara, 1981), all fungal mitochondrial genomes examined are physically circular.

B. Organizational Features

1. Noncoding Regions

The large increase in size of many fungal mitochondrial genomes over that needed to code for their known functions raises the question of the function of the additional DNA.

Several fungal mitochondrial genomes have been largely or completely sequenced, and it is clear that few additional open reading frames (ORFs) exist.

Thus, in some cases, 50 to over 80% of fungal mitochondrial DNA (mtDNA) sequences lack a coding function. Although the lack of a coding function for these sequences does not of itself argue that they are predominantly nonfunctional, or serve a function that is independent of sequence, several observations do point to that conclusion. One is the wide variation observed in mitochondrial genome size among related organisms. For instance, in the genus *Saccharomyces*, *S. cerevisiae* mtDNA is 72–78 kb (Sanders *et al.*, 1977) while *S. exiguus* is 23.7 kb (Clark-Walker *et al.*, 1983). In our own recent work with several basidiomycetes, *Coprinus cinereus* mtDNA size is 43 kb, whereas that of *C. stercorarius* is 91 kb (C. Weber, unpublished observations). Such observations are consistent with the suggestion (Butow *et al.*, 1985) that deletion of large blocks of noncoding sequences takes place in an infrequent, punctuated manner, causing mitochondrial genomes to vary widely in size, and to approach the minimum size necessary to encode the present-day mitochondrial functions.

A second observation is the nature of some of the sequence patterns in the noncoding regions. In *N. crassa* mtDNA, all coding regions are flanked by members of the 50–100 conserved GC-rich palindromic sequences containing double *Pst*I sites (Yin *et al.*, 1981). Although their location is suitable for a role such as that of signaling sites for transcript processing, recent mapping of processing sites does not involve these clusters (Breitenberger *et al.*, 1985). Similar distinctive repeated elements have not been recognized in the sequencing of *A. nidulans*, but possibly analogous GC-rich clusters are well known to be present more than 100 times on *S. cerevisiae* mtDNA (Prunell and Bernardi, 1977). One type of yeast cluster is associated with *Hae*III–*Hpa*II site clusters. Their presence has been particularly well studied in the var1 region, where three of the up to five clusters seen in some strains have turned out to be optional, and thus unlikely to be associated with any fundamental function (Zassenhaus *et al.*, 1983; Hudspeth *et al.*, 1984). We have suggested, rather, that these clusters may be analogous to nuclear *Alu*I or *Kpn*I families in primates (Singer, 1982), which may be species-specific promoters of genomic rearrangement (Butow *et al.*, 1983, 1985). The finding of these clusters so far in only the two largest fungal mitochondrial genomes studied in detail may indicate their involvement in the proposed recombination events leading to genome size reduction, as previously postulated for petite deletion mutants (Bernardi *et al.*, 1980a,b).

2. Repeated Sequences

Repeat sequences increase genome size without affecting coding capacity. Several short direct repeats have been seen in *Aspergillus niger* (Brown *et al.*, 1983a), *N. crassa* (Yin *et al.*, 1982), and in the chloroplast genomes of *Euglena* (Rawson *et al.*, 1978; Helling *et al.*, 1979). They may be uncommon because of their instability to recombination; for instance, a stopper mutant (see Section VIII) of *N. crassa* appears to involve cyclical recombination events at directly

TABLE I.
Sizes of Fungal Mitochondrial Genomes

Taxon	Mitochondrial genome size (kb)	Reference
Myxomycetes		
Physarum polycephalum	69	Kawano et al. (1982)
Oomycetes		
Achlya	49.8–50.7	Hudspeth et al. (1983); Boyd et al. (1984)
Phytophthora infestans	36.2	Klimczak and Prell (1984)
Pythium ultimum	57	Hudspeth (unpublished)
Saprolegnia sp.	44.5	Clark-Walker and Gleason (1973)
Saprolegnia ferax	46.5	Hudspeth (unpublished)
Zygomycetes		
Phycomyces blakesleeanus	25.6	Hudspeth (unpublished)
Ascomycetes		
Aspergillus nidulans	31.5	Stepien et al. (1978)
Brettanomyces custersii	108	Clark-Walker et al. (1981)
Cephalosporium acremonium	27	Minuth et al. (1982)
Claviceps purpurea	45	Tudzynski et al. (1983)

Cochliobolus heterostrophus	115	Garber and Yoder (1985)
Neurospora crassa	60	Bernard et al. (1976)
Podospora anserina	94	Cummings et al. (1979)
Penicillium chrysogenum	48.3–49.2	Saunders et al. (1984); Smith et al. (1984)
Selected Yeasts		
Hansenula mrakii	55	Weslowski and Fukuhara (1981)
Hansenula petersonii	42	Falcone (1984)
Kloeckera africana	27.1	Clark-Walker et al. (1981)
Kluyveromyces lactis	37	Groot and Van Harten-Loosbroek (1980)
Saccharomyces cerevisiae	68–78	Sanders et al. (1977)
Saccharomyces exiguus	23.7	Clark-Walker et al. (1983)
Saccharomycopsis (Yarrowia) lipolytica	44–48	Kuck et al. (1980); Weslowski et al. (1981)
Schizosaccharomyces pombe	19	Del Giudice et al. (1981)
Torulopsis glabrata	18.9	Clark-Walker et al. (1981)
Basidiomycetes		
Coprinus cinereus	43	Weber (unpublished)
Coprinus stercorarius	91	Weber (unpublished)
Schizophyllum commune	50.3–52.2	Specht et al. (1983)
Suillus grisellus	121	Bruns (unpublished)
Ustilago cynodontis	76.5	Mery-Drugeon et al. (1981)

repeated tRNA^{met} genes (Gross et al., 1984). Inverted repeats, however, which are a routine feature of chloroplast DNAs from higher plants (Whitfeld and Bottomley, 1983), have now been found in mtDNA of most oomycetes examined (Hudspeth et al., 1983; Boyd et al., 1984, and unpublished observations) and in an ascomycete (Clark-Walker et al., 1981).

Mitochondrial DNA from the oomycetous water mold *Achlya ambisexualis* was found to contain an approximately 12-kb inverted repeat in its 50-kb genome (Hudspeth et al., 1983). This inverted repeat, like those of all other organellar sources examined, contains the genes for ribosomal RNAs. Examination of the mtDNA population showed the presence of two isomers that differed in the orientation of one unique region with respect to the other. Furthermore, the midpoint of the proposed "flip" event, deduced from restriction analysis, was indistinguishable from the physical midpoints of the unique regions. The results are a predicted consequence of homologous recombination between the repeat arms and suggest that these mtDNAs, in analogy with the yeast 2μ plasmid system (Gerbaud et al., 1979; Broach and Hicks, 1980; Cox, 1983), encode a flipase function.

C. A Note on DNA Isolation

Examination of novel mitochondrial genomes is often hampered by the need to first devise an isolation protocol. We previously introduced a rapid method for isolating yeast mtDNA from a crude mitochondrial pellet by taking advantage of the significant enhancement of buoyant separation in cesium chloride gradients provided by the fluorescent dye bisbenzimide (Hudspeth et al., 1980). We have since found that the original protocol, with relatively minor modifications, works satisfactorily for all fungal mtDNAs to which it has been applied. These include the Oomycetes, Zygomycetes, Ascomycetes, and Basidiomycetes. The major modification has been elimination of enzymatic digestion with Zymolyase as an adjunct to mechanical cell breakage.

III. GENES AND THE GENETIC CODE

A. Protein Structural Genes

All known mitochondrial genomes contain in common a number of genes coding for components of the electron transport chain. These are cytochrome *c* oxidase subunits I, II, and III (COI–COIII), ATPase subunit 6, and apocytochrome *b* (Cyt *b*) (Gray, 1982; Wallace, 1982; Clayton, 1984). In addition, other structural gene products have been identified in particular organisms. In yeast, the Var1 gene codes (Hudspeth et al., 1982) for a protein of the small mitochondrial ribosome subunit (Terpstra et al., 1979). A biochemically cognate but structurally dissimilar protein, S-5, is found in Neurospora mitochondria (Lambowitz et al., 1979; LaPolla and Lambowitz, 1981) and its gene has been provisionally identified and sequenced (Burke and RajBhandary, 1982). A second additional gene, found thus far only in yeast, is ATPase 9 (Macino and Tzagoloff, 1979; Hensgens et al., 1979); in Neurospora and Aspergillus ATPase 9 is a nuclear gene (Sebald et al., 1979; Turner et al., 1979). That result is made more provocative, however, by the finding that an intact, but nonfunctioning, ATPase 9 is also present on Neurospora (Van den Boogaart et al., 1982) and Aspergillus mtDNAs (Brown et al., 1984) (see Section VI, below). Most recently, the gene for ATPase subunit 8 has been identified on yeast mtDNA (Macreadie et al., 1983) and, by sequence similarity, on other fungal mtDNAs, as well as on mammalian mtDNAs as the previously unidentified open reading frame URF A6L.

B. Structural RNA Genes

A second category of genes consists of those for the structural RNAs. All genomes examined contain a small (S-rRNA) and large (L-rRNA) ribosomal RNA gene and at least 22 transfer RNAs (Gray, 1982; Wallace, 1982; Clayton, 1984).

Both rRNA genes have been sequenced in S. cerevisiae (Sor and Fukuhara, 1980, 1983) and A. nidulans (Köchel and Küntzel, 1981; Netzker et al., 1982) and are slightly larger than those in Escherichia coli. Size differences are comprised primarily of contiguous nucleotides operationally considered domains on the basis of discrete regions of secondary structure generated by minimum energy considerations. The yeast L-rRNA gene, like the E. coli but unlike the Aspergillus gene, contains sequences similar to eukaryotic 5.8 S rRNA and 3' sequences similar to chloroplast 4.5 S rRNA. In all cases, significant homology exists between the mitochondrial and E. coli rRNAs, as well as between their derived secondary structures. Based on these and other comparisons, Köchel and Küntzel (1982) have argued for a eubacterial origin of both fungal and animal mitochondria, thus supporting the endosymbiont hypothesis for the origin of organelles (but see Gray and Doolittle, 1982).

At least 24 mitochondrial tRNA genes have been identified in fungi (see, e.g., Köchel et al., 1981; Netzker et al., 1982). This number would be insufficient to recognize all codons in the cytoplasm. In mitochondria two factors reduce the requirement for tRNAs. One is the altered genetic code (see Section III,D), which increases the symmetry of the genetic code table. The second is the absence of a modified U in the first anticodon position of some tRNAs, allowing them to read all four codons of a four-codon family (Heckman et al., 1980).

C. Introns and Unidentified Reading Frames

In addition to the identified protein genes, open but unidentified reading frames (URFs) are present in all of the fungal mtDNAs examined by sequence analysis, as well as in mammalian mitochondria. In *A. niger*, which is nearly completely sequenced, a total of 15 URFs have been found. Three of these are present in introns—the second and third introns of COI and the single intron in Cyt *b*—and may well be maturases (see Section V,B,1). A fourth, URFx, shares 48% amino acid homology (Brown *et al.*, 1983a) with the yeast gene for ATPase subunit 8 (Macreadie *et al.*, 1983) and is located in the same relative position, just preceding the ATPase subunit 6 gene. Four reading frames, URFs 1, 3, 4, and 5, are named in analogy with the four URFs on human mtDNA (Anderson *et al.*, 1981), with which they share significant homology (Netzker *et al.*, 1982; Brown *et al.*, 1983b; Scazzocchio *et al.*, 1983). The absence on yeast mtDNA of ORFs homologous to the human mitochondrial URFs had raised the question of whether the human URFs were not universal, but limited to mammals. The finding of these "mammalian" URFs on fungal mtDNA eliminates this possibility and favors the alternative explanation that at least some mammalian URFs not present on other mtDNAs will be found in the nucleus. Six of the remaining URFs, A1–A6, are thus far unique to *A. nidulans*. However, two of these (A5 and A6) span a region not yet sequenced and may, in fact, be part of the same gene.

The final URF in *Aspergillus* is present as an intron in the L-rRNA gene (Netzker *et al.*, 1982). The L-rRNA gene contains an intron also in *S. cerevisiae* (Dujon, 1980) and *N. crassa* (Burke and RajBhandary, 1982) and all three introns contain an open reading frame not contiguous with the exons and located in the same relative position in each gene. In *S. cerevisiae* the intron, ω, is optional (Dujon, 1980). The derived proteins, however, are not homologous. While the nature of the yeast protein is unclear (but see below), the *N. crassa* intron product has been provisionally assigned as S-5 (Burke and RajBhandary, 1982), a protein of the small mitochondrial ribosome subunit that is functionally analogous to var1 in yeast. The *Aspergillus* intron protein shows considerable homology to S-5, although S-5 shows only limited similarity to var1 in DNA sequence (Burke and RajBhandary, 1982; Jacquier and Dujon, 1982) and both amino acid composition and polarity (Hudspeth *et al.*, 1982). In an extensive search for other yeasts containing the *S. cerevisiae* L-rRNA intron, Jacquier and Dujon (1983) established its presence in all *Kluyveromyces* species tested and in 4 of 25 *Saccharomyces* isolates. As with *S. cerevisiae*, *S. carlsbergensis* isolates contain the intron optionally. Thus, either of two introns, or none, can be present in different fungi at the same position in the large rRNA gene.

In *S. pombe*, which is also completely sequenced, one freestanding reading frame, URFa, has been found (Lang *et al.*, 1983). This URF is not similar to any

3. Fungal Mitochondrial Genomes

other known mitochondrial reading frame, and appears to be a mutator locus unique to *S. pombe* (Seitz-Mayr and Wolf, 1982). In addition, Cyt *b* and COI genes contain a total of three intron ORFs that may be analogous to maturases in other fungal mitochondrial genes.

In *N. crassa*, besides the L-rRNA gene discussed above, ORFs are found in introns of Cyt *b* (Helmer Critterich *et al.*, 1983) and, uniquely, ATPase 6 (Morelli and Macino, 1984). In addition, although not present in laboratory strains, introns in COI have been found in strains isolated from nature (Collins and Lambowitz, 1983). In Cyt *b*, intron 2 (I2) contains an ORF not homologous to any other fungal ORF examined. Furthermore, both it and I1 are located in different positions in the Cyt *b* gene from the introns in *A. nidulans* and *S. cerevisiae*, which are found at the same position. In the ATPase 6 gene, I2 contains a large ORF.

Saccharomyces cerevisiae contains 6 to 14 ORFs, with all but 4 of the 14 present in the introns of identified genes. In addition to the L-rRNA intron ORF, intron reading frames are found in the COI and Cyt *b* genes. Both COI and Cyt *b* are found in laboratory strains as so-called short or long forms, differing in the number of introns they contain. COI contains seven introns in the long form (Bonitz *et al.*, 1980b; Hensgens *et al.*, 1983a); the first six contain ORFs, of which the first five are contiguous and in phase with their upstream exons. The short form lacks two introns that split exon 5 and thus contain four ORFs. Cyt *b* contains either five introns, of which I2–I4 contain long ORFs contiguous with their upstream exons, or two introns. In the short form, I2 is equivalent to I4 of the long form and thus contains an ORF.

Four freestanding ORFs are also present in some strains of *S. cerevisiae*—one near COII (Coruzzi *et al.*, 1981), one near COIII (Michel, 1984), and two near ATPase 6 (Macino and Tzagoloff, 1980; Grivell, 1982; Baldacchi *et al.*, 1984). Each of the ORFs near COIII (Thalenfeld *et al.*, 1983) and the ORF closest to ATPase 6 (Cobon *et al.*, 1982) is known to be lacking in at least one strain. Transfer of mitochondria lacking the ORF near ATPase 6 into a ρ° strain of *S. cerevisiae* produced respiratory competent cells, showing that ORF is not required for oxidative metabolism (P. S. Perlman, personal communication).

Current evidence indicates that at least 11 of the 13 *S. cerevisiae* introns are optional. Together with the observations discussed above that the nature, number, and location of introns can differ in different organisms, it is not difficult to speculate that the presence of introns is, in general, unrelated to gene function. Nevertheless, we can ask what functions are carried out by intron ORF products. These fall into several categories. One consists of the *Aspergillus* and *Neurospora* L-rRNA intron products, which appear not to be optional, and to be a mitochondrial ribosomal protein. A second category is the L-rRNA intron product in *S. cerevisiae* and other yeasts. Recent evidence (Zinn and Butow, 1985) has provided a good indication that it has a transposition function, acting in the conversion of ω^- to ω^+ in a manner similar to the involvement of HO endonuclease in the initiation of mating type switching in yeast (Kostriken et al., 1983). A third category, and the largest, is that of maturases (Lazowska et al., 1980; De La Salle et al., 1982; Anziano et al., 1982; Carignani et al., 1983; Weiss-Brummer et al., 1982, 1983; Mahler, 1983), *trans*-acting proteins encoded by some of the intron ORFs involved in splicing primary transcripts. Finally, a number of ORFs, such as the mammalian homologues found in *A. nidulans*, have no function associated with them at present.

D. Genetic Code Variation

Variations from the "universal" genetic code in mitochondria, first demonstrated in mammals (Barrell et al., 1979), are now well established. In fungi, examination of the genetic code has taken place only in *Saccharomyces* and *Neurospora*. The results, however, have been startling. As in all other mitochondrial genomes, TGA encodes tryptophan rather than (opal) translation termination. In *Neurospora*, that represents the only difference from the universal code (Browning and RajBhandary, 1982). In yeast two other differences are present: (1) the CTN family represents threonine rather than leucine (Bonitz et al., 1980a), an assignment found nowhere else, and (2) ATA represents methionine rather than isoleucine (Hudspeth et al., 1982), as in animal mitochondria. Unlike animal mitochondria (Barrell et al., 1980), the fungal genomes studied do use TGA and TGG as sense codons, apparently with the universal meaning of arginine (Bonitz et al., 1980a). TGA is a serine codon in *Drosophila* mitochondria (de Bruijn, 1983).

IV. GENE ORDER

Gene order among fungal mitochondrial genomes is surprisingly diverse. This is reflected by Table II, which presents the gene order for some known fungal mitochondrial DNAs. The table, which consists of members of the class Ascomycetes, shows that, exclusive of tRNA cistrons and URFs, the order of genes is quite plastic, even at the euascomycete and hemiascomycete subclass levels. It is difficult, however, to assume that gene orders can be random, as the features of coding strand bias and polycistronic transcripts are common to both the yeasts and the filamentous fungi and presumably must be maintained. Furthermore, the order of known protein-encoding and rRNA genes of all organisms listed in Table II can be interconverted by postulating four or fewer recombinational events. For example, in the most complex case, involving nine genes, the gene order of the yeast *Torulopsis* can be converted into the order of the distantly related *S. cerevisiae* as follows: (1) inversion of the COII–COIII–S-rRNA block;

TABLE II.

Gene Order of Selected Mitochondrial DNAs

Species	Gene order	Reference
A. nidulans	L-rRNA COIII S-rRNA ATPase 6 ATPase 8 Cyt b COII ATPase 9 COI	Scazzocchio et al. (1983)
N. crassa	L-rRNA COIII S-rRNA COII ATPase 9 ATPase 6 COI Cyt b	Macino (1980); Van den Boogaart et al. (1982)
P. anserina	L-rRNA S-rRNA COIII COI Cyt b COII ATPase 6	Wright et al. (1982)
S. cerevisiae	L-rRNA COII COIII S-rRNA COI ATPase 8 ATPase 6 Cyt b ATPase 9 Var1	Dujon (1981)
S. exiguus	L-rRNA S-rRNA COIII Cyt b COI ATPase 6 ATPase 9 COII	Clark-Walker et al. (1983)
T. glabrata	L-rRNA Var1 S-rRNA COIII COII ATPase 9 ATPase 6 COI Cyt b	Clark-Walker and Sriprakash (1983)
S. pombe	L-rRNA S-rRNA COI COIII Cyt b ATPase 6 ATPase 8 ATPase 9 COII	Ahne et al. (1984)

(2) inversion of the Var1–L-rRNA block or transposition of either gene; (3) inversion of the COI–ATPase 6 block or transposition of either gene; and (4) transposition of ATPase 9.

The key to the flexibility of ascomycete mitochondrial gene order may lie in the overall organization of the genomes. As noted above, stretches of relatively homologous GC-rich nucleotide sequences are dispersed throughout the genomes of both the Euascomycetes and the Hemiascomycetes. These sequences, many of which are optional in different laboratory strains, appear either in tandem or inverse orientation and frequently occur flanking genes in the relatively long (vis-á-vis animal mtDNA) intergenic sequences. These homologous sequences may serve as sites for intra- or intermolecular recombination in a manner that can preserve strand bias. Should the recombinational event occur in which transcription takes place from a new promoter, then the new gene order should be viable. Indeed, it has already been demonstrated in a rearranged *S. cerevisiae* petite that cytochrome *b* is transcribed from the ATPase 9 promoter (Dieckmann *et al.*, 1984a).

A disturbing feature of the proposed explanation of the plastic fungal mitochondrial genome is the apparent absence (other than in *S. cerevisiae* petites) of intraspecific gene order rearrangements. This may simply reflect the fact that laboratories studying mitochondrial genomes usually study fewer than five strains within any particular species. Furthermore, unless a rearrangement conferred a competitive advantage, it would remain diluted within its original population until the formidable task of analyzing significant numbers of individual isolates was undertaken. This task is made even more formidable upon consideration of the known genome homogenization mechanism, which occurs between the multiple genomes in each *S. cerevisiae* mitochondrion. To be detectable, a rearranged genome would have to escape or dominate the homogenization mechanism. Thus, until a great number of individual isolates has been examined we will be unable to assess the rate of gene order change and its true plasticity.

V. TRANSCRIPTION

A. Promoters

Unlike higher animal mitochondrial genomes, where a single primary transcript of the entire genome is subsequently processed (Clayton, 1984), fungal mitochondria use multiple promoters. That was first suggested by the observation *in vitro* that the yeast ribosomal RNA genes are preferentially transcribed, although they are located in nearly opposite positions on the 75-kb mitochondrial genome (Levens *et al.*, 1981a). It is now clear that at least 19 promoters exist

(Edwards et al., 1983a); however, each gene is not separately promoted, and primary transcripts of several genes are processed, as in animal cells, to yield mature messages or stable RNAs (Zassenhaus et al., 1984; Cobon et al., 1982; Christianson et al., 1983; Thalenfeld et al., 1983; Locker and Rabinowitz, 1981).

Two complementary approaches, both developed in Rabinowitz's laboratory, have defined most of the primary transcripts in *S. cerevisiae* mitochondria. One was the introduction of the use of vaccinia virus guanylyltransferase. This enzyme transfers GMP to the di- or triphosphate terminus of a polynucleotide. If such ends on mitochondrial RNA represent the initiating triphosphate, then use of [γ-^{32}P]GTP allows guanylyltransferase to selectively label primary transcripts (Levens et al., 1981b).

The second approach was *in vitro* transcription using purified mitochondrial RNA polymerase (Edwards et al., 1982). Since a number of primary transcripts could be mapped by S1 protection experiments to long AT-rich regions, *in vitro* transcription of sequenced regions could be used to determine the initiation point of transcription. This was done by omitting CTP or GTP and counting back from the stopping point by the length of the abbreviated transcript.

Study of the two rRNA genes (Osinga and Tabak, 1982), three protein genes [COI (Osinga et al., 1984), Var1 (Zassenhaus et al., 1984), and ATPase 9 (Edwards et al., 1983b)], four tRNA genes [thr1 (Osinga et al., 1984), phe (Christianson and Rabinowitz, 1983), fmet (Miller et al., 1983), and glu (Christianson et al., 1983)], and four replication origins (Osinga et al., 1982, 1984) has shown that the start of transcription of each takes place immediately after a common consensus sequence element, the nonanucleotide ATATAAGTA. For instance, for COI messenger, nuclease protection places the initiating nucleotide 540 bp from the AUG start. When a cloned fragment containing this start site is used as the template for *in vitro* transcription with mitochondrial RNA polymerase, transcript sizes are predictable in both a runoff assay and an assay performed in the absence of CTP. Both approaches show that transcription initiation occurs within this nonanucleotide sequence.

The four tRNA transcripts all contain a nonanucleotide initiation site that differs by one base from the consensus, whereas the ribosomal and protein genes and the replication origins all match it exactly. The variant initiation sequence, if it holds for other tRNA genes, may offer a way to regulate tRNA gene transcription from that of other transcripts. Indeed, the initiation sequence found in front of the tRNA^{cys} gene (Bos et al., 1979) does not appear to be used (Frontali et al., 1982).

That the nonanucleotide sequence is used in promotion of transcripts is further supported by the finding that it is found in the same position with respect to the start of the rRNAs in the distantly related yeast *Kluyveromyces lactis* (Osinga et al., 1982). Furthermore, the finding in *S. cerevisiae* of the consensus nonanucleotide in front of tRNA, rRNA, and protein genes, as well as replication origins, argues that, in contrast to the nucleus, a single RNA polymerase is responsible for all transcriptional classes.

Despite the correlative and *in vitro* transcription evidence that the conserved nonanucleotide sequence functions as a promoter, several preliminary observations indicate that the situation is, in reality, more complex (Tabak *et al.*, 1983). One indication comes from *in vitro* transcriptional studies of mutated templates. In one case analyzed, a G → T substitution within the nonanucleotide box had no effect on the specificity or the efficiency of transcription, whereas a double substitution at +10 and +13 with respect to the end of the box abolished transcription. A second observation was that cloned rRNA genes of *K. lactis* do not initiate correctly with an *S. cerevisiae* enzyme preparation, although *S. cerevisiae* rRNA genes do initiate correctly. This observation held even when the two rRNA genes were combined on the same plasmid. Since each contains the identical nonanucleotide box, other specific DNA sequences or protein factors may exist. Lastly, some genes contain tandem copies of the nonanucleotide box, yet only the upstream box is used at significant levels.

B. Message Processing and Splicing

Two types of transcript processing take place in fungal mitochondria. The first is the processing of primary transcripts to produce single gene messages and stable RNAs. The second is splicing—the removal of introns from the transcripts in which they are present.

1. Processing of Primary Transcripts

In *S. cerevisiae*, where transcription has been studied in most detail, a number of genes are known to be processed from multigene primary transcripts. The major large transcripts identified contain L-rRNA–tRNA^{thr} (Locker and Rabinowitz, 1981), tRNA^{val}–COIII (Thalenfeld *et al.*, 1983), ATPase 9–tRNA^{ser}–Var1 (Zassenhaus *et al.*, 1984), tRNA^{glu}–Cyt *b* (Christianson *et al.*, 1983), tRNA^{fmet}–9 S RNA–tRNA^{pro} (Miller *et al.*, 1983), and COI–ATPase 8–ATPase 6 (Cobon *et al.*, 1982). In most cases tRNA genes are part of these large primary transcripts and their release as mature tRNAs requires processing at both their 5′ and 3′ ends.

Processing of the 5′ ends of tRNAs requires the presence of a 9 S RNA encoded by mtDNA between tRNA^{fmet} and tRNA^{pro} (Miller and Martin, 1983; Underbrink-Lyon *et al.*, 1983; Miller *et al.*, 1983). Petite mutants lacking this locus can produce mitochondrial tRNAs whose 3′ ends are correctly processed and can be charged, but whose 5′ ends are extended (Frontali *et al.*, 1982; Miller *et al.*, 1983). A number of analogies have been drawn between the 9 S RNA of this "tRNA processing locus" and the RNA moiety of *E. coli* RNase P, including the ability of RNase P to process the precursor found in petite mutants lacking the 9 S locus (Miller *et al.*, 1983; Miller and Martin, 1983). If the analogy is correct, the protein portion of the tRNA processing complex would presumably be a nuclear product; so too, clearly, would be the functions needed to produce mature 3' ends.

Perhaps the most detailed study thus far of the processing of mitochondrial primary transcripts has been carried out in the ATPase 9–tRNA^{ser}–Var1 region (Farrelly *et al.*, 1982; Zassenhaus *et al.*, 1984). In that case, a polycistronic primary transcript (as judged by *in vitro* capping) of the three genes is initiated from one of two copies of the nonanucleotide promoter, located about 550 and 630 bp 5' to the ATPase 9 gene; the most distal one is probably the predominant site. Processing releases ATPase 9, a precursor to tRNA^{ser}, and a 19 S species that is further processed to the 16 S Var1 mRNA, as well as poorly understood 13 S and 14 S species that do not contain complete Var1 coding sequences.

The processing sites for all of those species have been determined. One type appears sequence-specific and examples are found at the 3' ends of the mature ATPase 9 and Var1 messages. A different sequence-specific processing site is found at the 5' ends of mature Var1 message (16 S). The second specific sequence, unlike the first, is also found, but not used, a number of other times in the Var1 region, indicating that the sequence information alone is not adequate to serve as a processing site.

The second type of processing sites involve the tRNA^{ser} locus and appear to depend on secondary structure rather than specific sequence. This conclusion for the 3' end rests on the observations that no consensus sequence is apparent for the 3' ends of mitochondrial rRNAs and that correct processing takes place in petite mutants with non-wild-type sequences 3' to the cleavage site. However, the non-wild-type sequences start some distance away, and the hypothesis has not been critically tested in a case where such sequences start immediately 3' to the processing site. Similar observations hold for the 5' end, although a mature end is not generated in the absence of the tRNA synthesis locus. The steps that do occur can do so with non-wild-type sequences as close as 180 bp.

2. Intron Splicing

The mitochondrial genes for L-rRNA, Cyt *b*, COI, and ATPase 6 can contain introns in various fungi. These introns have a number of unusual features compared to traditional introns in nuclear protein-coding genes. First, they do not obey the GT..AG rule, a feature found also in the nuclear tRNA genes of *S. cerevisiae,* in chloroplast tRNA and rRNA genes, and in protozoan rRNA genes (reviewed in Michel and Dujon, 1983). Second, some of the introns have open reading frames that code for *trans*-acting factors (Lazowska *et al.*, 1980), called maturases (Jacq *et al.*, 1980), essential for RNA splicing. Third, at least in *S. cerevisiae*, a minimum of 11 of 13 possible introns is optional (COI introns 3 and 5 have not been critically tested) and therefore cannot be required for any function except that of splicing other introns.

Introns have been divided into two classes based on their sequence homologies and potential secondary structures (Michel *et al.*, 1982; Hensgens *et al.*, 1983a; Michel and Dujon, 1983). Both groups contain members that do and do not have ORFs. The most common class, group I, includes all known mitochondrial introns except I1, I2, and I5 (I5 in the long form) of *S. cerevisiae* COI, I1 of *S. cerevisiae* Cyt *b*, and the *Zea mays* COII intron. Group I introns contain the conserved sequence elements P, Q, R, and S, which have been proposed to direct the folding of these introns into conserved secondary structures (Davies *et al.*, 1982, 1983; Michel and Dujon, 1983) by pairing of R and S and of P and Q. The 5' and 3' splice junctions are brought near to each other in this conserved or "core" structure. This model is supported by the isolation of cis-dominant mutations and their revertants that disrupt and restore, respectively, base pairing in the core structure (Michel and Dujon, 1983; Rödel *et al.*, 1983). Not all mitochondrial introns, however, are processed by mitochondrial maturases. For instance, the *S. cerevisiae* L-rRNA intron can be processed in petite mutants, indicating that its translation product is not required for splicing (Faye *et al.*, 1974, 1975; Tabak *et al.*, 1981). Apocytochrome *b* I1 and I5 can also be processed in petites and therefore do not require mitochondrial gene products (Halbreich *et al.*, 1980; Bonitz *et al.*, 1982).

Group I intron elements are also found in some nuclear and chloroplast stable RNA genes. The most unusual example is the large rRNA gene in *Tetrahymena*, where the intron has been shown *in vitro* to be spliced nonenzymatically by a proposed transesterification mechanism (Cech *et al.*, 1981; Kruger *et al.*, 1982; Zaug *et al.*, 1983). Cech *et al.* (1983) have pointed out that the secondary structure of the *Tetrahymena* rRNA intron shares conserved sequence elements with mitochondrial group I introns, and, recently, a transcript of the *N. crassa* mitochondrial large rRNA gene has been shown to undergo self-splicing *in vitro* (Gariga and Lambowitz, 1985). Whether these reactions are enzyme-catalyzed *in vivo* is not at all clear. It is possible that maturases function to aid the correct folding of group I introns for nonenzymatic splicing. In any case, the conservation of both secondary structure and sequence elements in these introns suggests that they share a common splicing mechanism.

The group II introns are also related. All four fungal introns are excised as circular RNAs (Arnberg *et al.*, 1980; Hensgens *et al.*, 1983b; Halbreich *et al.*, 1980), presumably through a similar mechanism. They contain a consensus sequence homology at their 3' ends (past the ORF, when present) that has the potential for a distinct secondary structure (Michel and Dujon, 1983).

VI. PSEUDOGENES

The nuclear and mitochondrial genomes interact to provide the subunits of functional mitochondrial complexes. This finding has implied, but not, of course, proved, that each organelle contains different coding sequences that, in aggregate, specify the mitochondrial complexes. In the past year and a half, however, it has become clear that this view is too simple, and that nucleus, mitochondria, and chloroplasts can all provide sequence information to each other.

Given the popularity of the endosymbiont hypothesis for the origin of organelles, it is not, in retrospect, surprising that cells contain mechanisms for transferring genetic information between compartments, since such transfer presumably accounts for the nuclear location of genes whose products interact with organelle products. Reports have now appeared documenting the presence in nuclear DNA of mitochondrial sequences in yeast (Farrelly and Butow, 1983), locust (Gellissen *et al.*, 1983), sea urchin (Jacobs *et al.*, 1983), maize (Kemble *et al.*, 1983), rat (Hadler *et al.*, 1983) and *Podospora* (Wright and Cummings, 1983). In addition, two other pairwise combinations have been observed: chloroplast DNA is found in the nucleus (in spinach) (Timmis and Scott, 1983) and in mitochondria (in maize) (Stern and Lonsdale, 1982).

The question of the direction of transfer of sequences between nucleus and mitochondria is difficult to approach. On the one hand, mitochondrial sequences may have been transposed to the nucleus, possibly by a mechanism first used in the establishment of proposed endosymbionts. On the other hand, any particular nuclear sequence could have been transferred to mitochondria. In either case, if the sequence movement is a copy operation, the copy that is not used would not be under selective pressure and its resemblance to the functional copy ought to diminish with time. This expectation appears true in some cases and untrue in others.

Two clear cases of pseudogenes have been analyzed at the sequence level. In yeast, 289 bases of nuclear DNA contained the total nuclear homology to mtDNA. That cloned fragment was composed of sequences homologous to a replication origin region, and to regions of Cyt *b* and Var1, and was shown by blot hybridization to contain yeast transposable (Ty) elements (Farrelly and Butow, 1983). These observations led to the suggestions that the organization of these sequence fragments was achieved by the reductional recombination that commonly forms petite mtDNA, and that the transfer of the proposed petite to the nucleus may have involved the direct participation of the Ty elements. The time of transfer was estimated from the nuclear and mitochondrial sequence divergence to have been recent—about 25 million years ago. This number probably represents an upper limit, since two factors about which little information exists—the degree of polymorphism of present-day yeast mitochondrial genes, and the possible rapid evolution of yeast mtDNA, analogous to that found in animal cells (Brown et al., 1979)—could not be considered. Independent of the time and mechanism of transfer, the presence in the nucleus of rearranged fragments of functional mitochondrial genes indicates that the direction of transfer was from mitochondria to nucleus. A similar conclusion about the direction of transfer is possible for a rearranged nuclear homologue of sea urchin mtDNA (Jacobs et al., 1983).

The direction of transfer is far less clear in perhaps the most interesting situation, the presence of the gene for ATPase subunit 9 in both the nucleus and mitochondria of *Neurospora* (Van den Boogaart et al., 1982) and *Aspergillus* (Brown et al., 1984). The nuclear gene is the one expressed exclusively, at least under active growth conditions, based on several lines of evidence: (1) in both *Aspergillus* and *Neurospora*, synthesis of ATPase subunit 9 is blocked by cycloheximide but not chloramphenicol (Turner et al., 1979; Sebald et al., 1979); (2) oligomycin resistance shows nuclear segregation in *Neurospora;* (3) the derived amino acid sequence of the *Neurospora* mitochondrial reading frame does not correspond to the sequence derived from purified subunit 9 protein or its mRNA (Viebrock et al., 1982); and (4) no transcript could be found in *Neurospora* for the mtDNA open reading frame for subunit 9 (Van den Boogaart et al., 1982).

Despite the strength of the evidence discussed above, the presence of an undamaged reading frame for subunit 9 in both *Neurospora* and *Aspergillus* mtDNA (whose derived proteins are 84% homologous) is too remarkable to dismiss yet. The possibility that these reading frames are undamaged because of a very recent transfer from the nucleus appears unlikely on two counts: (1) the *Neurospora* protein and derived mitochondrial amino acid sequence differ at 29 of the 73 positions where they can be aligned, and (2) unless the transfer of nuclear sequences to *Neurospora* and *Aspergillus* took place as independent events, it must have occurred before their divergence from a common ancestor.

Indeed, the mitochondrially derived amino acid sequences resemble expressed subunit 9 proteins more closely than does the *Neurospora* nuclear protein. When the amino acid sequences for the derived *Neurospora* and *Aspergillus* proteins are aligned, along with the yeast, bovine, and expressed *Neurospora* protein sequences, 28 positions are identical in all cases. At an additional 19 positions, four of the five sequences agree. Interestingly, the predicted *Neurospora* mitochondrial sequence does not differ from the consensus at any of these 19 positions and the predicted *Aspergillus* sequence does so only once, whereas the expressed *Neurospora* protein differs eight times.

An additional point of interest, noted by Van den Boogaart et al. (1982), is that the complementary strand of the *Neurospora* mitochondrial ATPase 9 reading frame also contains an open reading frame of 61 amino acids, starting with methionine. We thought it interesting to evaluate whether the potential exists for nonsense mutations in either reading frame (by a single base change) that would

3. Fungal Mitochondrial Genomes

be silent in the other. We found that, of 17 potential stop codons in the ATPase 9 reading frame, 8 would be silent for the complementary strand reading frame. Similarly, 5 of 28 in the latter would be silent for ATPase 9. A similar analysis holds for the *Aspergillus* mtDNA sequences presented, although the analogous complementary strand reading frame protein, which is 57% homologous, does not start with methionine.

More recent work (de Vries *et al.*, 1983; E. Agsteribbe, personal communication) indicates that minor transcripts of the ATPase 9 region exist and that their size is related to growth phase. Thus, the function of this mitochondrial coding region is still an open question. In any case, the significantly greater homology between the *Neurospora* mitochondrial ATPase reading frame and the bovine protein raises the question of whether the ATPase 9 gene in the line of descent leading to animals came from an independent transfer of the mitochondrial gene like the one seen in *Neurospora* and *Aspergillus* nuclei.

Although the presence of mitochondrial sequences in the nucleus has been widely—almost commonly—reported, there has not, in general, been any reason to suppose that this movement of DNA serves any short-term biological purpose; rather, one supposes that such gene transfers make infrequent and accidental use of existing cellular mechanisms. However, recent work with *Podospora* indicates that the transposition of specific mitochondrial sequences to the nucleus accompanies cellular senescence. *Podospora anserina* is characterized by race-specific timing of senescence, which is associated with excision and amplification of discrete mitochondrial sequences; the most common excised sequences contain the genes for subunits I and III of cytochrome *c* oxidase (see Section VIII,E). These mitochondrial sequences are absent from nuclei of young cells but present in the nuclei of senescent cells (Wright and Cummings, 1983). It is not clear, however, whether this transposition is mechanistically connected to senescence, or results only because the high concentration of autonomously replicating excised mitochondrial sequences provides an efficient substrate for transposition systems.

VII. MITOCHONDRIAL PLASMIDS

Among the increasing number of reports documenting the presence of plasmids in fungal cells, intramitochondrial plasmids have been demonstrated in only five genera. These are the euascomycetes *Neurospora, Claviceps, Cochliobolus, Aspergillus,* and *Podospora.* In all instances the presence of the plasmids is optional. The *Aspergillus* and *Podospora* plasmids, however, are directly derived from the mtDNA and are discussed in Section VIII. (See also Chapter 16 by Tudzynski and Esser.)

Three discrete *Neurospora* plasmids, isolated from *N. crassa* and *N. intermedia* strains, have been characterized to various extents. All exist in monomeric and head-to-tail multimeric forms. The *N. crassa* (Mauriceville 1c strain) plasmid monomeric length is 3.6 kb (Collins *et al.*, 1981) and the two *N. intermedia* plasmid monomeric lengths are 4.1–4.3 kb (Labelle strain) and 5.2–5.3 kb (Fiji strain) (Stohl *et al.*, 1982). Hybridization experiments failed to show any significant sequence homology between any of the plasmids or with the host mitochondrial genome. Sequencing of the Mauriceville plasmid, however, has revealed the presence of the canonical *N. crassa* mtDNA 18-nucleotide *Pst*I palindromic sequence (Nargang *et al.*, 1983, 1984). This same plasmid, unlike those in *N. intermedia*, also hybridizes to a specific transcript unique to the plasmid-containing strain. Analysis of the large ORF present (710 codons in the Mauriceville plasmid) shows a number of features typical of mitochondrial group I introns, such as type of codon usage and conserved sequence elements. On this basis, these plasmids have been suggested to be mobile elements that are or were the progenitors of mtDNA introns (Nargang *et al.*, 1984). The *Neurospora* plasmids are described in greater detail by Kinsey in Chapter 9.

Preliminary characterizations of plasmids have been reported in *Claviceps* and *Cochliobolus*. In the ergot fungus *Claviceps purpurea*, two linear plasmids of 5.3 and 6.6 kb have been detected (Tudzynski *et al.*, 1983). Hybridization experiments have indicated DNA sequence homology between the two plasmids and, at least in the larger plasmid, an absence of significant homology with the mitochondrial genome. A single isolate of *Cochliobolus heterostrophus* contains a multimeric series of a circular plasmid of 1.9-kb monomeric length (Garber *et al.*, 1984). This plasmid hybridizes to a specific site in the mitochondrial genome of both the plasmid-containing isolate and all non-plasmid-containing isolates. To date, there appears to be no effect of the plasmid on the integrity of the mitochondrial genome, nor does any sequence homology appear to exist between this plasmid and those found in *Neurospora* and *Podospora*.

VIII. EFFECTS OF ALTERATION OF mtDNA

Alterations in the DNA content of fungal mitochondria produce respiratory deficient phenotypes. The most studied of these are the petite mutation of *S. cerevisiae*, the *poky* and stopper mutations of *N. crassa*, the ragged mutation of *Aspergillus amstelodami*, and senescence in *Podospora anserina*. Morphologically, petites and pokys are characterized as small, slowly growing colonies, stoppers as undergoing cyclical or stop–start periods of growth, raggeds as showing uneven colonial morphology on solid medium, and senescence as limited vegetative growth.

A. Petite Mutations

The petite mutation in *S. cerevisiae*, originally described by Slonimski and Ephrussi (1949), is the most classical of the respiratory deficient phenotypes dealt with here. As petites have been recently reviewed (Dujon, 1981) and as they encompass a vast body of literature, we have limited ourselves to a brief discussion of petite genome organization and the utility of petites in genetic studies.

Mitochondrial genomes in petites result from large deletions of the wild-type genome such that less than one-third of the DNA is usually retained. These retained sequences, which may be as small as 66 base pairs, can represent any region of the genome and thus a series of overlapping petites of all mtDNA sequences can be found. Furthermore, the retained sequences are usually amplified—as either direct or inverted tandem repeats—to the extent that the concentration of mtDNA appears to be close to that of the wild-type DNA. The overall result is a "natural" cloning system providing a source of specific mtDNA sequences, whose utility has been exploited for genetic mapping and DNA sequencing.

The proposed mechanism of petite genome formation is a reductional recombination at short direct repeat sequences contained within the GC-rich clusters scattered on the genome, or at the AT-rich regions commonly found in intergenic regions (Faugeron-Fonty *et al.*, 1979; Bernardi *et al.*, 1980a,b).

B. *poky* Mutations

The *Neurospora poky* mutation, reviewed by Gilham (1978), is in part characterized by deficiencies of small ribosomal subunits (Rifkin and Luck, 1971). These deficiencies are suggestive of defects either in the mitochondrially encoded small ribosomal protein, S-5, or of the S-rRNA. Nevertheless, two-dimensional gel electrophoresis of *poky* S-5 and RNA fingerprinting and gel electrophoresis of *poky* S-rRNA revealed no apparent defects (Lambowitz *et al.*, 1979; Collins *et al.*, 1979). However, DNA sequence analysis of complementation group I mtDNA of several extranuclear mutants of the *poky* phenotype has now revealed that all mutants differ from the wild-type only in a 4-bp deletion (Akins and Lambowitz, 1984). The location of the deletion is just inside the 5' end of the S-rRNA and involves a putative processing site that destabilizes a potential hairpin structure by a factor of almost 4. This apparently results in the production of aberrantly processed S-rRNAs missing 38–45 nucleotides from the 5' end and thus substantially reduces mitochondrial translation efficiency.

C. Stopper Mutations

The stopper phenotype in *N. crassa* is a result of either insertions or deletions in the mitochondrial genome. Two well-characterized insertions of 1.9 kb (Collins and Lambowitz, 1981) and 2.2 kb (Mannella *et al.*, 1979a) contain overlapping sequences that appear as head-to-tail tandem repeats, with the larger having a reiteration frequency of up to eight. In the case of the 2.2-kb insert, defective molecules represent only one-half of the mtDNA population; the remaining molecules are apparently unaffected. A third insert of 4 kb, which does not overlap the others, has also been described (Bertrand *et al.*, 1980). All of these inserts have been localized to the approximately one-third of the genome containing the rRNA–tRNA genes.

Both small and large deletions have also been correlated with stopper mutants. Two small deletions of 7–8 bp and 75 bp have been located approximately 180° from the aforementioned insertions and appear to be genetically inseparable from the 2.2-kb insert (Mannella *et al.*, 1979b). However, a causal relationship between these deletions and the insertion has not been established. Other small deletions of 0.35 and 5 kb (Bertrand *et al.*, 1980) have been physically mapped as nonoverlapping, but, like the other deletions, lie outside the rRNA–tRNA region.

Large deletions representing less than one-half of the genome appear as a predominant feature of many stopper mtDNAs. Furthermore, the retained portion of the DNA invariably extends throughout the rRNA–tRNA region. In many instances in which large deletions are evident, the majority of the remaining portion of the genome is present but greatly underrepresented (Mannella *et al.*, 1979a; Bertrand *et al.*, 1980; de Vries *et al.*, 1981; Gross *et al.*, 1984). For two cases (de Vries *et al.*, 1981; Gross *et al.*, 1984) it has been shown that, although the molecular types complement each other, the full complement of sequences is not present. These totally deleted sequences lie in a common area outside the rRNA–tRNA region. The most recent work (Gross *et al.*, 1984) established the presence of both 21- and 43-kb circles, which account for 80% of the genome. The cyclical phase in this case is accompanied by the increased, but never predominant, appearance of the larger circle. It is postulated that the dual population of molecules is due to recombination between the molecules, each of which contains a copy of the direct repeat involving the tRNA^{met} locus.

D. Ragged Mutations

The ragged mutants of *A. amstelodami* are accompanied by the excision and head-to-tail amplification of mtDNA sequences (Lazarus *et al.*, 1980; Lazarus and Küntzel, 1981). The excision event originates from one of two regions of the mitochondrial genome, denoted region 1 and region 2, and results in a population of DNAs consisting of intact mitochondrial genomes, genomes with a deleted sequence, and plasmids containing head-to-tail amplifications of the deleted sequence. The single isolate from region 1 contained a plasmid with a 900-bp head-to-tail amplification derived downstream from the L-rRNA cistron. Five mutants contained plasmids with amplified head-to-tail sequences of 1.5–2.7 kb located within URFx and extending into ATPase 6. These latter mutants, which define region 2, all contain a 215-bp sequence that, in part, forms a hairpin structure in the intact mtDNA resembling the origins of replication of both *Saccharomyces* and mammalian mtDNAs. One of the region 2 plasmids arose from the original region 1 isolate and eventually displaced the former plasmid. Such displacement not only strengthens arguments of relative replicative efficiencies of plasmids but also supports speculation that the hairpin structure in region 2 is a primary replication origin in *Aspergillus* mtDNA.

E. Senescence in *Podospora*

Major inroads have been made toward elucidating the involvement of *Podospora* mtDNA in senescence. The senescence appears to involve the generation of several distinct mitochondrially derived plasmids (senDNAs). As these plasmids originate from different regions of the nonsenescing juvenile mitochondrial genome with different frequencies, they have been distinguished as α-event, β-event, etc., senDNAs.

Race-specific correlations have been established between the physiological onset and progression of senescence and the relative proportion of mitochondrially derived plasmids to intact mitochondrial genomes. In juvenile cultures the α-event plasmid ratios of 1 : 20 for the slowly senescing s^+ and 1 : 1 for the rapidly senescing A^+ increase to 1 : 1 for s^+ and 20 : 1 for A^+ in senescent cultures. Furthermore, the senescent state can be accelerated by the transformation of juvenile protoplasts with senescent plasmids (Tudzynski et al., 1980), and senescing cultures can be rejuvenated by the use of chemical or physical factors that affect the rate of mtDNA replication (ethidium bromide, chloramphenicol, low temperature, etc.) (Tudzynski and Esser, 1977, 1979; Belcour and Begel, 1980; Koll et al., 1984). The latter rejuvenation, at least in the case of ethidium bromide, results in a corresponding decrease in plasmid-to-intact-mtDNA ratios.

At least five plasmid types have been identified, and all hybridize to intact mtDNA within, or adjacent to, transcriptionally active regions (Wright et al., 1982). The most frequently encountered and most studied plasmid, α-event DNA, originates from the COI locus, whereas β-event, γ-event, and δ-event DNAs originate from the COIII, L-rRNA, and COII loci, respectively. The θ-event DNA hybridizes to a region of the mtDNA whose transcript is unidentified.

The nucleotide sequence of α-event DNA and its flanking regions within the mtDNA has been determined (Osiewacz and Esser, 1983). Within the 2500 bp of the plasmid a URF exhibiting reasonable homology to those encoded by introns 1 and 2 of the *S. cerevisiae* COI gene has been found. The flanking regions in the mtDNA indicate that the excision event occurs between two nonidentical 10-bp palindromes (Cummings and Wright, 1983). However, no information is available on the other plasmids to indicate whether similar palindromes are involved in the excision event.

The most intriguing observation related to senescence is the transfer of mitochondrial nucleotide sequences to the nuclear genome during senescence. In the two examples reported, both α-event and β-event sequences are undetectable in the juvenile nuclear genomes but appear as integrated sequences during senescence. The α-event transfer is race-specific, with the rapidly senescing A^+ race containing a higher copy number of multimeric head-to-tail nuclear sequences than the slowly senescing s^+—an observation consistent with the relative proportions of α-event DNAs in juvenile cultures. A unique exception to this α-event transfer involves the nonsenescing *mex-1* mutant, whose intact mtDNA lacks the α-event DNA, but whose nuclear genome contains an integrated single copy of the sequence in both juvenile and senescent cultures. Finally, the β-event transfer appears to be a monomeric integration in the s^+ race and may not be present in A^+.

IX. NUCLEAR–CYTOPLASMIC INTERACTIONS

Well over 200 nuclear genes contribute their products to produce functional mitochondria. Some of these are now being studied at the molecular level to learn their functions and the way in which they interact with mitochondrial gene products to produce a complete organelle. Two broad classes of nuclear genes may be defined. One consists of genes whose products are in general well known, such as those for polypeptides of the electron transport chain, mitochondrial ribosomal proteins, mitochondrial tRNA synthetases, and enzymes of the citric acid cycle. The second class consists of gene products that participate in what might be termed molecular housekeeping, whose existence has relatively recently been defined or surmised by genetic experiments, such as enzymes for processing mitochondrial transcripts and components of the machinery for transporting nuclear gene products to mitochondria.

Of the well-established gene class, whose products had generally been studied in detail before their genes, emphasis has been on genes whose products interact for function with mitochondrial products. As noted above, all known mitochondrial gene products are part of multigene complexes involving a majority of nuclear-coded components. In fungi thus far, nuclear genes have been cloned for subunit β of the yeast F_1-ATPase complex (Saltzgaber-Muller *et al.*, 1983), subunits V and VI of yeast cytochrome *c* oxidase (Cumsky *et al.*, 1983; Wright *et al.*, 1984), yeast cytochrome *c* peroxidase (Goltz *et al.*, 1982), *Neurospora* ATPase subunit 9 (Viebrock *et al.*, 1982), yeast elongation factor Tu (Nagata *et al.*, 1983), yeast cytochrome *c* (Montgomery *et al.*, 1978), yeast adenine nucleotide translocator (O'Malley *et al.*, 1982), subunits of the yeast cytochrome bc_1 complex (van Loon *et al.*, 1982), and a yeast mitochondrial outer membrane polypeptide (Riezman *et al.*, 1983). In addition, some evidence has been presented for the identification of mitochondrial ribosomal protein genes in yeast by the isolation of suppressors to rRNA mutations (Bolotin-Fukuhara et al., 1983). Since all mitochondrial ribosomal proteins in yeast except var1 are nuclear-coded, this approach has great potential.

Genes of the second class have been made accessible by the isolation of mutants. As noted above, genetic and biochemical evidence indicates that the Cyt b I4 maturase splices both Cyt b I4 and COI I4. The *NAM2* nuclear locus was recently identified by isolating suppressors to maturase mutants in Cyt b I4 (Groudinsky et al., 1981). Such mutants are pleiotropic and lead to a deficiency in both Cyt b and COI (Church et al., 1979). While intron 4 of COI contains an ORF whose derived protein shows a substantial sequence homology to the Cyt b intron 4 product (Bonitz et al., 1980b), it is not able to function as a maturase unless mutated (Dujardin et al., 1982). The *NAM2* product appears to function by activating the latent COI intron 4 maturase (Dujardin et al., 1983), although the process by which it does so remains to be elucidated.

A number of nuclear loci that function in the processing of mitochondrial transcripts have been identified by analyzing pet$^-$ mutants, a general class of nuclear mutants that abolish respiration in mitochondria. Some of these loci have been obtained as plasmid clones by transforming mutants with a yeast nuclear gene bank. The pet$^-$ mutants affecting the accumulation of cytochrome b and COI have been studied most closely. A single nuclear gene (MSS51) has been cloned and analyzed (Faye and Simon, 1983; Simon and Faye, 1984) whose product is needed for the correct maturation of the precursor to COI mRNA at the level of at least one intron splicing event. Since normal splicing takes place for cytochrome b and L-rRNA precursors, the MSS51 product appears to be gene-specific, providing a formal mechanism for regulation of expression of an individual gene.

For cytochrome b expression, several complementing groups of pet$^-$ mutants have been studied in some detail (Dieckmann et al., 1982a,b, 1984a,b; McGraw and Tzagoloff, 1983; Pillar et al., 1983a). Most nuclear products are required for specific intron processing, including those specific to the long form of the gene (Pillar et al., 1983a,b). One mutation, CBPI, is more unusual in being involved in 5' end-processing of pre-mRNA for Cyt b (Dieckmann et al., 1984a,b). In this mutant, although transcription of Cyt b pre-mRNA can be shown to occur, nucleolytic degradation takes place. Interestingly, petite mutants containing a rearrangement that fuses the 5' untranslated leader of the ATPase 9 gene to the 5' side of the Cyt b coding sequence allow the transcription of stable but novel species of Cyt b.

Although a number of nuclear genes whose products participate in mitochondrial transcript processing have been cloned and sequenced (all the ones examined have ORFs), their mode of action is unclear. A suggestion for those involved in intron splicing is that the nuclear product interacts with mitochondrial maturases for function, in keeping with the division of labor seen for all other mitochondrial products. However, no case yet exists where both a nuclear and a mitochondrial gene are required to excise the same intron.

In contrast to the splicing mutants, the nuclear amber mutation *PET494* specifically blocks the accumulation of COIII (Müller et al., 1985; Müller and Fox, 1984), although normal amounts of transcript are present. A mitochondrial partial suppressor contained, in addition to ρ^+ mtDNA, ρ^- mtDNA that had rearranged to replace the 5' nontranslated region of COIII mRNA. This interpretation provides another example of potential specific regulation of mitochondrial function.

All nuclear-coded proteins destined for mitochondria must be imported from their site of synthesis. Most such proteins are synthesized with an N-terminal presequence (cytochrome *c* is a well-known exception) and then imported. Precursors to be imported bind to the mitochondrial outer membrane (Hennig and Neupert, 1981); in an energy-requiring reaction (Gasser et al., 1982; Schleyer et al., 1982), precursors then enter the mitochondria and the presequence is removed as the limit reaction of a matrix protease (Bohni et al., 1980; Miura et al., 1982; McAda and Douglas, 1982; Cerletti et al., 1983). Although the outlines of this process are becoming clear (Schatz and Butow, 1983), little molecular detail is available, and it is not yet known whether the presequence is necessary and sufficient to direct a protein to mitochondria. The recent redirection of a fused β subunit of F_1-ATPase–β-galactosidase hybrid protein to mitochondria (Douglas et al., 1984) defined a useful system, but involved as many as the first 350 amino-terminal residues of the ATPase portion; further experiments will be needed to isolate the role of the presequence. Two nuclear mutants in yeast that block mitochondrial protein import (Yaffe and Schatz, 1984) should begin to allow a more detailed analysis. These temperature-sensitive mutants, *mas1* and *mas2*, were isolated by screening for the accumulation of β-subunit precursor of F_1-ATPase. The mutants are pleiotropic, however, inhibiting the import of at least the precursors to citrate synthase, cytochrome b_2 and an unidentified 90-kilodalton matrix protein; furthermore, the effect on each precursor differed for the two mutants.

As this brief summary indicates, the role of the nucleus in the assembly and function of mitochondria, which was always appreciated, is now an area that is increasingly amenable to experimentation, and one that is becoming a real focus.

ACKNOWLEDGMENTS

We are grateful to the many investigators who provided us with manuscripts and information prior to publication. Karla Barton and Janet Lee provided dedicated secretarial assistance. The work in our laboratory was supported by NIH grant GM-26546 to L.I.G.; Christine A. Weber and Thomas D. Bruns kindly allowed us to quote from their unpublished data.

REFERENCES

Ahne, F., Merlos-Lange, A., Lang, B. F., and Wolf, K. (1984). The mitochondrial genome of the fission yeast *Schizosaccharomyces pombe*. 5. Characterization of mitochondrial deletion mutants. *Curr. Genet.* **8**, 517–524.

Akins, R. A., and Lambowitz, A. M. (1984). The [*poky*] mutant of *Neurospora crassa* contains a 4-base-pair deletion at the 5′ end of the mitochondrial small rRNA. *Proc. Natl. Acad. Sci. U.S.A.* **81**, 3791–3795.

Anderson, S., Bankier, A. T., Barrell, B. G., de Bruijn, M. H. L., Coulson, A. R., Drouin, J., Eperon, I. C., Nierlich, D. P., Roe, B. A., Sanger, F., Schreier, P. H., Smith, A. J. H., Staden, R., and Young, I. G. (1981). Sequence and organization of the human mitochondrial genome. *Nature (London)* **290**, 457–465.

Anziano, P. Q., Hanson, D. K., Mahler, H. R., and Perlman, P. S. (1982). Functional domains in introns: *trans*-acting and *cis*-acting regions of intron 4 of the *cob* gene. *Cell (Cambridge, Mass.)* **30**, 925–932.

Arnberg, A. C., VanOmmen, G. B. J., Grivell, L. A., Van Bruggen, E. F. J., and Borst, P. (1980). Some yeast mitochondrial RNAs are circular. *Cell (Cambridge, Mass.)* **19**, 313–319.

Baldacchi, G., Cherif-Zahar, B., and Bernardi, G. (1984). The initiation of DNA replication in the mitochondrial genome of yeast. *EMBO J.* **3**, 2115–2120.

Barrell, B. G., Bankier, A. T., and Drouin, J. (1979). A different genetic code in human mitochondria. *Nature (London)* **282**, 189–194.

Barrell, B. G., Anderson, S., Bankier, A. T., de Bruijn, M. H. L., Chen, E., Coulson, A. R., Drouin, J., Eperson, I. C., Nierlich, D. P., Roe, B. A., Sanger, F., Schreier, P. H., Smith, A. J. H., Staden, R., and Young, I. G. (1980). Different pattern of codon recognition by mammalian mitochondrial tRNAs. *Proc. Natl. Acad. Sci. U.S.A.* **77**, 3164–3166.

Belcour, L., and Begel, O. (1980). Life-span and senescence in *Podospora anserina*: effect of mitochondrial genes and functions. *J. Gen. Microbiol.* **119**, 505–515.

Bernard, U., Goldthwaite, C., and Küntzel, H. (1976). Physical map of *Neurospora crassa* mitochondrial DNA and its transcription unit for ribosomal RNA. *Nucleic Acids Res.* **3**, 3101–3108.

Bernardi, G., Baldacci, G., Bernardi, G., Faugeron-Fonty, G., Gaillard, C., Goursot, R., Huyard, A., Mangin, M., Marotta, R., and de Zamoroczy, M. (1980a). The petite mutation: excision sequences, replication origins and suppressivity. *In* "The Organization and Expression of the Mitochondrial Genome" (A. H. Kroon and C. Saccone, eds.), pp. 21–31. Elsevier/North-Holland, Amsterdam.

Bernardi, G., Baldacci, G., Culard, F., Faugeron-Fonty, G., Gaillard, C., Goursot, R., Strauss, R., and de Zamoroczy, M. (1980b). Excision and replication sequences in the mitochondrial genome of yeast. *In* "Mobilization and Reassembly of Genetic Information" (W. A. Scott, R. Werner, D. R. Joseph, and J. Schultz, eds.), pp. 119–132. Academic Press, New York.

Bertrand, H., Collins, R. A., Stohl, L. L., Goewert, R. R., and Lambowitz, A. M. (1980). Deletion mutants of *Neurospora crassa* mitochondrial DNA and their relationship to the "stop–start" growth phenotype. *Proc. Natl. Acad. Sci. U.S.A.* **11**, 6032–6036.

Bohni, P. C., Gasser, S., Leaver, C., and Schatz, G. (1980). A matrix-located mitochondrial protease processing cytoplasmically-made precursors to mitochondrial proteins. *In* "The Organization and Expression of the Mitochondrial Genome'' (A. M. Kroon and C. Saccone, eds.), pp. 423–433. Elsevier/North-Holland, Amsterdam.

Bolotin-Fukuhara, M., Sor, F., and Fukuhara, H. (1983). Mitochondrial ribosomal RNA mutations and their nuclear suppressors in yeast. *In* ''Mitochondria 1983: Nucleo-Mitochondrial Interactions'' (R. J. Schweyen, K. Wolf, and F. Kaudewitz, eds.), pp. 455–467. de Gruyter, Berlin.

Bonitz, S. G., Berlani, R., Coruzzi, G., Li, M., Macino, G., Nobrega, F. G., Nobrega, M. P., Thalenfeld, B. E., and Tzagoloff, A. (1980a). Codon recognition rules in yeast mitochondria. *Proc. Natl. Acad. Sci. U.S.A.* **77**, 3167–3170.

Bonitz, S. G., Coruzzi, G., Thalenfeld, B. E., Tzagoloff, A., and Macino, G. (1980b). Assembly of the mitochondrial membrane system. Structure and nucleotide sequence of the gene coding for subunit 1 of yeast cytochrome oxidase. *J. Biol. Chem.* **255**, 11927–11941.

Bonitz, S. G., Homison, G., Thalenfeld, B. E., Tzagoloff, A., and Nobrega, F. G. (1982). Assembly of the mitochondrial membrane system. Processing of the apocytochrome *b* precursor RNAs in *Saccharomyces cerevisiae* D273-10B, *J. Biol. Chem.* **257**, 6268–6274.

Bos, J. L., Osinga, K. A., Van der Horst, G., and Borst, P. (1979). Nucleotide sequence of the mitochondrial structural genes for cysteine-tRNA and histidine-tRNA of yeast. *Nucleic Acids Res.* **6**, 3255–3266.

Boyd, D. A., Hobman, T. C., Gruenke, S. A., and Klassen, G. R. (1984). Evolutionary stability of mitochondrial DNA organization in *Achlya*. *Can. J. Biochem. Cell Biol.* **62**, 571–576.

Breitenberger, C. A., Browning, K. S., Alzner-DeWeerd, B., and RajBhandary, U. L. (1985). RNA processing in *Neurospora crassa* mitochondria: use of transfer RNA sequences as signals. *EMBO J.* **4**, 185–195.

Broach, J. R., and Hicks, J. B. (1980). Replication and recombination functions associated with the yeast plasmid, 2μ circle. *Cell (Cambridge, Mass.)* **21**, 501–508.

Brown, T. A., Davies, R. W., Waring, R. B., Ray, J. A., and Scazzocchio, C. (1983a). DNA duplication has resulted in transfer of an amino-terminal peptide between two mitochondrial proteins. *Nature (London)* **302**, 721–723.

Brown, T. A., Davies, R. W., Ray, J. A., Waring, R. B., and Scazzocchio, C. (1983b). The mitochondrial genome of *Aspergillus nidulans* contains reading frames homologous to human URFs 1 and 4. *EMBO J.* **2**, 427–435.

Brown, T. A., Ray, J. A., Waring, R. B., Scazzocchio, C., and Davies, R. W. (1984). A mitochondrial reading frame which may code for a second form of ATPase subunit 9 in *Aspergillus nidulans*. *Curr. Genet.* **8**, 489–492.

Brown, W. M., George, M., Jr., and Wilson A. C. (1979). Rapid evolution of animal mitochondrial DNA. *Proc. Natl. Acad. Sci. U.S.A.* **76**, 1967–1971.

Browning, K. S., and RajBhandary, U. L. (1982). Cytochrome oxidase subunit III gene in *Neurospora crassa* mitochondria. Location and sequence. *J. Biol. Chem.* **257**, 5253–5256.

Burke, J. M., and RajBhandary, U. L. (1982). Intron within the large rRNA gene of *N. crassa* mitochondria: a long open reading frame and a consensus sequence possibly important in splicing. *Cell (Cambridge, Mass.)* **31**, 509–520.

Butow, R. A., Ainley, W. M., Zassenhaus, H. P., Hudspeth, M. E., and Grossman, L. I. (1983). The unusual organization of the yeast mitochondrial *var1* gene. *In* ''Manipulation and Expression of Genes in Eukaryotes'' (P. Nagley, A. W. Linnane, A. W. Peacock, and J. A. Pateman, eds.), pp. 269–277. Academic Press, New York.

Butow, R. A., Perlman, P. S., and Grossman, L. I. (1985). The unusual *var1* gene of yeast mitochondrial DNA. *Science* **228**, 1496–1502.

Carignani, G., Groudinsky, O., Frezza, D., Schiavon, E., Bergantino, E., and Slonimski, P. P. (1983). An mRNA maturase is encoded by the first intron of the mitochondrial gene for the subunit I of cytochrome oxidase in *S. cerevisiae*. *Cell (Cambridge, Mass.)* **35**, 733–742.

Cech, T. R., Zaug, A. J., and Grabowski, P. J. (1981). *In vitro* splicing of the ribosomal RNA precursor of *Tetrahymena:* involvement of a guanosine nucleotide in the excision of the intervening sequence. *Cell (Cambridge, Mass.)* **27,** 487–496.

Cech, T. R., Tanner, N. K., Tinoco, I., Jr., Weir, B. R., Zucker, M., and Perlman, P. S. (1983). Secondary structure of the *Tetrahymena* ribosomal RNA intervening sequence: structural homology with fungal mitochondrial intervening sequences. *Proc. Natl. Acad. Sci. U.S.A.* **80,** 3903–3907.

Cerletti, N., Bohni, P. C., and Suda, K. (1983). Import of proteins into mitochondria. Isolated yeast mitochondria and a solubilized matrix protease correctly process cytochrome *c* oxidase subunit V precursor at the NH_2 terminus. *J. Biol. Chem.* **258,** 4944–4949.

Christianson, T., and Rabinowitz, M. (1983). Identification of multiple transcriptional initiation sites on the yeast mitochondrial genome by *in vitro* capping with guanylyltransferase. *J. Biol. Chem.* **258,** 14025–14033.

Christianson, T., Edwards, J. C., Mueller, D., and Rabinowitz, M. (1983). Identification of a single transcription initiation site for glutamic tRNA and *cob* genes in yeast mitochondria. *Proc. Natl. Acad. Sci. U.S.A.* **80,** 5564–5568.

Church, G. M., Slonimski, P. P., and Gilbert, W. (1979). Pleiotropic mutations within two yeast mitochondrial cytochrome genes block mRNA processing. *Cell (Cambridge, Mass.)* **18,** 1209–1215.

Clark-Walker, G. D., and Gleason, F. H. (1973). Circular DNA from the water mold *Saprolegnia*. *Arch. Mikrobiol.* **92,** 209–216.

Clark-Walker, G. D., and Sriprakash, K. S. (1983). Map location of transcripts from *Torulopsis glabrata* mitochondrial DNA. *EMBO J.* **2,** 1465–1472.

Clark-Walker, G. D., Sriprakash, K. S., McArthur, C. R., and Azad, A. A. (1980). Mapping of mitochondrial DNA from *Torulopsis glabrata:* location of ribosomal and transfer RNA genes. *Curr. Genet.* **1,** 209–217.

Clark-Walker, G. D., McArthur, C. R., and Sriprakash, K. S. (1981). Partial duplication of the large ribosomal RNA sequence in an inverted repeat in circular mitochondrial DNA from *Kloeckera africana*. *J. Mol. Biol.* **147,** 399–415.

Clark-Walker, G. D., McArthur, C. R., and Sriprakash, K. S. (1983). Order and orientation of genic sequences in circular mitochondrial DNA from *Saccharomyces exiguus:* implications for evolution of yeast mtDNAs. *J. Mol. Evol.* **19,** 333–341.

Clayton, D. A. (1984). Transcription of the mammalian mitochondrial genome. *Annu. Rev. Biochem.* **53,** 573–594.

Cobon, G. S., Beilharz, M. W., Linnane, A. W., and Nagley, P. (1982). Mapping of transcripts from the *oli2* region of the mitochondrial DNA in two genomic strains of *Saccharomyces cerevisiae*. *Curr. Gent.* **5,** 97–107.

Collins, R. A., and Lambowitz, A. M. (1981). Characterization of a variant *Neurospora crassa* mitochondrial DNA which contains tandem reiterations of a 1.9 kb sequence. *Curr. Genet.* **4,** 131–133.

Collins, R. A., and Lambowitz, A. M. (1983). Structural variations and optional introns in the mitochondrial DNAs of *Neurospora* strains isolated from nature. *Plasmid* **9,** 53–70.

Collins, R. A., Bertrand, H., LaPolla, R. J., and Lambowitz, A. M. (1979). Mitochondrial ribosome assembly in *Neurospora crassa:* Mutants with defects in mitochondrial ribosome assembly. *Mol. Gen. Genet.* **177,** 73–84.

Collins, R. A., Stohl, L. L., Cole, M. D., and Lambowitz, A. M. (1981). Characterization of a novel plasmid DNA found in mitochondria of *N. crassa*. *Cell (Cambridge, Mass.)* **24,** 443–452.

Coruzzi, G., Bonitz, S. G., Thalenfeld, B. E., and Tzagoloff, A. (1981). Assembly of the mitochondrial membrane system. Analysis of the nucleotide sequence and transcripts in the *oxi1* region of yeast mitochondrial DNA. *J. Biol. Chem.* **256,** 12780–12787.

Cox, H. M. (1983). The FLP protein of the yeast 2 micron plasmid: expression of a eukaryotic genetic recombination system in *Escherichia coli*. *Proc. Natl. Acad. Sci. U.S.A.* **80,** 4223–4227.

Cummings, D. J., and Wright, R. M. (1983). DNA sequence of the excision sites of a mitochondrial plasmid from senescent *Podospora anserina*. *Nucleic Acids Res.* **11,** 2111–2119.

Cummings, D. J., Belcour, L., and Grandchamp, C. (1979). Mitochondrial DNA from *Podospora anserina*. Isolation and characterization. *Mol. Gen. Genet.* **171,** 229–238.

Cumsky, M. G., McEwen, J. E., Ko, C., and Poyton, R. O. (1983). Nuclear genes for mitochondrial proteins. Identification and isolation of a structural gene for subunit V of yeast cytochrome *c* oxidase. *J. Biol. Chem.* **258,** 13418–13421.

Davies, R. W., Waring, R. B., Ray, J. A., Brown, T. A., and Scazzocchio, C. (1982). Making ends meet: a model for RNA splicing in fungal mitochondria. *Nature (London)* **300,** 719–724.

Davies, R. W., Waring, R. B., Brown, T. A., and Scazzocchio, C. (1983). Implications of fungal mitochondrial intron RNA structure for the mechanisms of RNA splicing. *In* "Mitochondria 1983: Nucleo-Mitochondrial Interactions" (R. J. Schweyen, K. Wolf, and F. Kaudewitz, eds.), pp. 179–189. de Gruyter, Berlin.

de Bruijn, M. H. L. (1983). *Drosophila melanogaster* mitochondrial DNA, a novel organization and genetic code. *Nature (London)* **304,** 234–241.

De La Salle, H., Jacq, C., and Slonimski, P. P. (1982). Critical sequences within mitochondrial introns: pleiotropic mRNA maturase and *cis*-dominant signals of the *box* intron controlling reductase and oxidase. *Cell (Cambridge, Mass.)* **28,** 721–732.

Del Giudice, L., Wolf, K., Buono, C., and Manna, F. (1981). Nucleo-cytoplasmic interactions in the petite negative yeast *Schizosaccharomyces pombe*. Inhibition of nuclear and mitochondrial DNA syntheses in the absence of cytoplasmic protein synthesis. *Mol. Gen. Genet.* **181,** 306–308.

de Vries, H., de Jonge, J. C., van't Sant, P., Agsteribbe, E., and Amberg, A. (1981). A "stopper" mutant of *Neurospora crassa* containing two populations of aberrant mitochondrial DNA. *Curr. Genet.* **3,** 205–211.

de Vries, H., de Jonge, J. C., Arnberg, A., Peijnenburg, A. A. C. M., and Agsteribbe, E. (1983). The expression of the mitochondrial genes for subunit 1 of cytochrome *c* oxidase and for an ATPase proteolipid in *Neurospora crassa:* nucleotide sequence and transcript analysis. *In* "Mitochondria 1983: Nucleo-Mitochondrial Interactions" (R. J. Schweyen, K. Wolf, and F. Kaudewitz, eds.), pp. 343–356. de Gruyter, Berlin.

Dieckmann, C. L., Pape, L. K., and Tzagoloff, A. (1982a). Identification and cloning of a yeast nuclear gene (CBP1) involved in the expression of mitochondrial cytochrome *b*. *Proc. Natl. Acad. Sci. U.S.A.* **79,** 1805–1809.

Dieckmann, C. L., Bonitz, S. G., Hill, J., Homison, G., McGraw, P., Pape, L., Thalenfeld, B. E., and Tzagoloff, A. (1982). Structure of the apocytochrome-*b* gene and processing of apocytochrome-*b* transcripts in *Saccharomyces cerevisiae*. *In* "Mitochondrial Genes" (P. Slonimski, P. Borst, and G. Attardi, eds.), pp. 213–223. Cold Spring Harbor Lab. Press, Cold Spring Harbor, New York.

Dieckmann, C. L., Koerner, T. J., and Tzagoloff, A. (1984a). Assembly of the mitochondrial membrane system. CBP1, a yeast nuclear gene involved in 5′ end processing of cytochrome *b* pre-mRNA. *J. Biol. Chem.* **259,** 4722–4731.

Dieckmann, C. L., Homison, G., and Tzagoloff, A. (1984b). Assembly of the mitochondrial membrane system. Nucleotide sequence of a yeast nuclear gene (CBP1) involved in 5′ end processing of cytochrome *b* pre-mRNA. *J. Biol. Chem.* **259,** 4732–4738.

Douglas, M. G., Geller, B. L., and Emr, S. D. (1984). Intracellular targeting and import of an F_1-ATPase β-subunit–β-galactosidase hybrid protein into yeast mitochondria. *Proc. Natl. Acad. Sci. U.S.A.* **81,** 3983–3987.

Dujardin, G., Jacq, C., and Slonimski, P. P. (1982). Single base substitution in an intron of oxidase gene compensates splicing defects of the cytochrome *b* gene. *Nature (London)* **298**, 628–632.

Dujardin, G., Labouesse, M., Netter, P., and Slonimski, P. P. (1983). Genetic and biochemical studies of the nuclear suppressor *NAM2:* extraneous activation of a latent pleiotropic maturase. *In* "Mitochondria 1983: Nucleo-Mitochondrial Interactions" (R. J. Schweyen, K. Wolf, and F. Kaudewitz, eds.), pp. 232–250. de Gruyter, Berlin.

Dujon, B. (1980). Sequence of the intron and flanking exons of the mitochondrial 21S rRNA gene of yeast strains having different alleles at the *omega* and *rib-1* loci. *Cell (Cambridge, Mass.)* **20**, 185–187.

Dujon, B. (1981). Mitochondrial genetics and functions. *In* "The Molecular Biology of the Yeast Saccharomyces: Life Cycle and Inheritance" (J. N. Strathern, E. W. Jones, and J. R. Broach, eds.), pp. 505–635. Cold Spring Harbor Lab. Press, Cold Spring Harbor, New York.

Edwards, J. C., Levens, D., and Rabinowitz, M. (1982). Analysis of transcriptional initiation of yeast mitochondrial DNA in a homologous *in vitro* transcription system. *Cell (Cambridge, Mass.)* **31**, 337–346.

Edwards, J. C., Christianson, T., Mueller, D., Biswas, T. K., Levens, D., Li, D., Wettstein, J., and Rabinowitz, M. (1983a). Initiation and transcription of yeast mitochondrial DNA. *In* "Mitochondria 1983: Nucleo-Mitochondrial Interactions" (R. J. Schweyen, K. Wolf, and F. Kaudewitz, eds.), pp. 69–78. de Gruyter, Berlin.

Edwards, J. C., Osinga, K. A., Christianson, T., Hensgens, L. A. M., Janssens, P. M., Rabinowitz, M., and Tabak, H. F. (1983b). Initiation of transcription of the yeast mitochondrial gene coding for ATPase subunit 9. *Nucleic Acids Res.* **11**, 8269–8282.

Falcone, C. (1984). The mitochondrial DNA of the yeast *Hansenula petersonii:* genome organization and mosaic genes. *Curr. Genet.* **8**, 449–455.

Farrelly, F., and Butow, R. A. (1983). Rearranged mitochondrial genes in the yeast nuclear genome. *Nature (London)* **301**, 296–301.

Farrelly, F., Zassenhaus, H. P., and Butow, R. A. (1982). Characterization of transcripts from the *var1* region on mitochondrial DNA from *Saccharomyces cerevisiae. J. Biol. Chem.* **257**, 6581–6587.

Faugeron-Fonty, G., Culard, F., Baldacci, G., Goursot, R., Prunell, A., and Bernardi, G. (1979). The mitochondrial genome of wild-type yeast cells. VIII. The spontaneous cytoplasmic "petite" mutation. *J. Mol. Biol.* **134**, 493–537.

Faye, G., and Simon, M. (1983). Analysis of a yeast nuclear gene involved in the maturation of mitochondrial pre-messenger RNA of cytochrome oxidase subunit I. *Cell (Cambridge, Mass.)* **32**, 77–87.

Faye, G., Kujawa, C., and Fukuhara, H. (1974). Physical and genetic organization of petite and grande mitochondrial DNA. IV. *In vitro* translation products and localization of 23S ribosomal RNA in petite mutants of *Saccharomyces cerevisiae. J. Mol. Biol.* **88**, 185–203.

Faye, G., Kujawa, C., Dujon, B., Bolotin-Fukuhara, M., Wolf, K., Fukuhara, H., and Slonimski, P. P. (1975). Localization of the gene coding for the 16S ribosomal RNA using rho^- mutants of *Saccharomyces cerevisiae. J. Mol. Biol.* **99**, 203–217.

Frontali, L., Palleschi, C., and Francisci, S. (1982). Transcripts of mitochondrial tRNA genes in *Saccharomyces cerevisiae. Nucleic Acids Res.* **10**, 7283–7293.

Garber, R. C., and Yoder, O. C. (1985). Mitochondrial DNA of the filamentous ascomycete *Cochliobolus heterostrophus.* Characterization of the mitochondrial chromosome and population genetics of a restriction enzyme polymorphism. *Curr. Genet.* **8**, 621–628.

Garber, R. C., Turgeon, B. G., and Yoder, O. C. (1984). A mitochondrial plasmid from the plant pathogenic fungus *Cochliobolus heterostrophus. Mol. Gen. Genet.* **196**, 301–310.

Garriga, G., and Lambowitz, A. (1985). RNA splicing in *Neurospora* mitochondria. Self-splicing of a mitochondrial intron *in vitro. Cell (Cambridge, Mass.)* **39**, 631–641.

Gasser, S. M., Daum, G., and Schatz, G. (1982). Import of proteins into mitochondria. Energy-dependent uptake of precursors by isolated mitochondria. *J. Biol. Chem.* **257,** 13034–13041.

Gellissen, G., Bradfield, J. Y., White, B. N., and Wyatt, G. R. (1983). Mitochondrial DNA sequences in the nuclear genome of a locust. *Nature (London)* **301,** 631–634.

Gerbaud, C., Fournier, P., Blanc, H., Aijle, M., Heslot, H., and Guerineau, M. (1979). High frequency of yeast transformation by plasmids carrying part or entire 2-μm yeast plasmid. *Gene* **5,** 233–253.

Gilham, N. W. (1978). "Organelle Heredity" pp. 268–275. Raven, New York.

Goddard, J. M., and Cummings, D. J. (1975). Structure and replication of mitochondrial DNA from *Paramecium aurelia*. *J. Mol. Biol.* **97,** 593–609.

Goldbach, R. W., Bollen-DeBoer, J. E., Van Bruggen, E. F. J., and Borst, P. (1979). Replication of the linear mitochondrial DNA of *Tetrahymena pyriformis*. *Biochim. Biophys. Acta* **562,** 400–417.

Goltz, S., Kaput, J., and Blobel, G. (1982). Nucleotide sequence of the yeast nuclear gene for cytochrome *c* peroxidase precursor. Functional implications of the presequence for protein transport into mitochondria. *J. Biol. Chem.* **257,** 15054–15058.

Gray, M. W. (1982). Mitochondrial genome diversity and the evolution of mitochondrial DNA. *Can. J. Biochem.* **60,** 157–171.

Gray, M. W., and Doolittle, W. F. (1982). Has the endosymbiont hypothesis been proven? *Microbiol. Rev.* **46,** 1–42.

Grivell, L. A. (1982). Restriction and genetic maps of yeast mitochondrial DNA. In "Genetic Maps: A Compilation of Linkage and Restriction Maps of Genetically Studied Organisms" (S. J. O'Brien, ed.), Vol. 2, pp. 221–235. Cold Spring Harbor Lab. Press, Cold Spring Harbor, New York.

Groot, G. S. P., and Van Harten-Loosbroek, N. (1980). The organization of the genes for ribosomal RNA on mitochondrial DNA of *Kluyveromyces lactis*. *Curr. Genet.* **1,** 133–135.

Gross, S. R., Hsiah, J.-S., and Levine, P. H. (1984). Intramolecular recombination as a source of mitochondrial chromosome heteromorphism in *Neurospora*. *Cell (Cambridge, Mass.)* **38,** 233–239.

Groudinsky, O., Dujardin, G., and Slonimski, P. P. (1981). Long range control circuits within mitochondria and between nucleus and mitochondria. II. Genetic and biochemical analyses of suppressors which selectively alleviate the mitochondrial intron mutations. *Mol. Gen. Genet.* **184,** 493–503.

Hadler, H. I., Dimitrijevis, B., and Mahalingam, R. (1983). Mitochondrial DNA and nuclear DNA from normal rat liver have a common sequence. *Proc. Natl. Acad. Sci. U.S.A.* **80,** 6495–6499.

Halbreich, A., Pajot, P., Foucher, M., Grandchamp, C., and Slonimski, P. P. (1980). A pathway of cytochrome *b* mRNA processing in yeast mitochondria: specific splicing steps and an intron-derived circular RNA. *Cell (Cambridge, Mass.)* **19,** 321–329.

Heckman, J. E., Sarnoff, J., Alzner-DeWeerd, B., Yin, S., and RajBhandary, U. L. (1980). Novel features in the genetic code and codon reading patterns in *Neurospora crassa* mitochondria based on sequences of six mitochondrial tRNAs. *Proc. Natl. Acad. Sci. U.S.A.* **77,** 3159–3163.

Helling, R. B., ElGewely, M. R., Lomax, M. I., Baumgartner, H. E., Schwartzbach, S. D., and Barnett, W. E. (1979). Organization of the chloroplast ribosomal RNA genes of *Euglena gracilis bacillaris*. *Mol. Gen. Genet.* **174,** 1–4.

Helmer Critterich, M., Morelli, G., and Macino, G. (1983). Nucleotide sequence and intron structure of the apocytochrome *b* gene in *Neurospora crassa* mitochondria. *EMBO J.* **2,** 1235–1242.

Hennig, B., and Neupert, W. (1981). Assembly of cytochrome *c*. Apocytochrome *c* is bound to specific sites on mitochondria before its conversion to holocytochrome *c*. *Eur. J. Biochem.* **121,** 203–212.

Hensgens, L. A. M., Grivell, L. A., Borst, P., and Bos, J. L. (1979). Nucleotide sequence of the mitochondrial structural gene for subunit 9 of the yeast ATPase complex. *Proc. Natl. Acad. Sci. U.S.A.* **76,** 1663-1667.

Hensgens, L. A. M., Bonen, L., de Haan, M., Van der Horst, G., and Grivell, L. A. (1983a). Two intron sequences in yeast mitochondrial COXI gene: homology among URF-containing introns and strain-dependent variations in flanking exons. *Cell (Cambridge, Mass.)* **32,** 379-389.

Hensgens, L. A. M., Arnberg, A. C., Roosendaal, E., Van der Horst, G., Van der Veen, R., Van Ommen, G. J. B., and Grivell, L. A. (1983b). Variation, transcription and circular RNAs of the mitochondrial gene for subunit I of cytochrome *c* oxidase. *J. Mol. Biol.* **164,** 35-38.

Hudspeth, M. E. S., Shumard, D. S., Tatti, K. M., and Grossman, L. I. (1980). Rapid purification of yeast mitochondrial DNA in high yield. *Biochim. Biophys. Acta* **610,** 221-228.

Hudspeth, M. E. S., Ainley, W. M., Shumard, D. S., Butow, R. A., and Grossman, L. I. (1982). Location and structure of the *var1* gene on yeast mitochondrial DNA: nucleotide sequence of the *40.0* allele. *Cell (Cambridge, Mass.)* **30,** 617-626.

Hudspeth, M. E. S., Shumard, D. S., Bradford, C. J. R., and Grossman, L. I. (1983). Organization of *Achlya* mtDNA: A population with two orientations and a large inverted repeat containing the rRNA genes. *Proc. Natl. Acad. Sci. U.S.A.* **80,** 142-146.

Hudspeth, M. E. S., Vincent, R. D., Perlman, P. S., Treisman, L. O., Shumard, D. S., and Grossman, L. I. (1984). The expandable *var1* gene of yeast mitochondrial DNA: in-frame insertions can explain the strain-specific protein size polymorphisms. *Proc. Natl. Acad. Sci. U.S.A.* **81,** 3148-3152.

Jacobs, H. T., Posakony, J. W., Grula, J. W., Roberts, J. W., Xin, J. H., Britten, R. J., and Davidson, E. H. (1983). Mitochondrial DNA sequences in the nuclear genome of *Strongylocentrotus purpuratus*. *J. Mol. Biol.* **165,** 609-632.

Jacq, C., Lazowska, J., and Slonimski, P. P. (1980). Sur un noveau méchanisme de la régulation de l'expression génétique. *C. R. Hebd. Seances Acad. Sci., Ser. D* **290,** 89-92.

Jacquier, A., and Dujon, B. (1983). The intron of the mitochondrial 21S rRNA gene: distribution in different yeast species and sequence comparison between *Kluyveromyces thermotolerans* and *Saccharomyces cerevisiae*. *Mol. Gen. Genet.* **192,** 487-499.

Kawano, S., Suzuki, T., and Kuroiwa, T. (1982). Structural homogeneity of mitochondrial DNA in the mitochondrial nucleoid of *Physarum polycephalum*. *Biochim. Biophys. Acta* **696,** 290-298.

Kemble, R. J., Mans, R. J., Gabay-Laughnan, S., and Laughnan, J. R. (1983). Sequences homologous to episomal mitochondrial DNAs in the maize nuclear genome. *Nature (London)* **304,** 744-747.

Klimczak, L. J., and Prell, H. H. (1984). Isolation and characterization of mitochondrial DNA of the oömycetous fungus *Phytophthora infestans*. *Curr. Genet.* **8,** 323-326.

Köchel, H. G., and Küntzel, H. (1981). Nucleotide sequence of the *Aspergillus nidulas* mitochondrial gene coding for the small ribosomal subunit RNA: homology to *E. coli* 16S rRNA. *Nucleic Acids. Res.* **9,** 5689-5696.

Köchel, H. G., and Küntzel, H. (1982). Mitochondrial L-rRNA from *Aspergillus nidulans:* potential secondary structure and evolution. *Nucleic Acids Res.* **10,** 4795-4801.

Köchel, H. G., Lazarus, C. M., Basak, N., and Küntzel, H. (1981). Mitochondrial tRNA gene clusters in *Aspergillus nidulans:* organization and nucleotide sequence. *Cell (Cambridge, Mass.)* **23,** 625-633.

Koll, F., Begel, O., Keller, A.-M., Vierny, C., and Belcour, L. (1984). Ethidium bromide rejuvenation of senescent cultures of *Podospora anserina:* loss of senescence-specific DNA and recovery of normal mitochondrial DNA. *Curr. Genet.* **8,** 127-134.

Kostriken, R., Strathern, J. N., Klar, A. J. S., Hicks, J. B., and Heffron, F. (1983). A site-specific endonuclease essential for mating-type switching in *Saccharomyces cerevisiae*. *Cell (Cambridge, Mass.)* **356,** 167-174.

Kruger, K., Grabowski, P. J., Zaug, A. J., Sands, J., Gottschling, D. E., and Cech, T. R. (1982). Self-splicing RNA: autoexcision and autocyclization of the ribosomal RNA intervening sequence of *Tetrahymena*. *Cell (Cambridge, Mass.)* **31,** 147–157.

Kuck, U., Stahl, U., Shermitte, A., and Esser, K. (1980). Isolation and characterization of mitochondrial DNA from the alkane yeast *Saccharomycopsis lipolytica*. *Curr. Genet.* **2,** 97–101.

Lambowitz, A. M., LaPolla, R. J., and Collins, R. A. (1979). Mitochondrial assembly in *Neurospora*. Two-dimensional gel electrophoretic analysis of mitochondrial ribosomal proteins. *J. Cell Biol.* **82,** 17–31.

Lang, B. F., Ahne, F., Distler, S., Zrinkl, H., Kaudewitz, F., and Wolf, K. (1983). Sequence of the mitochondrial DNA, arrangement of genes and processing of their transcripts in *Schizosaccharomyces pombe*. In "Mitochondria 1983: Nucleo-Mitochondrial Interactions" (R. J. Schweyen, K. Wolf, and F. Kaudewitz, eds.), pp. 313–329. de Gruyter, Berlin.

LaPolla, R. J., and Lambowitz, A. M. (1981). Mitochondrial ribosome assembly in *Neurospora crassa*. Purification of the mitochondrially synthesized ribosomal protein, S-5. *J. Biol. Chem.* **256,** 7064–7067.

Lazarus, C. M., and Küntzel, H. (1981). Anatomy of amplified mitochondrial DNA in "ragged" mutants of *Aspergillus amstelodami*: excision points within protein genes and a common 215 bp segment containing a possible origin of replication. *Curr. Genet.* **4,** 99–107.

Lazarus, C. M., Earl, A. J., Turner, G., and Küntzel, H. (1980). Amplification of a mitochondrial DNA sequence in the cytoplasmically inherited "ragged" mutant of *Aspergillus amstelodami*. *Eur. J. Biochem.* **106,** 633–641.

Lazowska, J., Jacq, C., and Slonimski, P. P. (1980). Sequence of intron and flanking exons in wild-type and *box3* mutants of cytochrome *b* reveals an interlaced splicing protein coded by an intron. *Cell (Cambridge, Mass.)* **22,** 333–348.

Levens, D., Morimoto, R., and Rabinowitz, M. (1981a). Mitochondrial transcription complex from *Saccharomyces cerevisiae*. *J. Biol. Chem.* **256,** 1466–1473.

Levens, D., Ticho, B., Ackerman, E., and Rabinowitz, M. (1981b). Transcriptional initiation and 5′ termini of yeast mitochondrial RNA. *J. Biol. Chem.* **256,** 5226–5232.

Locker, J., and Rabinowitz, M. (1981). Transcription in yeast mitochondria: analysis of the 21S rRNA region and its transcripts. *Plasmid* **6,** 302–314.

Lonsdale, D. M. (1985). A review of the structure and organization of the mitochondrial genome of higher plants. *Plant Mol. Biol.* **3,** 201–206.

McAda, P. C., and Douglas, M. G. (1982). A neutral metallo endoprotease involved in the processing of an F_1-ATPase subunit precursor in mitochondria. *J. Biol. Chem.* **257,** 3177–3182.

McGraw, P., and Tzagoloff, A. (1983). Assembly of the mitochondrial membrane system. Characterization of a yeast nuclear gene involved in the processing of the cytochrome *b* pre-mRNA. *J. Biol. Chem.* **258,** 9459–6468.

Macino, G. (1980). Mapping of mitochondrial structural genes in *Neurospora crassa*. *J. Biol. Chem.* **255,** 10563–10565.

Macino, G., and Tzagoloff, A. (1979). Assembly of the mitochondrial membrane system. The DNA sequence of a mitochondrial ATPase gene in *Saccharomyces cerevisiae*. *J. Biol. Chem.* **254,** 4617–4623.

Macino, G., and Tzagoloff, A. (1980). Assembly of the mitochondrial membrane system: sequence analysis of a yeast mitochondrial ATPase gene containing the *oli2* and *oli4* loci. *Cell (Cambridge, Mass.)* **20,** 507–517.

Macreadie, I. G., Novitski, C. E., Maxwell, R. J., John, U., Ooi, B. G., McMullen, G. L., Lukins, H. B., Linnane, A. W., and Nagley, P. (1983). Biogenesis of mitochondria: the mitochondrial gene (*aap1*) coding for ATPase subunit 8 in *Saccharomyces cerevisiae*. *Nucleic Acids Res.* **11,** 4435–4451.

Mahler, H. (1983). Exon–intron structure of genes. *Int. Rev. Cytol.* **82,** 1–98.

Mannella, C. A., Goewert, R. R., and Lambowitz, A. M. (1979a). Characterization of variant

Neurospora crassa mitochondrial DNAs which contain tandem reiterations. *Cell (Cambridge, Mass.)* **18,** 1197–1207.
Mannella, C. A., Pittenger, T. H., and Lambowitz, A. M. (1979b). Transmission of mitochondrial deoxyribonucleic acid in *Neurospora crassa* sexual crosses. *J. Bacteriol.* **137,** 1449–1451.
Mery-Drugeon, E., Crouse, E. J., Schmitt, J. M., Bohnert, H. J., and Bernardi, G. (1981). The mitochondrial genomes of *Ustilago cynodontis* and *Acanthamoeba costellanii. Eur. J. Biochem.* **114,** 577–583.
Michel, F. (1984). A maturase-like coding sequence downstream of the *oxi2* gene of yeast mitochondrial DNA is interrupted by two GC clusters and a putative end-of-messenger signal. *Curr. Genet.* **8,** 307–317.
Michel, F., and Dujon, B. (1983). Conservation of RNA secondary structures in two intron families including mitochondrial-, chloroplast-, and nuclear-encoded members. *EMBO J.* **2,** 33–38.
Michel, F., Jacquier, A., and Dujon, B. (1982). Comparison of fungal mitochondrial introns reveals extensive homologies in RNA secondary structure. *Biochimie* **64,** 867–881.
Miller, D. L., and Martin, N. C. (1983). Characterization of the yeast mitochondrial locus necessary for tRNA biosynthesis: DNA sequence analysis and identification of a new transcript. *Cell (Cambridge, Mass.)* **34,** 911–917.
Miller, D. L., Underbrink-Lyon, K., Najarian, D. R., Krupp, J., and Martin, N. C. (1983). Transcription of yeast mitochondrial tRNA genes and processing of tRNA gene transcripts. *In* "Mitochondria 1983: Nucleo-Mitochondrial Interactions" (R. J. Schweyen, K. Wolf, and F. Kaudewitz, eds.), pp. 151–164. de Gruyter, Berlin.
Minuth, W., Tudzynski, P., and Esser, K. (1982). Extrachromosomal genetics of *Cephalosporium acremonium:* I. Characterization and mapping of mitochondrial DNA. *Curr. Genet.* **5,** 227–231.
Miura, S., Mori, M., Amaya, Y., and Tatibani, M. (1982). A mitochondrial protease that cleaves the precursor to ornithine carbamoyltransferase. Purification and properties. *Eur. J. Biochem.* **122,** 641–647.
Montgomery, D. L., Hall, B. D., Gillam, S., and Smith, M. (1978). Identification and isolation of the yeast cytochrome *c* gene. *Cell (Cambridge, Mass.)* **14,** 673–680.
Morelli, G., and Macino, G. (1984). Two intervening sequences in the ATPase subunit 6 gene of *Neurospora crassa:* a short intron (93 bp) and a long intron that is stable after excision. *J. Mol. Biol.* **178,** 491–508.
Müller, P. P., and Fox, T. D. (1984). Molecular cloning and genetic mapping of the *PET494* gene of *Saccharomyces cerevisiae. Mol. Gen. Genet.* **195,** 275–280.
Müller, P. P., Reif, M. K., Zonghou, S., Sengstag, C., Mason, T. L., and Fox, T. D. (1985). A nuclear mutation that post-transcriptionally blocks accumulation of a yeast mitochondrial gene product can be suppressed by a mitochondrial gene rearrangement. *J. Mol. Biol.* **175,** 431–452.
Nagata, S., Tsunetsuga-Yokota, Y., Naito, A., and Kazibo, Y. (1983). Molecular cloning and sequence determination of the nuclear gene coding for mitochondrial elongation factor Tu of *Saccharomyces cerevisiae. Proc. Natl. Acad. Sci. U.S.A.* **80,** 6192–6196.
Nargang, F. E., Bell, J. B., Stohl, L. L., and Lambowitz, A. M. (1983). A family of repetitive palindromic sequences found in *Neurospora* mitochondrial DNA is also found in a mitochondrial plasmid DNA. *J. Biol. Chem.* **258,** 4257–4260.
Nargang, F. E., Bell, J. B., Stohl, L. L., and Lambowitz, A. M. (1984). The DNA sequence and genetic organization of a *Neurospora* mitochondrial plasmid suggests a relationship to introns and mobile elements. *Cell (Cambridge, Mass.)* **38,** 441–453.
Netzker, R., Köchel, H. G., Basak, N., and Küntzel, H. (1982). Nucleotide sequence for *Aspergillus nidulans* mitochondrial genes coding for ATPase subunit 6, cytochrome oxidase subunit 3, seven unidentified proteins, four tRNAs and L-rRNA. *Nucleic Acids Res.* **10,** 4783–4794.

O'Malley, K., Pratt, P., Roberton, J., Lilly, M., and Douglas, M. G. (1982). Selection of the nuclear gene for the mitochondrial adenine nucleotide translocator by genetic complementation of the *op1* mutation in yeast. *J. Biol. Chem.* **257,** 2097–2103.

Osiewacz, H. D., and Esser, K. (1983). DNA sequence analysis of the mitochondrial plasmid of *Podospora anserina*. *Curr. Genet.* **1,** 219–223.

Osinga, K. A., and Tabak, H. F. (1982). Initiation of transcription of genes for mitochondrial ribosomal RNA in yeast: comparison of the nucleotide sequence around the 5'-ends of both genes reveals a homologous stretch of 17 nucleotides. *Nucleic Acids Res.* **10,** 3617–3626.

Osinga, K. A., De Haan, M., Christianson, T., and Tabak, H. F. (1982). A nonanucleotide sequence involved in promotion of ribosomal RNA synthesis and RNA priming of DNA replication in yeast mitochondria. *Nucleic Acids Res.* **10,** 7993–8006.

Osinga, K. A., De Vries, E., Van der Horst, G. T. J., and Tabak, H. F. (1984). Initiation of transcription in yeast mitochondria: analysis of origins of replication and of genes coding for a messenger RNA and a transfer RNA. *Nucleic Acids Res.* **12,** 1889–1900.

Pillar, T., Kreike, J., Lang, B. F., and Kaudewitz, F. (1983a). Nuclear mutants defective in processing of *cob* and *oxi3* transcripts in *Saccharomyces cerevisiae*. *In* "Mitochondria 1983: Nucleo-Mitochondrial Interactions" (R. J. Schweyen, K. Wolf, and F. Kaudewitz, eds.), pp. 411–423. de Gruyter, Berlin.

Pillar, T., Lang, B. F., Steinberger, I., Vogt, B., and Kaudewitz, F. (1983b). Expression of the "split gene" *cob* in yeast mtDNA. Nuclear mutations specifically block the excision of different introns from its primary transcript. *J. Biol. Chem.* **258,** 7954–7959.

Prunell, A., and Bernardi, G. (1977). The mitochondrial genome of wild-type yeast cells, VI. Genome organization. *J. Mol. Biol.* **110,** 53–74.

Rawson, J. R. Y., Kushner, S. R., Vapnek, D., Kirby, N., Boerma, A., and Boerma, C. L. (1978). Chloroplast ribosomal RNA genes in *Euglena gracilis* exist as three clustered tandem repeats. *Gene* **3,** 191–209.

Riezman, H., Hase, T., van Loon, A. P. G. M., Grivell, L. A., Suda, K., and Schatz, G. (1983). Import of proteins into mitochondria: a 70 kilodalton outer membrane protein with a large carboxy-terminal deletion is still transported to the outer membrane. *EMBO J.* **2,** 2161–2168.

Rifkin, M. R., and Luck, D. J. L. (1971). Defective production of mitochondrial ribosomes in the *poky* mutant of *Neurospora crassa*. *Proc. Natl. Acad. Sci. U.S.A.* **68,** 287–290.

Rödel, G., Holl, J., Schmelzer, C., Schmidt, C., Schweyen, R., Weiss-Brummer, B., and Kaudewitz, F. (1983). Cob intron 1 and 4: studies of mutants and revertants uncover functional intron domains and test the validity of predicted RNA-secondary structures. *In* "Mitochondria 1983: Nucleo-Mitochondrial Interactions" (R. J. Schweyen, K. Wolf, and F. Kaudewitz, eds.), pp. 191–201. de Gruyter, Berlin.

Saltzgaber-Muller, J., Kunapuli, S. P., and Douglas, M. G. (1983). Nuclear genes coding the yeast mitochondrial adenosine triphosphate complex. Isolation of *ATP2* coding the F_1-ATPase β subunit. *J. Biol. Chem.* **258,** 11465–11470.

Sanders, J. P. M., Heyting, C., Verbeet, M. P., Meijlink, C. P. W., and Borst, P. (1977). The organization of genes in yeast mitochondrial DNA. III. Comparison of the physical maps of the mitochondrial DNAs from three wild-type *Saccharomyces* strains. *Mol. Gen. Genet.* **157,** 239–261.

Saunders, G., Rogers, M. E., Adlard, M. W., and Holt, G. (1984). Chromatographic resolution of nucleic acids extracted from *Penicillium chrysogenum*. *Mol. Gen. Genet.* **194,** 343–345.

Scazzocchio, C., Brown, T. A., Waring, R. B., Ray, J. A., and Davies, R. W. (1983). Organization of the *Aspergillus nidulans* mitochondrial genome. *In* "Mitochondria 1983: Nucleo-Mitochondrial Interactions" (R. J. Schweyen, K. Wolf, and F. Kaudewitz, eds.), pp. 303–312. de Gruyter, Berlin.

Schatz, G., and Butow, R. A. (1983). How are proteins imported into mitochondria? *Cell (Cambridge, Mass.)* **32,** 316–318.

3. Fungal Mitochondrial Genomes

Schleyer, M., Schmidt, B., and Neupert, W. (1982). Requirements of a membrane potential for the posttranslational transfer of proteins into mitochondria. *Eur. J. Biochem.* **125**, 109–116.

Sebald, W., Hoppe, J., and Wachter, E. (1979). Amino acid sequence of the ATPase proteolipid from mitochondria, chloroplasts and bacteria (wild type and mutants). *In* "Function and Molecular Aspects of Biomembrane Transport" (E. Quagliariello, F. Palmieri, S. Papa, and M. Klingenberg, eds.), pp. 63–74. Elsevier/North-Holland, Amsterdam.

Seitz-Mayr, G., and Wolf, K. (1982). Extrachromosomal mutator inducing point mutations and deletions in mitochondrial genome of fission yeast. *Proc. Natl. Acad. Sci. U.S.A.* **79**, 2618–2622.

Simon, M., and Faye, G. (1984). Steps in processing of the mitochondrial cytochrome oxidase subunit I pre-mRNA affected by a nuclear mutation in yeast. *Proc. Natl. Acad. Sci. U.S.A.* **81**, 8–12.

Singer, M. F. (1982). Highly repeated sequences in mammalian genomes. *Int. Rev. Cytol.* **76**, 67–112.

Slonimski, P. P., and Ephrussi, B. (1949). Action de l'acriflavine sur les leuvres. V. Le système des cytochromes des mutants "petite colony." *Ann. Inst. Pasteur, Paris* **77**, 47–63.

Smith, T. M., Saunders, G., Stacey, L. M., and Holt, G. (1984). Restriction endonuclease map of mitochondrial DNA from *Penicillium chrysogenum*. *J. Biotechnol.* **1**, 37–46.

Sor, F., and Fukuhara, H. (1980). Séquence nucléotidique du gène de l'ARN ribosomique 15S mitochondrial de la leuvre. *C. R. Hebd. Seances Acad. Sci. Ser. D* **291**, 933–936.

Sor, F., and Fukuhara, H. (1983). Complete DNA sequence coding for the large ribosomal RNA of yeast mitochondria. *Nucleic Acids Res.* **11**, 339–348.

Specht, C. A., Novotny, C. P., and Ulrich, R. C. (1983). Isolation and characterization of mitochondrial DNA from the basidiomycete *Schizophyllum commune*. *Exp. Mycol.* **1**, 336–343.

Stepien, P. P., Bernard, U., Cooke, H. J., and Küntzel, H. (1978). Restriction endonuclease cleavage map of mitochondrial DNA from *Aspergillus nidulans*. *Nucleic Acids Res.* **5**, 317–330.

Stern, D. B., and Lonsdale, D. M. (1982). Mitochondrial and chloroplast genomes of maize have a 12-kilobase DNA sequence in common. *Nature (London)* **299**, 698–702.

Stohl, L. L., Collins, R. A., Cole, M. E., and Lambowitz, A. M. (1982). Characterization of two new plasmid DNAs found in mitochondria of wild-type *Neurospora intermedia* strains. *Nucleic Acids Res.* **10**, 1439–1458.

Tabak, H. F., VanderLaan, J., Osinga, K. A., Schooten, J. P., Van Boom, J. H., and Veeneman, G. H. (1981). Use of a synthetic DNA oligonucleotide to probe the precision of RNA splicing in a yeast mitochondrial petite mutant. *Nucleic Acids Res.* **9**, 4475–4483.

Tabak, H. F., Osinga, K. A., DeVries, E., Van der Bliek, A. M., Van der Horst, G. T. J., Groot Koerkamp, M. J. A., Van der Horst, G., Zwarthoff, E. C., and Macdonald, M. E. (1983). Initiation and transcription of yeast mitochondrial DNA. *In* "Mitochondria 1983: Nucleo-Mitochondrial Interactions" (R. J. Schweyen, K. Wolf, and F. Kaudewitz, eds.), pp. 79–93. de Bruyter, Berlin.

Terpstra, P., Zanders, E., and Butow, R. A. (1979). The association of var1 with the 38S mitochondrial ribosomal subunit in yeast. *J. Biol. Chem.* **254**, 12653–12661.

Thalenfeld, B. E., Hille, J., and Tzagoloff, A. (1983). Assembly of the mitochondrial membrane system. Characterization of the *oxi2* transcript and localization of its promoter in *Saccharomyces cerevisiae*. *J. Biol. Chem.* **258**, 610–615.

Timmis, J. N., and Scott, N. S. (1983). Sequence homology between spinach nuclear and chloroplast genomes. *Nature (London)* **305**, 65–67.

Tudzynski, P., and Esser, K. (1977). Inhibitors of mitochondrial function prevent senescence in the ascomycete *Podospora anserina*. *Mol. Gen. Genet.* **153**, 111–113.

Tudzynski, P., and Esser, K. (1979). Chromosomal and extrachromosomal control of senescence in the ascomycete Podospora anserina. Mol. Gen. Genet. **173**, 71–84.

Tudzynski, P., Stahl, U., and Esser, K. (1980). Transformation to senescence with plasmid like DNA in the ascomycete *Podospora anserina*. *Curr. Genet.* **2**, 181–184.
Tudzynski, P., Duvell, A., and Esser, K. (1983). Extrachromosomal genetics of *Claviceps purpurea*. I. Mitochondrial DNA and mitochondrial plasmids. *Curr. Genet.* **1**, 145–150.
Turner, G., Imam, G., and Küntzel, H. (1979). Mitochondrial ATPase complex of *Aspergillus nidulans* and the dicyclohexylcarbodiimide-binding protein. *Eur. J. Biochem.* **97**, 565–571.
Underbrink-Lyon, K., Miller, D. L., Ross, N. A., Fukuhara, H., and Martin, N. C. (1983). Characterization of a yeast mitochondrial locus necessary for tRNA biosynthesis. Deletion mapping and restriction enzyme studies. *Mol. Gen. Genet.* **191**, 512–518.
Van den Boogaart, P., Samallo, J., and Agsteribbe, E. (1982). Similar genes for a mitochondrial ATPase subunit in the nuclear and mitochondrial genomes of *Neurospora crassa*. *Nature (London)* **298**, 187–189.
van Loon, A. P. G. M., de Groot, R. J., van Eyk, E., Van der Horst, G. T. J., and Grivell, L. A. (1982). Isolation and characterization of nuclear genes coding for subunits of the yeast ubiquinol–cytochrome *c* reductase complex. *Gene* **20**, 323–337.
Viebrock, A., Perz, A., and Sebald, W. (1982). The imported preprotein of the proteolipid subunit of the mitochondrial ATPase synthase from *Neurospora crassa*. Molecular cloning and sequencing of the RNA. *EMBO J.* **1**, 565–571.
Wallace, D. C. (1982). Structure and evolution of organelle genomes. *Microbiol. Rev.* **46**, 208–240.
Ward, B. L., Anderson, R. S., and Bendich, A. J. (1981). The mitochondrial genome is large and variable in a family of plants (Cucurbitaceae). *Cell (Cambridge, Mass.)* **25**, 793–803.
Weiss-Brummer, B., Rödel, G., Schweyen, R. J., and Kaudewitz, F. (1982). Expression of the split gene *cob* in yeast: evidence for a precursor of a "maturase" protein translated from intron 4 and preceding exons. *Cell (Cambridge, Mass.)* **29**, 527–536.
Weiss-Brummer, B., Holl, J., Schweyen, R. J., Rödel, G., and Kaudewitz, F. (1983). Processing of yeast mitochondrial RNA: involvement of intramolecular hybrids in splicing of *cob* intron 4 RNA by mutation and reversion. *Cell (Cambridge, Mass.)* **33**, 195–202.
Weslowski, M., and Fukuhara, H. (1981). Linear mitochondrial deoxyribonucleic acid from the yeast *Hansenula mrakii*. *Mol. Cell. Biol.* **1**, 387–383.
Whitfeld, P. R., and Bottomley, W. (1983). Organization and structure of chloroplast genes. *Annu. Rev. Plant Physiol.* **34**, 279–310.
Wright, R. M., and Cummings, D. J. (1983). Integration of mitochondrial gene sequences within the nuclear genome during senescence in a fungus. *Nature (London)* **302**, 86–88.
Wright, R. M., Horrum, M. A., and Cummings, D. J. (1982). Are mitochondrial structural genes selectively amplified during senescence in *Podospora anserina*? *Cell (Cambridge, Mass.)* **29**, 505–515.
Wright, R. M., Ko, C., Cumsky, M. G., and Poyton, R. O. (1984). Isolation and sequence of the structural gene for cytochrome *c* oxidase subunit VI from *Saccharomyces cerevisiae*. *J. Biol. Chem.* **259**, 15401–15407.
Yaffe, M. P., and Schatz, G. (1984). Two nuclear mutations that block mitochondrial protein import in yeast. *Proc. Natl. Acad. Sci. U.S.A.* **81**, 4819–4823.
Yin, S., Heckman, J., and RajBhandary, U. L. (1981). Highly conserved GC-rich palindromic DNA sequences flank tRNA genes in *Neurospora crassa* mitochondria. *Cell (Cambridge, Mass.)* **26**, 325–332.
Yin, S., Burke, J., Chang, D. D., Browning, K. S., Heckman, J. E., Alzner-DeWeerd, B., Potter, M. J., and RajBhandary, U. L. (1982). *Neurospora crassa* mitochondrial tRNAs and rRNAs: structure, gene organization and DNA sequences. *In* "Mitochondria 1983: Nucleo-Mitochondrial Interactions" (R. J. Schweyen, K. Wolf, and F. Kaudewitz, eds.), pp. 361–373. de Gruyter, Berlin.
Zassenhaus, H. P., Farrelly, F., Hudspeth, M. E. S., Grossman, L. I., and Butow, R. A. (1983).

3. Fungal Mitochondrial Genomes

Transcriptional analysis of the *Saccharomyces cerevisiae* mitochondrial *var1* gene: anomalous hybridization of RNA from AT-rich regions. *Mol. Cell. Biol.* **3**, 1615–1624.

Zassenhaus, H. P., Martin, N. C., and Butow, R. A. (1984). Origins of transcripts of the yeast mitochondrial *var1* gene. *J. Biol. Chem.* **259**, 6019–6028.

Zaug, A. J., Grabowski, P. J., and Cech, T. R. (1983). Autocatalytic cyclization of an excised intervening sequence RNA in a cleavage–ligation reaction. *Nature (London)* **301**, 578–583.

Zinn, A., and Butow, R. A. (1985). Kinetics and intermediates of yeast mitochondrial DNA recombination. *Cold Spring Harbor Symp. Quant. Biol.* **49**, 115–121.

4

Modeling the Environment for Gene Expression

BARBARA E. WRIGHT
Department of Microbiology
University of Montana
Missoula, Montana

I.	Introduction	105
II.	Modeling Metabolism under Steady-State Conditions	106
	A. Illustrating the Approach Using Simple Theoretical Models	107
	B. Discovering Metabolic Compartments	108
	C. Discovering Rate-Limiting Steps	112
III.	Modeling Metabolism during Product Accumulation	115
	A. Illustrating the Approach Using a Complex Transition Model	117
	B. Discovering How to Make More of a Product or How to Make It Sooner	119
	References	121

I. INTRODUCTION

To understand selective gene expression we must eventually understand how the inducers and repressors that orchestrate genetic activity are distributed within a cell and within an organism having different cell types. We must also understand the intracellular and intercellular distribution of gene products relative to the distribution of the substrates, effectors, and other cellular components with which gene products interact and upon which gene expression depends. Complicated metabolic processes such as differentiation, aging, industrial fermentations, and dimorphism in fungi clearly involve the interaction and interdependence of many different kinds of cellular structures and biochemical events. An in-depth understanding of any such process will ultimately require a dynamic systems analysis, that is, an analysis under *in vivo* conditions of the physical and biochemical relationships of the relevant organelles, enzymes, genes, mRNA, inducers, precursors, energy sources, and metabolites within the living systems. As cell rupture alters or destroys metabolic organization, analytical approaches to understanding this aspect of living systems are very limited. Kinetic modeling represents a powerful technique in the analysis of metabolism *in vivo*.

II. MODELING METABOLISM UNDER STEADY-STATE CONDITIONS

In the analysis of complicated metabolic processes, it is logical first to examine the dynamics of metabolism at single points in time, and then proceed to an understanding of the additional complexities of a metabolic transition from one point in time to another. To construct a steady-state model, data must be obtained over a relatively brief period of time, during which the concentrations of metabolites do not change significantly, and the flux into each "pool" is equal to the flux out. Overall flux through the metabolic pathway should be known (e.g., from O_2 consumption, the rate of product accumulation or substrate utilization), as well as the total cellular concentration of each metabolite. To assess the rate of flux of reactions *in vivo*, cells can be exposed to tracer (nonperturbing) levels of a radioactive metabolite over brief periods of time under steady-state conditions. Cell samples are then removed at successive points in time, the metabolites isolated, and their specific radioactivities (SRs) determined. A computer program called TFLUX has been developed to simulate the changes in SR of each metabolite pool as a function of time (Sherwood *et al.*, 1979).

The systems that TFLUX can simulate consist of chemically distinct species (pools) with time-varying SRs. Chemical reactions or transport mechanisms provide an interchange of tracers among pools. The metabolic system must be in steady state; that is, all pool sizes must be constant (the pool size is the sum of labeled and unlabeled material), and the total (labeled plus unlabeled) fluxes between pools must also be constant. TFLUX solves a set of simultaneous first-order differential equations to simulate the SRs for all pools as a function of time. To use TFLUX, the user specifies a number of pools, giving the name, size (millimolar cell volume), and initial SR for each pool. The interconnections between pools are specified by giving the (constant) flux between any pair of pools. Flux is expressed as the millimolar concentration per minute, which is equivalent to the number of micromoles per minute per milliliter cell volume. Unknown fluxes and compartmentation ratios of certain metabolites are varied until the SR curves generated by the model match the data. It is also possible to specify "external" pools, whose properties include infinite size and constant SR.

A. Illustrating the Approach Using Simple Theoretical Models

A hypothetical metabolic pathway, called model I, is depicted in Fig. 1. Pools ^{14}C-A, A, B, and C are all 0.01 mM in concentration, while the end product is 200 times larger. Overall flux through the system is 0.005 mM/min. Pools ^{14}C-A and A equilibrate so rapidly that, under these conditions with the time scale

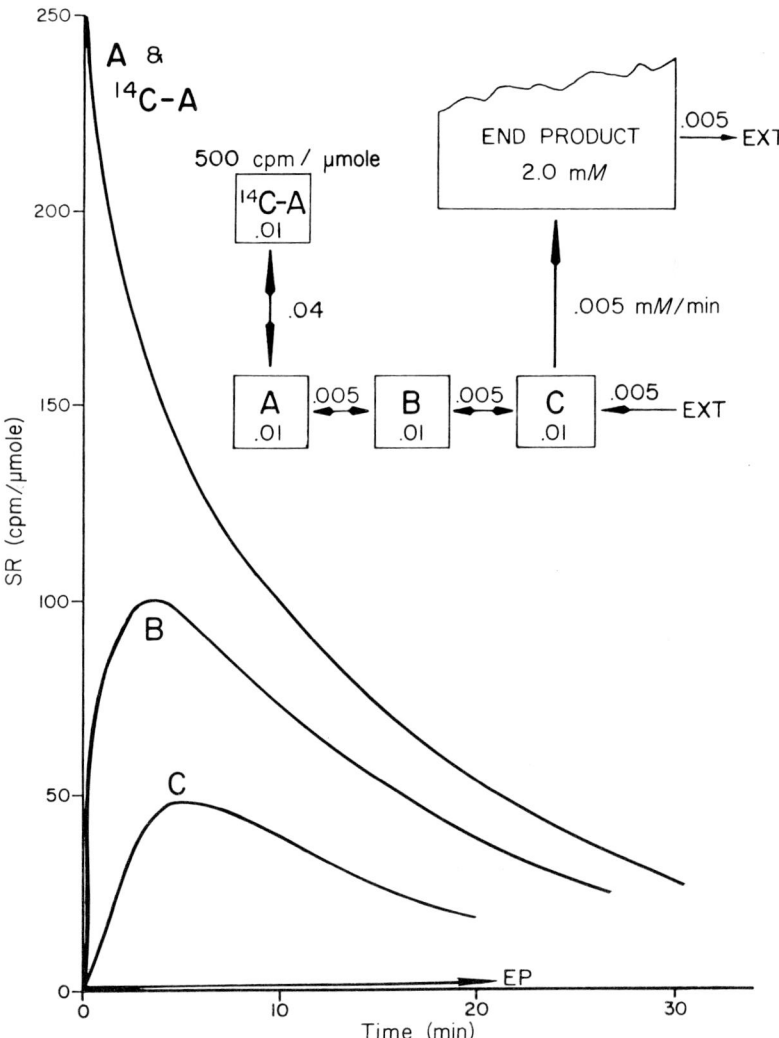

Fig. 1. A theoretical model, called model I, showing the specific radioactivity labeling patterns of each metabolite pool, given the pool sizes, fluxes, and specific radioactivity of ^{14}C-A indicated in the model. EXT, external pool. See text for details.

shown, they fall in SR at virtually the same rate. Pool B peaks at less than half the maximum A SR, and C peaks at about half the maximum B SR. The SR of the end product is very low, and the kinetics of labeling is relatively linear compared to the small, rapidly turning over pools. To maintain the steady-state conditions required by the TFLUX program, pool C must receive a flux of 0.005 mM/min from an external (EXT) unlabeled pool, and flux out of the end product pool must equal flux into it. This latter efflux can simply be thought of as the rate at which the end product accumulates. The cold flux from the external pool into C can be thought of as arising from endogenous metabolism, e.g., the diphosphates of the bases of RNA resulting from RNA turnover, or the hexose phosphates generated in the turnover of glycogen (Wright and Kelly, 1981).

Depending upon the area of metabolism under analysis, a metabolic pool in the position of B may be the one receiving a flux of unlabeled material from endogenous metabolism. In that event, model II (Fig. 2) may be more appropriate. Indeed, the experimental data available on metabolite levels and SRs should suggest the kind of model to consider. It is interesting to compare the characteristics of models I and II. The dashed lines indicate only those parameters of model II that differ from those of I (solid lines). Thus, since a flux of 0.005 from an external pool is entering B, a unidirectional flux of 0.005 must go to C from B. Furthermore, a flux must not go to C from an external pool, or steady-state conditions would not exist. The difference in metabolite SR patterns (dashed lines for model II, solid for I) is striking. Pool A' falls faster than A, as the cold external flux is nearer to it; B' peaks earlier and has a lower SR than B, as it now receives cold endogenous material directly. Pool C', on the other hand, now has a higher SR than C, as it is not cooled directly by an external pool.

Let us consider one more type of model—model III in Fig. 3. In this case, a source of cold endogenous material mixes directly with the radioactive tracer used—e.g., glutamate generated from protein turnover, mixing rapidly with the tracer [^{14}C]glutamate added to label intermediates of the citric acid cycle (Kelly *et al.*, 1979a,b). In such a situation, as might be expected, label is lost very quickly from the system (dotted lines).

B. Discovering Metabolic Compartments

Model I will be used to illustrate how simulation analyses can yield information about metabolic compartments and rate-limiting steps *in vivo*.

Let us suppose that pool C is larger than we estimate, based on the amount we isolate, e.g., part of it is enzyme-bound, and only the unbound material is isolated. However, *in vivo*, the entire poor is metabolically homogeneous. In this case, if C is actually 0.04 mM instead of 0.01 mM (dashed lines, Fig. 4), it will take longer to label it, the SR will be lower, and it will peak later (C' compared to C). Since B is being cooled by a larger pool, it will also have a lower SR, as

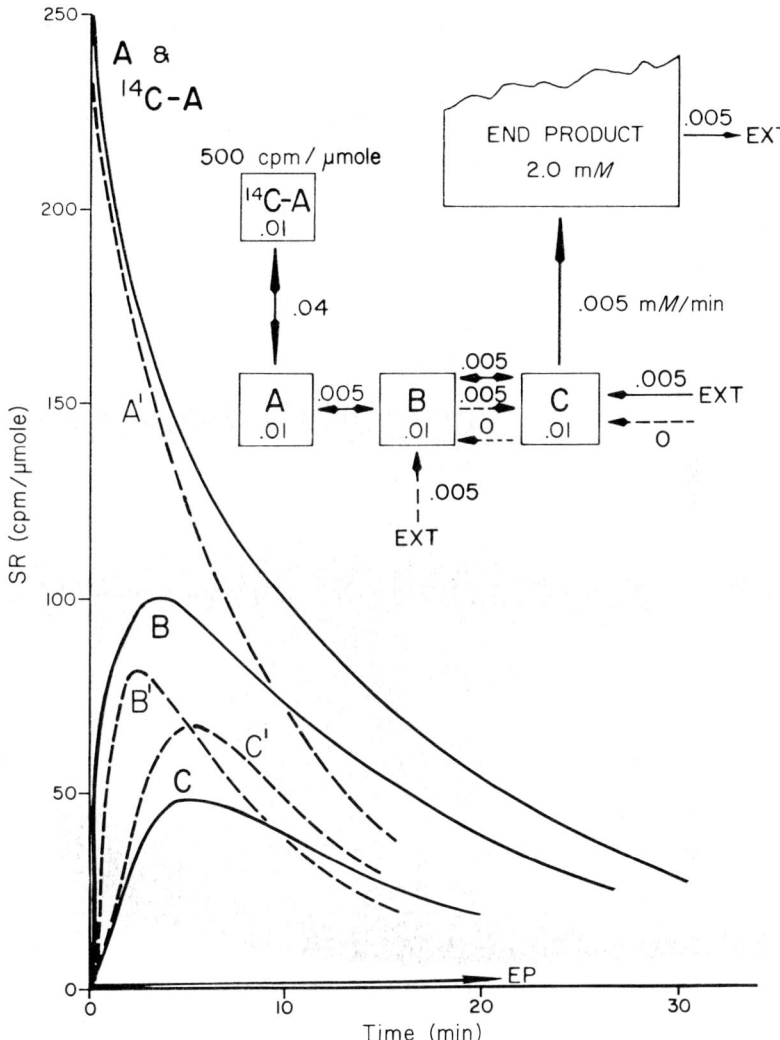

Fig. 2. A theoretical model, called model II, indicated by the dashed lines, A', B', and C', compared to model I (solid lines). See Fig. 1 and text for further details.

will A (B' and A' compared to B and A). Could an increased flux between B and C (dotted line) compensate in such a way as to mask the presence of a larger pool of C? Yes, with respect to the SR of C, increasing this flux from 0.005 to 0.05 does compensate (compare C″ to C' and C). However, B″ does not approach B, but approaches the behavior of C″, because of their increased rate of equilibration. Thus, the behavior of the B pool SR may indicate which circumstances prevail *in vivo*.

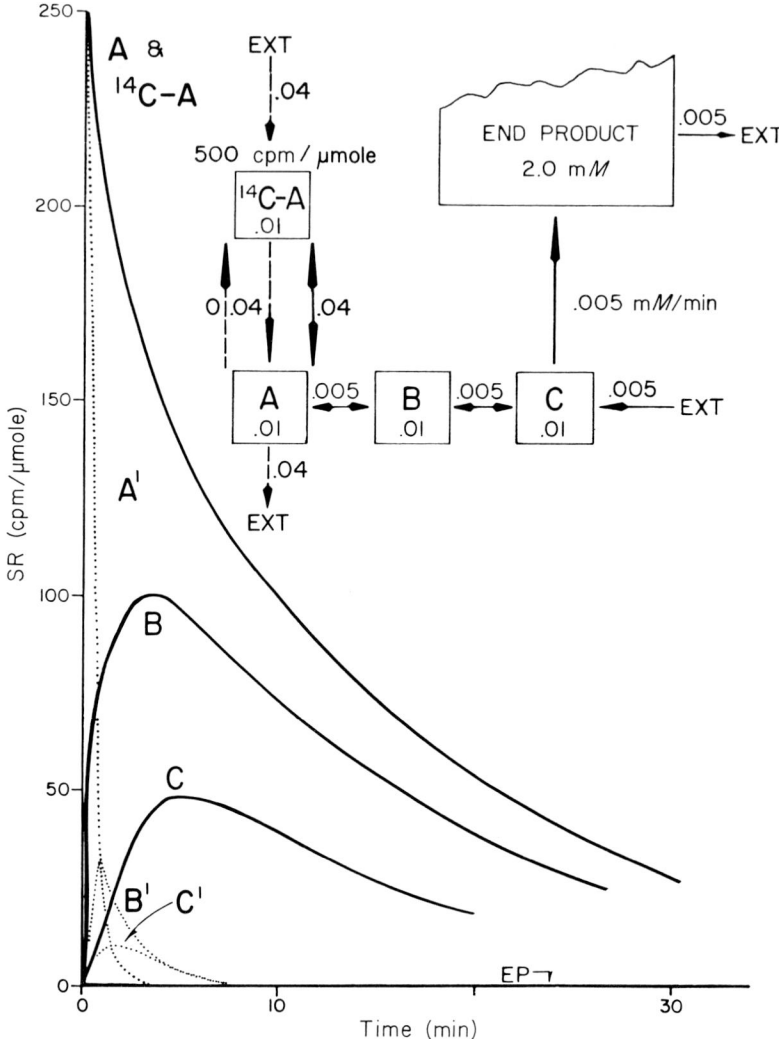

Fig. 3. A theoretical model, called model III, indicated by the dashed lines, A′, B′, and C′, compared to model I (solid lines). See Fig. 1 and text for further details.

Now consider a different kind of metabolic compartmentation, in which there are two physically separate pools of C, called C and CC (Fig. 5). When cells are ruptured, C and CC mix, so the

$$SR = \frac{C + CC \text{ total counts per minute}}{C + CC \text{ total micromoles}}$$

4. Modeling the Environment for Gene Expression

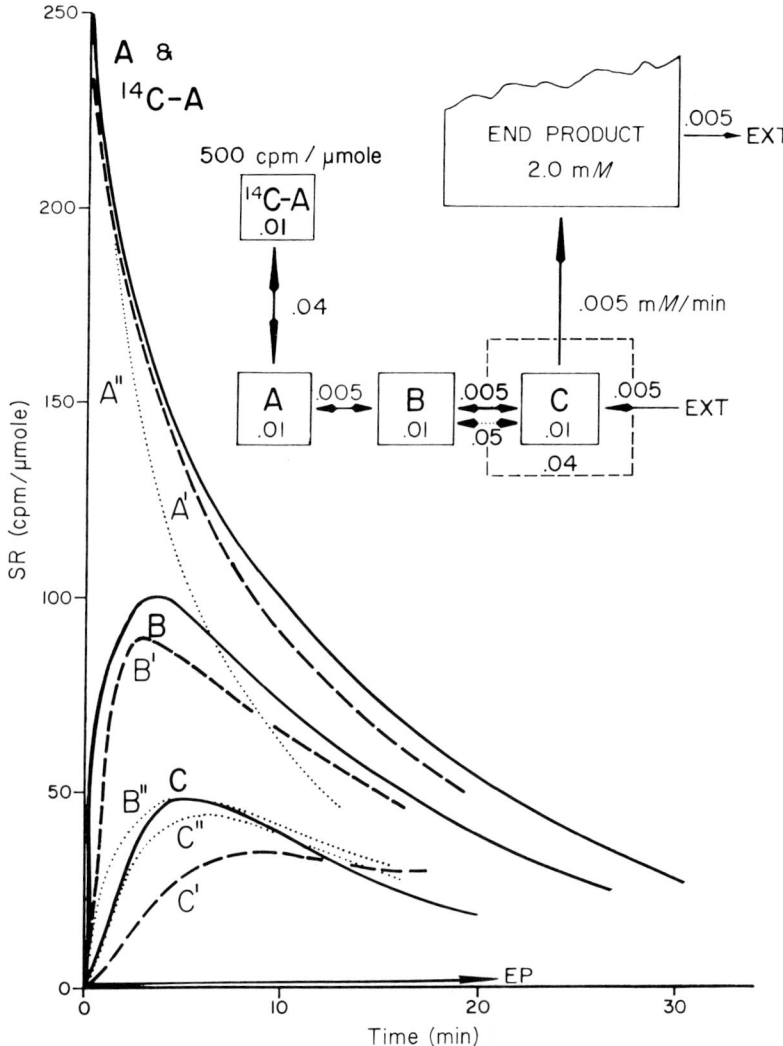

Fig. 4. A variation on model I, showing the specific radioactivity patterns assuming that pool C is 4 times larger (dashed lines, A′, B′, and C′), or that pool C is 4 times larger and the rate of exchange between C and B is 10 times higher (dotted lines, A″, B″, and C″). See text for further details.

If no exchange occurs between C and CC, the SR will be exactly half and the peak time will be identical (lower dashed line). As the rate of exchange between C and CC increases, the SR curve approaches that of C alone. The SR of B is affected very little with this kind of compartmentation, compared to that discussed previously (Fig. 4).

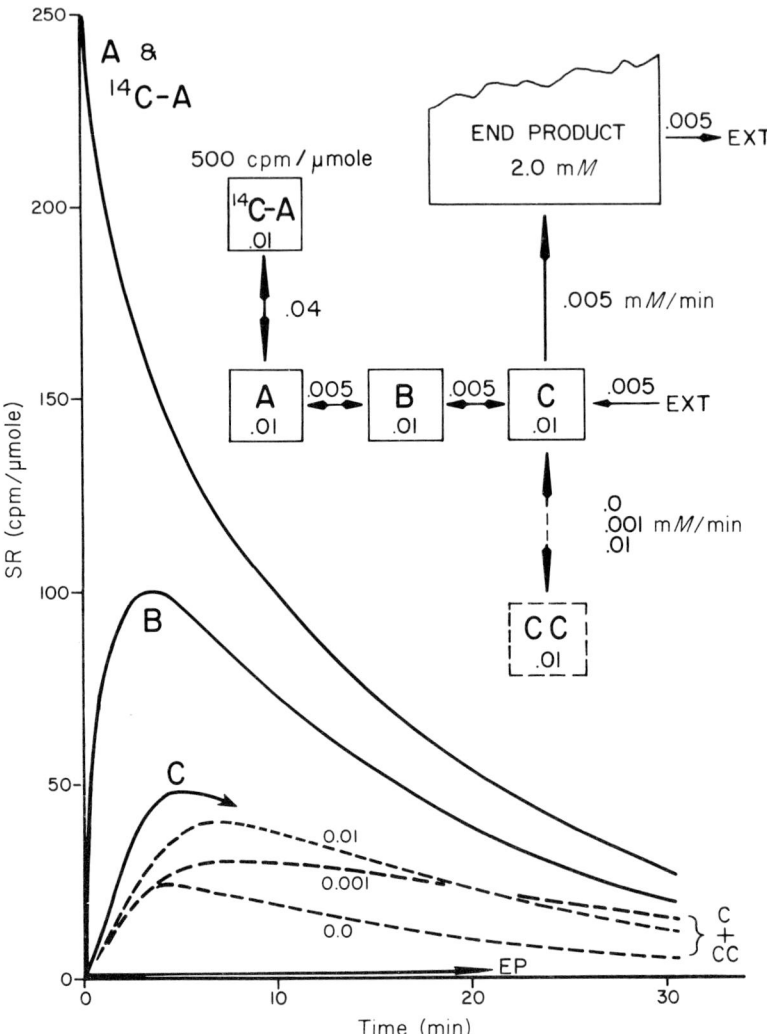

Fig. 5. A variation on model I, assuming that metabolite C exists in two separate compartments (C and CC), which mix on cell rupture. Specific radioactivity patterns for the mixed pools are shown (dashed lines) at different rates of exchange between them.

C. Discovering Rate-Limiting Steps

Perhaps it is important to determine which step in a pathway is rate-limiting, in order to try and increase overall flux and hence the rate of end product accumulation. Figure 6 indicates the pattern of SR curves if the exchange between A and B is 6 times more than between B and C (dotted lines). Pool A is cooled more

4. Modeling the Environment for Gene Expression

rapidly by B and C (dotted line, A′), but B and, to a lesser extent, C have higher SRs and peak earlier (B′ and C′). Contrast this situation with one in which the exchange between A and B is 6 times less than between B and C (Fig. 7, dotted lines). In this case, B is lower in SR (B′), and the peak time is not affected; A′ and C′ are qualitatively similar to Fig. 6. It was interesting to examine the effect of the type of model on this analysis of rate-limiting steps. That is, using model

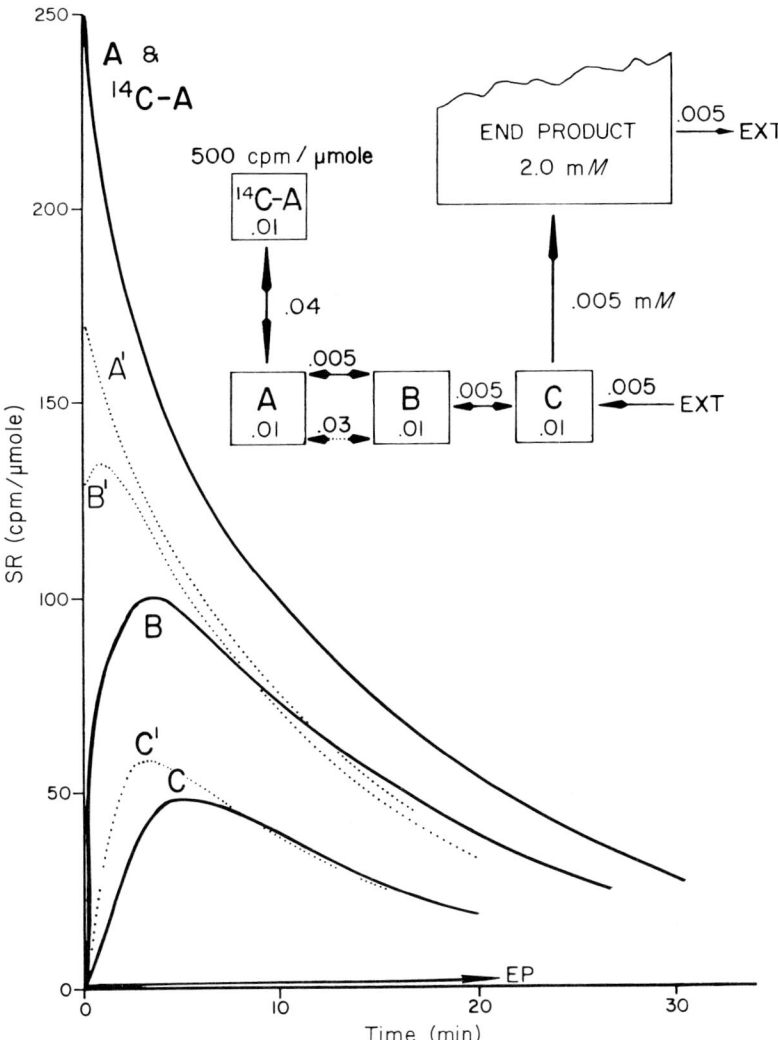

Fig. 6. A variation on model I showing specific radioactivity patterns when the exchange between A and B is 6 times higher (dotted lines, A′, B′, and C′).

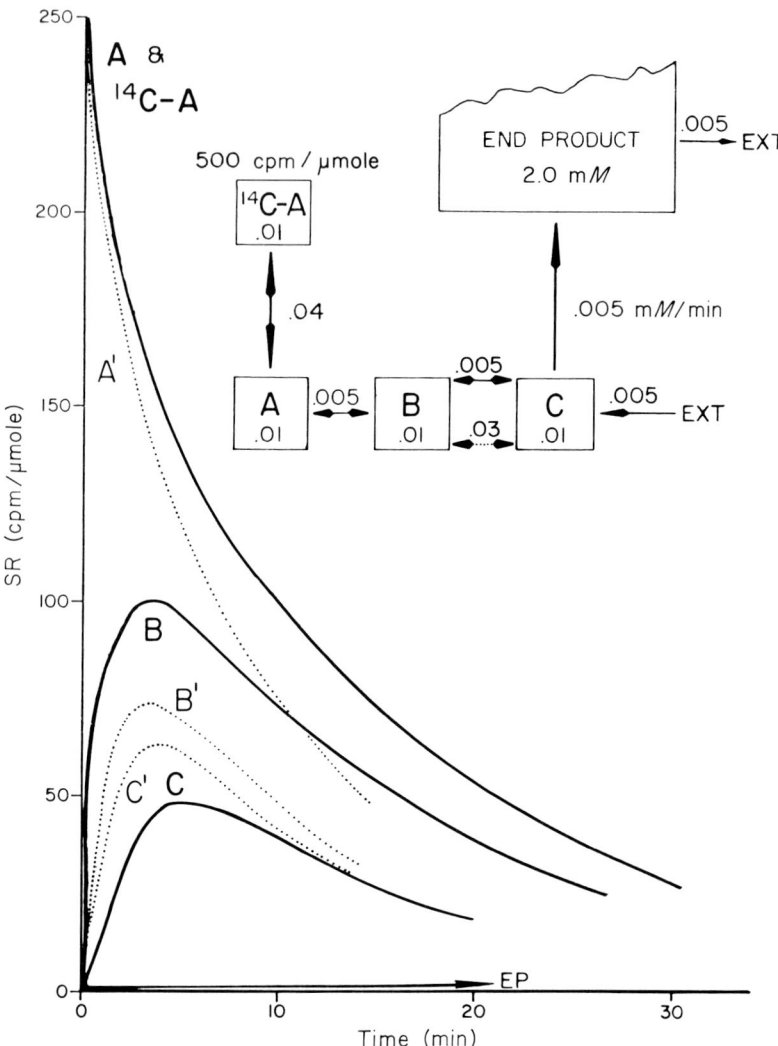

Fig. 7. A variation on model I showing specific radioactivity patterns when the exchange between B and C is 6 times higher (dotted lines, A′, B′, and C′).

II or III, would the relative SR patterns be similar? Surprisingly, yes. Table I indicates that for all three models: (1) as either exchange rate increases, SR peak times are the same or earlier; (2) as A ↔ B increases, SR values for B and C increase; (3) as B ↔ C (or B → C) increases, the SR value for B decreases and for C increases.

This analysis serves to illustrate that, knowing the overall flux through a

4. Modeling the Environment for Gene Expression

TABLE I.

Comparing Models I, II, and III in the Analysis of Rate-Limiting Steps

Flux (mM/min)	Metabolite	Peak time (min)			Value (SR)		
		I	II	III	I	II	III
A ↔ B = 0.005 ⎫	B	4	2.5	1	99	83	31
B ↔ C = 0.005^a ⎭	C	5	5	2	48	69	10
A ↔ B = 0.03	B	1	1	1	136	132	58
	C	3	4	1	60	90	21
B ↔ C = 0.03^a	B	4	1	1	74	28	21
	C	4	3.5	1	64	109	17

^{a} Or B → C = 0.005 for model II.

pathway, total pool sizes, and the maximum values and shapes of the SR curves, it is possible to obtain information about the compartmentation of metabolites, the relative sizes of the compartments, and the rates of exchange between them. It is also possible to gain some insight into rate-limiting steps, regardless of a detailed knowledge of the metabolic relationships operative *in vivo*. Clearly, metabolite SR data cannot be interpreted without a model, even in very simple pathways. Using the TFLUX program, steady-state models have been constructed for the analysis of intra- and extramitochondrial pools in *Dictyostelium* (Kelly *et al.*, 1979a,b), as well as for the analysis of intercellular metabolic compartments in that organism (Wright *et al.*, 1982).

III. MODELING METABOLISM DURING PRODUCT ACCUMULATION

As mentioned in Section I, an understanding of metabolic compartments and flux relationships at one point in time is a necessary starting point for the construction of a transition model simulating, for example, the accumulation of a specific end product over a period of hours. Such models are much more complex, requiring a computer program such as METASIM (Park and Wright, 1973). Table II compares some characteristics of the two kinds of model analysis.

Enzyme kinetic mechanisms and constants determined *in vitro* are assumed to be valid *in vivo* and are used as input in the transition models [(e) and (f)]. However, enzyme specific activities or enzyme "tissue contents" determined *in vitro* are not assumed to be valid *in vivo*. Enzyme activity in crude extracts rarely reflects activity in the intact cell or organism. Artifacts can arise during the

TABLE II.

A Comparison of Steady-State and Transition Models

Type of model:	Steady-state	Transition
Computer program:	TFLUX	METASIM
Input required:	(a) Metabolic map	(a) Metabolic map
	(b) Overall flux	(b) Initial fluxes
	(c) Metabolite concentrations	(c) Initial metabolite concentrations
	(d) SR tracer	(d) Compartments
		(e) Enzyme mechanisms
		(f) Kinetic constants
		(g) Calculated enzyme activities
Parameters varied:	(a) Unknown compartmentation ratios	(a) Enzyme activity/time
	(b) Unknown fluxes	(b) Enzyme mechanisms
		(c) Metabolite compartments
		(d) Flux relationships
		(e) External perturbations
Output matched with:	(a) Experimental SR curves	(a) Experimental metabolite concentration/time
		(b) Experimental flux/time

preparation of cell extracts: enzyme activators or inhibitors may be diluted out, concentrated, lost, or created; compartmentation at the level of organelles, enzyme complexes, or enzyme–substrate complexes may be destroyed. Enzyme-to-substrate ratios *in vivo* are usually orders of magnitude higher than those used for *in vitro* assays. Finally, enzymes usually are assayed under nonphysiological conditions of temperature, pH, substrate, and effector concentration. Indeed, in our experience, there is a very poor correlation between enzyme activities *in vivo* and *in vitro*. Thus, in the enzyme kinetic expressions the enzyme activities [(g), comparable to V_{max}], are *calculated* as the only unknown, based on the other parameters in the rate expression, i.e., the rate of the reaction, metabolite levels, and enzyme kinetic constants. The computer program METASIM integrates all of the input information for the metabolic network to be simulated. The differential equations describing each reaction are solved simultaneously, and output from the model consists of changes in reaction rates and metabolite concentrations as a function of time. This output must be consistent with experimental data obtained during the metabolic transition under analysis.

A. Illustrating the Approach Using a Complex Transition Model

Over the past 25 years about 40 investigators, working for various periods of time, obtained the data necessary for the construction of transition models of carbohydrate metabolism (reactions R1–R26) and energy metabolism (reactions 1–13) in *Dictyostelium discoideum*. This organism undergoes a simple differentiation process in which one cell type (an amoeba) becomes two (stalk and spore). These models, which continuously evolve in complexity as data become available and predictions are tested, are shown in Fig. 8.

Ninety percent of the parameters composing these models are data. These data are of four kinds: (1) evidence for the existence of all the reactions and metabolic compartments depicted in the metabolic map; (2) the cellular concentrations (millimolar packed cell volume) over the course of differentiation and aging of all the metabolites shown; (3) reaction rates determined *in vivo* with isotope tracers (all of the reaction rates represented by solid lines have been so determined); and (4) enzyme kinetic mechanisms and constants (all enzymes catalyzing the numbered reactions within circles have been purified and kinetically characterized). Approximately 45 flux patterns, 100 metabolic accumulation profiles (for normal and perturbed differentiation), 20 enzyme mechanisms, and 60 kinetic constants have been determined. As mentioned above, the enzyme activities are calculated, as the only unknown, and are expressed as a function of time (Wright and Kelly, 1981). The double boxes represent metabolites in stalk cells, which become very permeable as they are dying, but still retain a vestigial type of metabolism. The broken boxes on the edge of the model represent exogenous metabolites with which the organism and the model have been perturbed, to test the model under new conditions. Approximately one-fourth of the ATP generated from protein degradation can be accounted for by known metabolic demands: protein synthesis, RNA synthesis, and the ATP-dependent reactions indicated in the model. The contribution of carbohydrate metabolism to energy metabolism in this system is not significant; i.e., R24 is 0.01 of the flux from protein into the citric acid cycle. The more complicated, constrained, and unique a model becomes, the more nearly it approaches the complex and interdependent relationships characteristic of metabolic networks in living cells.

We believe these models reflect many aspects of metabolism *in vivo,* because predictive value frequently has been demonstrated. This, of course, is the ultimate test of any model. Of the many predictions made thus far, about 30 have been tested, and 90% of these substantiated. These predictions have been concerned with the flux and turnover of metabolites *in vivo;* the activity patterns of enzymes both *in vivo* and *in vitro;* the kinetic mechanisms and constants of specific reactions; the presence of enzyme inhibition *in vivo;* the effects of

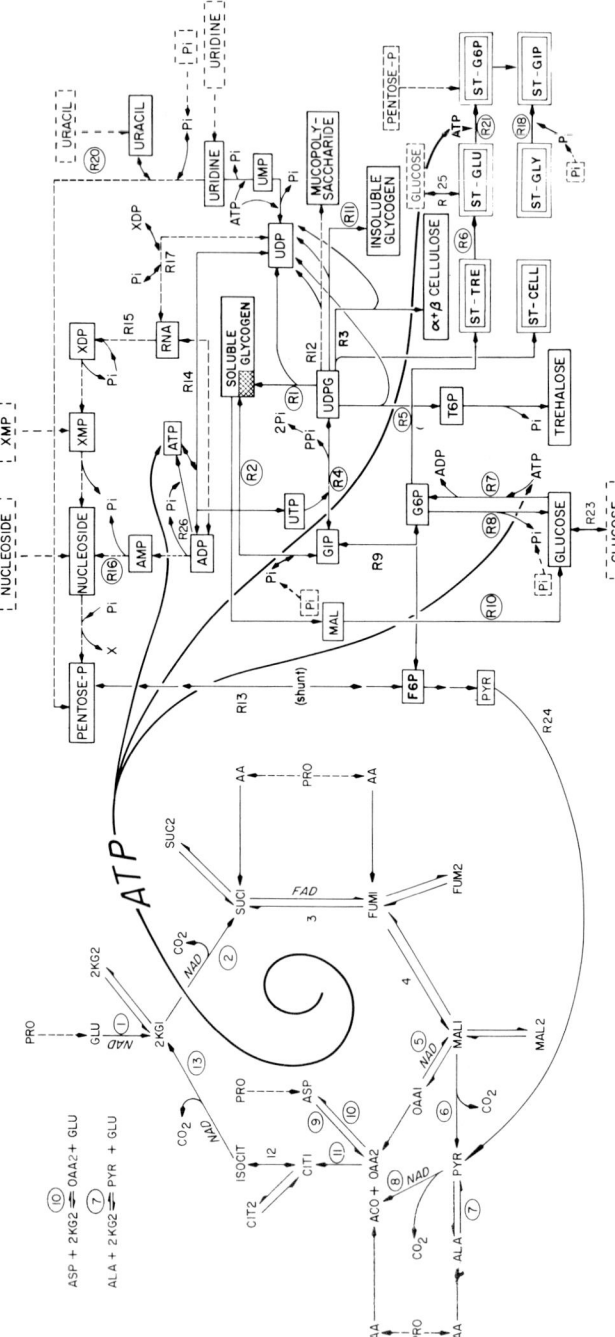

Fig. 8. The metabolic pathways and reactions representing kinetic models of energy and carbohydrate metabolism in *Dictyostelium discoideum*. Abbreviations: PRO, protein; GLU, glutamate; ALA, alanine; ASP, aspartate; AA, amino acids; 2KG1 and 2KG2, intra- and extramitochondrial 2-ketoglutarate; SUC1 and SUC2, intra- and extramitochondrial succinate; FUM1 and FUM2, intra- and extramitochondrial fumarate; MAL1 and MAL2, intra- and extramitochondrial malate; OAA1 and OAA2, two pools of oxaloacetate; PYR, pyruvate; ACO, acetyl-CoA; CIT1 and CIT2, intra- and extramitochondrial citrate; ISOCIT, isocitrate; CO_2, carbon dioxide; P_i, inorganic phosphate; PP_i, inorganic pyrophosphate; UTP, uridine triphosphate; G1P, glucose 1-phosphate; G6P, glucose 6-phosphate; ATP, adenosine triphosphate; T6P, trehalose 6-phosphate; UMP, uridine monophosphate; UDP, uridine diphosphate; UDPG, uridine diphosphoglucose; ADP, adenosine diphosphate; AMP, adenosine monophosphate; XMP and XDP, a mixture of the mono- and diphosphates of the bases in RNA. The double boxes represent stalk (ST)- specific metabolites. The metabolites in broken boxes are those used to perturb the organism and the model. See text for further details.

4. Modeling the Environment for Gene Expression

external metabolites or altered enzyme activities on respiration and on endogenous metabolite accumulation patterns; and permeability patterns, intra-, and intercellular compartmentation of metabolites and enzymes (Wright and Kelley, 1981; Wright et al., 1982; Wright and Emyanitoff, 1983; and unpublished observations). We have identified over 100 rate-limiting substrates, inhibitors, effectors, and enzymes controlling 40 interdependent reactions essential to biochemical differentiation in this organism.

B. Discovering How to Make More of a Product or How to Make It Sooner

The accumulation of one metabolite in the *Dictyostelium* system will be used to illustrate how model analysis can provide information allowing us to influence metabolite levels.

Trehalose accumulates from a level of $0.2-10$ mM (expressed in terms of cell volume) during the metabolic transition from the undifferentiated to the differentiated organism. Model analysis helped us understand the rate-limiting events controlling the accumulation of this disaccharide. Although the model represents an integrated system, with all pathways interacting and interdependent, the focus of the following discussion will concern those predictions and insights especially relevant to trehalose accumulation (for references, see Wright and Kelly, 1981).

1. *In vitro* evidence indicated that the specific activity of trehalose-6-phosphate (T6P) synthase (catalyzing R5) increased over the course of differentiation. However, simulations predicted that, because G6P and UDPG accumulated to maximum levels during the same period of time, this enzyme could not increase similarly in activity *in vivo*, as such an increase would prevent the accumulation of the two substrates. More than a year after this prediction was made, it was demonstrated that observed changes *in vitro* in the activity of the T6P synthase do not reflect activity *in vivo:* using [^{14}C]glucose, the rate of trehalose synthesis was determined, based on the specific radioactivity of its precursors. This rate was negligible from aggregation until late in the culmination process, at which time a 100-fold burst in the rate of synthesis occurred. It was also demonstrated that T6P synthase was "masked" *in vitro*, and that artifacts were present that interfered with the assessment of enzyme activity in crude extracts.

2. It was predicted that increasing the activity of glycogen phosphorylase during differentiation would enhance the levels of trehalose, G6P, and UDPG, in that order, but have little effect on cellulose levels. This prediction was unknowingly substantiated by Hames and Ashworth (1974), who determined the levels of glucose, G6P, UDPG, trehalose, and cellulose in an axenic mutant with higher levels of glycogen phosphorylase.

3. From observations of model behavior following perturbation by a flux of glucose, it was predicted that glucose perturbation of *Dictyostelium* should result

in an increase in the levels of trehalose and glycogen but not cellulose, and that P_i perturbation should increase trehalose but not glycogen levels. These predictions were substantiated.

4. It is apparent from the kinetic constants for T6P synthase and the cellular concentrations of UDPG and G6P that the latter substrate is in all probability the most rate-limiting *in vivo*. Analysis of the model suggested that uracil should inhibit trehalose synthesis by lowering G6P levels, primarily by utilizing pentose phosphate (a source of G6P) to form uridine (R20). Based on the kinetic constants for glucokinase, G6P phosphatase, and uridine phosphorylase, glucose should overcome this inhibition by increasing G6P levels via R7. Inorganic phosphate should enhance G6P accumulation and should overcome uracil inhibition by stimulating R20 (and hence R13) and by inhibiting R8. These predictions were substantiated.

5. Our dynamic understanding of the metabolic network leading to trehalose accumulation indicated that T6P synthase does not limit the rate of trehalose synthesis or the cessation of trehalose accumulation. This prediction was substantiated by the observation that exogenous glucose could enhance trehalose levels fivefold in mature (30-hr) sorocarps, many hours after the maximum levels of trehalose had accumulated. Thus, glucose equivalents are not only rate-limiting during differentiation, but are also critical variables in the cessation of biochemical differentiation.

Most of the above predictions and insights regarding the regulation of trehalose synthesis and accumulation resulted from the construction and analysis of kinetic models of this system. Although the primary purpose of these analyses was to understand the mechanisms involved, this understanding could be used to enhance the levels of trehalose. We may then end this analysis with one more prediction: the most efficient conversion of exogenous glucose to trehalose should occur in aged sorocarps in the presence of high levels of P_i. Although P_i accumulates during development, it does so only in stalk cells (Rutherford, 1976), which would no longer contain trehalose or be metabolically active. In the spore cells, however, G6P and trehalose levels would be enhanced by the stimulation of R7 by glucose, by the stimulation of R20 (R13) by P_i, and by the inhibition of R8 (and perhaps R4) by P_i. Glucose added at this time, over relatively brief (e.g., 1-hr) periods, should primarily be converted to trehalose; glycogen phosphorylase is undetectable, and glycogen turnover should have decreased to a very low level.

Every organism must have its own unique model, and all the data in the model should be derived from that organism. Moreover, to construct a realistic model with predictive value, the model should consist primarily of data. While this is no small task, an understanding of the complexities of metabolism in living cells is surely the most direct route to the identification of rate-limiting steps and, ultimately, to influencing the course of metabolism *in vivo*.

ACKNOWLEDGMENTS

This work was supported by Public Health Service grants AG03821 and AG03884 from the National Institutes of Health.

REFERENCES

Hames, B. D., and Ashworth, J. M. (1974). The metabolism of macromolecules during the differentiation of myxamoebae of the cellular slime mould *Dictyostelium discoideum* containing different amounts of glycogen. *Biochem. J.* **142,** 301–315.

Kelly, P. J., Kelleher, J. K., and Wright, B. E. (1979a). The tricarboxylic acid cycle in *Dictyostelium discoideum*. Metabolite concentrations, oxygen uptake and ^{14}C-amino acid labeling patterns. *Biochem. J.* **184,** 581–588.

Kelly, P. J., Kelleher, J. K., and Wright, B. E. (1979b). The tricarboxylic acid cycle in *Dictyostelium discoideum*. A model of the cycle at preculmination and aggregation. *Biochem. J.* **184,** 589–597.

Park, D. J. M., and Wright, B. E. (1973). METASIM, a general purpose metabolic simulator for studying cellular transformations. *Comp. Programs Biomed.* **3,** 10–26.

Rutherford, C. L. (1976). Cell specific events occurring during development, *J. Embryol. Exp. Morphol.* **35,** 335–343.

Sherwood, P., Kelly, P. J., Kelleher, J. K., and Wright, B. E. (1979). TFLUX: A general purpose program for the interpretation of radioactive tracer experiments. *Comp. Programs Biomed.* **10,** 66–74.

Wright, B. E., and Emyanitoff, R. (1983). Metabolic organization during differentiation in the slime mold. *In* "Fungal Differentiation" (J. E. Smith, ed.), pp. 19–41. Dekker, New York.

Wright, B. E., and Kelly, P. J. (1981). Kinetic models of metabolism in intact cells, tissues or organisms. *Curr. Top. Cell. Regul.* **19,** 103–158.

Wright, B. E., Thomas, D. A., and Ingalls, D. J. (1982). Metabolic compartments in *Dictyostelium discoideum*. *J. Biol. Chem.* **257,** 7587–7594.

II
Yeasts

5

Saccharomyces cerevisiae as a Paradigm for Modern Molecular Genetics of Fungi

JASPER RINE
Department of Biochemistry
University of California
Berkeley, California

MARIAN CARLSON
Department of Human Genetics and Development
College of Physicians and Surgeons
Columbia University
New York, New York

I. Introduction	126
II. Neoclassical Genetics	127
A. Isolation of Mutations	127
B. Construction of a Genetic Map	129
III. Transformation	133
A. General Methods	133
B. Requirements for Selectable Markers	134
C. Vectors	135
D. Manipulation of Vectors	137
IV. Cloning a Gene	141
A. Construction of a Library	142
B. Isolation of the Desired Gene	143
C. Criteria for Demonstrating the Identity of a Gene	145
V. Manipulation of a Cloned Gene	147
A. Use of a Cloned Gene to Facilitate Genetic Mapping	147
B. Manipulation of a Cloned Gene for Analysis of Chromosomal Mutations	148
C. Use of the Cloned Gene to Introduce Mutations into Yeast	150
D. Use of Gene Fusions for Studying Expression of a Cloned Gene	151

E. Use of Promoter Fusions for High-Level Expression of Cloned Genes in Yeast	151
VI. Other Uses of Cloned Genes	152
A. Making Defined Chromosomal Rearrangements	152
B. Complementation Tests in Heterokaryons	152
VII. An Agenda for Progress	153
A. Development of Recombinant Genetics	153
B. Potential for Surrogate Genetics	154
References	155

I. INTRODUCTION

As a group, the fungi include organisms with diverse ecological roles. Some fungi are of obvious economic importance due to useful secondary metabolites, as described elsewhere in this volume. Others, such as *Candida,* are important pathogens of humans. In order to exploit or combat the biological properties of these fungi efficiently, one must have a thorough understanding of such parameters as the life cycle of the organism, the nutritional requirements, and the mechanisms of gene expression. Although no single experimental approach is sufficient to unravel all of the complexities of an organism, the experience gained over the past 35 years with the yeast *Saccharomyces cerevisiae* indicates that a genetic approach has a number of unique virtues.

Saccharomyces cerevisiae is an economically important organism used in brewing, baking, and industrial fermentations. This yeast has received the widest scientific attention of any fungus because of its substantial contribution to our understanding of the basic biology of eukaryotic cells. *Saccharomyces cerevisiae* is without doubt one of the premier organisms of all time for the study of genetics. Classical genetics of *Saccharomyces* has evolved to a high level, with each of the 17 chromosomes heavily labeled with mapped genetic markers. The application of recombinant DNA methods, which we shall refer to as recombinant genetics, has had an obvious impact on genetic research on yeast and all other organisms. However, two qualities of *S. cerevisiae* have enabled recombinant genetics to have a particularly strong impact: (1) transformation, i.e., the ability to add exogenous DNA to yeast cells and have it propagated and expressed, and (2) homologous interactions between DNA molecules, which allow the replacement of chromosomal sequences with a mutated version constructed *in vitro* (Scherer and Davis, 1979; Rothstein, 1983; Winston *et al.,* 1983).

In this chapter our *cerevisiae* chauvinism will be apparent as we refer to *S. cerevisiae* simply as yeast. We have no intention of reviewing all or even most of the refinements of classical and recombinant genetics in *Saccharomyces,* as recent reviews have covered this topic (Botstein and Davis, 1982). Instead, our goal is to consider those developments that we judge to have the greatest potential for being successfully exported to the analysis of other fungi. We will deal primarily with approaches that have a proven record of success. We will, however, describe a few techniques that have received little attention, yet display a strong potential for widespread application. We wish to emphasize clearly here at the beginning that recombinant DNA per se is not the driving force of modern yeast molecular genetics. Rather, it is the combination and synergism of classical and recombinant genetics that has given *Saccharomyces* its preeminent position as a powerful and flexible experimental organism. This crucial interplay between classical and modern approaches will be a recurring theme throughout this chapter.

II. NEOCLASSICAL GENETICS

An important foundation for either classical or recombinant genetics is a well-defined and -described genetic map. The map of the yeast genome satisfies this requirement since there are well over 300 mapped loci including centromere-linked markers for all but one chromosome (Mortimer and Schild, 1980). The nuclear cytology of yeast is too crude to be useful in making a cytogenetic map of yeast chromosomes. Therefore the map has been derived by measuring map distances between genetic markers as a function of recombination frequencies. The yeast haploid genome consists of 14,000 kilobase pairs (kb) of DNA (Lauer *et al.*, 1977) and is genetically defined as containing approximately 4600 centimorgans (cM). The correspondence between physical distance and map distance ranges from 2.7 kb/cM on average for chromosome III (Strathern *et al.*, 1979) to approximately 10 kb/cM for a rather limited region of the genome (Shalit *et al.*, 1981). Thus, yeast is endowed with a rather large amount of recombination per base pair in comparison to *Drosophila melanogaster*, which has tenfold more DNA and a total of 284 cM in the entire genome (Lindsley and Grell, 1968). This relatively high level of recombination may contribute to the ease with which homologous DNA interactions are observed in yeast in comparison to organisms with larger genomes.

A. Isolation of Mutations

Construction of a sophisticated map requires large numbers of easily scorable genetic markers. For organisms capable of growth on a defined medium, auxotrophic mutations are easily isolated and characterized both genetically and biochemically. Auxotrophic mutations can be identified (1) by screening, (2) by enrichment plus screening, or in a few cases (3) by direct selection. Screening is accomplished by growing many individual colonies from a mutagenized culture on rich medium, and then replica-plating the colonies to a variety of supplemented minimal media each lacking a specific component. Colonies unable to grow on the supplemented minimal medium are auxotrophic for the nutrient(s) lacking in that medium. Enrichment for auxotrophic mutants or other conditionally lethal mutants is accomplished by treating cells in a way that causes death in cells that try to grow. A classic example of an enrichment method in yeast is based upon inositol starvation (Henry *et al.*, 1975). Inositol auxotrophs can form colonies on medium supplemented with inositol, an essential membrane lipid. If inositol auxotrophs are provided with all essential nutrients except for inositol, the cells begin to grow and divide, yet cannot form functional membranes and soon die. Those rare cells in a population that have acquired an additional auxotrophy (or conditional lethal mutation) will become limited for the new requirement, will not attempt to grow, and will thus survive a period of inositol starvation. Therefore, a population of cells surviving a brief period of inositol starvation will become enriched in those containing additional mutations. An analogous selection is the fatty acid-less death selection (Henry and Horowitz, 1975), which, like the inositol-less death selection, requires use of a strain auxotrophic for a particular nutrient. An improved variation on the nystatin selection (Snow, 1966) is the use of echinocandin B, a recently discovered inhibitor of cell wall biosynthesis (McCammon and Parks, 1982). Cells treated with echinocandin B respond in the same fashion as bacteria treated with penicillin. These cells form faulty cell walls and lyse in hypotonic medium. This inhibitor appears to be the most efficient at enriching for auxotrophs (20-fold enrichment with one round of treatment) and has the added virtue of working on wild-type cells.

1. Selection of Specific Mutations

One approach to the selection of specific mutations is based upon the fact that organisms can metabolize nontoxic compounds to toxic compounds. An example of this approach is the use of medium containing methyl mercury to select for mutations in the *MET15* locus (Singh and Sherman, 1975). Similarly, mutations in the arginine permease gene, *CAN1*, may be selected by growth in the presence of canavanine, an arginine analog that is incorporated into proteins, rendering them nonfunctional (Grenson *et al.*, 1966; Whelan *et al.*, 1979). Bach and Lacroute (1972) reported a method for selection for *ura3* mutations in cells grown in the presence of ureidosuccinic acid. However, this selection cannot be applied to all yeast strains, possibly due to lack of a cytoplasmically inherited factor influencing uptake of ureidosuccinic acid (Lacroute, 1971). An improved method for selection of *ura3* mutations has been developed based upon the toxicity of 5-fluoro-orotic acid in cells containing OMP decarboxylase, the product of the *URA3* locus; *ura3* mutants are resistant to 5-fluoro-orotic acid (F.

Lacroute, personal communication). Yeast cells that lack the *LYS2* or *LYS5* gene product are capable of growing on medium containing α-aminoadipate as the sole nitrogen source, whereas wild-type cells are not (Chattoo *et al.*, 1979). Further discussion of the uses of *lys2* selections is found in Chapter 7 by Barnes and Thorner. Several other selections are described in Littlewood (1975). However, since selections are available for only a small percentage of all genes, the more arduous routes are necessary to identify enough genes to make a map. Nevertheless, the existence of even a single gene for which there is both positive and negative selection would make that gene a logical choice upon which to establish recombinant vector systems (see below).

2. Potential Problems in Isolating Auxotrophic Mutations

Sometimes expected classes of auxotrophic mutations are not found. An example in yeast is the absence for many years of mutations conferring proline auxotrophy. It was then discovered that proline auxotrophs are incapable of growth on conventional yeast-rich medium and hence were selected against in mutant hunts (Brandriss, 1979). Duplicated genes present an occasional problem in isolating some mutations, even in an organism such as yeast in which 95% of the genome is unique sequence. For example, there are two copies of the genes encoding asparagine synthetase, carbamyl-phosphate synthetase, and glyceraldehyde-phosphate dehydrogenase (Ramos and Wiame, 1980; Lacroute *et al.*, 1965; Holland and Holland, 1980). Thus, although most potential auxotrophic mutations should be found for any fungus capable of growth on a defined medium, the repeated failure to identify specific mutations may be symptomatic of either a medium problem or a gene dosage problem.

B. Construction of a Genetic Map

For molecular genetics of an organism to be useful, nothing is more important than a reliable and well-described genetic map. Ascomycetes are particularly appealing for mapping studies because each ascus contains all of the haploid products of a single meiosis. The more genetic markers that are located on a map, the easier it becomes to map a new gene. Therefore, the effort spent in mapping genes becomes more valuable over time. The most difficult part of mapping a new gene in yeast is the first step: determining the chromosome on which a particular gene resides. Once mapped to its chromosome, a gene's position can be located precisely by conventional tetrad analysis. Several fast and efficient methods have been developed for mapping genes to chromosomes in yeast (e.g., Klapholtz and Esposito, 1982). Below we describe a few of these techniques, the first of which is applicable only to organisms that already possess a relatively well-developed map.

1. Superstrains

Because of the many mutations already mapped in yeast, it has been possible to construct a set of nine strains that together contain a genetic marker on average every 50 cM for the entire genome (R. K. Mortimer, Yeast Genetics Stock Center, Berkeley, California). By crossing a strain with an unmapped mutation to each of these strains, a mutation may be mapped to within 50 cM of its location by analyzing no more than nine crosses. One significant drawback to this technique is that some yeast strains do not grow well even on supplemented medium when they contain many auxotrophies. Thus, the segregation of all markers in these crosses may be difficult to monitor. In addition, relatively large numbers of tetrads are required to establish significant linkage relationships at distances near 50 cM.

2. Chromosome Loss Mapping

Chromosome loss mapping refers to a class of mapping techniques that cause the loss of random individual chromosomes from a diploid, thus uncovering recessive markers that were heterozygous in the original diploid. To illustrate the method of mapping by chromosome loss, consider how a dominant mutation X would be mapped. First, the X mutant is mated to several strains, which together contain recessive auxotrophic mutations on each chromosome. In practice, several strains would be used because strains with many genetic markers often grow poorly. The resulting diploid is prototrophic for each auxotrophy present in the tester strain. After induction of chromosome loss (see below), the diploid cells are plated onto rich medium and allowed to grow into colonies. The colonies are then replica-plated to various diagnostic media, each lacking one nutritional supplement, to determine which recessive markers are revealed by loss of the chromosome containing the X mutation. In the case of dominant mutations, colonies losing the chromosome containing the dominant mutation will become auxotrophic for the traits carried on the homologue. Although more than one recessive marker may be revealed if multiple chromosomes are lost, only one marker will consistently be associated with loss of the X mutation.

Mapping a recessive mutation, x, by chromosome loss mapping is somewhat more subtle than mapping a dominant mutation. The procedure employed is the same as that described above, but the analysis differs. Since the recessive x trait is never expressed unless the homologue containing the marker chromosome is lost, the x mutation is mapped by a process of elimination. That is, whichever recessive markers are *never* expressed in colonies that do express the x mutation define the chromosome that contains the X gene. Figure 1 illustrates the application of chromosome loss mapping to a dominant and a recessive drug resistance marker. Table I contains data that illustrate an application of chromosome loss mapping. These data were obtained by the *rad52* chromosome loss mapping

5. Saccharomyces cerevisiae as a Paradigm for Molecular Genetics of Fungi

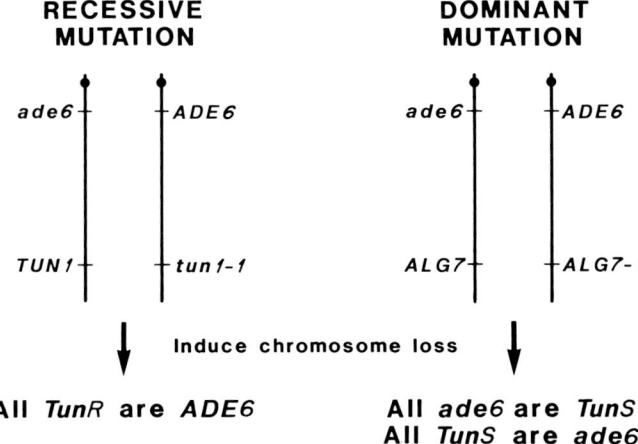

Fig. 1. Mapping mutations by chromosome loss methods. The figure illustrates mapping either a recessive mutation (*tun1-1*) or a dominant mutation (*ALG7-1*), either of which can confer resistance to tunicamycin (TunR). Wild-type cells are TunS. Diploid cells are shown that are heterozygous at the *ADE6* locus and are heterozygous at either *TUN1* or *ALG7*. In the case of the recessive mutation, the diploid is TunS and in the case of the dominant mutation the diploid is TunR.

method (see below) and were used to map both dominant *ALG7* and recessive *tun1* mutations, each of which confers resistance to tunicamycin (Barnes *et al.*, 1984).

The application of chromosome loss to genetic mapping requires that the frequency of chromosome loss be elevated sufficiently to make it practical to identify individual colonies that, among them, have lost each of the chromosomes of yeast. Chemical, physical, and genetic methods have been developed to raise the frequency of random chromosome loss. Acute treatment of diploid cells with methylbenzimidazole-2-yl carbamate (MBC), an inhibitor of microtubule polymerization, causes a high rate of chromosome loss and can be used to facilitate mapping (Wood, 1982). Several experimenters have reported difficulty in using MBC in this manner and advocate the use of benomyl, a commercial fungicide containing MBC and other ingredients, instead of MBC. Nevertheless, at least with some preparations of MBC, the frequency of chromosome loss can be as high as one or more losses in 50% of treated cells.

There are also genetic approaches to elevating chromosome loss. Diploids homozygous for the mutation *cdc6*, a temperature-sensitive lethal mutation, lose chromosomes after brief exposure to elevated temperatures. Similarly, diploids homozygous for the *rad52* mutation lose chromosomes at a high frequency when treated with either methyl methanesulfonate or ionizing radiation. (D. Schild, personal communication; Mortimer *et al.*, 1981). Therefore a strain carrying an unmapped mutation and either *cdc6* or *rad52* is mated to a variety of well-

TABLE I.

Chromosome Loss Mapping of *TUN1* and *ALG7*[a]

Chromosome	1	2	3	4	5	6	7	8	9	10	11	12	13	14	15	16
tun1-1																
violations	10	18	—	46	6	7	0	10	6	28	6	—	5	20	12	3
number of losses	16	30	—	91	19	25	48	29	11	38	10	—	17	28	28	12
number of colonies	120	119	—	239	117	117	240	240	117	240	117	—	119	117	117	117
tun1-2																
violations	4	34	—	52	12	10	0	53	12	41	7	—	17	0	9	12
number of losses	16	54	—	113	34	24	55	82	27	98	27	—	37	20	76	35
number of colonies	120	213	—	333	125	120	456	576	243	456	243	—	213	243	243	120
ALG7-1																
violations	5	14	—	22	12	7	0	49	31	46	24	—	33	32	11	8
number of losses	3	13	—	23	12	6	27	40	23	41	11	—	15	16	26	5
number of colonies	120	175	—	375	120	120	401	601	226	401	226	—	401	226	226	120
ALG7-2																
violations	30	60	—	82	27	20	1	82	20	43	12	—	27	15	6	31
number of losses	7	46	—	56	14	21	35	57	21	36	15	—	5	11	23	14
number of colonies	122	243	—	243	122	183	303	425	182	303	182	—	121	182	182	122

[a] This mitotic mapping technique employs diploids heterozygous for the mutation of interest and heterozygous for additional mutations marking various chromosomes. Chromosome loss events are measured in which particular marked chromosomes are lost and the mutation of interest is lost. We refer to violations as loss events that are inconsistent with the mutation of interest being on the chromosome being monitored. Specifically, violations represent the sum of the losses of the mutation of interest (but not the marked chromosome) plus the losses of the marked chromosome (but not the mutation of interest). Number of losses is the number of colonies in which the chromosome was lost. Number of colonies refers to the total number of colonies that were screened.

marked *cdc6* or *rad52* strains. After induction of chromosome loss, the map position is determined as described above.

A second class of mapping methods are based upon the properties of endogenous yeast plasmids when integrated into the genome. These methods are described in Section V,A.

III. TRANSFORMATION

Transformation is the *sine qua non* of recombinant genetics. The full potential of research with purified genes can by realized only when those genes can be faithfully and functionally reintroduced into the genome of the organism from which they came. Without this capability, the analysis of recombinant clones is little more than molecular anatomy. Although transformation is now possible in *Drosophila* and cultured mammalian cells, this procedure has achieved its highest level of sophistication in yeast.

A. General Methods

We will operationally define transformation as the acquisition by yeast cells of exogenously added DNA and the subsequent inheritance and expression of that DNA. Current procedures can be divided into those that use spheroplasts (Hinnen *et al.*, 1978; Beggs, 1978) and those that use intact cells (Hisao *et al.*, 1983). Spheroplasts are prepared by enzymatic removal of the yeast cell wall in medium containing 1 M sorbitol for osmotic support of the otherwise fragile spheroplasts. The spheroplasts are incubated briefly with DNA in the presence of Ca^{2+} ions and are then treated with polyethylene glycol, at which point DNA is taken up into the cell. Typically, fewer than 0.1% of the viable cells take up the exogenous DNA and subsequently express it. Therefore, the transformed cells must be selected from the nontransformed cells by plating the spheroplasts in top agar and selecting for the expression of a gene carried by the exogenously added DNA. The frequency of cotransformation is quite high; most transformed cells take up multiple DNA molecules (Stinchcomb *et al.*, 1979). With many strains it is common to obtain a thousand or more transformants per microgram of DNA for most replicating plasmid vectors. However, the efficiency of transformation can vary over four orders of magnitude from strain to strain, and it would therefore be prudent to screen a variety of strains in attempting to develop transformation in another fungus.

The other popular method of transformation relies upon the observation that intact yeast cells treated with Li^+ ions will take up exogenous DNA. The counter ion in these experiments has a large effect on efficiency, with acetate being the preferred counter ion. Generally, we have experienced a 10-fold reduction in transformation efficiency with LiAc protocols versus conventional spheroplast transformation. Strains that transform by one procedure seem to transform by the other procedure as well. Further discussion of yeast transformation may be found in Chapter 6, by Boguslawski.

B. Requirements for Selectable Markers

To facilitate selection of the transformed cells from the nontransformed cells, the exogenously added DNA, generally a plasmid vector, must carry a gene whose expression in yeast offers a selective growth advantage. To date, the two approaches used for selection have been the complementation of recessive mutations in the yeast strain and, more recently, the use of markers whose presence can be selected for in wild-type cells, so-called dominant transformation markers. Both approaches offer specific advantages and disadvantages.

1. Complementation of Recessive Mutations

Any wild-type gene encoding an essential biosynthetic enzyme should serve as a selectable marker in cells bearing a mutation in this gene. The earliest efforts in yeast recombinant genetics were directed at obtaining a functional clone of any conventional biosynthetic gene. This goal was readily accomplished by isolating yeast genes capable of complementing auxotrophic mutations in *Escherichia coli* (see Section IV,B,2); an example is the yeast *LEU2* gene, which complements *E. coli leuB* mutations (Ratzkin and Carbon, 1977). Recombinant plasmids containing the LEU2 gene were capable of transforming yeast *leu2* mutants to leucine prototrophy (Hinnen *et al.*, 1978). The success of these pioneering experiments relied upon the ability to construct, by *in vivo* recombination in yeast, a nonreverting allele of *leu2* containing two independent point mutations. A virtue of using a yeast gene to complement a yeast mutation is that no foreign gene products, which could have unpredictable secondary effects, are being added to the cell. Furthermore, the yeast gene on the vector provides a region of homology to the yeast genome, so that it is possible to integrate the entire vector at a defined location where it will be stably inherited as part of a yeast chromosome. The disadvantage of this approach is that it requires that a strain have a recessive mutation in the chromosomal copy of the gene carried on the plasmid.

2. Dominant Selectable Markers

It is desirable to have vectors containing genes whose expression can be selected for in wild-type cells. These genes could be used in ways analogous to the drug resistance transposons of bacteria. To date, no naturally occurring drug resistance plasmids or transposons have been discovered in yeast. Therefore prokaryotic drug resistance determinants have been tested for expression and function in yeast. G418 is an aminoglycoside antibiotic that is active against a wide variety of organisms including fungi. The bacterial transposon Tn601 encodes an aminoglycoside phosphotransferase that phosphorylates G418 and thus inactivates it. Fortuitously, the phosphotransferase encoded by Tn601 is expressed in yeast and provides resistance to otherwise lethal doses of G418 (Davies and Jimenez, 1980). Therefore Tn601 promises to be useful as a dominant selectable marker for yeast transformation. The *E. coli* chloramphenicol acetyltransferase gene is also expressed in yeast and offers a selective growth advantage in the presence of high levels of chloramphenicol (Cohen *et al.*, 1980), but it is still unclear whether the level of resistance is sufficient to allow use as a selectable marker.

C. Vectors

A panoply of yeast vectors are available for a variety of specific tasks. The more useful vectors consist of all or most of pBR322, to which has been added other DNA fragments. All of these vectors are shuttle vectors capable of being maintained and selected for in either *E. coli* or yeast. A widely accepted vector nomenclature has been established and is illustrated by the following examples of some plasmid vectors: YIp5, YEp24, YCp50, and YRp7. The Y indicates that the vector has a marker that can be selected in yeast; p that the vector is a plasmid; I that the plasmid is an integrating plasmid; R that the plasmid is capable of autonomous replication due to the existence of particular chromosomal sequences described below; E that the plasmid contains sequences derived from the endogenous plasmid of yeast and can be maintained episomally; and C that the plasmid contains a centromere. Numbers distinguish different vectors of the same type (Botstein *et al.*, 1979).

1. Integrative Vectors

Integrative vectors are all modeled after the prototype YIp5, which consists of pBR322 plus the yeast *URA3* gene (Scherer and Davies, 1979). Since neither pBR322 nor *URA3* contains sequences that can function as a yeast origin of replication, YIp5 is incapable of autonomous replication in yeast cells. Therefore YIp5 can transform yeast only by integrating into the genome, which occurs in yeast through homologous recombination. When a fragment of yeast DNA is cloned into YIp5, the plasmid can then integrate either at the *URA3* locus on the chromosome or at the locus defined by the cloned segment. The property of integration by homologous recombination is exploited in Section V,A.

2. High-Frequency Transformation

In several laboratories it was discovered simultaneously that certain segments of the yeast genome, when present on circular plasmids, are capable of causing transformation to occur with several orders of magnitude greater efficiency than integrative transformation (Beggs, 1978; Struhl et al., 1979; Hsiao and Carbon, 1979; Stinchcomb et al., 1979; Tschumper and Carbon, 1980). One class of these segments contains a sequence referred to as an ARS, for autonomously replicating sequence, that provides any plasmid with the ability to replicate as an episome in yeast. Evidence is accumulating that ARS1 is an authentic origin of replication. Statistically, ARSs are encountered every 40 kb in the yeast genome (Beach et al., 1980).

Although ARS plasmids replicate once per generation (Fangman et al., 1983), they are not evenly distributed in a population of cells. Even when the cells are grown under conditions that select for the presence of the plasmid, as many as 90% of the cells lack the plasmid. The explanation for this paradoxical behavior is the asymmetric distribution of plasmids between mother cells and daughter cells. The ARS plasmids have a strong segregational bias toward staying with the mother cell rather than the bud. The number of copies in the mother cells can be 50 or more (Murray and Szostak, 1983).

The endogenous plasmid of yeast, known as the 2-μm circle, has been used as a source of a replication origin for yeast transformation vectors (Beggs, 1978; Broach et al., 1979; Broach and Hicks, 1980). Plasmids containing a 2-μm origin of replication are replicated at multiple copies per cell. The endogenous 2-μm plasmid encodes both cis- and trans-acting replication functions, whereas some vectors containing the 2-μm plasmid origin, such as YEp24 (Botstein et al., 1979), contain only the cis-acting functions. Thus, these recombinant plasmids are dependent on the endogenous plasmids for replication functions. Although the endogenous plasmid is present at approximately 80 copies per cell (Gerbaud and Guerineau, 1980), recombinant plasmids containing only the 2-μm origin appear to be present at approximately 10 copies per cell. The YEp24 plasmids are relatively stable even when cells are grown for a short time in the absence of selection. This stability may be due to a plasmid-specific segregation system for the 2-μm plasmid (Kikuchi, 1983).

3. Centromere Plasmids

The stable mitotic and meiotic inheritance of eukaryotic chromosomes is due to the function of centromeres. The addition of a centromere to an ARS plasmid should confer both mitotic and meiotic stability to the plasmid. The centromeres of chromosomes III, IV, VI, and XI have been isolated both by chromosome walking and by selection for mitotic stabilization of otherwise unstable plasmids (Stinchcomb et al., 1982; Clarke and Carbon, 1980; Panzeri and Philippsen, 1982; Fitzgerald-Hayes et al., 1982a; Hsiao and Carbon, 1981). Subclones of approximately 1 kb have successfully conferred mitotic stability, replication at a single copy per cell, and meiotic segregation. In fact, incorporation of a centromere into an otherwise multicopy YEp plasmid reduces the copy number of the plasmid to approximately one copy per cell (Tschumper and Carbon, 1983).

The mechanism for copy number control in this case remains obscure. In general, centromere plasmids behave as authentic minichromosomes, although the mitotic stability of centromere plasmids is still two orders of magnitude below the stability of authentic chromosomes. Therefore, there must be more elements contributing to the stability of a chromosome than simply a centromere. Size may play a role in enhancing stability since 50-kb CEN plasmids are more mitotically stable than are 10-kb CEN plasmids (C. Mann, personal communication).

A subtle but important feature of centromere plasmids is that they are restricted to autonomous replication and are not found integrated into chromosomal DNA. Undoubtedly, integration can occur but would result in dicentric chromosomes that would be mitotically unstable (McClintock, 1939; Mann and Davis, 1983; Haber *et al.*, 1984). This instability of dicentric chromosomes will be exploited in Section V as a way of recovering chromosomal mutations.

4. Linear Plasmids

Linear plasmids have been constructed in yeast by ligating the telomeres of the extrachromosomal copies of *Tetrahymena* rDNA to the ends of a linearized plasmid vector. The *Tetrahymena* telomeres function in yeast, allowing these plasmids to replicate as autonomous linear plasmids. These plasmids then served as vectors to allow the cloning of authentic yeast telomeres (Szostak and Blackburn, 1982).

Linear ARS plasmids are more stable than circular ARS plasmids, yet are less stable than circular centromere plasmids. In addition, short linear plasmids with a centromere are less stable than short linear plasmids without a centromere (Murray and Szostak, 1983). These observations led the authors to propose that both the telomeres and centromeres contribute to the mitotic stability and that the mechanisms of mitotic stabilization are antagonistic. Linear plasmids are proving indispensable for analysis of chromosome structure and function, but the difficulty in using these plasmids *in vitro* has limited their general utility.

D. Manipulation of Vectors

1. Recovery of Episomal Vectors

Plasmids that replicate autonomously in yeast may be recovered by isolating total yeast DNA from a small culture and transforming *E. coli*, with selection for a marker on the vector. The amount of plasmid DNA from a small yeast culture is generally too small to allow direct transformation of another yeast strain.

2. Cytoduction

The transfer of numerous plasmids into a large number of strains is a tedious and time-consuming task by conventional procedures of transformation or genetic crosses. A recent innovation in transferring plasmids between strains is a cytoduction method, similar in spirit to bacterial matings, that uses yeast *kar1* mutants, which are deficient in nuclear fusion. In matings between a wild-type cell and a *kar1* mutant, a transient heterokaryon is formed from which haploid nuclei bud after cytoplasmic mixing occurs (Conde and Fink, 1976). During the time in which both nuclei reside in the same cytoplasm, whole chromosomes move from one nucleus to the other at a low frequency. Furthermore, the frequency of chromosome transfer is inversely proportional to chromosome size. Those with the smallest linkage group are transferred most frequently and those with the largest linkage group are transferred least frequently (Dutcher, 1981). Sigurdson *et al.* (1981) demonstrated that 2-μm plasmid transferred from one nucleus to the other in heterokaryons, suggesting that matings between a *kar1* and a *KAR1* strain could be used to transfer plasmids between strains. Recent results suggest that plasmid transfer between nuclei in heterokaryons is, in fact, a practical way in which to move many plasmids from strain to strain (J. Rine, unpublished observations).

The experimental protocol for plasmid transfer by mating is illustrated in Fig. 2. Consider a plasmid containing a *URA3* gene in a *ura3 kar1* host or donor strain. The recipient strain is constructed with a recessive drug resistance marker such as *can1* (resistance to canavanine) and a *ura3* mutation. The donor strain is replica-plated onto a lawn of the recipient strain and the two are allowed to grow

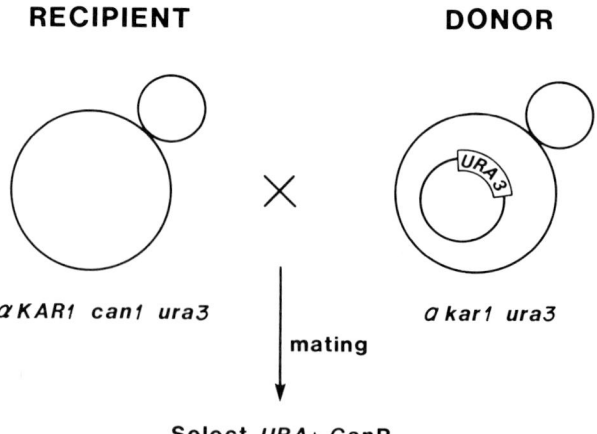

Fig. 2. Interstrain plasmid transfer in *KAR1* × *kar1* matings. Both strains are Ura⁻; the recipient strain contains the *can1* mutation that renders the cell resistant to canavanine (CanR). The cells are allowed to mate under permissive conditions and are then transferred to selective media. Under selective conditions, the only cells that can grow are those with the recipient cell's nucleus and the donor cell's plasmid. Further details are given in the text.

together and mate on rich medium. The cells on rich medium are then replicaplated onto minimal medium containing canavanine. Growth on canavanine selects against the donor cells and against any rare diploids that may form. Growth on minimal medium selects for the plasmid. Thus, the colonies that survive the selection are recipient cells that have acquired the plasmid from the donor nucleus. The quantitative transfer efficiency is presented in Table II for a variety of yeast plasmids and a qualitative demonstration of this method is presented in Fig. 3. This method has been used to transfer 100 random plasmids from a plasmid library in a *kar1* donor strain into a recipient strain with only three Petri plates: a master plate, a rich-medium plate, and a selective plate. All plasmids transferred with similar efficiency. Therefore, this technique may prove to be a useful way around the difficulties encountered in moving many plasmids between many strains. In addition, this approach offers a solution to introducing plasmids into nontransformable strains since the mating efficiency of yeast is essentially unity. Although intact chromosomes also can move from nucleus to nucleus as well as the plasmids, the frequency at which chromosomes move relative to the plasmids is too low to cause problems.

TABLE II.

Quantitative Measure of Plasmid Transfer in *kar* × *KAR* Matings

Plasmid	Transfer efficiency[a]
ARS	4×10^{-4}[b]
	7×10^{-2}
CEN	1.2×10^{-3}
YEP	4.6×10^{-3}

[a] The transfer efficiency refers to the number of *kar* donor cells that were able to transfer their plasmid to a *KAR* recipient strain during a period of nonselective growth on rich media.
[b] The lower transfer efficiency of the *ARS* plasmid is calculated based on the number of cells from this culture used in the experiment. The higher transfer efficiency is calculated based on the number of cells in the culture that actually contained a plasmid.

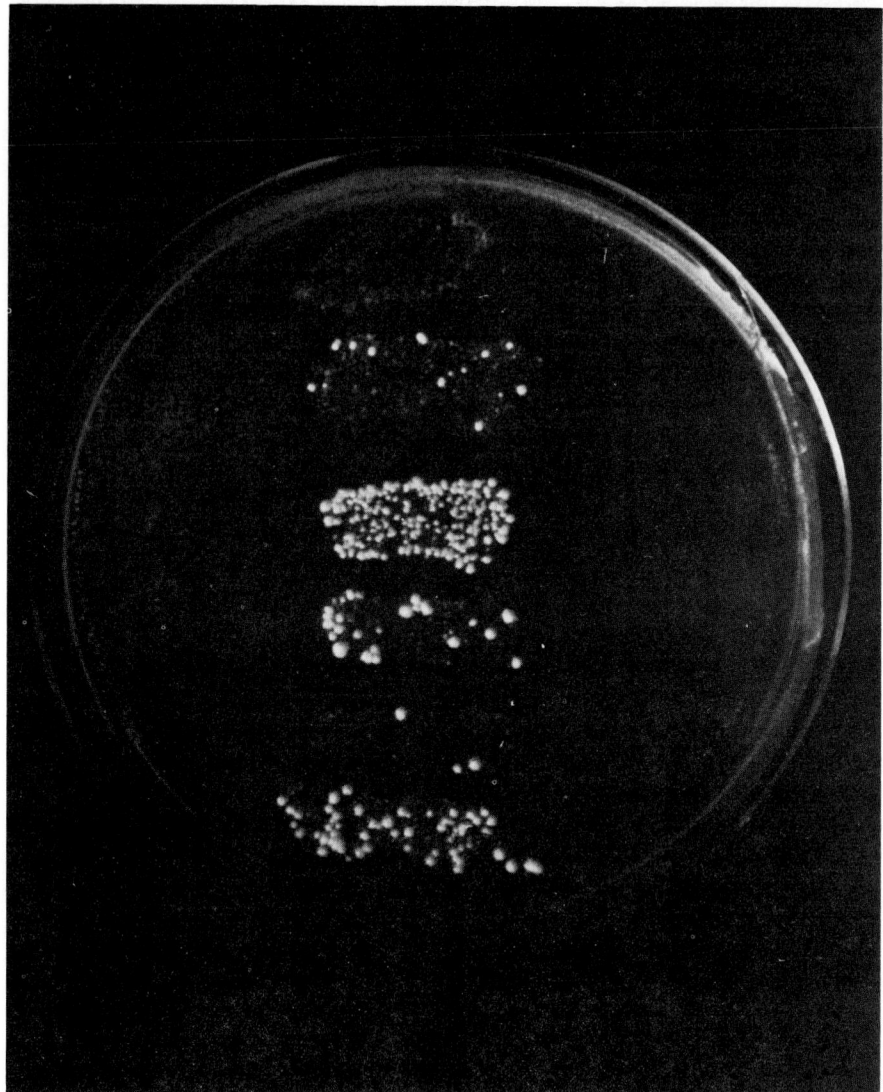

Fig. 3. Qualitative demonstration of plasmid transfer technique. The colonies growing on this selective plate are recipient cells that have received a plasmid from a donor strain. The top patch is a control mating in which the donor contained no plasmid. The other patches contain cells that have received the following plasmids in order from top to bottom: ARS, YEp, CEN, ARS, CEN.

3. Targeting Plasmid Integration to Specific Sites

Since plasmids integrate into the yeast genome by homologous recombination, the position in the genome at which a plasmid integrates is determined by the yeast sequences present on the plasmid. Most plasmids used to transform yeast have two different sequences homologous to different sites in the genome; one is the selectable marker and the other is the cloned gene of interest. A mechanism for targeting the site of integration has come from a study of the behavior of circular plasmids that have been linearized *in vitro* prior to transformation. The linear ends are recombinogenic and direct integration to the site in the genome homologous to the ends. Linearization also increases the efficiency of integrative transformation from 10- to 50-fold (Orr-Weaver *et al.*, 1981, 1983). Therefore, if a plasmid has two sequences, each homologous to a different site in the genome, integration can be targeted to either site by digesting the plasmid with a restriction enzyme that cleaves only in the sequence of interest.

4. Recovery of Integrated Plasmids

Plasmids that have integrated in the genome are recovered by isolating genomic DNA from cells containing the integrated plasmid. The genomic DNA is then digested *in vitro* with a restriction enzyme that cleaves genomic DNA at sites flanking the vector but not in the vector itself. The DNA is then ligated at a low DNA concentration to favor intramolecular ligation and used to transform *E. coli* (Roeder and Fink, 1980; Stiles, 1983).

Perhaps less obvious than the procedure for recovering integrated plasmids is the reason for wanting to recover an integrated plasmid. A reasonably small, cloned segment of DNA can direct integration of a vector into the genome, and large regions of flanking DNA can then be cloned by recovering the integrated vector using a genetic selection in *E. coli*. This procedure is, in general, much easier than constructing a complete library and identifying overlapping clones by hybridization. Although this method could also be used to clone chromosomal mutations, transformation with gapped molecules is probably more efficient for this purpose (see Section V,B).

IV. CLONING A GENE

One of the primary benefits derived from the application of recombinant DNA genetics to yeast has been the vastly increased scope of molecular genetic methods available for analysis of the structure, regulation, and function of particular cloned genes. In this section we will discuss procedures for cloning a gene: constructing a library, isolating a particular gene, and rigorously establishing the identity of the cloned gene. Emphasis is on methods that take advantage of yeast transformation.

A. Construction of a Library

The first step in cloning a gene is constructing a library that includes the gene of interest. The usual approach is to construct a library containing cloned DNA segments representing the entire yeast genome; however, in some cases a library can be prepared from DNA that has been enriched for the gene. For some specific applications a cDNA library may be desirable. We will consider the general case in which a genomic library is desired. Two factors are important in constructing a library: choice of vector and means of generating DNA segments for cloning.

1. Choice of Vector

The choice of vector depends on several factors. The most important is the method available for identifying the cloned gene of interest. In yeast the most commonly used methods involve transformation of yeast (see below), and therefore a vector providing high-efficiency transformation is usually chosen. Such vectors include the YRp, YEp, and YCp plasmid vectors described in Section III,C. These are all shuttle vectors able to be maintained and selected for in both yeast and *E. coli*. The YRp and YEp vectors differ from the YCp vectors with respect to the copy number at which they are maintained in yeast; the YCp vectors are maintained in one or two copies per cell. A stable, low-copy vector may be necessary if the presence of multiple copies of the gene of interest is lethal to the host cell. Each vector carries a marker allowing selection in yeast, and choice of a vector may be influenced by preference for a particular selectable marker. Finally, different vectors contain different restriction sites suitable for insertion of yeast DNA segments. If a method not requiring yeast transformation will be used to identify the cloned gene (for example, hybridization or complementation in *E. coli*), then any one of a variety of bacterial plasmid or phage vectors may serve (Maniatis *et al.*, 1982).

2. Preparation of DNA Segments for Cloning

Yeast DNA segments appropriate for insertion into a vector are conveniently generated by digestion of genomic DNA with a restriction endonuclease. Unless one has prior information regarding the disposition of restriction sites within and around a gene, it is prudent to prepare a partial digest of genomic DNA for cloning. A close to random distribution of fragments can be obtained by partial digestion with an enzyme that recognizes a site present in the genome at high frequency. For example, yeast DNA libraries have been constructed by inserting fragments generated by partial digestion with *Sau*3AI into the *Bam*HI site of a plasmid vector (Nasmyth and Reed, 1980; Carlson and Botstein, 1982). Although one could use an enzyme that cleaves the genome infrequently and generates a set of partial fragments of nonrandom size that contain the gene of interest, it is preferable to generate a random series of genomic fragments for cloning because fragment size and the nature of the other sequences included on the cloned fragment can affect cloning efficiency.

One final point to consider when constructing a library is size. It is obviously important that the library be sufficiently large to include the gene of interest, bearing in mind that some clones will inevitably be underrepresented in a library. Even the best identification scheme will not result in recovery of the desired cloned gene if it is not present in the library. An appropriate size for the library can be calculated from the size of the organism's genome and the size of the cloned DNA segments (Maniatis *et al.*, 1982). In addition, the library should be constructed using DNA from the canonical wild-type strain for any given fungus, unless special circumstances apply.

B. Isolation of the Desired Gene

1. Complementation in Yeast

The easiest and most common method for isolating a particular gene from a yeast DNA library is by complementation for function in yeast. Complementation is feasible provided that a strain carrying a recessive mutation in that gene is available and that a selection or convenient screen for the wild-type phenotype exists. The mutant strain is simply transformed with DNA from the library and selection for the gene is applied. Although this approach is straightforward, in many cases a two-step procedure is advisable: the mutant strain is first transformed with the library DNA and transformants are selected by virtue of the selectable marker carried by the vector; then the transformants are pooled and transformants containing the gene of interest are identified by selection or screen. This latter approach minimizes the likelihood of recovering revertants rather than transformants and avoids the possible problem that the gene of interest may not be expressed well enough in regenerating spheroplasts to ensure survival of the transformant. In addition, this approach is particularly useful if the selection for the gene is not stringent or if a screen must be used.

2. Complementation in E. coli or Related Fungi

Another method that has been successfully used to isolate cloned yeast genes is complementation for function in *E. coli* or in related fungal species. A number of yeast genes have been cloned by virtue of their ability to complement a defect in the corresponding *E. coli* function; an estimated one-third of the attempts to clone yeast genes in this manner met with success (Struhl *et al.*, 1976; Ratzkin and Carbon, 1977; Bach *et al.*, 1979). The *cdc28* gene derived from *Saccharomyces cerevisiae* has been shown to complement the *cdc2* mutation in the fission yeast *Schizosaccharomyces pombe* (Beach *et al.*, 1982). Complementation of defects in transformable yeast species may prove useful for isolation of genes from fungi for which transformation is either impossible or inefficient.

3. Copy Number Effects

A relatively obscure and overlooked aspect of both prokaryotic and eukaryotic genetics is the phenomenon of gene dosage effects. Although the elegant mechanisms that have evolved to regulate autogenously the expression of some genes are obviously interesting and important, part of the reason these mechanisms are so interesting is that they are the exceptions. Strong evidence for the fact that most genes do not dosage-compensate comes from the fact that genes can be mapped by measuring the relative changes in the level of a gene product as the copy number of specific chromosomal regions is altered (Lindsley et al., 1972). Since plasmid vectors of the YEp variety replicate at multiple copies per cell, transformation of yeast with recombinant libraries constructed with these vectors, yields a population of cells, each containing a small segment of the yeast genome amplified about tenfold. These partially aneuploid cells can be used for the direct selection of specific recombinant clones as described in the next section.

a. Use of Inhibitors. Only 10% of the yeast genes have been identified by mutations. Although it is simple to clone genes for which a chromosomal mutation exists, it is more difficult to clone hitherto unidentified genes. The use of competitive inhibitors of specific enzymes allows the targeted selection of clones containing either the structural gene or regulators of these enzymes. To use this method, it is necessary to determine the amount of inhibitor necessary to block the growth of cells with a single copy of the gene. A collection of transformed cells is then plated onto medium containing the inhibitor. The colonies that grow are generally those carrying recombinant clones whose products act to elevate the level of the enzyme blocked by the inhibitor. This approach has been used to clone genes involved in glycoprotein biosynthesis, sterol biosynthesis, and protein synthesis (Rine et al., 1983, 1984; Fried and Warner, 1981). The major strengths of this approach are: (1) wild-type cells can be used; no mutations are required in the genes of interest; (2) the approach is general as there are thousands of inhibitors that block the growth of yeast; (3) the method is a selection and not a screen; (4) the method selects for the gene(s) that is most limiting in the synthesis of an enzyme, whether it is the structural gene or a regulator of the structural gene; and (5) the genes cloned by this method by definition have a dominant phenotype and can be used as dominant selectable markers in yeast transformation (Rine et al., 1983).

b. Compensatory and Regulatory Interactions. Gene dosage effects can also allow the identification of interesting compensatory or regulatory interactions. For example, in the case of the *ste13* mutations, two different clones have been identified that complement the mutations. One clone, namely the *STE13* gene, can complement *ste13* mutations when the clone is present at high or low copy number. The other clone complements *ste13* only when the clone is present at high copy number. It is likely that the *STE13* gene encodes a heat-stable dipeptidyl aminopeptidase that processes alpha factor, a yeast mating pheromone, from an inactive precursor to an active form (Julius *et al.*, 1983). Yeast cells contain a second, heat-labile dipeptidyl aminopeptidase, and it is possible that the second clone that complements *ste13* is the structural gene for the heat-labile enzyme.

As a second example of unexpected gene dosage effects, consider the four *SIR* genes whose products are required to repress expression of the silent mating type genes of yeast (Rine, 1979). Through use of multicopy plasmids, it was discovered that overproduction of *SIR3* gene product compensates for the complete absence of *SIR4* gene product even though the two genes are unlinked. In wild-type cells all four *SIR* gene products are required concurrently for proper regulation. Observations such as these have led to the proposal that *SIR4* gene product acts as a positive effector or regulator of *SIR3* (R. Schnell, W. Kimmerly, and J. Rine, unpublished observations).

In addition, the entire spectrum of methods devised for cloning genes from other eukaryotes can be applied to yeast and other fungi. These methods will not be addressed here because they do not benefit from the genetic attributes manifested by fungi.

C. Criteria for Demonstrating the Identity of a Gene

Unfortunately, the cloned gene isolated by a selection or screening procedure may not, in fact, be the gene desired. Even genetic complementation for function is not foolproof: another gene may be recovered that is capable of suppressing the mutation in the desired gene when present in multiple copies in the mutant cell (see Section IV,B,3). It is therefore necessary to provide a convincing demonstration that the correct gene has been cloned.

1. Gene Disruption

The most elegant method available in yeast is to use an *in vitro*-generated mutation in the cloned DNA to cripple the homologous chromosomal sequence, thereby creating a mutation. If this mutation confers the expected phenotype, the cloned gene is probably the correct one, that is, a wild-type copy of the mutant gene used to isolate the clone; however, rigorous proof of its identity demands an allelism test. A variety of techniques for gene disruption have been described. A mutation, such as a deletion or insertion, can be introduced *in vitro* into the cloned gene on an integrating plasmid vector. The mutant gene can then be substituted for the chromosomal gene by transplacement: the plasmid is inserted into the corresponding chromosomal locus by integrative transformation and colonies are then screened for recombination events that excise the wild-type copy of the gene and the vector sequences, leaving the mutated gene in the chromosome (Scherer and Davis, 1979). Another method for gene disruption involves integration of a plasmid carrying an internal segment of the cloned gene at the chromosomal locus of that gene; the integration event results in two mutant copies of the cloned gene, one with a deletion of the 5' end and one with a deletion of the 3' end (Shortle *et al.*, 1982). A third method, called one-step gene disruption, entails the *in vitro* insertion of a selectable yeast gene into the cloned gene. A linear fragment carrying the disrupted cloned gene, flanked by sequences homologous to the gene's chromosomal locus, can be used to transform yeast cells, with selection for the inserted marker. Colonies in which the wild-type chromosomal copy of the cloned gene has been replaced with the disrupted gene will be recovered among the transformants (Rothstein, 1983). The success of this method is due to the recombinogenic property of linear DNA fragments in yeast.

In the case of essential genes, such as actin (Shortle *et al.*, 1982), disruption of the gene will result in a recessive lethal phenotype. Disruption must then be carried out in a diploid, using a method that results in linkage of a genetic marker to the disrupted gene. Genetic analysis can then be employed to demonstrate linkage of the genetic marker to the lethal mutation.

2. Cloning a Mutant Allele

A different approach to the problem of proving the identity of the cloned gene is to clone a mutant allele. The most rigorous proof requires cloning a nonsense allele and demonstrating that the cloned mutant allele is capable of providing a functional product or phenotype only when the host cell carries the appropriate nonsense suppressor (Carlson and Botstein, 1982). Cloning an undefined mutant allele is less satisfactory because it is difficult to control for the possibility that failure to complement results from some problem other than the presence of the expected mutation.

3. Linkage

Another genetic approach that is correlative, but not definitive, is to map the sequences in the genome that are homologous to the cloned DNA (Hicks and Fink, 1977). Operationally, a plasmid carrying the cloned DNA and a genetic marker can be integrated into the chromosomal locus homologous to the cloned DNA by transformation. Genetic analysis can then be carried out to assess linkage of the integrated genetic marker to the desired gene. Such data cannot, however, rule out the possibility that a suppressor gene tightly linked to the gene of interest has been cloned.

4. Other Methods

Other techniques can also be used to establish the identity of a cloned gene if the necessary information and reagents are available. For example, if the sequence of the protein encoded by the gene is known, then nucleotide sequence analysis of the cloned DNA should suffice to prove that the correct gene has been cloned. If antibody to the protein is available, the identity of a cloned gene can be confirmed by cell-free protein synthesis using hybrid-selected RNA or by hybrid arrest of translation (Miller *et al.*, 1983). Alternatively, one can express the cloned genes in *E. coli* and show that a protein with the expected properties is produced.

V. MANIPULATION OF A CLONED GENE

Having a cloned gene in hand makes possible a great variety of endeavors: analysis of coding potential, identification and mapping of RNAs, evaluation of homology to other genes, studies of gene evolution, engineering high-level synthesis of the gene product, and so on. Because our primary interest in this chapter is the application of recombinant DNA methodology to the molecular genetics of yeast, we will focus here on methods for manipulating a cloned gene to facilitate genetic and molecular genetic studies of the gene's structure, regulation, and function.

A. Use of a Cloned Gene to Facilitate Genetic Mapping

To determine the genetic map position of a gene, an allele conferring an easily scored phenotype must be obtained; this may be difficult, particularly if phenotypic expression of mutations in this gene is dependent on genetic background. If the gene has been cloned, an alternative approach is feasible. As described in Section IV,C, a plasmid carrying the cloned gene X and a convenient genetic marker (e.g., the yeast *URA3* gene) can be integrated by homologous recombination at gene X's chromosomal locus. The *URA3* gene then provides a closely linked marker for use in mapping analysis.

A recent chromosomal mapping method relies directly on the availability of a cloned gene (Falco and Botstein, 1983). The gene is inserted into a vector carrying 2-μm circle DNA sequences and a selectable marker, and this plasmid is integrated at the gene's chromosomal locus by transformation. The integrant is then mated to a second, untransformed strain to construct a diploid heterozygous for multiple markers. The presence of the integrated 2-μm circle DNA sequence induces chromosome loss, thereby uncovering recessive markers on the other homologue and allowing identification of the chromosome on which the cloned gene resides.

A new technique for chromosome separation can also be applied to chromosomal mapping of a cloned gene. Successful separations of yeast chromosomes by using pulsed field gradient electrophoresis have been reported (Schwartz et al., 1982). The chromosomal DNA can be transferred to a filter and the chromosome exhibiting homology to the cloned gene can be identified by hybridization (Southern, 1975).

B. Manipulation of a Cloned Gene for Analysis of Chromosomal Mutations

Many yeast genes have been extensively studied by the classical genetic methods of mutant isolation and characterization. However, in most cases the exact nature and position of the mutation within the gene are not known. The cloned gene can be used to recover mutant alleles from the chromosome for further analysis, in particular for analysis of the changes in DNA sequence responsible for the mutant phenotype.

One method for cloning chromosomal mutations has been termed "eviction" (Roeder and Fink, 1980; Winston et al., 1983). To evict a chromosomal mutation, a plasmid carrying the wild-type cloned gene is integrated at the corresponding chromosomal locus; this event results in a duplication of the cloned segment with the wild-type and mutant copies of the gene flanking the integrated vector sequences on either side. Genomic DNA isolated from the transformant is then cleaved with a restriction endonuclease, ligated, and used to transform bacteria; if an appropriate restriction enzyme is chosen, plasmids carrying the mutant copy of the gene, which has been evicted from the genome, can be recovered. If sufficient information regarding the nature or position of the mutation is available, it may be possible to design the experiment to ensure recovery of the mutant allele. If not, the presence of the mutation on the recovered plasmid can be tested by transforming the plasmid into the original mutant strain and determining whether or not wild-type recombinants are generated; however, because such negative evidence cannot be definitive, the presence of the mutation must be verified by nucleotide sequence analysis.

It is worth noting that with the appropriate choice of restriction enzyme for digestion of the genomic DNA, this same technique can be used to recover sequences flanking the genomic copy of the cloned segment but not included in the cloned DNA.

A second method for recovery of chromosomal mutations involves transformation with a gapped plasmid (Orr-Weaver et al., 1981, 1983) (Fig. 4). A double-strand gap is generated by cleavage at two restriction sites within the cloned DNA segment, and the gapped plasmid is used to transform a strain containing a mutation that maps to a portion of the chromosomal gene that is

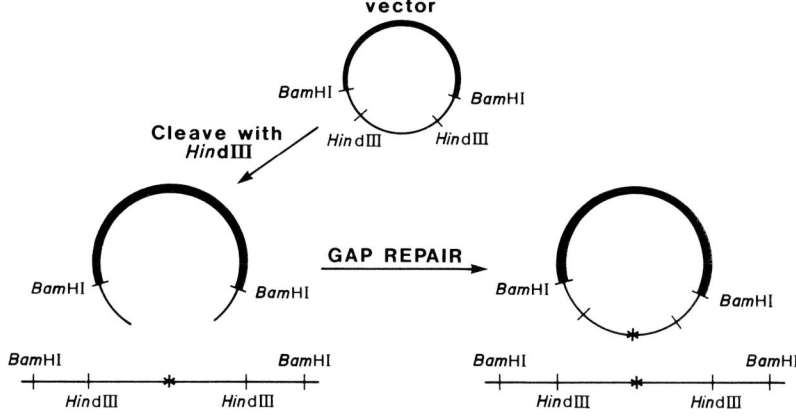

Fig. 4. Gap repair method for cloning mutant alleles. In this example a genomic *Bam*HI fragment containing the gene of interest is cloned into a yeast plasmid. The clone is cleaved with an enzyme that will produce a gapped molecule leaving homology to the yeast chromosome on both sides of the gap. The gapped molecule is introduced into a yeast strain containing a mutation in the gene of interest. The recombinogenic nature of the ends causes the gap on the plasmid to be repaired by sequences on the chromosome.

removed in making the gap. Interaction of the gapped plasmid with the homologous chromosomal region results in repair of the plasmid from chromosomal information. If an integrating plasmid vector is used, the integration event will result in a duplication of the cloned segment in which both copies carry the mutation; then recovery of the plasmid by eviction will always result in recovery of the mutation. If a replicating vector is used, the gap will be similarly repaired from the chromosomal allele. The result of this process can be either an integrated, repaired plasmid or a replicating, repaired plasmid in the case of YRp and YEp vectors. Use of a YCp vector, which contains a centromere sequence, will prevent integration; thus gap repair with a YCp vector will always ensure transfer of a chromosomal mutation directly to an autonomous plasmid. It is particularly easy to recover a replicating plasmid containing the chromosomal mutation by transformation of bacteria with genomic DNA from cells in which the gapped plasmid has been repaired. The extent of the sequence homology to the chromosome that must be present on either side of the gap to allow gap repair has not been precisely determined; as little as 100 bp appears to suffice, but a larger region of homology probably increases the efficiency of the process.

Another use of a cloned gene is to facilitate fine-structure mapping of chromosomal mutations. One technique involves transformation of a mutant strain with a gapped plasmid; wild-type transformants will be recovered only if the gap does not include the mutated site (Orr-Weaver et al., 1983). A second method relies on homologous recombination in yeast between chromosomes and plasmids carrying 2-μm circle DNA (Falco et al., 1983). A series of mapping deletions constructed on a YEp plasmid is introduced into strains carrying point mutations in the gene. The frequency with which wild-type recombinants arise allows the deduction of a map position for the mutation relative to the deletion end points.

C. Use of the Cloned Gene to Introduce Mutations into Yeast

Once one has cloned a gene, the most efficient way to obtain mutations in that gene is by *in vitro* mutagenesis of the cloned DNA. Such mutations can then be introduced into yeast by transformation to examine their effects *in vivo*. A variety of different types of mutations have proved useful in studies of gene function and regulation: deletions, small insertions (for example, linker insertions), small substitutions, and single base changes. Many sophisticated methods developed for the *in vitro* construction of such mutations and for site-directed mutagenesis are described in recent reviews (Shortle *et al.*, 1981).

The great advantage of yeast in these studies is that the effects of mutations created *in vitro* can be analyzed *in vivo*. The easiest approach is simply to transform yeast with a replicating plasmid carrying the mutation of interest. Usually a yeast strain carrying a recessive mutation in the gene under study is chosen as the recipient. This approach has, however, a potentially serious drawback: the mutant gene is present in multiple copies on a plasmid and is flanked by sequences different from those found at its normal chromosomal locus. There is no way to predict whether these factors will affect the regulation of expression of the gene or have unexpected effects on the cell (e.g., abnormally high levels of a gene product may prove lethal to the cell). The problem to copy number can, of course, be overcome by using a YCp vector, which is maintained in one or two copies per cell, or by using an integrating vector.

In many cases, a better approach is to replace the chromosomal copy of the gene with the mutated copy. This procedure eliminates copy number problems and ameliorates potential position effects. Replacement of the chromosomal gene with the mutated gene can be accomplished by either transplacement (Scherer and Davis, 1979) or a variation of one-step gene disruption (Rothstein, 1983) (see Section IV,C). These methods are most useful when the introduction of the mutation into the chromosome can be easily verified by physical means, as is the case for insertions, deletions, and mutations affecting a restriction site. The introduction of a single base change that produces a phenotypic alteration can also be easily monitored.

D. Use of Gene Fusions for Studying Expression of a Cloned Gene

Gene fusions have been used to facilitate studies of the expression of cloned yeast genes in yeast, such as *URA3* (Rose *et al.*, 1981), *CYC1* (Guarente and Ptashne, 1981), *HIS4* (Silverman *et al.*, 1982), and histone H2A (Osley and Hereford, 1982). Fusions of the cloned gene to the *E. coli lacZ* gene can be constructed in such a way that a hybrid protein with β-galactosidase activity is produced; the hybrid protein is composed of the amino terminus of the cloned gene attached to an enzymatically active portion of β-galactosidase. Such hybrid proteins display activity in yeast cells and can be detected easily by assay; moreover, yeast colonies containing such hybrid proteins turn blue when grown on medium containing the chromogenic substrate X-Gal (5-bromo-4-chloro-3-indolyl-β-D-galactoside). These properties greatly facilitate studies of the expression of cloned genes that are not themselves easily assayable. In addition, analysis of gene fusions that include varying portions of the cloned yeast gene can allow identification of regions of that gene required for regulation of gene expression, transport within the cell, or secretion. Methods for the construction of such gene fusions and their use to study gene expression and regulation have been reviewed (Casadaban *et al.*, 1983; Rose and Botstein, 1983; Guarente, 1983). A new application of gene fusions has been in the cellular localization of gene products; antibody to β-galactosidase can be used to localize a fusion protein in the intact yeast cell (L. Hereford, personal communication) by indirect immunofluorescent staining (Kilmartin *et al.*, 1982).

Yet another application of gene fusion technology is in the identification and study of the protein product encoded by a cloned gene. A variety of vectors are available for construction of gene fusions that produce hybrid proteins at high levels in *E. coli* (Reed, 1982; Gray *et al.*, 1982; Weinstock *et al.*, 1983; Spindler *et al.*, 1984; Young and Davis, 1983). Antisera prepared against such a hybrid protein can be used to identify the protein encoded in yeast by the yeast gene used in the fusion. Studies of the regulation of synthesis, cellular localization, and modification or processing of this protein are then possible.

E. Use of Promoter Fusions for High-Level Expression of Cloned Genes in Yeast

A convenient way to obtain high-level expression of a cloned homologous or heterologous gene in yeast is to construct a fusion between that gene and a strong yeast promoter. Some of the promoters that have been successfully used are the *GAL10* (UDP-galactose epimerase) promoter (Guarente *et al.*, 1982), the *ADH1* (alcohol dehydrogenase I) promoter (Valenzuela *et al.*, 1982; Hitzeman *et al.*, 1981; Ammerer, 1983), the *PHO5* (acid phosphatase) promoter (Miyanohara *et al.*, 1983; Kramer *et al.*, 1984), and the *PGK* (3-phosphoglycerate kinase) promoter (Hitzeman *et al.*, 1983a,b). Although the application of promoter fusion technology has thus far been directed primarily toward the synthesis in yeast of commercially important products, promoter fusions can clearly also be exploited in addressing basic problems of gene expression.

VI. OTHER USES OF CLONED GENES

We can expect the application of recombinant genetics to increase for some time and surely there will be a cornucopia of new uses. In this section we describe a few specific examples that may prove not only specific to fungi but also of significant importance to fungal genetics.

A. Making Defined Chromosomal Rearrangements

The ability to rearrange the organization of genes on chromosomes allows the reexamination of many classical issues of chromosome mechanics. As alluded to earlier, it has been possible to construct plasmid minichromosomes with two copies of a yeast centromere. Upon introduction of these plasmids into yeast, it has been possible to catalog most or all of the routes that cells have for dealing with chromosomes having two centromeres (Mann and Davis, 1983).

Studies of mitotic and meiotic pairing have been aided by a specific set of chromosome rearrangements (Clarke and Carbon, 1983). In these studies the centromere or chromosome III was replaced by the centromere of chromosome XI, and vice versa. Further rearrangements were made in which the orientation of the centromeres relative to both chromosome arms was inverted. None of these rearrangements affected the ability of these chromosomes to pair and disjoin. Thus these studies clearly and elegantly separate the role of the centromeres from the phenomenon of chromosome pairing.

B. Complementation Tests in Heterokaryons

A further use of cloned genes and chromsome rearrangements will be to enable geneticists working with certain fungi to perform an unambiguous complementation test. The limitation of complementation testing in some fungi is well illustrated by an example from *Aspergillus* genetics. *Aspergillus* can be propagated as either a heterokaryon (a cell with two or more genetically distinguishable nuclei) or a diploid (a cell with one nucleus). Mutations conferring resistance to fluoroacetate have been identified in three genes. When these mutations are tested for their ability to complement each other in diploids, the mutations are recessive and those mutations that are in different genes complement. However, when the same mutations are tested in heterokaryons, the mutations all appear to be dominant and do not complement (Apirion, 1966). Thus, it appears to be important to have two mutations in the same nucleus to obtain a clear complementation test. In fungi such as *Neurospora,* it is not possible to grow cells as stable diploids. Thus it is difficult to interpret certain regulatory mutations that appear to be cis-dominant when there is uncertainty over whether the complementation test is valid. These problems are discussed at length in Metzenberg (1979). Recombinant genetics should allow a more direct test of dominance or recessiveness of mutations in these fungi by transforming cells with a wild-type copy of the gene of interest.

VII. AN AGENDA FOR PROGRESS

The development of *S. cerevisiae* as an organism for genetic research has not always been an orderly progression of achievements, nor is it likely that genetics can be developed in any organism without some purely empirical research. Our hope is that lessons from *Saccharomyces* will speed the development of genetics in other fungi. To this end, we will first discuss what we judge to be a prudent and effective strategy for developing the genetics of a new fungus *de novo,* and then discuss the potential development of yeast as a host for surrogate genetics.

A. Development of Recombinant Genetics

1. The first objective is development of a defined solid medium that will permit both the sexual and asexual stages of the life cycle.
2. Many auxotrophic mutations should be obtained, with a particular focus on mutations of the purine and pyrimidine biosynthetic pathways as these pathways appear to be conserved enough to allow some cross-species complementation (Henikoff *et al.,* 1981). With these mutants, crosses should be performed to establish the ploidy and segregation properties of the chromosomes, assuming the organism possesses a sexual phase. The auxotrophs should be screened for mutations that either do not revert or revert at a very low frequency. The construction of a genetic map should begin but need not be extensively developed before the next steps.
3. The fungus should be examined for the presence of endogenous plasmids. These plasmids will be used as a source of an origin of replication for construction of recombinant plasmids.
4. A wild-type gene must be obtained and identified as a recombinant clone. The most efficient routes in *Saccharomyces* research have been complementation

of mutations in *E. coli* (approximately one-third of yeast genes are expressed in *E. coli*) and cross-species hybridization with cloned genes from related organisms. *Saccharomyces* genes should serve this purpose well.

5. Given a nonreverting mutation and a wild-type gene linked to an origin of replication, the time is right to develop transformation. In attempting transformation, several different strains should be tested as the frequency of transformation varies tremendously among strains of yeast. If no nonreverting mutations are found, strains mutant in two genes can be used in cotransformation experiments with two cloned genes. When a transformable strain is found, it is important to establish one strain as the canonical wild-type strain for that species and use that strain for all further developments. The importance of a wild-type strain cannot be overemphasized as it provides the necessary base line for all subsequent comparisons. DNA from the wild-type strain should then be used to build a recombinant library. The recombinant library must be a thorough representation of the genome, as described earlier.

6. Development of promoter fusion vectors has proved useful in overproducing a variety of gene products in yeast. The genes for glycolytic enzymes promise to have some of the strongest promoters due to the levels of these enzymes in the cell. However, should one want a highly regulated promoter, it is worth giving some thought to the ecology of the particular fungus. For example, in *S. cerevisiae,* there are no amino acid biosynthetic genes with the induction ratios of these genes in *E. coli,* perhaps because in nature yeast probably never has to cope with widely fluctuating levels of single amino acids. Yeast does, however, have to cope with dramatic shifts in the sugars that are available for growth, and it is therefore not surprising that genes involved in sugar metabolism have very high induction ratios in yeast and are efficiently regulated.

B. Potential for Surrogate Genetics

In addition to homologous genetics, that is, studying yeast genes in yeast, efforts are being made to use yeast as a surrogate host for the genetic analysis of sequences from other organisms. The premise behind these experiments is that the function a sequence provides in yeast will be an accurate reflection of the function provided in the organism from which that sequence was derived. The evidence for this view is promising but not yet compelling. For example, the functioning of *Tetrahymena* telomeres in yeast (Szostak and Blackburn, 1982) argues that the structural requirements for telomeres from any eukaryote may be very similar. Yeast integrating plasmid vectors such as YIp5 have been used to select for sequences from other organisms that will function as ARS elements in yeast. Such sequences may be bona fide origins of replication from the original organism, but this has yet to be demonstrated in a single case (Stinchcomb *et al.,* 1980).

An additional way in which *Saccharomyces* genetics could aid development of genetics in other fungi rests upon the assumption that genes from many fungi will be expressed in yeast. If this assumption bears out, genes from fungus X can be cloned by complementation of mutations in yeast. Furthermore, if a clone can complement mutations in two or more different genes, then these genes in fungus X must be physically linked regardless of their relative map position in yeast. The construction of a cosmid library from fungus X in a *kar1* yeast donor strain might allow the efficient acquisition of this type of mapping data.

The wholesale application of surrogate genetics is difficult to justify in a conventional sense because often some steps require assumptions that cannot be tested ahead of time. Nevertheless, we believe that the ease of genetic manipulation as it is practiced in yeast is a sufficient justification to try experiments that seem reasonable and have a significant potential benefit even though the probability of success is, at times, unpredictable.

ACKNOWLEDGMENTS

We thank Kerrie Rine for preparation of the figures and Peggy McCuthchan for help in preparation of the manuscript.

The research in the laboratory of J. R. is supported by grant GM31105 from the National Institutes of Health. Research in the laboratory of M. C. is supported by grant NP358a from the American Cancer Society and grant GM32065 from the National Institutes of Health.

REFERENCES

Ammerer, G. (1983). Expression of genes in yeast using the *ADC1* promoter. *In* "Recombinant DNA," Part C (R. Wu, L. Grossman, and K. Moldave, eds.), Methods in Enzymology, Vol. 101, pp. 192–201. Academic Press, New York.

Apirion, D. (1966). Recessive mutations at unlinked loci which complement in diploids but not in heterokaryons of *Aspergillus nidulans*. *Genetics* **53,** 935–941.

Bach, M. L., and Lacroute, F. (1972). Direct selection techniques for the isolation of pyrimidine auxotrophs in yeast. *Mol. Gen. Genet.* **115,** 126–130.

Bach, M. L., Lacroute, F., and Botstein, D. (1979). Evidence for transcriptional regulation of orotidine-5′-phosphate decarboxylase in yeast by hybridization of mRNA to the yeast structural gene cloned in *Escherichia coli. Proc. Natl. Acad. Sci. U.S.A.* **76,** 368–390.

Barnes, G., Hansen, W. J., Holcomb, C. L., and Rine, J. (1984). Asparagine linked glycosylation in *Saccharomyces cerevisiae:* genetic analysis of an early step. *Mol. Cell. Biol.* **4,** 2381–2388.

Beach, D., Piper, M., and Shall, S. (1980). Isolation of chromosomal origins of replication in yeast. *Nature (London)* **284,** 185–186.

Beach, D., Durkacz, B., and Nurse, P. (1982). Functionally homologous cell cycle control genes in budding and fission yeast. *Nature (London)* **300,** 706–709.

Beggs, J. D. (1978). Transformation of yeast by a replicating hybrid plasmid. *Nature (London)* **275,** 104–108.

Botstein, D., and Davis, R. W. (1982). Principles and practices of recombinant DNA with yeast. *In* "The Molecular Biology of the Yeast *Saccharomyces: Metabolism and Gene Expression*" (J. N. Strathern, E. W. Jones, and J. R. Broach, eds.), pp. 607–636. Cold Spring Harbor Lab. Press, New York.

Botstein, D., Falco, S. C., Stewart, S. E., Brennan, M., Scherer, S., Stinchcomb, D. T., Struhl, K., and Davis, R. W. (1979). Sterile host yeasts (SHY): a eucaryotic system for biological containment for recombinant DNA experiments. *Gene* **8,** 17–24.

Brandriss, M. (1979). Isolation and preliminary characterization of *Saccharomyces cerevisiae* proline auxotrophs. *J. Bacteriol.* **138,** 816–822.

Broach, J. R., and Hicks, J. B. (1980). Replication and recombination functions associated with the yeast 2 micron circle. *Cell (Cambridge, Mass.)* **21,** 501–508.

Broach, J. R., Strathern, J. N., and Hicks, J. B. (1979). Transformation in yeast: development of a hybrid cloning vector and isolation of the *CAN1* gene. *Gene* **8,** 121–133.

Carlson, M., and Botstein, D. (1982). Two differentially regulated mRNAs with different 5' ends encode secreted and intracellular forms of yeast invertase. *Cell (Cambridge, Mass.)* **28,** 145–154.

Casadaban, M. J., Martinez-Arias, A., Shapira, S. K., and Chou, J. (1983). β-Galactosidase gene fusions for analyzing gene expression in *Escherichia coli* and yeast. *In* "Recombinant DNA," Part B (R. Wu, L. Grossman, and K. Moldave, eds.), Methods in Enzymology, Vol. 100, pp. 293–308. Academic Press, New York.

Chattoo, B. B., Sherman, F., Azubalis, D. A., Fjellstedt, T. A., Mehnert, D., and Ogur, M. (1979). Selection of *lys2* mutants of the yeast *Saccharomyces cerevisiae* by the utilization of alpha-aminoadipate. *Genetics* **93,** 51–65.

Clarke, L., and Carbon, J. (1980).Isloation of a yeast centromere and construction of functional small circular chromosomes. *Nature (London)* **287,** 504–509.

Clarke, L., and Carbon, J. (1983). Genomic substitutions of centromeres in *Saccharomyces cerevisiae*. *Nature (London)* **305,** 23–28.

Cohen, J. D., Eccleshall, T. R., Needleman, R. B., Federoff, H., Buchferer, B. A., and Marmur, J. (1980). Functional expression in yeast of the *E. coli* plasmid gene coding for chloramphenicol acetyl transferase. *Proc. Natl. Acad. Sci. U.S.A.* **77,** 1078–1082.

Conde, J., and Fink, G. R. (1976). A mutant of *Saccharomyces cerevisiae* defective for nuclear fusion. *Proc. Natl. Acad. Sci. U.S.A.* **73,** 3651–3655.

Davies, J., and Jimenez, J. (1980). A new selective agent for eucaryotic cloning vectors. *Am. J. Trop. Med. Hyg.* **295,** 1089–1092.

Dutcher, S. K. (1981). Internuclear transfer of genetic information in *kar1-1/KAR1* heterokaryons in *Saccharomyces cerevisiae*. *Mol. Cell. Biol.* **1,** 245–253.

Falco, S. C., and Botstein, D. (1983). A rapid chromosome mapping method for cloned fragments of yeast DNA. *Genetics* **105,** 857–872.

Falco, S. C., Rose, M., and Botstein, D. (1983). Homologous recombination between episomal plasmids and chromosomes in yeast. *Genetics* **105,** 843–856.

Fangman, W. L., Hice, R. H., and Chlebowicz-Sledziewska, E. (1983). ARS replication during the yeast S phase. *Cell (Cambridge, Mass.)* **32,** 831–838.

Fitzgerald-Hayes, M., Buhler, J. M., Cooper, T. G., and Carbon, J. (1982a). Isolation and subcloning analysis of functional centromere DNA (Cen11) from *Saccharomyces cerevisiae* chromosome XI. *Mol. Cell. Biol.* **2,** 82–87.

Fitzgerald-Hayes, M., Clarke, L., and Carbon, J. (1982b). Nucleotide sequence analysis of yeast centromere DNAs. *Cell (Cambridge, Mass.)* **29,** 235–244.

Fried, H. M., and Warner, J. R. (1981). Cloning of yeast gene for trichodermin resistance and ribosomal protein L3. *Proc. Natl. Acad. Sci. U.S.A.* **78,** 238–242.

Gerbaud, C., and Guerineau, M. (1980). 2μm plasmid copy number in different yeast strains and

repartition of endogenous and 2um chimeric plasmids in transformed strains. *Curr. Genet.* **1**, 219–228.
Gray, M. R., Colot, H. V., Guarente, L., and Rosbach, M. (1982). Open reading frame cloning: identification, cloning and expression of open reading frame DNA. *Proc. Natl. Acad. Sci. U.S.A.* **79**, 6598–6602.
Grenson, M., Mousset, M., Wiame, J. M., and Bechet, J. (1966). Multiplicity of the amino acid permeases in *Saccharomyces cerevisiae*. I. Evidence for a specific arginine transporting system. *Biochim. Biophys. Acta* **127**, 325–338.
Guarente, L. (1983). Yeast promoters and *lacZ* fusions designed to study expression of cloned genes in yeast. *In* "Recombinant DNA," Part C (R. Wu, L. Grossman, and K. Moldave, eds.), Methods in Enzymology, Vol. 101, pp. 181–191. Academic Press, New York.
Guarente, L., and Ptashne, M. (1981). Fusion of *Escherichia coli lacZ* to the cytochrome *c* gene of *Saccharomyces cerevisiae*. *Proc. Natl. Acad. Sci. U.S.A.* **78**, 2199–2203.
Guarente, L., Yocum, R. R., and Gifford, P. (1982). A *GAL10–CYC1* hybrid yeast promoter identifies the *GAL4* regulatory region as an upstream site. *Proc. Natl. Acad. Sci. U.S.A.* **79**, 7410–7414.
Haber, J. E., Thorburn, P. C., and Rogers, D. (1984). Meiotic and mitotic behavior of dicentric chromosomes in *Saccharomyces cerevisiae*. *Genetics* **106**, 185–205.
Henikoff, S., Tatchell, K., Hall, B. D., and Nasmyth, K. A. (1981). Isolation of a gene from *Drosophila* by complementation in yeast. *Nature (London)* **289**, 33–37.
Henry, S. A., and Horowitz, B. (1975). A new method for mutant selection in *Saccharomyces cerevisiae*. *Genetics* **79**, 175–186.
Henry, S. A., Donahue, T. F., and Culbertson, M. R. (1975). Selection of spontaneous mutants by inositol starvation in yeast. *Mol. Gen. Genet.* **145**, 5–11.
Hicks, J., and Fink, G. R. (1977). Identification of chromosomal location of yeast DNA from hybrid plasmid pYeleu10. *Nature (London)* **269**, 265–272.
Hinnen, A., Hicks, J. B., and Fink, G. R. (1978). Transformation of yeast. *Proc. Natl. Acad. Sci. U.S.A.* **75**, 1929–1933.
Hisao, I., Fukuda, Y., Murata, K., and Kumura, A. (1983). Transformation of intact yeast cells treated with alkali cations. *J. Bacteriol.* **153**, 163–168.
Hitezman, R. A., Hagie, F. E., Levine, H. L., Goeddel, D. V., Ammerer, G., and Hall, B. D. (1981). Expression of a human gene for interferon in yeast. *Nature (London)* **293**, 717–722.
Hitezman, R. A., Leung, D. W., Perry, L. J., Kohr, W. J., Levine, H. L., and Goeddel, D. V. (1983a). Secretion of human interferons by yeast. *Science (Washington, D.C.)* **219**, 620–625.
Hitezman, R. A., Chen, C. Y., Hagie, F. E., Patzer, E. J., Liu, C. C., Estell, D. A., Miller, J. V., Yaffe, A., Kleid, D. G., Levinson, A. D., and Oppermann, H. (1983b). Expression of hepatitis B virus surface antigen in yeast. *Nucleic Acids Res.* **11**, 2745–2763.
Holland, J. P., and Holland, M. J. (1980). Structural comparison of two nontandemly repeated yeast glyceraldehyde-3-phosphate dehydrogenase genes. *J. Biol. Chem.* **255**, 2596–2605.
Hsiao, C. L., and Carbon, J. (1979). High frequency transformation of yeast by plasmids containing the cloned yeast *ARG4* genes. *Proc. Natl. Acad. Sci. U.S.A.* **76**, 3829–3833.
Hsiao, C. L., and Carbon, J. (1981). Direct selection procedure for the isolation of functional centromeric DNA. *Proc. Natl. Acad. Sci. U.S.A.* **78**, 3760–3764.
Julius, D., Blair, L., Brake, A., Sprague, G., and Thorner, J. (1983). Yeast alpha factor is processed from a larger precursor polypeptide: the essential role of a membrane-bound dipeptidyl aminopeptidase. *Cell (Cambridge, Mass.)* **32**, 839–852.
Kikuchi, Y. (1983). Yeast plasmid requires a *cis*-acting locus and two plasmid proteins for its stable maintenance. *Cell (Cambridge, Mass.)* **35**, 487–493.
Kilmartin, J. V., Wright, B., and Milstein, C. (1982). Rat monoclonal antitubulin antibodies derived by using a new nonsecreting rat cell line. *J. Cell Biol.* **93**, 567–572.

Klapholtz, S., and Esposito, R. (1982). A new mapping method employing a meiotic rec^- mutant of yeast. *Genetics* **100**, 387–412.

Kramer, R. A., DeChiara, T. M., Schaber, M. D., and Hilliker, S. (1984). Regulated expression of a human interferon gene in yeast: control by phosphate concentration or temperature. *Proc. Natl. Acad. Sci. U.S.A.* **81**, 367–370.

Lacroute, F. (1971). Non-Mendelian mutation allowing ureidosuccinic acid uptake in yeast. *J. Bacteriol.* **106**, 519–522.

Lacroute, F., Pierard, A., Grenson, M., and Wiame, J. M. (1965). The biosynthesis of carbamoyl phosphate in *Saccharomyces cerevisiae*. *J. Gen. Microbiol.* **40**, 127–142.

Lauer, G. O., Roberts, T. M., and Klotz, L. C. (1977). Determination of the nuclear DNA content of *Saccharomyces cerevisiae* and implications for the organization of DNA in yeast chromosomes. *J. Mol. Biol.* **114**, 507–526.

Lindsley, D. L., and Grell, E. H. (1968). Genetic variations of *Drosophila melanogaster*. *Carnegie Inst. Wash., Publ.* No. 627.

Lindsley, D. L., Sandler, L., Baker, B. S., Carpenter, A. T. C., Denell, R. E., Hall, J. C., Jacobs, P. A., Miklos, G. L. G., Davis, B. K., Gethmann, R. C., Hardy, R. W., Hessler, A., Miller, S. M., Nozawa, H., Parry, D. M., and Gould-Somero, M. (1972). Segmental aneuploidy and the genetic gross structure of the *Drosophila Genetics* **71**, 157–184.

Littlewood, B. S. (1975). Methods for selecting auxotrophic and temperature sensitive mutants in yeasts. *Methods Cell Biol.* **11**, 273–285.

McCammon, M., and Parks, L. W. (1982). Enrichment for auxotrophic mutants in *Saccharomyces cerevisiae* using the cell wall inhibitor echinocandin B. *Mol. Gen. Genet.* **186**, 295–297.

McClintock, B. (1939). The behavior of successive nuclear divisions of a chromosome broken at meiosis. *Proc. Natl. Acad. Sci. U.S.A.* **25**, 405–416.

Maniatis, T., Fritsch, E. F., and Sambrook, J. (1982). "Molecular Cloning: A Laboratory Manual." Cold Spring Harbor, New York.

Mann, C., and Davis, R. W. (1983). Instability of dicentric plasmids in yeast. *Proc. Natl. Acad. Sci. U.S.A.* **80**, 228–232.

Metzenberg, R. L. (1979). Implications of some genetic control mechanisms in *Neurospora*. *Microbiol. Rev.* **43**, 361–383.

Miller, J. S., Paterson, B. M., Ricciardi, R. P., Cohen, L., and Roberts, B. E. (1983). Methods utilizing cell-free protein-synthesizing systems for the identification of recombinant DNA molecules. *In* "Recombinant DNA," Part C (R. Wu, L. Grossman, and K. Moldave, eds.), Methods in Enzymology, Vol. 101, pp. 650–674. Academic Press, New York.

Miyanohara, A., Toh-e, A., Nozaki, C., Hamada, F., Ohtomo, N., and Matsubara, K. (1983). Expression of hepatitis B surface antigen in yeast. *Proc. Natl. Acad. Sci. U.S.A.* **80**, 1–5.

Mortimer, R. K., and Schild, D. (1980). Genetic map of *Saccharomyces cerevisiae*. *Microbiol. Rev.* **44**, 519–571.

Mortimer, R. K., Contopoulou, R., and Schild, D. (1981). Mitotic chromosome loss in a radiation sensitive strain of the yeast *Saccharomyces cerevisiae*. *Proc. Natl. Acad. Sci. U.S.A.* **78**, 5778–5782.

Murray, A. W., and Szostak, J. W. (1983). Pedigree analysis of plasmid segregation in yeast. *Cell (Cambridge, Mass.)* **34**, 961–970.

Nasmyth, K. A., and Reed, S. I. (1980). Isolation of genes by complementation in yeast: molecular cloning of a cell cycle gene. *Proc. Natl. Acad. Sci. U.S.A.* **77**, 2119–2123.

Orr-Weaver, T., Szostak, J. W., and Rothstein, R. J. (1981). Yeast transformation: a model system for the study of recombination. *Proc. Natl. Acad. Sci. U.S.A.* **78**, 6354–6358.

Orr-Weaver, T., Szostak, J. W., and Rothstein, R. J. (1983). Genetic applications of yeast transformation with linear and gapped plasmids. *In* "Recombinant DNA," Part C (R. Wu, L.

Grossmann, and K. Moldave, eds.), Methods in Enzymology, Vol. 101, pp. 228–245. Academic Press, New York.
Osley, M. A., and Hereford, L. (1982). Identification of a sequence responsible for periodic synthesis of yeast histone 2A mRNA. *Proc. Natl. Acad. Sci. U.S.A.* **79,** 7689–7693.
Panzeri, L., and Philippsen, P. (1982). DNA sequences in the centromere from yeast chromosome VI. *EMBO J.* **1,** 1605–1611.
Ramos, F., and Wiame, J. M. (1980). Two asparagine synthetases in *Saccharomyces cerevisiae*. *Eur. J. Biochem.* **108,** 373–377.
Ratzkin, B., and Carbon, J. (1977). Functional expression of cloned yeast DNA in *Escherichia coli*. *Proc. Natl. Acad. Sci. U.S.A.* **74,** 487–491.
Reed, S. I. (1982). Preparation of product specific antisera by gene fusion: antibodies specific for the product of the yeast cell-division-cycle gene *CDC28*. *Gene* **20,** 255–265.
Rine, J. (1979). Regulation and transposition of cryptic mating type genes in *Saccharomyces cerevisiae*. Ph.D. Thesis, Univ. of Oregon, Eugene.
Rine, J., Hansen, W., Hardeman, E., and Davis, R. W. (1983). Targeted selection of recombinant clones through gene dosage effects. *Proc. Natl. Acad. Sci. U.S.A.* **80,** 6750–6754.
Rine, J., Barnes, G., Hansen, W., and Holcomb, C. (1984). Cloning *Saccharomyces cerevisiae* genes through use of segmental aneuploidy. *Microbiology (Washington, D.C.)* pp. 140–143.
Roeder, G. S., and Fink, G. R. (1980). DNA rearrangements associated with a transposable element in yeast. *Cell (Cambridge, Mass.)* **21,** 239–249.
Rose, M., and Botstein, D. (1983). Construction and use of gene fusions to *lacZ* (β-galactosidase) that are expressed in yeast. *In* "Recombinant DNA," Part C (R. Wu, L. Grossman, and K. Moldave, eds.), Methods in Enzymology, Vol. 101, pp. 167–181. Academic Press, New York.
Rose, M., Casadaban, M. J., and Botstein, D. (1981). Yeast genes fused to β-galactosidase in *Escherichia coli* can be expressed normally in yeast. *Proc. Natl. Acad. Sci. U.S.A.* **78,** 2460–2464.
Rothstein, R. J. (1983). One step gene disruption in yeast. *In* "Recombinant DNA," Part C (R. Wu, L. Grossman, and K. Moldave, eds.), Methods in Enzymology, Vol. 101, pp. 202–211. Academic Press, New York.
Scherer, S., and Davis, R. W. (1979). Replacement of chromosome segments with altered DNA sequences constructed *in vitro*. *Proc. Natl. Acad. Sci. U.S.A.* **76,** 4951–4955.
Schwartz, D. C., Saffran, W., Welsh, J., Haas, R., Goldenberg, M., and Cantor, C. R. (1982). New techniques for purifying large DNAs and studying their properties and packaging. *Cold Spring Harbor Symp. Quant. Biol.* **47,** 189–195.
Shalit, P., Loughney, K., Olson, M., and Hall, B. D. (1981). Physical analysis of the *CYC1–sup4* interval in *Saccharomyces cerevisiae*. *Mol. Cell. Biol.* **1,** 228–236.
Shortle, D., DiMaio, D., and Nathans, D. (1981). Directed mutagenesis. *Annu. Rev. Genet.* **15,** 265–294.
Shortle, D., Haber, J. E., and Botstein, D. (1982). Lethal disruption of the yeast actin gene by integrative DNA transformation. *Science (Washington, D.C.)* **217,** 371–373.
Sigurdson, D. C., Gaarder, M. E., and Livingston, D. M. (1981). Characterization of the transmission during cytoductant formation of the 2 micron DNA plasmid from *Saccharomyces*. *Mol. Gen. Genet.* **183,** 59–65.
Silverman, S. J., Rose, M., Botstein, D., and Fink, G. R. (1982). Regulation of *HIS4–lacZ* fusions in *Saccharomyces cerevisiae*. *Mol. Cell. Biol.* **2,** 1212–1219.
Singh, A., and Sherman, F. (1975). Genetic and physiological characterization of *MET15* mutants of *Saccharomyces cerevisiae*: a selective system for forward and reverse mutations. *Genetics* **81,** 75–97.

Snow, R. (1966). An enrichment method for auxotrophic yeast mutants using the antibiotic, nystatin. *Nature (London)* **211,** 206–207.
Southern, E. M. (1975). Detection of specific sequences among DNA fragments separated by gel electrophoresis. *J. Mol. Biol.* **98,** 503–517.
Spindler, K. R., Rosser, D. S. E., and Berk, A. J. (1984). Analysis of adenovirus transforming proteins from early regions 1A and 1B with antisera to inducible fusion antigens produced in *Escherichia coli*. *J. Virol.* **49,** 132–141.
Stiles, J. I. (1983). Use of integrative transformation of yeast in the cloning of mutant genes and large segments of contiguous chromosomal sequences. *In* "Recombinant DNA," Part C (R. Wu, L. Grossman, and K. Moldave, eds.), Methods in Enzymology, Vol. 101, pp. 290–300. Academic Press, New York.
Stinchcomb, D. T., Struhl, K., and Davis, R. W. (1979). Isolation and characterization of a yeast chromosomal replicator. *Nature (London)* **282,** 39–43.
Stinchcomb, D. T., Thomas, M., Kelley, J., Selker, E., and Davis, R. W. (1980). Eucaryotic DNA sequences capable of autonomous replication in yeast. *Proc. Natl. Acad. Sci. U.S.A.* **77,** 4559–4563.
Stinchcomb, D. T., Mann, C., and Davis, R. W. (1982). Centromeric DNA from *Saccharomyces cerevisiae*. *J. Mol. Biol.* **158,** 157–179.
Strathern, J. N., Newlon, C. S., Herskowitz, I., and Hicks, J. B. (1979). Isolation of a circular derivative of yeast chromosome III: implications for the mechanism of mating type interconversion. *Cell (Cambridge, Mass.)* **18,** 309–319.
Struhl, K., Cameron, J. R., and Davis, R. W. (1976). Functional genetic expression of eucaryotic DNA in *E. coli*. *Proc. Natl. Acad. Sci. U.S.A.* **73,** 1471–1475.
Struhl, K., Stinchcomb, D. T., Scherer, S., and Davis, R. W. (1979). High frequency transformation of yeast: autonomous replication of hybrid DNA molecules. *Proc. Natl. Acad. Sci. U.S.A.* **76,** 1035–1039.
Szostak, J. W., and Blackburn, E. H. (1982). Cloning yeast telomeres on linear plasmid vectors. *Cell (Cambridge, Mass.)* **29,** 245–255.
Tschumper, G., and Carbon, J. (1980). Sequence of a yeast DNA fragment containing a chromosomal replicator and the *TRP1* gene. *Gene* **10,** 157–166.
Tschumper, G., and Carbon, J. (1983). Copy number control by a yeast centromere. *Gene* **23,** 221–232.
Valenzuela, P., Medina, A., Rutter, W. J., Ammerer, G., and Hall, B. D. (1982). Synthesis and assembly of hepatitis B virus surface antigen particles in yeast. *Nature (London)* **298,** 347–350.
Weinstock, G. M., Ap Rhys, C., Berman, M. L., Hampar, B., Jackson, D., Silhavy, T. J., Weisemann, J., and Zweig, M. (1983). Open reading frame expression vectors: a general method for antigen production in *Escherichia coli* using fusion proteins to β-galactosidase. *Proc. Natl. Acad. Sci. U.S.A.* **80,** 4432–4436.
Whelan, W. L., Gocke, E., and Manney, T. (1979). The *CAN1* locus of *Saccharomyces cerevisiae*: fine structure analysis and forward mutation rates. *Genetics* **91,** 35–51.
Winston, F., Chumley, F., and Fink, G. R. (1983). Eviction and transplacement of mutant genes in yeast. *In* "Recombinant DNA," Part C (R. Wu, L. Grossman, and K. Moldave, eds.), Methods in Enzymology, Vol. 101, pp. 211–228. Academic Press, New York.
Wood, J. S. (1982). Mitotic chromosome loss induced by methyl benzimidazole-2yl carbamate as a rapid mapping method in *Saccharomyces cerevisiae*. *Mol. Cell. Biol.* **2,** 1080–1087.
Young, R. A., and Davis, R. W. (1983). Efficient isolation of genes using antibody probes. *Proc. Natl. Acad. Sci. U.S.A.* **80,** 1194–1198.

6

Yeast Transformation

GEORGE BOGUSLAWSKI
Biosynthesis Research Laboratory
Biotechnology Group
Miles Laboratories, Inc.
Elkhart, Indiana

I.	Introduction	161
	A. General Principles	162
	B. Methodology	162
II.	Specific Transformation Systems in Yeast	167
	A. Types of Yeast Plasmid Vectors	167
	B. Artificial Chromosomes (Yeast Linear Plasmids - YLp)	177
III.	Applications	178
	A. Isolation and Characterization of Genes	178
	B. Expression of Heterologous Genes in Yeast	180
	C. Genetic Recombination Models	182
IV.	Conclusions	187
	References	188

I. INTRODUCTION

The recent astounding progress in the molecular genetics of yeast makes it difficult to remember that one of the major reasons for the progress, the development of an efficient transformation system, became a reality only a few years ago (Hinnen *et al.*, 1978). This system, combined with gene cloning, has permitted the analysis of chromosome structure and behavior, and the study of gene regulation and expression, in ways not previously possible. Among the many exciting achievements, several stand out prominently: the elucidation of the structure and function of the mating-type locus (Herskowitz and Oshima, 1981; Nasmyth *et al.*, 1981), the construction of artificial chromosomes, either in a circular or a linear form (Clarke and Carbon, 1980; Murray and Szostak, 1983a), and the discovery and characterization of transposable elements in yeast (Cameron *et al.*,

1979; Fink et al., 1980). In all these areas, molecular and genetic analyses were facilitated by transformation of cells with DNA molecules constructed in vitro. The knowledge and understanding of the molecular structure and function of the yeast hereditary material, gained from transformation and the associated techniques, fully justify the concept of a "new yeast genetics" (Struhl, 1983a).

The early reports on transformation in yeast (Oppenoorth, 1962; Tuppy and Wildner, 1965; Khan and Sen, 1974) were poorly documented, and none demonstrated that the transforming DNA became a heritable component of the recipient genome. The first successful demonstration of genetic transformation and integration of the exogenous DNA into the genome was reported by Hinnen et al. (1978), and was followed rapidly by the development of a variety of transforming DNA vectors (see below).

Although several methods for DNA uptake by yeast cells have been described (Hinnen et al., 1978; Ito et al., 1983; Sherman et al., 1983), understanding of the specifics of the uptake is still lacking. Therefore, this review will not deal so much with the mechanistic aspects of transformation, but rather will emphasize the new vistas opened by the development of novel plasmid vectors and in vitro construction of new chromosomes. I hope to be able to summarize the salient aspects of the field without provoking the opinion that "this is the kind of reading matter that you can't pick up once you put it down."

A. General Principles

In most transformation experiments the particular yeast DNA segment is inserted into a plasmid vector such as ColE1 or pBR322 which can be easily amplified in *Escherichia coli* and thus prepared in large quantities. The resulting chimeric plasmid DNA is then taken up by the yeast cells and transformants are examined. The essential aspect of transformation is the presence in the DNA of an easily selectable genetic marker. Because transformation efficiency is relatively low, the genetic markers of both the donor and the recipient DNA should be stable so that spontaneous mutations do not complicate the analysis and/or lead the investigator astray. As yeast cells are not naturally competent for DNA uptake, the permeability barrier must be sufficiently modified, usually by enzymatic treatment and the addition of calcium or lithium salts. Once the DNA enters the cell, it may become integrated into one of the chromosomes, or it may remain in an autonomous state, depending on the nature of the DNA molecules presented to the cell.

B. Methodology

Two basic methods of transformation of yeast are commonly used. The first one (Hinnen et al., 1978; Beggs, 1978; Hsiao and Carbon, 1979; Sherman et al.,

1983) involves the removal of cell wall by treatment with digestive enzymes (Glusulase, Zymolyase). The spheroplasts, maintained intact by osmotic stabilization, are mixed with DNA in the presence of calcium ions, followed by polyethylene glycol (PEG). The resulting aggregated spheroplasts are allowed to regenerate on appropriate selective media and then are analyzed for the acquisition and stability of the desired trait(s).

The second method (Ito *et al.*, 1983; Iimura *et al.*, 1983) eschews the spheroplast preparation step and instead relies on the action of alkali cations (especially lithium) on cells to induce DNA intake. However, PEG is still required. This procedure, although much simpler, is sometimes less efficient. The two methods are described in detail below. The reagents and media used in transformation are listed in Table I.

Efficiency of transformation is strongly dependent on the strain and on the plasmid used. Some strains do not transform at all, and the usefulness of a particular yeast strain in transformation must be determined experimentally. In the author's laboratory, strains such as AH2 (*a his4-519, leu2-3. leu2-112, can1;* obtained from G. R. Fink, Massachusetts Institute of Technology) and YNN27 (α *trp1-289, ura3-52, gal2;* obtained from R. W. Davis, Stanford University) have been used with excellent results.

1. Spheroplast Transformation

Yeast cells are grown overnight to stationary phase in YEPD at 30°C, and diluted 25-fold into fresh YEPD. The growth continues for 3–3.5 hr until the culture reaches the density of 2×10^7 cells/ml. [Hicks and co-workers (1982) have suggested that YEP be used in place of YEPD. This may result in 5- to 10-fold increase in transformation efficiency. The plateau of growth in this medium is at the cell density optimal for harvesting, and the cells remain in a transformable state for several hours. The drawback is that some strains cannot grow in the absence of an additional carbon source.]

The steps outlined below are for a 200-ml culture. This can be scaled up or down as desired.

1. Harvest cells by a 5-min centrifugation (2000 g) and suspend in 20 ml of SED.
2. Incubate at 30°C without aeration for 10–60 min. The presence of EDTA and DTT enhances the action of Glusulase.
3. Harvest cells at room temperature in a tabletop centrifuge (e.g., IEC clinical centrifuge).
4. Suspend cells by gentle swirling in 20 ml of SCE. Remove a 50-μl aliquot and mix with 0.5 ml of 5% SDS. This will serve as control for spheroplast formation.

TABLE I.
Media and Reagents for Transformation Procedures

Spheroplast transformation	

YEPD:
  1% yeast extract
  2% bactopeptone
  2% glucose, added after autoclaving separately
  2% agar for solid medium
YEP:
  Same as YEPD but without glucose
SD:
  0.67% Difco yeast nitrogen base without amino acids
  2% glucose
  2% agar
  0.8 M Sorbitol
SED:
  1 M sorbitol
  25 mM EDTA, pH 8.0
  50 mM dithiothreitol (DTT)
SCE:
  1 M sorbitol
  0.1 M sodium citrate, pH 5.8
  0.01 M EDTA
SY:
  1.0 M sorbitol in 67% YEPD
	SYTC:
	  0.9 M sorbitol
	  67% YEPD
	  10 mM Tris-HCl, pH 7.5
	  10 mM CaCl$_2$
	PEG:
	  20% polyethylene glycol 4000
	  10 mM Tris-HCl, pH 7.5
	  10 mM CaCl$_2$
	Regeneration agar:
	  0.67% yeast nitrogen base without amino acids
	  0.9 M sorbitol
	  2% glucose
	  3% agar
	Top agar:
	  same as regeneration agar but kept at 45–50°C
	Glusulase (Endo Laboratories):
	  use undiluted
	SDS:
	  5% sodium dodecyl sulfate

Whole-cell transformation	

YEPD
TE:
  10 mM Tris-HCl, pH 7.5
  1 mM EDTA
1 M Lithium acetate
PEG:
  50% (w/v) polyethylene glycol 4000
	Carrier DNA in TE (3 mg/ml):
	  salmon testes or calf thymus DNA.
	  Just prior to use, boil for 10 min; cool rapidly on ice

5. Add 0.2 ml of Glusulase, mix, and incubate at 30°C with occasional swirling. From time to time remove 50-μl samples into 0.5 ml of SDS to observe clearing. When over 90% of cells lyse (almost complete clearing), the spheroplast formation is complete. This usually takes about 15 min at 30°C. Sometimes, especially with overgrown cultures, Glusulase treatment may require up to 1 hr. Avoid overdigestion as it may decrease regeneration of spheroplasts.

6. Spin down the spheroplasts (1000 g, 5 min) and wash the pellet twice with 0.8 M sorbitol (5 ml each wash). Allow sorbitol solution to run along the side of

the tube and suspend the cells by gentle swirling. Centrifuge and discard supernatant.

7. Resuspend pellet in 10 ml of SY and incubate at 30°C for 1 hr without shaking. This step allows the cell wall regeneration to begin.

8. Centrifuge as above at room temperature; discard the supernatant.

9. Resuspend pellet in 1 ml of SYTC by gentle swirling. The cells may be stored at this point for at least 2–4 days at 4°C. Alternatively, the competent spheroplasts can be frozen at $-70°C$ after dimethyl sulfoxide is added to 15% concentration. However, a 5- to 10-fold loss in transformation efficiency may result (Orr-Weaver et al., 1983).

10. Distribute 0.1-ml samples of spheroplast suspension into disposable screw cap tubes (Falcon 2047) and add DNA as required (usually 0.1–5 µg in 1–20 µl). Incubate at room temperature for 15 min.

11. To each tube add 1.0 ml PEG; allow the solution to run along the side and mix in gently.

12. Incubate at 30°C for 20–30 min. The PEG will cause aggregation of spheroplasts and DNA. The mixture of cells and PEG can be stored at 4°C for 2–4 days.

13. Transfer 0.2 ml of the PEG suspension into a fresh Falcon 2047 tube and add 7 ml of top agar (held at 50°C) containing the necessary supplements (e.g., if AH2 strain is being transformed to Leu$^+$, include histidine but not leucine in the agar). Mix by rapid inversion and pour on regeneration agar plate previously warmed to room temperature.

14. Incubate at 30°C for 2–5 days. If transformants are too numerous, go back and dilute the suspension with 0.8 M sorbitol. Then repeat plating.

15. Controls: (1) no DNA; (2) top agar supplemented with all required nutrients (histidine and leucine in the example shown above); (3) dilution of spheroplast suspension with 0.8 M sorbitol and plating on a nonselective regeneration medium to estimate regeneration efficiency (a sample should also be plated without osmotic stabilization).

16. Transformants should be purified by streaking on selective media (e.g., SD) supplemented with appropriate nutrients. The stability of a marker under study should be tested by streaking transformants on nonselective medium, replica-plating on selective agar, and checking for growth. This could also be done in liquid media in which growth can be monitored quantitatively.

Following these steps, the transformants may be analyzed for the presence of the transforming DNA in either integrated or autonomous form. This analysis can be done in a variety of ways. Colony hybridization with an appropriate probe, combined with tetrad analysis, allows a simple and rapid verification of the transformation event and a test of linkage of the transforming DNA to chromosomal sequences (Ilgen et al., 1979). Unintegrated plasmids will show no linkage to any particular chromosomal locus ($4^+:0^-$ or $0^-:4^-$ segregation of the plasmid-borne markers in meiosis). The Southern blotting technique (Southern, 1975) also distinguishes between integrated and unintegrated transforming DNA because of differences in size between plasmid DNA and chromosomal DNA and because integration frequently changes the restriction pattern of the genetic locus in question (Orr-Weaver *et al.*, 1983; Winston *et al.*, 1983).

2. Whole-Cell Transformation

Cells are grown as described in Section I,B,1 and harvested by centrifugation when the culture density reaches 2×10^7 cells/ml. The steps described below are for a 100-ml culture.

1. Resuspend cells in 5 ml of TE; pellet by centrifugation.
2. Resuspend cells in 5 ml of 0.1 M lithium acetate (LiAc) in TE; pellet by centrifugation.
3. Resuspend cells in 1 ml of 0.1 M LiAc in TE; incubate at 30°C for 60 min; vortex gently at the end of incubation. The cells can be stored at this stage for as long as 10 days at 4°C but the efficiency of transformation will decrease about fivefold.
4. Pipette 0.1-ml aliquots of the cell suspension into 1.5-ml Eppendorf tubes. To each tube add 15 µl of carrier DNA and up to 15 µl of the transforming plasmid DNA. Incubate for 30 min at 30°C.
5. To each tube add 0.7 ml of 40% PEG in TE containing 0.1 M LiAc, vortex briefly, and incubate for 60 min at 30°C.
6. Transfer tubes into a 42°C water bath. Incubate for 5 min.
7. Centrifuge the suspensions for 4–5 sec in an Eppendorf microfuge. Shake off the supernatants.
8. Wash the pellets twice (2 sec) with 0.5 ml of TE. Each time vortex vigorously to resuspend pellets completely.
9. Suspend the washed pellets in 0.1 ml of TE and plate the entire suspension on an appropriate solid medium.
10. Incubate for 3–4 days at 30°C.

Further manipulations and analysis of transformants are the same as described in the section on spheroplast transformation (I,B,1). The whole-cell transformation technique is simple and efficient. Ito *et al.* (1983) observed only a few transformants when plasmids with a 2-µm origin of replication (see below) were used. This has not been the case in the author's laboratory, and the yields of transformants are very high (10^3–10^4 per microgram of plasmid DNA). However, treatment of *S. cerevisiae* with lithium acetate may induce deletions of unselected markers within transforming DNA molecules (Clancy *et al.*, 1984).

II. SPECIFIC TRANSFORMATION SYSTEMS IN YEAST

In principle, any DNA sequence that is functional in the yeast cell can be used to transform a suitable recipient. The only requirements are an adequate supply of DNA and of competent cells. In practice, the DNA supply is assured by constructing chimeric plasmids capable of replicating in *E. coli*. DNA from such plasmids grown in *E. coli* can be easily prepared on a large scale, which is usually not the case with plasmids, especially of the integrative type, grown in yeast. In fact, the difficulty of preparing plasmids directly from yeast is one of the disadvantages of this transformation system.

A. Types of Yeast Plasmid Vectors

Regardless of the associated genetic markers, the yeast plasmids can be categorized as autonomous or nonautonomous. The autonomous plasmids contain, in addition to various genetic markers, the yeast replicator sequences derived from chromosomal, mitochondrial, or 2-μm DNA (see below). A logical and clear system of classification has been described by Struhl *et al.* (1979) and Botstein and Davis (1982), and is summarized in Table II and in Fig. 1.

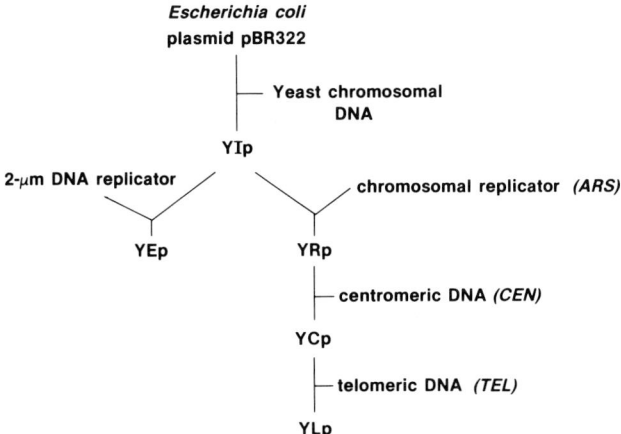

Fig. 1. Schematic pedigree of plasmids used in yeast transformation. See text for details.

TABLE II.

Classification of Yeast Plasmids

Plasmid type (and example)	Mode of maintenance	DNA required for maintenance	Genetic markers for selection in bacteria and/or yeast	Transformation frequency (colonies/μg DNA)	Plasmid stability Mitosis	Plasmid stability Meiosis	Reference
Integrating (YIp5)	Integrative, one or few copies	Homologous to chromosomal	Ap^r Tc^r URA3$^+$	<10	+^{a}	+^{a}	Scherer and Davis (1979)
Episomal (YEp13)	Autonomous, multicopy	2-μm	Ap^r Tc^r LEU2$^+$	10^3–10^4	–^{b}	–^{c}	Broach et al. (1979)
Replicating (YRp7)	Autonomous, multicopy	ARS^d	Ap^r Tc^r TRP1$^+$	10^3–10^4	–^{b}	–^{c}	Stinchcomb et al. (1979)
Centromeric (YCp19)	Chromosome-like, usually single copy	ARS, CEN^e	Ap^r TRP1$^+$ URA3$^+$	10^3–10^4	+^{f}	+^{f}	Stinchcomb et al. (1982)
Linear (YLp21)	Chromosome-like, usually single copy	ARS, CEN, TEL^g	TRP1$^+$ HIS3$^+$	10^3–10^4	+^{h}	+	Murray and Szostak (1983a)

^{a} About 1% excision per cell division.
^{b} Of transformants, 10–30% lose the plasmid after 15–20 generations on selective medium.
^{c} Non-Mendelian segregation (predominantly 4$^+$: 0$^-$; occasionally 0$^+$: 4$^-$).
^{d} Autonomously replicating sequence. In YRp7, this is derived from chromosome IV together with the TRP1$^+$ gene.
^{e} Centromeric DNA. In YCp19, this DNA and the ARS DNA are derived from chromosome IV. See text for further details.
^{f} Stability varies (3–14% loss in mitosis; 3% nondisjunction in meiosis).
^{g} Telomeric DNA.
^{h} Less than 1% loss of plasmid on selective medium.

6. Yeast Transformation

1. Yeast Integrating Plasmids—YIp

These plasmids cannot replicate by themselves and must integrate into homologous yeast DNA. Integration of transforming circular plasmid DNA is carried out by the yeast recombination systems. A rare single crossover event between the chromosomal marker and the homologous DNA of the plasmid results in the λ-style insertion of the entire molecule (Campbell, 1971), so that the yeast DNA sequence flank the vector sequences. In contrast to this "additive" integration, an even less frequent double crossover (or gene conversion) may result in "substitution" of the donor allele for the host allele. Integration has also been noted to occur at chromosomal locations unlinked to the allele in question due to the presence of *Ty* and δ sequences (Kingsman *et al.*, 1981). All these alternatives have been observed (Hinnen *et al.*, 1978) and are summarized in Fig. 2.

Some transformants arising from addition integration may segregate out one or the other region of duplication at a low frequency (Ilgen *et al.*, 1979). Also, nonreplicating YIp plasmids may integrate as a tandem multicopy array. Orr-Weaver and Szostak (1983b) have shown that the multimers arise by sequential integration of the plasmid DNA into the same chromosomal location.

2. Yeast Episomal Plasmids—YEp

The episomal plasmids owe their autonomous existence to the presence of certain sections of the 2-μm plasmid (also known as Scp 1). The 6300-bp 2-μm DNA circle, originally described by Clark-Walker and Miklos (1974), exists in 60–100 copies per haploid cell (Futcher and Cox, 1984). Even though the plasmid resides in the nucleus, its mode of inheritance is non-Mendelian. The double-stranded DNA of the 2-μm plasmid has been analyzed extensively (Broach, 1981), and found to contain several genes concerned with its own maintenance, including replication and interconversion to alternative topological forms (Beggs, 1978; Broach and Hicks, 1980; Broach *et al.*, 1982).

Very few phenotypic traits are associated with the presence of the 2-μm plasmid in the yeast cell [cir^+]. Certain strains of yeast carrying the *nib* mutation exhibit lethal colony sectoring resulting from the uncontrolled replication of the plasmid (Holm, 1982). However, there are no obvious physiological or metabolic differences between isogenic strains of yeast with or without the 2-μm circle.

Transforming YEp plasmids may undergo recombination with endogenous 2-μm circles. Under proper selective conditions, the recombinant plasmids may displace the resident 2-μm DNA completely because of replicative advantage (Hollenberg, 1982). Furher growth of cells under nonselective conditions may then lead to a loss of the plasmid and generation of a [cir^0] strain.

Although the 2-μm DNA is usually maintained autonomously, it is possible under certain conditions to integrate segments of this DNA into some chromosomal locations (Falco *et al.*, 1982). The integration occurs by recombination

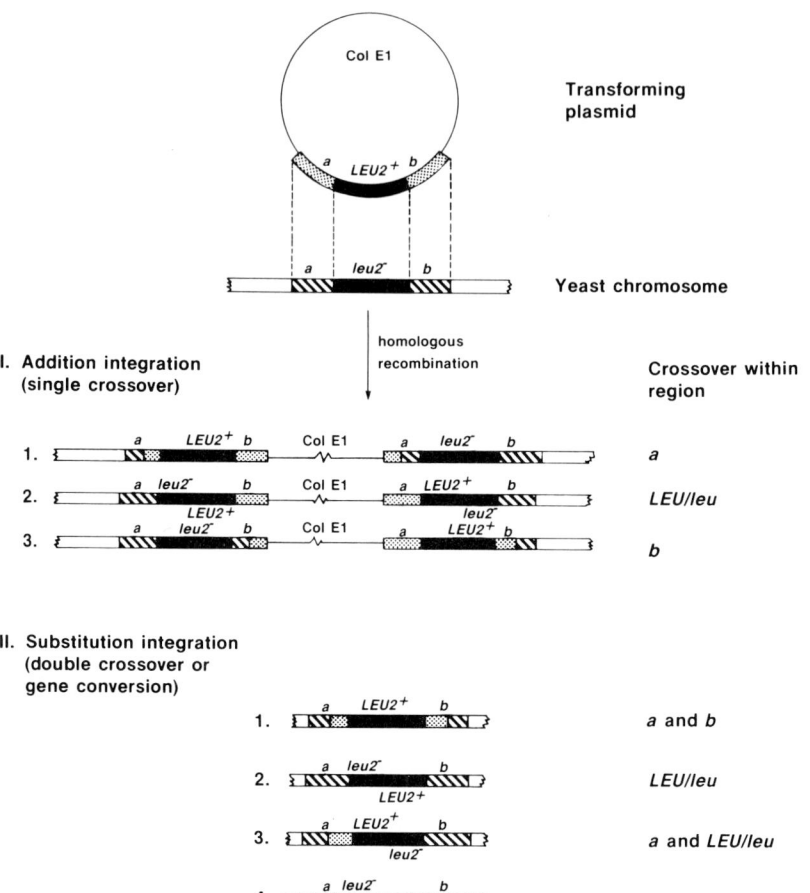

Fig. 2. Possible integration events involving a chimeric yeast–*E. coli* plasmid. The plasmid in this example [based on data of Hinnen *et al.* (1978)] is a hybrid of *Col*El (thin line) and yeast DNA sequences (boxed regions). The filled boxes represent the selectable marker ($LEU2^+$ on the plasmid, $leu2^-$ on the chromosome). The stippled and hatched boxes indicate yeast DNA sequences flanking the *LEU* marker. These sequences are differentiated to make it easier to trace the individual crossover events. Addition integrations result in duplication; substitution events evict the resident alleles. Not shown are integration events due to the presence of unrelated sequences homologous to other regions of the genome (e.g., *Ty* elements; see text).

within the regions of homology between a chromosomal locus and the corresponding segment borne by the plasmid. For example, if the YEp plasmid carries $LEU2^+$ $ura3^-$ markers, and the recipient is $leu2^-$ $ura3^-$, prototrophic LEU^+ transformants can be selected which contain portions of the 2-μm DNA integrated at the *URA3* locus. These transformants are unstable for markers residing on the chromosome with the 2-μm DNA insertion. In heterozygous diploids, the entire chromosome may be lost, but often the loss of markers is due to homozygote formation by mitotic recombination. Falco and Botstein (1983) took advantage of this instability to develop a new method for mapping cloned yeast genes (see Rine and Carlson, Chapter 5). The instability is observed only when an intact site-specific recombination gene (FLP, *flip-flop*) is present in the resident 2-μm DNA circles. This allows multiple copies of the plasmid to recombine into the chromosome bearing the integrated segments. The resulting inverted repeated sequences are the sites where FLP can promote unequal sister chromatid exchanges leading to formation of dicentric and acentric chromosomes. The loss of these rearranged chromosomes is the cause of the instability of transformants (Falco *et al.*, 1982). It strains devoid of resident 2-μm DNA plasmids, the integrated sequences remain reasonably stable.

3. Yeast Replicating Plasmids—YRp

Interest in the mechanics of chromosome replication in yeast led to the isolation and identification of a DNA fragment capable of replication without integration into the genome (Stinchcomb *et al.*, 1979). A 1.4-kb fragment obtained by *Eco*RI endonuclease digestion contained a centromere-linked $TRP1^+$ gene. The DNA was cloned into the *Eco*RI cleavage site of the bacterial plasmid pBR322, and the resulting chimeric plasmid (YRp7) was used to transform $trp1^-$ yeast cells. The efficiency of transformation to Trp^+ phenotype was very high but the stability of transformants was poor (18% loss after one generation in nonselective media). The transforming DNA was recovered as supercoiled circular molecules not integrated into the cell genome. This was in striking contrast to the results obtained with YIp plasmids (Hinnen *et al.*, 1978; Struhl *et al.*, 1979), and immediately suggested that YRp plasmids replicate autonomously. Since other yeast sequences (e.g., $LEU2^+$ or $HIS3^+$ structural genes) do not exhibit such behavior when cloned into pBR322, the element responsible for autonomous replication was associated with the 1.4-kb DNA fragment carrying the $TRP1^+$ gene (Stinchcomb *et al.*, 1979). The element was named *ARS1* (autonomous replicating sequence). A number of similar elements, both in yeast and in other eukaryotes, have since been described (Hsiao and Carbon, 1979; Chan and Tye, 1980; Stinchcomb *et al.*, 1980; Beach *et al.*, 1980; Kiss *et al.*, 1981; Davison and Thi, 1982; Banks, 1983). The ARS regions may normally serve as DNA replication origins in the intact chromosome (Hsiao and Carbon, 1979; Newlon and Lipchitz, 1983; Zakian and Scott, 1982; Broach *et al.*, 1983).

a. Stability of ARS-bearing Plasmids. As mentioned above, the stability of YRp transformants is relatively low. Stinchcomb *et al.* (1979) observed that in the presence of the *ARS1* element, the cell doubling time on minimal medium increased from 2.5 hr to 3.5–4 hr. Other *ARS* plasmids have a similar effect

(Clarke and Carbon, 1980; Tschumper and Carbon, 1982). In addition to decreased growth rate, there is a very high frequency of plasmid loss during vegetative growth, so that in a logarithmically growing culture perhaps only 30–50% of the cells (Stinchcomb et al., 1979) or even as few as 5% of the cells (Tschumper and Carbon, 1982) exhibit the selected prototrophic phenotype. One reason for such mitotic instability may be the competition for the initiator factor(s) necessary for DNA replication. Another reason is that the segregation of *ARS* plasmids during mitotic division deviates from random (Stinchcomb et al., 1981; Futcher and Cox, 1984). Particularly instructive in this regard have been the data obtained from pedigree analysis of plasmid segregation. Murray and Szostak (1983b) have shown that circular *ARS* plasmids have a high segregation frequency and a strong bias to segregate to the mother cell at mitosis. The presence of sequences involved in chromosomal segregation served to stabilize the YRp plasmids. Indeed, that was the rationale originally used by Clarke and Carbon (1980) in isolating centromeric DNA from yeast chromosomes (see below).

b. Structure and Activity of ARS. Several chromosomal replicators have been isolated and their complete DNA sequence determined (Tschumper and Carbon, 1980, 1982; Feldmann et al., 1981; Broach et al., 1983; Kearsey, 1983). Given the haploid yeast genome size of 15,000 kb and the fact that, on average, DNA replication is initiated once every 32–36 kb of DNA, there should be about 400 chromosomal origins of replication per genome (Newlon and Burke, 1980; Chan and Tye, 1980). The *ARS* elements have been shown to act only in *cis* configuration, i.e., they allow autonomous replication of a plasmid only when they are located on the same DNA molecule (Stinchcomb et al., 1979; Broach et al., 1983). The presence of several *ARS* sequences in the same plasmid does not disturb its replication pattern but, as mentioned above, the *ARS* plasmids are mitotically unstable. Stable transformants can be obtained at low frequency as a consequence of integration into an *ARS* locus at a chromosomal location.

Sequence analysis of *ARS1* (Tschumper and Carbon, 1980), *ARS2* (Tschumper and Carbon, 1982), *ARS3* (Stinchcomb et al., 1981), *ARS* elements associated with mating-type loci (Broach et al., 1983), and 2-μm origin of replication (Hartley and Donelson, 1980) revealed a limited homology between the elements. As shown in Table III, an 11-bp consensus sequence has been identified within a 100-bp region of each of the replicators (average A + T content of 79%). In addition to this common structural feature, the known replicator sequences share the property of being transcriptionally inactive. It is not clear what prevents the transcription of these sites. For example, even though the silent mating-type cassettes, *HML* and *HMR*, contain the consensus sequences, they

TABLE III.

Homology between Some Known Replicator Sequences

Site	Sequence	Reference
ARS1	TTTTATGTTTA	Stinchcomb et al. (1981)
ARS2	ATTTATATTTA	Tschumper and Carbon (1982)
ARS3	ACTTATATTTA	Stinchcomb et al. (1981)
2-μm circle origin	TTTTATGTTTA	Broach et al. (1983)
HMR left	TTTTATATTTA	Broach et al. (1983)
HML left	TTTTATGTTTT	Broach et al. (1983)
HML right	TTTTATATTTT	Broach et al. (1983)
HO	TTTAATATTTT	Kearsey (1983)
Consensus	^A_TTTTAT^A_GTTT^A_T	Broach et al. (1983)

remain repressed and inactive until transposed into the *MAT* (mating type) locus (Nasmyth, 1982; Strathern *et al.*, 1982).

Analysis of the *ARS1* element by digestion with restriction endonucleases revealed the bipartite structure of the element (Stinchcomb *et al.*, 1981; Stritch and Scott, 1983; Woontner and Scott, 1983). One domain (A) allows a high frequency of transformation but only limited autonomous replication. The other domain (B) facilitates replication of domain A but cannot replicate on its own.

Most *ARS* elements, or putative origins of replication, exist as individual units in the yeast genome. However, two repetitive families of *ARS*s have been recognized. The *ARS*s of one family are found within every unit of tandemly arranged ribosomal DNA sequences (Szostak and Wu, 1979). Recent studies by Chan and Tye (1983a) revealed that another family of *ARS* elements is found in close association with a repetitive sequence called *131* (1–1.5 kb). Restriction mapping and DNA hybridization demonstrated that the *ARS*s of this family are located not within sequence *131*, but next to it, and are themselves embedded within a moderately conserved sequence X (0.3–3.75 kb) or a highly conserved sequence Y (5.2 kb). Further analyses (Chan and Tye, 1983b) showed that these sequences (*131*, X, and Y) are associated with many yeast telomeres and map in the order: telomere end — (Y-*131*)$_n$—X, where *n* may range from 1 to 4, and Y-*131* represents a unit of 6.7 kb. The density of replication origins in this region is very high, with *ARS* elements spaced 6.7 kb apart in contrast to an average of 36 kb.

c. Yeast Acentric Replicating Plasmids—YARp. The 1.4-kb *Eco*RI fragment carrying the *TRP1* gene and the *ARS1* sequence can function independently of its original YRp-associated state. Zakian and Scott (1982) have found that when the fragment is self-ligated and introduced into yeast cells, it promotes transformation at a high frequency. The resulting transformants are very stable, and the copy number of the plasmid (also known as *TRP1* RI circle) is 100–200 per cell. The stability and high copy number are apparently a direct consequence of the small plasmid size. Unlike the centromeric plasmids (see below), which segregate $2^+:2^-$ in meiosis, the YARp plasmids segregate $4^+:0^-$. The usefulness of these plasmids resides in the fact that their DNA is organized in nucleosomes whose size and spacing are exactly as those in the bulk chromatin, and also in that they do not contain any DNA foreign to yeast. Thus, YARp plasmids may be an excellent minichromosome system for the study of chromosome replication and regulation of transcription. The main drawback is in preparing enough YARp DNA for further work since the plasmids do not replicate in *E. coli*.

4. Yeast Centrometric Plasmids—YCp

In searching for DNA sequences that could stabilize *ARS*-containing plasmids, Clarke and Carbon (1980) isolated a 1.6-kb fragment of DNA, from the yeast chromosome III, which allowed Mendelian segregation of plasmid markers in meiosis and conferred relative plasmid stability in mitosis. The presence of this DNA also brought the copy number down from about 60 to about 1. This DNA fragment thus behaved in a manner expected of a centromere. Since the original discovery, other centromeric DNAs have been identified and characterized (Fitzgerald-Hayes *et al.*, 1982a; Stinchcomb *et al.*, 1982; Panzeri and Philippsen, 1982; Maine *et al.*, 1984a; Blackburn and Szostak, 1984). The overall stability of plasmids containing centromeric DNA approaches but does not equal that of normal chromosomes in yeast (Tschumper and Carbon, 1983). Perhaps not surprisingly, when two centromeric DNA fragments are introduced into yeast on the same plasmid, the resulting transformants contain a heterogeneous mixture of constantly rearranging molecules with poor stability (Mann and Davis, 1983).

a. Structure of Centromeric (*CEN*) DNA. The overall DNA sequences of different centromeres vary enough so that there is no cross-hybridization between them, but they do share certain common features (Fig. 3) (Fitzgerald-Hayes *et al.*, 1982b; Bloom and Carbon, 1982; Panzeri and Philippsen, 1982). The centromeres contain an extremely A + T-rich core region of 87–89 bp (element II), which is flanked by a closely related 140-bp-long element I on the 5′ side and an 11-bp element III on the 3′ side. There is also element IV, 200–250 bp downstream from element III. Elements I, III, and IV of *CEN3* show perfect homology with the corresponding elements of *CEN11* DNA and considerable homology with *CEN6*. Bloom and Carbon (1982) have shown that the *CEN3* and *CEN11* DNAs are organized into nucleosomes. In each case a specific centromere region

6. Yeast Transformation

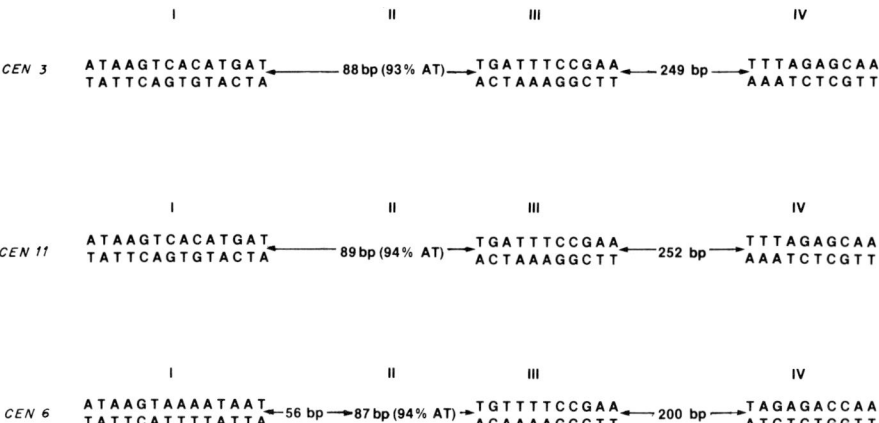

Fig. 3. Sequence similarities between centromeric DNA regions of yeast. The sequences [data of Fitzgerald-Hayes et al. (1982b), Bloom and Carbon (1982), and Panzeri and Philippsen (1982)] have been aligned to emphasize structural analogies. In CEN6, the 56-bp region also shows a moderate homology to element I, and the 200-bp region contains at least two sequences analogous to element IV.

of 220–250 bp is protected from micrococcal nuclease digestion, and is flanked by highly nuclease-sensitive sites. The protected region encompasses the elements I, II, and III, and thus includes the most highly conserved (71% homology) segment of 130 bp from element I to about 20 bp beyond element III. Identical organization of chromatin in CEN DNA is observed in the genome and on autonomous plasmids (minichromosomes) in yeast. The 130-bp region may be the functional unit of the centromere.

b. Functional Aspects of Centromeric DNA. By definition, centromeres serve as attachment sites for mitotic and meiotic spindle tubules. The presence in the CEN DNA of the unique spatial arrangements of the four elements, and of a nuclease-protected region, suggests that these structural features may be responsible for interaction with proteins involved in spindle formation. Whether the centromeric DNA directs synthesis of specific proteins is not known, but open reading frames have been detected in all centromeres analyzed to date (Fitzgerald-Hayes *et al.*, 1982b; Panzeri and Philippsen, 1982) and several RNA species have been found to hybridize to CEN4 DNA (Stinchcomb *et al.*, 1982).

c. Maintenance of Minichromosomes. As mentioned above, the stability of CEN plasmids approaches that of normal chromosomes. Recently, Maine *et al.* (1984b) demonstrated that the maintenance of these plasmids depends on the activity of yeast chromosomal genes. Mutants defective in minichromosome maintenance (Mcm$^-$) have been assigned to 16 complementation groups, and can be divided into two classes depending on the type of *ARS* elements present on the minichromosomes. One class of mutations affects the stability of all *ARS*-bearing plasmids, regardless of the source of *ARS* or the presence of a particular *CEN* DNA. The 2-μm plasmid is also destabilized. These mutants are termed nonspecific and most likely perturb a general facet of maintenance of circular plasmids, be it replication or segregation.

The other class of Mcm$^-$ mutants shows decreased stability of some, but not all, *ARS*-containing minichromosomes. Again, this is not dependent on the *CEN* DNA present on the plasmid. The unstable minichromosomes can be stabilized by the addition of a "functional" *ARS* onto the plasmid carrying a "nonfunctional" *ARS*. This epistatic effect demonstrates that all *ARS*s are not equal and that the Mcm$^-$ mutations can discriminate between them by a mechanism that still remains to be understood. These results suggest that besides the consensus sequence present in most *ARS* elements (see Table III) there may exist other structural features that allow interactions between *ARS* and Mcm gene products, interactions presumably necessary for initiation of DNA replication.

5. *Other Forms of Transforming DNA*

Singh *et al.* (1982) reported that single-stranded circular (ssc) DNA molecules transformed yeast cells 10- to 30-fold more efficiently than the double-stranded circles of identical sequence. The ssc DNA molecules were constructed by ligating the appropriate fragments of yeast DNA into the RF1 DNA of bacteriophage *fd*. The hybrid DNA was used to infect *E. coli*, and the single-stranded DNA was isolated from the phage particles. Upon transformation of yeast cells with an ssc DNA containing *ARS1* or the 2-μm plasmid replicator, double-stranded circular molecules were produced and maintained in the normal autonomous fashion.

Linear DNA has been used by Imai *et al.* (1983) and Suzuki *et al.* (1983) to transform yeast cells at high frequency. The plasmid vectors (pJDB219 = pMB9 + 2-μm DNA *LEU2*$^+$; YIp5 = *Col*E1 + *URA3*$^+$) were cut outside the yeast marker DNA sequences and introduced into yeast. Circular molecules were recovered, indicating the presence of a ligation system in *Saccharomyces cerevisiae*. Even DNA fragments containing noncohesive ends (e.g., *Hind*III and *Eco*RI) could be ligated. The resulting plasmids exhibited meiotic distribution and mitotic stability similar to those of YEp plasmids.

Mitochondrial DNA has been employed in yeast transformation (Vaughan *et al.*, 1980). A respiratory-deficient (*rho*^{0}) strain of yeast requiring leucine for growth (*leu2*$^-$) was transformed with a mixture of mitochondrial DNA from a *rho*$^+$ (*oli2*$^+$) strain and the YEp13 (*LEU2*$^+$) plasmid DNA. The resulting Leu$^+$ transformants were crossed with *mit*$^-$ tester strains (deficient in mitochondrial functions, lacking respiratory chain), and respiratory-competent diploids were obtained. However, no biochemical analysis was carried out to elucidate the fate of the transforming mitochondrial DNA. Woo et al. (1982) also showed that composite plasmids consisting of YEp13 DNA and $oli2^+$ DNA could not restore respiratory proficiency on the recipient mit^- cells. They suggested that because YEp13 replicates in the nucleus, the composite plasmid could not interact with the mitochondrial DNA of the tester strain.

Mitochondrial DNA (mtDNA) segments from a petite mutant (ρ^-) have been inserted into YIp5 plasmid (Hyman et al., 1982). The resulting hybrids transformed $ura3^-$ cells to URA^+ and replicated autonomously as unstable circular molecules (YRMp—yeast replicating mit DNA plasmids). The efficiency of transformation was somewhat reduced in ρ^0 hosts compared to the wild-type isogenic parent (ρ^+), but the stability and copy number of YRMp were similar in both, and were comparable to those of a control YRp12 plasmid. There is some evidence that YRMp is maintained in the nucleus and not in the mitochondria.

B. Artificial Chromosomes (Yeast Linear Plasmids—YLp)

Natural chromosomes possess the following major classes of functional elements: genes, origins of replication, centromeres, and telomeres (chromosomal ends). Telomeres received considerable attention recently, when Kiss et al. (1981) and Szostak and Blackburn (1982) demonstrated that plasmids containing terminal regions of extrachromosomal ribosomal DNA from *Tetrahymena* are capable of autonomous replication in yeast. The structure of the chromosome ends is of interest because of their involvement in successful replication and chromosome stability. The DNA ends created by artificially induced chromosome breakage are very unstable and lethal if not repaired (Resnick and Martin, 1976). When plasmid DNA is cleaved by restriction endonucleases, the resulting ends are also unstable, very recombinogenic, and susceptible to degradation or ligation (Orr-Weaver et al., 1981, 1983; Suzuki et al., 1983). True telomeres are stable; they contain a repetitive sequence; they seem to terminate in a hairpin loop; most remarkably, they appear to be conserved through evolution with respect to both general structure and function (Szostak, 1982; Blackburn and Challoner, 1984; Blackburn and Szostak, 1984).

The isolation and cloning of yeast telomeric DNA opened the way for combining the individual functional elements into an artificial chromosome. Murray and Szostak (1983a) constructed a series of linear plasmids, all of which contained a selectable gene marker (e.g., $TRP1^+$ $HIS3^+$, $URA3^+$), *ARS1* replicator, *CEN3* DNA, and telomeres from *Tetrahymena* ribosomal DNA. The plasmids varied in length from 9.8 to 55 kb.

Several interesting conclusions emerged from this work. First, the addition of telomeric DNA onto small (<16 kb) plasmids drastically *reduces* their mitotic stability and increases their copy number even though the *ARS* and *CEN* sequences remain intact. In fact, the stability of such plasmids is as poor as that of acentric linear molecules (Dani and Zakian, 1983). Thus, it appears that the presence of telomeric DNA has a suppressing effect on the *CEN* DNA function when the two are placed close to each other.

Second, very large (~50 kb) linear plasmids containing *ARS, CEN,* and telomeric DNA are present at one or two copies per cell and are more mitotically stable than their circular *ARS, CEN* counterparts. This stability contrasts with the behavior of the short linear plasmids described above, and suggests that the distance between the *CEN* and telomeres may be crucial for proper partitioning of chromosomes. However, even with this improved stability and faithful segregation, the YLp plasmids are still lost more frequently than the normal chromosomes, and their copy number is less rigidly controlled.

Third, the artificial chromosomes behave properly in meiosis. In diploids containing an artificial chromosome pair (YLp21, TRP^+URA^-; and YLp22, TRP^-URA^+), the segregation of markers is predominantly $4^+ : 0^-$; $2^+ : 2^-$; $0^+ : 4^-$. Among the tetrads showing $2^+ : 2^-$ segregation for *TRP* and *URA,* the two spores which contain the particular plasmid are sisters; that is, the two copies of the plasmid segregate from each other in the second meiotic division as expected of natural chromosomes. Furthermore, the two plasmids segregate from each other relatively cleanly, yielding two TRP^+URA^- and two TRP^-URA^+ spores (first division segregation), again showing typical chromosomal behavior.

III. APPLICATIONS

Several recent reviews have analyzed the transformation system of yeast as a tool with which to study gene structure and function (Ilgen *et al.,* 1979; Hicks *et al.,* 1982; Botstein and Davis, 1982; Struhl, 1983a). In addition, hundreds of papers made a reference to transformation as a means of arriving at a better understanding of a variety of biochemical and genetic activities of the yeast cells. To list, and comment on, all such papers would be a nearly impossible task whose only assured outcome could be a writer's cramp. Instead, what follows is a not-all-inclusive listing of examples of recent experimental advances in yeast molecular biology and genetics. These advances are chosen to exemplify the resolving power of recombinant DNA and transformation technology. Further examples can be found in Rine and Carlson (Chapter 5).

A. Isolation and Characterization of Genes

The principle of gene isolation is simple: obtain a strain mutated in a particular gene, transform it with yeast plasmids bearing random segments of the total yeast DNA, plate transformants on a selective medium, and pick the clone(s) showing the restored wild-type phenotype. This ''shotgun'' approach works well with almost any gene for which phenotypically distinct mutations are known and whose defect can be complemented by the functional allele of the gene. Either integrating or autonomous yeast plasmids can be used for the purpose of cloning. The integration of a plasmid-borne allele by homologous recombination (see Fig. 2) has the advantage of localizing the allele to a fixed position within the genome; thus, confirmation of the location and linkage by genetic mapping is feasible. Integration also allows: (1) the introduction of *in vitro*-modified alleles (gene transplacement, Scherer and Davis, 1979; Struhl, 1983b), (2) the cloning of mutant forms of genes (gene eviction, Winston *et al.*, 1983), and (3) the determination of whether the cloned gene is essential (gene disruption, Rothstein, 1983). The autonomously replicating plasmids offer the convenience of multiple gene copies, therefore simplifying the study of complementation, cis–trans dominance, and gene dosage effects, and permitting an easy reisolation of the cloned gene.

The *in vitro* DNA manipulations are particularly useful in the study of regulatory mutations, whose effects may be subtle and difficult to select for. The isolated DNA segments can be modified in a systematic way, and even mutations that do not have a phenotypic effect can be introduced into the genome. This is especially important in determining which portions of a particular sequence are essential for a given function, and which are dispensable. *In vitro* mutagenesis procedures are not burdened with the bias of *in vivo* genetic selection; thus, any mutational change can be introduced and studied in this manner (Struhl, 1983a).

In addition to transplacement, eviction, and disruption, gene fusion has been an important tool in modern yeast genetics (Rose and Botstein, 1983a,b; Guarente, 1983). In most cases, the gene or portion of the gene under study is fused to a structural gene whose activity is simple to measure, both by plate assay and in cell-free extracts (e.g., the *E. coli* β-galactosidase gene, *lacZ*), but whose own regulatory sequences are missing. Thus, the expression of β-galactosidase is made dependent on the presence of regulatory regions in the gene sequence in question. An excellent example of how successful this approach can be has been provided by the work of Guarente and co-workers (Guarente and Mason, 1983; Guarente *et al.*, 1984). Their studies on *CYC1–lacZ* fusions have revealed that heme regulates the expression of the *lacZ* gene when that gene is fused to the regulatory sequences of the *CYC1* gene. This role for the heme would have been difficult to prove using classical methods since, in addition to its regulatory effect, heme is necessary for the activity of cytochrome *c*. Thus, the dissection of the regulatory circuit was made possible by disassociating the structural gene from its controlling elements. With gene fusions of this kind, "the gene product that is assayed does not affect the regulation being analysed" (Struhl, 1983a).

Silverman *et al.* (1982) made use of transformation methodology and of gene fusion to investigate the general amino acid control of the expression of the *HIS4* gene. They showed that the *lacZ* gene fused to the *HIS4* regulatory sequences responded to amino acid starvation and to regulatory mutations (*aas 1-11* and *tra 3-1*) in the same manner as the normal *HIS4* gene. The controlling region of *HIS4* was localized within the 700-bp sequence upstream of the ATG start codon of the structural gene. Further work narrowed the region to 235-bp upstream and demonstrated the presence of a sequence 5'-TGACTC-3', repeated three times (at positions -194, -182, and -138) (Donahue *et al.*, 1983). The removal of all three repeats resulted in the loss of *HIS4* activity. When the two sequences at -194 and -182 were removed, the activity was much reduced but was still regulated normally by the general amino acid control. This work defined the promoter region (between -235 and -173) as well as the regulatory region around nucleotide -136 responsible for the reaction to amino acid starvation. A similar situation obtains for other genes under general amino acid contol, all of which possess the highly conserved core (TGACTC) within a 9-bp consensus sequence (A^A_TGTGACTC) repeated two to four times in their 5' flanking regions (Hinnebusch and Fink, 1983).

B. Expression of Heterologous Genes in Yeast

Saccharomyces cerevisiae is an attractive organism for cloning, expression, and synthesis of a number of gene products of commercial interest. It is safe in large-scale operations, can be adapted to continuous fermentations, grows easily on a variety of media, and has been very well characterized biochemically and genetically. In addition, as a eukaryote, it is expected to be able to utilize the specifically eukaryotic means of protein processing and transport, and thus to serve adequately as a vehicle for preparation of various mammalian gene products. Some prokaryotic genes also function in yeast when they are fused to yeast promoters (e.g., β-galactosidase gene, *lacZ*). Table IV lists some of the foreign genes that have been cloned and expressed in *S. cerevisiae*.

Several examples illustrate the successes and failures of yeast as a host for foreign gene expression. The thymidine kinase gene from *Herpes simplex* virus, originally cloned in yeast by McNeil and Friesen (1981), has been found to complement the *cdc8* conditionally lethal mutation in *S. cerevisiae* (Sclafani and Fangman, 1984). The observation is an interesting one because yeast cells, being normally deficient in thymidine kinase (TK), use thymidylate synthetase (TS) for production of dTMP (dUMP $\overset{TS}{\to}$ dTMP rather than deoxythymidine $\overset{TK}{\to}$ dTMP). The dTMP is subsequently converted to dTDP by the action of thymidilate kinase (dTMP kinase). The lesion in the *CDC8* gene prevents phosphorylation of dTMP to dTDP, suggesting that the $CDC8^+$ gene is responsible for the formation or expression of dTMP kinase. As the viral TK^+ gene overcomes the *cdc8* defect, it appears that the TK^+ protein is able to catalyze both reactions. Thus, the importance of this study lies not only in the fact that the role of the $CDC8^+$ gene has been elucidated, but also in the discovery of dual enzymatic activity for the single protein species.

TABLE IV.

Expression of Foreign Genes in Yeast

Gene	Source	Expression RNA	Expression Protein	Export	Reference
Chymosin	Calf	N.T.[a]	+	No	Mellor et al. (1983)
Hepatitis B surface antigen	Hepatitis B virus	N.T.	+	No	Miyanohara et al. (1983)
Epidermal growth factor	Human	N.T.	+	No	Urdea et al. (1983)
Leukocyte interferon α	Human	+	In vitro	N.A.[b]	Chandra and Kung (1983)
Interferon α and γ	Human	N.T.	+	Yes	Hitzeman et al. (1983)
Thymidine kinase	Herpes simplex virus	N.T.	+	No	McNeil and Friesen (1981)
Growth hormone	Rat	+	No	No	Ammerer et al. (1981)
Thaumatin	*Thaumatococcus daniellii* Benth	N.T.	+	N.T.	Edens et al. (1984)
β-Globin	Rabbit	+, not spliced	−	N.A.	Beggs et al. (1980)
Proinsulin	Human	N.T.	+	No	Stepien et al. (1983)
Adenine-8+	*Drosophila*	+	Presumably	No	Henikoff et al. (1981)
α-Antitrypsin	Human	N.T.	+	No	Kawasaki et al. (1983)
Ovalbumin	Chicken	N.T.	+	No	Mercereau-Puijalon et al. (1980)
β-Lactamase	*E. coli* (pBR325)	N.T.	+	Perhaps	Roggenkamp et al. (1981, 1983)
Chloramphenicol acetyltransferase	*E. coli* (pBR325)	N.T.	+	No	Cohen et al. (1980)
Aminoglycoside 3′-phosphotransferase	*E. coli* (Tn903)	N.T.	+	No	Jimenez and Davies (1980)
β-Galactosidase	*E. coli*	N.T.	+	No	Panthier et al. (1980)
Outer membrane protein (*ompA*)	*E. coli*	N.T.	+	Yes	Janowicz et al. (1982)

[a] Not tested.
[b] Not applicable.

Hitzeman et al. (1981) and Ammerer et al. (1981) described the cloning of the rat growth hormone gene in yeast and showed that the gene is faithfully transcribed. However, no immunoreactive material could be detected. Therefore, an efficient and accurate transcription of a particular heterologous gene is not sufficient for proper translation and synthesis of a functional protein. Preferences in codon usage (Bennetzen and Hall, 1982) and the sequence context around the initiator codon (Kozak, 1981; Dobson et al., 1982) may be responsible for this failure. Placing the gene of interest under the control of a strong yeast promoter and replacing the -1 to -10 sequence with a DNA segment acceptable to yeast may allow a high expression level. An example of this is the production of prochymosin directed by the yeast phosphoglycerokinase promoter (Mellor et al., 1983). The sequences at the 3' end of the gene that are responsible for termination of transcription may also be important (Mellor et al., 1983; Zaret and Sherman, 1982).

Beggs et al. (1980) described the cloning and transcription of the rabbit β-globin gene in yeast. However, the primary transcript was not spliced and was terminated prematurely. Also, some 20–40 nucleotides normally present at the 5' end of mature mRNA were missing, as were the poly(A) tails. These results suggest that yeast lacks a splicing system capable of recognizing intron-exon boundaries in RNA transcripts derived from higher eukaryotes. Similarly, yeast cells were unable to remove introns from RNA transcripts of *Acanthamoeba* actin I gene or duck α^D-globin gene (Langford et al., 1983). On the other hand, the introns present in the yeast gene transcripts are excised efficiently (Langford et al., 1983; Pikielny et al., 1983). This indicates that in addition to the consensus intron–exon boundary (exon–AG/GTAAGTA–intron–AG/G, Flint, 1982), the yeast system requires a recognition sequence away from the junction [ICS (internal conserved sequence, UACUAAC)] (Pikielny et al., 1983).

C. Genetic Recombination Models

1. Gene Conversion

Mechanism of recombination has been one of the most interesting and difficult topics in fungal genetics (Stahl, 1979; Whitehouse, 1982). In particular, an aspect of recombination, gene conversion, remains an object of intense studies because of its importance in elucidation of fine gene structure (Fogel et al., 1982a, 1983). Gene conversion is formally defined as nonreciprocal transfer of genetic information between homologous chromatids. Thus, for a particular marker, instead of the expected $2^+:2^-$ meiotic segregation, a $3^+:1^-$ or $3^-:1^+$ segregation of alleles is observed. Gene conversion is postulated to result from correction of a heteroduplex joint that is created during crossing-over between chromatids. Since gene conversion is a relatively rare event, its study by

6. Yeast Transformation

classical methods is laborious and time-consuming and requires the analysis of thousands of tetrads (Fogel *et al.*, 1983). The advances in yeast molecular biology, aided by the efficient transformation methods, lessened the amount of labor and permitted new insights into the mechanism of recombination at the DNA level.

Linearization or formation of gaps in yeast plasmids within the yeast DNA sequences increases the frequency of integrative transformation (Ilgen *et al.*, 1979; Orr-Weaver *et al.*, 1981). This observation gave an impetus to the development of a model of genetic recombination and gene conversion in yeast based on the formation of double-stranded breaks in chromosomal DNA (Orr-Weaver *et al.*, 1981, 1983; Orr-Weaver and Szostak, 1983a; Szostak *et al.*, 1983). The model postulates the repair of a double-strand break or gap by two rounds of single-strand repair synthesis, which results in the formation of regions of heteroduplex DNA on each side of the repaired gap. Unlike the Meselson–Radding model (Meselson and Radding, 1975), in which DNA strands form a single Holliday junction (Holliday, 1964), the present scheme proposes the existence of two such junctions (Fig. 4). (A Holliday junction is a postulated site of crossing-over in DNA recombination models.) When both are resolved by cutting the same pair of strands (inner or outer), there is no recombination of flanking markers; when each junction is resolved in the opposite sense, crossing-over between the outside markers is observed. The substantial difference between the two models is that one (Meselson and Radding, 1975) postulates that the initial nick occurs on the DNA strand that is the donor of the genetic information in a conversion event; the other model assumes the initial nick to occur on the recipient strand (Fig. 4).

Orr-Weaver *et al.* (1981) and Orr-Weaver and Szostak (1983a,b) analyzed the fate of integrative and autonomous yeast plasmids linearized or gapped within regions of homology with chromosomal DNA. Gapped molecules are repaired by using the resident DNA as a template. This is true whether transformation leads to integration or not. As shown earlier (Fig. 2), integration into the chromosome requires crossovers while autonomous plasmids are maintained independently. When nonintegrating gapped plasmids are used in transformation, approximately equal numbers of integrated and autonomous plasmids are recovered. Thus, this type of gene conversion, due to double-strand gap repair, occurs with or without crossing-over. This is analogous to meiotic gene conversion, which also may or may not be accompanied by recombination between flanking markers (Fogel *et al.*, 1982a). It remains to be seen whether double-strand gap repair is affected in correction-deficient mutants, *cor* (Fogel *et al.*, 1983). These mutants exhibit reduced conversion frequency for some alleles and the DNA repair tracks seem much shorter than in COR^+ strains.

Furthermore, meiotic recombination, meiotic gene conversion, and the repair of single- and double-strand gaps require the activity of the *RAD52* gene (Game

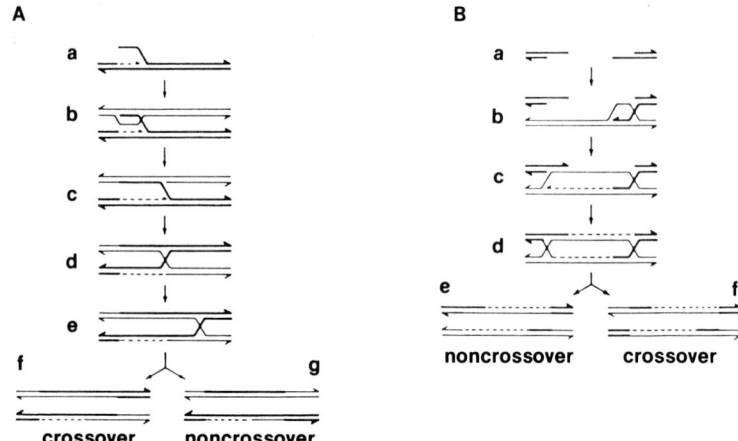

Fig. 4. Models of genetic recombination. The Meselson–Radding model (A) and the double-strand-break model (B) are presented side by side. In (A), recombination is initiated (a) by a single-strand nick on one of the two interacting duplexes. The 3' end of the nicked strand acts as a primer for DNA synthesis, which displaces the strand ahead of it. The displaced single strand invades the other duplex at a homologous site, displacing a D loop (b) and forming a small region of asymmetric heteroduplex DNA. The single-stranded D loop is degraded, and the invading strand is ligated in place. The limited region of asymmetric heteroduplex DNA is expanded (c) by concerted DNA synthesis on the first (donor) duplex, and by exonucleolytic degradation on the second (recipient) duplex. After the enzymatically driven production of asymmetric heteroduplex DNA stops, either branch migration or isomerization can bring the 5' and 3' single-stranded ends into apposition so that they can be ligated. The resulting Holliday junction (d) can move along the duplex by the process of branch migration, generating symmetric heteroduplex DNA (e). Resolution can yield either the crossover (f) or the noncrossover (g) configuration. In (B), a double-strand cut is made in one duplex (a), and a gap flanked by 3' single strands is formed by the action of exonucleases. (b) One 3' end invades a homologous duplex, displacing a D loop. (c) The D loop is enlarged by repair synthesis until the other 3' end can anneal to complementary single-stranded sequences. (d) Repair synthesis from the second 3' end completes the process of gap repair, and branch migration results in the formation of two Holliday junctions. Resolution of two junctions by cutting either inner or outer strands leads to two possible noncrossover (e) and two possible crossover (f) configurations. In the illustrated resolutions, the right-hand junction was resolved by cutting the inner, crossed strands. [Reproduced from Szostak *et al.* (1983); by permission. ©MIT, 1983.]

et al., 1980; Resnick and Martin, 1976; Resnick *et al.*, 1981). Strains bearing a *rad52-1* mutation are unable to integrate linear or gapped plasmids (Orr-Weaver *et al.*, 1981). However, integration of circular plasmids takes place in these strains. The same is true in *in vitro* recombination assays (Symington *et al.*, 1983). In the former case, DNA repair is required; in the latter it is not. Therefore, the *RAD52* gene must be involved in DNA repair functions of the cell, and some aspects of the integration process must be different for circular and linear plasmids. Earlier, Resnick and Martin (1976) showed that *RAD52* gene activity was necessary for repair of double-strand breaks in nuclear DNA. Such breaks are lethal in *rad52-1* mutants. The mutation also results in chromosome instability (Mortimer *et al.*, 1981), and decreases mitotic recombination (Malone and Esposito, 1980) and mitotic gene conversion not associated with reciprocal exchanges. The less frequent mitotic conversion events that are accompanied by exchange of flanking markers apparently are not affected (Jackson and Fink, 1981).

The contribution—indeed, indispensability—of yeast transformation technology to the understanding of recombination and gene conversion at the DNA level has been vividly illustrated by the studies on genetic behavior of nontandem duplications in yeast (Klein and Petes, 1981; Jackson and Fink, 1981; Fogel *et al.*, 1982b). Such duplications are easily obtained by using integrating yeast plasmids. The difference between interchromosomal and intrachromosomal exchanges is readily detected by quantitative Southern blotting and hybridization analysis of the recombinants for retention of the duplications and of vector sequences and/or changes in gene copy number. Either type of exchange may be reciprocal or nonreciprocal (gene conversion).

Intrachromosomal transfer of genetic information has been observed in both meiosis (Klein and Petes, 1981; Fogel *et al.*, 1982b) and mitosis (Jackson and Fink, 1981). Gene conversion is the predominant recombination event in mitotic exchanges involving nontandem duplications. In most cases the conversion is not accompanied by reciprocal recombination of the flanking markers. The *rad52-1* mutation decreases the conversion frequency 30- to 200-fold but, as mentioned above, it has little effect on reciprocal recombination involving intrachromosomal exchanges. This indicates that gene conversion and reciprocal recombination can occur through different pathways in mitosis. In meiosis, the *rad52-1* mutation eliminates reciprocal recombination and gene conversion.

Intrachromosomal gene conversion and unequal sister chromatid exchanges have been shown to result in gene amplification, and to be important in the maintenance of sequence homogeneity of repeated DNA units (Petes, 1980; Szostak and Wu, 1980; Jackson and Fink, 1981; Fogel and Welch, 1982). All these studies were facilitated by integration of cloned DNA into plasmids and subsequent transformation of yeast. In the case of the *CUP1* locus of yeast (Fogel and Welch, 1982), the resistance to high concentrations of copper was found to be the result of an increase in the number of *CUP1* alleles in the chromosome. Whereas copper-sensitive haploid cells contain a single copy of the gene, the highly resistant strain bears some 15 copies of the same gene, tandemly aligned on chromosome VIII.

Tandem repeats, in general, are known to exhibit a high frequency of unequal sister chromatid exchanges in meiosis and mitosis (Szostak and Wu, 1980; Petes, 1980). In the case of the ribosomal DNA array in yeast these exchanges generate deletions and duplications of six to eight repeat units. Szostak and Wu (1980) indicate that the rate of mitotic sister chromatid exchange within the ribosomal

DNA cluster is sufficiently high to maintain the homogeneity of the region. On the other hand, the studies of Klein and Petes (1981) and of Jackson and Fink (1981) on duplications of the *LEU2* and *HIS4* genes, respectively, suggest that intrachromosomal gene conversion may be a more likely means of rectification of repeated sequences in the genome.

2. Control of Ty Elements of Yeast

Transposable elements of yeast, originally described by Cameron *et al.* (1979) and Fink *et al.* (1980), have been studied extensively but many questions concerning their functions remain unanswered. In particular, the regulation of transposition and the mechanism of *Ty* recombination are poorly understood.

The *Ty* elements associated with the *HIS4* locus of yeast have been studied by Roeder *et al.* (1980). When two such elements, called *Ty912* and *Ty917*, are inserted into the *HIS4* 5' noncoding regulatory region, the transcription of the *HIS4* gene stops and the cells acquire a His$^-$ phenotype. Like other *Ty* elements, these two consist of a central region of about 5600 bp of DNA flanked by direct repeat sequences (about 330 bp) named delta (δ). Recombination between the δ sequences (δ–δ recombination) results in excision of the *Ty* element, with a single δ remaining in place (*his4-912δ*). Excision of *Ty912* leads to a cold-sensitive His$^+$ phenotype.

Analysis of His$^+$ revertants of the *his4-912* and *his4-917* mutations revealed that in several cases the reversion was caused by secondary mutations at loci unlinked to each other or to *HIS4* (Roeder *et al.*, 1980; Roeder and Fink, 1983; Winston *et al.*, 1984). These mutations, originally named *spm* (suppressor–mutator) and later renamed *spt*, have been assigned to genes *SPT1* through *SPT7* (suppressor of *Ty*). The *spt1* suppresses the *Ty* insertion effect in both *his4-912* and *his4-917*; *spt2*, *spt3*, and *spt6* suppress the *his4-917* but not the *his4-912* mutation. All Spts suppress the His$^-$ phenotype of *his4-912δ* but none affects the δ–δ recombination or *Ty* excision. Only one mutation (*spt4*) suppresses the His$^-$ phenotype caused by a non-*Ty* insertion into the *HIS4* locus. Such an insertion was created by transforming a *HIS4$^+$* strain with YIp5 vector containing a 274-bp fragment from the 5' end of the *HIS4* gene (Winston *et al.*, 1984). Additive integration at the *HIS4* locus resulted in a duplication similar in size and structure to that to *Ty912* (5600 bp of YIp5 flanked by 274-bp direct repeats). The transformants containing this insertion in the 5' noncoding region were phenotypically His$^-$ and could be tested for suppressibility by the various *spt* mutations. This is yet another example of the application of transformation and plasmid constructions to the analysis of gene structure and behavior.

The molecular basis of *spt* suppression of *Ty* and δ insertions remains undefined. It is thought that in some instances the suppression occurs at the transcriptional level (G. R. Fink, personal communication). Whether this is due to a change in the specificity of transcription initiation or promoter recognition is not known. Many *spt* mutations have pleiotropic effects on cells resulting in temperature sensitivity (*spt6, spt7-159*), mating defect (*spt3*), sensitivity to methylmethane sulfonate (*spt4-3*), and other defects (Winston *et al.*, 1984).

IV. CONCLUSIONS

Even though the methodology of yeast transformation has been developed only recently, the combination of this methodology and molecular biology has already enabled workers in the field to understand in detail the basic phenomena regulating the life cycle, physiology, and genetics of *S. cerevisiae*. However, to maintain a sense of proportion, it must be pointed out that the time-honored and well-established transformation system of *E. coli* should not be abandoned completely in favor of the yeast system. It is useful, therefore, to enumerate the several advantages and limitations of yeast and *E. coli* as transformation models. (To limit the comparison to yeast and *E. coli* does not imply that other organisms are not worthy of study.)

1. It is much easier to isolate plasmids in large quantity from *E. coli* than from yeast; plasmid amplification is possible in *E. coli*.

2. Transformation efficiency is usually higher in *E. coli*, and the recovery of plasmids is easier.

3. Excellent selective markers are available for bacterial plasmids; there is a paucity of good, universal drug resistance markers in yeast (Gritz and Davies, 1983; Webster and Dickson, 1983).

4. Gene replacement is simpler and more efficient in yeast than in *E. coli*. This facilitates the analysis of gene mutations induced *in vitro* (Struhl, 1983b).

5. Efficient cloning and expression of mammalian genes, and especially protein processing, is probably more likely in yeast than in bacteria.

These respective limitations notwithstanding, it is obvious that both transformation systems have contributed enormously to the progress in our understanding of gene structure and function. Many apparent puzzles have been solved, and simple explanations are available for previously unyielding problems. In view of this progress it seems fitting to conclude this summary with the words of T. H. Morgan (1919):

> That the fundamental aspects of heredity should have turned out to be so extraordinarily simple supports us in the hope that nature may, after all, be entirely approachable. Her much-advertised inscrutability has once more been found to be an illusion due to our ignorance. This is encouraging, for, if the world in which we live were as complicated as some of our friends would have us believe, we might well despair that biology could ever become an exact science.

NOTE ADDED IN PROOF

Boeke et al. (1985) have shown that Ty transposition occurs via an RNA intermediate. The presence of reverse transcriptase activity in cells competent for transposition has also been established (G. R. Fink, personal communication). These findings demonstrate a great similarity between transposition of Ty elements and of retroviruses.

ACKNOWLEDGMENTS

I wish to thank G. R. Fink for reading the manuscript, for sharing results before publication, and for useful advice. I also thank Ms. Terri Collins for the cheerful handling of numerous retypings of the manuscript.

REFERENCES

Ammerer, G., Hitzeman, R., Hagie, E., Barta, A., and Hall, B. D. (1981). The functional expression of mammalian genes in yeast. In "Recombinant DNA" (A. G. Walton, ed.), pp. 185 197. Elsevier, Amsterdam.

Banks, G. R. (1983). Chromosomal DNA sequences from *Ustilago maydis* promote autonomous replication of plasmid in *Saccharomyces cerevisiae*. Curr. Genet. **7**, 79–84.

Beach, D., Piper, M., and Shall, S. (1980). Isolation of chromosomal origins of replication in yeast. *Nature (London)* **284**, 185–187.

Beggs, J. D. (1978). Transformation of yeast by a replicating hybrid plasmid. *Nature (London)* **275**, 104–109.

Beggs, J. D., van den Berg, J., van Ooyen, A., and Weissmann, C. (1980). Abnormal expression of chromosomal rabbit β-globin gene in *Saccharomyces cerevisiae*. *Nature (London)* **283**, 835–840.

Bennetzen, J. L., and Hall, B. D. (1982). Codon selection in yeast. J. Biol. Chem. **257**, 3026–3031.

Blackburn, E. H., and Challoner, P. B. (1984). Identification of a telomeric DNA sequence in *Trypanosoma brucei*. *Cell (Cambridge, Mass.)* **36**, 447–457.

Blackburn, E. H., and Szostak, J. W. (1984). The molecular structure of centromeres and telomeres. Annu. Rev. Biochem. **53**, 163–194.

Bloom, K. S., and Carbon, J. (1982). Yeast centromere DNA is in a unique and highly ordered structure in chromosomes and small circular minichromosomes. *Cell (Cambridge, Mass.)* **29**, 305–317.

Boeke, J. D., Garfinkel, D. J., Styles, C. A., and Fink, G. R. (1985). Ty elements transpose through an RNA intermediate. *Cell* **40**, 491–500.

Botstein, D., and Davis, R. W. (1982). Principles and practice of recombinant DNA research with yeast. In "The Molecular Biology of the Yeast Saccharomyces. II. Metabolism and Gene Expression" (J. N. Strathern, E. W. Jones, and R. J. Broach, eds.), pp. 607–636. Cold Spring Harbor Lab. Press, Cold Spring Harbor, New York.

Broach, J. R. (1981). The yeast plasmid 2μ circle. In "The Molecular Biology of the Yeast Saccharomyces. I. Life Cycle and Inheritance" (J. N. Strathern, E. W. Jones, and J. R. Broach, eds.), pp. 445–470. Cold Spring Harbor Lab. Press, Cold Spring Harbor, New York.

Broach, J. R., and Hicks, J. B. (1980). Replication and recombination functions associated with the yeast plasmid, 2μ circle. *Cell (Cambridge, Mass.)* **21**, 501–508.

Broach, J. R., Strathern, J. N., and Hicks, J. B. (1979). Transformation in yeast: development of a hybrid cloning vector and isolation of the *CAN1* gene. *Gene* **8,** 121–133.

Broach, J. R., Guarascio, V. R., and Jayaram, M. (1982). Recombination within the yeast plasmid 2μ circle is site-specific. *Cell (Cambridge, Mass.)* **29,** 227–234.

Broach, J. R., Li, Y.-Y., Feldman, J., Jayaram, M., Abraham, J., Nasmyth, K. A., and Hicks, J. B. (1983). Localization and sequence analysis of yeast origins of DNA replication. *Cold Spring Harbor Symp. Quant. Biol.* **47,** 1165–1173.

Cameron, J. R., Loh, E. Y., and David, R. W. (1979). Evidence for transposition of dispersed repetitive DNA families in yeast. *Cell (Cambridge, Mass.)* **16,** 739–751.

Campbell, A. (1971). Genetic structure. *In* "The Bacteriophage Lambda" (A. D. Hershey, ed.), pp. 13–44. Cold Spring Harbor Lab. Press, Cold Spring Harbor, New York.

Chan, C. S. M., and Tye, B.-K. (1980). Autonomously replicating sequences in *Saccharomyces cerevisiae*. *Proc. Natl. Acad. Sci. U.S.A.* **77,** 6329–6333.

Chan, C. S. M., and Tye, B.-K. (1983a). A family of *Saccharomyces cerevisiae* repetitive autonomously replicating sequences that have very similar genomic environments. *J. Mol. Biol.* **168,** 505–523.

Chan, C. S. M., and Tye, B.-K. (1983b). Organization of DNA sequences and replication origins at yeast telomeres. *Cell (Cambridge, Mass.)* **33,** 563–573.

Chandra, P. K., and Kung, H.-F. (1983). *In vitro* synthesis of biologicaly active human leukocyte interferon in a RNA-dependent system from *Saccharomyces cerevisiae*. *Proc. Natl. Acad. Sci. U.S.A.* **80,** 2569-2573.

Clancy, S., Mann, C., Davis, R. W., and Calos, M. P. (1984). Deletion of plasmid sequences during *Saccharomyces cerevisiae* transformation. *J. Bacteriol.* **159,** 1065–1067.

Clarke, L., and Carbon, J. (1980), Isolation of a yeast centromere and construction of functional small circular chromosomes. *Nature (London)* **287,** 504–509.

Clark-Walker, G. D., and Miklos, G. L. G. (1974). Localization and quantification of circular DNA in yeast. *Eur. J. Biochem.* **41,** 359–365.

Cohen, J. D., Eccleshall, T. R., Needleman, R. B., Fedoroff, H., Buchferer, B. A., and Marmur, J. (1980). Functional expression in yeast of the *Escherichia coli* plasmid gene coding for chloramphenicol acetyltransferase. *Proc. Natl. Acad. Sci. U.S.A.* **77,** 1078–1082.

Dani, G. M., and Zakian, V. A. (1983). Instability of dicentric plasmids in yeast. *Proc. Natl. Acad. Sci. U.S.A.* **80,** 228–232.

Davison, J., and Thi, V. H. (1982). The *Trypanosoma brucei* maxi-circle DNA contains *ars* elements active in *Saccharomyces cerevisiae*. *Curr. Genet.* **6,** 19–20.

Dobson, M. J., Trite, M. F., Roberts, N. A., Kingsman, A. J., Kingsman, S. M., Perkins, R. E., Conroy, S. C., Dunbar, B., and Fothergill, L. A. (1982). Conservation of high efficiency promoter sequences in *Saccharomyces cerevisiae*. *Nucleic Acids Res.* **10,** 26125–26137.

Donahue, T. F., Daves, R. S., Lucchini, G., and Fink, G. R. (1983). A short nucleotide sequence required for regulation of *HIS4* by the general control system of yeast. *Cell (Cambridge, Mass.)* **32,** 89–98.

Edens, L., Bom, I., Ledeboer, A. M., Matt, J., Tonen, M. Y., Visser, C., and Verrips, C. T. (1984). Synthesis and processing of the plant protein thaumatin in yeast. *Cell (Cambridge, Mass.)* **37,** 629–633.

Falco, S. C., and Botstein, D. (1983). A rapid chromosome-mapping method for cloned fragments of yeast DNA. *Genetics* **105,** 857–872.

Falco, S. C., Li, Y., Broach, J. R., and Botstein, D. (1982). Genetic properties of chromosomally integrated 2μ plasmid DNA in yeast. *Cell (Cambridge, Mass.)* **29,** 585–594.

Feldmann, H., Olah, J., and Friedenreich, H. (1981). Sequence of a yeast DNA fragment containing a chromosomal replicator and a tRNA$_3^{Glu}$ gene. *Nucleic Acids Res.* **9,** 2949–2959.

Fink, G. R., Farabaugh, P., Roeder, G., and Chaleff, D. (1980). Transposable elements (*Ty*) in yeast. *Cold Spring Harbor Symp. Quant. Biol.* **45,** 575–580.

Fitzgerald-Hayes, M., Buhler, J.-M., Cooper, T. G., and Carbon, J. (1982a). Isolation and subcloning analysis of functional centromere DNA (*CEN11*) from *Saccharomyces cerevisiae* chromosome XI. *Mol. Cell. Biol.* **2,** 82–87.

Fitzgerald-Hayes, M., Clarke, L., and Carbon, J. (1982b). Nucleotide sequence comparisons and functional analysis of yeast centromere DNA. *Cell (Cambridge, Mass.)* **29,** 235–244.

Flint, S. J. (1982). RNA processing. *Fed. Proc., Fed. Am. Soc. Exp. Biol.* **41,** 2781–2789.

Fogel, S., and Welch, J. W. (1982). Tandem gene amplification mediates copper resistance in yeast. *Proc. Natl. Acad. Sci. U.S.A.* **79,** 5342–5346.

Fogel, S., Mortimer, R. K., and Lusnak, K. (1982a). Mechanism of meiotic gene conversion, or "Wanderings on a foreign strand." *In* "The Molecular Biology of the Yeast Saccharomyces. I. Life Cycle and Inheritance" (J. Strathern, E. W. Jones, and J. R. Broach, eds.), pp. 289–339. Cold Spring Harbor Lab. Press, Cold Spring Harbor, New York.

Fogel, S., Choi, T., Kilgore, D., Lusnak, K., and Williamson, M. (1982b). The molecular genetics of non-tandem duplications at *ADE8* in yeast. *Recent Adv. Yeast Mol. Biol.* **1,** 269–288.

Fogel, S., Mortimer, R. K., and Lusnak, K. (1983). Meiotic gene conversion in yeast: molecular and experimental perspectives. *In* "Yeast Genetics. Fundamental and Applied Aspects" (J. F. T. Spencer, D. M. Spencer, and A. R. W. Smith, eds.), pp. 65–107. Springer-Verlag, Berlin and New York.

Futcher, A. B., and Cox, B. S. (1984). Copy number and the stability of 2-μm circle-based artificial plasmids of *Saccharomyces cerevisiae*. *J. Bacteriol.* **157,** 283–290.

Game, J. C., Zamb, T. J., Brown, R. J., Resnick, M., and Roth, R. M. (1980). The role of radiation (*rad*) genes in meiotic recombination in yeast. *Genetics* **94,** 51–68.

Gritz, L., and Davies, J. (1983). Plasmid-encoded hygromycin B resistance: the sequence of hygromycin B phosphotransferase gene and its expression in *Escherichia coli* and *Saccharomyces cerevisiae*. *Gene* **25,** 179–188.

Guarente, L. (1983). Yeast promoters and *lacZ* fusions designed to study expression of cloned genes in yeast. *In* "Recombinant DNA," Part C (R. Wu, L. Grossman, and K. Moldave, eds.), Methods in Enzymology, Vol. 101, pp. 181–191. Academic Press, New York.

Guarente, L., and Mason, T. (1983). Heme regulates transcription of the *CYC1* gene of *S. cerevisiae* via an upstream activation site. *Cell (Cambridge, Mass.)* **32,** 1279–1286.

Guarente, L., Lalonde, B., Gifford, P., and Alani, E. (1984). Distinctly regulated tandem upstream activation sites mediate catabolite repression of the *CYC1* gene of *Saccharomyces cerevisiae*. *Cell (Cambridge, Mass.)* **36,** 503–511.

Hartley, J. L., and Donelson, J. E. (1980). Nucleotide sequence of the yeast plasmid. *Nature (London)* **286,** 860–864.

Henikoff, S., Tatchell, K., Hall, B. D., and Nasmyth, K. A. (1981). Isolation of a gene from *Drosophila* by complementation in yeast. *Nature (London)* **289,** 33–37.

Herskowitz, I., and Oshima, Y. (1981). Control of cell type in *Saccharomyces cerevisiae:* mating type and mating-type interconversion. *In* "The Molecular Biology of the Yeast Saccharomyces. I. Life Cycle and Inheritance" (J. Strathern, E. W. Jones, and J. R. Broach, eds.), pp. 181–209. Cold Spring Harbor Lab. Press, Cold Spring Harbor, New York.

Hicks, J. B., Strathern, J. N., Klar, A. J. S., and Dellaporta, S. L. (1982). Cloning by complementation in yeast: the mating type genes. *In* "Genetic Engineering" (J. K. Setlow and A. Hollaender, eds.), Vol. 4, pp. 219–248. Plenum, New York.

Hinnebusch, A. G., and Fink, G. R. (1983). Repeated DNA sequences upstream from *HIS1* also occur at several other co-regulated genes in *Saccharomyces cerevisiae*. *J. Biol. Chem.* **258,** 5238–5247.

Hinnen, A., Hicks, J. B., and Fink, G. R. (1978). Transformation of yeast. *Proc. Natl. Acad. Sci. U.S.A.* **75**, 1929–1933.

Hitzeman, R. A., Hagie, F. E., Levine, H. L., Goeddel, D., Ammerer, G., and Hall, B. D. (1981). Expression of a human gene for interferon in yeast. *Nature (London)* **293**, 717–722.

Hitzeman, R. A., Leung, D. W., Perry, L. J., Kohr, W. J., Levine, H. L., and Goeddel, D. V. (1983). Secretion of human interferons by yeast. *Science (Washington, D.C.)* **219**, 620–625.

Hollenberg, C. P. (1982). Cloning with 2-μm DNA vectors and the expression of foreign genes in *Saccharomyces cerevisiae*. *In* "Gene Cloning in Organisms Other than *E. coli*" (P. H. Hofschneider and W. Goebel, eds.), pp. 119–144. Springer-Verlag, Berlin and New York.

Holliday, R. (1964). A mechanism for gene conversion in fungi. *Genet. Res.* **5**, 282–304.

Holm, C. (1982). Clonal lethality caused by the yeast plasmid 2μ DNA. *Cell (Cambridge, Mass.)* **29**, 585–594.

Hsiao, C.-L., and Carbon, J. (1979). High-frequency transformation of yeast by plasmids containing the cloned yeast *ARG4* gene. *Proc. Natl. Acad. Sci. U.S.A.* **76**, 3829–3833.

Hyman, B. C., Cramer, J. H., and Rownd, R. H. (1982). Properties of a *Saccharomyces cerevisiae* mtDNA segment conferring high-frequency yeast transformation. *Proc. Natl. Acad. Sci. U.S.A.* **79**, 1578–1582.

Iimura, Y., Gotoh, K., Ouchi, K., and Nishiya, T. (1983). Yeast transformation without the spheroplasting process. *Agric. Biol. Chem.* **47**, 897–901.

Ilgen, C., Farabaugh, P. J., Hinnen, A., Walsh, J. M., and Fink, G. R. (1979). Transformation of yeast. *In* "Genetic Engineering" (J. K. Setlow and A. Hollaender, eds.), Vol. 1, pp. 117–132. Plenum, New York.

Imai, Y., Suzuki, K., Yamashita, I., and Jukui, S. (1983). High-frequency transformation of *Saccharomyces cerevisiae* with linear deoxyribonucleic acid. *Agric. Biol. Chem.* **47**, 915–918.

Ito, H., Fukuda, Y., Murata, K., and Kimura, A. (1983). Transformation of intact yeast cells treated with alkali cations. *J. Bacteriol.* **153**, 163–168.

Jackson, J. A., and Fink, G. R. (1981). Gene conversion between duplicated genetic elements in yeast. *Nature (London)* **292**, 306–311.

Janowicz, Z. A., Henning, U., and Hollenberg, C. P. (1982). Synthesis of *Escherichia coli* outer membrane *Omp A* protein in yeast. *Gene* **20**, 347–358.

Jimenez, A., and Davies, J. (1980). Expression of a transposable antibiotic resistance element in *Saccharomyces*. *Nature (London)* **287**, 869–871.

Kawasaki, G. H., Woodbury, R. C., and Forstrom, J. F. (1983). Production of human α1-antitrypsin in yeast. *Abstr. Mol. Biol. Yeast Meet., Cold Spring Harbor Lab., New York* p. 230.

Kearsey, S. (1983). Analysis of sequences conferring autonomous replication in baker's yeast. *EMBO J.* **2**, 1571–1575.

Khan, N. C., and Sen, S. P. (1974). Genetic transformation in yeasts. *J. Gen. Microbiol.* **83**, 237–250.

Kingsman, A. J., Gimlich, R. L., Clarke, L., Chinault, A. C., and Carbon, J. (1981). Sequence variation in dispersed repetitive sequences in *Saccharomyces cerevisiae*. *J. Mol. Biol.* **145**, 619–632.

Kiss, G. B., Amin, A. A., and Pearlman, R. E. (1981). Two separate regions of the extrachromosomal ribosomal deoxyribonucleic acid of *Tetrahymena thermophilia* enable autonomous replication of plasmids in *Saccharomyces cerevisiae*. *Mol. Cell. Biol.* **1**, 535–543.

Klein, H. L., and Petes, T. D. (1981). Intrachromosomal gene conversion. *Nature (London)* **289**, 144–148.

Kozak, M. (1981). Possible role of flanking nucleotides in recognition of the AUG initiator codon by eukaryotic ribosomes. *Nucleic Acids Res.* **9**, 5233–5252.

Langford, C., Nellen, W., Niessing, J., and Gallwitz, D. (1983). Yeast is unable to excise foreign

intervening sequences from hybrid gene transcirpts. *Proc. Natl. Acad. Sci. U.S.A.* **80,** 1496–1500.
McNeil, J. B., and Friesen, J. D. (1981). Expression of the *Herpes simplex* virus thymidine kinase gene in *Saccharomyces cerevisiae. Mol. Gen. Genet.* **184,** 386–393.
Maine, G. T., Surosky, R. T., and Tye, B.-K. (1984a). Isolation and characterization of the centromere from chromosome V (*CEN5*) of *Saccharomyces cerevisiae. Mol. Cell. Biol.* **4,** 86–91.
Maine, G. T., Sinha, P., and Tye, B.-K. (1984b). Mutants of *S. cerevisiae* defective in the maintenance of minichromosomes. *Genetics* **106,** 365–385.
Malone, R. E., and Esposito, R. E. (1980). The *RAD52* gene is required for monothallic interconversion of mating types and spontaneous mitotic recombination in yeast. *Proc. Natl. Acad. Sci. U.S.A.* **77,** 503–507.
Mann, C., and Davis, R. W. (1983). Instability of dicentric plasmids in yeast. *Proc. Natl. Acad. Sci. U.S.A.* **80,** 228–232.
Mellor, J., Dobson, M. J., Roberts, N. A., Tuite, M. F., Entage, J. S., White, S., Lowe, P. A., Patel, T., Kingsman, A. J., and Kingsman, S. M. (1983). Efficient synthesis of enzymatically active calf chymosin in *Saccharomyces cerevisiae. Gene* **24,** 1–14.
Mercereau-Puijalon, O., Lacroute, F., and Kourilsky, P. (1980). Synthesis of a chicken ovalbumin-like protein in the yeast *Saccharomyces cerevisiae. Gene* **11,** 163–167.
Meselson, M. S., and Radding, C. M. (1975). A general model for genetic recombination. *Proc. Natl. Acad. Sci. U.S.A.* **75,** 358–361.
Miyanohara, T., Toh-e, A., Nozaki, C., Hamada, F., Ohtomo, N., and Matsubara, K. (1983). Expression of hepatitis B surface antigen gene in yeast. *Proc. Natl. Acad. Sci. U.S.A.* **80,** 1–5.
Morgan, T. H. (1919). "The Physical Basis of Heredity." Lippincott, Co., Philadelphia, Pennsylvania.
Mortimer, R. K., Contopoulou, R., and Schild, D. (1981). Mitotic chromosome loss in radiation-sensitive strain of yeast, *Saccharomyces cerevisiae. Proc. Natl. Acad. Sci. U.S.A.* **78,** 5778–7782.
Murray, A. W., and Szostak, J. W. (1983a). Construction of artificial chromosomes in yeast. *Nature (London)* **305,** 189–193.
Murray, A. W., and Szostak, J. W. (1983b). Pedigree analysis of plasmid segregation in yeast. *Cell (Cambridge, Mass.)* **34,** 961–970.
Nasmyth, K. A. (1982). The regulation of yeast mating-type chromatin structure by *SIR:* an action at a distance affecting both transcription and transposition. *Cell (Cambridge, Mass.)* **30,** 567–578.
Nasmyth, K. A., Tatchell, K., Hall, B. D., Astell, C., and Smith, M. (1981). Physical analysis of mating type loci in *Saccharomyces cerevisiae. Cold Spring Harbor Symp. Quant. Biol.* **45,** 961–981.
Newlon, C. S., and Lipchitz, L. R. (1983). Cloning, mapping, and localization of *ARS*'s on a circular derivative of yeast chromosome III. *Abstr. Mol. Biol. Yeast Meet., Cold Spring Harbor Lab., New York* p. 16.
Oppenoorth, W. F. F. (1962). Transformation in yeast: evidence of a genetic change by the action of DNA. *Nature (London)* **193,** 706.
Orr-Weaver, T. L., and Szostak, J. W. (1983a). Yeast recombination: the association between double-strand gap repair and crossing-over. *Proc. Natl. Acad. Sci. U.S.A.* **80,** 4417–4421.
Orr-Weaver, T. L., and Szostak, J. W. (1983b). Multiple, tandem plasmid integration in *Saccharomyces cerevisiae. Curr. Genet.* **6,** 19–20.
Orr-Weaver, T. L., Szostak, J. W., and Rothstein, R. J. (1981). Yeast transformation: a model system for the study of recombination. *Proc. Natl. Acad. Sci. U.S.A.* **78,** 6354–6358.
Orr-Weaver, T. L., Szostak, J. W., and Rothstein, R. J. (1983). Genetic applications of yeast

6. Yeast Transformation

transformation with linear and gapped plasmids. *In* "Recombinant DNA," Part C (R. Wu, L. Grossman, and K. Moldave, eds.), Methods in Enzymology, Vol. 101, pp. 228–245. Academic Press, New York.

Panthier, J.-J., Fournier, P., Heslot, H., and Rambah, A. (1980). Cloned β-galactosidase gene of *Escherichia coli* is expressed in the yeast *Saccharomyces cerevisiae*. *Curr. Genet.* **2**, 109–113.

Panzeri, L., and Philippsen, P. (1982). Centromeric DNA from chromosome VI in *Saccharomyces cerevisiae* strains. *EMBO J.* **1**, 1605–1611.

Petes, T. D. (1980). Unequal meiotic recombination within tandem arrays of yeast ribosomal DNA genes. *Cell (Cambridge, Mass.)* **19**, 765–774.

Pikielny, C. W., Teeru, J. L., and Rosbash, M. (1983). Evidence for the biochemical role of an internal sequence in yeast nuclear mRNA introns: implications for U1 RNA and metazoan mRNA splicing. *Cell (Cambridge, Mass.)* **34**, 395–403.

Resnick, M. A., and Martin, P. (1976). The repair of double-stranded breaks in the nuclear DNA of *Saccharomyces cerevisiae* and its genetic control. *Mol. Gen. Genet.* **143**, 119–129.

Resnick, M. A., Kasimos, J. N., Game, J. C., Braun, R. J., and Roth, R. M. (1981). Changes in DNA during meiosis in a repair-deficient mutant (*rad52*) of yeast. *Science (Washington, D.C.)* **212**, 543–545.

Roeder, G. S., and Fink, G. R. (1983). Transposable elements in yeast. *In* "Mobile Genetic Elements" (J. A. Shapiro, ed.), pp. 299–328. Academic Press, New York.

Roeder, G. S., Farabaugh, P. J., Chaleff, D. T., and Fink, G. R. (1980). The origins of gene instability in yeast. *Science (Washington, D.C.)* **209**, 1375–1380.

Roggenkamp, R., Kustermann-Kuhn, B., and Hollenberg, C. P. (1981). Expression and processing of bacterial β-lactamase in the yeast *Saccharomyces cerevisiae*. *Proc. Natl. Acad. Sci. U.S.A.* **78**, 4466–4470.

Roggenkamp, R., Hoppe, J., and Hollenberg, C. P. (1983). Specific processing of the bacterial β-lactamase precursor in *Saccharomyces cerevisiae*. *J. Cell. Biochem.* **22**, 141–149.

Rose, M., and Botstein, D. (1983a). Construction and use of gene fusions to *lacZ* (β-galactosidase) that are expressed in yeast. *In* "Recombinant DNA," Part C (R. Wu, L. Grossman, and K. Moldave, eds.), Methods in Enzymology, Vol. 101, pp. 167–180. Academic Press, New York.

Rose, M., and Botstein, D. (1983b). Structure and function of the yeast *URA3* gene. Differentially regulated expression of hybrid β-galactosidase from overlapping coding sequences in yeast. *J. Mol. Biol.* **170**, 883–904.

Rothstein, R. J. (1983). One-step gene disruption in yeast. *In* "Recombinant DNA," Part C (R. Wu, L. Gro-sman, and K. Moldaver, eds.), Methods in Enzymology, Vol. 101, pp. 202–211. Academic Press, New York.

Scherer, S., and Davis, R. W. (1979). Replacement of chromosome segments with altered DNA sequences constructed *in vitro*. *Proc. Natl. Acad. Sci. U.S.A.* **76**, 4951–4955.

Sclafani, R. A., and Fangman, W. L. (1984). Yeast gene *CDC8* encodes thymidylate kinase and is complemented by herpes thymidine kinase gene TK. *Proc. Natl. Acad. Sci. U.S.A.* **81**, 5821–5825.

Sherman, F., Fink, G. R., and Hicks, J. B. (1983). "Methods in Yeast Genetics," pp. 106–120. Cold Spring Harbor Lab. Press, Cold Spring Harbor, New York.

Silverman, S. J., Rose, M., Botstein, D., and Fink, G. R. (1982). Regulation of *HIS4–lacZ* fusions in *Saccharomyces cerevisiae*. *Mol. Cell. Biol.* **2**, 1212–1219.

Singh, H., Bieker, J. J., and Dumas, L. B. (1982). Genetic transformation of *Saccharomyces cerevisiae* with single-stranded circular DNA vectors. *Gene* **20**, 441–449.

Southern, E. H. (1975). Detection of specific sequences among DNA fragments separated by gel electrophoresis. *J. Mol. Biol.* **98**, 503–517.

Stahl, F. W. (1979). "Genetic Recombination: Thinking About It in Phage and Fungi." Freeman, San Francisco, California.
Stepien, P. P., Brousseau, R., Wu, R., Narang, S., and Thomas, D. Y. (1983). Synthesis of a human insulin gene. 6. Expression of the synthetic proinsulin gene in yeast *Saccharomyces cerevisiae*. *Gene* **24**, 289–298.
Stinchcomb, D. T., Struhl, K., and Davis, R. W. (1979). Isolation and characterization of a yeast chromosomal replicator. *Nature (London)* **282**, 39–43.
Stinchcomb, D. T., Thomas, M., Kelly, J., Selker, E., and Davis, R. W. (1980). Eukaryotic DNA segments capable of autonomous replication in yeast. *Proc. Natl. Acad. Sci. U.S.A.* **77**, 4559–4563.
Stinchcomb, D. T., Mann, C., Selker, E., and Davis, R. W. (1981). DNA sequences that allow the replication and segregation of yeast chromosomes. *ICN–UCLA Symp. Mol. Cell. Biol.* **22**, 473–488.
Stinchcomb, D. T., Mann, C., and Davis, R. W. (1982). Centromeric DNA from *Saccharomyces cerevisiae*. *J. Mol. Biol.* **158**, 157–179.
Strathern, J. N., Klar, A. J. S., Hicks, J. B., Abraham, J. A., Ivy, J. M., Nasmyth, K. A., and McGill, C. (1982). Homothallic switching of yeast mating type cassettes is initiated by a double-strand cut in the *MAT* locus. *Cell (Cambridge, Mass.)* **31**, 183–192.
Stritch, R., and Scott, J. (1983). Insertion mutagenesis and subclones of *ARS1*. *Abstr. Mol. Biol. Yeast Meet., Cold Spring Harbor Lab., New York* p. 56.
Struhl, K. (1983a). The new yeast genetics. *Nature (London)* **305**, 391–397.
Struhl, K. (1983b). Direct selection for gene replacement events in yeast. *Gene* **26**, 231–242.
Struhl, K., Stinchcomb, D. T., Scherer, S., and Davis, R. W. (1979). High frequency transformation of yeast: autonomous replication of hybrid DNA molecules. *Proc. Natl. Acad. Sci. U.S.A.* **76**, 1035–1039.
Suzuki, K., Imai, Y., and Yamashita, I. (1983). In vivo ligation of linear DNA molecules to circular forms in the yeast *Saccharomyces cerevisiae*. *J. Bacteriol.* **155**, 747–754.
Symington, L. S., Fogarty, L. M., and Kolodner, R. (1983). Genetic recombination of homologous plasmids catalysed by cell-free extracts of *Saccharomyces cerevisiae*. *Cell (Cambridge, Mass.)* **35**, 805–813.
Szostak, J. W. (1982). Replication and resolution of telomers in yeast. *Cold Spring Harbor Symp. Quant. Biol.* **47**, 1187–1194.
Szostak, J. W., and Blackburn, E. H. (1982). Cloning yeast telomeres on linear plasmid vectors. *Cell (Cambridge, Mass.)* **29**, 245–255.
Szostak, J. W., and Wu, R. (1979). Insertion of a genetic marker into the ribosomal DNA of yeast. *Plasmid* **2**, 536–554.
Szostak, J. W., and Wu, R. (1980). Unequal crossing over in ribosomal DNA of *Saccharomyces cerevisiae*. *Nature (London)* **284**, 426–430.
Szostak, J. W., Orr-Weaver, T. L., Rothstein, R. J., and Stahl, F. W. (1983). The double-strand-break repair model for recombination. *Cell (Cambridge, Mass.)* **33**, 25–35.
Tschumper, G., and Carbon, J. (1980). Sequence of a yeast DNA fragment containing a chromosomal replicator and the *TRP1* gene. *Gene* **10**, 157–166.
Tschumper, G., and Carbon, J. (1982). Delta sequences and double symmetry in a yeast chromosomal replicator region. *J. Mol. Biol.* **156**, 293–307.
Tschumper, G., and Carbon, J. (1983). Copy number control by a yeast centromere. *Gene* **23**, 221–232.
Tuppy, H., and Wildner, G. (1965). Cytoplasmic transformation: mitochondria of wild type brewer's yeast restoring respiratory capacity in respiratory deficient "petite" mutant. *Biochem. Biophys. Res. Commun.* **20**, 733–738.

6. Yeast Transformation

Urdea, M. S., Merryweather, J. P., Mullenbach, G. T., Coit, D., Heberlein, V., Valenzuela, P., and Barr, P. J. (1983). Chemical synthesis of a gene for human epidermal growth factor, urogastrone, and its expression in yeast. *Proc. Natl. Acad. Sci. U.S.A.* **80,** 7461–7465.

Vaughan, P. R., Woo, S. W., Novitski, C. E., Linnane, A. W., and Nagley, P. (1980). Cotransformation of yeast with mitochondrial DNA and a 2-μ based cloning vector. *Biochem. Int.* **1,** 610–619.

Webster, T. D., and Dickson, R. C. (1983). Direct selection of *Saccharomyces cerevisiae* resistant to the antibiotic G418 following transformation with a DNA vector carrying the kanamycin-resistance gene of Tn903. *Gene* **26,** 243–252.

Whitehouse, H. L. K. (1982). "Genetic Recombination: Understanding the Mechanism of Heredity." Wiley, New York.

Winston, F., Chumley, F., and Fink, G. R. (1983). Eviction and transplacement of mutant genes in yeast. *In* "Recombinant DNA," Part C (R. Wu, L. Grossman, and K. Moldave, eds.), Methods in Enzymology, Vol. 101, pp. 211–228. Academic Press, New York.

Winston, F., Chaleff, D. T., Valent, B., and Fink, G. R. (1984). Mutations affecting Ty-mediated expression of the *HIS*4 gene of *Saccharomyces cerevisiae*. *Genetics* **107,** 179–197.

Woo, S. W., Linnane, A. W., and Nagley, P. (1982). Genetic changes in yeast cells cotransformed with mitochondrial DNA and plasmid YEp13 are not elicted by recombinant molecules made by covalent insertion of mitochondrial DNA into YEp13. *Biochem. Biophys. Res. Commun.* **109,** 455–462.

Woontner, M., and Scott, J. (1983). Deletion analysis of *ARS*1 in *TRP*1 RI circle. *Abstr. Mol. Biol. Yeast Meet., Cold Spring Harbor Lab., New York* p. 57.

Zakian, V. A., and Scott, J. F. (1982). Construction, replication, and chromatin structure of *TRP1* RI circle, a multiple copy synthetic plasmid derived from *Saccharomyces cerevisiae* chromosomal DNA. *Mol. Cell. Biol.* **2,** 221–232.

Zaret, K. S., and Sherman, F. (1982). DNA sequence required for efficient transcription termination. *Cell (Cambridge, Mass.)* **28,** 563–573.

7

Use of the *LYS2* Gene for Gene Disruption, Gene Replacement, and Promoter Analysis in *Saccharomyces cerevisiae*

DEBRA A. BARNES AND JEREMY THORNER
Department of Microbiology and Immunology
University of California
Berkeley, California

I.	Introduction	197
	A. Basic Concepts in the Use of Plasmid Vectors for Yeast Transformation	197
	B. Genetic Advantages of the *LYS2* Locus as a Selectable Marker	200
II.	Characterization of the *LYS2* Gene and Its Product	203
	A. Biochemistry of Lysine Biosynthesis in Fungi	203
	B. Isolation and Properties of the *LYS2* Gene	205
	C. Transcriptional Regulation of *LYS2* Gene Expression	208
III.	Genetic Manipulations Utilizing the *LYS2* Gene	210
	A. Targeted Integration of Linear and Gapped Plasmids	210
	B. Gene Replacement by Plasmid Integration and Excision	210
	C. One-Step Gene Disruption and Gene Replacement	213
	D. Construction of Promoter Fusions in Yeast	218
	E. Other Applications	221
IV.	Conclusions and Prospectus	222
	References	223

I. INTRODUCTION

A. Basic Concepts in the Use of Plasmid Vectors for Yeast Transformation

The yeast *Saccharomyces cerevisiae* has become an increasingly popular organism of study over the past several years. One major reason for this popularity is that a wide variety of genetic manipulations are readily available in the yeast system. Many of these manipulations are based on the ability to introduce genes into and remove genes from the yeast genome at will. This new feature of yeast genetics became possible with the development of dependable methods, devised originally by Beggs (1978) and by Hinnen *et al.* (1978), for transformation of yeast with DNA molecules. These techniques are described in detail in this volume by Boguslawski (Chapter 6) and by Rine and Carlson (Chapter 5). The purpose of this chapter is to highlight the special advantages of the *LYS2* gene. The genetic utility of this locus was first recognized by Chattoo *et al.* (1979). We have developed this system further by cloning the *LYS2* gene, by constructing convenient vectors carrying this gene, and by devising procedures for the genetic manipulation of yeast cells utilizing these tools.

Transformation techniques depend on the use of plasmid vectors that can be shuttled from *Escherichia coli* to *S. cerevisiae*. These vectors contain sequences that allow replication in *E. coli* so that sufficient amplification is achieved to permit the purification of enough plasmid DNA to work with *in vitro*. To permit their propagation as episomes in yeast, the vectors usually carry sequences that promote autonomous replication. Such sequences are derived either from chromosomal DNA (so-called ARS elements, Struhl *et al.*, 1979) or from the endogenous 2-μm DNA plasmid (Broach, 1981). The vectors also carry markers selectable in both bacterial and yeast cells. Bacterial markers are generally genes encoding enzymes that confer resistance to specific antibiotics, for example, ampicillin, tetracycline, and kanamycin. Therefore, bacterial transformants can be selected on media containing such drugs. Selectable markers in yeast, on the other hand, are usually genes encoding enzymes of essential biosynthetic pathways. Thus, when introduced into this organism, such cloned genes will complement the nutritional requirement caused by a mutation in the chromosomal copy of the same gene. For example, a plasmid carrying the cloned *URA3* gene will transform a *ura3* mutant that requires uracil for growth (Ura$^-$ phenotype) to uracil prototrophy (Ura$^+$ phenotype); hence, successful transformants will survive on medium lacking uracil.

One problem with cloning by complementation in yeast is that the recipient strain to be transformed must carry a recessive, and preferably nonreverting, mutation in the chromosomal locus corresponding to the cloned gene. In most cases, introduction of this kind of mutation is done by a genetic cross using preexisting mutations. In such a cross, a haploid strain carrying the recessive mutation (and perhaps not much else to recommend it) is mated with a strain of interest containing an assortment of desirable traits. Diploids are selected on minimal medium and then sporulated. The meiotic products (tetrads) are dissected by micromanipulation. Colonies arising from single haploid spores are then tested, by replica-plating, for their ability to grow on various media. In this way, the phenotype of a cell may be determined and its genotype inferred.

7. Use of *LYS2* for Genetic Manipulation of Yeast

Unfortunately, due to Mendelian segregation and the fact that *S. cerevisiae* has 17 discrete chromosomes, it may be tedious, and often difficult, to recover all of the markers of interest in a single genetic background.

A distinct improvement for introducing recessive mutations into a strain would be provided by direct selections for useful mutations. Positive selections for suitable recessive mutations are available for only a few loci.

One may directly select for *ura3* (and *ura5*) mutants by plating on minimal medium containing 5-fluoro-orotic acid and uracil (Boeke *et al.*, 1984). This pyrimidine analog is metabolized to toxic compounds (F-UMP and F-dUMP) by cells containing functional *URA3* and *URA5* genes. Therefore, wild-type cells plated on a medium containing 5-fluoro-orotic acid are killed, whereas spontaneously arising *ura3* or *ura5* mutants grow to form colonies. The utility of this approach is limited because the key ingredient of the medium, 5-fluoro-orotic acid, is rather expensive ($32.00 for 50 mg from P-L Biochemicals) and a high concentration of this compound is required in the medium (500–870 μg/ml).

It has been shown recently that resistance to 2-deoxygalactose provides a positive selection for *gal1* mutants. However, to be effective, the cells must be expressing the galactose-metabolizing enzymes constitutively (due to *gal80* or *GAL4^c* mutations) and selections must be done on medium containing a nonfermentable (nonhexose) carbon source so that uptake and phosphorylation of the sugar analog are not competitively inhibited (Platt, 1984). Furthermore, although 50–75% of the 2-deoxygalactose-resistant colonies have *gal1* mutations, galactose permease-negative (*gal2*), regulatory defects (*gal4*), and unknown galactose-negative mutants (*rdg*) appear. Nonetheless, since the cloned bacterial galactokinase gene (*galK*) can be expressed and produces functional galactokinase in yeast cells (Rymond *et al.*, 1983; Heusterspreute *et al.*, 1984), the 2-deoxygalactose selection furnishes (1) a direct way to isolate *gal1* recipients for such vectors and (2) a counterselection for the loss of such vectors from transformed *gal1* cells.

There are, of course, other recessive markers for which direct selections exist, for example, *cyh2* (cycloheximide resistance), *tcm1* (trichodermin resistance), and *can1* (canavanine resistance) (Mortimer and Schild, 1981). There are, however, no obvious methods for directly selecting for the presence of the normal alleles of these genes because they would confer drug sensitivity.

One way to avoid the necessity for construction of appropriate recipient strains carrying recessive chromosomal markers is to have available plasmid vectors that contain a dominant genetic marker conferring resistance to a particular compound that is inhibitory to yeast growth. One such marker is the kanamycin resistance gene of Tn601 (Tn6) or Tn903 (Tn55). These transposons encode an aminoglycoside phosphotransferase that inactivates aminoglycoside antibiotics (Davies and Smith, 1978). Although kanamycin itself is not fungistatic, a related drug, G418 (a gentamycin derivative), will inhibit yeast growth. When Tn601 is carried on a yeast episomal (2-μm) vector, yeast cells transformed with this plasmid become resistant to drug G418 at a concentration of 1 mg/ml (Jimenez and Davies, 1980). However, as pointed out by Hollenberg (1982), the optimal concentration of G418 for selection of transformants is lower, about 600 μg/ml. At this lower concentration, spontaneously arising antibiotic-resistant variants appear at very high frequency and may constitute 50% or more of the total G418-resistant colonies on a transformation plate; this level of false positives would severely compromise many recombinant DNA procedures, for example, "shotgun" screening of a library for complementation of a yeast mutation of interest. To circumvent this problem, elaborate strategies involving newer vector constructions and special culture conditions were devised (Webster and Dickson, 1983). More recently, however, vectors have been developed in which the G418-resistance gene is expressed from strong eukaryotic promoters that function well in yeast, for example, the 5'-flanking region of *ENO1*, the structural gene for the glycolytic enzyme enolase (McAlister and Holland, 1982). Such plasmids do provide a reproducible and sufficiently high level of drug resistance to permit the direct selection of antibiotic-resistant transformants (D. Gelfand, personal communication).

Another dominant locus that shows some promise as a selectable marker is *CUP1*, which was cloned by Fogel and Welch (1982). Most yeast strains carry the recessive *cup1* allele and are unable to grow on plates containing 3 m*M* $CuSO_4$. When strains are transformed by the cloned *CUP1* allele, which contains multiple copies of the metallothionein-like copper-binding protein encoded at this locus, cells become resistant to this concentration of copper. Although transcription of *CUP1* is Cu^{2+}-inducible and hence might provide a useful regulatable promoter (Karin *et al.*, 1984), the main advantage of *CUP1* as a selectable marker is limited to its use in vectors for gene isolation. Because there are no positive selections for the loss of *CUP1* or the G418-resistance gene, these loci may not be useful for performing certain types of genetic manipulations that can be applied to a yeast gene once it has been isolated. The importance of the ability to select for the loss, as well as gain, of a gene function is explained further in subsequent sections.

B. Genetic Advantages of the *LYS2* Locus as a Selectable Marker

In contrast to the genes mentioned in the preceding section, one advantage of using the *LYS2* gene as a selectable marker is that there are convenient methods for directly selecting both forward and reverse mutations at this locus.

As first described by Chattoo *et al.* (1979), mutations in the *LYS2* gene may be readily obtained by plating wild-type cells on minimal medium containing 2 mg/ml DL-α-aminoadipic acid as the sole nitrogen source and 30 μg/ml lysine [which cannot serve as a nitrogen source for *S. cerevisiae* (see Cooper, 1982)]. Although a relatively high concentration of α-aminoadipate is required, this compound is inexpensive ($48.70 for 5 g from Aldrich). Only spontaneously arising *lys2* and *lys5* mutants can grow on this medium. About 97% of the total number of colonies appearing on such a plate are *lys2* mutants. In *lys2* (and *lys5*) mutants, the pool of α-aminoadipate remains sufficiently high that the reversible α-aminoadipate aminotransferase reaction (Fig. 1, step 6) can proceed in the nonbiosynthetic direction. In this way, α-aminoadipate serves to transaminate α-ketoglutarate to glutamate, the normal cellular nitrogen donor made in abundance when free NH_4^+ is the nitrogen source. In normal $LYS2^+$ cells, the α-aminoadipate pool is much lower due to its conversion to subsequent pathway intermediates and hence, in the absence of ammonia in the medium, cells cannot make enough glutamate to support growth. Under the conditions of selection, normal cells arrest their cell division cycle as either large unbudded cells or singly budded cells (Winston and Bhattacharjee, 1982). It should be noted that even for *lys2* and *lys5* mutants the growth rate is not very high in this medium, as it takes 5–7 days for visible colonies to appear. In addition, it has been observed that even in minimal medium containing NH_4^+, the presence of α-aminoadipate is inhibitory to the growth of wild-type cells (cited in Bhattacharjee, 1983). The fungistatic action of high levels of α-aminoadipate in normal yeast cells is not completely understood, but may be due to overaccumulation of α-aminoadipate-δ-semialdehyde (Cooper, 1982; Zaret and Sherman, 1985).

Because the *lys2* mutants are auxotrophic for lysine, there is a strong positive selection for the restoration of a functional *LYS2* gene. Only cells that have regained *LYS2* function will be prototrophs and will grow on minimal medium lacking lysine.

These genetic features of the *LYS2* locus provide several powerful advantages in establishing a new vector–host system for cloning and manipulation of yeast genes. If the *LYS2* gene were cloned and available as the selectable marker on a plasmid, any strain could be made a potential recipient for transformation by such a vector by using the α-aminoadipate procedure to select a *lys2* mutant. Successful transformation of the *lys2* mutant by the plasmid would restore lysine prototrophy; therefore, transformants could be selected readily on medium lacking lysine. Finally, for several reasons, it is often useful to be able to remove a plasmid from a transformed cell, regardless of whether the vector exists unstably as an episome or has integrated stably into the genome. Derivatives of transformants that have lost the *LYS2* sequences are functionally *lys2* mutants and should regain the ability to grow on α-aminoadipate medium, and hence can be directly selected on that medium.

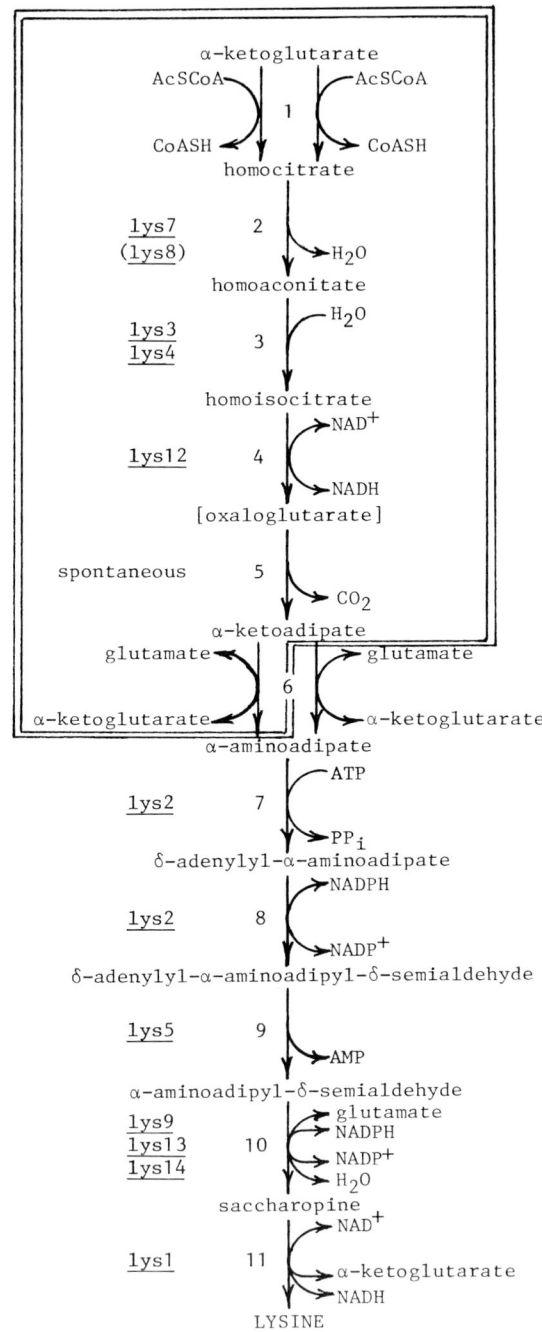

II. CHARACTERIZATION OF THE *LYS2* GENE AND ITS PRODUCT

A. Biochemistry of Lysine Biosynthesis in Fungi

Unlike bacteria and green plants, which synthesize lysine from aspartate via the diaminopimelate pathway, fungi have evolved a unique route for biosynthesis of lysine. In fungi, lysine is derived from α-ketoglutarate, a TCA cycle intermediate, by a series of nearly a dozen steps (Fig. 1). At least 10 of these reactions are known to be enzyme-catalyzed, and genetic loci presumably representing the structural genes for many of these enzymes have been identified (Table I).

One essential intermediate in this pathway, α-aminoadipate, is converted to the next metabolite, α-aminoadipate-δ-semialdehyde, by a series of three reactions (Sinha and Bhattacharjee, 1971). The δ-carboxyl is activated through formation of an adenylyl derivative at the expense of ATP. This derivative is reduced by NADPH and, finally, AMP is eliminated with concomitant formation of the semialdehyde. The three steps collectively are referred to as "α-aminoadipate reductase." The overall reaction requires Mg^{2+}.

The products of at least two genes, *LYS2* and *LYS5*, are required for formation of α-aminoadipate-δ-semialdehyde. Extracts from *lys5* mutants are capable of carrying out the first two steps of the α-aminoadipate reductase reaction, but are incapable of resolving the adenylylated intermediate by releasing α-aminoadipate-δ-semialdehyde and AMP (Sinha and Bhattacharjee, 1971). Therefore, the *LYS5* gene product is probably involved only in the hydrolysis step. Extracts from *lys2* mutants, on the other hand, cannot perform the first two steps of the reaction. One *lys2* mutant (*aau*) has an enzyme with an altered K_m for α-aminoadipate (Chattoo *et al.*, 1979) and all *lys2* mutants tested fail to catalyze α-aminoadipate-dependent NADPH oxidation, which is conveniently followed by a spectrophotometric assay measuring the disappearance of absorbance at 340 nm (Sinha and Bhattacharjee, 1971; Barnes and Thorner, 1985). Hence, the *LYS2* gene product is likely to be responsible for at least the initial activation step. Whether the *LYS2* protein actually participates in the reduction step is likely, but has not yet been proven definitively. Mattoon and co-workers (1961) described a Lys⁻ mutant that excreted what appeared to be δ-adenylyl-α-aminoadipate when starved for lysine. This mutant could represent a strain with a

---

Fig. 1. Lysine biosynthesis in *S. cerevisiae*. The pathway presented was compiled from information that can be found in Bhattacharjee (1983), Jones and Fink (1982), and Mortimer and Schild (1981). The steps enclosed by double solid lines are thought to occur in the mitochondrion; the other steps presumably occur in the cytosol. The reactions prevented by particular genetic blocks are indicated. Individual enzymes are listed in Table I.

TABLE I.
Gene–Enzyme Relationships for the Lysine Biosynthetic Pathway in *Saccharomyces cerevisiae*[a]

Reaction	Enzyme	EC number	Locus	Map position
1	Homocitrate synthase, two isozymes [3-hydroxy-3-carboxyadipate 2-oxoglutarate-lyase (CoA-acetylating)]	4.1.3.21	?	?
2	Homocitrate dehydrase (3-hydroxy-3-carboxyadipate hydro-lyase)	?	*lys7*	13R
3	Homoaconitate hydratase, homoaconitase (2-hydroxy-3-carboxyadipate hydro-lyase)	4.2.1.36	*lys3*	?
			lys4	4R
4	Homoisocitrate dehydrogenase [2-hydroxy-3-carboxyadipate : NAD oxidoreductase (decarboxylating)]	1.1.1.87	*lys12*	?
5	Spontaneous process			
6	α-Aminoadipate aminotransferase, two isozymes (L-2-aminoadipate : 2-oxoglutarate aminotransferase)	2.6.1.39	?	?
7–9	α-Aminoadipate reductase (L-2-aminoadipate-6-semialdehyde : NADP oxidoreductase)	1.2.1.31	*lys2*	2R
			lys5	7L
10	Saccharopine reductase [N^5-(1,3-dicarboxylpropyl)-L-lysine : NADP oxidoreductase (L-glutamate-forming)]	1.5.1.10	*lys9*	14R
			lys13	?
			lys14	?
11	Saccharopine dehydrogenase [N^5-(1,3-dicarboxylpropyl)-L-lysine : NAD oxidoreductase (L-lysine-forming)]	1.5.1.7	*lys1*	9R

[a] Several other mutations affecting lysine biosynthesis have been isolated, including *lys6* (?), *lys8* (?), *lys10* (14L), *lys11* (9L), *lys15* (?), and *lys16* (?). The role of these gene products, whether regulatory or enzymatic, in lysine biosynthesis is currently unknown. In addition, the map positions of many of these genes, as well as of several of the genes listed above that disrupt a known enzymatic reaction, are presently unknown (?). The *lys6*, *lys8*, *lys10*, *lys11*, and *lys15* complementation groups are blocked in the pathway prior to α-aminoadipate. To date, no mutants affecting homocitrate synthase and α-aminoadipate aminotransferase have been isolated, presumably because a double mutation would be required to eliminate both of the isozymes that catalyze each of these steps. The *lys6* and *lys8* mutants are also pleiotropic in the sense that they impose a requirement for glutamate as well as lysine, are respiratory-incompetent, and are deficient for the TCA cycle enzyme aconitate hydratase (4.2.1.3). Recent studies indicate that *lys9* and *lys13* are allelic, or very tightly linked (Borell *et al.*, 1984).

lesion specifically in the reduction step. Unfortunately, this lys^- mutation was not genetically characterized and hence its allelism to $lys2$ mutations is unknown.

B. Isolation and Properties of the LYS2 Gene

We isolated the *LYS2* gene from a library of yeast genomic DNA in the vector YEp24, also called pFL1 (Fasiolo *et al.*, 1981). This library was generously provided by M. Carlson and D. Botstein and was used to transform a *ura3 lys2* strain selecting initially for Ura$^+$ transformants. The transformants were then screened for any that also possessed a Lys$^+$ phenotype. Potential candidates were obtained and tested to determine if the apparent Ura$^+$Lys$^+$ phenotype was due to the presence of a plasmid. This test was accomplished by removing the selective pressure for plasmid maintenance by growing independently isolated transformants in rich medium for several generations. When the broth-grown cells were plated and replicas on media lacking either uracil or lysine were made, for certain transformants many colonies had become simultaneously Ura$^-$ and Lys$^-$, presumably due to the loss of a plasmid carrying both the *URA3* and *LYS2* genes. DNA was isolated from several of these Lys$^+$Ura$^+$ transformants and was used to transform *E. coli* HB101 to ampicillin resistance so that large amounts of plasmid could be purified. One plasmid initially isolated by this complementation procedure was pDA6200. A physical map of the restriction endonuclease cleavage sites in the insert contained in this plasmid is given in Fig. 2.

That pDA6200 contained the *LYS2* gene of *S. cerevisiae* was confirmed by several complementary genetic and biochemical methods (Barnes and Thorner, 1985). First, this plasmid conferred lysine prototrophy to strains carrying five independently derived *lys2* alleles, including both missense and nonsense mutations. Second, spectrophotometric assay of the α-aminoadipate-dependent oxidation of NADPH showed that the presence of this plasmid not only restored α-aminoadipate reductase activity in extracts of transformed *lys2* mutants, but resulted in a seven- to ninefold overproduction of the enzyme. Third, subcloned segments of the insert carried by pDA6200, which were unable to confer a Lys$^+$ phenotype by themselves, were nonetheless able to recombine with and restore *LYS2* function to chromosomal *lys2* alleles in marker rescue experiments. Finally, and most conclusively, it was determined that the insert contained in pDA6200 corresponded to the *LYS2* region of chromosome II by integration of this plasmid into the genome. This was accomplished in the following way.

A *lys2 ura3* recipient was transformed with pDA6200 that had been linearized by cleavage at the unique *Xho*I site within the insert (Fig. 2). This approach was taken because it has been shown that double-stranded ends dramatically promote homologous recombination in yeast (Orr-Weaver *et al.*, 1981, 1983). In this way, integration of pDA6200 was "targeted" to the region of the genome homologous to the DNA sequences carried by the plasmid insert. Transformants

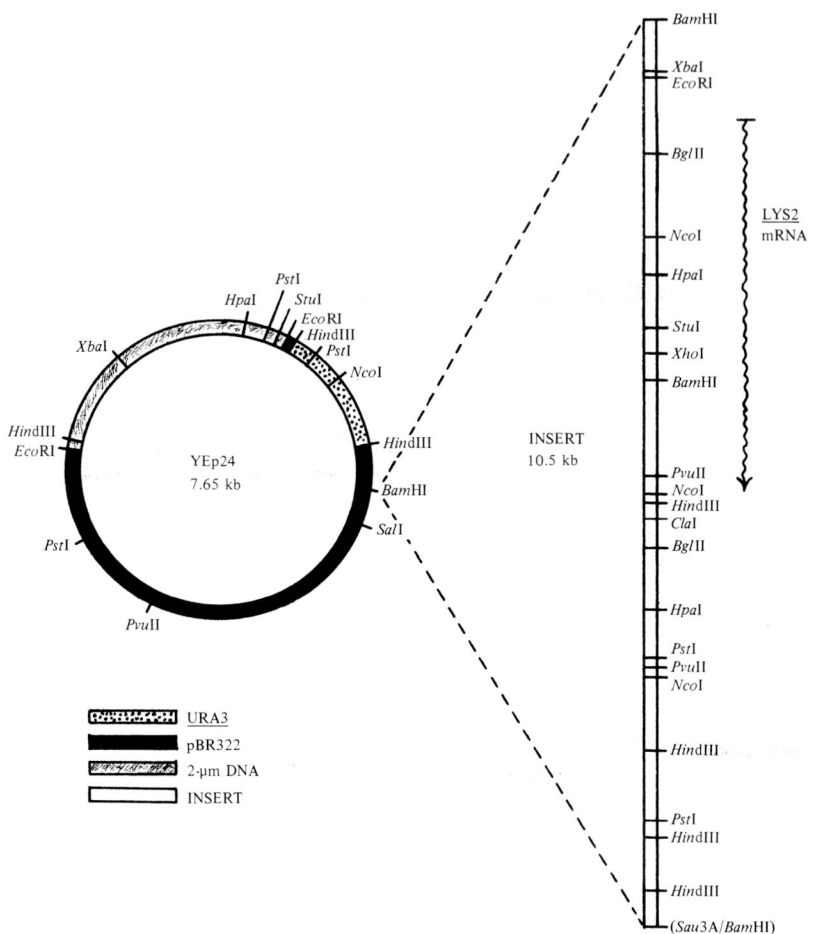

Fig. 2. Physical map of the *LYS2* gene of *S. cerevisiae*. Plasmid pDA6200 (Barnes and Thorner, 1985) was digested with a variety of different restriction endonucleases, singly and in various combinations, and subjected to electrophoresis in agarose gels, all following standard methods (Maniatis *et al.*, 1982). The plasmid contains a yeast genomic DNA segment of 10.5 kb. Within the insert *Xho*I is a unique site. The vector YEp24 (also called pFL1) is described in Fasiolo *et al.* (1981) and its complete restriction endonuclease cleavage site map is available because the entire nucleotide sequences are known for all of the elements of this plasmid: pBR322 (Sutcliffe, 1978); the small *Eco*RI fragment of the B form of yeast 2-μm circle (Hartley and Donelson, 1980); and the *Hin*dIII fragment containing the yeast *URA3* gene (Rose *et al.*, 1984). A functional *LYS2* gene is contained in the 5-kb *Eco*RI–*Hin*dIII segment of the insert and encodes a single transcript of about 4 kb (Eibel and Phillipsen, 1983; Barnes and Thorner, 1985).

were obtained that appeared to be stable integrants of the entire plasmid, as judged by the fact that both the Ura$^+$ and Lys$^+$ phenotypes were retained even after extensive passage of the cells under nonselective conditions and by the fact that sequences homologous to the bacterial portion (pBR322) of the vector were present in the chromosomal DNA isolated from the transformants and examined by the Southern blotting procedure. To confirm that pDA6200 had integrated at a single site, the stable transformants were backcrossed to a *ura3 lys2* strain and the resulting diploids were subjected to genetic analysis. All tetrads dissected segregated two viable and two nonviable spores. Thus, integration of pDA6200, which contains 2-μm DNA (Fig. 2), into chromosomal DNA appeared to result in chromosome loss due, presumably, to nondisjunction during meiosis. In every tetrad, the two viable spores were always either Lys$^-$ and Ura$^-$ or Lys$^+$ and Ura$^+$, indicating that the *LYS2* and *URA3* genes were tightly linked, as expected if they were inserted into the genome by integration of the same DNA fragment. To determine if the integration had occurred on chromosome II, where the *LYS2* locus has been shown to map (Mortimer and Schild, 1981), advantage was taken of the genetic instability of the chromosome into which the 2-μm DNA-containing plasmid had integrated. As first noted by Falco *et al.* (1982), in diploids a chromosome containing 2-μm DNA is also mitotically unstable and is lost during growth (or undergoes homozygosis with its homologue) at high frequency. This instability can be exploited for the purpose of genetic mapping if the haploid 2-μm-containing integrant is mated to a set of haploid strains bearing well-defined recessive markers on all the known linkage groups. The resulting diploids will frequently become monosomic (or homozygous), but only for the chromosome of interest, which is revealed by the uncovering of the recessive mutation(s) on the marked homologue (Falco and Botstein, 1983). One collection of well-characterized multiply marked strains that we have found particularly useful for this purpose is that of Klapholz and Esposito (1982). When our stable transformants were mated with *ura3* strains carrying markers on all yeast chromosomes (including *his7* and *tyr1* on chromosome II), the resulting diploids frequently became auxotrophic for uracil, presumably because the homologue into which pDA6200 (*URA3*$^+$) had integrated was lost. Most important, we found that all of these cells, without exception, had simultaneously become auxotrophic for histidine and tyrosine, but not auxotrophic for any other marker tested. These results are readily explained if the chromosome that is lost at high frequency is indeed chromosome II. To prove that pDA6200 had integrated at the *lys2* locus on chromosome II, an allelism test was performed by backcrossing the stable transformants (*ura3 lys2* :: *LYS2*$^+$ *URA3*$^+$) to a *LYS2*$^+$ *ura3* strain. Sporulation of the resulting diploids again produced tetrads containing two viable and two nonviable spores. In any given tetrad, the two viable spores were either both Ura$^+$ or both Ura$^-$. Most significantly, however, the viable spores were always Lys$^+$. The lack of separation of the *lys2* and *LYS2* markers proved that the DNA segment carried by pDA6200 had integrated at the *lys2* locus and therefore was indeed derived from the genomic region that includes the *LYS2* gene.

The *LYS2* gene has been isolated independently by at least two other groups of investigators (C. Falco, personal communication; Eibel and Philippsen, 1983). To prove that the DNA segment obtained corresponded to the *LYS2* gene, and was not a phenotype suppressor of *lys2* mutations, Eibel and Philipssen (1983) used a different, but equally valid, approach from the one described above. They first determined the transcribed region contained in the DNA segment which they found could complement a known *lys2* mutation when carried on a multicopy plasmid. They then subcloned an internal fragment (lacking the putative amino-terminal and carboxyl-terminal ends of the presumptive coding sequence) into an integrating vector, YIp5. Transformation of cells with this plasmid, which carries an internal segment of a yeast gene but possesses no origin for replication in yeast, produces an integrant containing two tandem, but inactivated, copies of the homologous chromosomal gene (Shortle *et al.*, 1982). Such transformants had indeed become lysine auxotrophs and were shown to be *lys2* mutants by complementation tests in which the transformants were backcrossed against tester strains carrying mutations in each of the known *lys* loci (see Table I and its footnote). The physical map of the *LYS2* DNA obtained by Eibel and Philippsen is identical to that obtained by us and shown in Fig. 2.

Extensive subcloning of pDA6200 revealed that the 5-kb *Eco*RI-*Hin*dIII fragment within the insert (Fig. 2) was sufficient to confer a Lys$^+$ phenotype to *lys2* cells. This fragment and the slightly larger *Eco*RI–*Cla*I segment were used to construct a variety of useful yeast cloning vehicles (Fig. 3). The plasmids we constructed include: (1) high-copy-number vectors containing either chromosomal or 2-μm circle-derived origins of replication (YRp610 and YEp620, respectively); (2) a low-copy-number centromere-containing vector (YCp630); and (3) an integrative plasmid (YIp600). At least two similar cloning vehicles, an integrative plasmid (YIp333) and an autonomously replicating plasmid (YRp31), have been prepared by Eibel and Philippsen (1983). However, these plasmids utilize the significantly larger *Eco*RI–*Pst*I (7-kb) segment of the *LYS2* region (see Fig. 2) and hence are more cumbersome and less convenient to use than the more compact vectors we constructed.

C. Transcriptional Regulation of *LYS2* Gene Expression

Many of the amino acid biosynthetic enzymes of *S. cerevisiae* are under so-called general amino acid control (Jones and Fink, 1982). This type of regulation is manifest as the simultaneous derepression of the synthesis of many enzymes from several unrelated amino acid biosynthetic pathways when cells are limited for any one of the amino acids involved in this control. It was anticipated that the

7. Use of *LYS2* for Genetic Manipulation of Yeast

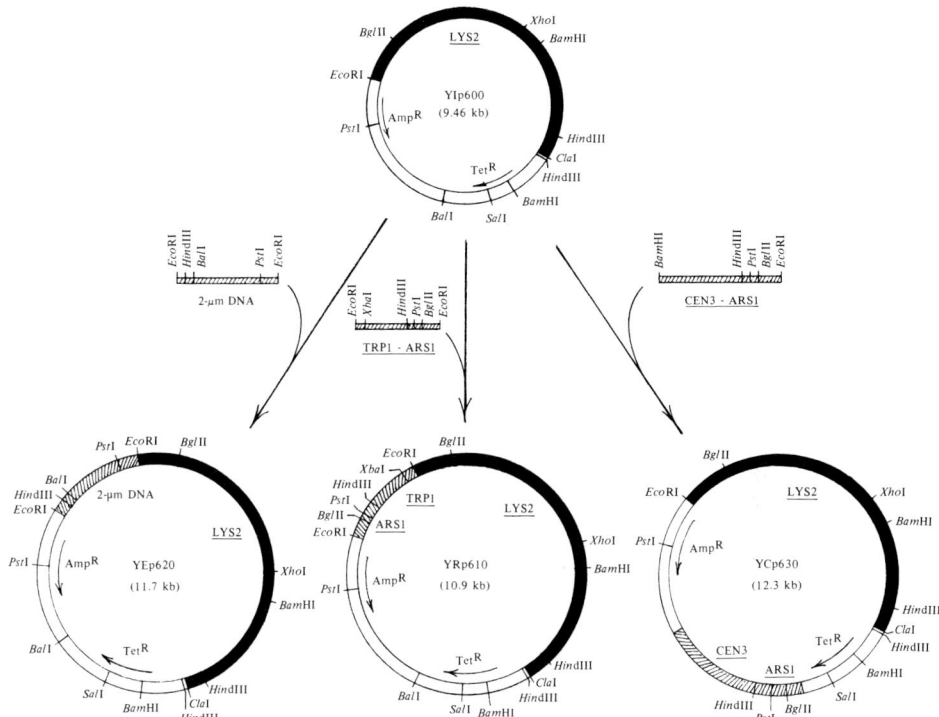

Fig. 3. Yeast cloning vectors utilizing the *LYS2* gene. The constructions of the yeast–*E. coli* shuttle plasmids YIp600, YRp610, YEp620, and YCp630 are described in detail in Barnes and Thorner (1985).

LYS2 gene might be regulated in this manner because it has been observed that cells starved for histidine exhibit an apparently marked increase in the activity of several enzymes of the lysine biosynthetic pathway, including α-aminoadipate reductase (Wolfner *et al.*, 1975).

When the *Hin*dIII fragment of plasmid pDA6200 (see Fig. 2), which contains the entire *LYS2* gene, was recloned into pBR322 and used as a hybridization probe, only a single transcript of 4.2 kb was observed by blotting poly A^+ mRNA species from wild-type yeast cells (Barnes and Thorner, 1985). A transcript of similar size was also detected by Eibel and Philippsen (1983). It was found that the *LYS2* transcript was present at significantly higher levels (5- to 10-fold) when cells were grown in minimal medium (or even in minimal medium containing lysine) than when they were grown in rich broth (or in minimal medium supplemented with all 20 amino acids) (Barnes and Thorner, 1985). This finding is in agreement with the measurements of enzyme levels made previously (Wolfner *et al.*, 1975) and strongly suggests that general amino acid control of α-aminoadipate reductase is exerted at the level of transcription of the *LYS2* gene. Indeed, DNA sequence analysis of the 5'-flanking region of the *LYS2* gene (D. Pridmore and P. Philippsen, personal communication) has revealed at least one copy of the sequence (A^A_TGTGACTC) that has been implicated in general amino acid control of transcription. This nucleotide sequence has been found, usually in multiple copies, upstream of every other gene encoding an amino acid biosynthetic enzyme that has been cloned and analyzed to date (Donahue *et al.*, 1983; Hinnebusch and Fink, 1983).

III. GENETIC MANIPULATIONS UTILIZING THE *LYS2* GENE

A. Targeted Integration of Linear and Gapped Plasmids

Many of the genetic manipulations that will be described in the subsequent sections of this chapter are dependent upon integration of a plasmid or fragment of DNA into a chromosomal locus. As first observed by Hicks *et al.* (1979) and extended in great detail by Orr-Weaver *et al.* (1981, 1983), integration of cloned DNA is greatly facilitated if a double-strand break, generated by digestion with a restriction endonuclease, is present in the cloned segment within the region that is homologous to the yeast genome. This discovery, that double-stranded DNA ends are highly recombinogenic in yeast (Szostak *et al.*, 1983), permits the efficient "targeting" of the integration event. As mentioned above, we used such a procedure to demonstrate that pDA6200 (Fig. 2) carries a DNA segment that directs its integration to the *LYS2* locus of chromosome II.

As additionally demonstrated by Orr-Weaver *et al.* (1981, 1983), when yeast cells are transformed with a plasmid DNA molecule that has been cut with restriction endonucleases at two sites within the insert, leaving a gap, the missing sequences are repaired by the cognate chromosomal information during integration. If the gapped plasmid contains a functional origin for replication in yeast (ARS or 2-μm origin), then the repaired plasmid can remain as a replicating episome. Therefore, this method can be used to recover mutant chromosomal alleles of genes that already have been cloned, provided a second selectable marker is also carried by the plasmid.

B. Gene Replacement by Plasmid Integration and Excision

One of the most important developments in the area of yeast molecular biology is the ability to test the effects of alterations of genes and their regulatory regions by inserting them in single copy at their normal chromosomal location in place of the wild-type gene. This process, termed "transplacement," was first devised by Scherer and Davis (1979).

7. Use of *LYS2* for Genetic Manipulation of Yeast

The method developed by Scherer and Davis (1979) originally involved the use of a yeast integrative vector (YI plasmid) that carried some selectable yeast marker, and an altered cloned gene of interest. Since the plasmid cannot replicate as an episome in yeast, transformants can arise under this circumstance only if the plasmid has integrated into homologous chromosomal DNA. To prevent the plasmid from integrating at the locus corresponding to the selectable marker and to direct integration to the site of the gene of interest, either a strain carrying a deletion or rearrangement of the marker locus is used as the recipient or integration of the plasmid is targeted by linearizing the plasmid by restriction enzyme cleavage within the insert (or both). The latter method has now become the most useful and most widely used one since it may be applied to any type of vector (YE, YR, YC, as well as YI, plasmids).

Integration of the plasmid generates a tandem duplication consisting of a copy of the altered gene and its wild-type homologue separated by vector sequences. Such duplications are genetically unstable because homologous recombination can occur at a detectable frequency between the duplicated sequences; therefore, removal of the selective pressure allows growth of cells that have lost the vector sequences and either the altered gene sequence or the wild-type sequence. After the selective pressure has been removed, one must screen for those cells that have lost the vector and then test such cells for the presence of the altered gene by Southern analysis, if some distinguishing physical characteristic is available (for example, a restriction enzyme cleavage site polymorphism).

Use of the *LYS2* gene as the selectable marker for gene transplacement vehicles would facilitate several aspects of this procedure (Fig. 4). First, any strain of interest could be made the recipient for a *LYS2* vector by application of the direct selection method for *lys2* mutants. Second, integration events are readily selected as restoration of stable lysine prototrophy. Third, one could select directly for subsequent loss of the integrated vector by demanding growth on α-aminoadipate medium. This latter step would save considerable time and effort since screening becomes unnecessary.

As a direct test for the utility of this approach, we attempted to use the cloned *TRP1* region to replace a mutant *trp1-289* allele in the chromosome with a normal gene. Integration of the vector YRp610 (Fig. 3) was targeted to the *TRP1* locus on chromosome IV by linearizing the plasmid DNA by digestion with XbaI and transforming strain DA100 (*MATa ura3-52 leu2-3,112 trp1-289 lys2 his4-580 ade2-1*), selecting for $LYS2^+$ cells on medium lacking lysine. Almost all of the Lys^+ transformants were also Trp^+. (In the transformants that were not, the *lys2* locus on chromosome II appeared to have undergone simple gene conversion to $LYS2^+$ at the expense of the plasmid sequence.) Because YRp610 contains an autonomous replicating sequence (ARS), the stability of the Lys^+Trp^+ phenotype was examined to ensure that the plasmid had, in fact, integrated. As episomes, ARS-containing plasmids are notoriously unstable dur-

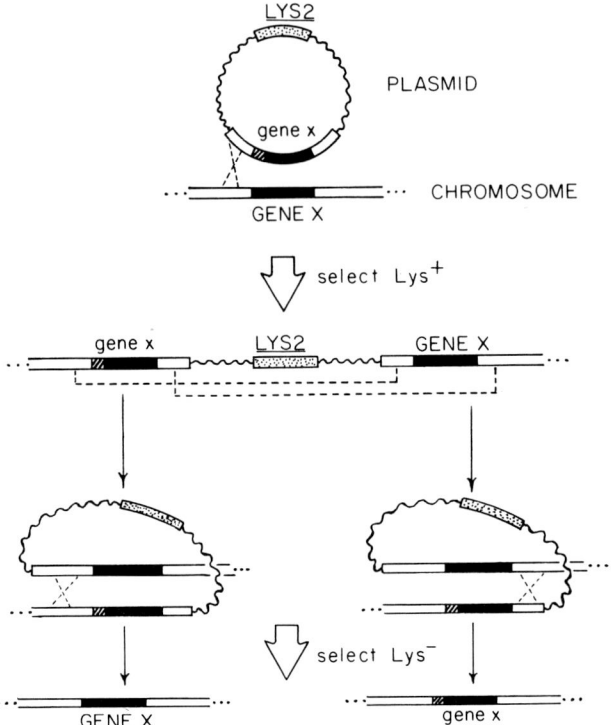

Fig. 4. Gene replacement by integration and excision of a plasmid carrying the *LYS2* gene. A vector carrying the *LYS2* gene, for example, YIp600 (see Fig. 3), that contains a cloned gene of interest that has been altered *in vitro* (gene x) will integrate by homologous recombination into the region of the yeast genome where the normal copy of this gene (GENE X) resides. The integration event can be directly selected by demanding the formation of lysine prototrophs from *lys2* recipients. Integration can be directed exclusively to the GENE X locus either by linearizing the plasmid by restriction enzyme cleavage within the gene x insert or by utilizing a recipient strain that carries a *lys2* deletion mutation. The tandem duplication created by integration is relatively unstable and the vector can be removed by homologous recombination between the gene x and GENE X regions. Depending upon which sequences of the gene x and GENE X segment participate in the recombination, either the former or the latter will be excised from the genome along with the vector. Eviction of the vector can be directly selected by demanding growth on α-aminoadipate medium, which is permissive only for cells that have lost the *LYS2* gene. If the length of the homologous DNA flanking GENE X on either side is sufficiently great, up to 50% of the *lys2* mutants selected from the integrant should contain the altered gene x.

ing mitosis and are lost at high frequency even under selective conditions, as often as once per cell per every two or three divisions (Murray and Szostak, 1983). Several of the Lys$^+$Trp$^+$ transformants were picked and grown in broth (YPD medium*) under nonselective conditions overnight. The overnight cultures were then plated on YPD plates. Colonies on the YPD plates were

*Yeast extract (1%), bactopeptone (2%), and dextrose (2%).

replica-plated to medium lacking either lysine or tryptophan. Even though Lys⁻ or Trp⁻ cells appeared at low frequency ($\sim 10^{-3}$), most cells retained a Lys⁺Trp⁺ phenotype, as expected for an integrative transformant. To purposefully evict the integrated vector, transformants were streaked onto α-aminoadipate medium. The *lys2* colonies appearing on these plates were then tested for their ability to grow on medium lacking tryptophan. About 1% of the *lys2* colonies had become Trp⁺, presumably due to loss of the mutant *trp1-289* allele upon excision of the vector sequences resulting from homologous recombination between the *TRP1* gene brought in by the plasmid and the formerly resident *trp1* locus on the genome. The fact that only 1% of the *lys2* cells were *TRP1*⁺, and not 50%, as would be expected if recombination could occur extensively on either side of the site of the *trp1-289* lesion, suggests that the *trp1-289* mutation is reasonably close to the 5' end of the coding region, which, in physical terms, must be near the *Eco*RI site at the right boundary of the cloned *TRP1* segment in YRp610 (Fig. 3).

C. One-Step Gene Disruption and Gene Replacement

If one's primary goal is to test the effect of eliminating gene function, a simple and direct gene transplacement method can be used which involves transformation by linear cloned fragments homologous to the yeast genome into which a selectable marker has been inserted (Rothstein, 1983). All that is required is detailed knowledge of the restriction enzyme cleavage sites within the cloned yeast segment or individual gene of interest. First, a DNA fragment containing a selectable yeast marker gene is inserted by ligation *in vitro* into a convenient restriction site in the gene of interest. As examples, the *URA3* gene is carried on a *Hin*dIII fragment and can be inserted into *Hin*dIII site(s); the *LEU2* gene is carried on a *Sal*I–*Xho*I segment and can be inserted into *Sal*I or *Xho*I sites; and the *HIS3* gene is carried on a *Bam*HI fragment and can be inserted into *Bam*HI, *Bgl*II, *Bcl*I, or *Sau*3A sites. Although such insertional mutations should, like transposon insertions, disrupt gene function, it is probably safer also to delete as much of the gene of interest as possible prior to insertion of the marker gene. Second, the gene of interest is excised from its vector by digestion with a restriction enzyme(s) that cuts on either side of, but not within, the selectable marker sequences that have been inserted. Third, as long as the sequences homologous to the yeast genome that remain on either side of the selectable marker are at least 100 bp or so, transformation of yeast with such a fragment replaces the cognate locus with the mutant insertion-containing copy. Formally, it could be considered that selection for the marker gene demands that the double-stranded ends of the fragment promote a highly efficient double-recombination event, like that which occurs for chromosomal fragments in bacterial transduction. More recent evidence suggests that the replacement process is more akin to a gene conversion event (Szostak *et al.*, 1983).

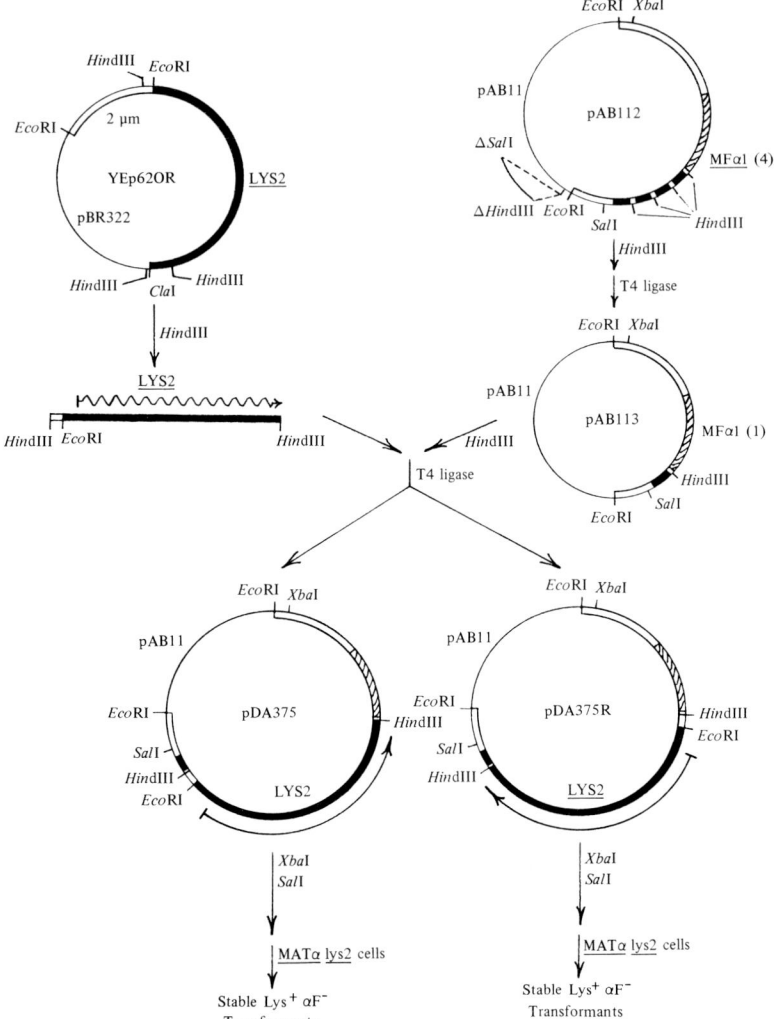

Fig. 5. One-step disruption of the α-factor structural gene (*MFα1*) using a *LYS2* fragment. A 1.75-kb genomic *Eco*RI fragment containing the *MFα1* gene was cloned into pAB11, a vector derived from pBR322 by removal of the *Hin*dIII and *Sal*I region, producing pAB112 (Brake *et al.*, 1983). The normal coding region contains four repeats of the amino acid sequence corresponding to the mature α-factor pheromone which are separated by *Hin*dIII sites (Kurjan and Herskowitz, 1982; Brake *et al.*, 1983; Singh *et al.*, 1983). A shorter *MFα1* gene containing just one repeat of pheromone coding sequence was prepared by digestion of pAB112 to completion with *Hin*dIII and religation forming pAB113. A segment of YEp620R containing the intact *LYS2* gene and only 100 bp of 2-μm circle DNA was prepared by digestion with *Hin*dIII and ligated to pAB112, which had been cleaved with the same enzyme to disrupt the *MFα1* gene. Because the sites are identical, derivatives

7. Use of *LYS2* for Genetic Manipulation of Yeast

The first advantage of using the *LYS2* gene as the selectable marker for one-step gene disruptions is that any strain could be made a recipient simply by plating cells on α-aminoadipate medium to obtain *lys2* mutants. Such *lys2* strains then could be transformed with linear fragments in which the gene of interest has been disrupted by insertion of the LYS2 gene by selecting on plates lacking lysine.

To test the utility of the *LYS2* gene for one-step gene inactivation, we inserted the *LYS2* gene into the major structural gene for the mating pheromone α-factor, *MFα1*, and then replaced the normal genomic copy of this gene (Figs. 5 and 6). The coding region for the α-factor precursor contains four *Hin*dIII sites. A functional *LYS2* gene can be excised intact from certain vectors by cleavage with *Hin*dIII. Hence, it was possible to insert the *LYS2* gene into *MFα1* to disrupt the coding region.

Two plasmids were isolated: pDA375, in which the direction of *LYS2* transcription goes against the direction of *MFα1* transcription (so-called ANTI orientation), and pDA375R, in which *LYS2* transcription is in the same direction as that of *MFα1* (so-called SYN orientation) (Fig. 6). Both pDA375 and pDA375R were cut to completion with *Xba*I and *Sal*I, which generated fragments containing homology to the *MFα1* gene flanking the *LYS2* sequence at both ends. The fragments were used to transform strain DA2102 (*MATα ura3-52 leu2-3,112 his4 lys2*) to Lys$^+$. These Lys$^+$ transformants were tested for their ability to make α-factor by use of a bioassay. The assay consists of replica-plating patches of cells to be tested onto an indicator lawn of *MAT*a cells supersensitive to the effects of the α-factor pheromone (Julius *et al.*, 1983). If the patches are secreting α-factor, then the lawn of *MAT*a haploids surrounding the patch are arrested in the G$_1$ phase of their cell cycle, and hence do not grow. The result is a clear zone or "halo" around the α-factor-secreting patches. As expected, stable Lys$^+$ transformants that were α-factor negative were readily obtained. Two of these transformants (DA2102::375 and DA2102::375-R) representing the two different orientations of *LYS2* within the *MFα1* gene were selected for further study.

To show conclusively that the fragments had integrated at the *MFα1* locus and had replaced the normal chromosomal *MFα1* gene, we isolated total yeast DNA from the transformants and from the untransformed parent DA2102. The DNA was digested with *Eco*RI, subjected to electrophoresis in an agarose gel, and then blotted to nitrocellulose filter paper. The filter was hybridized with a radioactive

---

in which the *LYS2* gene was inserted with the same (SYN) direction of transcription as the *MFα1* gene and with the opposite (ANTI) direction of transcription were recovered, namely, pDA375R and pDA375, respectively. Because the *LYS2* segment inserted lacks *Xba*I and *Sal*I sites, the insertion-containing *MFα1* segments were excised from pDA375 and pDA375R by digestion to completion with these two restriction enzymes and were used to transform *MAT*α *lys2* recipients to lysine prototrophy.

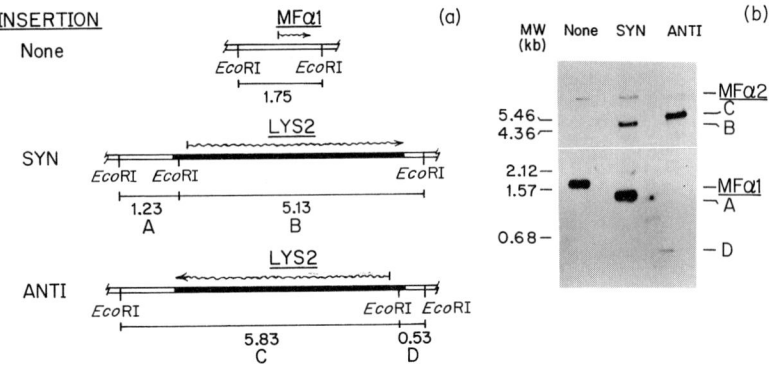

Fig. 6. Hybridization analysis of one-step disruption of the *MFα1* locus by the *LYS2* gene. (a) Predicted *Eco*RI fragment lengths of the normal *MFα1* gene and of the two *LYS2* insertion mutations constructed and integrated as described in the legend of Fig. 5. (b) Genomic DNAs were isolated from the untransformed recipient (strain DA2102) and from the transformants containing the *LYS2* gene in either the SYN (strain DA2102::375R) or the ANTI (strain DA2102::375) orientation. The DNA preparations were digested to completion with *Eco*RI, subjected to electrophoresis in a 0.8% agarose gel, and then transferred to nitrocellulose filter paper by the procedure of Southern (1975). The 1.75-kb *Eco*RI fragment of pAB112 (Fig. 5) was purified, radioactively labeled by nick-translation (Rigby *et al.*, 1977), and then hybridized to the filter. The molecular weight standards used were a mixture of two separate digests (*Rsa*I and *Eco*RI) of pBR322 and an *Eco*RI digest of YIp5. The faint high-molecular-weight band present in all three lanes is the minor α-factor structural gene (*MFα2*), which shares some limited homology with *MFα1*. It should be noted that the band corresponding to the normal *MFα1* region in the untransformed recipient is missing in both the SYN and ANTI transformants and has been replaced by two new fragments (A and B or C and D, respectively) in the transformants.

probe containing *MFα1* sequences. In each case, fragments of the expected size (Fig. 6a) were found to be present in the yeast genome (Fig. 6b), confirming that the chromosomal *MFα1* gene had been replaced by the versions containing *LYS2* insertions.

It has been shown that expression of the *MFα1* gene is tightly regulated at the transcriptional level, depending on the cell type (Brake *et al.*, 1983; Emr *et al.*, 1983). The *MFα1* mRNA is made in *MAT*α haploids, but not in *MAT*a haploids or in *MAT*a/*MAT*α diploids. The *LYS2* gene inserted into *MFα1* carried its own promoter and was expressed in *MAT*α cells regardless of its orientation since Lys$^+$ transformants were recovered at the same frequency for both insertions. It was of interest to know, therefore, if the control of *MFα1* expression might override the *LYS2* gene promoter and prevent *LYS2* gene expression from the insertions in *MAT*a haploids or in *MAT*a/*MAT*α diploids. To examine this question, the integrative transformants were mated with strain DA320 (*MAT*a *lys2 ade2 ura3-11 leu1 rme1 can1-11 cyh2-2*). Both the diploid cells and *MAT*a spores carrying the *mfα1* :: *LYS2* insertion derived by sporulation and tetrad

7. Use of *LYS2* for Genetic Manipulation of Yeast

dissection of the diploid were phenotypically Lys⁺ and were able to grow just as well on solid medium lacking lysine as the *MAT*α integrants. Clearly, the lack of expression of the *MF*α*1* promoter in *MAT***a** and *MAT***a**/*MAT*α cells is not due to long-range effects in chromatin structure that completely inactivate any other promoter placed in this region. However, because we do not yet know what level of *LYS2* gene expression is sufficient to confer a Lys⁺ phenotype *in vivo*, it is still possible that the level of expression of the *LYS2* gene inserted at *MF*α*1* is reduced significantly in *MAT***a** and *MAT***a**/*MAT*α cells.

A second advantage of the *LYS2* gene for gene disruptions is that once the *LYS2* gene has been inserted into the yeast genome in place of the normal chromosomal copy of a gene of interest, one has a convenient recipient for subsequent one-step gene replacements. For example, we constructed a fusion of the *MF*α*1* promoter to a convenient indicator enzyme, *SUC2*, to permit ready monitoring of the level of gene expression (Emr *et al.*, 1983). A similar fusion was constructed in such a way as to also retain homology to the 3'-flanking region of the *MF*α*1* gene (M. C. Flessel and J. Thorner, unpublished results). By transforming a *MAT*α *mf*α*1* : : *LYS2* strain with a linear fragment containing only the *MF*α*1*–*SUC2* fusion gene, integrants in which the fusion replaced the *LYS2* insertion were readily obtained by plating on α-aminoadipate medium because the selection demanded the loss of the *LYS2* sequences. Alterations of the *MF*α*1* promoter (deletions, point mutations, linker insertions) have been constructed *in vitro* in the *MF*α*1*–*SUC2* fusion gene (M. C. Flessel and J. Thorner, unpublished observations). In a similar way, each of these alterations of the *MF*α*1* promoter can be placed in single copy at the normal *MF*α*1* locus by using the appropriate fragments of the *MF*α*1*–*SUC2* fusion to transform the *mf*α*1* : : *LYS2* insertion strain and selecting for *lys2* mutants on α-aminoadipate medium. The effects of each of the alterations on *MF*α*1* gene expression can then be quantitatively assessed.

Several other strategies for creating recipient strains that provide positive selections for one-step gene replacement have been developed recently. One system utilizes the cloned *ochre*-suppressing mutant tyrosine-inserting tRNA genes, *SUP4* or *SUP11* (Guthrie and Abelson, 1982), and recipient strains that carry *ochre*-suppressible markers both in a convenient biosynthetic pathway (e.g., *trp5-2°*) and in the arginine permease gene (e.g., *can1-100°*). If a cloned *SUP* gene is inserted *in vitro* into a gene of interest, replacements of the genomic copy of the gene can be directly selected in *trp5* hosts as Trp⁺ cells on medium lacking tryptophan, since presence of the suppressor tRNA will restore tryptophan prototrophy. Conversely, a *SUP*-containing insertional mutation can be directly replaced subsequently by any other version of the gene of interest by counterselecting for canavanine-resistant cells, because loss of the suppressor tRNA will result in a nonfunctional permease and hence an inability to take up canavanine (a toxic arginine analog) (K. Nasmyth, personal communication).

Alternatively, even in the absence of a $can1^{ochre}$ mutation in the background, cells that have lost the suppressor can be selected on medium of high osmolarity (St. John et al., 1981) because, for as yet unknown reasons, the presence of efficient *ochre* suppressors often makes cells sensitive to growth inhibition by a medium of high osmotic strength (Sherman, 1982).

Another system for one-step gene replacement utilizes the cloned *CYH2* gene (Kaufer et al., 1983), which encodes a ribosomal protein. In $CYH2^+$ strains, yeast protein synthesis is blocked by the drug cycloheximide. The *CYH2* locus can mutate, however, to confer resistance to this translation inhibitor. The cloned *CYH2* gene can be inserted *in vitro* into a gene of interest and replacements of the genomic copy of the gene isolated. If the original recipient strain carried a $cyh2^R$ allele on chromosome VII, then such *CYH2*-containing insertional mutations can be directly replaced subsequently by any other version of the gene of interest by selecting for cycloheximide-resistant cells, because cells that have lost the inserted $CYH2^+$ (cycloheximide-sensitive) gene will express only the $cyh2^R$ chromosomal allele and hence will be drug-resistant. This strategy has been used to insert altered sequences into the genome to study various aspects of the promoters and terminators of the *HIS3* (Struhl, 1983) and *CYC1* genes (F. Sherman, personal communication).

A third promising system for one-step gene replacement utilizes the cloned *CAN1* (arginine permease) gene (Broach et al., 1979). Yeast cells that are arg^-can1^- are nonviable on minimal medium containing NH_4^+ as the nitrogen source because ammonia represses the general amino acid permease (Jones and Fink, 1982) and, in the absence of a functional arginine-specific permease, sufficient exogenous arginine cannot enter the cell to spare the auxotrophy. On rich medium (lacking NH_4^+), however, the general amino acid permease is derepressed, and arg^-can1^- cells can grow, albeit poorly. If the cloned *CAN1* gene is inserted into a gene of interest, then replacements of the chromosomal copy by the *CAN1*-containing segment can be directly selected on minimal medium containing arginine and NH_4^+. Such *CAN1*-containing insertions are useful recipients for one-step gene replacements because eviction of the *CAN1*-containing copy by another incoming DNA segment can be directly selected as canavanine-resistant cells on rich medium containing this analog (J. Shuster, personal communication).

D. Construction of Promoter Fusions in Yeast

Several genes are potentially useful for promoter isolation and analysis in yeast. The features that make a gene attractive for this application are (1) the ability to separate facilely the coding region from its normal promoter and (2) some readily selectable phenotype or assay for successful expression of the gene. These ''headless'' genes can be either bacterial or yeast in origin. Among the convenient bacterial genes that can function in yeast are *cat,* which encodes the chloramphenicol acetyltransferase gene of Tn9, and, of course, the *lacZ* gene of *E. coli,* which encodes β-galactosidase. To date, there have not been many yeast genes that have been used in promoter fusions. However, one gene that has shown some promise recently is the *SUC2* gene encoding invertase (Emr *et al.,* 1983).

Use of the *cat* gene for constructing promoter fusions in yeast is based on the observation by Cohen *et al.* (1980) that yeast cells are sensitive to chloramphenicol at a concentration of 500 μg/ml, but only when grown on a nonfermentable carbon source such as glycerol or ethanol. Chloramphenicol binds to and inhibits the 70 S ribosome of mitochondria; therefore, to test for sensitivity to the drug, cell growth must be strictly dependent on respiratory metabolism. The *cat* gene is extremely easy to use and is now even sold commercially as a promoterless gene on convenient restriction fragments or as the "cartridges" originally constructed by Close and Rodriguez (1982). These cartridges contain the complete CAT polypeptide coding sequence, with an upstream *E. coli* ribosome binding site but no promoter sequence. The CAT cartridges have been fused to yeast promoters and have been found to function in yeast (R. Rodriguez, personal communication). Rodriguez and collaborators have also fused such a CAT cartridge to three different DNA fragments containing the *CYC1* promoter and the initiating ATG of *CYC1* followed by 1, 2, or 3 base pairs. Although the three protein fusions should have been in three different reading frames, only one of which should have been able to produce a functional CAT protein, yeast cells transformed with vectors containing all three fusions were resistant to chloramphenicol. This result suggests that there is some secondary translational initiation site which is efficiently recognized in yeast as long as mRNA is produced (R. Rodriguez, personal communication). Hence, CAT protein is probably a poor indicator enzyme for detecting the construction of in-frame fusion proteins.

Modified versions of the G418-resistance gene of Tn601 appear more useful than CAT for identifying functional promoters and for constructing in-frame fusions (D. Gelfand, personal communication).

One of the most popular genes for creating, simultaneously, promoter and protein fusions is *lacZ*. Extremely well characterized, *lacZ* is expressed and relatively stable in yeast, and its product (β-galactosidase) is readily detected in yeast colonies on indicator plates and/or easily quantitated in extracts by direct enzyme assay. It has been determined that the first 27 amino acids of the amino terminus of the protein are dispensable (Brickman *et al.,* 1979). Therefore, in-frame fusions may be made anywhere within this region. Plasmids are available that contain poly-linkers in this N-terminal region of *lacZ*. Thus, yeast promoter fragments containing a portion of the coding region of a yeast gene can be fused to *lacZ in vitro. Saccharomyces cerevisiae* is naturally Lac$^-$. To our knowledge, expression of *lacZ* from yeast promoters is not sufficient to confer on yeast the ability to grow on lactose as a carbon source. Therefore, it is necessary to screen colonies for those which contain a functional fusion. This is not inconvenient, however, because yeast cells that contain an in-frame fusion have the ability to hydrolyze the indicator dye X-gal and will turn plates containing this compound blue (Rose et al., 1981; Guarente and Ptashne, 1981). Unlike the CAT cartridges, the yeast translational machinery seems to recognize the initial ATG of the yeast segment of protein fusions and not internal ATGs within the lacZ portion of such hybrid genes.

The yeast SUC2 gene has been used successfully in a fusion to the $MF\alpha1$ promoter and prepro segment of the α-factor precursor (Emr et al., 1983). The SUC2 gene provides an advantage for the study of secreted gene products. Because invertase can cross yeast membranes, fusions of SUC2 to a signal sequence allow secretion of invertase into the periplasmic space. If invertase activity has been secreted, one can screen colonies for the presence of successful SUC2 fusions on fermentation indicator plates (Schaefler, 1967). In this assay, patches of cells that produce invertase split sucrose, grow on the glucose and fructose so released, and thus generate acids that lower the pH of the medium. The pH change can be monitored with indicator dyes (e.g., bromcresol purple turns from royal purple to canary yellow). Even intracellular invertase production can be detected in the same way if the colonies are first permeabilized by exposure to chloroform vapor. Alternatively, successful fusions to SUC2 can be selected directly by growth on sucrose as sole carbon source (S. Emr, personal communication).

Our preliminary results suggest that the LYS2 gene may be another good candidate for promoter manipulation in yeast. First, it may be possible to construct promoter indicator plasmids using the LYS2 gene. Such a vector would carry the LYS2 gene lacking its own promoter sequences, but bearing convenient restriction enzyme sites immediately 5′ to the LYS2 coding region. Insertion of yeast DNA segments that contain a functional promoter could be directly selected by transformation of lys2 recipients. Conversely, if one were interested in determining the importance of certain nucleotides in a known promoter, one could select alterations of the promoter by fusing the promoter to LYS2 in vitro, introducing the fusion into lys2 hosts, and then selecting for loss of LYS2 expression by plating on α-aminoadipate medium. Since the LYS2 gene is probably a bigger target for mutagenesis than the promoter, most $lys2^-$ mutants obtained in this way would probably be in the LYS2 gene itself rather than in the promoter of interest. To avoid this problem, one could use targeted mutagenesis in vitro of the promoter sequence of such a lys2 fusion. Alterations of the promoter which destroyed LYS2 expression could then be directly isolated out of the pool of mutagenized plasmids by transformation of lys2 cells on medium containing α-aminoadipate, provided the vector contained another marker (e.g., URA3) to select for successful transformants.

7. Use of *LYS2* for Genetic Manipulation of Yeast

Second, marker rescue experiments using several subcloned segments of the *LYS2* gene indicate that every *lys2* allele tested to date lies in the promoter-distal portion of the gene. This finding suggests that, like β-galactosidase, the N-terminal region of the *LYS2* protein may not be essential for its enzymic function. Hence, it is possible that the *LYS2* gene may be useful for creating protein fusions as well as promoter fusions. Again, successful in-frame fusions of another yeast gene to *LYS2* lacking its promoter and initiator methionine codon could be directly selected by transforming a *lys2* recipient to lysine prototrophy on medium lacking lysine.

Finally, if the *LYS2* gene has been inserted downstream from a yeast gene of interest in a *lys2* host, it also should be possible to select for spontaneous deletions that remove the intervening information and fuse the promoter and/or coding region to the *LYS2* gene. The success of this technique would depend on several factors. First, the directions of transcription of the gene in question and the inserted *LYS2* gene must be the same. Second, the region to be deleted cannot be essential for cell viability. Third, the upstream promoter must be under some regulatory control, such that it is turned off under certain conditions. For example, the *MFα1* promoter is "off" in **a** and **a**/α diploids but is "on" in α cells. The *LYS2* promoter inserted into *MFα1* is still functional in **a**/α diploid, cells, however, as discussed in a previous section. Therefore, it should be possible to select for spontaneous deletions between the *MFα1* promoter region and the *LYS2* gene inserted in the SYN orientation (see Figs. 5 and 6) because in-frame fusions should prevent *LYS2* expression. To accomplish this, *MAT*a *CRY1*s/*MAT*α *cry1*R diploid cells carrying the *mfα1* : : *LYS2* insertion would be plated on α-aminoadipate-containing medium to select for *lys2* cells. Of course, *lys2* mutations could also arise by lesions in the *LYS2* sequence itself. However, diploid cells containing a successful *MFα1*–*LYS2* fusion should show a Lys+ phenotype, but only when the fusion is segregated into an α haploid background. Hence, the *lys2*⁻ diploids obtained on α-aminoadipate medium would be patched onto sporulation plates to re-form haploid spores. The sporulated patches would then be replica-plated to minimal medium lacking lysine but containing cryptopleurine to select for the α haploids. Patches that give Lys+ papillae presumably contain a fusion between the *MFα1* promoter region and the *LYS2* gene, which could be analyzed by Southern blotting analysis to confirm that a deletion had, in fact, occurred.

E. Other Applications

Because of the efficient selection for mutations at the *LYS2* locus, it is extremely easy to select a large number of spontaneous alterations of the gene and therefore possible to examine what might otherwise be rare events. Eibel and Philippsen (1984) took advantage of this fact to isolate a large number of *lys2* mutants. They found, as expected from previous work, that a significant fraction (2%) of spontaneously arising *lys2* mutations were due to insertions of the yeast transposable element, Ty1. Interestingly, most of the insertions that eliminated *LYS2* function were found preferentially in the transcription initiation region of the *LYS2* gene and not in the coding region itself. Excisions of those Ty1 elements that restore *LYS2* expression could be selected on medium lacking lysine and then characterized.

Because the protein encoded by the *LYS2* gene appears to be a multifunctional enzyme, as discussed in an earlier section, this protein is interesting in its own right. A large number of *lys2* alleles have been isolated and genetically well characterized (F. Sherman, personal communication). It should now be possible to correlate the genetic and biochemical properties of such *lys2* lesions with their exact location in the polypeptide by using the cloned *LYS2* gene to recover and characterize these mutations. The mutated sites can first be localized by marker rescue experiments with cloned subfragments of the *LYS2* gene and then recovered by transformation of the mutants with an appropriate episomal vector linearized by creation of a gap in the region of interest.

One very clever use of the powerful α-aminoadipate selection against a functional *LYS2* gene is being applied by J. O'Rear and J. Rine (personal communication). They have constructed a linear yeast plasmid containing the *URA3* gene, a centromere, and telomere ends like those first described by Szostak and Blackburn (1982). They have sandwiched the *LYS2* gene between one telomere end and a known telomere-linked DNA segment from *Drosophila*, which they have also cloned into the vector. If yeast spheroplasts carrying the linear plasmid are transformed with random fragments of total *Drosophila* DNA, recombination between *Drosophila* DNA and the linear plasmid can be demanded by selection for loss of the *LYS2* gene and for maintenance of the *URA3* marker. In this way, they hope to replace the yeast telomere end with an authentic *Drosophila* telomere. If successful, this will permit the detailed analysis of a chromosomal telomere from a complex eukaryote.

Finally, even though *Saccharomyces* strains of industrial importance may be polyploid for chromosome II, it is possible that the α-aminoadipate selection is sufficiently powerful to obtain *lys2* mutants of these cells. If so, our *LYS2* vectors could become potentially useful tools for transformation of yeast strains of commercial value.

IV. CONCLUSIONS AND PROSPECTUS

The *LYS2* gene appears to be a very useful tool for genetic manipulations of yeast cells. Libraries of yeast DNA in both the YEp620 and YCp630 vectors have been prepared and are available to other investigators upon request. One disadvantage of the *LYS2* gene for certain applications is its length and the concomitant number of sites for commonly used restriction enzymes within the gene. To increase the utility of the *LYS2* gene, we are currently attempting to determine which of these sites can be eliminated or modified without deleteriously perturbing *LYS2* function. Our descriptions of how the *LYS2* gene can be used in applying recombinant DNA methodology to the genetic analysis of yeast have provided a showcase for many techniques that were developed by others. We are certain that as other strategies are devised, the *LYS2* gene will prove equally useful. Just as important, we hope that the availability of the *LYS2* DNA, the information we have obtained about it, and the vehicles we have constructed containing it will permit the evolution of novel techniques that would not otherwise have been possible.

REFERENCES

Barnes, D. A., and Thorner, J. (1985). Convenient vectors for genetic manipulation of *Saccharomyces cerevisiae* utilizing the *LYS2* gene. *Mol. Cell. Biol.* Submitted for publication.

Beggs, J. D. (1978). Transformation of yeast by a replicating hybrid plasmid. *Nature (London)* **275**, 104–109.

Bhattacharjee, J. K. (1983). Lysine biosynthesis in eukaryotes. In "Amino Acids: Biosynthesis and Genetic Regulation" (K. M. Herrmann and R. L. Somerville, eds.), pp. 229–244. Addison-Wesley, Reading, Massachusetts.

Boeke, J. D., LaCroute, F., and Fink, G. R. (1984). A positive selection for mutants lacking orotidine-5′-phosphate decarboxylase activity in yeast: 5-fluoroorotic acid-resistance. *Molec. Gen. Genet.* **197**, 345–346.

Borell, C. W., Urrestarazu, L. A., and Bhattacharjee, J. K. (1984). Two unlinked lysine genes (*LYS9* and *LYS14*) are required for the synthesis of saccharopine reductase in *Saccharomyces cerevisiae*. *J. Bacteriol.* **159**, 429–432.

Brake, A. J., Julius, D. J., and Thorner, J. (1983). A functional prepro-α-factor gene in *Saccharomyces* yeasts can contain three, four, or five repeats of the mature pheromone sequence. *Mol. Cell Biol.* **3**, 1440–1450.

Brickman, E., Silhavy, T. J., Shuman, H. A., and Beckwith, J. R. (1979). Sites within gene *lacZ* of *E. coli* for formation of active hybrid β-galactosidase molecules. *J. Bacteriol.* **139**, 13–18.

Broach, J. R. (1981). The yeast plasmid 2μ circle. In "Molecular Biology of the Yeast *Saccharomyces*: Life Cycle and Inheritance" (J. N. Strathern, E. W. Jones, and J. R. Broach, eds.), pp. 445–470. Cold Spring Harbor Lab. Press, Cold Spring Harbor, New York.

Broach, J. R., Strathern, J. N., and Hicks, J. B. (1979). Transformation in yeast: development of a hybrid cloning vector and isolation of the *CAN1* gene. *Gene* **3**, 121–133.

Chattoo, B. B., Sherman, F., Fjellstedt, T. A., Mehnert, D., and Ogur, M. (1979). Selection of *lys2* mutants of the yeast *Saccharomyces cerevisiae* by the utilization of α-aminoadipate. *Genetics* **93**, 51–65.

Close, T. J., and Rodriguez, R. (1982). Construction and characterization of the chloramphenicol-resistance gene cartridge: a new approach to the transcriptional mapping of extrachromosomal elements. *Gene* **20**, 305–316.

Cohen, J. D., Eccleshall, T. R., Needleman, R. B., Federoff, N., Buchferer, B. A., and Marmur, J. (1980). Functional expression in yeast of the *E. coli* plasmid gene coding for chloramphenicol acetyltransferase. *Proc. Natl. Acad. Sci. U.S.A.* **77**, 1078–1082.

Cooper, T. G. (1982). Nitrogen metabolism in *Saccharomyces cerevisiae*. *In* "Molecular Biology of the Yeast *Saccharomyces:* Metabolism and Gene Expression" (J. N. Strathern, E. W. Jones, and J. Broach, eds.), pp. 39–99. Cold Spring Harbor Lab. Press, Cold Spring Harbor, New York.

Davies, J., and Smith, D. I. (1978). Plasmid-determined resistance to antimicrobial agents. *Annu. Rev. Microbiol.* **32**, 469–518.

Donahue, T. F., Daves, R. S., Lucchini, G., and Fink, G. R. (1983). A short nucleotide sequence required for regulation of *HIS4* by the general control system of yeast. *Cell (Cambridge, Mass.)* **32**, 89–98.

Eibel, H., and Philippsen, P. (1983). Identification of the cloned *Saccharomyces cerevisiae LYS2* gene by an integrative transformation approach. *Mol. Gen. Genet.* **191**, 66–73.

Eibel, H., and Philippsen, P. (1984). Preferential integration of yeast transposable element Ty into promoter region. *Nature (London)* **307**, 386–388.

Emr, S. D., Schekman, R., Flessel, M. C., and Thorner, J. (1983). An *MFα1–SUC2* (α-factor-invertase) gene fusion for study of protein localization and gene expression in yeast. *Proc. Natl. Acad. Sci. U.S.A.* **80**, 7080–7084.

Falco, S. C., and Botstein, D. (1983). A rapid chromosome mapping method for cloned fragments of yeast DNA. *Genetics* **105**, 857–872.

Falco, S. C., Li, Y., Broach, J. R., and Botstein, D. (1982). Genetic properties of chromosomally integrated 2μ plasmid DNA in yeast. *Cell (Cambridge, Mass.)* **29**, 573–584.

Fasiolo, F., Bonnet, J., and Lacrout, F. (1981). Cloning of the yeast methionyl-tRNA synthetase gene. *J. Biol. Chem.* **256**, 2324–2328.

Fogel, S., and Welch, J. W. (1982). Tandem gene amplification mediates copper resistance in yeast. *Proc. Natl. Acad. Sci. U.S.A.* **79**, 5342–5346.

Guarente, L., and Ptashne, M. (1981). Fusion of *Escherichia coli lacZ* to the cytochrome *c* gene of *Saccharomyces cerevisiae*. *Proc. Natl. Acad. Sci. U.S.A.* **78**, 2199–2203.

Guthrie, C., and Abelson, J. (1982). Organization and expression of tRNA genes in *Saccharomyces cerevisiae*. *In* "Molecular Biology of the Yeast *Saccharomyces:* Metabolism and Gene Expression" (J. N. Strathern, E. W. Jones, and J. Broach, eds.), pp. 487–528. Cold Spring Harbor Lab. Press, Cold Spring Harbor, New York.

Hartley, J. L., and Donelson, J. E. (1980). Nucleotide sequence of the yeast plasmid. *Nature (London)* **286**, 860–864.

Heusterspreute, M., Thi, V. H., and Davison, J. (1984). Expression of the *E. coli* galactokinase gene as a fusion protein in *E. coli* and *Saccharomyces cerevisiae*. *DNA* **3**, 377–386.

Hicks, J. B., Hinnen, A., and Fink, G. R. (1979). Properties of yeast transformation. *Cold Spring Harbor Symp. Quant. Biol.* **43**, 1305–1313.

Hinnebusch, A. G., and Fink, G. R. (1983). Repeated DNA sequences upstream from *HIS1* also occur at several other co-regulated genes in *Saccharomyces cerevisiae*. *J. Biol. Chem.* **258**, 5238–5247.

Hinnen, A., Hicks, J., and Fink, G. R. (1978). Transformation in yeast. *Proc. Natl. Acad. Sci. U.S.A.* **75**, 1929–1933.

Hollenberg, C. P. (1982). Cloning with 2μ DNA vectors and the expression of foreign genes in *S. cerevisiae*. *Curr. Top. Microbiol. Immunol.* **96**, 119–144.

Jimenez, A., and Davies, J. (1980). Expression of a transposable antibiotic resistance element in *Saccharomyces cerevisiae;* a potential selection for eukaryotic cloning vectors. *Nature (London)* **287**, 869–871.

Jones, E. W., and Fink, G. R. (1982). Regulation of amino acid and nucleotide biosynthesis in yeast. *In* "Molecular Biology of the Yeast *Saccharomyces cerevisiae:* Metabolism and Gene Expression" (J. N. Strathern, E. W. Jones, and J. Broach, eds.), pp. 81–299. Cold Spring Harbor Lab. Press, Cold Spring Harbor, New York.

7. Use of *LYS2* for Genetic Manipulation of Yeast

Julius, D. J., Blair, L., Brake, A., Sprague, G., and Thorner, J. (1983). Yeast α-factor is processed from a larger precursor polypeptide: the essential role of a membrane-bound dipeptidyl aminopeptidase. *Cell (Cambridge, Mass.)* **32**, 839–852.

Karin, M., Najarian, R., Haslinger, A., Valenzuela, P., Welch, J., and Fogel, S. (1984). Primary structure and transcription of an amplified genetic locus: the *CUP1* locus of yeast. *Proc. Natl. Acad. Sci. U.S.A.* **81**, 337–341.

Kaufer, N. F., Fried, H. M., Schwindinger, W. F., Jasin, M., and Warner, J. R. (1983). Cycloheximide resistance in yeast: the gene and its protein. *Nucleic Acids Res.* **11**, 3123–3135.

Klapholz, S., and Esposito, R. (1982). A new mapping method employing a meiotic Rec⁻ mutant of yeast. *Genetics* **100**, 387–412.

Kurjan, J., and Herskowitz, I. (1982). Structure of a yeast pheromone gene (*MFα1*): a putative α-factor precursor contains four tandem repeats of mature α-factor. *Cell (Cambridge, Mass.)* **30**, 933–943.

McAlister, L., and Holland, M. (1982). Targeted deletion of a yeast enolase structural gene: Identification and isolation of yeast enolase isozymes. *J. Biol. Chem.* **257**, 7181–7188.

Maniatis, T., Fritsch, E. F., and Sambrook, J. (1982). "Molecular Cloning." Cold Spring Harbor Lab. Press, Cold Spring Harbor, New York.

Mattoon, J. R., Moshier, T. A., and Kreiser, T. H. (1961). Separation and partial characterization of a lysine precursor produced by a yeast mutant. *Biochim. Biophys. Acta* **51**, 615–621.

Mortimer, R. K., and Schild, D. (1981). Genetic map of *Saccharomyces cerevisiae*. In "Molecular Biology of the Yeast *Saccharomyces*: Life Cycle and Inheritance" (J. N. Strathern, E. W. Jones, and J. Broach, eds.), pp. 641–727. Cold Spring Harbor Lab. Press, Cold Spring Harbor, New York.

Murray, A. W., and Szostak, J. W. (1983). Pedigree analysis of plasmid segregation in yeast. *Cell (Cambridge, Mass.)* **34**, 961–970.

Orr-Weaver, T. L., Szostak, J. W., and Rothstein, R. J. (1981). Yeast transformation: a model system for the study of recombination. *Proc. Natl. Acad. Sci. U.S.A.* **78**, 6354–6358.

Orr-Weaver, T. L., Szostak, J. W., and Rothstein, R. J. (1983). Genetic applications of yeast transformation with linear and gapped plasmids. In "Recombinant DNA," Part C (R. Wu, L. Grossman, and K. Moldave, eds.), Methods in Enzymology, Vol. 101, pp. 228–245. Academic Press, New York.

Platt, T. (1984). Toxicity of 2-deoxygalactose to *Saccharomyces cerevisiae* cells constitutively synthesizing galactose-metabolizing enzymes. *Mol. Cell. Biol.* **4**, 994–996.

Rigby, P. W. J., Dieckmann, M., Rhodes, C., and Berg, P. (1977). Labeling deoxyribonucleic acid to high specific activity *in vitro* by nick translation with DNA polymerase I. *J. Mol. Biol.* **113**, 237–257.

Rose, M., Casadaban, M. J., and Botstein, D. (1981). Yeast genes fused to β-galactosidase in *Escherichia coli* can be expressed normally in yeast. *Proc. Natl. Acad. Sci. U.S.A.* **78**, 2460–2464.

Rose, M., Grisafi, P., and Botstein, D. (1984). Structure and function of the yeast *URA3* gene: Expression in *E. coli*. *Gene* **29**, 113–124.

Rothstein, R. (1983). One-step gene disruption in yeast. In "Recombinant DNA," Part C (R. Wu, L. Grossman, and K. Moldave, eds.), Methods in Enzymology, Vol. 101, pp. 202–211. Academic Press, New York.

Rymond, B. C., Zitomer, R. S., Schumperli, D., and Rosenberg, M. J. (1983). The expression in yeast of the *Escherichia coli* galactokinase gene on *CYC1 : galK* fusion plasmids. *Gene* **25**, 249–262.

St. John, T. P., Scherer, S., McDonald, M. W., and Davis, R. W. (1981). Deletion analysis of the *Saccharomyces cerevisiae GAL* gene cluster: Transcription from three promoters. *J. Mol. Biol.* **152**, 317–334.

Schaefler, S. (1967). Inducible system for the utilization of β-galactosides in *E. coli. J. Bacteriol.* **93,** 254–263.

Scherer, S., and Davis, R. W. (1979). Replacement of chromosome segments with altered DNA sequences constructed *in vitro. Proc. Natl. Acad. Sci. U.S.A.* **76,** 4951–4955.

Sherman, F. (1982). Suppression in the yeast *Saccharomyces cerevisiae. In* "Molecular Biology of the Yeast *Saccharomyces:* Metabolism and Gene Expression" (J. N. Strathern, E. W. Jones, and J. Broach, eds.), pp. 463–486. Cold Spring Harbor Lab. Press, Cold Spring Harbor, New York.

Shortle, D., Haber, J. E., and Botstein, D. (1982). Lethal disruption of the yeast actin gene by integrative DNA transformation. *Science (Washington, D.C.)* **217,** 371–373.

Singh, A., Chen, E. Y., Lugovoy, J., Chang, C. N., Hintzman, R. A., and Seeburg, P. H. (1983). *Saccharomyces cerevisiae* contains two discrete genes coding for the α-factor pheromone. *Nucleic Acids Res.* **11,** 4049–4063.

Sinha, A. K., and Bhattacharjee, J. K. (1971). Lysine biosynthesis in *Saccharomyces. Biochem. J.* **125,** 743–749.

Southern, E. (1975). Detection of specific sequences among DNA fragments separated by gel electrophoresis. *J. Mol. Biol.* **98,** 503–517.

Struhl, K. (1983). Direct selection for gene replacement events in yeast. *Gene* **26,** 231–242.

Struhl, K., Stinchcomb, D. T., Scherer, S., and Davis, R. W. (1979). High frequency transformation of yeast: autonomous replication of hybrid DNA molecules. *Proc. Natl. Acad. Sci. U.S.A.* **76,** 1035–1039.

Sutcliffe, J. G. (1978). pBR322 restriction map derived from the DNA sequence: accurate DNA size markers up to 4361 nucleotide pairs long. *Nucleic Acids Res.* **5,** 2721–2728.

Szostak, J. W., and Blackburn, E. H. (1982). Cloning yeast telomeres on linear plasmid vectors. *Cell (Cambridge, Mass.)* **29,** 245–255.

Szostak, J., Orr-Weaver, T. W., Rothstein, R., and Stahl, F. (1983). The double-stranded break repair model for recombination repair. *Cell (Cambridge, Mass.)* **33,** 25–35.

Webster, T. D., and Dickson, R. C. (1983). Direct selection of *Saccharomyces cerevisiae* resistant to the antibiotic G418 following transformation with a DNA vector carrying the kanamycin resistance gene of Tn 903. *Gene* **26,** 243–252.

Winston, M. K., and Bhattacharjee, J. K. (1982). Growth inhibition by α-aminoadipate and reversal of the effect of specific amino acid supplements in *Saccharomyces cerevisiae. J. Bacteriol.* **152,** 874–879.

Wolfner, M., Yep, D., Messenguy, F., and Fink, G. (1975). Integration of amino acid biosynthesis into the cell cycle of *Saccharomyces cerevisiae. J. Mol. Biol.* **96,** 273–290.

Zaret, K. S., and Sherman, F. (1985). α Aminoadipate as a primary nitrogen source for *Saccharomyces cerevisiae* mutants. *J. Bacteriol.* **162,** 579–583.

III
Molds

8

Molecular Biology of the *qa* Gene Cluster of *Neurospora*

LAYNE HUIET[1] AND MARY CASE
Department of Genetics
University of Georgia
Athens, Georgia

I.	Introduction	229
II.	Molecular Analysis of the *qa* Cluster	230
	A. Background	230
	B. Physical Organization of the *qa* Gene Cluster	231
	C. Transcriptional Mapping and Regulation of the *qa* Genes	232
	D. DNA Sequence Analysis	234
	E. Activator-Independent Mutants of *qa-2*	234
	F. Regulation of the *qa* Genes	235
III.	Transformation of the *qa* Gene Cluster in *Neurospora*	236
	A. Utilization of Recombinant Plasmid DNA for Transformation	236
	B. General Outline for Transformation	237
	C. Factors Affecting *Neurospora* Transformation	239
	D. Genetic Characteristics of *N. crassa* Transformants	240
	E. Intergeneric Transformation and Gene Expression	241
	F. Expression of Bacterial Genes in *N. crassa*	241
IV.	Summary	242
	References	242

I. INTRODUCTION

One of the best characterized systems of gene regulation in lower eukaryotes is the quinate/shikimate catabolic pathway of *Neurospora crassa*. This chapter will provide an overview of the *qa* system, concentrating on recent molecular studies that have become possible with the cloning of the *qa* gene cluster. The first

[1]Present address: CSIRO Division of Plant Industry, Canberra City, A.C.T. 2601, Australia.

section will discuss the system in general, while the latter section will discuss the use of cloned *qa* genes in *Neurospora* transformation.

II. MOLECULAR ANALYSIS OF THE *qa* CLUSTER

A. Background

Previous genetic analysis indicated that the *qa* cluster was comprised of three structural genes (*qa-2*, *qa-3*, and *qa-4*) and a regulatory gene (*qa-1*), which were required for the catabolism of quinic acid and shikimic acid to protocatechuic acid (Fig. 1) (Giles *et al.*, 1973). Mutants in any of these genes were identified by their inability to utilize quinic acid as a sole carbon source. The genes have been mapped and the order determined to be centromere VII, *qa-1*, *qa-3*, *qa-4*, *qa-2*, *me-7* (Case and Giles, 1976). The three structural genes encode catabolic dehydroquinase (*qa-2*), quinate dehydrogenase (*qa-3*), and dehydroshikimate dehydratase (*qa-4*), and these three enzymes are coordinately induced by quinic acid. Induction does not occur in the presence of the protein synthesis inhibitor cycloheximide and has been shown by heavy-isotope labeling to be due to the *de novo* synthesis of these enzymes rather than enzyme activation or assembly (Reinert and Giles, 1977). More recently, experiments using an *in vitro* transla-

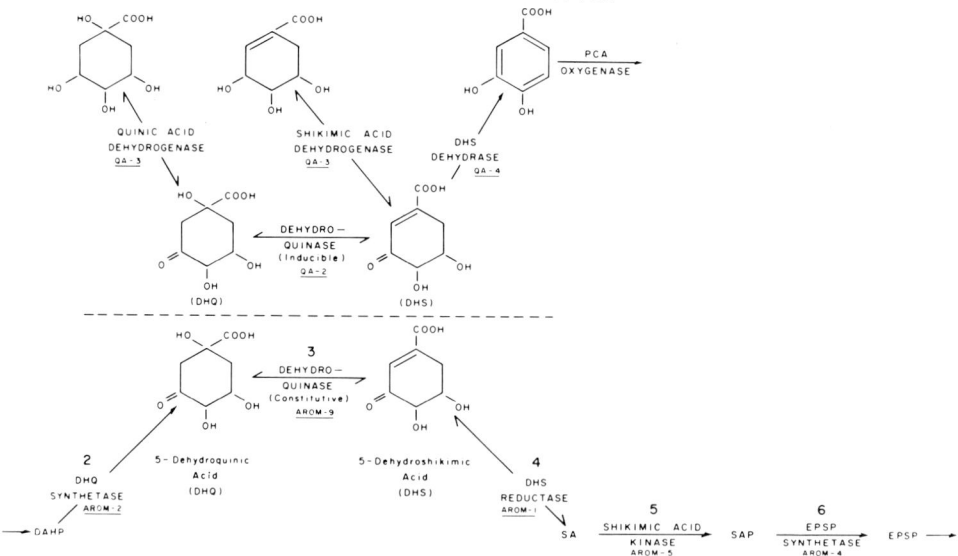

Fig. 1. Pathway for catabolism of quinic acid and shikimic acid to protocatechuic acid.

tion system (rabbit reticulocyte) indicated that *qa-3* mRNA is present only in quinate-induced cultures and not in uninduced cultures (Reinert *et al.*, 1981).

Initial genetic analysis suggested that the fourth gene in the *qa* cluster (*qa-1*) encoded a regulatory protein that exerted positive control over the coordinate production of the enzymes encoded by the structural genes. The evidence that *qa-1* was a regulatory protein came from genetic and biochemical data (Case and Giles, 1975). Both pleiotropic negative (noninducible) and constitutive *qa-1* mutants have been isolated which affect the synthesis of the three structural genes. Most *qa-1* mutants are negative types, being noninducible for the three genes, whereas *qa-1* constitutive mutants produce high levels of all three enzymes when grown in the absence of inducer. The existence of temperature-sensitive mutants indicated that the *qa-1* gene product was probably a protein and that the gene was not a controlling element (Case and Giles, 1975).

The characterization of *qa-1* mutants demonstrated two distinct and nonoverlapping groups, mapping at opposite ends of the gene (Case and Giles, 1975). These groups have been designated by the ability of a mutant to complement *qa-2* mutants as *qa-1*^{S} (slow or poorly complementing) and *qa-1*^{F} (fast or strongly complementing) (Rines, 1969). Constitutive *qa-1*^{c} mutants can be obtained directly from wild-type and from *qa-1*^{S} mutants by secondary mutations. However, *qa-1*^{F} types do not give rise to constitutives. On the basis of these results, it was suggested that the two classes of mutants represent two domains of a regulatory protein, one for inducer binding (*qa-1*^{F}) and the other for DNA binding (*qa-1*^{S}). The occurrence of pairs of *qa-1*^{S} and *qa-1*^{F} alleles which did not complement each other led to the proposal that the two mutant types comprised a single cistron. However, because *qa-1*^{S} exhibited negative complementation (semidominance), it could not be ruled out that *qa-1*^{S} and *qa-1*^{F} defined separate cistrons.

Many of the questions arising from the initial genetic and biochemical characterization of the *qa* system are now being resolved by the analysis of this region at the molecular level. Vapnek *et al.* (1977) demonstrated that it was possible to select for clones containing the *qa-2*$^+$ gene by complementation of *Escherichia coli* mutants (*aroD*) which lack the biosynthetic dehydroquinase of the polyaromatic amino acid pathway. Expression of the catabolic dehydroquinase of *Neurospora* (*qa-2*$^+$ gene) in *E. coli* transformants allows the bacteria to grow in the absence of aromatic amino acids. This fortuitous result made it possible to clone the entire *qa* gene cluster and led to the development of a transformation system for *Neurospora*.

B. Physical Organization of the *qa* Gene Cluster

Utilizing the positive selection procedure described previously, it has been possible to clone a 36.6-kb (kilobase pairs) fragment containing the complete *qa* gene cluster and its surrounding region of linkage group VII (Schweizer et al., 1981) via cosmid cloning. Due to the development of a transformation system for *Neurospora* it was possible to detect and localize the other *qa* structural genes (*qa-3, qa-4*) and the regulatory gene *qa-1* (Schweizer et al., 1981). Results from transformation experiments using subclones of the original plasmid confirmed the gene order as previously determined by genetic analysis. DNA–RNA hybridization experiments demonstrated that there were separate quinate-inducible mRNAs for *qa-2, qa-3,* and *qa-4,* plus two additional quinic acid-inducible mRNAs of unknown function, *qa-x* and *qa-y* (Patel et al., 1981). These results established that the structural genes were not transcribed as a single polycistronic mRNA, as in prokaryotic operons. The experiments also demonstrated that the regulation of *qa* enzyme synthesis was primarily at the transcriptional level and that this control directly involved the regulatory gene *qa-1*. In *qa-1* noninducible mutants (*qa-1^F* or *qa-1^S* types) induction of mRNAs for the *qa* structural genes does not occur, and in a *qa-1* constitutive strain, mRNA synthesis occurs under both inducing and noninducing conditions (Patel et al., 1981). Recent detailed analysis of the mRNA products of the *qa* structural genes has shown that multiple transcripts are produced and that this is the result of differences in both the 5' initiation sites and 3' termination sites of mRNA synthesis (Tyler et al., 1984; B. M. Tyler, unpublished observations).

Due to the isolation of a recombinant plasmid containing the entire *qa* cluster, it has been possible to analyze the region containing the regulatory gene *qa-1*. Utilizing both *Neurospora* transformation and DNA–RNA hybridization it was demonstrated that the *qa-1* region consists of two divergently transcribed genes, designated *qa-1S* and *qa-1F*, corresponding to the two original mutational types *qa-1^S* and *qa-1^F* (Huiet, 1984). The mRNA species for the two genes have been mapped and all seven genes are shown in Fig. 2. The seven genes comprise 17.5 kilobases and are all contiguous. The sizes of the transcripts are indicated, including the directions of transcription (Tyler et al., 1984, and unpublished observations; Huiet, 1984). Two of the genes, *qa-1S* and *qa-x,* contain intervening sequences.

C. Transcriptional Mapping and Regulation of the *qa* Genes

As mentioned in the previous section, multiple mRNA species exist for all of the *qa* structural genes, with the exception of *qa-y*. Subsequent 5' and 3' end-mapping of the mRNAs by nuclease digestion of RNA–DNA hybrids and by primer extension has revealed that the multiple *qa-x* and *qa-2* mRNAs result from heterogeneity at the 3' ends of the mRNAs, while the multiple *qa-3* and *qa-4* mRNA species are the result of heterogeneity at the 5' ends (Tyler et al., 1984; Tyler, unpublished observations). These experiments also revealed that the *qa-2* mRNAs have 5' heterogeneity which was not detectable by Northern blot hybridization.

8. The *qa* Gene Cluster of *Neurospora*

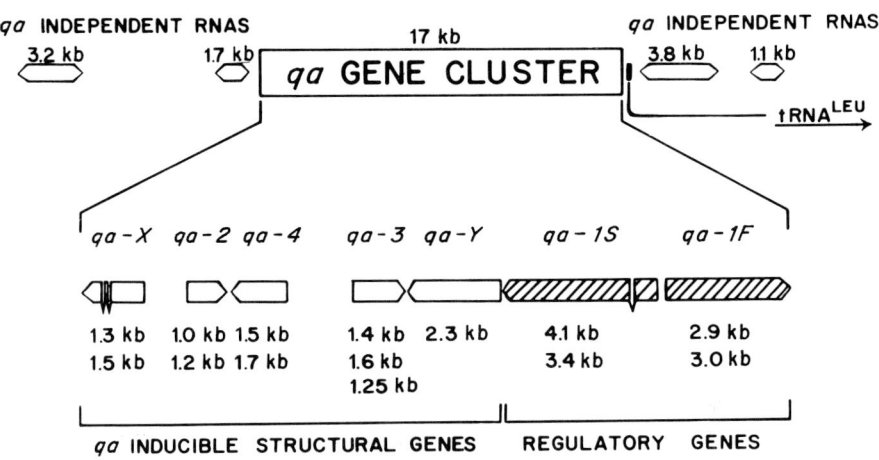

Fig. 2. The *qa* gene cluster.

Recent experiments have confirmed that in the case of *qa-2*, *qa-3*, *qa-4*, and *qa-y* mRNAs induction is abolished by mutations in either *qa-1S* or *qa-1F*. In *qa-1^c* strains the induction of these mRNAs does not require the presence of the inducer (QA). Also, catabolite repression appears to have little effect on the transcription of these four genes. When the constitutive strain is grown on sucrose and then shifted to carbon-free or glycerol media there is only an approximately twofold increase in the levels of the four mRNAs. However, *qa-x* appears to be regulated differently from the other four structural genes. The presence of a mutation in *qa-1F* reduces *qa-x* transcription only two- to threefold compared to a *qa-1F$^+$* strain. In contrast, a *qa-1S$^-$* mutation does block the induction of *qa-x*. Also, the transcription of *qa-x* is increased 5- to 10-fold by shifting a constitutive strain from sucrose to carbon-free media. It therefore appears that transcription of *qa-x* is controlled primarily by *qa-1S* and by catabolite repression.

Surprisingly, the mRNAs corresponding to the two regulatory genes *qa-1F* and *qa-1S* are also inducible. Both genes appear to be transcribed at low levels in cultures grown under noninducing conditions, but when the cultures are shifted to quinic acid the mRNA levels of both genes increase approximately 50-fold. However, results obtained with *qa-1F* and *qa-1S* mutants that are noninducible for the structural genes indicate that the induction of both *qa-1S* and *qa-1F* mRNA is dependent on the presence of a wild-type *qa-1F* product and also a wild-type *qa-1S* product. Therefore both genes appear to exhibit autogeneous regulation (Huiet, 1984; Patel and Giles, 1985). However, in the case of *qa-1S*, it is not clear whether this apparent autogenous regulation is due to a direct interaction of the *qa-1S* gene product with its own promoter or to an indirect effect such as its possible role in controlling *qa-1F* transcription.

D. DNA Sequence Analysis

The entire sequence of the *qa* structural genes and the regulatory genes has been determined, and the coding regions identified (Alton *et al.*, 1982; Rutledge, 1984; R. F. Geever, unpublished observations; L. Huiet, unpublished observations). In the case of the enzymes encoded by *qa-2*, *qa-3*, and *qa-4*, the molecular weight predicted from the DNA sequence is in good agreement with the previously published results (Barea and Giles, 1978; Strøman *et al.*, 1978). Also, where available, the amino acid sequences of these enzymes are in fair agreement with DNA sequences. The sequences of the two "unknown" genes, *qa-x* and *qa-y*, reveal that both contain long open reading frames. The *qa-y* sequence would encode a protein of molecular weight 60,000, while the *qa-x* sequence would encode a protein of molecular weight approximately 37,000. The two intervening sequences in *qa-x* are 69 and 74 base pairs (bp) in length and both of these introns conform to the consensus donor and acceptor sequences for eukaryotes (R. F. Geever, unpublished observations). The regulatory genes *qa-1F* and *qa-1S* encode the largest proteins. The *qa-1F* sequence contains a single open reading frame which would encode a protein of approximately 88,000 daltons, while the *qa-1S* sequence can encode a protein of approximately 100,000 daltons which contains a single intervening sequence of 66 bp. However, this intervening sequence does not have a consensus donor sequence. The donor sequence in this intron as determined by sequencing the mRNA begins with the dinucleotide GC rather than the consensus GT. It has been shown directly that GC can act as an efficient donor sequence (Fischer *et al.*, 1984).

The codon usage of the *qa* genes does not exhibit the very strong biases seen in the *Neurospora* glutamate dehydrogenase gene (Kinnaird and Fincham, 1983) and the *Neurospora* histone genes (Woudt *et al.*, 1983). However, the structural genes show a stronger bias than do the regulatory genes. In general, codons with C in the third position are preferred, while codons with A in this position are the least preferred. The nucleotide sequence analysis also revealed the presence of a leucine tRNA 300 bp from the 3' end of the *qa-1F* gene (Huiet *et al.*, 1984).

E. Activator-Independent Mutants of *qa-2*

Insights into the function of the *qa-1F* gene product have been provided by *qa-1F*-independent transcriptional mutants of *qa-2* (qa-2^{ai} mutants). These mutants were selected by demanding *qa-2* expression as "reversion" to prototrophy in a stable qa-$1F^-$ $arom$-9^- strain. A number of these strains have been characterized, the mutations cloned, and their DNA sequences determined. All of the mutations characterized occur in a region flanking the 5' end of the *qa-2* gene and include three types of alterations. Of 12 mutants analyzed, 8 are large reciprocal rearrangements, 2 are tandem direct duplications, and 2 have small base pair changes (Geever *et al.*, 1983). The mutants produce from 2 to 45% of induced wild-type levels of catabolic dehydroquinase and of *qa-2* mRNA. The mutations also stimulate to a lesser extent the transcription of surrounding *qa* genes such as *qa-2, qa-4,* and *qa-3* independently of the orientation or position (upstream or downstream) of the mutation relative to the gene (Tyler *et al.*, 1984). Thus the *qa-2*^{ai} mutations appear to create sequences with the properties of viral enhancers. Surprisingly, however, the *qa-2*^{ai} mutations stimulate *qa-1F*-independent transcription from only a subset of the multiple promoters of each *qa* gene (see above). For instance, *qa-2* transcription is initiated primarily from the −45 site in the *qa-2*^{ai} mutants rather than from the +1 site as in the wild type. Certain *qa-1F* temperature-sensitive mutations distinguish the same two subsets of initiation sites as the *qa-2*^{ai} mutations. For example, at the permissive temperature, transcription from the +1 site of *qa-2* is diminished 80% while transcription from the −45 site is unaffected. No transcription at all occurs at the nonpermissive temperature. These results have been interpreted as implying that two types of promoters occur in the *qa* gene cluster. Type II promoters, e.g., the −45 promoter of *qa-2,* would require only RNA polymerase II access or an active chromatin configuration for activity, while type I promoters, e.g., the +1 *qa-2* promoter, would also require direct binding of the *qa-1F* gene product. In the wild-type strain both of these requirements would be supplied directly or indirectly by *qa-1F*. In *qa-2*^{ai} mutants, only the requirement of the type II promoters would be provided by the enhancer sequences created by the *qa-2*^{ai} mutations, while the *qa-1F*^{ts} mutations would selectively impair the ability of *qa-1F* to stimulate type I promoters. Individual genes can have only type I promoters (e.g., *qa-y*), only type II promoters (e.g., *qa-4*), or both (e.g., *qa-2* and *qa-3*). The reason for having two types of promoters and the precise mechanisms by which *qa-1F* regulates the two promoter types remain unclear at this point.

F. Regulation of the *qa* Genes

The recent identification of two distinct regulatory genes has made it necessary to revise the earlier model of *qa* regulation, which involved only a single regulatory protein. It is apparent that these two genes must be involved in the expression of *qa* structural genes and that the different *qa-1* mutant phenotypes observed are the result of different mutations in these two genes.

All *qa-1F* mutants exhibit a recessive noninducible phenotype which is epistatic to *qa-1S*^{c} mutations. On the basis of these genetic data, it seems likely that the *qa-1F* gene product may encode a positive activator protein required for the induction of transcription of all of the *qa* genes, including itself and *qa-1S*. At the molecular level, however, it has been shown that several temperature-sensitive *qa-1F* mutants show alterations in the sites of transcription initiation for several of the *qa* structural genes (Tyler et al., 1984).

The noninducible *qa-1S* mutants are semidominant and revert to constitutivity. These phenotypes suggest that the role of *qa-1S* in *qa* regulation may be a negative one. Due to the fact that *qa-1F* is only transcribed at low levels in a *qa-1S* noninducible mutant it appears that the *qa-1S* gene product may regulate the *qa* structural genes by preventing transcription of the *qa-1F* gene. This would invoke a Jacob and Monod type model with the *qa-1S* gene encoding a repressor, which would block the transcription of the *qa-1F* gene under noninducing conditions by binding to an operator sequence adjacent to the promoter. In the presence of the inducer, quinic acid, the *qa-1S* gene product would be removed from the operator by interacting with the inducer, thus permitting the transcription of the *qa-1F* gene. The *qa-1S^c* mutants would therefore be interpreted, as in the lac system, as producing "inactive" repressor molecules which are defective in their ability to bind the operator binding site. Similarly, the *qa-1S* noninducible mutants would produce a "superrepressor" which cannot be removed from the operator by interaction with quinic acid.

This model of the interaction of the *qa-1S* and *qa-1F* genes does not require that the two gene products interact with one another. It is possible that the two genes do interact with one another to form a multimeric protein with both positive and negative regulatory functions, similar to the *araC* protein of *E. coli*. Since the genetic data suggest a negative role for the *qa-1S* gene product, it is surprising that the *qa-1S* gene is subject to induction. Perhaps this induction of *qa-1S* may have a feedback role in controlling *qa-1F* transcription. However, the possibility that *qa-1S* may also act positively cannot be excluded.

At this point the ideas and concepts of how the two regulatory genes interact to control *qa* gene expression are based mainly upon earlier genetic analysis. The physical localization of *qa-1S* and *qa-1F* mutations by DNA sequence analysis will determine whether the interpretation of their phenotypes is correct. The ultimate characterization of the gene products themselves may elucidate their interactions and role in *qa* gene expression.

III. TRANSFORMATION OF THE *qa* GENE CLUSTER IN NEUROSPORA

A. Utilization of Recombinant Plasmid DNA for Transformation

A molecular analysis of a gene cannot be totally accomplished without the cloning of the gene and the subsequent development of a suitable transformation system for the organism. Transformation of *Neurospora* was first described for the inositol gene by Mishra and Tatum (1973), utilizing total *N. crassa* DNA. This type of analysis could characterize the transformants only genetically. The cloning of the *qa-2* gene (Vapnek *et al.*, 1977) in *E. coli* permitted the development of an efficient transformation system for *N. crassa* (Case *et al.*, 1979). Subsequently, the entire *qa* gene cluster was cloned in *E. coli* and the localization of the genes in the *qa* gene cluster (Schweizer *et al.*, 1981; Huiet, 1984) was determined by the transformation procedure described here. The molecular analysis of the *qa* gene cluster has been presented in Section II. This part of the discussion will describe the basic transformation procedures, factors affecting transformation, genetic characteristics of the transformants, and attempts at intergeneric transformation and gene expression.

B. General Outline for Transformation

1. Utilization of Glusulase

The first method developed to transform *Neurospora* was based upon the transformation procedure described for yeast (Hinnen *et al.*, 1978). The procedure that will be described here has been used for selection of transformants of *qa* mutant strains (e.g., *qa-2* and *qa-1* recipient strains, Schweizer *et al.*, 1981). The utilization of this procedure for other *N. crassa* strains may require changes in growth conditions and in supplementations to the regeneration agar, e.g., *am* gene and *trp-1* gene (Kinnaird *et al.*, 1982; Schechtman and Yanofsky, 1983). A conidial suspension (5×10^{-6} conidia/ml) in 150 ml of one-half concentration Fries salt sucrose medium with a normal aromatic supplement is germinated for 4 hr at 25°C in a 250-ml Erlenmeyer flask on a rotary shaker (Schweizer *et al.*, 1981). The salt concentration of the minimal medium is reduced to half strength with distilled water to lower the calcium concentration. Spheroplasts are formed by incubation of 4-hr germinated conidia with 3% Glusulase (Endo Laboratories, Inc.) in 1 *M* sorbitol for 60–70 min at 30°C. The Glusulase is removed by washing twice with 1 *M* sorbitol and once with 1 *M* sorbitol in buffer A [10 m*M* morpholinepropanesulfonic acid (Sigma Chemical Co.), pH 6.3, 50 m*M* CaCl$_2$] in a tabletop centrifuge at 1000–2000 rpm. The spheroplast pellet (ca. 1×10^{-8} cells) is resuspended in 0.4 ml of 1 *M* sorbitol in buffer A, 5 µl of dimethyl sulfoxide, and 50 µl of 40% (wt/vol) polyethylene glycol (PEG) 4000 (Sigma Chemical Co.) in buffer A. Plasmid DNA (5–20 µg in 10 m*M* Tris, 1 m*M* EDTA, pH 8) is pretreated for 20 min on ice with 50 µl of heparin (5 mg/ml, Sigma Chemical Co., No. 3125) in buffer A to inhibit nucleases. The spheroplast suspension is added to the plasmid DNA and incubated on ice for 30 min. Then 2.5 ml of 40% (wt/vol) PEG (in buffer A) is added and incubated at room temperature (ca. 25°C) for 20 min. Regeneration of spheroplasts takes place in minimal medium containing 1 *M* sorbitol and 3% agar in order to maintain osmotic stability until the cell wall has regenerated. Transformants are visible after 2–3 days incubation at 30°C.

Transformants of qa-2, qa-3, or qa-4 recipient strains have been obtained in a liquid medium by selecting for growth on quinic acid as a carbon source following regeneration of the cell wall overnight in an osmotically stable medium (Case, 1983). The qa-1 $arom$-9 and qa-2 $arom$-9 double mutants require an aromatic supplement for growth. Such double mutants are blocked in the conversion of DHQ to DHS (Fig. 1) both in the aromatic biosynthetic pathway and in the quinic catabolic pathway. The localization of the qa-3 and qa-4 genes to plasmid clones was originally determined by transformation and selection for qa-2^+ expression in a multiple mutant recipient strain (qa-2^- qa-3^-, qa-4^-) (Schweizer et al., 1981). After selection for qa-2^+ transformants, complementation tests with appropriate qa-3^- and qa-4^- tester strains on liquid quinic acid permitted the localization of the qa-3^- and qa-4^- genes. To be able to localize the qa-3 and qa-4 genes to plasmid subclones that did not contain a qa-2^+ gene, the liquid transformation procedure was developed. The qa-3 and qa-4 mutants will grow on a minimal agar medium utilizing the residual carbon source in the agar. Thus, transformants could not be selected on the usual regeneration agar medium. The following modifications to the transformation procedure were made to allow the cell wall to regenerate prior to incubation on liquid quinic medium. After the addition of donor DNA, the spheroplasts were incubated overnight in liquid regeneration medium, either 20% sucrose (wt/vol) or 20% PEG (wt/vol) in sorbose–fructose–glucose minimal Fries medium. The regenerated cells were washed with water, added to 400 ml of 0.3% quinic acid (wt/vol) Fries minimal medium, and apportioned at 2 ml/tube into 13 × 100 mm test tubes to permit growth. Transformants were detected as cultures able to grow on the quinic acid medium after 2 weeks at 25°C. Positive growth was obtained with plasmids known to contain either the qa-4^+ or qa-3^+ genes; no transformants were obtained with plasmids that do not contain these genes. Transformation frequencies are low on liquid medium; nevertheless, this procedure does permit the localization of these genes to specific regions of the restriction map of the qa gene cluster.

2. Lithium Acetate Procedure

An alternative transformation procedure utilizing lithium acetate has been developed by Dhawale et al. (1984). This procedure is similar to the yeast transformation procedure where log-phase yeast cells treated with alkali cations (e.g., Li^+, Cs^+, Rb^+, and Na^+) took up plasmid DNA (Ito et al., 1983). In the N. crassa procedure, germinated conidia are treated with lithium acetate, followed by treatment with polyethylene glycol, and heat-shocked briefly at 37°C prior to plating on selective medium. Transformants are observed 5–6 days after plating at 30°C. The nature of the transformants as determined by the type of integration event (Section III,D) remains to be determined.

C. Factors Affecting *Neurospora* Transformation

Various experimental conditions that will be listed here have been examined to increase transformation frequencies. Alterations in pH indicated that pH 6.3 (buffer A) gave a higher transformation frequency than pH 7.5 Tris used in the yeast transformation procedure (Hinnen *et al.*, 1978). The optimal spheroplast concentration was found to be 1×10^{-8} to 2×10^{-8} with 5–10 μg plasmid DNA. The optimal PEG concentration was 40% when compared to 20 and 30%. Also, PEG 4000 was better than PEG 1500 and PEG 6000. The optimal $CaCl_2$ concentration was 50 mM when compared to 10 and 100 mM. The optimal time for incubation of spheroplast DNA on ice was found to be 20–30 min, although transformants have been observed after only 5 min (Case, 1982). The optimal time for incubation with PEG is 20 min, although only 2–3 min incubation is necessary to obtain some transformants.

The purification procedure used to isolate plasmid DNA does affect transformation. The best transformation frequencies are obtained with plasmid that has been banded two to three times on cesium chloride–ethidium bromide gradients (Case, 1982). Transformation in *E. coli* will occur with "dirty" DNA. Transformation in *N. crassa* with "dirty" DNA will be 100- to 1000-fold less than with clean DNA. The contaminant is probably RNA, as adding RNA to plasmid DNA inhibits transformation in *N. crassa* (M. E. Case, unpublished observations).

In yeast transformation experiments, digesting the plasmid with a restriction endonuclease at a site in the yeast DNA insert will increase the transformation frequency. A similar experiment with *N. crassa* did not show an increase in the transformation frequency.

The size of the *N. crassa* insert (from ca. 1 kb to 37 kb) into the plasmid does not grossly alter transformation frequencies. In fact, the transformation frequency is higher (ca. 500 to 1000 transformants per microgram of DNA) than with the smaller plasmids (70 transformants per microgram of DNA) (Case, 1982). The differences in transformation frequencies probably reflect the other genes and sequences that are present on the cloned *N. crassa* DNA, e.g., autonomously replicating Li sequences (see Section III,D).

Several different mutant strains, nuclease mutant *nuc-1* as well as UV-sensitive *uvs-2* and *uvs-3* strains, were crossed into the standard recipient strain *qa-2 arom-9* to determine whether any of these mutations would enhance transformation frequencies. None of these strains showed any alterations in the transformation frequency. However, the *uvs-3 qa-2 arom-9* strain was very sensitive to spheroplast formation and had only a 10% survival compared to the standard *qa-2 arom-9* strain.

D. Genetic Characteristics of *N. crassa* Transformants

1. Types of Transformants

After transformation two types of colonies have been observed on selective medium: colonies that continue to grow on selective medium on transfer and colonies that fail to grow on selective medium on transfer. These results suggest that those transformants which fail to grow on transfer on selective medium, termed "abortive transformants," may contain a slowly self-replicating plasmid that is unstable and ultimately lost. Not all plasmid subclones containing the *qa-2* gene give rise to abortive transformants (M. E. Case, unpublished observations). These results suggest that plasmids that give rise to abortive transformants may contain an autonomously replicating sequence that is not capable of prolonged replication in *N. crassa*.

The transformants that are stable and continue to grow on transfer have been characterized both genetically and by Southern blots (Case *et al.*, 1979; M. E. Case, unpublished observations). These data showed that the donor DNA can (1) integrate and replace the recipient gene, (2) integrate adjacent to the recipient gene as a linked insertion type, or (3) integrate at some other site within the genome. Southern blot analyses now indicate that essentially all nonhomokaryotic isolates contain additional *N. crassa* DNA and plasmid pBR322 sequences as well. When transformants from pVK88 (Case, 1982) are made homokaryotic, none of the replacement types and only half of the linked and unlinked duplication types contain pBR322 sequences integrated along with the *N. crassa* donor DNA. When transformants from the smaller plasmids pRC57 and pMEC1 (M. E. Case, unpublished observations) are made homokaryotic, most transformants are categorized as unlinked duplication types that contain pBR322 sequences integrated along with the *N. crassa* donor DNA. Both replacement types and linked insertion types are rare. In addition, certain transformants contain two copies of the plasmid inserted as a tandem repeat resulting from transformation with a dimeric form of the plasmid and integration into *N. crassa* DNA by a single crossover event (M. E. Case, unpublished observations).

2. Evidence for Self-Replicating Plasmids

Recent results suggest that a stable self-replicating plasmid has been constructed for *N. crassa* which contains the Labelle mitochondrial plasmid and the *qa-2* gene inserted into pBR325 (Stohl and Lambowitz, 1983). This chimeric plasmid can be recovered at a low frequency from DNA isolated from *N. crassa* transformants by selecting for antibiotic resistance in *E. coli*. When these transformants from *E. coli* are examined further, various types of plasmids are observed: (1) plasmids identical to the original donor DNA, (2) plasmids that have lost either the *qa-2* gene or the mitochondrial plasmid insert, and (3) plasmids that contain an additional insert of *N. crassa* chromosomal DNA not present in the original donor plasmid (Stohl and Lambowitz, 1983; M. E. Case, unpublished observations). Whether this chimeric plasmid or any of its derivatives will be useful as shuttle vectors for cloning *Neurospora* genes remains to be seen.

E. Intergeneric Transformation and Gene Expression

Chimeric plasmids have been constructed containing the qa-2^+ gene from *N. crassa* combined with each of these four genes from yeast: $trp1^+$, $ura3^+$, leu^+, $his3^+$. Transformation of the appropriate qa-2 trp, ura, leu, or his strains of *N. crassa* has indicated that transformation was obtained only with the qa-2^+ gene and none of these yeast genes was expressed in *N. crassa* (L. Huiet, unpublished observations). However, the pyr-4 gene from *N. crassa* has been used successfully to transform the ura gene of yeast (Ballance et al., 1983).

Successful intergeneric transformations have been obtained between *N. crassa* and *Aspergillus nidulans*. The pyr-4 gene from *N. crassa* has been successfully used to transform *A. nidulans* (Ballance et al., 1983). In addition, the cloned ornithine transcarbamylase gene from *A. nidulans* (OTCase) has been used successfully as donor DNA to transform an OTCase-deficient strain of *N. crassa* (arg-12) (Berse et al., 1983; Chapter 11, this volume). The results with intergeneric transformation between *N. crassa* and *A. nidulans* suggest that these two genera are more closely related than *N. crassa* is to yeast.

F. Expression of Bacterial Genes in *N. crassa*

Two antibiotic resistance genes from *E. coli*—the chloramphenicol acetyltransferase gene (chloramphenicol resistance) and the aminoglycoside phosphotransferase gene (2-deoxystreptamine antibiotic G418 resistance)—were tested for gene expression in *N. crassa*. Both of these genes are functionally expressed in yeast (Cohen et al., 1980; Jiminez and Davis, 1980). Since *N. crassa* reverts spontaneously at a high frequency for both chloramphenicol and G418 resistance, a plasmid was constructed carrying both of these genes as well as the qa-2^+ gene from *N. crassa*. Utilizing a chloramphenicol-sensitive qa-2^- recipient strain of *N. crassa* and selecting for qa-2^+ transformants, no qa-2^+ transformants in over 100 tested were able to grow in the presence of both antibiotics. Southern blot analyses indicated that the qa-2^+ gene as well as the entire plasmid had been integrated into the *N. crassa* genome (M. E. Case, unpublished observations). Expression of these genes in *N. crassa* might be obtained by inserting an *N. crassa* promoter in the chimeric plasmid at a site prior to the G418 or *cam* gene.

IV. SUMMARY

At present, regulation of the *qa* gene cluster in *Neurospora* is one of the best characterized systems in lower eukaryotes. The cloning of the entire *qa* gene cluster and the development of an efficient transformation system have permitted an in-depth study of gene regulation; this may be a useful model system for comparison with gene regulation in other fungal systems.

REFERENCES

Alton, N. K., Buxton, F., Patel, V., Giles, N. H., and Vapnek, D. (1982). 5′ untranslated sequences of two structural genes in the *qa* gene cluster of *Neurospora crassa*. *Proc. Natl. Acad. Sci. U.S.A.* **79**, 1955–1959.

Ballance, D. J., Buxton, F. P., and Turner, G. (1983). Transformation of *Aspergillus nidulans* by the orotidine-5′-phosphate decarboxylase gene of *Neurospora crassa*. *Biochem. Biophys. Res. Commun.* **112**, 284–289.

Barea, J. L., and Giles, N. H. (1978). Purification and characterization of quinate (shikimate) dehydroquinase, an enzyme in the inducible quinic acid catabolic pathway of *Neurospora crassa*. *Biochim. Biophys. Acta* **524**, 1–14.

Berse, B., Dmochowska, A., Skrzypek, M., Weglénski, P., Bates, M. A., and Weiss, R. L. (1983). Cloning and characterization of the ornithine carbamoyltransferase gene from *Aspergillus nidulans*. *Gene* **25**, 109–117.

Case, M. E. (1982). Transformation on *Neurospora crassa*. *In* "Genetic Engineering of Microorganisms" (A. Hollaender, R. D. DeMoss, S. Kaplan, J. Konisky, D. Savage, and R. S. Wolfe, eds.), pp. 87–100. Plenum, New York.

Case, M. E. (1983). Gene organization and regulation in *Neurospora crassa*. Evidence from cloning and transformation of the *qa* gene cluster. *In* "Genetic Engineering in Eukaryotes" (P. F. Lurquin and A. Kleinhofs, eds.), pp. 7–20. Plenum, New York.

Case, M. E., and Giles, N. H. (1975). Genetic evidence on the organization and action of the *qa-1* gene product: A protein regulating the induction of three enzymes in quinate catabolism in *Neurospora crassa*. *Proc. Natl. Acad. Sci. U.S.A.* **72**, 553–557.

Case, M. E., and Giles, N. H. (1976). Gene order in the *qa* gene cluster of *Neurospora crassa*. *Mol. Gen. Genet.* **147**, 83–89.

Case, M. E., Schweizer, M., Kushner, S. R., and Giles, N. H. (1979). Efficient transformation of *Neurospora crassa* utilizing hybrid plasmid DNA. *Proc. Natl. Acad. Sci. U.S.A.* **76**, 5259–5263.

Cohen, J. D., Ecclesall, T. R., Needleman, R. B., Federoff, H., Buchferer, B. A., and Marmur, J. (1980). Functional expression in yeast of the *Escherichia coli* plasmid gene coding for chloramphenicol acetyltransferase. *Proc. Natl. Acad. Sci. U.S.A.* **77**, 1078–1082.

Dhawale, S. S., Paletta, J. V., and Marzluf, G. A. (1984). A new rapid and efficient transformation procedure for Neurospora. *Curr. Genet.* **8**, 77–80.

Fischer, H. D., Dodgson, J. B., Hughes, S., and Engel, J. D. (1984). An unusual 5′ splice sequence is efficiently utilized *in vivo*. *Proc. Natl. Acad. Sci. U.S.A.* **81**, 2733–2737.

Geever, R. F., Case, M. E., Tyler, B. M., Buxton, F., and Giles, N. H. (1983). Point mutations and DNA rearrangements 5′ to the inducible *qa-2* gene of Neurospora allow activator-independent transcription. *Proc. Natl. Acad. Sci. U.S.A.* **80**, 7298–7302.

8. The *qa* Gene Cluster of *Neurospora*

Giles, N. H., Case, M. E., and Jacobson, J. W. (1973). Genetic regulation of quinate-shikimate catabolism in *Neurospora crassa*. In "Molecular Cytogenetics" (B. A. Hamkalo and J. Papaconstantinou, eds.), pp. 309-314. Plenum, New York.

Hinnen, A., Hicks, J. B., and Fink, G. R. (1978). Transformation of yeast. *Proc. Natl. Acad. Sci. U.S.A.* **75**, 1929-1933.

Huiet, L. (1984). Molecular analysis of the *Neurospora qa-1* regulatory region indicates that two interacting genes control *qa* gene expression. *Proc. Natl. Acad. Sci. U.S.A.* **81**, 1174-1178.

Huiet, L., Tyler, B. M., and Giles, N. H. (1984). A leucine tRNA gene adjacent to the *qa* gene cluster of *Neurospora crassa*. *Nucleic Acids Res.* **12**, 5757-5765.

Ito, J., Yasuki, F., Murata, K., and Kimura, A. (1983). Transformation of intact yeast cells treated with alkali cations. *J. Bacteriol.* **153**, 163-168.

Jimenez, A., and Davis, J. (1980). Expression of a transposable antibiotic resistance element in *Saccharomyces*. *Nature (London)* **287**, 869-871.

Kinnaird, J. H., and Fincham, J. R. S. (1983). The complete nucleotide sequence of the *Neurospora crassa am* gene. *Gene* **26**, 253-260.

Kinnaird, J. H., Keighren, M. A., Kinsey, J. A., Eaton, M., and Fincham, J. R. S. (1982). Cloning of the *am* (glutamate dehydrogenase) gene of *Neurospora crassa* through the use of a synthetic DNA probe. *Gene* **20**, 387-396.

Mishra, N. C., and Tatum, E. L. (1973). Non-Mendelian inheritance of DNA induces inositol independence in *Neurospora crassa*. *Proc. Natl. Acad. Sci. U.S.A.* **70**, 3873-3879.

Patel, V. B., and Giles, N. H. (1985). In preparation.

Patel, V. B., Schweizer, M., Dykstra, C. C., Kushner, S. R., and Giles, N. H. (1981). Genetic organization and transcriptional regulation in the *qa* gene cluster of *Neurospora crassa*. *Proc. Natl. Acad. Sci. U.S.A.* **78**, 5783-5787.

Reinert, W. R., and Giles, N. H. (1977). Proof of *de novo* synthesis of the *qa* enzymes of *Neurospora crassa* during induction. *Proc. Natl. Acad. Sci. U.S.A.* **74**, 4256-4260.

Reinert, W. R., Patel, V. B., and Giles, N. H. (1981). Genetic regulation of the *qa* gene cluster in *Neurospora crassa:* Induction of *qa* mRNA and dependency on *qa-1* function. *Mol. Cell. Biol.* **1**, 829-835.

Rines, H. W. (1969). Genetical and biochemical studies on the inducible quinic acid catabolic pathway in *Neurospora crassa*. Ph.D. Thesis, Yale Univ., New Haven, Connecticut.

Rutledge, B. J. (1984). Molecular characterization of the *qa-4* gene of *Neurospora crassa*. Ph.D. Thesis, Univ. of Georgia, Athens.

Schechtman, M., and Yanofsky, C. (1983). Structure of the trifunctional *trp-1* gene from *Neurospora crassa* and its aberrant expression in *Escherichia coli*. *J. Mol. Appl. Genet.* **2**, 83-99.

Schweizer, M., Case, M. E., Dykstra, C. C., Giles, N. H., and Kishner, S. R. (1981). Identification and characterization of recombinant plasmids carrying the complete *qa* gene cluster from *Neurospora crassa*, including the *qa-1*$^+$ regulatory gene. *Proc. Natl. Acad. Sci. U.S.A.* **78**, 5086-5090.

Stohl, L. L., and Lambowitz, A. M. (1983). Construction of a shuttle vector for the filamentous fungus *Neurospora crassa*. *Proc. Natl. Acad. Sci. U.S.A.* **80**, 1058-1062.

Strøman, P., Reinert, W. R., and Giles, N. H. (1978). Purification and characterization of dehydroshikimate dehydratase, an enzyme in the inducible quinic acid catabolic pathway of *Neurospora crassa*. *J. Biol. Chem.* **253**, 4593-4598.

Tyler, B. M., Geever, R. F., Case, M. E., and Giles, N. H. (1984). Cis-acting and trans-acting regulatory mutations define two types of promoters controlled by the *qa-1F* gene of *Neurospora*. *Cell (Cambridge, Mass.)* **36**, 493-502.

Vapnek, D., Hautala, J. A., Jacobson, J. W., Giles, N. H., and Kushner, S. R. (1977). Expression in *Escherichia coli* K-12 of the structural gene for catabolic dehydroquinase of *Neurospora crassa*. *Proc. Natl. Acad. Sci. U.S.A.* **74,** 3508–3512.

Woudt, L. P., Pastink, A., Kempers-Veenstra, A. E., Jansen, A. E. M., Mager, W. H., and Planta, R. J. (1983). The genes coding for histone H3 and H4 in *Neurospora crassa* are unique and contain intervening sequences. *Nucleic Acids Res.* **11,** 5347–5360.

9

Neurospora Plasmids

JOHN A. KINSEY

Department of Microbiology
University of Kansas School of Medicine
Kansas City, Kansas

I.	Introduction	245
II.	Naturally Occurring Plasmids	245
III.	Replicating Plasmids Constructed in the Laboratory	248
	A. Plasmids Constructed Using the qa-2^+ Gene	248
	B. Plasmids Constructed Using the am^+ Gene	250
IV.	Overview and Prospects for the Future	253
	References	256

I. INTRODUCTION

The subject of fungal plasmids has recently taken on a new importance with the application of molecular biological techniques to fungal systems. The paradigm of the yeast 2-μm circle and its usefulness in the construction of replicating vectors (see reviews in Broach, 1982; Botstein and Davis, 1982; Chapter 5, this volume) has prompted a renewed interest in the search for useful, naturally occurring plasmids of *Neurospora* and other filamentous fungi. Similarly, the demonstrated ability to construct plasmids capable of replicating in yeast and bacteria by inserting a selectable yeast marker and a yeast autonomously replicating sequence (ARS) into a bacterial plasmid such as pBR322 (see review in Botstein and Davis, 1982) has prompted a number of attempts to construct similar vectors for filamentous fungi. This chapter covers the small but emerging literature on natural and constructed *Neurospora* plasmids.

II. NATURALLY OCCURRING PLASMIDS

To date, all of the naturally occurring plasmids found in *Neurospora* have been mitochondrial in origin. Recent work with artificially constructed plasmids capable of persisting in the nucleus, discussed in Section III below, suggests that naturally occurring nuclear plasmids could exist. The availability of the large collection of wild-type *Neurospora* strains collected from around the world by Perkins and co-workers (1976) makes one hopeful that as more and more of these strains are screened, a true nuclear plasmid will turn up.

Two distinct kinds of plasmids or plasmid-like DNAs (PLDs) are found in mitochondria of *Neurospora*. The PLDs, defective deletion variants of segments of the normal mitochondrial chromosome, have also been reported from a number of other microbial eukaryotes (Cummings *et al.*, 1979; Locker *et al.*, 1979; Mannella *et al.*, 1979; Bertrand *et al.*, 1980; Jamet-Vierny *et al.*, 1980; Lazarus *et al.*, 1980). Although these PLDs, or parts thereof, may at a later time be useful in the construction of cloning vectors, at the moment their usefulness is somewhat limited by the fact that the ultimate outcome of transforming PLDs into healthy cells is to produce a weakened or dying strain.

The true plasmids, on the other hand, are unrelated, or only distantly related, to mitochondrial chromosomal sequences. They appear to be normal constitutents of mitochondria from healthy cells. These plasmids have been found in mitochondria of a few plants and other fungi; however, the plasmids from *Neurospora* have been the most extensively characterized (Collins *et al.*, 1981; Stohl *et al.*, 1982, 1983; Nargang *et al.*, 1983, 1984; Natvig *et al.*, 1984). The mitochondrial plasmids of *Neurospora* were discovered by Collins *et al.* (1981) in a screen for structural variants of the mitochondrial chromosome of wild-type *Neurospora* strains. They described a plasmid that was found in mitochondria of the strain Mauriceville-1c and referred to it as the Mauriceville plasmid. Initially, the plasmid DNA was found in *Eco*RI digests of mitochondrial DNA as an unexpected band of about 2.6 kilobase pairs (kb) that was present in a two- to threefold excess over other bands. A more detailed analysis of mitochondrial DNA from the Mauriceville strain indicated that in addition to the normal mitochondrial chromosome there was a closed-circular DNA molecule with a monomeric length of about 3.6 kb. By electron microscopy Collins *et al.* (1981) were able to show that in addition to monomer-sized circles there was a series of oligomeric forms, up to a hexamer.

By means of Southern analysis they found that Mauriceville plasmid hybridized neither to DNA from the mitochondrial chromosome nor to DNA from nuclear chromosomes. Using Southern analysis as well as nuclease protection experiments, they were able to conclude that the plasmid was probably confined to the mitochondrial compartment. As expected, the plasmid showed a strict maternal inheritance pattern.

Stohl *et al.* (1982) have identified two additional mitochondrial plasmids from *Neurospora intermedia* strains. One was from the P405 Labelle strain, the other from the Fiji N6-6 strain. The Labelle plasmid had a monomer length of 4.1–4.3 kb and Fiji plasmid, 5.2–5.3 kb. Both of these plasmids were present as monomers as well as oligomers arranged in head-to-tail repeats. Neither of these plasmids has substantial sequence homology with mitochondrial DNA, nuclear DNA, each other, or with the Mauriceville plasmid. As was found with the Mauriceville plasmid, these plasmids were shown to be mitochondrial in origin and to segregate with a maternal inheritance pattern. Stohl et al. (1983) identified a fourth mitochondrial plasmid from the *N. intermedia* strain Varkud-1c. This strain contains a 3.8-kb plasmid that is closely related to the Mauriceville plasmid.

Natvig et al. (1984) have identified related mitochondrial plasmids in four of seven isolates of *Neurospora tetrasperma* that were tested. Three of these, Hanalei, Lihue, and Waimea Falls, were from Hawaii. The fourth strain was from Surinam. The three Hawaiian plasmids appeared to be identical, while the Surinam plasmid was only slightly different. When tested against the group of plasmids isolated by Lambowitz and associates, these *N. tetrasperma* plasmids proved to be related to the *N. intermedia* strain Fiji plasmid, although they differed from Fiji plasmid at several restriction sites.

Clearly, these novel mitochondrial plasmids are a widespread feature of *Neurospora* strains, with related plasmids sometimes shared between two (or more?) species. What role do these plasmids play in the natural history of *Neurospora?* It seems unlikely that they have any essential function for the strains that carry them, since other strains of the same species lacking these plasmids do quite well. Stohl et al. (1982) point out that they may have some useful, but nonessential, function such as drug resistance. Alternatively, they may represent parasitic DNA.

One interesting point is that Mauriceville plasmid *in vivo* produces a unit length transcript with the 5' and 3' ends adjacent to each other (Nargang et al., 1984). This transcript has an open reading frame that could code for a 710-amino-acid protein, assuming normal mitochondrial codon usage, or two open reading frames (256 amino acids and 234 amino acids), assuming translation using the standard genetic code (Nargang et al., 1984). It is still unclear whether this transcript directs the synthesis of a protein product; however, Nargang et al. (1984) speculate that it might code for a protein required for propagation or expression of the plasmid. Neither Labelle nor Fiji plasmid appears to produce a major transcription product (Stohl et al., 1982); therefore, if the putative Mauriceville product is required for replication, the mechanism of replication of Mauriceville must differ from that of Fiji and Labelle.

Another particularly interesting feature of the Mauriceville plasmid is that it contains a cluster of *Pst*I sites that cover a 400-base-pair (bp) region of plasmid DNA. Nargang et al. (1983, 1984) sequenced this region and found that the cluster consisted of eight *Pst*I sites organized in five palindromic elements closely resembling the *Pst*I palindromes characteristic of *Neurospora* mitochondrial chromosomal DNA (Yin et al., 1981). In fact, the two outer palindromes of the cluster have the exact canonical sequence for mitochondrial DNA PstI palindromes (5'CCCTGCAGTACTGCAGGG3'). This cluster of PstI palindromes occurs about 120 bp downstream from the termination codon for the putative 710-amino-acid protein (Nargang et al., 1984). The PstI palindromes of mitochondrial DNA are frequently found at the boundaries of rRNA, tRNA, and mitochondrial protein genes (Yin et al., 1981). These features, along with conserved sequences characteristic of type I introns of mitochondrial genes, suggested to Stohl et al. (1983) and Nargang et al. (1983, 1984) that these sequences might be related to mobile elements. They postulated that such elements might propagate in Neurospora mitochondria and may have been the progenitors of mitochondrial introns. Full-length transcripts, as observed for the Mauriceville plasmid (Nargang et al., 1984), are also characteristic of certain eukaryotic mobile elements such as Ty1 of yeast (Elder et al., 1980).

III. REPLICATING PLASMIDS CONSTRUCTED IN THE LABORATORY

To date, chimeric plasmids containing two cloned Neurospora genes, $qa-2^+$ and am^+, have been constructed that appear to replicate independently in Neurospora. The recent cloning of additional selectable Neurospora nuclear genes (Schectman and Yanofsky, 1983; Buxton and Radford, 1983) suggests that it may soon be possible to construct additional replicating plasmids, perhaps with advantages over the two current plasmid sets.

A. Plasmids Constructed Using the $qa-2^+$ Gene

The first Neurospora gene to be cloned was $qa-2^+$, the structural gene for catabolic dehydroquinase (Vapnek et al., 1977). It was originally isolated on a chimeric plasmid, pVK55, capable of replicating in Escherichia coli, where it could complement an aro D6 mutant strain to yield prototrophic transformants. Case et al. (1979) utilized this cloned $qa-2^+$ gene, along with a stable qa-2 arom-9 double mutant strain of Neurospora, to develop the first efficient transformation system for Neurospora. The initial results with this system (Case et al., 1979; Schweizer et al., 1981; Case, 1981) demonstrated that transformation could occur by integration in the form of gene replacements, linked gene duplications, or unlinked integration. This was similar to the picture seen in the initial yeast transformation system (Hinnen et al., 1978). No evidence for transformation involving a replicating plasmid was seen in these studies.

Stohl and Lambowitz (1983a) utilized the qa system in experiments designed to test the possible usefulness of the mitochondrial plasmids (discussed in Section II above) for construction of nonintegrating vectors. They constructed a hybrid vector consisting of a 3.0-kb HindIII fragment containing the $qa\text{-}2^+$ gene, the Labelle mitochondrial plasmid, and pBR325. This construction, called pALS1, was then used to transform a *qa-2 arom-9* double mutant strain. The initial results indicated that hybrid plasmids containing Labelle plasmid sequences were more efficient at transformation than were equivalent plasmids lacking the Labelle plasmid sequences. In a more recent paper, Stohl *et al.* (1984) suggested that the addition of Labelle plasmid sequences might be beneficial to some plasmids and not to others with respect to increased transformation frequency. As they point out, transformation efficiency of *Neurospora* is susceptible to a number of factors, varying not only from one preparation of DNA to the next, but also with a number of factors, including differences in batches of Glusulase used to make the spheroplasts. The question of relative efficiency remains somewhat up in the air; however, Stohl and Lambowitz (1983a) were able to demonstrate not only that the pALS1 plasmid could transform *Neurospora*, but that it could be reisolated from *Neurospora* and the reisolated plasmid used to transform *E. coli* to chloramphenicol resistance. The yield of plasmid from *Neurospora* was low (approximately 10 transformants per microgram of DNA) but was consistently obtained. Thus, pALS1 met the primary requirement for a *Neurospora* shuttle vector, namely that it could replicate in and be reisolated from both *Neurospora* and *E. coli*.

An unexpected finding of Stohl and Lambowitz (1983a) was that, in addition to recovery of the original pALS1 plasmid, a smaller derivative plasmid, called pALS2, was recovered. In fact, in the initial transformation experiments, 90% of the *E. coli* transformed with DNA from *Neurospora* transformants contained pALS2 rather than pALS1. Restriction analysis of pALS2 plasmid indicated that most or all of the Labelle plasmid sequences had been deleted. In a subsequent paper, Stohl *et al.* (1984) sequenced the pertinent region of five independently isolated pALS2 plasmids. All five had completely lost the Labelle plasmid sequences and, in addition, were missing the four internal bases of the *Pst*I site of pBR325.

Stohl and Lambowitz (1983a) also found that sequential passages of pALS1 through *Neurospora* and *E. coli* seemed to produce a more stable version of the pALS1 which produced pALS2 less frequently.

Although these results are open to other interpretations, they fit quite well with the idea that both pALS1 and pALS2 are chimeric vectors capable of replicating independently in *Neurospora*. Differential fractionation of cellular components suggests that the plasmids are present in the nuclear compartment. Heterokaryon studies (Stohl and Lambowitz, cited in Grant *et al.*, 1984) support this conclusion. The fact that pALS2 appears capable of replication, although devoid of Labelle plasmid sequences, strongly suggests that the sequences used for the origin of replication are fortuitously present either in pBR325 or in the *qa-2* fragment. It thus appears that the combination of pBR325 and the 3.0-kb HindIII fragment containing the *qa-2* gene constitutes a replicating nuclear plasmid for *Neurospora* that can be recovered at low yield and used to retransform *E. coli*.

B. Plasmids Constructed Using the *am*$^+$ Gene

The *am* (glutamate dehydrogenase, GDH) gene of *Neurospora* was cloned in the lambda replacement vector λ-L47 (Kinnaird *et al.*, 1982). Using the deletion strain *am*$_{132}$ as a recipient (Kinsey and Hung, 1981), the clone, called C10, was shown to transform *am* mutants to prototrophy (Kinnaird *et al.*, 1982; Kinsey and Rambosek, 1984). All of the transformants resulted from integration, either within the normal resident chromosome, linkage group V (LGV), or within other chromsomes (Kinsey and Rambosek, 1984). A plasmid subclone was constructed with the 2.7-kb *Bam*HI fragment of the C10 clone (which contains the entire coding sequence for GDH) and inserted into the single *Bam*HI site of the bacterial plasmid pUC8. When this plasmid (pJR2) was used for transformation of the deletion strain the transformants appeared not to be integrated (Kinsey and Rambosek, 1984). This prompted a more detailed study of the nature of pJR2 transformation, which led to the conclusion that pJR2 was probably capable of autonomous replication in *Neurospora* (Grant *et al.*, 1984). An alternative explanation of the data involving unstable integration of tandem repeats could not be completely eliminated, but seemed much less likely. The primary evidence supporting autonomous replication came from genetic studies of the transformed deletion strain. The *am*$^+$ character could be maintained in mitotically dividing cells by selection; however, very few (fewer than 1%) *am*$^+$ spores were obtained when the transformants were backcrossed to the *am* deletion parent. When the rare *am*$^+$ spores were backcrossed or intercrossed they continued to yield <1% *am*$^+$ spores. Moreover, continuous mitotic selection by single conidial colony isolation for as many as 11 generations failed to produce strains that were stable in meiosis. When pJR2 was used to transform a point mutant, rather than the deletion, the surprising result was that most of the transformants were still apparently nonintegrated. Only 1 transformant of 19 tested could definitely be shown to be integrated in chromosomal DNA. That transformant had *am*$^+$ integrated into LGV, as was expected on the basis of the provided homology.

In an additional test for the effect of chromosomal homology upon integration, Grant *et al.* (1984) constructed a new plasmid vector (called pAAQ8) consisting of pJR2 with one *Bam*HI site deleted and a 1.2-kb *Bam*HI, *Bgl*II fragment containing the *qa-2*$^+$ gene inserted in the remaining *Bam*HI site. This plasmid was used to transform the *am*$_{132}$ deletion strain, with the same results obtained with the pJR2 plasmid itself (i.e., none of 14 transformants tested appeared to be integrated, as indicated by their meiotic instability). Thus, in spite of significant homology with, in one case LGV, and in the other case LGVII, most of the transformants appear to have obtained the am^+ gene in a nonintegrated form.

When the pJR2 transformants were tested for mitotic stability, all were found to be unstable, although not to the extreme degree seen in meiosis. The degree of mitotic stability varied considerably from one transformed strain to another. After a conservatively estimated 25–26 nuclear generations of growth on nonselective medium, the fraction of am^+ conidia ranged from 6 to 67% in different strains.

It seemed possible that mitotic instability was an artifact due to heterokaryosis. This could arise if there were an integration of am^+ DNA that caused recessive lethal damage to the recipient chromosome. If this were so, then all transformants would be heterokaryons between am^+ nuclei containing a recessive lethal mutation and *am* nuclei without the lethal mutation. Such a heterokaryon would superficially resemble an unstable transformant. Using diagnostic heterokaryons, Grant *et al.* (1984) demonstrated that this was not the case and that the apparent instability was real. They also demonstrated, by means of heterokaryon tests, that the am^+ character was limited to the nuclear compartment.

Grant *et al.* (1984) analyzed the plasmid DNA content of randomly selected transformed strains by means of genomic Southern analysis. In some uncut DNA preparations of some transformants, hybridizing bands were found where one would expect monomer-sized plasmid; however, the bulk of the DNA that hybridized to plasmid DNA was present as high-molecular-weight DNA. When the DNA was cut with restriction endonucleases, fragments expected for monomer plasmid were obtained in most cases; however, in most transformants other prominent bands also appeared. In some transformants no band appeared where monomer would have been expected, although a characteristic 0.72-kb *Hin*dIII fragment derived from the am^+ portion of the plasmid was always present. Colony filter hybridization (Stohl and Lambowitz, 1983b) and Southern analysis of *am* progeny from crosses of transformants to the am_{132} parent revealed that some *am* progeny had plasmid sequences (although not functional am^+ sequences) incorporated into chromosomal DNA. The Southern analysis data together with the genetic evidence for instability suggested that pJR2 is capable of replication in *Neurospora,* probably in a large oligomeric form similar to the situation described by Sakaguchi and Yamamoto (1982) for a *ura-1*-containing plasmid in *Schizosaccharomyces pombe.* Grant *et al.* (1984) estimated that the average oligomer size of pJR2 might be as large as 9–26 monomer units in various transformants.

The presence in restriction endonuclease-treated DNA preparations of prominent bands other than that expected for monomer plasmid and the presence of plasmid DNA sequences in *am* progeny spores suggested that the plasmid DNA was undergoing rearrangement and/or recombination with chromosomal sequences at some point during the transformation process. This idea was supported by the fact that plasmid was recovered from one transformant by retransformation of *E. coli*. This plasmid had undergone some kind of recombination event which had added a 1.1- to 1.2-kb fragment of normally unlinked *Neurospora* chromosomal DNA to the plasmid. Rearrangements are frequently seen with the pALS1, *qa-2* transformation system as well (Stohl and Lambowitz, 1983a; Stohl *et al.*, 1984). Thus, autonomously replicating DNA in *Neurospora* may be particularly prone to these kinds of events, as appears to be the case for mammalian SV40-based shuttle vectors (Razzaque *et al.*, 1983).

Grant *et al.* (1984) also made plasmid constructs that had the Labelle mitochondrial plasmid sequence inserted into the *Pst*I site of pJR2. These constructs were tested for the effect of Labelle sequences on transformation frequencies, using the am_{132} deletion strain as recipient. There was no indication that added Labelle sequences improved the transformation frequency; however, there was a pronounced increase in chromosomal integration with plasmids containing Labelle plasmid sequences (14–23% of the transformants had integrated am^+ sequences). Labelle plasmid has no homology with chromosomal DNA (Stohl *et al.*, 1982), but mitochondrial plasmids have been postulated to be related to mobile elements (Nargang *et al.*, 1983, 1984). It certainly seems that Labelle plasmid sequences acted in some way to enhance chromosomal integration.

All of the data taken together suggest that pJR2 is capable of autonomous replication in *Neurospora* as a head-to-tail oligomeric plasmid. This plasmid appears to have inherent instability in both meiosis and mitosis. In meiosis, where the plasmid is extremely unstable, the oligomeric plasmid sequences may behave in the same manner as acentric chromosomal fragments. Barry (1972) showed that acentric chromosomal fragments generated by chromosomal rearrangements are lost from the nuclei during meiosis. Grant *et al.* (1984) point out that Southern analysis of DNA from pALS1 transformants also showed a high-molecular-weight band that could be due to ologomeric plasmid. It may be that pALS1 simply has a higher monomer-to-oligomer ratio than does pJR2. This would presumably be the result of the different *Neurospora* DNA sequences contained in the two plasmids. A somewhat similar situation was recently reported for plasmids of *E. coli* (James *et al.*, 1983).

It is possible that transformants obtained with pJR2 contain plasmid as integrated tandem repeats in chromosomal DNA. This explanation is compatible with the Southern analysis data; however, it is a much less acceptable explanation for the genetic data for the following reasons:

1. No comparable instability has been observed for *Neurospora* integrated transformants obtained with the lambda C10 am^+ vector (Kinsey and Rambosek, 1984), the integrated transformants obtained by using pJR2 with an *am* point mutant as recipient, or the integrated transformants obtained with the derivatives of pJR2 that contain Labelle plasmid sequences. (Grant *et al.*, 1984).

2. Weiss *et al.* (Chapter 11, this volume) have obtained *Neurospora* transformants by using the cloned *Aspergillus* ornithine transcarbamylase gene and the

Neurospora arg-12 mutant strain. These transformants have *Aspergillus* sequences integrated as tandem repeats in *Neurospora* chromosomal DNA. These transformants are stable in both meiosis and mitosis.

3. Since only 1% (or fewer) of the spores produced by a backcross of transformants to am_{132} were am^+, the mechanism operating in these crosses would have had to reduce the copy number of am^+ sequences to zero in at least 98% of the spores that normally would have gotten an am^+ gene. No mechanism is known that routinely reduces the copy number of tandem repeats to zero in the majority of meioses.

Taken together, these arguments favor the explanation that pJR2 replicates in *Neurospora;* however, they do not eliminate the integrated tandem repeat hypothesis. Even if pJR2 is capable of replication, the paucity of monomers makes it less useful than pALS1 or pALS2 as a shuttle vector. The plasmid pJR2 may be useful in other ways, i.e., in searching for sequences that will increase the ratio of monomeric to oligomeric plasmid, or in looking for sequences that enhance GDH expression. Grant et al. (1984) found that primary transformants produced only 2–5% of normal GDH activity, whereas some passaged strains produced up to 55% of normal GDH activity. Part of this difference was almost certainly due to the heterokaryotic nature of primary transformants; however, there is the intriguing possibility that pJR2 was able to pick up sequences during replication that enhanced the promoter activity of the *am* gene. A previous report (Kinsey and Rambosek, 1984) had suggested that an upstream element might be required for full GDH expression.

IV. OVERVIEW AND PROSPECTS FOR THE FUTURE

The development of plasmids for molecular biological studies of *Neurospora* is still in a relatively early stage compared to the situation in yeast (Botstein and Davis, 1982). A naturally occurring nuclear plasmid with controlled replication and regular division between daughter nuclei, similar to the 2-μm circle of yeast, would provide the "ideal" shuttle plasmid; however, no nuclear plasmids have been found.

Mitochondrial plasmids are found in a number of *Neurospora* strains. They are extremely interesting from the standpoint of their biological origin, natural history, and possible functional/parasitic role in the organisms that they inhabit, but so far they have not been particularly useful in the construction of shuttle plasmids.

In yeast, a series of extremely useful plasmids has been constructed by taking advantage of what appears to be a rigid set of rules for the behavior of plasmids in these cells. Plasmids that contain a selectable yeast marker, such as the *LEU2* gene in the bacterial plasmid ColE1 (Hinnen et al., 1978), can transform a yeast auxotrophic strain to prototrophy at low efficiency. The transformants obtained are all integrated into chromosomal DNA and the integration appears to be exclusively determined by homology of yeast sequences in the plasmid with chromosomal DNA (these plasmids are referred to as YIp plasmids). If an autonomously replicating sequence is added to such a plasmid the plasmid will transform the yeast auxotrophic strain with a much higher frequency (about 10,000 times better than the YIp vector) and the transformants will contain autonomously replicating plasmid that is readily reisolated from the yeast transformant (Beggs, 1978; Stinchcomb et al., 1979). If the ARS used was all or part of the yeast 2-μm circle DNA, the plasmids are referred to as YEp plasmids. Transformants obtained with YEp plasmids are relatively stable and the plasmid they contain has a high copy number. If the ARS used was a cloned fragment of chromosomal DNA such as ars1, normally located near the *trp1* gene (Stinchcomb et al., 1979), the plasmids are referred to as YRp plasmids and the transformants obtained are very unstable and usually have a relatively low copy number. The occasional stable transformants obtained are due to integration.

A fourth category of yeast plasmid is obtained by adding a cloned centromere-containing fragment to YRp-type plasmids (Clarke and Carbon, 1980). These plasmids, called YCp plasmids, transform yeast strains with a high efficiency without integration and behave stably in mitosis and meiosis, showing approximately Mendelian segregation ratios (see Chapter 5 for a more detailed discussion of yeast plasmids).

In *Neurospora* the situation seems much less clear-cut. In the original transformation experiments with the pVK88 plasmid all of the qa-2^+ transformants contained integrated donor DNA, either as a duplication or replacement at the normal *qa-2* locus, or from integration events at some novel location in chromosomal DNA (Case et al., 1979). This is very similar to the original observations obtained by transformation of yeast with a YIp plasmid containing the *LEU2* gene (Hinnen et al., 1978). In the latter case, the nonhomologous transformants were due to non-*LEU2* sequences in the cloned DNA that are repeated a number of times in the yeast genome (Botstein and Davis, 1982). There is no indication that pVK88 contains repetitive DNA, so the nonhomologous integration obtained in *Neurospora* might be due to very short regions of homology. Alternatively, integration might occur by recombination between nonhomologous sequences. In order to know whether sequence homology is or is not required for integration, it will be necessary to sequence the DNA at the junction between integrated sequences and chromosomal DNA for a number of transformants with integrated donor DNA.

The situation in *Neurospora* with respect to replicating plasmids is again not nearly as clear as the yeast case. The plasmid pVK88 was reported to give transformation only by integration (Case et al., 1979; Case, 1981), whereas a similar but smaller plasmid, pALS2, appears to give transformation via a replicating plasmid (Stohl and Lambowitz, 1983a; Stohl et al., 1984). Possibly with pVK88 only the stable transformants were selected for further study (e.g., only transformants that came through crosses successfully). This would give the appearance that all of the transformation was via integration. If that were the case then probably either pBR322 or pBR325 plus *qa-2* sequences constitute a replicating plasmid similar to the YRp plasmids of yeast, capable of replicating independently but also able to give rise to stable integrants.

One difficulty in interpreting the results of transformation experiments with the *qa-2* system is the fact that homologous DNA is always present in the recipient cells; thus, at least in theory, the incoming DNA has the option of integration at the chromosomal *qa-2* locus. By using the *am* system, in which deletion mutants completely lacking homology with pJR2 are available, this difficulty is overcome. It seems that am^+ sequences present in a lambda vector transform *am* auxotrophs exclusively by integration (Kinsey and Rambosek, 1984). With the exception of the initial abortive transformants, no unstable transformants were found (J. A. Kinsey, unpublished observations). In contrast, when the am^+ sequences are present in the plasmid vector pJR2 and the deletion strain, am_{132}, is used as a recipient, all of the transformants appear to have the am^+ gene in a nonintegrated form. Furthermore, these transformants could be carried through 11 passages on selective medium without the am^+ DNA becoming stably associated with chromosomal DNA (Grant et al., 1984). Transformants with integrated am^+ were obtained when transformation was carried out using a point mutant as a recipient, or when the Labelle mitochondrial plasmid DNA was added to pJR2 and the deletion strain was used as a recipient. Even in these cases the am^+ sequences were not stably integrated in the majority of transformants. In the former case it appeared that integration occurred via the provided homology. In the latter case there was no major homology between the plasmid containing the Labelle plasmid sequences and chromosomal DNA, so the mechanism of integration remains unclear. Genetic analysis indicated that the integration of the am^+ sequences from the plasmid containing Labelle plasmid DNA occurred at sites other than the normal location for *am* (J. A. Kinsey, unpublished observations). Nargang et al. (1983, 1984) have suggested that *Neurospora* mitochondrial plasmids are related to mobile elements, so it is possible that the Labelle plasmid DNA has some positive effect on integration rather than merely providing a site of microhomology. In this regard, the apparently frequent precise deletion of Labelle sequences from pALS1 is suggestive that Labelle DNA could have some specific role in recombination. Thus pJR2 appears to transform *Neurospora* preferentially as a replicating plasmid; however, under certain conditions it can integrate into chromosomal DNA.

Superficially at least, pALS2 and pJR2 (and perhaps pVK88) resemble the YRp plasmids of yeast. It is not clear that there are any analogs for the YIp (integration only) plasmids in either of these two systems. One surprising result (in view of the relative efficiency of YIp and YRp plasmids in yeast) is that am^+ in the presumably nonreplicating lambda vector gives an efficiency of transformation as high as that of pJR2 (Kinsey and Rambosek, 1984). This suggests that in *Neurospora* the replicating plasmid might have no great advantage over a vector that transforms by integration only. On the other hand, there might be an inherent advantage for vectors with free ends that could promote recombination. Such an advantage was not observed for cut pVK88 (Case *et al.,* 1979).

If both pALS2 and pJR2 are replicating plasmids, the location of the autonomously replicating sequences is still an open question. Since neither of these plasmids contains very much extraneous *Neurospora* DNA, it is possible either that the ARSs are found within the genes themselves or that ARSs are found fortuitously within the pUC8 and pBR325 bacterial plasmids.

In the immediate future, attempts will undoubtedly be made to determine the exact location of the ARSs in pJR2 and pALS2 so that these sequences can be used in construction of other replicating plasmids. As new selectable *Neurospora* genes are cloned they will be compared to the *qa-2* and *am* systems for efficiency of transformation and ability to support a replicating plasmid. If these genes are advantageous in any way, but fail to support replication, then perhaps the ARSs from pALS2 or pJR2 could be used to advantage.

Because pALS2 can be reisolated consistently but at low efficiency from *Neurospora,* it may be useful as a shuttle vector for the isolation of *Neurospora* genes by complementation in *Neurospora*. The experience with am^+-containing plasmid suggests that some *Neurospora* sequences might tend to reduce or eliminate monomers, so that different sequences may be more or less amenable to isolation by use of this plasmid. For the future one hopes that either a naturally occurring nuclear plasmid or a chromosomal sequence that favors monomers can form the basis of a new shuttle plasmid series with a high yield of monomer plasmid. Another characteristic of such an "ideal" plasmid would be a dominant selectable marker, thus avoiding the need to build double or triple mutant strains for use as recipients in transformation experiments.

REFERENCES

Barry, E. G. (1972). Meiotic chromosome behavior of an inverted insertional translocation in *Neurospora. Genetics* **71,** 53–62.

Beggs, J. D. (1978). Transformation of yeast by a replicating hybrid plasmid. *Nature (London)* **275,** 104–108.

Bertrand, H., Collins, R. A., Stohl, L. L., Goewert, R. R., and Lambowitz, A. M. (1980). Deletion mutants of *Neurospora crassa* mitochondrial DNA and their relationship to the "stop–start" growth phenotype. *Proc. Natl. Acad. Sci. U.S.A.* **77,** 6032–6036.

Botstein, D., and Davis, R. W. (1982). Principles and practice of recombinant DNA research with yeast. *In* "The Molecular Biology of the Yeast *Saccharomyces:* Metabolism and Gene Expression'' (J. Strathern, E. Jones, and J. Broach, eds.), pp. 607–636. Cold Spring Harbor Lab. Press, Cold Spring Harbor, New York.

Broach, J. R. (1982). The yeast plasmid 2μ circle. *In* "The Molecular Biology of the Yeast *Saccharomyces:* Metabolism and Gene Expression'' (J. Strathern, E. Jones, and J. Broach, eds.), pp. 445–470. Cold Spring Harbor Lab. Press, Cold Spring Harbor, New York.

Buxton, F. P., and Radford, A. (1983). Cloning of the structural gene for orotidine 5' phosphate carboxylase of *Neurospora crassa* by expression in *Escherichia coli*. *Mol. Gen. Genet.* **190,** 403–405.

Case, M. E. (1981). Transformation of *Neurospora crassa* utilizing recombinant plasmid DNA. *In* "Genetic Engineering of Microorganisms for Chemicals'' (A. Hollaender, ed.), pp. 87–100. Plenum, New York.

Case, M. E., Schweizer, M., Kushner, S. R., and Giles, N. H. (1979). Efficient transformation of *Neurospora crassa* by utilizing hybrid plasmid DNA. *Proc. Natl. Acad. Sci. U.S.A.* **76,** 5259–5263.

Clarke, L., and Carbon, J. (1980). Isolation of a yeast centromere and construction of functional small circular chromosomes. *Nature (London)* **287,** 504–509.

Collins, R. A., Stohl, L. L., Cole, M. D., and Lambowitz, A. M. (1981). Characterization of a novel plasmid DNA found in mitochondria of *N. crassa*. *Cell (Cambridge, Mass.)* **24,** 443–452.

Cummings, D. J., Belcour, L., and Grandchamp, C. (1979). Mitochondrial DNA from *Podospora anserina*. II. Properties of mutant DNA and multimeric circular DNA from senescent cultures. *Mol. Gen. Genet.* **171,** 239–250.

Elder, R. T., St. John, T. P., Stinchcomb, D. T., and Davis, R. W. (1980). Studies on the transposable element Ty1 of yeast. I. RNA homologous to Ty1. *Cold Spring Harbor Symp. Quant. Biol.* **45,** 581–584.

Grant, D. M., Lambowitz, A. M., Rambosek, J. A., and Kinsey, J. A. (1984). Transformation of *Neurospora* with recombinant plasmids containing the cloned glutamate dehydrogenase (*am*) gene. Evidence for autonomous replication. *Mol. Cell. Biol.* **4,** 2041–2051.

Hinnen, A., Hicks, J. B., and Fink, G. R. (1978). Transformation of yeast. *Proc. Natl. Acad. Sci. U.S.A.* **75,** 1929–1933.

James, A. A., Morrison, P. T., and Kolodner, R. (1983). Isolation of genetic elements that increase frequencies of plasmid recombinants. *Nature (London)* **303,** 256–259.

Jamet-Vierny, C., Begel, O., and Belcour, L. (1980). Senescence in *Podospora anserina:* Amplification of a mitochondrial DNA sequence. *Cell (Cambridge, Mass.)* **21,** 189–194.

Kinnaird, J. H., Keighren, M. A., Kinsey, J. A., Eaton, M., and Fincham, J. R. S. (1982). Cloning the *am* (glutamate dehydrogenase) gene of *Neurospora crassa* through the use of a synthetic DNA probe. *Gene* **20,** 387–396.

Kinsey, J. A., and Hung, B. S. T. (1981). Mutation at the *am* locus of *Neurospora crassa*. *Genetics* **93,** 577–586.

Kinsey, J. A., and Rambosek, J. A. (1984). Transformation of *Neurospora crassa* with the cloned *am* (glutamate dehydrogenase) gene. *Mol. Cell. Biol.* **4,** 117–122.

Lazarus, C., Earl, A. J., Turner, G., and Kuntzel, H. (1980). Amplification of a mitochondrial DNA sequence in the cytoplasmically-inherited "ragged'' mutant of *Aspergillus amstelodami*. *Eur. J. Biochem.* **106,** 633–641.

Locker, J., Lewin, A., and Rabinowitz, M. (1979). The structure and organization of mitochondrial DNA from petite yeast. *Plasmid* **2,** 155–181.

Mannella, C. A., Goewert, R. R., and Lambowitz, A. M. (1979). Characterization of variant *Neurospora crassa* mitochondrial DNAs which contain tandem reiterations. *Cell (Cambridge, Mass.)* **18,** 1197–1207.

Nargang, F. E., Bell, J. B., Stohl, L. L., and Lambowitz, A. M. (1983). A family of repetitive palindromic sequences found in *Neurospora* mitochondrial DNA is also found in a mitochondrial plasmid DNA. *J. Biol. Chem.* **258,** 4257–4260.

Nargang, F. E., Bell, J. B., Stohl, L. L., and Lambowitz, A. M. (1984). The DNA sequence and genetic organization of a *Neurospora* mitochondrial plasmid suggests a relationship to the mitochondrial introns and mobile genetic elements. *Cell (Cambridge, Mass.)* **38,** 441–453.

Natvig, D. O., May, G., and Taylor, J. W. (1984). Distribution and evolutionary significance of mitochondrial plasmids in *Neurospora*. *J. Bacteriol.* **159,** 288–293.

Perkins, D. D., Turner, B. C., and Berry, E. G. (1976). Strains of *Neurospora* collected from nature. *Evolution* **30,** 281–313.

Razzaque, A., Mizusawa, H., and Seidman, M. M. (1983). Rearrangement and mutagenesis of a shuttle vector plasmid after passage in mammalian cells. *Proc. Natl. Acad. Sci. U.S.A.* **80,** 3010–3014.

Sakaguchi, J., and Yamamoto, M. (1982). Cloned *ura-1* locus of *Schizosaccharomyces pombe* propagates autonomously in this yeast assuming a polymeric form. *Proc. Natl. Acad. Sci. U.S.A.* **79,** 7819–7823.

Schechtman, M. G., and Yanofsky, C. (1983). Structure of the trifunctional *trp-1* gene from *Neurospora crassa* and its aberrant expression in *Escherichia coli*. *J. Mol. Appl. Genet.* **2,** 83–99.

Schweizer, M., Case, M. E., Dykstra, C. C., Giles, N. H., and Kushner, S. R. (1981). Identification and characterization of recombinant plasmids carrying the complete *qa* gene cluster from *Neurospora crassa* including the *qa-1*[+] regulatory gene. *Proc. Natl. Acad. Sci. U.S.A.* **78,** 5086–5090.

Stinchcomb, D. T., Struhl, K., and Davis, R. W. (1979). Isolation and characterization of a yeast chromosomal replicator. *Nature (London)* **282,** 39–43.

Stohl, L. L., and Lambowitz, A. M. (1983a). Construction of a shuttle vector for the filamentous fungus *Neurospora crassa*. *Proc. Natl. Acad. Sci. U.S.A.* **80,** 1058–1062.

Stohl, L. L., and Lambowitz, A. M. (1983b). A colony filter-hybridization procedure for the filamentous fungus *Neurospora crassa*. *Anal. Biochem.* **134,** 82–85.

Stohl, L. L., Collins, R. A., Cole, M. D., and Lambowitz, A. M. (1982). Characterization of two new plasmid DNAs found in mitochondria of wild-type *Neurospora intermedia* strains. *Nucleic Acids Res.* **10,** 1439–1458.

Stohl, L. L., Grant, D. M., Lambowitz, A. M., Bell, J. B., and Nargang, F. E. (1983). DNA sequence analysis of two *Neurospora* mitochondrial plasmids. *Fed. Proc., Fed. Am. Soc. Exp. Biol.* **42,** 1972.

Stohl, L. L., Akins, R. A., and Lambowitz, A. M. (1984). Characterization of deletion derivatives of an autonomously replicating *Neurospora* plasmid. *Nucleic Acids Res.* **12,** 6169–6178.

Vapnek, D., Hautala, J. A., Jacobson, J. W., Giles, N. H., and Kushner, S. R. (1977). Expression in *Escherichia coli* K12 of the structural gene for catabolic dehydroquinase of *Neurospora crassa*. *Proc. Natl. Acad. Sci. U.S.A.* **74,** 3508–3512.

Yin, S., Heckman, J., and RajBhandary, U. L. (1981). Highly conserved GC-rich palindromic DNA sequences flank tRNA genes in *Neurospora crassa* mitochondria. *Cell (Cambridge, Mass.)* **26,** 325–332.

10

Cloning and Transformation in *Aspergillus*

G. TURNER AND D. J. BALLANCE
Department of Microbiology
University of Bristol
Bristol, United Kingdom

I.	Introduction	259
II.	Transformation Methodology	260
	A. Entry of DNA	260
	B. Selectable Markers	264
III.	Fate of Transforming DNA after Entry	265
	A. Integration	265
	B. Marker Rescue	269
IV.	The Quest for Replicating Vectors	270
	A. Native Plasmids	271
	B. Yeast-Selected ARS	271
	C. Mitochondrial Sequences	271
V.	Development of Vectors	273
	A. Gene Isolation by Self-Cloning	273
	B. A Vector for Promoter Isolation	274
VI.	Future Prospects	275
	References	275

I. INTRODUCTION

The classical genetics of *Aspergillus nidulans* has been studied in some detail since Pontecorvo *et al.* (1953) first described the genetic system, and the main areas of study have been summarized (Smith and Pateman, 1977). In particular, there has been much interest in regulatory aspects of carbon and nitrogen source catabolism (Arst and Scazzocchio, Chapter 13). More recently, interest has grown in areas such as mitochondrial genetics (Turner and Rowlands, 1977;

Scazzocchio et al., 1983), mitosis (Bergen et al., 1984), and differentiation (Zimmerman et al., 1980).

The advent of techniques for gene isolation and analysis, first with bacteria and then baker's yeast, has resulted in rapid progress in our understanding of gene structure and expression at the molecular level in those organisms. It is clear that a similar understanding of gene expression in filamentous fungi requires the development and application of these new techniques in addition to the older, classical genetics, and a number of groups have been working to this end in the past few years. A transformation system for *Neurospora crassa* was reported some years ago (Case et al., 1979), but the transformation frequency has remained fairly low, making difficult the development of cloning vectors (Huiet and Case, Chapter 8). Although certain genes can be isolated from *A. nidulans,* as from other fungi, by shotgun cloning and selection for expression in *Escherichia coli* or yeast (Table I), the ideal is to develop a self-cloning system for *Aspergillus,* and good progress is now being made in this direction. The first step was the development of an efficient transformation system to reintroduce isolated genes into *A. nidulans,* and this has now been achieved, making possible, for instance, the study of expression of *in vitro*-mutated genes.

Many of the developments discussed below were unpublished at the time of writing, and are personal communications or were reported at the EMBO meeting on "Gene Expression in Filamentous Fungi" at Rhenen, Holland, April 1984.

II. TRANSFORMATION METHODOLOGY

A. Entry of DNA

1. Protoplast Formation

With organisms such as *Bacillus, Streptomyces,* yeast, and *Neurospora,* removal of the cell wall to form stable protoplasts is the best method of facilitating entry of DNA, though attempts have been made to generate "competent" cells by other means (Ito et al., 1983; Dhawale et al., 1984). Protoplast formation and regeneration of mycelium was perfected for *A. nidulans* some years ago (Peberdy and Ferenczy, 1985), and the commercially available enzyme Novozym 234 (Novo Industries) is most convenient for this purpose. We have used KCl (0.6 *M*) as an osmotic stabilizer for protoplast release and subsequent transformation and regeneration. A number of other inorganic salts can be used in place of KCl (Peberdy, 1979; Peberdy and Ferenczy, 1985). However, unlike the situation with yeast, use of sugars or sugar alcohols as the osmotic stabilizer results in rather poor yields of protoplasts.

$MgSO_4$ has been employed in the preparation of protoplasts for transformation

TABLE I.

Isolation of Genes from *Aspergillus*

Species	Gene locus	Gene product	Isolation method[a]	Genetic transformation	Reference
A. nidulans	*acuD*	Isocitrate lyase	EA	+	D. J. Ballance and G. Turner (unpublished)
	alcA	Alcohol dehydrogenase I	MH	+	R. A. Lockington et al. (unpublished)
	aldH	Aldehyde dehydrogenase	MH		R. A. Lockington et al. (unpublished)
	amdS	Acetamidase	MH	+	Hynes et al. (1983)
					Tilburn et al. (1983)
	argB	Ornithine carbamoyltransferase	ES	+	Berse et al. (1983)
	arom	Anabolic dehydroquinase	EE		Kinghorn and Hawkins (1982)
	gdhA	Anabolic NADH-dependent glutamate dehydrogenase	MH		J. R. Kinghorn (unpublished)
	oliC	Mitochondrial ATP-synthetase subunit 9	MH	+	M. Ward and G. Turner (unpublished)
	prnA,D,B,C (cluster)	Proline catabolizing enzymes	MH	+	P. M. Green et al. (unpublished)
	qutE	Catabolic dehydroquinase	MH		A. Hawkins et al. (unpublished)
	trpC	Phosphoribosyl anthranilate isomerase	EE	+	Yelton et al. (1983)
A. niger		Phosphoribosyl anthranilate isomerase	EE	Transforms *A. nidulans*	A. Kos et al. (unpublished)

[a] EA, expression in *A. nidulans*; EE, expression in *E. coli*; ES, expression in *S. cerevisiae*; MH, molecular hybridization (e.g., via cDNA, heterologous probing, differential hybridization, and other related methods).

(Tilburn et al., 1983) in order to facilitate separation of the protoplasts from mycelial debris (de Vries and Wessels, 1972; Tilburn et al., 1983). Purified protoplasts were then resuspended in buffer containing 1.2 M sorbitol for transformation (Table II).

2. DNA Uptake

Polyethylene glycol (PEG) plus Ca^{2+}-induced fusion of protoplasts has been used for some time in studies of interspecies hybrid formation (Peberdy and Ferenczy, 1985) and recombination of mitochondrial DNA (Earl et al., 1981), and since similar treatment of *Bacillus* (Chang and Cohen, 1979), *Streptomyces* (Bibb et al., 1978), yeast (Hinnen et al., 1978), and *Neurospora* (Case et al., 1979) is known to stimulate DNA uptake, this technique was modified and refined to obtain maximum transformation frequencies in *Aspergillus* once suitable selectable markers became available. Transformation protocols are shown in Table II. As far as comparison can be made, there does not seem to be a great deal of difference in transformation frequency between these methods, all achieving about 10–100 transformants per microgram of DNA. However, a number of interesting points are worth mentioning.

In transformation of *pyrG* auxotrophs to prototrophy, we have found that increasing the agar concentration in the overlay used for protoplast regeneration from 0.9 to 2% increases severalfold the number of transformants recovered. There is, however, no effect on the overall regeneration frequency, suggesting a differential effect on those protoplasts competent for transformation. This alteration is not possible in the acetamidase system because of the problem of utilization of N sources present in agar (Polkinghorne and Hynes, 1975).

Tilburn et al. (1983) found that increasing the concentration of PEG from 25 to 60% increased the frequency of transformation, though we have not found such an effect in our system. Time course studies have shown us that it is advantageous to reduce the time of exposure to PEG to a minimum and that subsequent dilution of the fusogenic medium is sufficient to alleviate these toxic effects.

Yelton et al. (1984) used the method of Tilburn et al. (1983) with a few modifications (Table II), the most significant of which is a period of incubation of the transformed protoplasts prior to spreading onto the surface of agar plates.

K. Wernars et al. (unpublished observations) have used protoplasts prepared from germinating conidia (Bos and Slakhorst, 1981) for transformation studies. Such protoplasts are more homogeneous than those made from mycelium, and tend to be uninucleate, but this latter feature is probably lost following PEG treatment because of the high frequency of fusion which occurs (Peberdy and Ferenczy, 1985).

TABLE II.

Comparison of Transformation Procedures[a]

	Ballance and Turner (pyr4)	Tilburn et al. (amdS)	Yelton et al. (trpC)
Growth of mycelium	Cellophane cultures	Liquid culture	Liquid culture
Cell wall digestion	Novozym 234 in 0.6 M KCl	Novozym 234 and helicase in 1.2 M MgSO$_4$	Novozym 234 and β-glucuronidase in 1.2 M MgSO$_4$
Purification of protoplasts	Filtration through sintered glass	Centrifugation	Centrifugation
Transformation buffer	0.6 M KCl, 50 mM CaCl$_2$	1.2 M sorbitol, 10 mM CaCl$_2$, 10 mM Tris-HCl, pH 7.5	1.2 M sorbitol, 10 mM CaCl$_2$, 10 mM Tris-HCl, pH 7.5
Incubation 1	Add DNA then 0.25 vol[b] PEG solution[c], ice, 20 min	Add DNA, room temperature, 20 min	Add DNA, room temperature, 25 min
Incubation 2	Add 10 vol PEG solution, room temperature, 20 min	Add 10 vol PEG solution[d], room temperature, 20 min	Add 12.5 vol PEG solution[d], room temperature, 20 min
Regeneration	Add 20 vol 0.6 M KCl, 50 mM CaCl$_2$, then add aliquots to medium containing 2% agar	Centrifuge, then wash protoplasts with transformation buffer; add to medium containing 0.25% agar	Centrifuge, then resuspend in 1.2 M sorbitol + yeast extract and glucose, 37°C, 2 hr before spreading onto surface of agar plates

[a] From Ballance et al. (1983), D. J. Ballance and G. Turner (unpublished observations), Tilburn et al. (1983), and Yelton et al. (1984).
[b] All volumes refer to the volume of transformation buffer.
[c] PEG solution of Ballance and Turner: 25% PEG 6000, 50 mM CaCl$_2$, 10 mM Tris-HCl, pH 7.5.
[d] PEG solution of Tilburn et al. and Yelton et al.: 60% PEG 4000, 10 mM CaCl$_2$, 10 mM Tris-HCl, pH 7.5.

B. Selectable Markers

1. Transformation of Nutritional Mutants

Transformation of nutritional mutants with cloned, wild-type genes has been the main approach to date, the initial problem having been the availability of an appropriate, cloned, prototrophic gene and stable, matching mutant. Earlier attempts at transformation with cloned yeast genes and the matching *A. nidulans* mutants, such as *ura3* (Bach *et al.*, 1979) and *pyrG* (Palmer and Cove, 1975), were unsuccessful. We have since introduced the yeast *ura3* gene into *Aspergillus* and failed to detect any expression. Similar failures have been reported for *N. crassa* (Case, 1982). However, the *pyr4* gene of *Neurospora* expresses well in *Aspergillus* and weakly in yeast (Ballance *et al.*, 1983), and has been used by us for the development of vectors. A number of *Aspergillus* genes which match available auxotrophic and other nutritional mutations have now been isolated and used in transformation studies and in the development of vectors (Table I). Analyses of transformation have been carried out with *pyr4* (Ballance *et al.*, 1983), *amdS* (Tilburn *et al.*, 1983), and *trpC* (Yelton *et al.*, 1984). The *amdS* mutant is unable to use acetamidase as an N source as a result of a deletion in the structural gene (Hynes, 1979). Because of a certain amount of background growth on N sources present in the agar, certain precautions have to be taken, including careful choice of agar and inclusion of Cs^+ ions in the medium (Tilburn *et al.*, 1983).

2. Antibiotic Resistance

An inconvenience of prototrophic selection as outlined above is that before any strain of *Aspergillus* can be transformed, the nutritional mutation has to be introduced by a sexual cross. Antibiotic resistance would avoid this, and such vectors have been developed for animal cells (Colbère-Garapin *et al.*, 1981) and yeast (Jimenez and Davies, 1980). The following antibiotics have been considered for *Aspergillus:*

a. Oligomycin. *Aspergillus nidulans* is sensitive to quite low levels (3 μg/ml) of this inhibitor of mitochondrial ATP synthetase, and resistance is conferred by mutations at mitochondrial and nuclear loci which alter the *in vitro* sensitivity of the enzyme complex (Rowlands and Turner, 1973, 1977). Mutations in the gene for subunit 9 are known to confer such resistance in yeast and *Neurospora* (Tzagoloff *et al.*, 1979). In yeast, this subunit is encoded in mitochondrial DNA (mtDNA), but in *Neurospora* and *Aspergillus,* subunit 9 is cytoplasmically translated and imported into mitochondria. This subunit is an 8000 M.W. hydrophobic polypeptide ("proteolipid") in the Fo (membrane) portion of the complex, specifically binds the inhibitor dicyclohexylcarbodiimide (DCCD), and is believed to be involved in proton translocation during oxidative phosphorylation.

In *Neurospora*, nuclear oligomycin resistance mutations occur at a single locus and alter the amino acid sequence of the proteolipid (Sebald and Hoppe, 1981). The cDNA for this gene has been cloned (Viebrock *et al.*, 1982). Since it had previously been demonstrated that subunit 9 is cytoplasmically translated in *Aspergillus* (Turner *et al.*, 1979), and all nuclear mutants reside at a single locus *oliC* (Rowlands and Turner, 1977), we assumed that *oliC* was the subunit 9 gene of *Aspergillus*. Using the cDNA clone from *Neurospora* as a probe against a lambda EMBL4 gene bank of an *Aspergillus* strain carrying the *oliC31* resistance mutation, we have isolated an homologous sequence and begun to analyze it (M. Ward and G. Turner, unpublished observations). The sequence has been subcloned into plasmid vectors carrying the *pyr4* gene, and these have been used to transform an *Aspergillus pyrG* mutant to prototrophy. These transformants showed various levels of resistance to oligomycin. It has also been possible to select directly for oligomycin resistance following transformation, and demonstrate the presence of *pyr4* in the transformants.

b. Chloramphenicol. This is an inhibitor of mitochondrial protein synthesis, but in general, eukaryotes are sensitive only to rather high concentrations of the drug. A mutant of *A. nidulans* is available which is relatively hypersensitive (Lazarus and Turner, 1977). Although both mitochondrial and nuclear chloramphenicol-resistant mutants can be isolated (Gunatilleke *et al.*, 1975; Lazarus and Turner, 1977), the mode of resistance is not yet known. In yeast, mitochondrially inherited resistance is conferred by an altered sequence of the mitochondrial large ribosomal subunit RNA (Dujon, 1980). Bacterial resistance, mediated by drug inactivation with chloramphenicol acetylase, is expressed weakly in yeast (Cohen *et al.*, 1980), but has not been useful as a selective marker. We have introduced the chloramphenicol resistance gene of pBR325 without selection into *Aspergillus* on a *pyr4* vector, and observed no consequent increased resistance to the drug. The same story can be told for bacterial kanamycin resistance and the antibiotic G418 (Jimenez and Davies, 1980). In both cases, it seems probable that a fungal promoter would first be essential to obtain expression, and such an approach remains a possibility. However, G418, like chloramphenicol, is not a particularly effective growth inhibitor.

III. FATE OF TRANSFORMING DNA AFTER ENTRY

A. Integration

Integration into the chromosome is, to date, the only way in which stable transformants can be obtained. The nature of the integration event can be deduced from analysis of the region of integration, and although further work is needed, a number of features can be described.

1. Abortive Transformants

Whatever the selective marker, over 90% of the initially observed colonies following transformation remain very small, and are incapable of further growth on subculture. A similar phenomenon has been reported for *Neurospora* (Case, 1982) and yeast (Hicks *et al.*, 1979), and may result from transient expression without stabilized integration of the transforming DNA. Tilburn *et al.* (1983) have observed that subculture of abortive colonies sometimes leads to outgrowth of stable transformant sectors.

2. Homologous Recombination

Transformation using the *Aspergillus amdS* (Tilburn *et al.*, 1983) and the *trpC* (Yelton *et al.*, 1984) genes is sometimes followed by integration at the host gene locus, and can be explained by a simple, single homologous recombination event. The bacterial plasmid region is flanked by wild-type and mutant genes (Fig. 1). This has been referred to as type I transformation by analogy with yeast (Hinnen *et al.*, 1978). More rarely, a double crossover event results in replacement of the mutant host gene by the transforming wild-type gene (type III transformation) (Fig. 1). Type II transformation refers to the integration of the transforming DNA at a site other than the host gene locus, and may result from the chance occurrence of some homology in another part of the host genome (Fig. 1). Just how much homology is required for recombination to occur is still a matter for speculation. However, it does appear that regions with greater homology are favored over those with less homology as sites for integration, and it

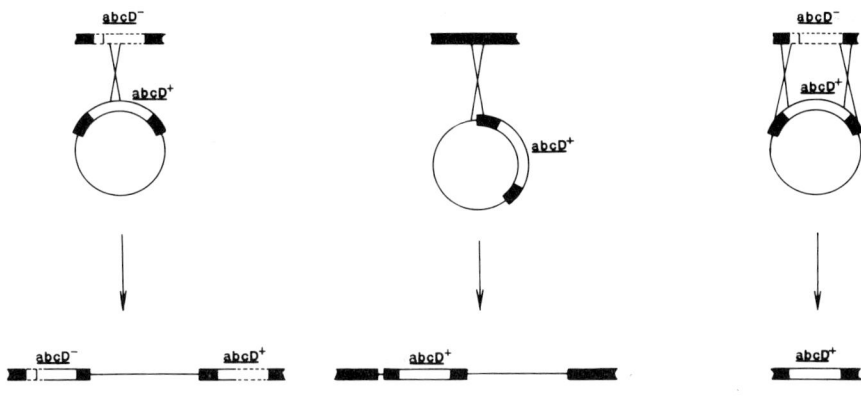

Fig. 1. Diagrammatic representation of possible integration events. [After Hinnen *et al.* (1978).] (———) bacterial plasmid DNA; (===) mutant gene sequences; (=) wild-type gene sequences; (▬) genomic DNA.

should be possible to exploit this when future attempts are made to target integration. Yelton *et al.* (1984) found that approximately half of their *trpC* transformants resulted from type I integration, and that type II were more common than type III. J. Tilburn and J. Kelly (unpublished observations), using the acetamidase system, found a lower proportion of type I transformants, perhaps reflecting the fact that the *amdS* recipient is deleted in that gene. Tilburn *et al.* (1983) were able to direct the *amdS* gene into the ribosomal repeat region by incorporating a fragment of the latter into their transforming plasmid.

Most of the work described has been carried out with circular plasmids, but some attempts have been made to examine the effect of linearization. In yeast, this is known to increase transformation frequency for integrating vectors, and to direct integration by the generation of recombinogenic ends (Orr-Weaver *et al.*, 1983). Yelton *et al.* (1984) cut their transforming plasmid within the *trpC* gene and observed some targeting effect (increase in proportion of type I and type III transformants), but no increase in transformation frequency.

3. Cotransformation

There is evidence from a number of laboratories for multiple integration in a single nucleus following transformation (Ballance *et al.*, 1983; Yelton *et al.*, 1984). These observations pointed to the feasibility of cotransformation with plasmid mixtures. If two marked plasmids are mixed prior to transformation, selection for one plasmid results in a high proportion of transformants containing the second plasmid. Experiments with *amdS*- and *prnC*-marked plasmids have shown that cotransformation can be very efficient (up to 80%, J. Tilburn and J. Kelly, unpublished observations). Similarly, we have cotransformed *Aspergillus* with a mixture of a plasmid carrying *pyr4* and the *oliC31* (resistant) gene carried in lambda. Selection for *pyr4* resulted in *oliC* cotransformation at a frequency of 4%.

4. Integration of a Foreign Gene

Transformation of *Aspergillus* using the *Neurospora pyr4* gene results in integration at more than one site on different linkage groups (i.e., equivalent to type II integration), and we have no evidence for a favored site. The equivalent *Aspergillus* gene, *pyrG*, is located on linkage group I, but no preference for this linkage group is apparent. This is not surprising in view of the very low homology between *pyr4* and *pyrG*, demonstrated by the failure of the former to hybridize to specific genomic DNA fragments of *A. nidulans*. This feature should make *pyr4* a useful selectable marker for transformation when attempts are made to replace wild-type host genes with *in vitro*-mutated genes, since it will increase the probability of integration at the required site.

One of the integration sites, on linkage group VI, has been analyzed in some detail by restriction mapping (Fig. 2). The point of recombination with the

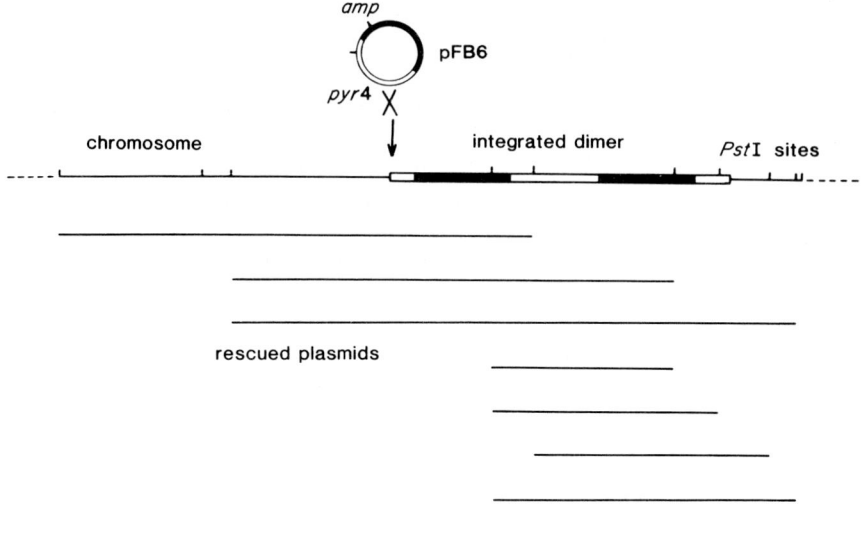

Fig. 2. Analysis of an integration event on chromosome VI (D. J. Ballance and G. Turner, unpublished observations). (■) bacterial plasmid DNA; (═) *Neurospora crassa* DNA carrying *pyr4*; (———) genomic DNA. Plasmids recovered in *E. coli* by marker rescue (selection for ampicillin resistance) following partial digestion and religation of total DNA from the transformant are shown.

Aspergillus chromosome can be located, and appears to be within the *Neurospora pyr4* region, since subcloning experiments have enabled us to delineate approximately the region required for expression of *pyrF* complementing activity in *E. coli*. This recombination interrupts the *pyr4* gene, generating a hybrid region which would not express. We have observed that a tandem repeat (Fig. 2) is generated in this transformant, which produces an intact *Neurospora* sequence between two bacterial plasmid sequences. Such a structure could be generated either by a single integration event involving a dimer, or successive events involving monomers.

Closer examination of the region of recombination shows that a part of the *Neurospora pyr4* in the region of recombination has been lost during integration. This may represent deletion or gene conversion, and can be further investigated when the corresponding *Aspergillus* sequence has been analyzed from an untransformed strain. Sequencing of the *pyr4* gene and the homologous *Aspergillus* sequence will provide the basis for an investigation of the recombination process itself.

It is interesting to note that the gene for phosphoribosylanthranilate isomerase has been isolated from *Aspergillus niger* and shown to express in an *A. nidulans trpC* mutant (Table I).

5. Genetic Analysis of Transformants

Genetic mapping can provide a convenient way of identifying the site of integration in any transformant, but experience has shown that the stability of the integrated DNA may vary between individual transformants at different stages in the life cycle.

a. Asexual Cycle. The phenotypic stability of transformants during normal vegetative growth, conidiation, and germination (mitotic stability) can be judged by plating a transformant on medium lacking selective pressure for the transforming marker. When individual colonies are then replicated onto selective medium, the rate of loss of the transforming marker can be assessed. The stability of a number of transformants in different laboratories with different genes has varied from 97 to >99.9%. However, it should be noted that transformants exhibit a wide variety of morphological phenotypes, and successive subculture of individual transformants often results in further morphological changes, indicative of some genetic instability even when the transforming marker remains.

b. Parasexual Cycle. In *Aspergillus*, markers can be easily located to particular chromosomes by constructing a diploid between the test strain and a master strain, the latter carrying at least one marker on each chromosome, then inducing haploidization, which generates haploids with chromosomes from either parent (McCully and Forbes, 1965). In experiments with the *Neurospora pyr4* gene, we have observed that some transformants retain the transforming marker during this process, and the marker can be assigned to a particular chromosome, while others lose the marker at some stage of the parasexual cycle.

c. Sexual Cycle. Reports from different laboratories have suggested variation in stability of markers through the sexual cycle. Tilburn *et al.* (1983) observed loss of the integrated *amdS* gene in a third of the progeny from a selfed cleistothecium. Yelton *et al.* (1984) also observed instability with some transformants (up to 12% loss of *trpC*) after selfing.

B. Marker Rescue

Since integration events by homologous recombination usually result in incorporation of the bacterial plasmid sequence into the chromosome, it is possible to rescue the region of integration in *E. coli*. This technique is important for detailed analysis of the region to deduce the mode of integration and for isolation of flanking chromosomal sequences. Any self-cloning method which relies on chromosomal integration will necessitate subsequent rescue to obtain the desired gene. If the integration event has generated a tandem repeat, as in Fig. 2, then complete digestion of total cell DNA with an enzyme which does not disrupt either ampicillin resistance or replication origin of the plasmid on subsequent religation will generate a viable plasmid which will transform *E. coli*. In order to analyze the whole region, it is necessary to religate a partial digest, generating a range of overlapping plasmids (Fig. 2). More surprisingly, it is possible to recover plasmids in *E. coli* at low frequency by transforming with untreated total cell DNA. Others (R. F. M. van Gorcom *et al.*, unpublished observations) have repeated the same observation with different plasmids employed in the initial transformation of *Aspergillus*, suggesting that it is a general phenomenon in *Aspergillus*, and cannot therefore be taken as evidence for the presence of freely replicating plasmids in the mycelium. This phenomenon, together with the relative ease of chromosomal integration of circular plasmids, creates a problem in the identification of autonomously replicating plasmids. This should be borne in mind when interpreting results for *Neurospora*, where claims for replicating plasmids have been made on the basis of ability to rescue without cutting (Stohl *et al.*, 1984). Marker rescue is not without problems. In some cases, plasmids rescued in *E. coli* show deletions and rearrangements. The mechanisms involved in these phenomena have not yet been fully investigated, but it is interesting to note that linear DNA can be recircularized by a recombination-dependent mechanism even in *recA* strains of *E. coli* (Conley and Saunders, 1984). When the linear DNA is monomeric, deletions often occur during the recombination process, but where dimeric structures are present, deletions do not occur. The first type of recombination may depend on chance homology between certain nonidentical sequences, or the presence of short repeated sequences, resulting in aberrant plasmids, whereas in the second type, precise homologous recombination is favored, producing an intact, monomeric plasmid. Consistent with this is the observation that rescue without cutting from the transformant shown in Fig. 2 has always yielded plasmids in *E. coli* identical with pFB6, the original plasmid.

IV. THE QUEST FOR REPLICATING VECTORS

In the case of *Saccharomyces cerevisiae*, where integration of circular plasmids is very inefficient, high-frequency transformation was obtained by inclusion of sequences permitting autonomous replication (Beggs, 1978). These include all or part of the native 2-μm plasmid, chromosomal fragments from yeast [putative replication origins or autonomously replicating sequences (ARSs)], chromosomal fragments from other eukaryotes, and mitochondrial sequences from yeast and other eukaryotes (see Chapters 5 and 6). Resulting transformants are mitotically unstable unless a centromere is added to generate a circular minichromosome. If linearized, telomeres are also required. The main advantages of autonomous vectors are the gain in transformation frequency, permitting isolation of yeast genes in yeast by shotgun cloning, and recovery from yeast without excision from the chromosome. Consequently, *Aspergillus* and *Neurospora* workers have tried similar approaches.

A. Native Plasmids

To date, no native plasmids have been found in *Aspergillus,* though mitochondrially located plasmids which show no homology with mtDNA have been isolated from some strains of *Neurospora* (Collins *et al.,* 1981; Kinsey, Chapter 9, this volume). Although attempts have been made to use these in the construction of replicating vectors, increases in transformation frequency have not been dramatic (5–10-fold, Stohl and Lambowitz, 1983), and it is not clear whether they are indeed replicating in *Neurospora.*

B. Yeast-Selected ARS

It is relatively easy to select for sequences with ARS behavior in yeast by using the nonreplicating yeast vector YIp5 (Stinchcomb *et al.,* 1980). Fragments of *Aspergillus* DNA with cohesive ends are ligated into a suitable site in YIp5 prior to transformation of a yeast strain carrying the *ura3-52* mutation, in which integration is negligible. Each transformant then carries a replicating plasmid, YIp5 plus "ARS", and this can be reisolated and incorporated into any *Aspergillus* vector. A number of such sequences have been tried in *Aspergillus* by ourselves and others, without any apparent effect on transformation frequency. However, one such sequence, a 3.5-kilobase (kb) *Eco*RI chromosomal fragment (Fig. 3) designated *ans1* (D. J. Ballance and G. Turner, unpublished observations) resulted in a 50- to 100-fold increase in transformation frequency when included in vectors carrying the *pyr4* as selective marker. A feature of such transformation is the drastic reduction in abortive transformants from >90 to <5%. Most transformants so obtained are mitotically stable without selective pressure being maintained, suggesting integration of transforming DNA. When the *ans1* sequence is used as a probe against *Eco*RI-digested *A. nidulans* DNA, in addition to the native *ans1* sequence, a number of other homologous bands can be detected. Whether this has any bearing on the high frequency of transformation remains to be answered. It cannot be ruled out that *ans1* replicates before integration, but we have no evidence to support this.

C. Mitochondrial Sequences

Yeast mitochondrial DNA contains a number of replication origins. Parts of the mitochondrial genome carrying such origins are occasionally excised, replicate faster than the intact genome, and give rise to respiratory-deficient, cyto-

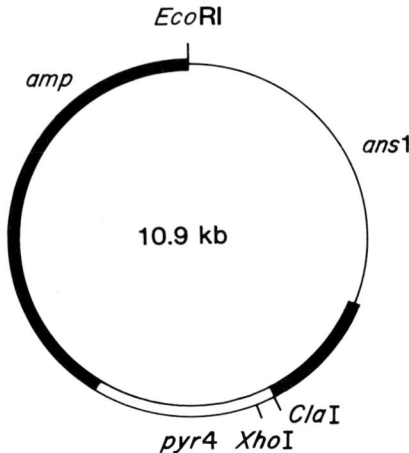

Fig. 3. High-frequency transformation vector pDJB3 and its cloning sites (D. J. Ballance and G. Turner, unpublished observations). (■) Sequences derived from pBR325; (═) *Neurospora crassa* DNA carrying *pyr4*; (──) *Aspergillus nidulans* DNA (*ans1*).

plasmically inherited petite mutations (Bernardi, 1982). Since the demonstration that such origins act as ARS in yeast vectors (i.e., outside the mitochondrion), there has been interest in using mitochondrial origins of filamentous fungi to construct vectors (see Tudzynski and Esser, Chapter 16). Since only a limited number of fungi exhibit petite-like phenomena (excision and independent replication of mitochondrial sequences), the method of Stinchcomb *et al.* (1980) as described above has been used to select ARS fragments from isolated mitochondrial DNA. A sequence of this type isolated from mitochondrial DNA of *A. nidulans* was included in a plasmid carrying *amdS*, but no increase in transformation frequency was observed, nor was there any evidence for autonomous replication of the plasmid in *Aspergillus* (Tilburn *et al.*, 1983).

Further experiments with yeast mitochondrial DNA have suggested that in addition to the very strong replication origins defined by their presence in suppressive petites, and by certain recognizable sequence features, it appears that many other sequences can behave as origins outside the mitochondrion when cloned in YIp5. Hyman *et al.* (1983) showed that such sequences occur every 1.7 kb around the mitochondrial genome. The specificity of sequence recognition by the yeast DNA polymerase thus appears to be rather low, and the value of this approach as a method for isolating specific replication origins from either mitochondrial or chromosomal DNA of other organisms is dubious. No conclusive results have yet emerged from attempts to construct replicative vectors for filamentous fungi using such sequences.

Petite-like behavior has been observed in a limited number of filamentous fungi, including *Aspergillus amstelodami* (Lazarus *et al.*, 1980; Lazarus and

Küntzel, 1981), but not *A. nidulans*, and in *Podospora anserina* (Cummings *et al.*, 1979; Esser *et al.*, 1980). *Aspergillus amstelodami* "ragged" mutants result from excision of either of two regions (region 1 and region 2). We have tested these sequences in yeast, and found that only region 1 has ARS behavior in the latter (J. Ham and G. Turner, unpublished observations). This sequence does not significantly improve transformation frequency in *A. nidulans*. A region of the mtDNA of *A. nidulans* which has homology with region 1 also acts as an ARS in yeast, and this sequence has also been tested in *A. nidulans* vectors. A slight improvement in transformation frequency was observed (R. Beri and D. J. Ballance, unpublished observations), but the basis for this apparent improvement is not yet understood. At present, there is no convenient way of isolating ARS by a functional test in *Aspergillus*, since integration occurs with a relatively high efficiency.

V. DEVELOPMENT OF VECTORS

A. Gene Isolation by Self-Cloning

Isolation of genes by direct shotgun cloning and selection in *Aspergillus* requires the development of vectors which transform at high frequency, contain suitable cloning sites, and can be recovered along with the desired sequence following transformation.

1. Plasmid Vectors

At present, we are experimenting with pDJB3, which we have recently constructed (Fig. 3). This plasmid transforms *A. nidulans* at a frequency of approximately 5000 transformants per microgram of DNA. Since 50 µg of DNA is not saturating in our system, it is possible to obtain 200,000 transformants without difficulty in a single experiment. The plasmid has single sites for *Eco*RI, *Cla*I, and *Xho*I. A gene bank of *Aspergillus* total DNA has been made by inserting fractionated (approximately 5–10 kb) quasi-random *Taq*I partial digest fragments into the *Cla*I site of the vector, and transforming *E. coli*.

This gene bank has recently been used successfully to isolate a sequence of DNA complementing an isocitrate lyase-deficient mutant, *acuD*, and a summary of the technique is given in Fig. 4. Since this sequence complements other alleles at the same locus, including three temperature-sensitive mutations (Armitt *et al.*, 1976), it is likely to carry the structural gene rather than a suppressor.

The vector pDJB3 is the first self-cloning vector for a filamentous fungus, and it should be possible to use it to isolate other genes by the same approach. Its limitation is only the size of DNA fragments which can be conveniently included

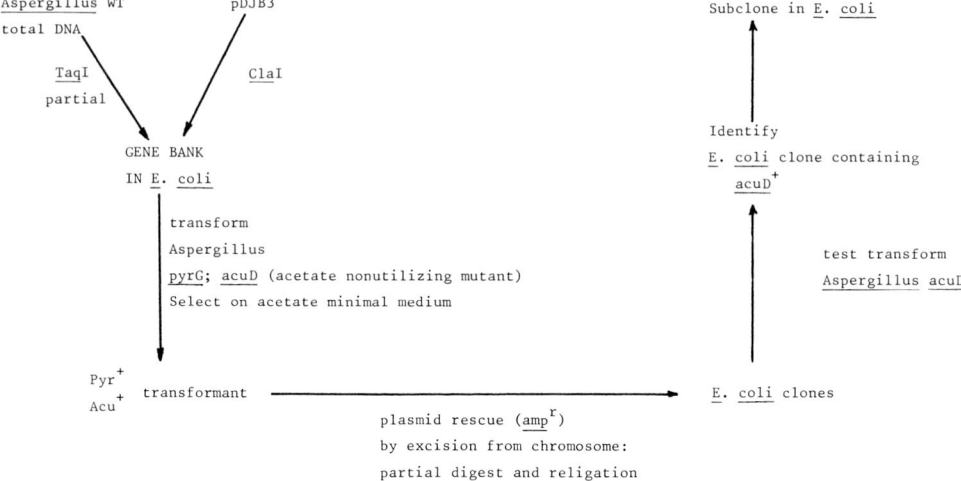

Fig. 4. Summary of the strategy employed in the isolation of the isocitrate lyase gene from *Aspergillus nidulans* (D. J. Ballance and G. Turner, unpublished observations).

in the gene bank. Attempts can now be made to isolate a number of genes of interest by expression in *Aspergillus*.

2. Cosmid Vectors

An alternative strategy to plasmid vectors would be to aim for a gene bank containing longer sequences, thus requiring fewer transformants to cover the genome. This would be valuable where there is no means of positive selection for the desired phenotype. Inclusion of a fungal selectable marker in a cosmid is the first step in this approach, and a combination of plasmid and cosmid vectors would give greater flexibility for gene isolation and reintroduction.

B. A Vector for Promoter Isolation

An interesting vector which has recently been constructed (R. F. M. van Gorcom *et al.*, unpublished observations) has potential for isolating fungal promoters. This vector carries *amdS* as a selective marker and part of the *trpC* gene. The C-terminal portion of the latter has been removed, and replaced by the *E. coli* β-galactosidase gene. Following transformation and integration in *Aspergillus*, the bacterial enzyme is expressed, and can be detected by incorporation of the indicator X-gal into the growth medium, with the consequent formation of blue *Aspergillus* colonies. Although *Aspergillus* itself produces a β-galactosidase and hence blue colonies on X-gal, this native enzyme can be repressed on glucose, while the bacterial enzyme is unaffected. Removal of the promoter region of the *trpC* gene from the vector has stopped the expression of the bacterial enzyme. This vector can now be used as a "promoter probe" to select fungal promoter sequences from *Aspergillus* DNA, and points the way to the construction of high-efficiency expression vectors.

VI. FUTURE PROSPECTS

It is now clear that the recent rapid advances in this field should make *A. nidulans* an ideal organism for basic studies on the molecular biology of filamentous fungi, and a starting point for applied work with commercially important, related organisms such as *A. niger, Penicillium,* and *Cephalosporium*. Furthermore, it is possible that *A. nidulans* might have potential as a cloning host for other fungal genes as a first step toward studies on genetically uncharacterized fungi.

REFERENCES

Armitt, S., McCullough, W., and Roberts, C. F. (1976). Analysis of acetate non-utilizing (*acu*) mutants in *Aspergillus nidulans. J. Gen. Microbiol.* **92,** 263–282.

Bach, M.-L., Lacroute, F., and Botstein, D. (1979). Evidence for transcriptional regulation of orotidine-5'-phosphate decarboxylase in yeast by hybridization of mRNA to the yeast structural gene cloned in *Escherichia coli. Proc. Natl. Acad. Sci. U.S.A.* **76,** 386–390.

Ballance, D. J., Buxton, F. P., and Turner, G. (1983). Transformation of *Aspergillus nidulans* by the orotidine-5'-phosphate decarboxylase gene of *Neurospora crassa. Biochem. Biophys. Res. Commun.* **112,** 284–289.

Beggs, J. D. (1978). Transformation of yeast by a replicating hybrid plasmid. *Nature (London)* **275,** 104–109.

Bergen, L. G., Upshall, A., and Morris, N. R. (1984). S-phase, G2 and nuclear division mutants of *Aspergillus nidulans. J. Bacteriol.* **159,** 114–119.

Bernardi, G. (1982). The origins of replication of the mitochondrial genome of yeast. *Trends Biochem. Sci.* **7,** 404–408.

Berse, B., Dmochowska, A., Skrzypek, M., Weglenski, P., Bates, M. A., and Weiss, R. L. (1983). Cloning and characterization of the ornithine carbamoyltransferase gene from *Aspergillus nidulans. Gene* **25,** 109–117.

Bibb, M. J., Ward, J. M., and Hopwood, D. A. (1978). Transformation of plasmid DNA into Streptomyces at high frequency. *Nature (London)* **274,** 398–400.

Bos, C. J., and Slakhorst, S. M. (1981). Isolation of protoplasts from *Aspergillus nidulans* conidiospores. *Can. J. Microbiol.* **27,** 400–407.

Case, M. E. (1982). Transformation of *Neurospora crassa. In* "Engineering of Microorganisms for Chemicals" (A. Hollaender, R. D. DeMoss, J. Koninsky, D. Savage, and R. S. Wolfe, eds.), Basic Life Sciences, Vol. 19, pp. 87–100. Plenum, New York.

Case, M. E., Schweizer, M., Kushner, S. R., and Giles, N. H. (1979). Efficient transformation of *Neurospora crassa* by utilizing hybrid plasmid DNA. *Proc. Natl. Acad. Sci. U.S.A.* **76,** 5259–5363.

Chang, S., and Cohen, S. N. (1979). High frequency transformation of *Bacillus subtilis* protoplasts by plasmid DNA. *Mol. Gen. Genet.* **168,** 111–115.

Cohen, J. D., Eccleshall, T. R., Needleman, R. B., Federoff, H., Buchferer, B. A., and Marmur, J. (1980). Functional expression in yeast of the *Escherichia coli* plasmid gene coding for chloramphenicol acetyltransferase. *Proc. Natl. Acad. Sci. U.S.A.* **77,** 1078–1082.

Colbère-Garapin, F., Horodniceanu, F., Kouritsky, P., and Colbère, A.-C. (1981). A new hybrid selective marker for higher eukaryotic cells. *J. Mol. Biol.* **150,** 1–14.

Collins, R. A., Stohl, L. L., Cole, M. D., and Lambowitz, A. M. (1981). Characterization of a novel plasmid DNA found in mitochondria of *N. crassa*. *Cell (Cambridge, Mass.)* **24,** 443–452.

Conley, E. C., and Saunders, J. R. (1984). Recombination-dependent recircularization of linearized pBR322 plasmid DNA following transformation of *Escherichia coli*. *Mol. Gen. Genet.* **194,** 211–218.

Cummings, D. J., Belcour, L., and Grandchamp, C. (1979). Mitochondrial DNA from *Podospora anserina*. Properties of mutant DNA and multimeric circular DNA from senescent cultures. *Mol. Gen. Genet.* **171,** 239–250.

de Vries, O. M. H., and Wessels, J. G. H. (1972). Release of protoplasts from *Schizophyllum commune* by a lytic enzyme preparation from *Trichoderma viride*. *J. Gen. Microbiol.* **73,** 13–22.

Dhawale, S. S., Paietta, J. V., and Marzluf, G. A. (1984). A new, rapid and efficient transformation procedure for Neurospora. *Curr. Genet.* **8,** 77–79.

Dujon, B. (1980). Sequence of the intron and flanking exons of the mitochondrial 21s rRNA gene of yeast strains having different alleles at the ω and *rib-1* loci. *Cell (Cambridge, Mass.)* **20,** 185–197.

Earl, A. J., Turner, G., Croft, J. H., Dales, R. B. G., Lazarus, C. M., Lünsdorf, H., and Küntzel, H. (1981). High frequency transfer of species specific mitochondrial DNA sequences between members of the Aspergillaceae. *Curr. Genet.* **3,** 221–228.

Esser, K., Tudzynski, P., Stahl, U., and Kück, U. (1980). A model to explain senescence in the filamentous fungus *Podospora anserina* by the action of plasmid-like DNA. *Mol. Gen. Genet.* **178,** 213–216.

Gunatilleke, I. A. U. N., Scazzocchio, C., and Arst, H. N., Jr. (1975). Cytoplasmic and nuclear mutations to chloramphenicol resistance in *Aspergillus nidulans*. *Mol. Gen. Genet.* **137,** 269–276.

Hicks, J. B., Hinnen, A., and Fink, G. R. (1979). Properties of yeast transformation. *Cold Spring Harbor Symp. Quant. Biol.* **43,** 1305–1313.

Hinnen, A., Hicks, J. B., and Fink, G. R. (1978). Transformation of yeast. *Proc. Natl. Acad. Sci. U.S.A.* **75,** 1929–1933.

Hyman, B. C., Cramer, J. H., and Rownd, R. H. (1983). The mitochondrial genome of *Saccharomyces cerevisiae* contains numerous, densely spaced autonomous replicating sequences. *Gene* **26,** 223–230.

Hynes, M. J. (1979). Fine structure mapping of the acetamidase structural gene and its controlling region in *Aspergillus nidulans*. *Genetics* **91,** 381–392.

Hynes, M. J., Corrick, C. M., and King, J. A. (1983). Isolation of genomic clones containing the *amdS* gene of *Aspergillus nidulans* and their use in the analysis of structural and regulatory mutations. *Mol. Cell. Biol.* **3,** 1430–1439.

Ito, H., Fukuda, Y., Murata, K., and Kimura, A. (1983). Transformation of intact yeast cells treated with alkali cations. *J. Bacteriol.* **153,** 163–168.

Jimenez, A., and Davies, J. (1980). Expression of a transposable antibiotic resistance element in Saccharomyces. *Nature (London)* **287,** 869–871.

Kinghorn, J. R., and Hawkins, A. R. (1982). Cloning and expression in *Escherichia coli* K-12 of the biosynthetic dehydroquinase function of the *arom* cluster gene from the eukaryote, *Aspergillus nidulans*. *Mol. Gen. Genet.* **186**, 145–152.

Lazarus, C. M., and Küntzel, H. (1981). Anatomy of amplified mitochondrial DNA in "ragged" mutants of *Aspergillus amstelodami:* excision points within protein genes and a common 215bp segment containing a possible origin of replication. *Curr. Genet.* **4**, 99–107.

Lazarus, C. M., and Turner, G. (1977). Extranuclear recombination in *Aspergillus nidulans:* Closely linked multiple chloramphenicol and oligomycin resistance loci. *Mol. Gen. Genet.* **156**, 303–311.

Lazarus, C. M., Earl, A. J., Turner, G., and Küntzel, H. (1980). Amplification of a mitochondrial DNA sequence in the cytoplasmically inherited "ragged" mutant of *Aspergillus nidulans*. *Eur. J. Biochem.* **106**, 633–641.

McCully, K. S., and Forbes, E. (1965). The use of *p*-fluorophenylalanine with "master strains" of *Aspergillus nidulans* for assigning genes to linkage groups. *Genet. Res.* **6**, 352–359.

Orr-Weaver, T. J., Szostak, J., and Rothstein, R. (1983). Genetic applications of yeast transformation with linear and gapped plasmids. *In* "Recombinant DNA," Part C (R. Wu, L. Grossman, and K. Moldave, eds.), Methods in Enzymology, Vol. 101, pp. 228–245. Academic Press, New York.

Palmer, L. M., and Cove, D. J. (1975). Pyrimidine biosynthesis in *Aspergillus nidulans*. Isolation and preliminary characterisation of auxotrophic mutants. *Mol. Gen. Genet.* **138**, 243–255.

Peberdy, J. F. (1979). Fungal protoplasts: isolation, reversion and fusions. *Annu. Rev. Microbiol.* **33**, 21–39.

Peberdy, J. F., and Ferenczy, L., eds. (1985). "Fungal Protoplasts: Applications in Biochemistry and Genetics." Dekker, New York.

Polkinghorne, M., and Hynes, M. J. (1975). Effect of L-histidine on the catabolism of nitrogen compounds in *Aspergillus nidulans*. *J. Gen. Microbiol.* **87**, 185–187.

Pontecorvo, G., Roper, J. A., Hemmons, L. M., Macdonald, K. D., and Bufton, A. W. J. (1953). The genetics of *Aspergillus nidulans*. *Adv. Genet.* **5**, 141–238.

Rowlands, R. T., and Turner, G. (1973). Nuclear and extranuclear inheritance of oligomycin resistance in *Aspergillus nidulans*. *Mol. Gen. Genet.* **126**, 201–216.

Rowlands, R. T., and Turner, G. (1977). Nuclear–extranuclear interactions affecting oligomycin resistance in *Aspergillus nidulans*. *Mol. Gen. Genet.* **154**, 311–318.

Scazzocchio, C., Brown, T. A., Waring, R. B., Ray, J. A., and Davies, R. W. (1983). Organisation of the *Aspergillus nidulans* mitochondrial genome. *In* "Mitochondria 1983" (R. J. Schweyen, K. Wolf, and F. Kaudewitz, eds.), pp. 303–312. de Gruyter, Berlin.

Sebald, W., and Hoppe, J. (1981). On the structure and genetics of the proteolipid subunit of the ATP synthase complex. *Curr. Top. Bioenerget.* **12**, 1–64.

Smith, J. E., and Pateman, J. A., eds. (1977). "Genetics and Physiology of Aspergillus." Academic Press, New York.

Stinchcomb, D. T., Thomas, M., Kelly, S., Selber, E. R., and Davies, R. W. (1980). Eukaryotic DNA segments capable of autonomous replication in yeast. *Proc. Natl. Acad. Sci. U.S.A.* **77**, 4559–4563.

Stohl, L. L., and Lambowitz, A. M. (1983). Construction of a shuttle vector for the filamentous fungus *Neurospora crassa*. *Proc. Natl. Acad. Sci. U.S.A.* **80**, 1058–1062.

Stohl, L. L., Akins, R. A., and Lambowitz, A. M. (1984). Characterisation of deletion derivatives of an autonomously replicating Neurospora plasmid. *Nucleic Acids Res.* **12**, 6169–6178.

Tilburn, J., Scazzocchio, C., Taylor, G. G., Zabicky-Zissman, J. H., Lockington, R. A., and Davies, R. W. (1983). Transformation by integration in *Aspergillus nidulans*. *Gene* **26**, 205–221.

Turner, G., and Rowlands, R. T. (1977). Mitochondrial genetics of *Aspergillus nidulans. In* "Genetics and Physiology of Aspergillus" (J. E. Smith and J. A. Pateman, eds.), pp. 319–337. Academic Press, New York.

Turner, G., Imam, G., and Küntzel, H. (1979). Mitochondrial ATPase complex of *Aspergillus nidulans* and the dicyclohexylcarbodiimide-binding protein. *Eur. J. Biochem.* **97,** 565–571.

Tzagoloff, A., Macino, G., and Sebald, W. (1979). Mitochondrial genes and translation products. *Annu. Rev. Biochem.* **48,** 419–441.

Viebrock, A., Perz, A., and Sebald, W. (1982). The imported preprotein of the proteolipid subunit of the mitochondrial ATP synthase from *Neurospora crassa*. Molecular cloning and sequencing of the mRNA. *EMBO J.* **1,** 565–571.

Yelton, M. M., Hamer, J. E., de Souza, E. R., Mullaney, E. J., and Timberlake, W. E. (1983). Developmental regulation of the *Aspergillus nidulans trpC* gene. *Proc. Natl. Acad. Sci. U.S.A.* **80,** 7576–7580.

Yelton, M. M., Hamer, J. E., and Timberlake, W. E. (1984). Transformation of *Aspergillus nidulans* by using a *trpC* plasmid. *Proc. Natl. Acad. Sci. U.S.A.* **81,** 1470–1474.

Zimmerman, C. R., Orr, W. C., Leclerc, R. F., Bernard, E. C., and Timberlake, W. E. (1980). Molecular cloning and selection of genes regulated in Aspergillus development. *Cell (Cambridge, Mass.)* **21,** 709–715.

11

Expression of *Aspergillus* Genes in *Neurospora*

RICHARD L. WEISS

Department of Chemistry and Biochemistry
University of California at Los Angeles
Los Angeles, California

DIANE PUETZ

Molecular Biology Institute
University of California at Los Angeles
Los Angeles, California

JAN CYBIS

Institute of Biochemistry and Biophysics
Polish Academy of Sciences
Warsaw, Poland

I. Transformation of *Neurospora crassa*	280
A. Techniques	280
B. Experimental Difficulties	280
C. Expression of Heterologous Genes	281
II. Isolation of the *Aspergillus nidulans* Gene for Ornithine Carbamoyltransferase	281
III. Transformation of the *Neurospora crassa arg-12* Mutant	282
A. Subcloning of the Gene	282
B. Transformation	283
C. Evidence for Expression of the *A. nidulans* Gene: Thermal Stability of Ornithine Carbamoyltransferase from *A. nidulans*, *N. crassa*, and the Transformants	284
D. Stability of the Transformants	285

E. Genetic Analysis of the Transformants 286
IV. Control of Gene Expression 288
V. Subcellular Localization of the Active Enzyme 289
VI. Conclusions 290
References 291

I. TRANSFORMATION OF *NEUROSPORA CRASSA*

A. Techniques

A variety of techniques have been described for the transformation of *Neurospora crassa* (see, e.g., Huiet and Case, Chapter 8). The first successful method involved the $CaCl_2$-dependent uptake of DNA into spheroplasts formed by mycelial wall-degrading enzymes (Case *et al.*, 1979). This method has now been modified somewhat (Schweizer *et al.*, 1981; Schechtman and Yanofsky, 1983) and has been successfully employed by a number of laboratories. Other methods have been reported, such as liposome-mediated genetic transformation (Radford *et al.*, 1981) and the use of $LiCl_2$ (Dhawale *et al.*, 1984), but have not yet found wide acceptance. Transformation frequencies vary widely, but 500–1000 transformants per microgram of plasmid DNA can usually be obtained if the plasmid contains a suitable selectable marker. These techniques provide a means of introducing foreign genes into *N. crassa*.

B. Experimental Difficulties

Progress in the application of recombinant DNA techniques to filamentous fungi has been slower than the progress in bacteria, yeasts, and even mammalian cells. The primary limitations appear to be the failure of most genes derived from filamentous fungi to be expressed in either bacteria or yeast and the lack of vectors capable of autonomous replication in filamentous fungi. The latter makes it difficult to recover cloned genes from transformants. Recent reports of autonomously replicating vectors derived from mitochondrial plasmids (Esser *et al.*, 1983; Stohl and Lambowitz, 1983) are promising, but these techniques have not yet been successfully applied to the isolation of specific genes.

Attempts to recover bacterial vector sequences from fungal transformants have been largely unsuccessful. In *N. crassa*, they are often lost during integration into the genome or appear to be modified so that they are not expressed when reintroduced into *Escherichia coli*. Although these limitations have prevented the large-scale identification and subsequent isolation of genes from filamentous fungi, they do not prevent investigation of the expression of cloned genes obtained in other ways.

C. Expression of Heterologous Genes

A number of genes from a variety of filamentous fungi have been cloned by complementation in *E. coli* or yeast. However, little has been done to examine the expression of such genes in heterologous filamentous fungi. The results of such experiments should be of interest for a variety of reasons. First, successful expression of cloned fungal genes in a number of filamentous fungi would suggest that the problem of obtaining expression of such genes in *E. coli* and yeast is the result of a common set of transcriptional and/or translational parameters which are shared by filamentous fungi. Second, the results should provide details of common mechanisms of regulation and organelle biogenesis which operate in this important group of microorganisms.

II. ISOLATION OF THE *ASPERGILLUS NIDULANS* GENE FOR ORNITHINE CARBAMOYLTRANSFERASE

Ornithine carbamoyltransferase is a central enzyme in the biosynthesis of arginine in bacteria and fungi and in the urea cycle of mammalian organisms. The enzyme catalyzes the condensation of ornithine and carbamoyl phosphate to form citrulline. This reaction takes place in the cytoplasm of some yeasts and in the mitochondria of most filamentous fungi, some yeasts, and mammals (Urrestarazu *et al.*, 1977; Davis, 1983). Both ornithine and carbamoyl phosphate are substrates for several enzymes. Ornithine is decarboxylated to form polyamines and is an intermediate in the pathway of arginine degradation in most lower eukaryotes. These metabolic relationships in *N. crassa* are illustrated in Fig. 1. Carbamoyl phosphate is also used in pyrimidine synthesis. Thus, ornithine carbamoyltransferase represents a metabolic branch point and would be expected to be subject to metabolic controls. A number of such controls have been described (reviewed in Davis, 1983).

The different localization of ornithine carbamoyltransferase in various eukaryotic microorganisms makes the enzyme particularly interesting from an evolutionary as well as a regulatory perspective. In addition, its mitochondrial localization in *N. crassa*, *A. nidulans,* and some yeasts makes it a candidate for investigations of mitochondrial biogenesis and the coordination between mitochondrial and cytoplasmic gene expression. The cytoplasmic enzyme of *Saccharomyces cerevisiae* has been cloned (Crabeel *et al.*, 1981). Neither the yeast nor the mammalian enzyme is immunologically related to the mitochondrial enzymes of *A. nidulans* or *N. crassa* (Bates, 1984).

The gene for ornithine carbamoyltransferase from *A. nidulans* has been cloned by Weglenski and his colleagues through its ability to complement mutants of *S. cerevisiae* (Berse *et al.*, 1983). The success of this approach is remarkable from two perspectives: the enzyme functions in the mitochondrion of *A. nidulans,*

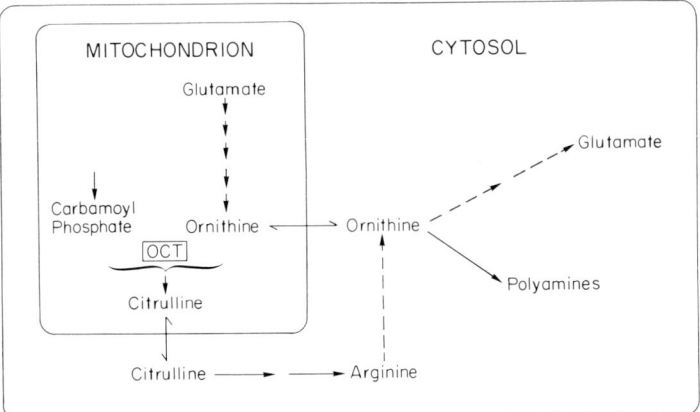

Fig. 1. Arginine metabolism and its organization in *Neurospora crassa*. Solid lines represent biosynthetic reactions, dashed lines represent catabolic reactions, and double-headed arrows represent movement of molecules across intramycelial membranes. Abbreviation: OCT, ornithine carbamoyltransferase.

whereas the corresponding enzyme is cytosolic in *S. cerevisiae;* and genes from filamentous fungi are rarely expressed in yeast. DNA–DNA hybridization and immunological tests demonstrated that the enzyme made in the transformants was the *A. nidulans* enzyme. The cloned gene was expressed in either orientation in the vector but was not expressed in *E. coli* without the occurrence of DNA rearrangements.

We have used the cloned gene to transform an ornithine carbamoyltransferase-deficient strain of *N. crassa* in order to investigate the expression of the gene, the processing of the cytosolic precursor, and the function of the enzyme. Our results indicate that the *A. nidulans* enzyme is expressed in *N. crassa*, that it is translocated into the mitochondrion, and that expression appears to respond to the same regulatory signals which control the expression of the arginine biosynthetic enzymes in *N. crassa*.

III. TRANSFORMATION OF THE *NEUROSPORA CRASSA* *ARG-12* MUTANT

A. Subcloning of the Gene

The ornithine carbamoyltransferase gene was originally isolated on a 6.2-kilobase (kb) *Bam*HI fragment in an *E. coli*—yeast chimeric vector (Berse *et al.*, 1983). The 6.2-kb insert of *A. nidulans* DNA from the resulting plasmid, pBB116, was excised with *Bam*HI, electrophoretically purified, and inserted into

11. Expression of *Aspergillus* Genes in *Neurospora*

the *Bam*HI site of pBR322. The resulting plasmid, pDP2, is pBB116 minus the yeast DNA (Fig. 2).

B. Transformation

The recombinant plasmid was used to transform an ornithine carbamoyltransferase-deficient strain of *N. crassa*, LA14 (*arg-12*). Transformation was performed by a modification of the spheroplast technique first described by Case *et al.* (1979) and modified as described by Schweizer *et al.* (1981) and Schechtman and Yanofsky (1983). Approximately 20 transformants were obtained per microgram of DNA. Thirty-six transformants were chosen for further study. Three of the 36 failed to grow when transferred to agar slants containing unsupplemented medium. In subsequent transformations, a large number of "abortive" transformants were observed. The growth rate and the amount of conidia formed varied widely among the stable transformants, suggesting a heterogeneous population of transformants.

Three of the transformants have been characterized further. The growth rates and enzyme levels of these transformants and wild-type strains of *N. crassa* and *A. nidulans* are shown in Table I. The activity of ornithine carbamoyltransferase in the transformants is very low. The activities are considerably less than those observed in wild-type strains of either *N. crassa* or *A. nidulans*. Transformant LA415 appeared to grow as fast as the wild-type strain despite a 20-fold reduction in ornithine carbamoyltransferase activity. This is not surprising since a strain of *N. crassa* having a mutation in the structural gene for ornithine carbamoyltransferase (*arg12^s*) is known with a similar wild-type growth rate but reduced ornithine carbamoyltransferase activity (Davis, 1962).

The results substantiate the conclusion that the transformants are hetero-

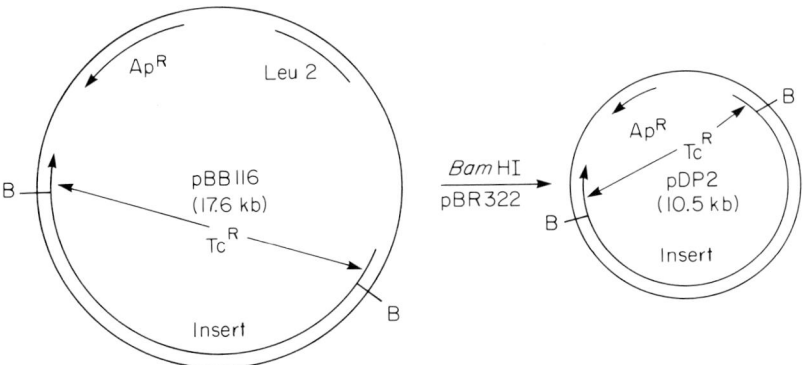

Fig. 2. Construction of plasmid pDP2. Abbreviations: Ap^R, ampicillin resistance; Tc^R, tetracycline resistance; B, *Bam*HI restriction site.

TABLE I.

Properties of *Neurospora crassa* Transformants[a]

Strain	Doubling time (hr)	Enzyme activity (units/mg protein)
Neurospora crassa		
LA1 (wild-type)	1.8	0.81
LA402 (transformant 2)	3.0	N.D.
LA415 (transformant 15)	1.8	0.036
LA430 (transformant 30)	2.5	0.037
Aspergillus nidulans		
LA1 (wild-type, FGSC 4)	N.D.	0.42

[a] Transformants were purified twice on selective medium. Growth rates were determined by the increase in turbidity (Basabe et al., 1979). Mycelia were grown to late log phase, harvested, and frozen until used for enzyme activity measurements. Mycelia were permeabilized and assayed under optimal conditions for the *N. crassa* enzyme (Davis, 1962).

geneous and suggest that the cloned gene is expressed poorly (or that defects exist in its processing; see Conclusions, Section VI). The reduced growth rates may be due to arginine limitation.

C. Evidence for Expression of the *A. nidulans* Gene: Thermal Stability of Ornithine Carbamoyltransferase from *A. nidulans*, *N. crassa*, and the Transformants

Although no "transformants" were obtained in the absence of added pDP2 DNA, it was possible that the apparent transformants were due to reversion or suppression of the parental *arg-12* mutation. Southern hybridization analysis (see below) detected the presence of DNA sequences homologous to the 6.2-kb *A. nidulans* insert in pDP2. No homologous sequences were detected in the recipient (host) strain or in wild-type strains.

In order to confirm the presence of the *A. nidulans* enzyme in the transformants, the thermal stability of the ornithine carbamolytransferase activity was determined in wild-type *N. crassa*, wild-type *A. nidulans*, and two transformants. The results are shown in Table II. The *A. nidulans* enzyme is considerably more sensitive to thermal inactivation than is the corresponding enzyme from *N. crassa*: 98% of the activity was lost after 10 min at 70°C, whereas only 23% of the *N. crassa* activity was lost after such treatment. The activities in the transformants have the thermal instability characteristic of the *A. nidulans* enzyme. This observation, coupled with the reappearance of the arginine auxotrophic phenotype in progeny of matings with wild-type strains (see below), suggests that the transformants express the *A. nidulans* gene and are not rever-

TABLE II.

Thermal Stability of Ornithine Carbamoyltransferase from *Neurospora crassa*, *Aspergillus nidulans*, and Transformants[a]

Strain	Enzyme activity (%)		
	25°C	60°C	70°C
Neurospora crassa			
LA1 (wild-type)	100	92	77
LA415 (transformant 15)	100	59	0
LA430 (transformant 30)	100	46	0
Aspergillus nidulans			
LA1 (wild-type, FGSC 4)	100	64	2

[a] Strains were grown to late log phase and then treated with 20 mM 3-amino-1,2,4-triazole for 3 hr. Mycelia were then harvested and frozen until used for the determination of enzyme activities. Three samples of each preparation of permeabilized mycelia were heated for 10 min at 25, 60, and 70°C before being assayed for ornithine carbamoyltransferase activity.

tants of the *arg-12* mutation. This is consistent with the absence of revertants in control transformations without pDP2 DNA.

D. Stability of the Transformants

1. Mitotic Stability

The original transformants were purified by several successive transfers on selective medium. The mitotic stability of two transformants was tested by plating conidia on nonselective medium (*Neurospora* culture agar). Microscopic examination of the plate revealed few nonviable conidia. Individual colonies were isolated and tested for growth on minimal medium. Twenty-four colonies of both LA415 and LA430 were tested. All the isolates were able to grow on unsupplemented medium, indicating that the transformation event was stable through mitosis.

2. Meiotic Stability

Numerous attempts were made to mate the transformants with the parental strain LA14 (*arg-12*). All such attempts were unsuccessful. This is reminiscent of the allelic infertility of arginine auxotrophs (Davis, 1979). Both transformants were fertile as either the male or female parent in matings to a wild-type strain. The outcome of these crosses (see below) was consistent with the stable transmission of the *A. nidulans* gene during meiosis.

E. Genetic Analysis of the Transformants

Tetrad analysis was performed on progeny of crosses between the transformants and a wild-type strain of *N. crassa*. The results (Table III) are consistent with the presence of a single functional genetic ornithine carbamoyltransferase locus in the transformants unlinked to the original *arg-12* mutation—that is, approximately 25% of the progeny were arginine auxotrophs. The significant number of tetratype tetrads suggests that this locus is not closely linked to the centromere of the chromosome to which the ornithine carbamoyltransferase gene had integrated. Further genetic analysis will be required to substantiate these conclusions. Although these results are consistent with a single functional genetic locus, the data cannot rule out expression of multiple copies of tandemly repeated genes.

Southern blot hybridization analysis was performed to determine the pattern of integration. Representative results are shown in Fig. 3. The *Sal*I/*Pst*I-digested transformant DNA in lanes 3 and 4 has bands corresponding to each of those in the *Sal*I digest of pDP2 (lane 1), except for the last band, which is cleaved by *Pst*I and runs off the gel. For transformant 2 (lane 3), the extra bands are fainter and may be those produced by the integration event(s). The intensity of the bands in lane 4 (transformant 30) suggests that the *A. nidulans* DNA sequences are present in high copy number in this transformant. We hypothesize that an initial integration event took place at one or two sites and was followed by several tandem integrations of pDP2. Similar tandemly repeated integration events have been observed for homologous transformation of *N. crassa* (A. M. Lambowitz, personal communication).

TABLE III.

Progeny from Crosses between Transformants and a Wild-Type Strain of *Neurospora crassa*[a]

Transformant	Parental ditypes	Nonparental ditypes	Tetratypes	ARG⁻ (%)
LA415	1	3	2	33
LA430	2	2	4	25

[a] Unordered tetrads were isolated and tested for their ability to grow in arginine-free medium. Four tetrad types were observed and classified as follows: 8 ARG⁺ (parental ditype); 4 ARG⁺ and 4 ARG⁻ (nonparental ditype); and 6 ARG⁺ and 2 ARG⁻ (tetratype). The assumptions concerning the presence of *A. nidulans* DNA sequences in the progeny are being tested by Southern hybridization.

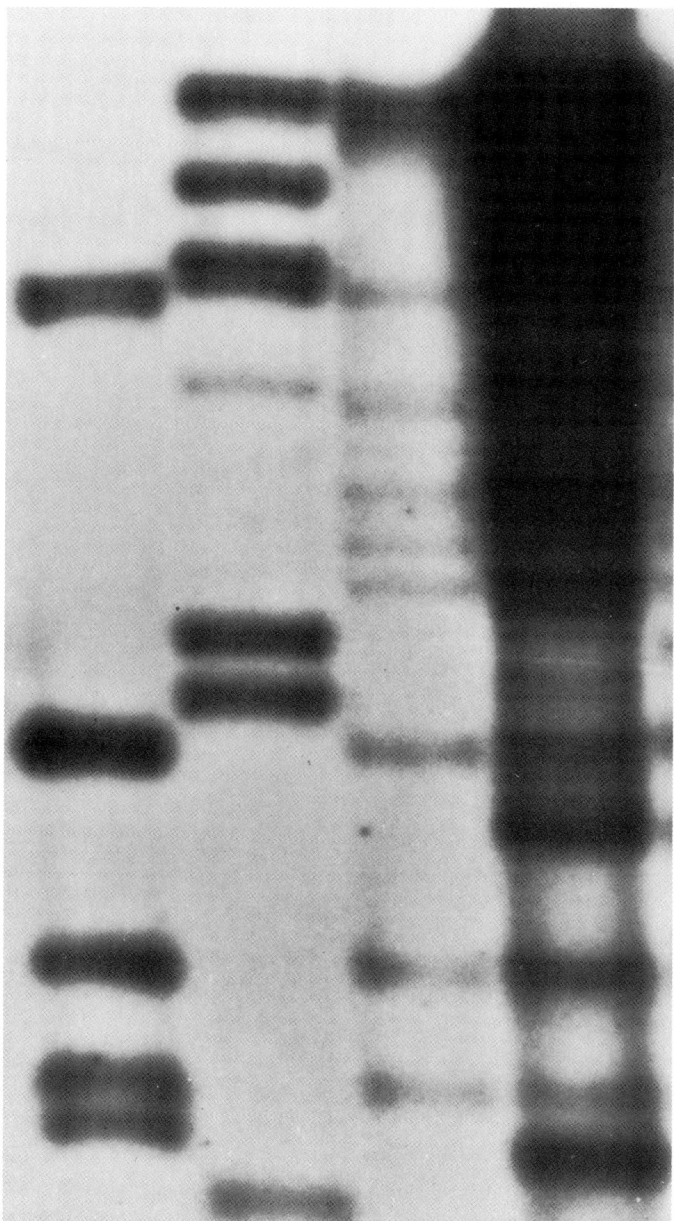

Fig. 3. Hybridization patterns of representative transformants. DNA preparations were probed with the 6-kb *Bam*HI fragment from pDP2. Lane 1, transforming plasmid pDP2 digested with *Sal*I; lane 2, lambda *Hin*dIII size standards; lane 3, 10 μg of DNA from LA402 digested with *Pst*I and *Sal*I; lane 4, 10 μg of DNA from LA430 digested with *Pst*I and *Sal*I.

IV. CONTROL OF GENE EXPRESSION

The expression of arginine biosynthetic enzymes in *N. crassa* is controlled by "cross-pathway" regulation: all the enzymes are derepressed in response to limitation (or starvation) for any of a number of amino acids (Carsiotis and Jones, 1974; Carsiotis *et al.*, 1974; Flint and Kemp, 1981; Kemp and Flint, 1982). None of the biosynthetic enzymes, with the exception of the small subunit of carbamoyl-phosphate synthetase, is repressed in the presence of exogenous arginine. Cross-pathway control is mediated by the product of the *cpc* locus (Barthelmess, 1982). The operation of the cross-pathway control system can be detected by observing the derepression of amino acid biosynthetic enzymes in response to histidine limitation. The latter can be brought about by treatment of mycelia of *N. crassa* with 3-amino-1,2,4-triazole.

Wild-type strains of both *A. nidulans* and *N. crassa* exhibit a three- to fivefold derepression of ornithine carbamoyltransferase in response to histidine limitation (Carsiotis *et al.*, 1974; Piotrowska, 1980; Flint and Kemp, 1981; Kemp and Flint, 1982). The results of such experiments on the transformants are shown in Table IV. The ornithine carbamoyltransferase activity in both transformants was derepressed upon treatment with 3-amino-1,2,4-triazole. The simplest explanation for these results is that the *A. nidulans* control sequences are present on the 6.2-kb insert in pDP2; these sequences integrate into the *N. crassa* genome along with the ornithine carbamoyltransferase structural gene; and these control sequences are recognized by the product of the *cpc* locus.

TABLE IV.

Control of Gene Expression in *Neurospora crassa*, *Aspergillus nidulans*, and Transformants[a]

Strain	Enzyme activity (units/mg protein)		Fold increase
	Minimal medium	+ Aminotriazole	
Neurospora crassa			
LA1 (wild-type)	0.45	2.3	5.1
LA415 (transformant 15)	0.016	0.066	4.1
LA430 (transformant 30)	0.019	0.055	2.9
Aspergillus nidulans			
LA1 (wild-type, FGSC 4)	0.42	0.82	1.9

[a] Germinated conidia were grown in minimal medium for 3 hr and then one-half of each culture was supplemented with 20 mM 3-amino-1,2,4-triazole. After 9 hr, the mycelia were harvested and frozen until used for determination of enzyme activities. The enzyme activities in older mycelia are lower than those observed in log-phase mycelia (Table I)

V. SUBCELLULAR LOCALIZATION OF THE ACTIVE ENZYME

The localization of the active enzyme was determined by subcellular fractionation of mycelia of the *N. crassa* transformants and a wild-type strain. The results are shown in Table V. Recoveries of a cytosolic marker enzyme (glucose-6-phosphate dehydrogenase) and a mitochondrial membrane marker (cytochrome oxidase) were measured to control for efficient separation of cytosolic and mitochondrial fractions. The low recoveries of ornithine carbamoyltransferase (a mitochondrial matrix enzyme) are largely due to mitochondrial breakage during fractionation.

A significant proportion of the ornithine carbamoyltransferase activity was recovered in the mitochondrial fraction. This suggests that the cytoplasmic precursor of the *A. nidulans* enzyme recognizes *N. crassa* mitochondria and can be processed by the *N. crassa* processing machinery. The lower recovery of ornithine carbamoyltransferase in this fraction in the transformants suggests that some functionally active molecules may be present in the cytoplasm in these strains. Such molecules might represent unprocessed precursors having reduced but significant catalytic activity. Such reduced processing efficiency may account for the low activities present in the transformants. These possibilities are being examined by immunoprecipitation using antibodies to the *N. crassa* enzyme.

TABLE V.

Subcellular Localization of Ornithine Carbamoyltransferase[a]

Strain	Mitochondrial enzyme activity (%)		
	Glucose-6-phosphate dehydrogenase	Cytochrome oxidase	Ornithine carbamoyltransferase
LA1 (wild-type)	<1	95	32
LA415	<1	95	20
LA430	<1	96	13

[a] Mycelia were grown to stationary phase in unsupplemented medium, harvested, and frozen until used. Mycelia were broken with glass beads in an osmotically stabilized buffer and the crude organellar fraction was collected by differential centrifugation. The mitochondria were then further purified on a sucrose step gradient. The cytosolic and mitochondrial fractions were then assayed for ornithine carbamoyltransferase activity. All three strains exhibited similar activities for glucose-6-phosphate dehydrogenase and cytochrome oxidase but the mutants had only 2% of the wild-type activity of ornithine carbamoyltransferase.

VI. CONCLUSIONS

The *A. nidulans* gene for ornithine carbamoyltransferase has been successfully used to transform an ornithine carbamoyltransferase-deficient strain of *N. crassa*. In at least one case, transformation appears to involve the tandem integration of several copies of the *A. nidulans* ornithine carbamoyltransferase gene at a single site unlinked to the *arg-12* locus. We have been unable to detect sequence homologies between the *A. nidulans* and *N. crassa* genes by DNA–DNA hybridization. Although these experiments were carried out under conditions of low stringency, it is possible that even less stringent conditions might detect nucleotide sequence homology. Alternatively, the nucleotide sequence homology might be very low despite immunological and possible amino acid sequence homology. However, these results are consistent with the observation that integration occurs at sites other than the *arg-12* locus. Such unlinked insertion events have also been observed with homologous genes (Case *et al.*, 1979). Tandem duplications of integrated homologous genes in *N. crassa* have been detected by others (A. M. Lambowitz, personal communication).

The ornithine carbamoyltransferase activity in the transformants is very low despite the presence of multiple copies of the gene. It is not clear, however, whether *all* the copies of the gene are expressed. The low activities suggest that the *A. nidulans* gene is expressed poorly and selection for prototrophs may have required multiple integration events. This would account for the low transformation frequency (compared to the *qa-2* gene). It is also not clear whether the low enzyme activities result from reduced transcription or inefficient processing of the cytoplasmic precursor of the mature, enzymatically active mitochondrial protein. These possibilities are being investigated. However, the successful expression of the *A. nidulans* gene in *N. crassa* and its derepression in response to histidine limitation suggest that both promoter and regulatory sequences are similar in the two organisms. The alternative explanation, that integration fortuitously placed the *A. nidulans* gene under the control of an *N. crassa* promoter and regulatory sequence, seems unlikely.

Ornithine carbamoyltransferase functions in the mitochondrion in both *N. crassa* and *A. nidulans*. Mitochondrial proteins are often synthesized as higher molecular weight precursors, which are processed to the mature form upon entry into the mitochondrion (reviewed in Hay *et al.*, 1984). Processing involves several recognition steps which are thought to be mediated by an amino terminal extension—the "signal" sequence. A higher molecular weight precursor of the *N. crassa* enzyme has been detected (Bates, 1984).

A significant fraction of the *A. nidulans* enzyme was found in the mitochondria of the transformants. This suggests that the *A. nidulans* gene product is processed so that it functions in its normal mitochondrial location. This indicates that the recognition machinery and signal sequences are similar in the two organisms. The lower recovery of enzyme activity in the mitochondrial fraction of the transformants versus the wild-type strain could arise in either of two possible ways: (1) the mitochondria of the transformants undergo more damage during isolation, with resultant loss of associated matrix enzymes, or (2) the transformants have enzymatically active enzyme molecules in the cytoplasm. The latter could arise from inefficient processing of the cytoplasmic precursor of the mitochondrial enzyme and might account for the low activities in the transformants.

The successful transformation of *N. crassa* with a gene from *A. nidulans* suggests that development of a reliable "gene recovery" system for either organism will be useful in the isolation of genes from both organisms. The results also suggest that both organisms share similarities in their transcriptional and regulatory machinery. Further attempts to transfer heterologous genes between species of filamentous fungi may prove to be a powerful tool in understanding their complex patterns of genetic regulation.

ACKNOWLEDGMENTS

This work was supported in part by U.S. Public Health Service grant GM28864 and National Science Foundation grant PCM82-04194 to R.L.W. D.P. was supported by U.S. Public Health Service National Research Service Award 07104 from the National Institutes of Health.

REFERENCES

Barthelmess, I. B. (1982). Mutants affecting amino acid cross-pathway control in *Neurospora crassa*. *Genet. Res.* **39**, 169–185.

Basabe, J. R., Lee, C. A., and Weiss, R. L. (1979). Enzyme assays using permeabilized cells of *Neurospora*. *Anal. Biochem.* **92**, 356–360.

Bates, M. A. (1984). Ornithine transcarbamylase of *Neurospora crassa:* purification, characterization and genetic analysis of synthesis and translocation. Ph.D. Thesis, Univ. of California, Los Angeles.

Berse, B., Dmochowska, A., Skrzypek, M., Weglenski, P., Bates, M. A., and Weiss, R. L. (1983). Cloning and characterization of the ornithine carbamoyltransferase gene from *Aspergillus nidulans*. *Gene* **25**, 109–117.

Carsiotis, M., and Jones, R. F. (1974). Cross-pathway regulation: tryptophan-mediated control of histidine and arginine biosynthetic enzymes in *Neurospora crassa*. *J. Bacteriol.* **119**, 889–892.

Carsiotis, M., Jones, R. F., and Wesseling, A. C. (1974). Cross pathway regulation: histidine-mediated control of histidine, tryptophan, and arginine biosynthetic enzymes in *Neurospora crassa*. *J. Bacteriol.* **119**, 893–898.

Case, M. E., Schweizer, M., Kushner, S. R., and Giles, N. H. (1979). Efficient transformation of *Neurospora crassa* by utilizing hybrid plasmid DNA. *Proc. Natl. Acad. Sci. U.S.A.* **76**, 5259–5263.

Crabeel, M., Messenguy, F., Lacroute, F., and Glansdorff, N. (1981). Cloning *arg3*, the gene for ornithine carbamoyltransferase from *Saccharomyces cerevisiae:* expression in *Escherichia coli* requires secondary mutations; production of plasmid β-lactamase in yeast. *Proc. Natl. Acad. Sci. U.S.A.* **78,** 5026–5030.

Davis, R. H. (1962). A mutant form of ornithine carbamoyltransferase found in a strain of *Neurospora* carrying a pyrimidine–proline supressor gene. *Arch. Biochem. Biophys.* **97,** 185–191.

Davis, R. H. (1979). Genetics of arginine biosynthesis in *Neurospora crassa. Genetics* **93,** 557–575.

Davis, R. H. (1983). Arginine synthesis in eukaryotes. *In* "Amino Acids—Biosynthesis and Genetic Regulation" (K. Herrmann and R. Somerville, eds.), pp. 81–101. Addison-Wesley, Reading, Massachusetts.

Dhawale, S. S., Paietta, J. V., and Marzluf, G. A. (1984). A new rapid and efficient transformation procedure for *Neurospora. Curr. Genet.* **8,** 77–80.

Esser, K., Kuck, U., Stahl, U., and Tudzynski, P. (1983). Cloning vectors of mitochondrial origin for eukaryotes: a new concept in genetic engineering. *Curr. Genet.* **7,** 239–243.

Flint, H. J., and Kemp, B. F. (1981). General control of arginine biosynthetic enzymes in *Neurospora crassa. J. Gen. Microbiol.* **124,** 129–140.

Hay, R., Bohni, P., and Gasser, S. (1984). How mitochondria import proteins. *Biochim. Biophys. Acta* **779,** 65–87.

Kemp, B. F., and Flint, H. J. (1982). Cross-pathway control of ornithine carbamoyltransferase synthesis in *Neurospora crassa. J. Gen. Microbiol.* **128,** 1503–1507.

Piotrowska, M. (1980). Cross-pathway regulation of ornithine carbamoyltransferase synthesis in *Aspergillus nidulans. J. Gen. Microbiol.* **116,** 335–339.

Radford, A., Pope, S., Sazci, A., Fraser, M. J., and Parish, J. H. (1981). Liposome-mediated genetic transformation of *Neurospora crassa. Mol. Gen. Genet.* **184,** 567–569.

Schechtman, M. G., and Yanofsky, C. (1983). Structure of the trifunctional *trp-1* gene from *Neurospora crassa* and its aberrant expression in *Escherichia coli. J. Mol. Appl. Genet.* **2,** 83–99.

Schweizer, M., Case, M. E., Dykstra, C. C., Giles, N. H., and Kushner, S. R. (1981). Identification and characterization of recombinant plasmids carrying the complete *qa* gene cluster from *Neurospora crassa* including the *qa-1*$^+$ regulatory gene. *Proc. Natl. Acad. Sci. U.S.A.* **78,** 5086–5090.

Stohl, L. L., and Lambowitz, A. M. (1983). Construction of a shuttle vector for the filamentous fungus *Neurospora crassa. Proc. Natl. Acad. Sci. U.S.A.* **80,** 1058–1062.

Urrestarazu, L. A., Vissers, S., and Wiame, J.-M. (1977). Change in location of ornithine carbamoyltransferase and carbamoylphosphate synthetase among yeasts in relation to the arginase/ornithine carbamoyltransferase regulatory complex and the energy status of the cells. *Eur. J. Biochem.* **79,** 473–481.

12

Gene Dosage Effects and Antibiotic Synthesis in Fungi

C. H. O'DONNELL[1]
A. UPSHALL[2]
Department of Biological Sciences
University of Lancaster
Lancaster, United Kingdom

K. D. MACDONALD
Chemical Defence Establishment
Salisbury, Wiltshire, United Kingdom

I. Introduction	293
II. Methods for the Amplification of Genetic Material in Microorganisms	295
III. Studies with Disomic Strains of *Aspergillus nidulans*	297
A. Isolation of a Strain with Two Copies of *penB2*	298
B. Isolation of a Strain with Two Copies of *penA1*	299
C. Isolation of a Strain with Extra Copies of *penA1* and *penB2*	299
References	305

I. INTRODUCTION

One of the more rewarding areas of research in biotechnology has been that concerned with strain improvement. Genetic manipulations in fungi and other microbes that produce antibiotics have led to substantial increases in the yields of these secondary metabolites. The most widely used method is mutagenic treat-

---

[1]Present address: Biosoft, Elsevier Biomedical Press, 68 Hills Road, Cambridge CB2 1LA, United Kingdom.
[2]Present address: Zymo Genetics, Inc., 2121 N. 35th Street, Seattle, Washington 98103.

ment of individual cells followed by the examination of clones grown from survivors. After adequate testing, a mutant with a better antibiotic yield than its parent is chosen as the new production strain and is also subjected to a further round of mutation and selection in a search for fresh mutants with still better yields.

The success of this approach has probably meant that less effort has gone into exploring methods of strain improvement concerned with the exchange of genetic information via sexual and parasexual processes. Also, the upsurge of interest in the practical applications of genetic engineering may have, to some extent, overshadowed work on the genetics of secondary metabolite production in industrial fungi. While compounds such as eukaryotic enzymes made by single genes are clearly candidates for manufacture in prokaryotes, it is more difficult to conceive genetically engineered prokaryotes making secondary metabolites normally produced by eukaryotes. Secondary metabolites are relatively small molecules and many genes are involved in their biosynthesis. Possibly, in the future, they may be synthesized on a sequential series of immobilized enzymes, but until then it would seem more sensible to investigate the genetics of secondary metabolite production, and ways of optimizing yield, in the fungi making them, rather than attempt to transfer all the genes concerned to another organism.

A number of studies on the genetics of fungi elaborating antibiotics have been concerned with the practical utilization of sexual or parasexual processes (Elander, 1967; Elander *et al.*, 1973; Ball, 1973; Calam *et al.*, 1976; Hamlyn and Ball, 1979) and there are reports of increases in antibiotic yields following what was presumed to have been the interchange of genetic information (Ball, 1973; Calam *et al.*, 1976; Hamlyn and Ball, 1979).

The effects of gene dosage on antibiotic production have not been studied extensively in fungi, but in *Penicillium chrysogenum* an increase in ploidy does not usually result in an increase in penicillin yield; the indications are that homozygous diploids do not normally produce more penicillin than their parental haploid (Macdonald, 1966; Sermonti, 1969), although one exception has been reported where the penicillin titer of a homozygous diploid exceeded that of the original haploid (Elander, 1967; Elander *et al.*, 1973). Here the haploid was found to throw off a considerable number of morphological variants and it was suggested that the haploid bore a number of recessive genes with a higher than normal mutation rate, including mutations reducing penicillin yield. Mutants could accumulate during growth in a production vessel and lower penicillin titer. If these mutations occurred at random then their effects could be partially hidden in the diploid so that its penicillin yield would be better than that of the haploid. There are occasions when a heterozygous diploid can elaborate more penicillin than the component haploid strains from which it was synthesized. This has sometimes been ascribed to heterosis (Sermonti, 1969). Each of the haploid parental strains may have borne deleterious, recessive, and nonallelic mutations reducing penicillin yield which would have been masked in the diploid. The further assumption has to be made that there were sufficient allelic, recessive mutations increasing penicillin yield in the haploids or that some of these positive penicillin mutations were dominant.

The outcome on antibiotic yield of increasing the number of positive penicillin mutations, by raising the ploidy of an organism, could be affected by modifying genes. During the derivation of a high-yielding mutant, after a series of mutation and selection steps, some cryptic mutations could accumulate which reduce the penicillin yield of the mutant to below its potential level. In a homozygous diploid there would be a double dose of these mutations injurious to penicillin production. However, if a partial diploid could be constructed where only positive penicillin mutations were duplicated, then their effects might be different and more beneficial than when they were present in a homozygous diploid.

Although it produces only small amounts of penicillin, the fungus *Aspergillus nidulans* has been employed as a model organism in studies of the genetics of penicillin production because its formal genetics is much better understood than that of *P. chrysogenum*, the industrial producer of penicillin. *Aspergillus nidulans* is normally haploid (n) but strains of this organism have been discovered in which one of the chromosomes has been duplicated; the latter are referred to as disomic strains ($n + 1$). Because *A. nidulans* has eight chromosomes there are eight different disomic strains which can be isolated. Upshall *et al.* (1979) have devised methods for making stable disomic strains of *A. nidulans*, and later in this chapter studies on the duplication of antibiotic genes in *A. nidulans* using their techniques are described.

II. METHODS FOR THE AMPLIFICATION OF GENETIC MATERIAL IN MICROORGANISMS

Most of the work in this area has been concerned with bacteria but there have been a few studies with fungi.

Continuous culture has been used for the isolation of mutants which are more efficient utilizers of a particular compound than their parent strain. If the microbe is supplied with limiting amounts of the compound as sole carbon or sole nitrogen source, then mutants which are better utilizers of the compound can be selected during growth in a continuous culture apparatus. Novick and Horiuchi (1961) isolated overproducers of β-galactosidase by growing cultures of *Escherichia coli* on low levels of lactose. Constitutive mutants were first produced, i.e., mutants which produced the enzyme in the absence of the sugar rather than as an adaptive response to its presence. Strains were then isolated where the appropriate region of the chromosome, the *lac* gene, had been duplicated in tandem. After about 100 generations, strains were found which yielded β-galactosidase in quantities of about 25% of the total cell protein. These had apparently several copies of the relevant gene but were found to be very unstable when removed from continuous culture (Horiuchi et al., 1962). Similarly, gene duplication has increased the production of ribitol dehydrogenase in *Klebsiella aerogenes* (Neuberger and Hartley, 1981).

In *E. coli* the replication of the chromosome begins from an initiator point and is independent of growth rate. Under conditions of fast growth rate, DNA replication may occur in less time than it takes for the replication complexes to divide the chromosome. The number of replication complexes can thus increase and lead to chromosomes having branched structures where genes replicated early will have multiple copies (Collins, 1976). However, increased gene dosage does not always lead to proportionally higher yields of the gene product. For example, Richmond (1968) found that the presence of two independent plasmids, each bearing a constitutive gene for penicillinase synthesis, did not result in an additive yield of the enzyme. Also, gene dosage effects will not be expected when there is autoregulation and the protein coded by a gene plays a negative regulatory part in the expression of the gene (Scaife, 1976).

Amplification of genes carried on plasmids can occur and individual cells may contain as many as 1000 or more plasmids under certain conditions (Birge, 1981). Also, recombination between a bacterial chromosome and plasmid DNA can lead to the occurrence of bacteria which are partially diploid for the portion of the chromosome borne on the plasmid (Collins, 1976).

Specialized transducing phage can also be utilized to increase gene dosage. For example, lambda, a temperate coliphage, yields specialized transducing phage when bacterial genes are integrated into its genome. These integrated genes are borne as part of the prophage when a bacterial cell is lysogenized and they render the host cell partially diploid for the host genes carried on the prophage. If the prophage is induced, it is excised from the bacterial chromosome and replicated to give around 50 copies of the phage genome in one bacterial cell. The 50 copies of a portion of the host chromosome integrated in each phage particle will be expressed until lysis occurs.

While not directly concerned with gene amplification, the structure of the promoter region of any gene is of considerable importance since the yield of a particular product can be raised by suitable alterations at the promoter site (Brammar, 1976). There are some general methods available for improving promoter activity. For example, from a temperature-sensitive mutant, selection can be made for mutants restoring the lost function at the restrictive temperature. A strain with restored function might carry a mutation that increased the rate of production of the mutant protein and thus elaborated sufficient for expression at the restrictive temperature. Also, revertants from leaky and slow-growing mutants which are able to grow normally could result from mutations which altered the rate of production of a protein.

Turning to fungi, the effects of gene dosage have been examined in *Saccharomyces cerevisiae*. Haploid, diploid, and tetraploid strains were compared which bore the gene responsible for tryptophan synthetase production singly, in double, or in quadruple copies, respectively. The amount of tryptophan synthetase activity was proportional to the copy number of the relevant gene (Cifferi *et al.*, 1968). Also, in *S. cerevisiae* resistance to a metabolic inhibitor can result from an increased copy number of specific genes whose product competitively inhibits the action of the toxin.

The occurrence of plasmids in fungi has been demonstrated in yeast and suggested in the case of a few other filamentous fungi. The so-called 2-μm plasmid exists in yeast usually at a level of about 100 copies per cell. When incorporated into *Escherichia coli* as a chimera, novel polypeptide fragments are elaborated (Gubbins *et al.*, 1977). What function it has other than its own preservation is as yet unknown. The employment of plasmids in yeasts and other industrial fungi, where they may be discovered, can be visualized for the future in at least some of the ways in which they have been utilized in bacteria for gene amplification.

As mentioned earlier, in *P. chrysogenum* an increase in ploidy does not usually result in an increase in penicillin yield. Therefore, the ability to produce a partial diploid could be useful. First, if such a strain could be constructed bearing positive penicillin mutations in duplicate, and cryptic deleterious mutations in single doses, then the former might act more beneficially than in a haploid or in a diploid strain. Second, an antibiotic such as penicillin is elaborated as the result of many different enzyme reactions, any one of which, presumably, can be rate-limiting. Doubling the dose of a gene responsible for a rate-limiting reaction while other genes concerned with antibiotic synthesis remained in single dosages might have an advantageous effect on antibiotic titer. The techniques devised by Upshall *et al.* (1979) for the production of disomic strains in *A. nidulans* are discussed in the next section with emphasis on antibiotic production.

III. STUDIES WITH DISOMIC STRAINS OF *ASPERGILLUS NIDULANS*

This section summarizes our work with *A. nidulans* on the effect of disomy on antibiotic production.

In *A. nidulans* three loci have been identified, *penA1*, *penB2*, and *penC3*, mutations at which can lead to an increase in penicillin titer (Ditchburn *et al.*, 1976). Genetic analysis has located *penA1* on a chromosome VIII, linked to *chaA* and *sE*, which determine, respectively, chartreuse conidial color and inability to utilize sulfate. The *penB2* mutation is located on chromosome III and *penC3* on chromosome IV (Macdonald and Holt, 1976; Makins *et al.*, 1983).

When compared to a nonmutant strain, *penA* and *penC* increased penicillin titer approximately fourfold and *penB*, twofold. The dominance relationships of the alleles and the epistatic relationships of the mutations have been determined (Ditchburn et al., 1976).

Since it has eight chromosomes, there are eight possible disomic genotypes in *A. nidulans*. Each disomic can be identified by its possessing a unique combination of growth rate, pigmentation, and conidiation abnormalities (Kafer and Upshall, 1973). In haploids, approximately 0.2% of the conidial progeny are aneuploid, primarily disomics, and although all chromosomes are capable of undergoing nondisjunction some appear to do so more frequently than others (Upshall, 1966, 1971). Mitotic breakdown in diploid strains generates a variety of aneuploid types usually disomic or trisomic for more than one chromosome; these abnormal types occur in a frequency of about 2% among diploid conidiophore progeny (Kafer, 1961). These low frequencies and lack of specificity make the recovery of a particular aneuploid a laborious process. However, a second and more serious complication of disomic strains is their instability resulting from the random loss of one of the disomic chromosomes at mitosis (Kafer and Upshall, 1973). To employ disomic strains in a study of gene dosage would require resolution of both of these problems, but particularly the latter. Thus, while it may not be straightforward to isolate a specific disomic strain, it is imperative that once isolated, it remain stable at least for the duration of any experiment where the properties of the disomic strain are being examined. Stabilization based on balanced lethal genotypes has been proposed by Ball (1973), but to be effective, the balancing mutations would need to be allelic or very closely linked to minimize the possibility of recombination producing a selectively advantaged haploid.

In studies of nuclear division in *A. nidulans*, when mutants of this fungus defective in anaphase segregation were sought, four conditional lethal, heat-sensitive mutants were identified. When cultured at the subrestrictive temperature of 37°C, these generated and stabilized specific disomies (Upshall et al., 1979). These mutants were designated *sod* (stabilization of disomy). Three of the *sod* mutants gave strains which were disomic for chromosome VI and the fourth mutant was disomic for chromosome III. The *sod* phenotype was determined by a single mutation in each case and the reversion rate was low (I. D. Mortimore and A. Upshall, unpublished observations).

The isolation of *pen* and *sod* mutations coupled with the knowledge available on the genetics of *A. nidulans* made it possible to consider a study examining the effects of gene dosage on penicillin production.

A. Isolation of a Strain with Two Copies of *penB2*

The mutation *penB2* is located on chromosome III and one of the *sod* mutations, $sod^{III}A$, which is located on chromosome VIII, induces stable disomy of chromosome III (I. D. Mortimer and A. Upshall, unpublished observations). The construction of a strain carrying both *penB2* and *sod^{III}A* should lead to disomy of chromosome III and hence two copies of *penB2* when the strain is cultured at 37°C.

B. Isolation of a Strain with Two Copies of *penA1*

In *A. nidulans* the translocation T2 III–VIII is unidirectional and relocates the distal segment of chromosome VIII to the distal end of the left arm of chromosome III (Clutterbuck, 1970). Since *penA1* maps near the distal end of chromosome VIII, a cross between a strain bearing this translocation and one bearing *penA1* might lead to the isolation of a recombinant carrying *penA1* both in the normal position on chromosome VIII and duplicated on chromosome III with the *sod^{III}A1* mutation from chromosome VIII.

C. Isolation of a Strain with Extra Copies of *penA1* and *penB2*

A strain with the T2 III–VIII translocation as described above could be isolated which carried not only *penA1* on chromosome VIII and duplicated on chromosome III but also *penB2* on chromosome III and *sod^{III}A1* on chromosome VIII. When grown at 37°C this should be disomic for chromosome III bearing two copies of *penA1* and two copies of *penB2*.

The lack of any well-characterized translocation complexes involving chromosome IV or a *sod* mutation specific for this chromosome precluded the involvement of *penC3* in experiments at this stage.

During the course of the work difficulties were encountered in obtaining recombinants bearing the *penA1* mutation. Some of these difficulties were probably the consequence of recombination between the *penA1* mutation and other genes in the crosses, which may have resulted in modifying the expression of *penA1*. It was not possible to confirm unambiguously the presence of the *penA1* mutation in potentially useful recombinants although further work is contemplated. However, suitable strains bearing the *penB2* mutation were isolated by utilizing as a recognition factor the abnormal morphology conferred by this mutation (Ditchburn *et al.*, 1976). The results reported here are therefore concerned with duplication of the *penB2* locus.

It was necessary to establish the stability of the disomic state under shake flask conditions of growth for penicillin production. Conidial suspensions taken from a stable *n* + III disomic strain heterozygous for *dilA1*, a mutation which results in dilute conidial color, were grown at 37°C in shake flask cultures containing penicillin production medium. After 2 days, samples were plated on complete medium agar and incubated at 30°C, at which temperature disomy is not maintained. The results are shown in Table I. In two experiments, over 95% of the

TABLE I.

Numbers and Types of Colonies from the Platings of Samples of Mycelium after 2 Days of Growth at 37°C in Shake Flasks Containing Penicillin Production Medium That Had Been Inoculated with Conidia of a Stable $n + $ III Disomic Strain Heterozygous for $dilA$[a]

Suspension	Number of colonies			
	Total	$n + $ III	Haploid	Others
1	962	926	1	35
2	9222	9054	127	41

[a] $dilA$ is borne on chromosome III and dilutes conidial color.

colonies recovered were phenotypically $n + $ III and the sectors arising on these colonies all showed segregation for dilute spore color. Colonies without sectors and phenotypically either $dilA1$ or $dilA1^+$ were presumed to be haploid. It was assumed that they had grown from revertants which arose in shake flask culture or perhaps by reversion of the $n + $ III type to euploidy in the early stages of colony formation at the permissive temperature of 30°C. The category listed under "Others" in Table I comprised an assortment of disomic types and may reflect a measure of instability during growth in shake flasks. The samples were taken after 2 days growth and similar patterns were observed in samples taken after 4 days growth. The results indicated that an $n + $ III disomic strain does not show a high degree of instability when grown at 37°C in a medium used for the production of penicillin.

A sexual cross was then set up as illustrated in Table II, and 12 recombinants were isolated with the abnormal morphological characteristics of strains bearing $penB2$ (Macdonald and Holt, 1976). The 12 recombinants, with suitable controls, were grown at 25°C in penicillin production medium as described by Ditchburn et al. (1974), with the results shown in Table II. When grown at 25°C, strains bearing $sod^{III}A$ were expected to be haploid.

An immediate observation was that the penicillin yields of strains G113 and L2 (bearing $penB2$ and $penB2^+$, respectively) were only half of those previously observed (Ditchburn et al., 1976). Also, the penicillin titers of strains bearing $penA1$ (strain COD69) and $penC3$ (strain G96) were both somewhat reduced. Minor variations in growth procedures or media can affect penicillin production in strains with different mutations enhancing penicillin titer (Macdonald, 1973). Despite the reductions observed, strain G113 ($penB2$) produced significantly more penicillin than strain L2 ($penB2^+$). Recombinants were selected bearing the pleiotropic character of the $penB2$ mutation, which resulted in reduced conidiation. Based on mean penicillin titer, the recombinants fell into two groups: one

12. Gene Dosage Effects and Antibiotic Synthesis

TABLE II.

Mean Penicillin Titers of Samples Taken after 5 Days Growth at 25°C in Penicillin Production Medium[a]

Strain	Genotype	Mean titre (units/ml)
L2 (control)	*pabaA1,yA2*	3.83
COD49 (control)	*pabaA1,yA2; penB2*	4.05
COD50 (control)	*yA2; penB2; riboB2,sod^{III}A1*	3.79
R/COD84	*penB2; nicB8*	3.86
R/COD85	*penB2; nicB8*	3.72
R/COD86	*penB2; nicB8; sod^{III}A*	3.99
R/COD87	*penB2; nicB8; sod^{III}A*	3.50
R/COD95	*penB2; riboB2,sod^{III}A*	3.77
R/COD96	*penB2; riboB2,sod^{III}A*	4.07
R/COD102	*yA2; penB2; nicB8; riboB2*	3.71
R/COD103	*yA2; penB2; nicB8; riboB2*	3.88
R/COD110	*yA2; penB2*	3.76
R/COD111	*yA2; penB2*	3.86
COD52 (control)	*pabaA1,yA2; penB2; nicB8; sod^{III}A1*	5.55
G113 (control)	*yA2; acrA1; penB2; riboB2*	5.05
R/COD106	*pabaA1,yA2; penB2; nicB8; sod^{III}A*	4.80
R/COD107	*pabaA1,yA2; penB2; nicB8; sod^{III}A*	5.58
COD69 (control)	*pabaA1,yA2; chaA1,penA1,nirA*	14.99
G92 (control)	*yA2,w-010; pyroA4,penC3*	13.81

[a] Twelve recombinants (prefix R) were isolated following a sexual cross between strain L113 (*proA1,pabaA1; nicB8*) and strain COD50. In this experiment strain L2 replaced strain L113 as control. Like strain L113, strain L2 bore *penB2*$^+$ and *sod^{III}A*$^+$.

All strains would grow as haploids at 25°C. The mean penicillin titers of the strains were from ten replicates, except strain L2, where there were nine replicates, and strain R/COD111, where there were five replicates. Apart from strain COD69 (bearing *penA1*) and strain G92 (bearing *penC3*), the control strains and recombinants were collected into two groups with similar penicillin yields.

For abbreviations see Clutterbuck (1981), except for *sod^{III}A* (see Upshall *et al.*, 1979) and *w-010* (see Ditchburn *et al.*, 1976). The *nirA* mutation in strain COD69 was selected by resistance to potassium chlorate (Cove, 1976).

in which the strains were similar in penicillin yield to strain G113 bearing *penB2*, and another in which the strains were similar in yield to strain L2 bearing *penB2*$^+$. It was concluded that the expression of *penB2* could be modified in certain genetic backgrounds, following recombination, when it was removed from its original genetic background.

In a pilot experiment a number of strains were then grown in penicillin produc-

TABLE III.

Mean Penicillin Titers of Samples Taken after 2 and 5 Days of Growth at 37°C in Penicillin Production Medium

		Mean titer (units/ml)[a]	
Strain	Genotype	Day 2	Day 5
L2	pabaA1,yA2	2.72	1.26
L11	Glasgow wild type (prototroph)	1.93	0.81
G113	yA2; acrA1; penB2; riboB2	3.51	3.58
L486	pabaA1,yA2; sod^{III}A1	1.75	0.57
COD69	pabaA1,yA2; chA1,penA1,nirA	11.18	4.89
R/COD107	pabaA1,yA2; penB2; nicB8; sod^{III}A	2.97	2.00

[a] The mean penicillin titers of the strains were from ten replicates on day 2 and from five replicates on day 5.

tion medium in shake flasks at 37°C and assayed after 2 and after 5 days (Table III). At 37°C, strains bearing sod^{III}A would grow as disomic for chromosome III which bears penB2. At 37°C, all strains yielded less penicillin than at 25°C. Also, although strains L2 and L11 both bore penB2$^+$, they apparently differed by an unselected marker or unselected markers increasing antibiotic titer in strain L2. Previous studies had shown that strain L2 (bearing sod^{III}A$^+$) and strain L486 (bearing sod^{III}A) had similar penicillin yields when grown at 25°C. However, at 37°C the penicillin yield of strain L486 was reduced relative to that of strain L2. Since strain L486 grows as disomic for chromosome III at 37°C, this suggested that disomy per se reduced penicillin titer.

To determine the effect on penicillin titer of duplicating penB2, comparisons ideally required isogenic haploid and disomic strains. Mitigating against this was first, the fact that strains bearing both penB2 and sod^{III}A were derived by recombination and might possibly exhibit a reassortment of other genetic factors from the parents, and second, and more important, any single strain can only be either haploid or disomic at 37°C. It was anticipated that strain L11 (bearing penB2$^+$) and strain G113 (bearing penB2) would be useful as controls. However, a complication was introduced by the differences in penicillin yield between strain L11 and strain L2 (both bearing penB2$^+$) when grown at 37°C. Strain L2 is probably the more appropriate control since it is an ancestor of one of the parents of the recombinants.

Further experiments were then performed at 37°C with ten of the recombinants grown previously in penicillin production medium at 25°C. Recombinants COD110 and COD111 were omitted. Neither carried the sod^{III}A mutation, and they would have grown as haploids at 37°C. The remaining recombinants, with

TABLE IV.

Mean Penicillin Titers of Samples Taken after 2 and 5 Days of Growth at 37°C in Penicillin Production Medium[a]

Strain	Ploidy[b]	Mean titer (units/ml) Day 2	Day 5
L2	h	3.21	1.83
L11	h	1.63	1.30
G113	h	3.33	3.27
L486	d	2.02	0.95
COD52	d	3.43	1.71
R/COD106	d	3.04	1.45
R/COD107	d	3.46	1.96
COD50	d	5.59	3.58
R/COD86	d	5.27	3.09
R/COD87	d	4.85	3.00
R/COD95	d	5.58	3.58
R/COD96	d	6.62	3.16
R/COD84	h	4.04	1.48
R/COD85	h	3.13	2.14
R/COD102	h	4.19	1.73
R/COD103	h	2.98	1.73

[a] On day 2 the mean penicillin titers of the strains were from 12 replicates, except strain R/COD95, where there were 8 replicates, and strain R/COD96, where there were 11 replicates. On day 5 the mean penicillin titers of the strains were from 8 replicates, except strain R/COD95, where there were 6 replicates, and strain R/COD96, where there were 7 replicates. The recombinant strains and two of the control strains, strain COD50 and strain COD52 (both bearing $penB2$ and $sod^{III}A$), were collected into groups with similar penicillin yields.

[b] h, haploid; d, disomic.

appropriate controls, were inoculated into penicillin production medium and grown at 37°C, penicillin assays being done after 2 and after 5 days (Table IV). All four control strains (L2, L11, G113, and L486) behaved as in the pilot experiment, again suggesting the existence of an unselected mutation or mutations influencing penicillin titer in strain L2, as well as suggesting that disomy per se reduces penicillin yield. With the exception of strain G113, all strains showed a decline in titer between day 2 and day 5. A comparison of the mean titers of strains which would grow as haploids with the mean titers of those which would grow as disomics showed them not to be significantly different. Also, the distribution of penicillin titers suggested that the variation between disomic strains was greater than that between haploid strains.

In Table IV, it was possible to divide the strains which bore *penB2* into three groups: (1) low-titer strains bearing $sod^{III}A1$, (2) high-titer strains bearing $sod^{III}A1$, and (3) strains bearing $sod^{III}A^+$.

As in previous studies (Merrick, 1976; Simpson and Caten, 1979), variation of penicillin titer was observed among progeny derived from a sexual cross between a high and a low penicillin yielder. The expression of mutations influencing penicillin titer may be dependent on a balanced combination of interactions (Simpson and Caten, 1979; Rowlands, 1983). In the work reported here it appeared that a number of such interactions were negative in that, in some cases, the full expression of *penB2* was prevented. However, there was no evidence that the $sod^{III}A$ mutation or any of the other genetic markers used in the present work significantly or consistently affected penicillin titer. All of the variation in penicillin titer observed after the growth of strains in penicillin production medium at 25°C could be attributed to differences between two groups of strains, one having strains with penicillin yields equivalent to that of strain L2 (bearing $penB2^+$) and the other having strains with penicillin yields equivalent to that of strain G113 (bearing *penB2*). It can be concluded tentatively from the results at 25°C that in those recombinants selected as carrying *penB2*, which had penicillin yields equivalent to that of strain L2, the expression of the *penB2* mutation was suppressed by deleterious interactions in their genotypes. However, different positive interactions have allowed the expression of *penB2* in recombinants with penicillin yields equivalent to that of strain G113.

After growth in penicillin production medium at 37°C, there was greater variation in penicillin titer among disomic than among haploid strains. While disomy per se apparently reduces penicillin titer, it appears that disomy involving duplication of the mutation *penB2* can compensate for the detrimental effect of disomy. The greater variance in penicillin yield among strains from the disomic group, after growth at 37°C, probably indicates wider interactions in disomics than in haploids, with the level of penicillin titer in disomics being dependent on the nature of the genotypic interactions. After 2 days at 37°C in penicillin production medium, certain disomic recombinants, bearing *penB2* in duplicate, have penicillin yields surpassing that of the haploid strain G113, bearing a single copy of *penB2*.

Another feature of the results was that the grouping of disomic strains with similar penicillin yields at 25°C, one with relatively low and one with relatively high penicillin titers, was maintained after growth in penicillin production medium at 37°C except that in terms of penicillin titer the relative positions were switched. The maintenance of this grouping suggested that it had a genetic basis, and the group switching suggested a differential interaction of genotypes with temperature.

Gene duplication by way of disomy can therefore be a way of influencing penicillin production and may increase penicillin titer. In the case of the *penB2* mutation, the effect appears to be dependent on appropriate interactions in a compatible genotype.

Antibiotics, like other secondary metabolites, are produced as the result of the activity of many different genes. Not only those genes concerned directly in their biosynthesis may influence the amounts which can be isolated from cultures, but others not directly involved with their biosynthesis can affect the final yields of antibiotics (Macdonald, 1983). While the amplification of the copy number of a particular gene may lead to increased production of a particular enzyme, to have an effect on the titer of a secondary metabolite the level of the substrate required for the enzyme may need to be increased simultaneously. A higher level of production of the secondary metabolite would thus be achieved only in the appropriately modified genotype.

Further developments in the use of gene duplication and other rational approaches for strain development in fungi of industrial importance await a fuller understanding of the biosynthesis of secondary metabolites and a greater knowledge of the formal genetics of the relevant microorganisms. Given the abnormal morphologies and vegetative instabilities of certain industrial microorganisms, it would not be surprising if certain of the steps in raising the yields of secondary metabolites have been the consequence of gene duplication.

ACKNOWLEDGMENTS

The work reported in this chapter was performed while C. H. O'Donnell was in receipt of a CASE award from the Science and Engineering Research Council in collaboration with the Procurement Executive, Ministry of Defence.

REFERENCES

Ball, C. (1973). The genetics of *Penicillium chrysogenum*. *Prog. Ind. Microbiol.* **12,** 47–72.
Birge, E. A. (1981). "Bacterial and Bacteriophage Genetics." Springer-Verlag, Berlin and New York.
Brammar, W. J. (1976). Genetic approaches to the stimulation of bacterial protein synthesis. *Proc.— Int. Symp. Genet. Ind. Microorg., 2nd, Sheffield, Engl., 1974* pp. 291–300.
Calam, C. T., Daglish, L. P., and McCann, E. P. (1976). Penicillin: tactics in strain improvement. *Proc.—Int. Symp. Genet. Ind. Microorg., 2nd, Sheffield, Engl., 1974* pp. 273–287.
Cifferi, O., Sara, S., and Tiboni, O. (1968). Effect of gene dosage on tryptophan synthetase activity in *Saccharomyces cerevisiae*. *Genetics* **61,** 567–576.

Clutterbuck, A. J. (1970). A variegated position effect in *Aspergillus nidulans. Genet. Res.* **16**, 303–316.

Clutterbuck, A. J. (1981). Loci and linkage map of *Aspergillus nidulans. Aspergillus News Lett.* **15**, 58–72.

Collins, J. F. (1976). Gene amplification in bacterial systems. *Proc.—Int. Symp. Genet. Ind. Microorg., 2nd, Sheffield, Engl., 1974* pp. 41–58.

Cove, D. J. (1976). Chlorate toxicity in *Aspergillus nidulans:* the selection and characterisation of chlorate resistant mutants. *Heredity* **36**, 191–203.

Ditchburn, P., Giddings, B., and Macdonald, K. D. (1974). Rapid screening for the isolation of mutants of *Aspergillus nidulans* with increased penicillin yields. *J. Appl. Bacteriol.* **37**, 515–523.

Ditchburn, P., Holt, G., and Macdonald, K. D. (1976). The genetic location of mutations increasing penicillin yield in *Aspergillus nidulans. Proc.—Int. Symp. Genet. Ind. Microorg., 2nd, Sheffield, Engl., 1974* pp. 213–227.

Elander, R. P. (1967). Enhanced penicillin biosynthesis in mutant and recombinant strains of *Penicillium chrysogenum. Abh. Dtsch. Akad. Wiss. Berlin, Kl. Med.* **4**, 403–423.

Elander, R. P., Espanshade, M. A., Pathac, S. G., and Pan, C. H. (1973). The use of parasexual genetics in an industrial strain selection programme with *Penicillium chrysogenum. In* "Genetics of Industrial Microorganisms" (Z. Vanek, Z. Hostalek, and J. Cudlin, eds.), Vol. 2, pp. 239–253. Academia, Prague.

Gubbins, E. J., Newlon, C. S., Kann, M. D., and Donelson, J. E. (1977). Sequence organisation and expression of a yeast plasmid DNA. *Gene* **1**, 185–207.

Hamlyn, P. F., and Ball, C. (1979). Recombination studies with *Cephalosporium acremonium. Genet. Ind. Microorg., Proc. Int. Symp., 3rd, Madison, Wis., 1978* pp. 185–191.

Horiuchi, T., Tomizawa, J., and Novick, A. (1962). Isolation and properties of bacteria capable of high rates of β-galactosidase synthesis. *Biochim. Biophys. Acta* **55**, 152–163.

Kafer, E. (1961). The process of spontaneous recombination in vegetative nuclei of *Aspergillus nidulans. Genetics* **46**, 1581–1609.

Kafer, E., and Upshall, A. (1973). The phenotypes of the eight disomics and trisomics of *Aspergillus nidulans. J. Hered.* **64**, 35–38.

Macdonald, K. D. (1966). Differences in diploids synthesised between the same parental strains of *Penicillium chrysogenum. Antonie van Leeuwenhoek* **32**, 431–441.

Macdonald, K. D. (1973). Genetics of penicillin production in *Penicillium chrysogenum* and *Aspergillus nidulans. In* "Genetics of Industrial Microorganisms" (Z. Vanek, Z. Hostalek, and J. Cudlin, eds.), Vol. 2, pp. 255–264. Academia, Prague.

Macdonald, K. D. (1983). Fungal genetics and antibiotic production. *In* "Biochemistry and Genetic Regulation of Commercially Important Antibiotics" (L. Vining, ed.), pp. 25–47. Addison-Wesley, Reading, Massachusetts.

Macdonald, K. D., and Holt, G. (1976). Genetics of biosynthesis and over-production of penicillin. *Sci. Prog. (Oxford)* **63**, 547–573.

Makins, J. E., Holt, G., and Macdonald, K. D. (1983). The genetic location of three mutations impairing penicillin production in *Aspergillus nidulans. J. Gen. Microbiol.* **129**, 3027–3033.

Merrick, M. J. (1976). Hybridisation and selection for penicillin production in *Aspergillus nidulans*—a biometrical approach to strain improvement. *Proc.—Int. Symp. Genet. Ind. Microorg., 2nd, Sheffield, Engl., 1974* pp. 229–242.

Neuberger, M. S., and Hartley, B. S. (1981). Structure of an experimentally evolved gene duplication encoding ribitol dehydrogenase in a mutant of *Klebsiella aerogenes. J. Gen. Microbiol.* **122**, 181–191.

Novick, A., and Horiuchi, T. (1961). Hyper-production of β-galactosidase by *Escherichia coli* bacteria. *Cold Spring Harbor Symp. Quant. Biol.* **26**, 239–245.

Richmond, M. H. (1968). The plasmids of *Staphylococcus aureus* and their relation to other extrachromosomal elements in bacteria. *Adv. Microb. Physiol.* **2,** 43–88.

Rowlands, R. T. (1983). Industrial fungal genetics and strain selection. *In* "The Filamentous Fungi, Vol. IV, Fungal Technology" (J. E. Smith, D. R. Berry, and B. Kristiansen, eds.), pp. 346–372. Arnold, London.

Scaife, J. (1976). Some observations on gene expression. *Proc.—Int. Symp. Genet. Ind. Microorg., 2nd, Sheffield, Engl., 1974* pp. 41–57.

Sermonti, G. (1969). "Genetics of Antibiotic-Producing Microorganisms." Wiley (Interscience), New York.

Simpson, I., and Caten, C. E. (1979). Recurrent mutation and selection for increased penicillin titre in *Aspergillus nidulans*. *J. Gen. Microbiol.* **113,** 209–217.

Upshall, A. (1966). Somatically unstable mutants of *Aspergillus nidulans*. *Nature (London)* **209,** 1113–1115.

Upshall, A. (1971). Phenotypic specificity of aneuploid states in *Aspergillus nidulans*. *Genet. Res.* **18,** 167–171.

Upshall, A., Giddings, B., Teow, C., and Mortimore, I. D. (1979). Novel methods of genetic analysis in fungi. *Genet. Ind. Microorg., Proc. Int. Symp., 3rd, Madison, Wis., 1978* pp. 197–204.

13

Formal Genetics and Molecular Biology of the Control of Gene Expression in *Aspergillus nidulans*

HERBERT N. ARST, JR.[1]
Department of Genetics
University of Newcastle upon Tyne
Claremont Place
Newcastle upon Tyne, United Kingdom

CLAUDIO SCAZZOCCHIO[2]
Department of Biology
University of Essex
Colchester, United Kingdom

I. Formal Genetic Methodology of *Aspergillus nidulans* as Applied to the Study of Control Systems	310
II. The Metabolic Versatility of *A. nidulans* and Its Exploitation	311
III. Regulatory Genes	313
A. Pathway-Specific Regulation	313
B. Wide Domain Regulation	319
C. Integrator Genes	325
IV. Putative Receptor Sites	327
A. *gabI* Mutations	328
B. *prn*^{d} Mutations	328
C. sB_0-90	329
D. *cis*-Acting Regulatory Mutations Affecting Expression of *amdS* . .	329
E. *cis*-Acting Regulatory Mutations Affecting Expression of *uapA* . .	329
V. The Spatial Organization of Functionally Related Genes	330
A. The Proline Catabolism Gene Cluster	332
B. The Nitrate Assimilation Gene Cluster	334
VI. At What Level Does Regulation of Gene Expression Occur?	335
References	337

[1]Present address: Department of Bacteriology, Royal Postgraduate Medical School, Hammersmith Hospital, Ducane Road, London W.12, United Kingdom.
[2]Present address: Bâtiment 409, Institut de Microbiologie, Université Paris XI, Centre d'Orsay, 91405-Orsay, France.

I. FORMAL GENETIC METHODOLOGY OF *ASPERGILLUS NIDULANS* AS APPLIED TO THE STUDY OF CONTROL SYSTEMS

The ability to perform sophisticated genetic manipulations has been of fundamental importance in establishing models of regulation of gene expression. On the whole, molecular analysis of various regulated systems has confirmed the formal models derived from purely genetic data. Among the microbial eukaryotes, only in *Saccharomyces cerevisiae* (e.g., reviews in the volume edited by Strathern *et al.*, 1982), *Aspergillus nidulans*, and, to a lesser extent, *Neurospora crassa* (see, e.g., Metzenberg, 1979; Huiet and Case, Chapter 8, this volume) have regulatory systems been submitted to a genetic analysis comparable in detail to that possible in some prokaryotic systems. But while the development of efficient transformation methodologies for *S. cerevisiae* (reviewed in Botstein and Davis, 1982; Struhl, 1983) has led to a flourishing of molecular analysis, research using *A. nidulans* as a model system has remained largely physiological and genetic. Extremely detailed genetic descriptions of a number of regulatory systems have accumulated, and in the past year the development of transformation techniques (Ballance *et al.*, 1983; Tilburn *et al.*, 1983; Yelton *et al.*, 1984) and of new cloning methodologies (see Chapter 14) has meant that now a second simple eukaryote is amenable to virtually all molecular and genetic manipulations. Here we describe genetic approaches to the study of control systems in *A. nidulans*, introduce molecular data that are largely unpublished, and review our knowledge about the regulation of a number of metabolic pathways in this organism.

The basic techniques of *A. nidulans* genetics are given by Pontecorvo *et al.* (1953) and Clutterbuck (1974), and a recent genetic map can be found in Clutterbuck (1982). *Aspergillus nidulans* is a homothallic ascomycete. Homothallism is actually an advantage as it allows crossing of any two of the many thousands of available strains, provided these carry a pair of complementing nutritional markers. Recently, homothallism has proved advantageous in another context. Work in various laboratories (see, e.g., Tilburn *et al.*, 1983) has indicated that transformation, using cloned sequences, for a number of markers results in integration of tandem repeats of all or part of the transforming DNA. Selfing of a transformant can result in both increases and decreases in the number of tandem copies of the cloned sequence, as expected from unequal crossover events (see, e.g., Tilburn *et al.*, 1983). Analysis of nonordered tetrads is possible in *A. nidulans*. For most work the less laborious analysis of random ascospores from crossed cleistothecia is sufficient (Clutterbuck, 1974). The parasexual cycle allows easy allocation of new mutations to linkage groups, one means of detection of extrachromosomal mutations, mapping by mitotic recombination, and identification of both reciprocal and nonreciprocal translocations (Käfer, 1958; McCully and Forbes, 1965; Clutterbuck, 1974). A detailed example of the mapping of a translocation is given by Arst *et al.* (1979), and Käfer (1977) gives a comprehensive survey of the genetics of translocations in *A. nidulans.* Exploitation of the ability to obtain and map translocations will be discussed further in the context of cloning methodologies (see Chapter 14).

Detailed fine-structure maps have been constructed for several genes or gene clusters in *A. nidulans.* Two methodologies have been employed: the use of flanking markers and deletion mapping. The use of flanking markers was first described by Pritchard (1955) and was used recently to map mutations within the regulatory gene *uaY* and the structural gene for xanthine dehydrogenase *hxA* (see Section III,A,2). The latter map is very detailed and shows an interesting clustering of mutations defining structural domains within the gene (H. M. Sealy-Lewis *et al.*, unpublished observations). Deletion mapping is far less laborious and has been used for the nitrate assimilation (Section V,B) and proline catabolism (where flanking markers were also used; Section V,A) gene clusters. For both clusters, positive selection techniques were used to obtain mutations in one of the genes of the cluster, followed by screening for the absence of expression of a second gene through the use of simple growth tests and/or quick heterokaryon complementation tests, and detailed fine-structure maps were constructed (Tomsett and Cove, 1979; Arst *et al.*, 1981; Sharma and Arst, 1985). The mutagen 1,2,7,8-diepoxyoctane, first employed in *N. crassa,* has proved useful for obtaining deletion mutations within genes in *A. nidulans* (*amdS*, Hynes, 1979; *uaY*, G. Ong and C. Scazzocchio, unpublished observations; *prnA*, Sharma and Arst, 1985). Deletion mapping of *uaY* has given results completely consistent with those obtained earlier with flanking markers (T. Sankarsingh and C. Scazzocchio, unpublished observations).

Both diploids and heterokaryons can be readily obtained in *A. nidulans.* This has been invaluable for developing the mitochondrial genetics of this organism (Turner and Rowlands, 1977; Waring and Scazzocchio, 1983). While this topic is outside the scope of this review, the ability to test complementation and dominance at these two different levels, with mutations within the same nucleus or in different nuclei in a common cytoplasm, is of considerable benefit in studies of gene regulation. Pontecorvo (1963) first postulated that regulatory gene mutations might show differences in complementation and dominance between diploids and heterokaryons. This will be discussed further in Section III,A,3.

II. THE METABOLIC VERSATILITY OF *A. NIDULANS* AND ITS EXPLOITATION

Aspergillus nidulans grows on extremely simple defined media and is able to utilize a remarkable range of nutrients. [For some measure of this metabolic versatility see the gene list compiled by Clutterbuck (1982).] Thus, a large

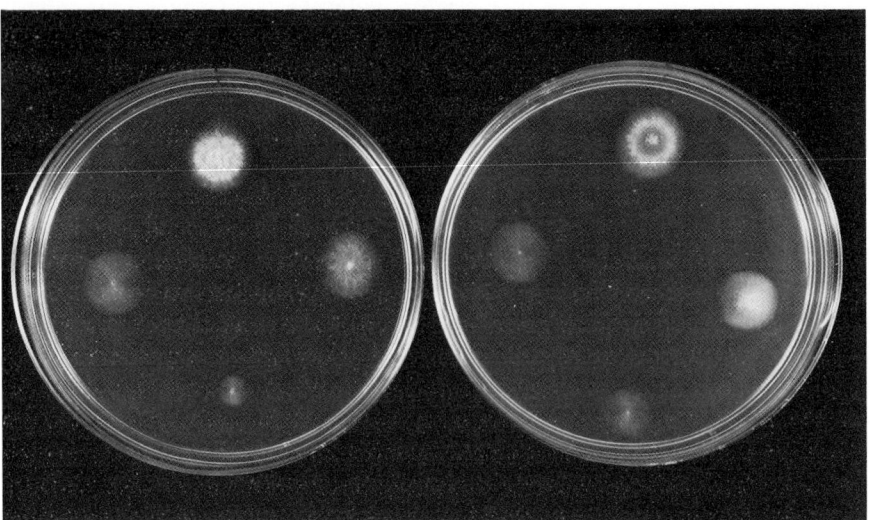

Fig. 1. Growth habits of strains of *Aspergillus nidulans* carrying mutations affecting the utilization of nitrogen sources. Plate on the left: proline as sole nitrogen source. Top, wild type; left, *prnA1*; bottom, *prnC61*; right, *prnB6*. Plate on the right: hypoxanthine as nitrogen source. Top, wild-type; left, *hxA18*; bottom, *uaZ11*; right, *azgA4*. *prnA1* is a mutation in the positive regulatory gene for the proline utilization pathway; *prnC61* is a mutation in the structural gene for the second enzyme of the pathway, Δ^1-pyrroline-5-carboxylate dehydrogenase; *prnB6* is a mutation in the gene coding for the major proline permease (see Section V,B); *hxA18* is a mutation in the structural gene for xanthine dehydrogenase (purine hydroxylase I); *uaZ11* is a mutation in the following enzyme of the purine degradation pathway, urate oxidase (see Section III,A,2); *azgA4* is a mutation affecting hypoxanthine and adenine (but not uric acid or xanthine) uptake (Darlington and Scazzocchio, 1967). Three different growth habits are evident: *prnA1* and *hxA18* show the typical growth of strains impaired in the utilization of a nitrogen source (proline and hypoxanthine, respectively). This ''nitrogen-starved growth'' is equivalent to the growth of all these strains when no nitrogen source is included in the medium. *prnC61* and *uaZ11* show the growth habit of strains that not only cannot utilize a nitrogen source but also are inhibited by an accumulated intermediate, Δ^1-pyrroline-5-carboxylate and uric acid, respectively. Finally, *prnB6* and *azgA4* show the typical leaky growth of permease mutations where uptake can also occur by an alternative uptake system (Arst *et al.*, 1980; D. Gorton and C. Scazzocchio, unpublished observations).

number of anabolic and catabolic pathways are amenable to study. The potential of *A. nidulans* for biotechnological use in this regard has yet to be appreciated. Extensive nutritional screening has, however, proved an invaluable tool for studying the control of gene expression in this organism (see, e.g., Darlington and Scazzocchio, 1967; Arst and Cove, 1973; Arst *et al.*, 1981). Few people who work with other organisms appreciate the sensitivity of growth testing that is possible with a colonial but differentiating and multicellular microorganism (Arst, 1981). To give but one example, there are four clustered genes where mutation can result in inability to utilize L-proline (to be discussed in Section

V,A). Not only does growth testing allow unambiguous classification of each of the four kinds of proline-nonutilizing single mutants, but it allows either complete or partial classification of all of the categories of multiple mutants and deletion mutants (Arst et al., 1981). Examples of phenotypes of some proline-nonutilizing mutations are shown in Fig. 1.

III. REGULATORY GENES

A number of trans-acting regulatory genes have been identified in *A.nidulans*. The criteria for assigning a direct regulatory function to a gene have been discussed (Arst and Bailey, 1977; Scazzocchio, 1980; Arst, 1981). Here we subdivide regulatory genes into three (somewhat arbitrary) categories. Pathway-specific regulatory genes mediate induction or repression of the synthesis of the permeases and enzymes of a single metabolic pathway. Wide domain regulatory genes control the expression of structural genes from a number of different pathways. They mediate responses to environmental factors having wide-ranging metabolic consequences. Integrator genes provide an alternative means of regulating the expression of structural genes from two or more apparently unrelated and independently regulated pathways. In this section, we will discuss some of the better characterized forms of control in each category. It should be noted that certain pathway-specific regulatory genes are briefly discussed in other sections, namely, *amdA* and *facB* (Section III,C), *alcR* and *aplA* (Section V), and *prnA* (Section V,A). Two other pathway-specific regulatory genes have been reviewed elsewhere; *galA*, involved in D-galactose utilization (Arst and Bailey, 1977), and *arcA*, involved in L-arginine catabolism (Bartnik et al., 1977).

A. Pathway-Specific Regulation

1. Induction by Nitrate and Nitrite

The positive-acting regulatory gene *nirA* mediates induction, by nitrate and nitrite, of the enzymes of nitrate assimilation, nitrate and nitrite reductases (reviewed in Cove, 1979). Several enzymes of the pentose phosphate pathway and related activities are also subject to nitrate induction mediated by *nirA*, presumably in order to ensure sufficient NADPH for nitrate assimilation (Hankinson and Cove, 1974, 1975). Loss-of-function mutations, designated $nirA^-$, lead to noninducibility of these activities and the inability to utilize nitrate or nitrite (Cove, 1979). Much rarer constitutive mutations, designated $nirA^c$, alleviate the requirement for a coinducer (Pateman and Cove, 1967; Rand and Arst, 1978). Another gain-of-function class of mutation, designated $nirA^d$, bypasses the need for the *areA* gene product and results in nitrogen metabolite derepressed expression of

TABLE I.

Levels of Nitrate Assimilation Pathway Enzymes in Wild-Type, *nirA^c*, *nirA^d*, and *nirA$^{c/d}$* Strains of *A. nidulans*[a]

		Relative activity	
Relevant genotype	Nitrogen source(s)	Nitrate reductase	Nitrite reductase
Wild-type (*nirA$^+$*)	Urea	6	2
	NO_3^-	100	100
	$NO_3^- + NH_4^+$	2	2
	NH_4^+	2	2
nirA^c-1	Urea	75	29
	NO_3^-	98	117
	$NO_3^- + NH_4^+$	20	21
	NH_4^+	3	6
nirA$^{c/d}$-101	Urea	85	27
	NO_3^-	143	111
	$NO_3^- + NH_4^+$	79	63
	NH_4^+	85	40
Wild-type (*nirA$^+$*)	L-Proline	1	2
	NO_3^-	100	100
	$NO_3^- + NH_4^+$	5	5
	NH_4^+	1	1
nirA^d-106	L-Proline	3	5
	NO_3^-	120	134
	$NO_3^- + NH_4^+$	70	61
	NH_4^+	10	6
nirA$^{c/d}$-113	L-Proline	93	47
	NO_3^-	155	81
	$NO_3^- + NH_4^+$	151	82
	NH_4^+	126	60

[a] Data above the line are taken from Rand and Arst (1978); those below the line are from Tollervey and Arst (1981). Growth and assay conditions differ so that data from the two papers cannot be compared directly. On each side of the line, activities are expressed as percentages of the induced, nitrogen metabolite derepressed wild-type activity.

activities under *nirA* control (Rand and Arst, 1978; Tollervey and Arst, 1981); *nirA^c nirA^d* double mutants (also designated *nirA$^{c/d}$*) can be constructed by mutation or intragenic recombination (Tollervey and Arst, 1981). This suggests that the *nirA* gene product contains two separate domains, a region affected by coinducer binding, defined by *nirA^c* mutations, and a region interacting with the *areA* gene product or with initiator sites adjacent to structural genes under *areA* and *nirA* control, defined by *nirA^d* mutations. Some effects of *nirA^c*, *nirA^d*, and *nirA*^{c/d} alleles are shown in Table I. The categories of mutations obtained in *nirA* bear a strong formal similarity to those obtained in the *araC* gene, which mediates induction of the L-arabinose regulon of *Escherichia coli* [discussed in Rand and Arst (1978) and Tollervey and Arst (1981)]. Like *nirA*, *araC* is a positive-acting pathway-specific regulatory gene (although it can also act negatively) whose regulon is also subject to regulation by a positive-acting wide domain regulatory gene.

The *nirA* gene product is apparently present in limiting amounts (Cove, 1969, 1979). The *nirA*⁻ mutations only partially complement in heterokaryons with mutations in structural genes under *nirA* control (although they do complement fully in diploids). This phenomenon is discussed in Section III,A,3.

It is relevant to point out here that another process affects the regulation of the synthesis of nitrate and nitrite reductases. Most mutations in *niaD*, the structural gene for nitrate reductase apoenzyme, and the *cnx* genes, which participate in the synthesis of a molybdenum-containing cofactor common to nitrate reductase and other molybdoenzymes (the purine hydroxylases; see Sections III,A,2 and V), result in constitutive synthesis of nitrite reductase and, where detectable, the inactive nitrate reductase protein (reviewed in Cove, 1979). It thus appears that nitrate reductase regulates its own synthesis along with that of nitrite reductase (and certain other related activities). This system provided one of the earliest examples of autogenous regulation. Three straightforward models can rationalize the data. In one, the product of *nirA* gene is active unless complexed with nitrate reductase. The coinducers nitrate and nitrite prevent complex formation. In the second, nitrate reductase, unless complexed with a coinducer, prevents synthesis of the *nirA* gene product. In the third, possibly less economical, each protein interacts with cis-acting regulatory sites adjacent to the cognate structural genes.

2. Induction by Uric Acid

Uric acid is the physiological inducer of the syntheses of a number of enzymes and permeases involved in purine degradation [i.e., adenine deaminase, xanthine dehydrogenase, urate oxidase, allantoinase, allantoicase, uric acid–xanthine permease (Scazzocchio and Darlington, 1968; Arst and Scazzocchio, 1975; Sealy-Lewis *et al.*, 1978)]. Mutations in one gene, *uaY*, result in noninducibility by uric acid as well as by several of its thioanalogs which act as gratuitous inducers for the wild type (Scazzocchio *et al.*, 1982). The structural genes have been identified rigorously for xanthine dehydrogenase (*hxA*) and urate oxidase (*uaZ*) (Scazzocchio and Sealy-Lewis, 1978; H. M. Sealy-Lewis *et al.*, unpublished observations) and tentatively for adenine deaminase (*nadA*), allantoinase (*alX*), allantoicase (*aaX*), and an inducible uric acid–xanthine permease (*uapA*) (Scazzocchio and Darlington, 1968; Arst and Scazzocchio, 1975; Scazzocchio *et al.*, 1982). There is a second uric acid–xanthine permease, possibly encoded by the *uapC* gene, which is also likely to be under *uaY* control (J. H. Zabicky-

Zissman *et al.*, unpublished observations). Indirect evidence indicates that two other genes, *xanA* and *hxB*, are also under *uaY* control (Scazzocchio, 1973a; Sealy-Lewis *et al.*, 1978; Scazzocchio *et al.*, 1982). The *xanA* gene product is necessary for an alternative pathway of oxidation of xanthine to uric acid, present in the absence of xanthine dehydrogenase. The *hxB* gene product is necessary for a probable posttranslational modification common to xanthine dehydrogenase (purine hydroxylase I) and the first enzyme of nicotinate catabolism, purine hydroxylase II (Scazzocchio, 1980). Three genes under *uaY* control are also subject to alternative induction mechanisms. Allantoinase and allantoicase (putative structural genes *alX* and *aaX*, respectively) are induced by uric acid and allantoin in an additive fashion. The *uaY*$^-$ mutations only affect uric acid induction (Scazzocchio and Darlington, 1968). Genetic evidence indicates that *hxB* is under the alternative control of at least one of the genes regulating nicotinate utilization, *aplA* (Scazzocchio, 1973b; see Section V). All of these genes are scattered throughout the genome; no clustering occurs (Scazzocchio and Darlington, 1968; Scazzocchio and Gorton, 1977; Scazzocchio *et al.*, 1982).

Two types of mutations mapping in the *uaY* gene have been described: a "nonleaky" type leading equally to loss of all activities under *uaY* control, and a "leaky" type reducing urate oxidase less than adenine deaminase, xanthine dehydrogenase, and transport of uric acid and xanthine. Fine-structure mapping indicates that these two types of mutations are interspersed and do not define discrete functional domains (Scazzocchio *et al.*, 1982; T. Sankarsingh and C. Scazzocchio, unpublished observations).

Attempts to select *uaY* alleles leading to constitutivity have yielded a class of "pseudoconstitutive" mutations hyperresponsive to low concentrations of intracellular inducers. At least some of these mutations map in the *oxpA* locus, closely linked (0.5 centimorgan) to *uaY* (Scazzocchio *et al.*, 1982). One *uaY*$^-$ allele, *uaY-205*, has especially interesting characteristics. It is the most centromere proximal (and thus the most *oxpA* proximal) *uaY* allele mapped. Unlike other *uaY*$^-$ alleles, which revert only to a wild-type phenotype, *uaY-205* reverts also to a number of intermediate phenotypes and, more strikingly, to a constitutive phenotype (T. Sankarsingh and C. Scazzocchio, unpublished observations). Moreover, a constitutive *uaY* allele has now been selected as a forward mutation from the wild type (J. M. Kelly and C. Scazzocchio, unpublished observations). It maps between *uaY-205* and the second most centromere proximal *uaY*$^-$ allele. The simplest hypothesis to accommodate these results is that *uaY-205* and the constitutive allele define the *uaY* promoter region, the intermediate revertant phenotypes from *uaY-205* resulting from partial promoter restoration and the constitutive phenotypes resulting from "up-promoter" mutations leading to higher concentrations of the *uaY* product.

The basis for this constitutivity is probably analogous to that for the derepression conferred by the *areA* allele *xprD-1* (see Section III,B,1) and to that for the apparent constitutivity of the *alcA* and *aldA* gene products in transformants carrying multiple copies of the regulatory gene *alcR* (R. A. Lockington *et al.*, unpublished observations; see also Section V). If an equilibrium exists between two conformers (two different interconvertible species) of a positive-acting regulatory protein, one of which is activating while the other is not, the active concentration of the inducing conformer can be increased in several ways, including (1) forming a complex with a low-molecular-weight effector (physiological regulation), (2) increasing the concentration of regulatory gene product (e.g., *xprD-1*, constitutive *uaY* alleles, multiple copies of regulatory genes), and (3) changing the affinity of a receptor site (e.g., *uap-100* for the *uaY* product; see Section IV,E).

A protein likely to be the product of *uaY* has been identified after chromatography on phosphocellulose and DNA–cellulose as binding with the natural (i.e., uric acid) or gratuitous (2-thiouric acid) effector. This protein peak is missing in strains carrying a null, nonrevertible mutation in the *uaY* gene and shows a modified elution profile in a strain carrying one of the leaky *uaY* mutations (Philippides and Scazzocchio, 1981). Further purification was achieved by affinity chromatography, using 8-thiouric acid bound to a Sepharose column (D. Philippides, unpublished observations). As this protein is present in very low amounts ($\leq 2 \times 10^{-6}$ μmol of binding sites per milligram of soluble protein in extracts; Philippides and Scazzocchio, 1981), further characterization and final proof that it is indeed the product of the *uaY* gene will depend on the molecular cloning of *uaY* and the production of larger amounts of the protein by using high-expression vectors.

3. Dose Effects and Nuclear Limitation of Regulatory Gene Products

Both uaY^- and $nirA^-$ (as well as $nirA^c$ and $nirA^{c/d}$) alleles are strictly codominant (or "semidominant") with their wild-type alleles (Cove, 1969; Rand and Arst, 1978; Scazzocchio *et al.*, 1982). Thus uaY^+/uaY^- diploids have enzyme levels intermediate between those of uaY^+/uaY^+ and uaY^-/uaY^- diploids (Table II). The same is true of $nirA^+/nirA^-$, $nirA^+/nirA^c$, and $nirA^-/nirA^c$ diploids (Table II). This is also the case for constitutive alleles of the regulatory gene *aplA* involved in nicotinate catabolism (Scazzocchio *et al.*, 1973; see also Sections III,A,2 and V). This indicates that there is a strict dose limitation, one copy of the regulatory gene being insufficient to regulate fully two sets of structural genes. As a corollary, these results illustrate that dominance relationships cannot be used to determine the mode of control (positive or negative) unless the regulatory gene product is present in excess. The mode of control must therefore be determined from other data such as the phenotype of deletion or polypeptide chain-termination mutations; or rearrangements such as translocations, inversions, or insertions (if a breakpoint or breakpoints occur

TABLE II.

Data Showing "Semidominance" of Mutations in Two Regulatory Genes, with Nitrate Reductase Levels Used to Illustrate the *nirA* System and Urate Oxidase Levels Used to Illustrate the *uaY* System[a]

Relevant genotype	Enzyme levels	
	Noninduced	Induced
nirA+/*nirA*+	3	100
nirA−/*nirA*−	2	2
*nirA*c/*nirA*c	137	137
nirA+/*nirA*−	3	62
nirA+/*nirA*c	50	88
*nirA*c/*nirA*−	36	57
uaY+ *oxpA*+/*uaY*+ *oxpA*+	3	100
uaY− *oxpA*+/*uaY*− *oxpA*+	3	2
uaY+ *oxpA*−/*uaY*+ *oxpA*−	41	130
uaY+ *oxpA*+/*uaY*− *oxpA*+	4	44
uaY+ *oxpA*+/*uaY*+ *oxpA*−	3	78
uaY− *oxpA*+/*uaY*+ *oxpA*−	3	36

[a] Data for the *nirA* system are taken from Cove (1969, 1979) and those for the *uaY* system are taken from Scazzocchio *et al.* (1982). Enzyme levels are expressed as a percentage of those for the fully induced wild-type diploid. Consult the original publications for comparable data for nitrite reductase for the *nirA* system and for adenine deaminase and xanthine dehydrogenase for the *uaY* system. Details of genotypes and conditions for growth, induction, and assay are given in the original papers. The values given for *uaY*− *oxpA*+/*uaY*− *oxpA*+, *uaY*− *oxpA*+/*uaY*+ *oxpA*−, and *uaY*+ *oxpA*+/*uaY*− *oxpA*+ diploids are averages of eight, ten, and ten diploids, respectively, using only extreme null *uaY*− alleles. The table shows both the dose effects revealed by the *nirA*−, *nirA*c, and *uaY*− mutations and the complete recessivity of a "pseudoconstitutive" mutation mapping at the *oxpA* locus. The dose effect for *uaY*− mutations is also apparent in the *uaY*− *oxpA*+/*uaY*+ *oxpA*− data.

within the gene); or the relative frequency of mutations to each phenotype. While it is clear that *uaY* and *nirA* are positive-acting (Cove, 1979; Scazzocchio *et al.*, 1982), the mode of control by *aplA*, originally considered to be positive-acting on the basis of fallacious reasoning (Scazzocchio and Darlington, 1967; Scazzocchio *et al.*, 1973), is more difficult to determine. As no definitively identified loss-of-function mutation in *aplA* is available, only the mutation frequency can give an indication of the mode of control. Mutations to the *aplA*c phenotype occur with a high frequency (10^{-4} among survivors of ultraviolet mutagenesis with 95% kill; J. M. Kelly and C. Scazzocchio, unpublished observations). It is therefore likely that the constitutive phenotype results from loss rather than modification of function. If so, *aplA* is negative-acting.

These three regulatory genes, all of whose products show stringent dose effects, also show an apparent limitation of their function to the nucleus (partial in the case of *nirA*). This is seen by comparing diploids with heterokaryons for complementation between regulatory and cognate structural gene mutations. The differences are most striking for the *uaY* control system. The *uaY*$^-$ mutations do not complement with *hxA*$^-$, *hxB*$^-$, or *uaZ*$^-$ mutations in heterokaryons while complementing fully in diploids (Scazzocchio *et al.*, 1982). Moreover, a functional *uaY* allele must be within the same nuclei as the up-promoter constitutive mutation *uap-100* for the latter to be expressed (Scazzocchio and Arst, 1978; see also Section IV,E). A similar relationship exists between *aplA*^{c} and *hxB*$^-$ mutations (indicating that *hxB* must be under *aplA* as well as *uaY* control; see above) and, to a less marked extent, between *nirA*$^-$ mutations and *niaD*$^-$ and *niiA*$^-$ structural gene mutations.

These differences in complementation patterns between heterokaryons and diploids might reflect limitation of regulatory gene products to the nucleus, or they might be a trivial consequence of a stringent dose effect (see above) when combined with a nonrandom distribution of nuclei in heterokaryons (Clutterbuck, and Roper, 1966). [For discussion see Scazzocchio *et al.* (1973, 1982); for similar phenomena in *N. crassa* see Burton and Metzenberg (1972) and Metzenberg (1979).] It is now possible to distinguish between these alternatives by using strains carrying tandem multiple copies of genes (including regulatory genes) in either their normal or other chromosomal locations arising by transformation (Tilburn *et al.*, 1983; H. N. Arst, Jr. and C. Scazzocchio, unpublished observations).

B. Wide Domain Regulation

1. Nitrogen Metabolite Repression

The syntheses of a large number of enzymes and permeases involved in nitrogen nutrition are subject to repression by preferred nitrogen sources, especially ammonium and L-glutamine. This repression is mediated by the positive-acting regulatory gene *areA* (Arst and Cove, 1973). Loss-of-function mutations in *areA*, designated *areA*^{r}, result in inability to utilize nitrogen sources other than ammonium and low, repressed levels (see Table III) of enzymes and permeases involved in nitrogen source utilization (Arst and Cove, 1973; Hynes, 1975a; Polya *et al.*, 1975; Rand and Arst, 1977; Arst *et al.*, 1980, 1982; Tollervey and Arst, 1982; Spathas *et al.*, 1982, 1983). The *areA* mutations were originally classed as loss of function on the basis of their frequency and recessivity (Arst and Cove, 1973), an interpretation strongly supported by selec-

TABLE III.

Levels of Certain Nitrogen-Metabolizing Enzymes in *A. nidulans* Wild Type and an *areA*^{r} Mutant

	Relative activity		
Relevant genotype	Nitrate reductase	Urease	L-Asparaginase
Wild-type (*areA*$^+$)	100	100	100
areA^r-*18*	0	20	2

^{a} Data are taken from Tollervey and Arst (1981) and Arst *et al.* (1982). Growth of mycelia was carried out in nitrogen metabolite derepressing conditions and, in the case of nitrate reductase, in the presence of a coinducer. Activities are expressed as percentages of the induced, nitrogen metabolite derepressed wild-type activities.

tion of an *areA*^{r} allele in which a translocation breakpoint apparently occurs within the *areA* gene (Rand and Arst, 1977; Arst, 1981) and of putative polypeptide chain-termination *areA*^{r} alleles (Al Taho *et al.*, 1984). Other, far rarer mutant *areA* alleles, designated *areA*^{d}, lead to nitrogen metabolite derepressed expression of one or more activities under *areA* control (Table IV). The *areA*^{d} alleles fall into two categories; all but one of the currently available *areA*^{d} alleles enhance expression of some activities while reducing or not affecting expression of others (Hynes and Pateman, 1970; Arst and Cove, 1973; Hynes, 1975a; Polkinghorne and Hynes, 1975, 1982; Arst and Scazzocchio, 1975; Arst, 1977; Shaffer and Arst, 1984). These probably alter the structure of the *areA* product. The one exceptional *areA*^{d} allele leads to nitrogen derepressed expression of nearly all activities under *areA* control (Cohen, 1972; Arst and Cove, 1973; Pateman *et al.*, 1973; Bartnik *et al.*, 1976), suggesting that it leads to increased production of wild-type or nearly wild-type *areA* product. Arst (1982) showed

TABLE IV.

Nitrate Reductase Levels in *A. nidulans* Wild Type and an *areA*^{d} Mutant^a

	Relative nitrate reductase activity		
Relevant genotype	Uninduced	Induced	Repressed (coinducer present)
Wild-type (*areA*$^+$)	8	100	13
xprD-1 (*areA*^{d})	1	179	141

^{a} Data are taken from Arst and Cove (1973). Activities are expressed as a percentage of the induced, nitrogen metabolite derepressed wild-type activity.

that this exceptional *areA*^{d} allele is an inversion and speculated that it might fuse the coding region of *areA* to a more efficient promoter and/or ribosome-binding site. Tables III and IV illustrate some effects of *areA* mutations on enzyme levels.

The existence of *areA*^{r} alleles subject to allele-specific suppression by putative polypeptide chain termination suppressors establishes that the *areA* product is a protein (Al Taho *et al.*, 1984). The nonhierarchical heterogeneity of phenotype of mutant *areA* alleles indicates that the *areA* product is directly involved in the regulation of gene expression and that receptor sites for the *areA* product differ in structure (Arst and Cove, 1973; Arst and Bailey, 1977). An explanation of this is given in the following simple example. Suppose that synthesis of the product of structural gene X is strongly nitrogen metabolite-repressible whereas expression of structural gene Y is only weakly subject to nitrogen metabolite repression. If the *areA* gene product were indirectly involved in the regulation of X and Y (e.g., as an enzyme involved in synthesis of a small molecule able to activate expression of X and Y) then partial loss-of-function mutations in *areA* might result in an X^-Y^+ phenotype but never in an X^+Y^- phenotype. No *areA* mutation could ever mask the hierarchy of structural genes reflecting their individual degrees of sensitivity to nitrogen metabolite repression. If, on the contrary, the *areA* gene product interacts directly with *cis*-acting regulatory sites adjacent to X and Y, then both X^-Y^+ and X^+Y^- phenotypes might be possible for different *areA* alleles, provided that the receptor site for the *areA* product adjacent to gene X differs in structure from that adjacent to gene Y. For *areA* this second situation applies. Such nonhierarchical heterogeneity is observed not only among leaky *areA*^{r} mutations but also when examining those whose phenotypes involve conditional (e.g., thermosensitive) expression of some structural genes but not others. Even more strikingly, some mutant *areA* alleles have opposing effects on expression of different structural genes, resulting in reduced (or no) expression of some and enhanced or nitrogen metabolite derepressed expression of others. This interpretation is strongly supported by the fact that some *areA*^{r} putative polypeptide chain-termination mutations result in thermosensitive utilization of some nitrogen sources but not others when suppressed by translational suppressors (Al Taho *et al.*, 1984). The product of the equivalent gene of *N. crassa, nit-2,* has been isolated as a DNA-binding protein located in the nucleus (Grove and Marzluf, 1981).

L-Glutamine is probably the effector for the *areA* gene product. Loss-of-function mutations in *gdhA,* the structural gene for NADP-linked glutamate dehydrogenase, and *glnA,* the structural gene for glutamine synthetase, led to nitrogen metabolite derepression (Arst and MacDonald, 1973; Pateman *et al.*, 1973; Kinghorn and Pateman, 1973, 1975; Arst *et al.,* 1982; MacDonald, 1982; Cornwell and MacDonald, 1984). The evidence that L-glutamine is the nitrogen metabolite corepressor in *N. crassa* is rather more direct in that glutamine specifically elutes the *nit-2* product from DNA–cellulose (Grove and Marzluf, 1981). Marzluf (1981) has proposed that in the presence of glutamine the *nit-2* gene product is unable to activate expression of structural genes under *nit-2* control. A similar model would be compatible with the available data for *A. nidulans*.

The selection of locus-specific extracistronic suppressor mutations restoring the ability of *areA*^{r} strains to utilize one or more nitrogen sources has proved a particularly fruitful method of selecting other kinds of regulatory mutations. The vast majority of these mutations [e.g., *intA*^{c} (Section III,C), *prn*^{d} (Section IV,B) *amdI* (Section IV,D), *gabI* (Section IV,A), *creA*^{d} (Section III,B,2), and *nis-5* (Section V,B) mutations and *nirA*$^{c/d}$ double mutations (Section III,A,1)] suppress *areA*^{r} mutations for utilization of a few nitrogen sources at most. The regulatory genes and cis-acting regulatory regions in which such mutations occur have a much more narrow regulatory domain than *areA*. However, Tollervey and Arst (1982) were able to select five extremely rare mutations at a locus designated *areB* which suppress *areA*^{r} mutations for the utilization of all nitrogen sources. All five mutations are associated with major chromosomal rearrangements, four with translocations (Tollervey and Arst, 1982) and one with an inversion (M. X. Caddick and H. N. Arst, Jr., unpublished observations; Wiame *et al.*, 1985). Remarkably, two of the translocations, although obtained in different parental strains in separate experiments and having distinct phenotypes, have nearly identical translocation breakpoints (Tollervey and Arst, 1982; Wiame *et al.*, 1985). It would seem, therefore, that *areB* mutations involve perforce one of a rather small number of chromosomal rearrangements. A plausible hypothesis is that *areB* might be an *areA*-related pseudogene, normally inactive but able to be activated by a chromosomal rearrangement that fuses it to a functional (but dispensable) promoter and/or ribosome-binding site. In support of this hypothesis, loss-of-function mutations, designated *areB*^{r}, have been obtained in *areB* as abolishing suppression of *areA*^{r} mutations without affecting the original *areB* translocation of the parental strain (H. N. Arst, Jr., unpublished observations; Wiame *et al.*, 1985); *areB* is probably the first pseudogene to be identified by classical genetic analysis.

Pateman and Kinghorn (1977) suggested that, in addition to *areA*, a second regulatory gene *tamA* is involved in mediating nitrogen metabolite repression. Results reported by Arst *et al.* (1982) indicate, however, that the apparent regulatory effects reported by Pateman and Kinghorn probably resulted from additional mutations in genes other than *tamA*. There is therefore no reason to suppose that *tamA* is involved in nitrogen metabolite repression in any direct way.

2. Carbon Catabolite Repression

Expression of many structural genes involved in carbon source utilization is subject to carbon catabolite repression. Such repression is mediated by a regulatory gene designated *creA* which is probably negative-acting (Arst and Cove,

1973; Bailey and Arst, 1975; Arst and MacDonald, 1975; Arst and Bailey, 1977). Putative loss-of-function mutations designated $creA^d$ lead to carbon catabolite derepression. The $creA^d$ mutations have been selected in several ways, but two methods are particularly powerful and instructive. One method involves reversion of $areA^r$ strains. The $areA^r$ mutants are able to utilize certain nitrogen sources [L-proline, acetamide, L-glutamine, and γ-amino-n-butyrate (GABA)] if these compounds also serve as sole carbon sources or if D-glucose, normally present in *A. nidulans* medium (Cove, 1966), is replaced by glycerol, L-arabinose, lactose, melibiose, or ethanol. The $creA^d$ mutations have been selected as allowing $areA^r$ strains to utilize these four nitrogen sources in the presence of a repressing carbon source such as D-glucose, D-xylose, or sucrose (Arst and Cove, 1973; Bailey, 1976). The other method involves the use of $pdhA^-$ strains lacking pyruvate dehydrogenase. As *A. nidulans* is an obligate aerobe, $pdhA^-$ mutants require acetate, but the acetate requirement can be met by certain precursors of acetate such as ethanol if they serve as sole carbon source or in the presence of other derepressing carbon sources such as glycerol, L-arabinose, lactose, or melibiose. Use of ethanol for supplementation of the acetate requirement is completely prevented by D-glucose, D-xylose, and sucrose, while a number of other carbon sources such as D-fructose, D-mannose, and D-galactose partially prevent it. The degree to which various carbon sources allow ethanol supplementation of $pdhA^-$ mutants closely parallels the degree to which they allow $areA^r$ mutants to utilize L-proline, acetamide, and so forth, indicating that both systems measure carbon catabolite repression. A number of $creA^d$ alleles have been obtained as allowing ethanol supplementation of $pdhA^-$ strains in the presence of a repressing carbon source (Bailey and Arst, 1975). Various $creA^d$ mutations show a nonhierarchical heterogeneity of phenotype, indicating that the *creA* gene product is directly involved in carbon catabolite repression (Arst and Bailey, 1977).

Some effects of $creA^{d}$-1 on enzyme levels are shown in Table V. The high degree of repressibility of alcohol dehydrogenase and its relief by $creA^{d}$-1 are consistent with the pattern of ethanol supplementation of $pdhA^-$ strains and the resulting method for selecting $creA^d$ mutations. Carbon catabolite repression of other activities such as NAD-linked glutamate dehydrogenase is only modestly (or even not at all) affected by $creA^d$ mutations (Table V). Little can be said about the identity of the effector(s) interacting with the *creA* product except that $3',5'$-cyclic AMP is probably not involved and uptake and some metabolism of repressing carbon sources are necessary for carbon catabolite repression to occur (Arst and Bailey, 1977).

3. Phosphorus Repression

The expression of a number of structural genes involved in phosphorus acquisition is subject to repression by phosphate (or a metabolite synthesized from phosphate) mediated by the product of the *palcA* gene. The $palcA^-$ mutants lack

TABLE V.

Relative Activities of Alcohol Dehydrogenase (ADH) and NAD-Linked Glutamate Dehydrogenase (GDH) in *A. nidulans* Wild Type (*creA*⁺) and a *creA*ᵈ Mutant[a]

Growth conditions		Relative activities			
		ADH		NAD–GDH	
Induction	Carbon catabolite repression	*creA*⁺	*creA*ᵈ⁻¹	*creA*⁺	*creA*ᵈ⁻¹
−	−	11	10	38	35
+	−	100	107	100	107
+	+	2	71	12	25

[a] Data are taken from Bailey and Arst (1975), where experimental details are given. Activities are expressed as percentages of the induced, carbon catabolite derepressed wild-type activity.

several phosphatases and at least one phosphate permease (Dorn, 1965a,b; Brownlee *et al.*, 1983; M. X. Caddick, A. G. Brownlee, and H. N. Arst, Jr., unpublished observations). As *palcA*⁻ mutations are likely to be a loss-of-function class, *palcA* is probably positive acting.

4. Sulfur Repression

A sulfate permease and at least some of the enzymes of L-cysteine biosynthesis are subject to repression by L-cysteine and L-homocysteine (Paszewski and Grabski, 1974; Paszewski *et al.*, 1977). This repression is mediated by the product of the *suAmeth* gene, likely to be a negative-acting regulatory gene (Paszewski and Grabski, 1975; Lukaszkiewicz and Paszewski, 1976). A putative loss-of-function mutation leads to derepressed expression of activities under *suAmeth* control (Paszewski and Grabski, 1975; Lukaszkiewicz and Paszewski, 1976; Paszewski *et al.*, 1977). Although the range of activities currently known to be subject to *suAmeth* control is sufficiently narrow to allow its classification as a pathway-specific regulatory gene (Section III,A), the breadth of the sulfur regulatory domain in *N. crassa* (reviewed in Metzenberg, 1979) makes it likely that *suAmeth* should be classified as a wide domain regulatory gene.

5. pH Regulation

The syntheses of certain permeases and secreted enzymes are dependent upon the pH of the growth medium. For example, *A. nidulans* wild type is able to synthesize high levels of alkaline phosphatase only at alkaline pH and high levels of acid phosphatase, GABA permease, and molybdate permease only at acid pH (M. X. Caddick *et al.*, unpublished observations). Similar pH regulation of secreted enzymes has been reported in other fungi (Cohen, 1980; Lindberg *et al.*, 1982; Nahas *et al.*, 1982). Mutations in any of several genes affect pH regulation; one of these genes is a likely candidate for a wide domain regulatory gene mediating pH regulation, and several of the others might be involved in the synthesis of a metabolite able to monitor pH and act as an effector (M. X. Caddick *et al.*, unpublished observations).

C. Integrator Genes

The concept of a regulatory gene able to integrate the expression of a number of structural genes which can also be expressed independently comes from a model for gene regulation developed by Britten and Davidson (1969, 1971; Davidson and Britten, 1973). Suppose that the product of a particular structural gene is required in several different contexts. For example, the same enzyme might catalyze steps necessary for (1) utilization of a nutrient, (2) transition from one developmental stage to another, and (3) biosynthesis of a secondary metabolite. The difficulty is that expression of the structural gene must be responsive to a different regulation signal or metabolite for each context. One solution would be continuous expression so that the structural gene product is synthesized constitutively irrespective of growth conditions or stage of growth. This is possibly wasteful and, if absence of the product is ever necessary, potentially harmful. An alternative is a separate copy of the structural gene for each context. However, the relative absence of repetitive DNA in the genome of *A. nidulans* (Timberlake, 1978) suggests that, at least for this organism, a rather severe constraint operates on genome size. A single copy of the structural gene will suffice if it is under the control of parallel-acting positive regulatory genes corresponding to each context and has a cis-acting receptor site for the product of each. As many "contexts" require the presence of more than one structural gene product, many of these positive-acting regulatory genes integrate the expression of a number of structural genes whose expression can also occur independently. These regulatory genes are termed integrator genes, and they occur on a "one integrator gene–one context" basis. A hypothetical regulatory circuit involving integrator genes is shown in Fig. 2.

The first identified integrator gene was *intA* of *A. nidulans* (Arst, 1976; Arst *et al.*, 1978; Hynes, 1978a; Makins *et al.*, 1981). *intA* controls the expression of the structural gene for acetamidase (*amdS*) and putative structural genes for GABA aminotransferase (*gatA*), GABA permease (*gabA*), and a lactamase (*lamA*), mediating induction by ω-amino acids. Acetamidase can be induced independently by acetate and by benzoate (Hynes, 1977, 1978a; Hynes *et al.*, 1983). The mechanism of benzoate induction remains obscure, but acetate induction seems to involve two independently acting positive regulatory genes, *facB* and *amdA* (Arst and Cove, 1973; Arst, 1976; Hynes, 1977, 1978a; Hynes *et al.*, 1983). *amdA* might also be an integrator gene because at least one other polypeptide (of unknown function) seems to be under its control (M. J. Hynes and P.

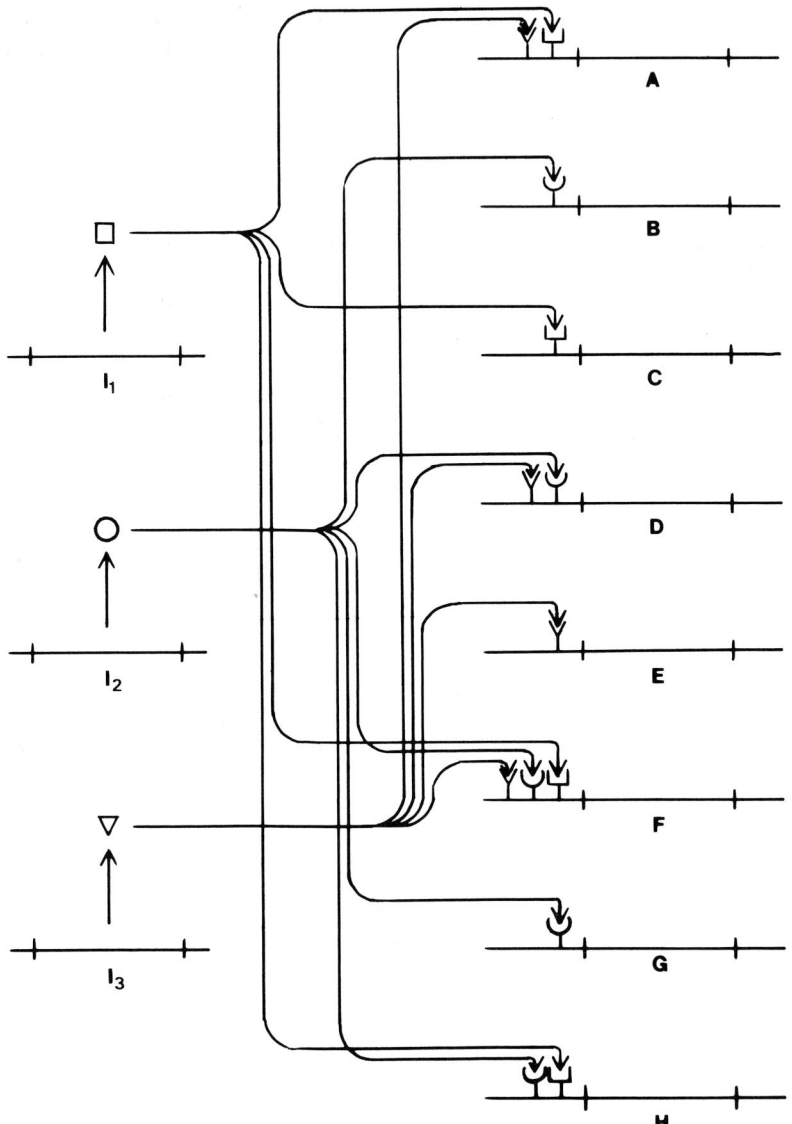

Fig. 2. Hypothetical scheme for a regulatory circuit. A–H are structural genes. I_1, I_2, and I_3 are integrator genes specifying activators (probably proteins, denoted □, ○, and ▽) for those structural genes appropriate to the particular context and active in the presence of a small coinducing molecule indicating the context.

TABLE VI.

Relative Activities of Acetamidase γ-Amino-*n*-butyrate (GABA) Aminotransferase and GABA Permease in *A. nidulans* Wild Type and *intA* Mutants[a]

	Relative activities					
	Acetamidase		GABA aminotransferase		GABA uptake	
Relevant genotype	Uninduced	Induced	Uninduced	Induced	Uninduced	Induced
Wild-type (*intA*⁺)	28	100	14	100	41	100
*intA*ᶜ-2	121	193	142	134	83	247
*intA*ᶜ-305	23	187	11	182	40	108
*intA*ᶜ-304	64	75	209	280	85	70
intA⁻-101	16	27	12	13	35	21
intA⁻-500	27	25	18	67	52	45

[a] Data are collated from Arst *et al.* (1978), Bailey *et al.* (1979, 1980), Penfold (1979), and C. R. Bailey (unpublished observations). β-Alanine (5 mM) was used as coinducer. Published references contain details of growth conditions and assays. Activities are expressed as percentages of the induced wild-type activity.

Atkinson, cited in Hynes *et al.*, 1983). The existence of these alternative parallel-acting modes of acetamidase induction is sufficient to establish *intA* as an integrator gene even if no alternative modes of induction of the *gatA, gabA,* or *lamA* products have been detected.

Loss-of-function *intA*⁻ alleles leading to noninducibility and *intA*ᶜ alleles leading to constitutivity have both been obtained. Table VI illustrates the effects of several *intA* mutations. The nonhierarchical heterogeneity of the phenotypes of *intA* alleles (seen, for example, in Table VI) establishes both that the *intA* gene product is directly involved in the regulation of gene expression and that *intA* product receptor sites adjacent to the structural genes under *intA* control differ in structure (Arst, 1976, 1981; Arst and Bailey, 1977; Arst *et al.*, 1978).

IV. PUTATIVE RECEPTOR SITES

Cis-acting regulatory mutations can, in principle, identify receptor sites for regulatory gene products. Wherever a structural gene is subject to control by more than one regulatory gene, there can be difficulties in determining which regulatory system is affected. There is also the possibility (realized in the case of *uap-100;* see below) that a cis-acting regulatory mutation might affect more than one form of regulation. One further caveat is necessary. Some cis-acting regulatory mutations, e.g., *nis-5* (Section V,B), the inversion-associated *areA*ᵈ allele (Section III,B,1), and rearrangement-associated *areB* mutations (Section III,B,1), result from chromosomal rearrangements. While such mutations are useful and interesting in their own right, they do not define (although they might help to locate) normal receptor sites. The frequency of cis-acting regulatory mutations associated with large-scale rearrangements is possibly less in *A. nidulans* than in *S. cerevisiae* because there is at present no evidence for a repetitive transposable element associated with cis-acting regulatory mutations (P. M. Green, P. Durrens, and C. Scazzocchio, unpublished observations; Hynes *et al.*, 1983). Insertion of such elements is responsible for many cis-acting regulatory mutations in *S. cerevisiae* (Errede *et al.*, 1981; Ciriacy and Williamson, 1981; Williamson *et al.*, 1981).

Beyond identification of receptor sites, cis-acting regulatory mutations can also yield information concerning the phenotype of other regulatory mutations. A clear example of this emerges from analysis of cis-acting regulatory mutations which supress *areA^r* mutations. If a cis-acting regulatory mutation is able to suppress an *areA^r* mutation for utilization of a particular nitrogen source, this establishes that the activity specified by the adjacent structural gene is growth-limiting for utilization of that nitrogen source by the *areA^r* mutant. This deduction is equally valid for receptor site mutations and chromosomal rearrangements. Thus *gabI* mutations (Bailey *et al.*, 1979) establish uptake as growth-limiting for utilization of GABA for *areA^r* strains, and *prn^d* mutations (Arst and MacDonald, 1975; Arst *et al.*, 1980) also establish uptake as growth-limiting for utilization of L-proline by *areA^r* strains. The *amdI* mutations (Hynes, 1975b, 1978b; Arst, 1978) establish acetamidase activity as limiting for acetamide utilization by *areA^r* strains, while *nis-5* (Rand and Arst, 1977) shows nitrite reductase to be limiting for nitrite utilization. The *uap-100* mutation (Arst and Scazzocchio, 1975) establishes that lack of uptake is responsible for the inability of *areA-102* strains (but not necessarily strains carrying other *areA* alleles) to utilize xanthine and uric acid. As *areA^r* mutations reduce levels of many enzymes and permeases, such mutations are often the only evidence that a particular activity is growth-limiting.

A. *gabI* Mutations

The *gabI* mutations were selected as suppressing *areA^r* mutations for GABA utilization and mapping adjacent to *gabA*, specifying GABA permease (Bailey *et al.*, 1979). Measurements of GABA permease levels tend to suggest that *gabI-1* relieves nitrogen metabolite repression and/or carbon catabolite repression, whereas *gabI-3* might have a general up-promoter effect.

B. *prn^d* Mutations

The *prn^d* mutations were selected as suppressing *areA^r* mutations for L-proline utilization and map adjacent to *prnB*, specifying L-proline permease, in the central cis-acting regulatory region of the *prn* gene cluster (see Fig. 4 and Section V,A) (Arst and MacDonald, 1975; Arst *et al.*, 1980, 1981). Their effects on permease levels suggest that they lead mainly to derepression, although it is unclear whether they relieve nitrogen metabolite repression, carbon catabolite repression, or both; *prn*^{d} mutations control expression of *prnB* in cis, but there is no evidence that they affect expression of *prnA*, *prnD*, or *prnC* (Arst *et al.*, 1980). Southern blotting, using the cloned *prn* cluster, has shown that none of the three *prn*^{d} mutations examined leads to a major DNA rearrangement (P. M. Green, P. Durrens, and C. Scazzocchio, unpublished observations).

C. sB_o-90

The *sB* gene is the structural gene for the sulfate permease (Arst, 1968; Lukaszkiewicz and Pieniazek, 1972); sB_o-90 is a likely operator mutation tightly linked to *sB* and results in hyperrepressibility of the permease due to enhanced affinity for the *suAmeth* repressor (see Section III,B,4) (Lukaszkiewicz and Paszewski, 1976).

D. cis-Acting Regulatory Mutations Affecting Expression of *amdS*

The laboratory of M. J. Hynes has studied the *amdS* gene, encoding acetamidase, in detail, using both classical and recombinant DNA techniques. An extensive fine-structure map has been constructed by recombination and physical mapping (Hynes, 1979; Hynes *et al.*, 1983). An extremely interesting collection of cis-acting regulatory mutations has been selected and analyzed. These include *amdI-18*, an up-promoter mutation (Hynes, 1978b); *amdI-9*, which leads to enhanced induction by acetate (probably mediated by *facB*) (Hynes, 1975b, 1977, 1978a); *amdI-93*, which prevents *intA*-mediated induction and has a partial down-promoter effect (Hynes, 1980); *amdI-66*, which magnifies the effect of a constitutive *amdA* allele (Hynes, 1982); and three extreme down-promoter alleles, *amd-406* (deletion), *amd-407* (short deletion), and *amd-205* (possibly a point mutation) (Hynes, 1979, 1982; Hynes *et al.*, 1983). Apart from the two deletions, none of these mutations involves a major rearrangement (Hynes *et al.*, 1983).

E. cis-Acting Regulatory Mutations Affecting Expression of *uapA*

The *uapA* gene specifies a permease for xanthine and uric acid (Darlington and Scazzocchio, 1967). The *areA-102* allele, although leading to derepressed expression of some structural genes under *areA* control, results in drastically reduced expression of *uapA* (Arst and Cove, 1973; Arst and Scazzocchio, 1975;

Gorton, 1983). Several cis-acting regulatory mutations for *uapA* have been selected as suppressing *areA-102* for utilization of xanthine and uric acid. The most extraordinary of these is *uap-100*, which, in addition to accommodating the mutant form of the *areA* gene product present in *areA-102* strains, results in initiator constitutivity and an up-promoter effect (Arst and Scazzocchio, 1975). (In this context an initiator is a receptor site for a positive-acting regulatory gene product.) This phenotype implies functional overlap between the promoter and two initiator sites. The *uap-100* phenotype is dependent upon a functional *uaY* allele, suggesting that it increases receptor site affinity for a particular conformer of the *uaY* product (Scazzocchio and Arst, 1978). The other cis-acting regulatory mutations for *uapA* seem to behave as simple, allele-specific suppressors of *areA-102* (Gorton, 1983).

V. THE SPATIAL ORGANIZATION OF FUNCTIONALLY RELATED GENES

A common (and possibly naive) prejudice is that clustering of genes involved in the same metabolic pathway occurs in bacteria (or at least in the well-characterized pathways of the Enterobacteriaceae) because it is a prerequisite for operon-type organization. In eukaryotes there is no definitive evidence for a functional operon organization, yet a number of metabolic pathways clearly show gene clustering.

The degree of clustering varies widely among different metabolic pathways in *A. nidulans*. In the purine degradation pathway (see Section III,A,2) all nine genes definitely or putatively under the control of *uaY* are meiotically unlinked to *uaY* and each other (Scazzocchio and Gorton, 1977; Scazzocchio *et al.*, 1982). The structural gene for alcohol dehydrogenase *alcA* is closely linked to its cognate positive-acting regulatory gene *alcR* but unlinked to the putative structural gene for aldehyde dehydrogenase *aldA,* also under the control of *alcR* (Pateman *et al.*, 1983; H. M. Sealy-Lewis, unpublished observations). Molecular cloning of the *alcA alcR* cluster has shown that approximately 2 kb separate *alcA* from *alcR* (Lockington *et al.*, 1985). This was shown by transformation of strains carrying $alcA^-$ or $alcR^-$ point mutations using subclones derived from a 13-kb genomic clone, recognized as hybridizing with an *alcA* cDNA clone (Lockington *et al.*, 1985).

In the nicotinate degradation pathway the structural gene for nicotinate hydroxylase (purine hydroxylase II) is tightly linked to both a positive control gene *hxnR* and a negative control gene *aplA* (Scazzocchio, 1980; see also Section III,A,3), probably in the order *hxnR aplA hxnS* (J. M. Kelly and C. Scazzocchio, unpublished observations). Another gene or gene cluster, specifying an enzyme(s) involved in a further step(s) of the pathway, although on the same

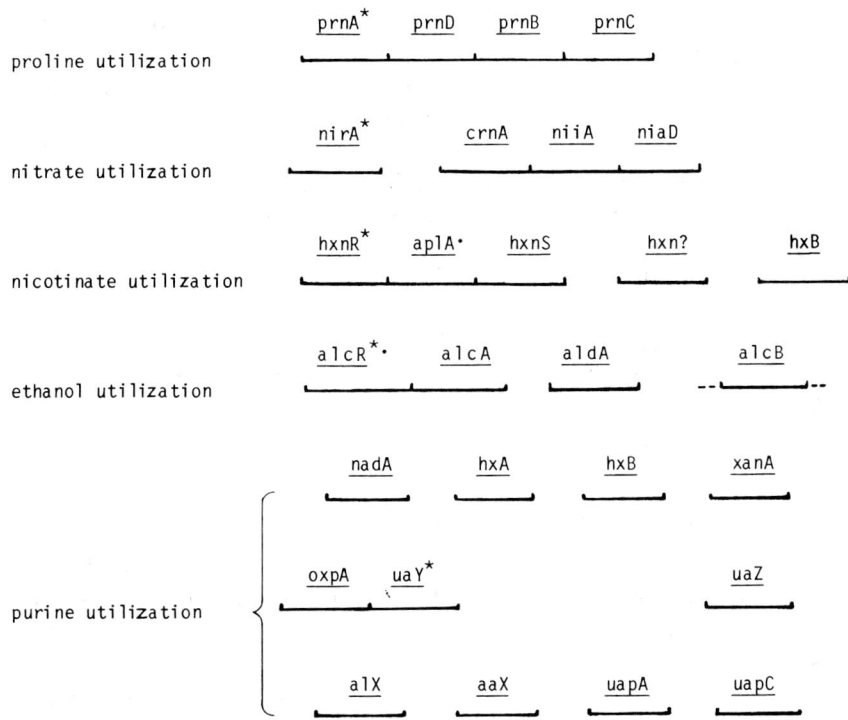

Fig. 3. Wide variation in the extent of clustering of genes involved in a number of metabolic pathways. Genes and pathways are described in the text. Positive regulatory genes are indicated with an asterisk (e.g., *prnA**), negative regulatory genes with a dot (*aplA*·). *hxn?* represents a number of closely linked mutations involved in nicotinate utilization and mapping on the same chromosome as the *hxnR aplA hxnS* cluster but recombining freely with mutations in the cluster. The mutations in *hxn?* might map in more than one gene, and one of these is possibly involved in the uptake of nicotinic acid. It is not clear whether these genes are under *hxnR aplA* control (J. M. Kelly, D. Lycan, S. Lee and C. Scazzocchio, unpublished data; see Section V). The ethanol utilization pathway merits further comment. A second, minor alcohol dehydrogenase (ADH II) that is repressed when the physiologically characterized alcohol dehydrogenase (ADH I) is induced is present in *A. nidulans*. All the evidence (Sealy-Lewis and Lockington 1984; Lockington *et al.*, 1985) is consistent with *alcR* acting as a positive regulatory gene for *alcA* and *aldA* (see text) but as a negative regulator for the gene coding for ADH II. This unidentified gene is labeled *alcB*. The fact that its position is unknown is indicated by the dotted lines in the drawing, but molecular data indicate that it is not within 26 kb of the *alcA alcR* genomic clone, as a large deletion encompassing both genes results in constitutivity for (not loss of) ADH II. We show next to *uaY* the closely linked gene *oxpA*. "Pseudoconstitutive" mutations map in this gene (see text and Table II). For a discussion of the possible role of this gene in the regulation of purine utilization see Scazzocchio *et al.* (1982). The genes shown do not represent all known genes participating in each pathway but only those under the control of the regulatory genes shown.

chromosome, does not show meiotic linkage to the *hxnR aplA hxnS* cluster (J. M. Kelly, C. Scazzocchio, S. Lee, and D. Lycan, unpublished observations). There is evidence that, as in the extensively studied *qa* cluster of *N. crassa* (Huiet and Case, Chapter 8), the regulatory and structural genes involved in quinate utilization comprise a gene cluster in *A. nidulans* (Hawkins *et al.*, 1984, 1985). Figure 3 shows examples of pathways illustrating different types of gene spatial organization. Below, two gene clusters, involved in proline catabolism and nitrate assimilation, respectively, are described in greater detail.

As in other eukaryotes, a number of "cluster genes" encoding polyfunctional polypeptides are present in *A. nidulans*. These genes have been reviewed elsewhere (Giles, 1978; Metzenberg, 1979) and are outside the scope of this chapter. They should not be confused with clusters of distinct and separate yet functionally related genes, discussed here, although there might be evolutionary connections between these two types of genetic organization.

A. The Proline Catabolism Gene Cluster

Four genes whose products are necessary and sufficient for conversion of exogenous L-proline to internal L-glutamate are clustered in linkage group VII (Fig. 4). *prnB* specifies the major proline permease; *prnD* and *prnC* are structural genes for proline oxidase and Δ^1-pyrroline-5-carboxylate (P5C) dehydrogenase, respectively; and *prnA* is a pathway-specific regulatory gene mediating proline induction (Arst *et al.*, 1980; Jones *et al.*, 1981). Some *prnA*⁻ mutations are suppressible by translational suppressors and deletion mutations in *prnA* have a non-inducible phenotype, showing that the *prnA* product is a protein and that it is positive-acting (Sharma and Arst, 1985). cis-Acting regulatory mutations, designated *prn*^{d}, map in the center of the cluster and control expression of *prnB* (see Section IV,B), suggesting that the direction of transcription of *prnB* is toward *prnC*. Genetic analysis of the *prn* cluster benefits from use of an unlinked mutation resulting in proline toxicity (to provide a positive selection method for

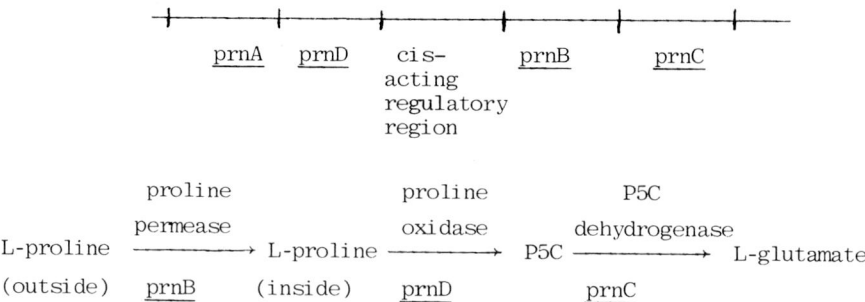

Fig. 4. The *prn* gene cluster (after Arst *et al.*, 1981) and the L-proline catabolic pathway.

obtaining a number of categories of *prn*⁻ mutations) and a battery of growth tests rendering all classes of single mutants and most classes of double mutants readily distinguishable (Arst *et al.*, 1981).

A particularly interesting feature of the *prn* gene cluster is that deletions that remove the central cis-acting regulatory region without extending into *prnC* nevertheless reduce *prnC* expression, as judged by both enzyme assays and complementation tests (Arst and MacDonald, 1978; Arst *et al.*, 1981). Originally this was interpreted as indicating that *prnC* is expressed via a *prnB prnC* dicistronic mRNA initiated in the central regulatory region plus an overlapping transcript initiated elsewhere. Cloning of the *prn* cluster has permitted a direct test of this hypothesis (P. M. Green and C. Scazzocchio, unpublished observations). RNA blot analysis suggests that *prnC* is expressed via a single mRNA of monocistronic size. Deletion of the central regulatory region reduces the amount of this transcript, indicating that regulation is at the level of transcription. It thus appears that the central regulatory region (or possibly an adjacent portion of the *prnB* or *prnD* genes removed by all ten of the available relevant deletions) contains an enhancer-like element for *prnC* expression. As the effect of the deletion mutations is much more pronounced in mycelia grown at 25°C than in those grown at 37°C (Arst and MacDonald, 1978; Arst *et al.*, 1981), this cis-acting element would seem to be thermosensitive.

The *prn* gene cluster is open to molecular analysis now that the whole of it has been cloned (P. M. Green and C. Scazzocchio, unpublished observations). Most of the cluster is contained in a 14-kb λ Charon 4a clone. Correlation of the restriction map of this clone with a genetic deletion map of the cluster has shown that the clone lacks part of the *prnA*. However, an overlapping Charon clone containing all of the *prnA* has been obtained. Northern blotting has identified messages for *prnB, prnC, prnD*, and possibly *prnA*. Another gene apparently lies between *prnA* and *prnD* because an additional transcript, proline-inducible, mapping in this region has been identified.

Transformation experiments, using the cloned *prn* cluster, have provided some interesting information (P. M. Green, P. Durrens, C. Scazzocchio, and H. N. Arst, Jr., unpublished observations). For example, integration of multiple copies of the structural genes can compensate for the absence of a functional copy of the positive-acting regulatory gene *prnA*. Apparently, basal levels of structural gene expression are sufficient for proline utilization, presumably provided that enough copies of the structural genes are present. In the presence of a functional *prnA* gene, the integration of multiple copies of the structural genes (probably *prnB* in particular) can apparently result in a *prn*^{d} phenotype. The integration of multiple copies in transformation experiments with *A. nidulans* was first observed for the acetamidase structural gene *amdS*, and it was noted that selfed cleistothecia from such transformants contain some ascospores from which all the integrated copies have been eliminated (Tilburn *et al.*, 1983; J.

Tilburn and C. Scazzocchio, unpublished observations). This was also found in the case of *prn* transformants, where, likewise, integration does not necessarily occur in the normal genome position. Selfing a *prn* transformant having a *prn*^{d} phenotype also yields *prn*$^+$ (i.e., nonderepressed, wild-type) progeny, presumably having a single functional copy (or very few functional copies) of the *prnB* gene. Most surprisingly, transformation followed by selfing can yield *prn*$^-$ strains differing in phenotype from the *prn*$^-$ recipient strain used for transformation. These include new deletions apparently generated by the transformation event(s), which are revealed when additional functional gene copies are eliminated in meiosis. Thus, transformation followed by selfing is likely to provide an efficient means of generating new deletion mutations.

B. The Nitrate Assimilation Gene Cluster

Linkage group VIII of *A. nidulans* contains a gene cluster comprised of three genes involved in nitrate assimilation (Tomsett and Cove, 1979): *niaD*, the structural gene for the apoenzyme of nitrate reductase (MacDonald and Cove, 1974); *niiA*, the structural gene for nitrite reductase (Rand and Arst, 1977); and *crnA*, whose product is involved in nitrate uptake (Brownlee and Arst, 1983) (see Fig. 5). Whereas expression of *niaD* and *niiA* is subject to control by *nirA* (see Section III,A,1), expression of *crnA* is probably not (Pateman and Cove, 1967; Rand and Arst, 1978; Cove, 1979; Tollervey and Arst, 1981; Brownlee and Arst, 1983). Expression of all three genes of the cluster is subject to control by *areA*, the positive-acting regulatory gene mediating nitrogen metabolite repression (see Section III,B,1) (Arst and Cove, 1973; Tollervey and Arst, 1981; Arst *et al.*, 1982; Brownlee and Arst, 1983). In addition, there is evidence (summarized in Cove, 1979) that nitrate reductase regulates its own synthesis as well as that of nitrite reductase (see Section III,A,1).

A cis-acting regulatory mutation for *niiA*, designated *nis-5*, partially bypasses the requirements for both the *nirA* and *areA* gene products (Rand and Arst, 1977). Regulation of *niaD* expression is, however, not affected (Rand and Arst,

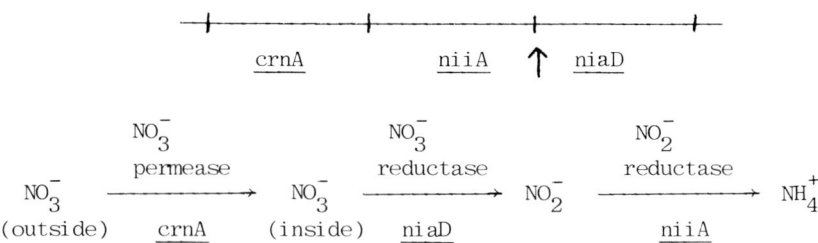

Fig. 5. The nitrate assimilation gene cluster (after Tomsett and Cove, 1979) and pathway. The vertical arrow indicates the site of the *nis-5* insertional translocation (Arst *et al.*, 1979).

1977), nor is that of *crnA* (Brownlee and Arst, 1983). Arst *et al.* (1979) showed that *nis-5* is an insertional translocation in which a considerable portion of the right arm of linkage group II is inserted between the *niiA* and *niaD* genes. They proposed that *nis-5* fuses *niiA* to a promoter normally in linkage group II. Because maximal *niiA* expression still requires nitrate or nitrite induction and the absence of nitrogen metabolite repression, the normal *niiA* promoter/initiator must still be present, presumably in tandem to the new promoter. Two important deductions follow: first, transcription of *niiA* must proceed from the *niaD* side, and second, expression of the three genes of the cluster cannot occur solely via a tricistronic mRNA, nor can expression of *niiA* occur solely via a *niaD niiA* dicistronic mRNA. However, the maximal level of nitrite reductase in *nis-5* strains is only about 40% of that in the wild type (Rand and Arst, 1977). This probably indicates that the translocation has impaired functioning of the normal promoter/initiator (as a consequence of the functioning of the new promoter or of the positioning of the translocation breakpoint) (Arst *et al.*, 1979). An alternative possibility is that expression of *niiA* involves at least two kinds of overlapping transcripts, a tri- or dicistronic mRNA initiated on the *niiA*-distal side of *niaD* (and abolished by the *nis-5* translocation) and a di- or monocistronic mRNA initiated between *niaD* and *niiA* (Arst *et al.*, 1979). These alternatives can be distinguished once the genes have been cloned, as done for the *prn* cluster (see Section V,A).

VI. AT WHAT LEVEL DOES REGULATION OF GENE EXPRESSION OCCUR?

It has usually been assumed that gene regulation in the pathways described above occurs at the level of transcription. Early studies using inhibitors (see, e.g., Cybis and Weglenski, 1972; summarized in Arst and Bailey, 1977) were consistent with this prejudice. Recently, studies using reticulocyte translation systems have shown that specific induction or repression alters the level of translatable mRNA for a number of enzymes including alcohol dehydrogenase I (Pateman *et al.*, 1983; Sealy-Lewis and Lockington, 1984), alcohol dehydrogenase II (Sealy-Lewis and Lockington, 1984), aldehyde dehydrogenase (Pateman *et al.*, 1983), acetamidase (Hynes *et al.*, 1983), purine hydroxylase I and urate oxidase (M. Winther *et al.*, unpublished observations), and purine hydroxylase II (F. Hanselman and C. Scazzocchio, unpublished observations).

Cloned genes allow direct quantitation of the relative abundance of transcripts in mycelia grown under inducing, noninducing, and repressing growth conditions. Hynes *et al.* (1983) have established that induction of actamidase affects the steady-state concentration of *amdS* transcripts, compatible with induction occurring at the level of transcription. Similar results were obtained in the labora-

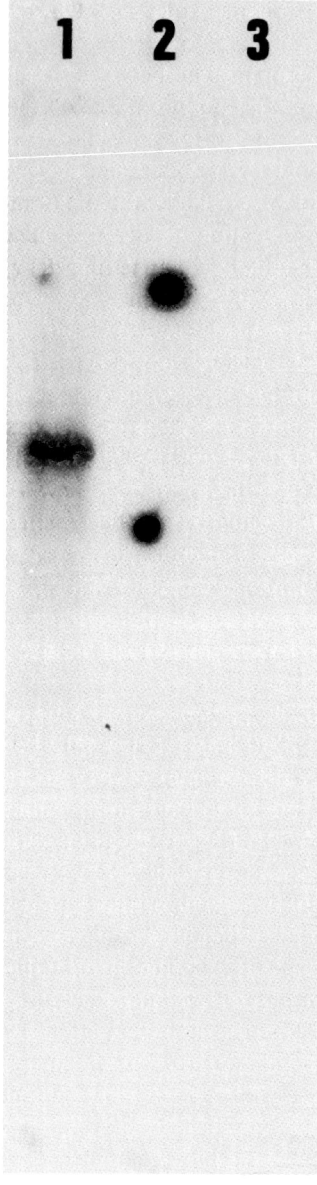

Fig. 6. Regulation at the level of transcription. A Northern blot of polyadenylated RNA extracted from: 1, wild-type strain grown under inducing conditions (0.1% fructose + 1% ethanol); 2, wild type grown under noninducing conditions (0.1% fructose as a carbon source); 3, a strain carrying the *alcR125* mutation, a putative chain termination mutation mapping in *alcR* (see Section V), grown under inducing conditions. The probe was an *alcA* genomic clone not containing any *alcR* sequences. [For experimental details see Lockington *et al.* (1985).]

tory of C. Scazzocchio for the specific induction of *prnB*, *prnC*, and *prnD* transcripts (P. M. Green *et al.*, unpublished observations; see Section V,A), for the specific induction of *aldA* transcripts, and for the induction and repression of *alcA* transcripts (Lockington *et al.*, 1985; Section V). An example of these results in given in Fig. 6.

Thus, the assumption that the regulatory gene products described in this chapter affect the transcriptional process seems fully justified, at least on the basis of present evidence. As, potentially, all *A. nidulans* genes can be cloned (see Turner and Ballance, Chapter 10; Green and Scazzocchio, Chapter 14), the next step should be a study of the interactions of the various regulatory gene products at the DNA and chromatin levels.

ACKNOWLEDGMENTS

We are extremely grateful to all those, cited as appropriate, who kindly allowed us to use their unpublished data. Over the years, work in our laboratories has been generously supported by the Royal Society (H.N.A.), the Science and Engineering Research Council (H.N.A. and C.S.), and the Medical Research Council (H.N.A. and C.S.), to whom we are very grateful.

NOTE ADDED IN PROOF

Northern blotting experiments in the C. Scazzocchio laboratory have shown that the *alcR* mRNA is subject to autogenous regulation and carbon catabolite repression (R. A. Lockington *et al.*, unpublished data). Sequencing of *alcA* and *aldA* has been almost completed in the R. W. Davies laboratory. Strictly conserved sequences 5' to the coding region of these two genes might define an *alcR* product binding site (D. I. Gwynne *et al.*, unpublished data). The *areA* gene has now been cloned, as has a portion of the region centromere distal to *areA*, establishing that the region between *areA* and the telomere of the right arm of chromosome III is at least 80 kb (M. X. Caddick, L. Taylor, R. I. Johnson, A. G. Brownlee, and H. N. Arst, Jr., unpublished results).

REFERENCES

Al Taho, N., Sealy-Lewis, H. M., and Scazzocchio, C. (1984). Suppressible alleles in a wide domain regulatory gene in *Aspergillus nidulans*. *Curr. Genet.* **8,** 245–251.
Arst, H. N., Jr. (1968). Genetic analysis of the first steps of sulphate metabolism in *Aspergillus nidulans*. *Nature (London)* **219,** 268–270.
Arst, H. N., Jr. (1976). Integrator gene in *Aspergillus nidulans*. *Nature (London)* **262,** 231–234.
Arst, H. N., Jr. (1977). Some genetical aspects of ornithine metabolism in *Aspergillus nidulans*. *Mol. Gen. Genet.* **151,** 105–110.
Arst, H. N., Jr. (1978). GABA transaminase provides an alternative route of β-alanine synthesis in *Aspergillus nidulans*. *Mol. Gen. Genet.* **163,** 23–27.
Arst, H. N., Jr. (1981). Aspects of the control of gene expression in fungi. *Symp. Soc. Gen. Microbiol.* **31,** 131–160.
Arst, H. N., Jr. (1982). A near terminal pericentric inversion leads to nitrogen metabolite derepression in *Aspergillus nidulans*. *Mol. Gen. Genet.* **188,** 490–493.

Arst, H. N., Jr., and Bailey, C. R. (1977). The regulation of carbon metabolism in *Aspergillus nidulans. In* "Genetics and Physiology of *Aspergillus*" (J. E. Smith and J. A. Pateman, eds.), pp. 131–146. Academic Press, New York.

Arst, H. N., Jr., and Cove, D. J. (1973). Nitrogen metabolite repression in *Aspergillus nidulans*. *Mol. Gen. Genet.* **126,** 111–141.

Arst, H. N., Jr., and MacDonald, D. W. (1973). A mutant of *Aspergillus nidulans* lacking NADP-linked glutamate dehydrogenase. *Mol. Gen. Genet.* **122,** 261–265.

Arst, H. N., Jr., and MacDonald, D. W. (1975). A gene cluster in *Aspergillus nidulans* with an internally located *cis*-acting regulatory region. *Nature (London)* **254,** 26–31.

Arst, H. N., Jr., and MacDonald, D. W. (1978). Reduced expression of a distal gene of the *prn* gene cluster in deletion mutants of *Aspergillus nidulans:* genetic evidence for a dicistronic messenger in an eukaryote. *Mol. Gen. Genet.* **163,** 17–22.

Arst, H. N., Jr., and Scazzocchio, C. (1975). Initiator constitutive mutation with an "up-promoter" effect in *Aspergillus nidulans*. *Nature (London)* **254,** 31–34.

Arst, H. N., Jr., Penfold, H. A., and Bailey, C. R. (1978). Lactam utilisation in *Aspergillus nidulans:* evidence for a fourth gene under the control of the integrator gene *int*A. *Mol. Gen. Genet.* **166,** 321–327.

Arst, H. N., Jr., Rand, K. N., and Bailey, C. R. (1979). Do the tightly linked structural genes for nitrate and nitrite reductases form an operon? Evidence from an insertional translocation which separates them. *Mol. Gen. Genet.* **174,** 89–100.

Arst, H. N., Jr., MacDonald, D. W., and Jones. S. A. (1980). Regulation of proline transport in *Aspergillus nidulans*. *J. Gen. Microbiol.* **116,** 285–294.

Arst, H. N., Jr., Jones, S. A., and Bailey, C. R. (1981). A method for the selection of deletion mutations in the L-proline catabolism gene cluster of *Aspergillus nidulans*. *Genet. Res.* **38,** 171–195.

Arst, H. N., Jr., Brownlee, A. G., and Cousen, S. A. (1982). Nitrogen metabolite repression in *Aspergillus nidulans:* a farewell to *tam*A? *Curr. Genet.* **6,** 245–257.

Bailey, C. R. (1976). Carbon catabolite repression in *Aspergillus nidulans*. Ph.D. Thesis, Univ. of Cambridge.

Bailey, C., and Arst, H. N., Jr. (1975). Carbon catabolite repression in *Aspergillus nidulans*. *Eur. J. Biochem.* **51,** 573–577.

Bailey, C. R., Penfold, H. A., and Arst, H. N., Jr. (1979). *Cis*-dominant regulatory mutations affecting the expression of GABA permease in *Aspergillus nidulans*. *Mol. Gen. Genet.* **169,** 79–83.

Bailey, C. R., Arst, H. N., Jr., and Penfold, H. A. (1980). A third gene affecting GABA transaminase levels in *Aspergillus nidulans*. *Genet. Res.* **36,** 167–180.

Ballance, D. J., Buxton, F. P., and Turner, G. (1983). Transformation of *Aspergillus nidulans* by the orotidine-5′-phosphate decarboxylase gene of *Neurospora crassa*. *Biochem. Biophys. Res. Commun.* **112,** 284–289.

Bartnik, E., Klimczuk, J., Kowalska, I., and Weglenski, P. (1976). Effect of the *are*A gene on regulation of arginine catabolism in *Aspergillus nidulans*. *Acta Microbiol. Pol.* **25,** 169–173.

Bartnik, E., Guzewska, J., Klimczuk, J., Piotrowska, M., and Weglenski, P. (1977). Regulation of arginine catabolism in *Aspergillus nidulans. In* "Genetics and Physiology of *Aspergillus*" (J. E. Smith and J. A. Pateman, eds.), pp. 243–254. Academic Press, New York.

Botstein, D., and Davis, R. W. (1982). Principles and practice of recombinant DNA research with yeast. *In* "The Molecular Biology of the Yeast *Saccharomyces*. Metabolism and Gene Expression" (J. N. Strathern, E. W. Jones, and J. R. Broach, eds.), pp. 607–636. Cold Spring Harbor Lab., Cold Spring Harbor, New York.

Britten, R. J., and Davidson, E. H. (1969). Gene regulation for higher cells: a theory. *Science (Washington, D.C.)* **165,** 349–357.

Britten, R. J., and Davidson, E. H. (1971). Repetitive and non-repetitive DNA sequences and a speculation on the origins of evolutionary novelty. *Q. Rev. Biol.* **46**, 111–138.
Brownlee, A. G., and Arst, H. N. Jr. (1983). Nitrate uptake in *Aspergillus nidulans* and involvement of the third gene of the nitrate assimilation gene cluster. *J. Bacteriol.* **155**, 1138–1146.
Brownlee, A. G., Caddick, M. X., and Arst, H. N., Jr. (1983). A novel phosphate-repressible phosphodiesterase in *Aspergillus nidulans*. *Heredity* **51**, 529.
Burton, E. G., and Metzenberg, R. L. (1972). Novel mutation causing derepression of several enzymes of sulfur metabolism in *Neurospora crassa*. *J. Bacteriol.* **109**, 140–151.
Ciriacy, M., and Williamson, V. M. (1981). Analysis of mutations affecting *Ty*-mediated gene expression in *Saccharomyces cerevisiae*. *Mol. Gen. Genet.* **182**, 159–163.
Clutterbuck, A. J. (1974). *Aspergillus nidulans*. *In* "Handbook of Genetics" (R. C. King, ed.), Vol. 1. pp. 447–510. Plenum, New York.
Clutterbuck, A. J. (1982). *Aspergillus nidulans*. *Genet. Maps* **2**, 209–217.
Clutterbuck, A. J., and Roper, J. A. (1966). A direct determination of nuclear distribution in heterokaryons of *Aspergillus nidulans*. *Genet. Res.* **7**, 185–194.
Cohen, B. L. (1972). Ammonium repression of extracellular protease in *Aspergillus nidulans*. *J. Gen. Microbiol.* **71**, 293–299.
Cohen, B. L. (1980). Transport and utilization of proteins by fungi. *In* "Microorganisms and Nitrogen Sources" (J. W. Payne, ed.), pp. 411–430. Wiley, New York.
Cornwell, E. V., and MacDonald, D. W. (1984). *gln*A mutations define the structural gene for glutamine synthetase in *Aspergillus*. *Curr. Genet.* **8**, 33–36.
Cove, D. J. (1966). The induction and repression of nitrate reductase in the fungus *Aspergillus nidulans*. *Biochim. Biophys. Acta* **113**, 51–56.
Cove, D. J. (1969). Evidence for a near limiting intracellular concentration of a regulator substance. *Nature (London)* **224**, 272–273.
Cove, D. J. (1979). Genetic studies of nitrate assimilation in *Aspergillus nidulans*. *Biol. Rev. Cambridge Philos. Soc.* **54**, 291–327.
Cybis, J., and Weglenski, P. (1972). Arginase induction in *Aspergillus nidulans*. The induction and decay of the coding capacity of messenger. *Eur. J. Biochem.* **30**, 262–268.
Darlington, A. J., and Scazzocchio, C. (1967). Use of analogues and the substrate-sensitivity of mutants in analysis of purine uptake and breakdown in *Aspergillus nidulans*. *J. Bacteriol.* **93**, 937–940.
Davidson, E. H., and Britten, R. J. (1973). Organization, transcription, and regulation in the animal genome. *Q. Rev. Biol.* **48**, 565–613.
Dorn, G. (1965a). Genetic analysis of the phosphatases of *Aspergillus nidulans*. *Genet. Res.* **6**, 13–26.
Dorn, G. (1965b). Phosphatase mutants in *Aspergillus nidulans*. *Science (Washington, D.C.)* **150**, 1183–1184.
Errede, B., Cardillo, T. S., Wever, G., and Sherman, F. (1981). Studies on transposable elements in yeast. I. ROAM mutations causing increased expression of yeast genes: their activation by signals directed toward conjugation functions and their formation by insertion of *Ty*1 repetitive elements. *Cold Spring Harbor Symp. Quant. Biol.* **45**, 593–607.
Giles, N. H. (1978). The organization, function and evolution of gene clusters in eukaryotes. *Am. Nat.* **112**, 641–657.
Gorton, D. J. (1983). Genetic and biochemical studies of the uptake of purines and their degradation products in *Aspergillus nidulans*. Ph.D. Thesis, Univ. of Essex.
Grove, G., and Marzluf, G. A. (1981). Identification of the product of the major regulatory gene of the nitrogen control circuit of *Neurospora crassa* as a nuclear DNA-binding protein. *J. Biol. Chem.* **256**, 463–470.
Hankinson, O., and Cove, D. J. (1974). Regulation of the pentose phosphate pathway in the fungus *Aspergillus nidulans*. The effect of growth with nitrate. *J. Biol. Chem.* **249**, 2344–2353.

Hankinson, O., and Cove, D. J. (1975). Regulation of mannitol-1-phosphate dehydrogenase in *Aspergillus nidulans*. *Can. J. Microbiol.* **21,** 99–101.

Hawkins, A. R., Da Silva, A. J. F., and Roberts, C. F. (1984). Evidence for two control genes regulating expression of the quinic acid utilization (*qut*) gene cluster in *Aspergillus nidulans*. *J. Gen. Microbiol.* **130,** 567–574.

Hawkins, A. R., Da Silva, A. J. F., and Roberts, C. F. (1985). Cloning and characterization of the three enzyme structural genes *qut*B, *qut*C and *qut*E from the quinic acid utilization gene cluster in *Aspergillus nidulans*. *Curr. Genet.* **9,** 305–312.

Hynes, M. J. (1975a). Studies on the role of the *are*A gene in the regulation of nitrogen catabolism in *Aspergillus nidulans*. *Aust. J. Biol. Sci.* **28,** 301–313.

Hynes, M. J. (1975b). A *cis*-dominant regulatory mutation affecting enzyme induction in the eukaryote *Aspergillus nidulans*. *Nature (London)* **253,** 210–212.

Hynes, M. J. (1977). Induction of the acetamidase of *Aspergillus nidulans* by acetate metabolism. *J. Bacteriol.* **131,** 770–775.

Hynes, M. J. (1978a). Multiple independent control mechanisms affecting the acetamidase of *Aspergillus nidulans*. *Mol. Gen. Genet.* **161,** 59–65.

Hynes, M. J. (1978b). An ''up-promoter'' mutation affecting the acetamidase of *Aspergillus nidulans*. *Mol. Gen. Genet.* **166,** 31–36.

Hynes, M. J. (1979). Fine-structure mapping of the acetamidase structural gene and its controlling region in *Aspergillus nidulans*. *Genetics* **91,** 381–392.

Hynes, M. J. (1980). A mutation, adjacent to gene *amd*S, defining the site of action of positive-control gene *amd*R in *Aspergillus nidulans*. *J. Bacteriol.* **142,** 400–406.

Hynes, M. J. (1982). A *cis*-dominant mutation in *Aspergillus nidulans* affecting the expression of the *amd*S gene in the presence of mutations in the unlinked gene, *amd*A. *Genetics* **102,** 139–147.

Hynes, M. J., and Pateman, J. A. J. (1970). The genetic analysis of regulation of amidase synthesis in *Aspergillus nidulans*. I. Mutants able to utilize acrylamide. *Mol. Gen. Genet.* **108,** 97–106.

Hynes, M. J., Corrick, C. M., and King, J. A. (1983). Isolation of genomic clones containing the *amd*S gene of *Aspergillus nidulans* and their use in the analysis of structural and regulatory mutations. *Mol. Cell. Biol.* **3,** 1430–1439.

Jones, S. A., Arst, H. N., Jr., and MacDonald, D. W. (1981). Gene roles in the *prn* cluster of *Aspergillus nidulans*. *Curr. Genet.* **3,** 49–56.

Käfer, E. (1958). An 8-chromosome map of *Aspergillus nidulans*. *Adv. Genet.* **9,** 105–145.

Käfer, E. (1977). Meiotic and mitotic recombination in *Aspergillus* and its chromosomal aberrations. *Adv. Genet.* **19,** 33–131.

Kinghorn, J. R., and Pateman, J. A. (1973). NAD and NADP L-glutamate dehydrogenase activity and ammonium regulation in *Aspergillus nidulans*. *J. Gen. Microbiol.* **78,** 39–46.

Kinghorn, J. R., and Pateman, J. A. (1975). The structural gene for NADP L-glutamate dehydrogenase in *Aspergillus nidulans*. *J. Gen. Microbiol.* **86,** 294–300.

Lindberg, R. A., Rhodes, W. G., Eirich, L. D., and Drucker, H. (1982). Extracellular acid proteases from *Neurospora crassa*. *J. Bacteriol.* **150,** 1103–1108.

Lockington, R. A., Sealy-Lewis, H. M., Scazzocchio, C., and Davies, R. W. (1985). Cloning and characterization of the ethanol utilization regulon in *Aspergillus nidulans*. *Gene* **33,** 137–149.

Lukaszkiewicz, Z., and Paszewski, A. (1976). Hyper-repressible operator-type mutant in sulphate permease gene of *Aspergillus nidulans*. *Nature (London)* **259,** 337–338.

Lukaszkiewicz, Z., and Pieniazek, N. J. (1972). Mutations increasing the specificity of the sulphate permease in *Aspergillus nidulans*. *Bull. Acad. Pol. Sci., Ser. Sci. Biol.* **20,** 833–836.

McCully, K. S., and Forbes, E. (1965). The use of *p*-fluorophenylalanine with ''master strains'' of *Aspergillus nidulans* for assigning genes to linkage groups. *Genet. Res.* **6,** 352–359.

MacDonald, D. W. (1982). A single mutation leads to loss of glutamine synthetase and relief of ammonium repression in *Aspergillus nidulans*. *Curr. Genet.* **6,** 203–208.

MacDonald, D. W., and Cove, D. J. (1974). Studies on temperature-sensitive mutants affecting the assimilatory nitrate reductase of *Aspergillus nidulans*. *Eur. J. Biochem.* **47,** 107–110.
Makins, J. F., Holt, G., and Macdonald, K. D. (1981). *lamA* mutants of *Aspergillus nidulans* show decreased β-lactamase activity but do not influence penicillin production. *Aspergillus News Lett.* **15,** 30–33.
Marzluf, G. A. (1981). Regulation of nitrogen metabolism and gene expression in fungi. *Microbiol. Rev.* **45,** 437–461.
Metzenberg, R. L. (1979). Implications of some genetic control mechanisms in *Neurospora*. *Microbiol. Rev.* **43,** 361–383.
Nahas, E., Terenzi, H. F., and Rossi, A. (1982). Effect of carbon source and pH on the production and secretion of acid phosphatase (EC 3.1.3.2) and alkaline phosphatase (EC 3.1.3.1) in *Neurospora crassa*. *J. Gen. Microbiol.* **128,** 2017–2021.
Paszewski, A., and Grabski, J. (1974). Regulation of S-amino acid biosynthesis in *Aspergillus nidulans*. Role of cysteine and/or homocysteine as regulatory effectors. *Mol. Gen. Genet.* **132,** 307–320.
Paszewski, A., and Grabski, J. (1975). Enzymatic lesions in methionine mutants of *Aspergillus nidulans:* role and regulation of an alternative pathway for cysteine and methionine synthesis. *J. Bacteriol.* **124,** 893–904.
Paszewski, A., Prazmo, W., and Landman-Balinska, M. (1977). Regulation of homocysteine metabolizing enzymes in *Aspergillus nidulans*. *Mol. Gen. Genet.* **155,** 109–112.
Pateman, J. A., and Cove, D. J. (1967). Regulation of nitrate reduction in *Aspergillus nidulans*. *Nature (London)* **215,** 1234–1237.
Pateman, J. A., and Kinghorn, J. R. (1977). Genetic regulation of nitrogen metabolism. *In* "Genetics and Physiology of *Aspergillus*" (J. E. Smith and J. A. Pateman, eds.), pp. 203–241. Academic Press, New York.
Pateman, J. A., Kinghorn, J. R., Dunn, E., and Forbes, E. (1973). Ammonium regulation in *Aspergillus nidulans*. *J. Bacteriol.* **114,** 943–950.
Pateman, J. A., Doy, C. H., Olsen, J. E., Norris, U., Creaser, E. H., and Hynes, M. (1983). Regulation of alcohol dehydrogenase (ADH) and aldehyde dehydrogenase (AldDH) in *Aspergillus nidulans*. *Proc. R. Soc. London, Ser. B* **216,** 243–264.
Penfold, H. A. (1979). Omega-amino acid catabolism in *Aspergillus nidulans*. Ph.D. Thesis, Univ. of Cambridge.
Philippides, D., and Scazzocchio, C. (1981). Positive regulation in a eukaryote, a study of the *ua*Y gene of *Aspergillus nidulans*. II. Identification of the effector binding protein. *Mol. Gen. Genet.* **181,** 107–115.
Polkinghorne, M., and Hynes, M. J. (1975). Mutants affecting histidine utilization in *Aspergillus nidulans*. *Genet. Res.* **25,** 119–135.
Polkinghorne, M. A., and Hynes, M. J. (1982). L-Histidine utilization in *Aspergillus nidulans*. *J. Bacteriol.* **149,** 931–940.
Polya, G. M., Brownlee, A. G., and Hynes, M. J. (1975). Enzymology and genetic regulation of a cyclic nucleotide-binding phosphodiesterase–phosphomonoesterase from *Aspergillus nidulans*. *J. Bacteriol.* **124,** 693–703.
Pontecorvo, G. (1963). Microbial genetics: retrospect and prospect. *Proc. R. Soc. London, Ser. B* **158,** 1–23.
Pontecorvo, G., Roper, J. A., Hemmons, L. M., Macdonald, K. D., and Bufton, A. W. J. (1953). The genetics of *Aspergillus nidulans*. *Adv. Genet.* **5,** 141–238.
Pritchard, R. H. (1955). The linear arrangement of a series of alleles of *Aspergillus nidulans*. *Heredity* **9,** 343–371.
Rand, K. N., and Arst, H. N., Jr. (1977). A mutation in *Aspergillus nidulans* which affects the regulation of nitrite reductase and is tightly linked to its structural gene. *Mol. Gen. Genet.* **155,** 67–75.

Rand, K. N., and Arst, H. N., Jr. (1978). Mutations in *nir*A gene of *Aspergillus nidulans* and nitrogen metabolism. *Nature (London)* **272**, 732-734.

Scazzocchio, C. (1973a). The genetic control of molybdoflavoproteins in *Aspergillus nidulans*. II: Use of the NADH dehydrogenase activity associated with xanthine dehydrogenase to investigate substrate and product induction. *Mol. Gen. Genet.* **125**, 147-155.

Scazzocchio, C. (1973b). Appendix: The induction of xanthine dehydrogenase II. In C. Scazzocchio, F. B. Holl, and A. I. Foguelman. The genetic control of molybdoflavoproteins in *Aspergillus nidulans*. Allopurinol resistant mutants constitutive for xanthine dehydrogenase. *Eur. J. Biochem.* **36**, 428-445.

Scazzocchio, C. (1980). The genetics of the molybdenum-containing enzymes. *In* "Molybdenum and Molybdenum-Containing Enzymes" (M. P. Coughlan, ed.), pp. 487-515. Pergamon, Oxford.

Scazzocchio, C., and Arst, H. N., Jr. (1978). The nature of an initiator constitutive mutation in *Aspergillus nidulans*. *Nature (London)* **274**, 177-179.

Scazzocchio, C., and Darlington, A. J. (1967). The genetic control of xanthine dehydrogenase and urate oxidase synthesis in *Aspergillus nidulans*. *Bull. Soc. Chim. Biol.* **49**, 1503-1508.

Scazzocchio, C., and Darlington, A. J. (1968). The induction and repression of the enzymes of purine breakdown in *Aspergillus nidulans*. *Biochim. Biophys. Acta* **166**, 557-568.

Scazzocchio, C., and Gorton, D. (1977). The regulation of purine breakdown. *In* "Genetics and Physiology of *Aspergillus*" (J. E. Smith and J. A. Pateman, eds.), pp. 255-265. Academic Press, New York.

Scazzocchio, C., and Sealy-Lewis, H. M. (1978). A mutation in the xanthine dehydrogenase (purine hydroxylase I) of *Aspergillus nidulans* resulting in altered specificity. Implications for the geometry of the active site. *Eur. J. Biochem.* **91**, 99-109.

Scazzocchio, C., Holl, F. B., and Foguelman, A. I. (1973). The genetic control of molybdoflavoproteins in *Aspergillus nidulans*. Allopurinol resistant mutants constitutive for xanthine dehydrogenase. *Eur. J. Biochem.* **36**, 428-445.

Scazzocchio, C., Sdrin, N., and Ong, G. (1982). Positive regulation in a eukaryote, a study of the *ua*Y gene of *Aspergillus nidulans:* I Characterisation of alleles, dominance and complementation studies and a fine structure map of the *ua*Y-*oxp*A cluster. *Genetics* **100**, 185-208.

Sealy-Lewis, H. M., and Lockington, R. A. (1984). Regulation of two alcohol dehydrogenases in *Aspergillus nidulans*. *Curr. Genet.* **8**, 253-259.

Sealy-Lewis, H. M., Scazzocchio, C., and Lee, S. (1978). A mutation defective in the xanthine alternative pathway of *Aspergillus nidulans*. Its use to investigate the specificity of *ua*Y mediated induction. *Mol. Gen. Genet.* **164**, 303-308.

Shaffer, P. M., and Arst, H. N., Jr. (1984). Regulation of pyrimidine salvage in *Aspergillus nidulans:* a role for the major regulatory gene *are*A mediating nitrogen metabolite repression. *Mol. Gen. Genet.* **198**, 139-145.

Sharma, K. K., and Arst, H. N., Jr. (1985). The product of the regulatory gene of the proline catabolism gene cluster of *Aspergillus nidulans* is a positive-acting protein. *Curr. Genet.* **9**, 299-304.

Spathas, D. H., Pateman, J. A., and Clutterbuck, A. J. (1982). Polyamine transport in *Aspergillus nidulans*. *J. Gen. Microbiol.* **128**, 557-563.

Spathas, D. H., Clutterbuck, A. J., and Pateman, J. A. (1983). Putrescine as a nitrogen source for wild type and mutants of *Aspergillus nidulans*. *FEMS Microbiol. Lett.* **17**, 345-348.

Strathern, J. N., Jones, E. W., and Broach, J. R., eds. (1982). "The Molecular Biology of the Yeast *Saccharomyces*. Metabolism and Gene Expression." Cold Spring Harbor Lab., Cold Spring Harbor, New York.

Struhl, K. (1983). The new yeast genetics. *Nature (London)* **305**, 391-397.

Tilburn, J., Scazzocchio, C., Taylor, G. G., Zabicky-Zissman, J. H., Lockington, R. A., and Davies, R. W. (1983). Transformation by integration in *Aspergillus nidulans*. *Gene* **26**, 205-221.

Timberlake, W. E. (1978). Low repetitive DNA content in *Aspergillus nidulans*. *Science (Washington, D.C.)* **202,** 973-975.

Tollervey, D. W., and Arst, H. N., Jr. (1981). Mutations to constitutivity and derepression are separate and separable in a regulatory gene of *Aspergillus nidulans*. *Curr. Genet.* **4,** 63-68.

Tollervey, D. W., and Arst, H. N., Jr. (1982). Domain-wide, locus-specific suppression of nitrogen metabolite repressed mutations in *Aspergillus nidulans*. *Curr. Genet.* **6,** 79-85.

Tomsett, A. B., and Cove, D. J. (1979). Deletion mapping of the *nii*A *nia*D gene region of *Aspergillus nidulans*. *Genet. Res.* **34,** 19-32.

Turner, G., and Rowlands, R. T. (1977). Mitochondrial genetics of *Aspergillus nidulans*. *In* "Genetics and Physiology of *Aspergillus*" (J. E. Smith and J. A. Pateman, eds.), pp. 319-337. Academic Press, New York.

Waring, R. B., and Scazzocchio, C. (1983). Mitochondrial four point crosses in *Aspergillus nidulans:* mapping of a suppressor of a mitochondrially inherited cold-sensitive mutation. *Genetics* **103,** 409-428.

Wiame, J.-M., Grenson, M., and Arst, H. N., Jr. (1985). Nitrogen catabolite repression in yeasts and filamentous fungi. *Adv. Microb. Physiol.* **26,** 1-87.

Williamson, V. M., Young, E. T., and Ciriacy, M. (1981). Transposable elements associated with constitutive expression of yeast alcohol dehydrogenase II. *Cell (Cambridge, Mass.)* **23,** 605-614.

Yelton, M. M., Hamer, J. E., and Timberlake, W. E. (1984). Transformation of *Aspergillus nidulans* using a *trpC* plasmid. *Proc. Natl. Acad. Sci. U.S.A.* **81,** 1470-1474.

14

A Cloning Strategy in Filamentous Fungi

P. M. GREEN[1] AND CLAUDIO SCAZZOCCHIO[2]

Department of Biology
University of Essex
Colchester, United Kingdom

Text	345
References	352

This brief chapter was conceived and originally written as an appendix to Chapter 13 of this book, "Formal Genetics and Molecular Biology of the Control of Gene Expression in *Aspergillus nidulans*," by Arst and Scazzocchio. We did not intend then to make a thorough analysis of cloning methodologies, and thus the discussion of certain methods is rather more cursory than if it had been written as an autonomous chapter. The editors thought it more appropriate to include this as an autonomous chapter in Part II of the book and kindly took it on themselves to make the necessary adaptations. However, a number of the genes mentioned here were defined in Arst and Scazzocchio (Chapter 13 and references therein). The reader is referred to appropriate sections of that chapter for these definitions.

Workers with filamentous fungi have been hampered by the lack of an efficient transformation system. Only recently have transformation frequencies sufficient to allow cloning of chosen genes by direct complementation of suitable mutations been obtained (Turner and Ballance, Chapter 10; Huiet and Case, Chapter 8). Methods involving heterologous expression in other organisms are

---

[1]Present address: Paediatric Research Unit, The Prince Philip Research Laboratories, Guy's Tower, London Bridge, London SE1 9RT, United Kingdom.

[2]Present address: Bâtiment 409, Institut de Microbiologie, Université Paris XI, Centre d'Orsay, 91405-Orsay, France.

TABLE I.
Cloned Genes from *Aspergillus nidulans*[a]

Gene	Function	Method of Cloning	Reference
arom	"Cluster gene" including coding for steps 2–6 of aromatic biosynthesis	Complementation in *E. coli* for the biosynthetic dehydroquinate hydrolase activity	Kinghorn and Hawkins (1982); A. Hawkins (personal communication)
argB	Ornithine carbamoyltransferase	Complementation in *S. cerevisiae*	Berse *et al.* (1983)
trypC	"Cluster gene" comprising glutamine aminohydrolase, phosphoribosylanthranilate isomerase and indoglycerolphosphate synthetase	Complementation in *E. coli*	Yelton *et al.* (1983)
amdS	Acetamidase	Screen of genomic library by differential hybridization	Hynes *et al.* (1983)
aldA and *alcA alcR*	Aldehyde dehydrogenase and alcohol utilization gene cluster comprising the structural gene for alcohol dehydrogenase I and its cognate regulatory gene	Differential hybridization of cDNA gene library, cDNA clones used to identify genomic clones. The *alcA alcR* gene cluster was also obtained subsequently by the chromosomal aberration brute force method	Lockington *et al.* (1985)
gdhA	Biosynthetic NADP-linked glutamate dehydrogenase	Hybridization with *N. crassa* probe containing isologous gene	A. Hawkins, S. Gurr, C. Drainas, and J. R. Kinghorn (personal communication)

oliC	Subunit 9 of the mitochondrial ATP synthetase	Hybridization with probe from *N. crassa* containing isologous gene	M. Ward and G. Turner (personal communication)
qutBCE	Gene cluster coding for the three enzymes of quinate utilization	Hybridization with probes containing parts of isologous cluster from *N. crassa*	Hawkins *et al.* (1984)
Undesignated locus	Phosphoglycerate kinase	Hybridization with probe from *S. cerevisiae* containing isologous gene	J. M. Clemens and C. F. Roberts (personal communication)
prnADBC	Gene cluster comprising all the activities involved in proline utilization (see Section V,B, Chapter 13)	Chromosomal aberration brute force method	P. M. Green, C. Scazzocchio, H. N. Arst, Jr., and R. W. Davies (unpublished)
Ya	Genomic segment including the yellow (yA) gene specifying a p-diphenoloxidase expressed only in conidiospores	Transformation of *A. nidulans* mutant strain with a cosmid gene library	M. M. Yelton, W. E. Timberlake, and C. A. M. J. J. Van den Hondel (personal communication)
pabaA	Gene coding for unspecified step in p-aminobenzoic acid biosynthesis	Transformation of *A. nidulans* mutant strain with a cosmid gene library	M. M. Yelton *et al.* (personal communication)
acuD	Isocitrate lyase	Transformation of *A. nidulans* mutant strain with a plasmid gene library	D. J. Ballance and G. Turner (personal communication)

[a] We have indicated all the genes involved in metabolic pathways which have been cloned to date to the best of our knowledge. Not included in the table are a number of developmentally regulated gene clusters (Zimmerman *et al.*, 1980) and a number of tubulin genes cloned in R. Morris's laboratory using isologous probes.

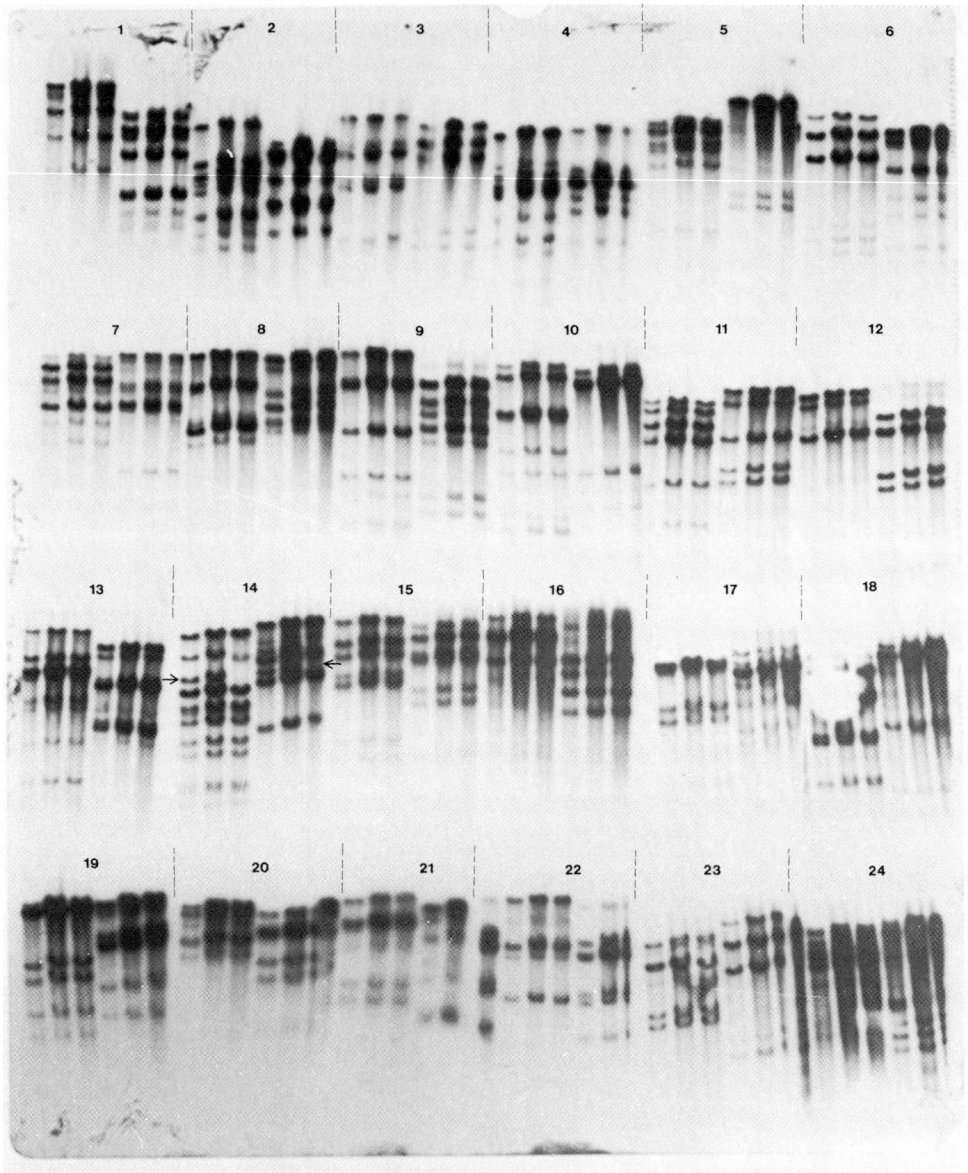

Fig. 1. Brute force cloning of the *A. nidulans alcA alcR* gene cluster. The autoradiogram shows 24 filters, representative of a batch of 100, each hybridized with a mixture of two clones from an *A. nidulans* genomic library made in a Charon 4 λ vector (Orr and Timberlake, 1982). DNA extracted from three strains (designated 1, 2, and 3) has been digested with *Hin*dIII (first three tracks of each filter) and *Xho*I (following three tracks). Strain 1 is *yA2 pyroA4 pantoB100 uaZ11 hxn10 nis5 xprD1*.

limited in that, at the very least, they require transformable organisms with mutations that can be complemented by the clone sought.

All fungal genes cloned heretofore by using mRNA or cDNA probes are expressed at relatively high levels. While in principle, methods based on immunological purification of specific polysomes (Kraus and Rosenberg, 1982) could be used, these methods are made less attractive by the difficulty of isolating high-order polysomes from fungi. Cascade experiments of the type performed by Timberlake and co-workers (Zimmerman et al., 1980), in which cDNA probes prepared from mRNA made from fully induced mycelia were enriched by removal of hybrids made with mRNA from translocation-associated or deletion null mutants, have failed to yield clones for the *prnADBC* or *hxnR aplA hxnS* gene clusters or the *uaZ* gene, all of which are represented by mRNAs that account for less than 0.05% of the total translatable mRNA (R. A. Lockington, F. Hanselman, P. M. Green, and C. Scazzocchio, unpublished observations; see also Sections III,A,2, V, and V,A of Chapter 13).

Strain 2 is *biA1 nia506 uaY27 creA^{d}30*. Strain 3 is *pabaA1 sB43 drkB5 areA^{r}18 areB403 alc500 brlA12*. The relevant mutations associated with chromosomal aberrations are: strain 1: *uaZ11*, a mutation in the structural gene for urate oxidase inseparable from a I–VIII translocation (Scazzocchio et al., 1982; Section III,A,2, Chapter 13, this volume); *hxn10*, a deletion that does not recombine with any point mutant in the *hxnS* and *hxnR* genes (see Section V, Chapter 13, this volume; J. M. Kelly, D. Lycan, and C. Scazzocchio, unpublished observations); *nis5*, an insertion of part of chromosome II between the *niaD* and *niiA* genes in chromosome VIII (see Sections IV and V,B, Chapter 13); *xprD1*, a pericentric inversion involving the *areA* gene (see Section III,B,1, Chapter 13). Strain 2: *nia506*, a very large deletion in the *crnA niiA niaD* gene cluster (Tomsett and Cove, 1979; see Sections I and V,B, Chapter 13, this volume); *uaY27*, a well-mapped deletion within the *uaY* gene (T. Sankarsingh and C. Scazzocchio, unpublished observations; see Section III,A,2, Chapter 13, this volume); *creA^{d}30*, a pericentric inversion resulting in loss of function of the *creA* gene (D. W. Tollervey and H. N. Arst, Jr., unpublished observations; see Section III,B,2, Chapter 13, this volume). Strain 3: *areA^{r}18*, a IV→III insertional translocation inseparable from an *areA^{r}* mutation (see Section III,B,1, Chapter 13); *areB403*, a I–VII translocation possibly activating a pseudogene and thus bypassing the *areA* function (see Section III,B,1, Chapter 13); *alc500*, a large deletion in the *alcA alcR* gene cluster (see Section V, Chapter 13, this volume; M. J. Hynes, unpublished observations; Lockington et al., 1985). *brlA12*, a VIII→III nonreciprocal translocation resulting in a variegated position effect at the *brlA* gene. Mutations in this gene drastically affect the development of the conidial apparatus (Clutterbuck, 1970). Other mutations are irrelevant to this table and are listed in Clutterbuck (1984).

There are 11 mutations but, as the translocations and inversions have more than one breakpoint, the three strains comprise 20 breakpoints (on the basis of current mapping data). With 20 breakpoints, one would need to screen 200 clones (15-kb insert) to have a 90% probability of finding a clone overlapping one of the breakpoints. With translocations and inversions only clones overlapping one of the breakpoints would lead directly to the desired gene. However, if clones overlapping the other breakpoint were obtained, one could easily "jump the genome" (see, e.g., Bender et al., 1983) using a gene library made from the strain carrying the translocation or inversion.

The arrow in filter 14 shows that a band is missing in both digests of strain 3. This is seen more clearly for the *Hin*dIII digest. The data in Fig. 1 show that this missing band together with a band of modified mobility, more clearly visible in Fig. 2, result from the *alc500* deletion.

Table I lists the genes from *A. nidulans* which have been cloned and the methods with which the clones have been recognized.

We have taken advantage of some features of the genetics of *A. nidulans* to develop a brute force cloning method (P. M. Green and C. Scazzocchio, unpublished) capable of allowing recognition of clones irrespective of gene expression. This method involves the use of chromosomal aberrations large enough to be detectable in Southern blots. As the method used to detect translocations in *A. nidulans* is potentially applicable to other organisms having a parasexual cycle (e.g., *Aspergillus niger,* Penicillia, slime molds) and as translocations and other chromosomal aberrations can be detected cytologically in a number of organisms, we believe this method might have widespread application and describe it below.

The principle of the method is the following: any large-scale chromosomal aberration will result in a change in the restriction pattern of a given region on the genome. Thus, if a mutation resulting from a chromosomal aberration (deletion, insertion, translocation, or inversion) can be identified genetically and clones from a gene library tested one by one in Southern transfers of restricted DNA from the wild type and a strain carrying a given chromosomal aberration, a clone overlapping the rearranged sequences will eventually be found. While this technique is *prima facie* extremely laborious, several modifications have enabled us to use it for routine cloning. For example, if ten chromosomal aberrations mapping in ten different genes can be screened simultaneously, 400 λ clones having *A. nidulans* DNA inserts of 15 kb average would have to be screened to have a 90% probability of finding sequences overlapping any one of the chromosomal aberrations. The number of Southern blots necessary can be halved as two λ clones can be screened in one blot. (Obviously, cosmid clones could be used rather than λ clones.) To screen for several genes simultaneously, three multiply mutant strains each carrying several chromosomal aberrations (but no two carrying aberrations affecting the same gene) were constructed. Thus, for any given gene of interest, two strains serve as wild-type controls for the third. Moreover, the DNA is restricted (separately) with two different enzymes to offset the possibility of a band having coincidentally the same mobility as a wild-type band in one of the digests [see Figs. 5 and 7 of Tilburn *et al.* (1983) for an example of this].

Figure 1 shows an example of the method. Tracks c and f of transfer 14 show an altered restriction pattern. This could be due to any of the four chromosomal aberrations carried by this strain. Figure 2 shows that this difference results from the *alc-500* deletion (a large deletion covering both genes of the *alcA alcR* cluster; Lockington *et al.,* 1985). Sequences containing the *alcA alcR* gene cluster had already been identified by conventional cDNA cloning, and thus the identity of the clone obtained by the "chromosomal aberration screening method" was easily confirmed.

14. A Cloning Strategy in Filamentous Fungi

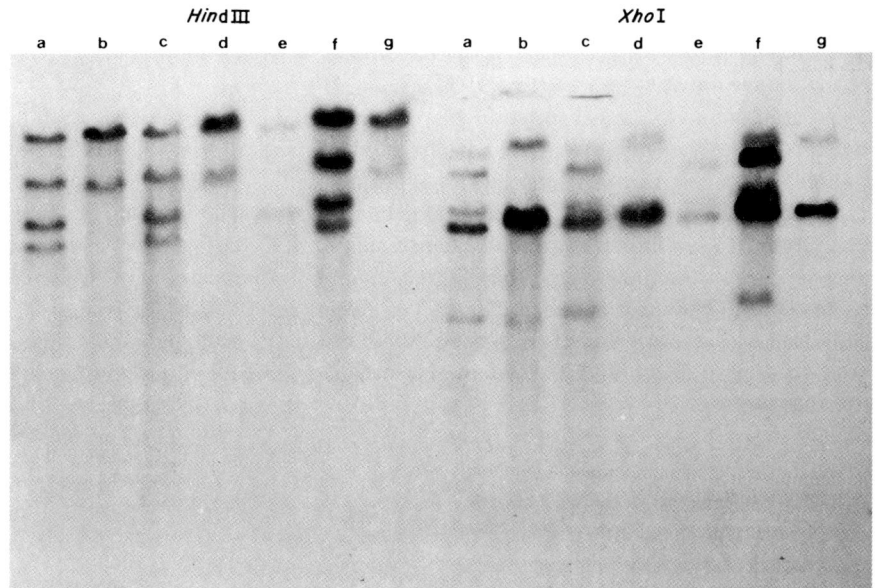

Fig. 2. Brute force cloning of the *A. nidulans alcA alcR* gene cluster. This autoradiogram shows *Hin*dIII and *Xho*I digests of DNA extracted from a number of *A. nidulans* strains and hybridized with one of the genomic clones of the mixture used to hybridize filter 14 in Fig. 1. The strains from which the DNAs were extracted carry the following markers: (a) *biA1* (wild type for all relevant markers, (b) from strain 3 of Fig. 1, (c) *puA2 areA^{r}18*, (d) *yA2 pyroA4 sB43 alc500*, (e) *biA1 brlA12 drkB5*, (f) *biA1 pabaA1 areB403*, (g) *sB43 alc500 areA^{r}18 brlA12 drkB5*. All relevant mutations have been defined in Fig. 1. It is clear that the alteration in the restriction pattern (two bands missing and the position of a third altered in the *Hin*dIII digest and a less straightforward modification in the *Xho*I digest) is associated with *alc500* and not with any other of the chromosomal aberrations originally present in strain 3. The identity of this clone was confirmed by hybridization with *A. nidulans alcA alcR* clones that had been obtained shortly before (see text and Section V, Chapter 13, this volume; Lockington *et al.*, 1984). The fact that two, rather than one, bands are missing in strains carrying *alc500* was obscured in the original screen by the presence of sequences hybridizing with the other Charon clone in the mixture. This second clone gives identical restriction patterns with all the different strains mentioned in Figs. 1 and 2.

We have used this method to identify clones containing the whole *prnADBC* cluster (see Section V,A in Chapter 13), and modifications of this procedure are now used routinely in our laboratories. Confirmation of the identity of clones is routinely achieved by transformation, as even very low transformation frequencies suffice for this purpose (Lockington *et al.*, 1985; P. M. Green and C. Scazzocchio, unpublished). While it is clear that this method is feasible for organisms with small genomes [2.6×10^7 base pairs in *A. nidulans* (Timberlake, 1978)], it might also be applied to organisms with larger genomes if gene libraries from individual chromosomes can be constructed [after separating chromosomes by flow cytometry (Davies et al., 1981) or pulsed field gradient electrophoresis (Schwartz and Cantor, 1984)].

Direct cloning by transformation can be achieved even with the medium transformation frequencies (~100 transformants per microgram of DNA) now routinely obtained in most laboratories working with *A. nidulans* by using a similar brute force approach, in which multimarked strains are transformed and then screened for complementation of individual nonselected markers. The combination of these methods with the high transformation frequencies now obtained by Turner and Ballance (Chapter 10) would allow the molecular cloning of any gene from *A. nidulans*, structural or regulatory, metabolic or developmental, and point the way to developing cloning methodologies applicable to fungi of industrial importance.

ACKNOWLEDGMENTS

We warmly thank the editors for their help, H. N. Arst for critical reading of the manuscript, and all our colleagues at Essex and elsewhere who allowed us to incorporate unpublished data.

NOTE ADDED IN PROOF

The cloning of the *yA* gene (Table I) has been published (Yelton et al., 1984). Since this chapter was written, several *A. nidulans* genes have been cloned by direct complementation. Among these is *brlA*, a gene involved at the onset of conidiation (Johnstone et al., 1984). The cloning of *areA*, the key gene of nitrogen metabolite repression, has been referred to in the note added in proof to Chapter 13.

REFERENCES

Bender, W., Spierer, P., and Hogness, D. S. (1983). Chromosomal walking and jumping to isolate DNA from the *Ace* and *rosy* loci and the bithorax complex in *Drosophila melanogaster*. *J. Mol. Biol.* **168**, 17–33.

Berse, B., Dmochowska, A., Skrzypek, Weglenski, P., Bates, M. A., and Weiss, R. L. (1983). Cloning and characterization of the ornithine carbamoyltransferase gene from *Aspergillus nidulans*. *Gene* **25**, 109–117.

Clutterbuck, A. J. (1970). A variegated position effect in *Aspergillus nidulans*. *Genet. Res.* **16**, 303–316.

Clutterbuck, A. J. (1984). *Aspergillus nidulans*. *Genet. Maps* **3**, 265–273.

Davies, K. E., Young, B. D., Elles, R. G., Hill, M. E., and Williamson, R. (1981). Cloning of a representative genomic library of the human X chromosome after sorting by flow cytometry. *Nature (London)* **293**, 374–376.

Hawkins, A. R., Da Silva, A. J. F., and Roberts, C. F. (1984). Cloning and characterization of the three enzyme structural genes *qutB*, *qutC* and *qutE* from the quininic acid utilization cluster in *Aspergillus nidulans*. *Curr. Genet.* **9**, 305–312.

14. A Cloning Strategy in Filamentous Fungi

Hynes, M. J., Corrick, C. M., and King, J. A. (1983). Isolation of genomic clones containing the *amdS* gene of *Aspergillus nidulans* and their use in the analysis of structural and regulatory mutations. *Mol. Cell. Biol.* **3**, 1430–1439.

Johnstone, J. L., Hughes, S. G., and Clutterbuck, A. J. (1984). Cloning an *Aspergillus nidulans* developmental gene by transformation. *EMBO J.* **4**, 1307–1311.

Kinghorn, J. R., and Hawkins, A. R. (1982). Cloning and expression in *Escherichia coli* K-12 of the biosynthetic dehydroquinase function of the *arom* cluster gene from the eukaryote *Aspergillus nidulans*. *Mol. Gen. Genet.* **186**, 145–152.

Kraus, J. P., and Rosenberg, L. E. (1982). Purification of low abundance messenger RNAs from rat liver by polysome immunoadsorption. *Proc. Natl. Acad. Sci. U.S.A.* **79**, 4015–4019.

Lockington, R. A., Sealy-Lewis, H. M., Scazzocchio, C., and Davies, R. W. (1985). Cloning and characterization of the ethanol utilization regulon in *Aspergillus nidulans*. *Gene* (in press).

Orr, W. C., and Timberlake, W. E. (1982). Clustering of spore-specific genes in *Aspergillus nidulans*. *Proc. Natl. Acad. Sci. U.S.A.* **79**, 5976–5980.

Scazzocchio, C., Sdrin, N., and Ong, G. (1982). Positive regulation in a eukaryote, a study of the *ua*Y gene of *Aspergillus nidulans:* I. Characterization of alleles, dominance and complementation studies and a fine structure map of the *ua*Y-*oxp*A cluster. *Genetics* **100**, 185–208.

Schwartz, D. C., and Cantor, C. R. (1984). Separation of yeast chromosome-sized DNA by pulsed field gradient electrophoresis. *Cell (Cambridge, Mass.)* **37**, 67–75.

Tilburn, J., Scazzocchio, C., Taylor, G. G., Zabicky-Zissman, J. H., Lockington, R. A., and Davies, R. W. (1983). Transformation by integration in *Aspergillus nidulans*. *Gene* **26**, 205–221.

Timberlake, W. E. (1978). Low repetitive DNA content in *Aspergillus nidulans*. *Science (Washington, D.C.)* **202**, 973–975.

Tomsett, A. B., and Cove, D. J. (1979). Deletion mapping of the *nii*A*nia*D gene region of *Aspergillus nidulans*. *Genet. Res.* **34**, 19–32.

Yelton, M. M., Hamer, J. E., de Souza, E. R., Mullaney, E. J., and Timberlake, W. E. (1983). Developmental regulation of the *Aspergillus nidulans trp*C gene. *Proc. Natl. Acad. Sci. U.S.A.* **80**, 7576–7580.

Yelton, M. M., Timberlake, W. E., and van den Hondel, C. A. M. J. J. (1984). A cosmid for selecting genes by complementation in *Aspergillus nidulans:* selection of the developmentally regulated *y*A locus. *Proc. Natl. Acad. Sci. U.S.A.* **82**, 834–838.

Zimmerman, C. R., Orr, W. C., Leclerc, R. F., Barnard, E. C., and Timberlake, W. E. (1980). Molecular cloning and selection of genes regulated during *Aspergillus* development. *Cell (Cambridge, Mass.)* **21**, 709–713.

IV
Applications

15

Primary Metabolism and Industrial Fermentations

RAMUNAS BIGELIS

Biotechnology Group
Miles Laboratories, Inc.
Elkhart, Indiana

I.	Primary Metabolites	358
II.	Genetic Approaches to the Production of Primary Metabolites	359
III.	Organic Acids	360
	A. Citric Acid	361
	B. Itaconic Acid	365
	C. Gluconic Acid	365
	D. Fumaric Acid	366
	E. Malic Acid	367
	F. Lactic Acid	367
	G. Other Organic Acids	368
IV.	Amino Acids	368
	A. Tryptophan	369
	B. Lysine	369
	C. Methionine	370
	D. Other Amino Acids	370
	E. Single-Cell Protein	371
V.	Polysaccharides	372
	A. Pullulan	372
	B. Scleroglucan	373
	C. Other Fungal Polysaccharides	374
VI.	Lipids	374
VII.	Nucleotides and Nucleic Acid-Related Compounds	376
	A. Nucleotides	376
	B. Nucleic Acids	377
VIII.	Vitamins	378
	A. Riboflavin	378
	B. Fungal Production of Vitamins	379
	C. Vitamins A and D	379
IX.	Polyols	380
	A. Glycerol	380

B. Other Polyols 381
X. Ethanol 382
XI. The Promise of Biotechnology 383
    References 385

Fermented foods and beverages were produced for centuries by unknowing exploitation of fungi and microorganisms. Only within the past 100 years or so has an understanding of microbiological principles allowed the culture of individual species and the development of reliable large-scale processes for the production of food, fuel, and chemicals. Modern industrial fermentations with genetically improved strains produce millions of tons of useful compounds yearly, often much more cheaply than chemical synthesis or extraction from biological materials. Four main classes of substances are produced by industrial fermentations: primary metabolites, secondary metabolites, enzymes, and the whole cells themselves (Demain, 1981). This review will concern the commercial production of primary metabolites by fermentations employing fungi, and will examine the role of genetics in improving the productivity and efficiency of such industrial fermentations.

I. PRIMARY METABOLITES

What are primary and secondary metabolites? The distinction between primary and secondary metabolism is readily apparent in most instances, but blurred in some. Bu'lock (1961) has considered secondary metabolites as compounds that have "no obvious function in general metabolism" and have a restricted distribution. Demain (1971) has noted that while secondary metabolites have no general function in cellular processes, they may play an important role in the organisms producing them. Secondary metabolism is usually species-specific and involves synthetic processes which appear to be restricted to microbes and green plants (Turner, 1971). Secondary metabolites are derived from primary metabolites (Drew and Demain, 1977), which are, on the other hand, cellular substances that are indispensable for growth and life processes. Furthermore, primary metabolism is universal among microorganisms and involves those biosynthetic and degradative pathways that sustain life (Bennett, 1983). This review considers primary metabolites in the broad sense of compounds essential to life. It regards cellular substances closely related to metabolic intermediates as primary metabolites and discusses a wide range of biological compounds of actual or potential commercial importance.

II. GENETIC APPROACHES TO THE PRODUCTION OF PRIMARY METABOLITES

Classical genetics, especially when coupled with biochemistry, has revealed much about the function and regulation of primary metabolism in fungi. But Roper's (1973) assertion over ten years ago is still true today: "fungal genetics has so far contributed little to programs of industrial strain improvement." Perhaps the reason rests partly on the difference between basic and industrial research. Basic microbiological research has focused to a large extent on convenient experimental organisms amenable to genetic manipulation for probing fundamental biological questions, while industrial and sometimes applied research have centered on commercially important microorganisms, often chosen after lengthy screening procedures and without regard to their willingness to cooperate with geneticists. Thus, genetic recombination in industrial microorganisms is quite rare. As a result, industrial strain improvement programs have relied largely on mutagenesis and screening and not classical genetics as a means of improving the economics of fermentation processes. Now, in the age of the new genetic technology, with the advent of interspecific gene transfer, there is a sharp increase in the exchange of information between basic and applied researchers. The gap between the favorite organisms of geneticists, such as *Escherichia coli* and yeast, and those prolific but stubborn industrial organisms has narrowed. Genes from industrial organisms can be introduced into a convenient host and, if desired, engineered and then returned to the donor, creating a better strain for commercial processes. Thus, the new genetic technology has caused a redefinition of the means of strain improvement. Recombinant DNA and cell fusion methods for increasing the efficiency of metabolite production by fermentation have also prompted a redefinition of the aims of strain improvement programs, leading to the search for new or superior genes or nucleotide sequences with commercial potential.

Nevertheless, mutagenesis and screening is still a valuable procedure for the improvement of industrial strains and the efficiency of commercial fermentations. The method is reliable and has a record of improving productivity in a short time period (Elander, 1982). It has allowed the development of better strains that are able to utilize cheaper raw materials and are adapted to specialized conditions of the fermentation process. Advances in the understanding of pathway regulation and the mechanisms of mutagenesis have contributed greatly to the cost-effectiveness of mutation and screening procedures. Technological advances in processing of microbial isolates have also contributed; automation now substantially increases the likelihood of obtaining the desired strain (Rowlands, 1984). Thus, when a genetic methodology is absent for an industrial microorganism, mutation and selection still remains an indispensable tool for the biotechnologist.

Other genetic approaches to the production of primary metabolites by fungi have involved a variety of procedures, many of which are referred to in this and accompanying chapters. Apart from the screening of natural isolates, the most successful methods have been based on mutant enrichment procedures, auxotrophy, reversion, resistance or sensitivity to metabolic analogs, resistance to inhibitors, tolerance of the metabolite that is overproduced, inability to assimilate a metabolite, temperature sensitivity, osmotic sensitivity, transport and permeability, colonial or morphological characteristics, continuous culture, and specialized selections, for example, those used in the brewing and wine industries. Additional genetic manipulation has included recombination via mating, parasexuality, hybridization, and polyploidization. These genetic approaches have been considered in depth elsewhere (Pontecorvo *et al.*, 1953a; Calam, 1964, 1970; Hopwood, 1970; Esser, 1974, 1978; Johnston, 1975; Malik, 1979; Elander and Chang, 1979; Ball, 1980; Elander, 1982; Spencer and Spencer, 1983). Viruses of some industrial fungi have been described but rarely used for genetic manipulation (Lemke *et al.*, 1976).

The chapters in this volume also discuss new genetic methods offering exciting opportunities for the improvement of industrial fungi. Protoplast fusion is one of these methods (Peberdy, 1979, 1980; Ferenczy, 1981). Interspecific protoplast fusion and recombination are now possible, raising the possibility of gene exchange between unrelated organisms. The technique may allow the generation of superior recombinants after fusion of "overmutagenized" industrial strains and related vigorous, but poorly producing, strains (Demain, 1981). Recombinant DNA technology will also revolutionize the exploitation of industrial fungi. For example, foreign genes introduced into industrial strains may allow growth on cheap, readily available polysaccharides; amplification of host genes via multicopy plasmids may increase the production rate or yield of a metabolite or protein; *in vitro* tailoring of nucleotide sequences may favorably alter the regulation of the synthesis of a desired product; or the combination of foreign and host genes may give rise to novel biochemical pathways for primary metabolites (Demain, 1981; Jackson, 1982). Advances in recombinant DNA chemistry, in conjunction with progress in fungal transformation and vector construction, will lead to improvements of industrial fungal strains and eventually increase the efficiency of large-scale fermentations.

III. ORGANIC ACIDS

A variety of fungi have been employed for the industrial production of organic acids, primarily citric, itaconic, and gluconic acids. The overproduction of other organic acids has been reported but remains to be commercially exploited. Organic acid production has been reviewed extensively over the years and the cited references provide a historical perspective on the development of the science and art of organic acid fermentation (Cochrane, 1948; Perlman, 1949; Johnson, 1954; Perlman and Sih, 1960; Martin, 1963; Smith *et al.*, 1974; Miall, 1978; Lockwood, 1979; Röhr *et al.*, 1983a; Abou-Zeid and Ashy, 1984).

Citric, itaconic, and gluconic acids, which will be considered first, are three organic acids related to primary metabolism that are produced commercially by fungal fermentations. However, considerable research is also being devoted to the formation of other organic acids by fungi. The availability of genetically engineered strains, greater demand for organic acids, and higher prices for precursors of chemical synthesis could make fungal processes for a number of these organic acids competitive with chemical manufacture.

A. Citric Acid

Citric acid is the most important organic acid produced by industrial fermentation and is widely used in the food and pharmaceutical industries (Atticus, 1975). It serves as an acidulant in soft drinks, a buffer in jellies, jams, and desserts, an antioxidant of fats, an emulsifying and stabilizing agent, a detergent, a blood anticoagulant, and a component of effervescent tablets and powders. Other industrial uses of citric acid involve electroplating processes, plasticizers, water conditioners, and secondary oil recovery (Kapoor *et al.*, 1982).

1. *Aspergillus niger*

Since Wehmer's (1903) first unsuccessful attempts to commercially exploit the fermentation of sugar to citric acid by penicillia, industrial citric acid production has evolved from a process dependent on the precipitation of 10,000 tons of the calcium salt from pressed citrus fruit per year to large-scale fungal fermentations which today generate more than 283,000 metric tons annually in the United States, Western Europe, and Japan (Strauss, 1981). The ability to form citric acid is not uncommon among fungi, especially the penicillia and aspergilli (Foster, 1949). A patent by Zahorsky (1913) first revealed *Aspergillus niger*, a conidial fungus first described in connection with gallic acid production (Van Tieghem, 1867), to be a citric acid producer and led to refinements that allowed yields of over 60% after several weeks of fermentation in surface culture (Thom and Currie, 1916). However, present-day fermentations with selected strains of *A. niger* employ submerged cultures which substantially reduce fermentation times, utilize cheap raw carbohydrate-based media with meticulously controlled phosphorus, metal ion, and nitrogen concentrations, and yield up to 90% (w/w) citric acid from sucrose (Kubicek and Röhr, 1982).

Though many *Aspergillus* species produce some citric acid, *A. niger*, which represents a morphologically and biochemically diverse group, is generally accepted as the most useful producer. The selection of desirable production strains of *A. niger* must contend with a number of critical variables: citric acid accumulation, by-product formation, growth rate, capacity for citric acid reutilization, metal sensitivity, response to acid pH, cellular morphology, ability to grow on the desired carbohydrate source, and susceptibility to contaminants. While the specific details of fermentation and extraction processes remain proprietary or patented information, it is the high-yielding strains themselves that are the most guarded secrets of commercial citric acid producers. Thus, little has been published on the improvement of citric acid-producing strains of *A. niger*.

Attempts to develop improved citric acid-producing strains of *Aspergillus* have been hampered by the absence of a sexual cycle (Fincham *et al.*, 1979). However, *A. niger* has been studied genetically using parasexual recombination after formation of balanced heterokaryons (Pontecorvo *et al.*, 1953b; Lhoas, 1967) and a diploid strain has even been reported in nature (Nga *et al.*, 1975). Early attempts to apply heterokaryons of *Aspergillus fonsecaeus* to commercial fermentations indicated that heterokaryons were less efficient producers of citric acid and were also unstable, segregating component mutants during fermentation (Ciegler and Raper, 1957). Model studies with *Aspergillus nidulans* suggest that the unstable constructed diploids of *A. niger* may have generated unstable segregants which were either aneuploid or partial duplication strains (Ball, 1967; Ball and Azevedo, 1976).

Since the early work of Ciegler and Raper (1957), other studies (Seichertova and Leopold, 1969; Ilczuk, 1971b; Das and Roy, 1978) have endeavored to augment citric acid production with diploid strains of *A. niger* but have not reported more than 20% enhancement of citric acid production by heterokaryons. Ikeda (1961), however, reported a 60% improvement in a tetraploid strain, Shcherbakova and Rezvaya (1977) a 50% increase using diploids induced by colchicine, and Ilczuk (1971a) up to a severalfold increase using diploids over low-yielding haploid strains. In no cases were the constructed strains superior to the most productive haploid strains. Based on crosses with low-yielding auxotrophic haploid strains of *A. niger*, Chang and Terry (1973) concluded that nuclear genes are primarily responsible for citric acid production and that gene dosage is not responsible for the observed increases in citric acid yield in somatic diploids. They proposed that the increased yields in diploids or heterokaryons were due to intergenic complementation which corrected for the effects of pleiotropic mutations causing not only the nutritional deficiencies but also impaired citric acid metabolism in the parental haploids.

Significant strain improvement has been achieved after mutagenesis of *A. niger* and then screening for elevated production of citric acid. This approach was first used successfully in 1935 (Kresling and Shtern, 1935) employing ultraviolet light and radium as mutagenic agents. Subsequent studies have also exposed spores to x rays, hertzian waves, and a variety of mutagenic chemicals (Perlman and Sih, 1960; Das, 1972; Azevedo and Bonatelli, 1982); screened for superior strains using organic acid detection methods (Miles Laboratories, 1951; James et al., 1956; Röhr et al., 1979); screened with shake flasks; and then developed optimal environmental conditions for citric acid overproduction, paying particular attention to the metal constituents of the medium. Mutagenesis procedures for improved citric acid production with wild-type A. niger have been optimized on the basis of their efficiency in generating morphological mutants (Das and Roy, 1981). In addition, productive mutants have been isolated that have altered morphological or cultural characteristics (Shcherbakova, 1964) or altered conidial color (Ilczuk, 1968), but no correlation between these morphological characteristics and the ability to synthesize large amounts of citric acid has been found. Pellet formation, a morphological trait related to citric acid overproduction (Whittaker and Long, 1973; Metz and Kossen, 1977; Röhr et al., 1983a), has never been exploited as the basis for selection of citric acid-producing mutants. Citric acid-producing strains have also been isolated that are more tolerant to metals (Bernhauer, 1929; Gardner et al., 1956; Trumpy and Millis, 1963).

Selective breeding after chemical mutagenesis and five stages of selection has generated A. niger strains that produce 70% more citric acid but half the amount of oxalic acid, an undesirable by-product (Golubtsova et al., 1979). However, it is surprising that after many decades of research no selection procedures for A. niger or genetic loci have been described that involve specific steps of citric acid metabolism in this organism (Röhr et al., 1983a). More recent approaches such as those taken by Bonatelli et al. (1982) with morphological, auxotrophic, and Benlate-resistant mutants and their revertants may eventually allow association of genetic loci with the citric acid excretion phenotype.

The advent of new genetic technology will undoubtedly have an impact on the exploitation of A. niger for industrial chemicals and enzymes. Genetic engineering approaches should eventually make citric acid production more economical. As suggested by Eveleigh (1981) and Demain (1981), the transfer to A. niger of genes involved in the breakdown of cellulose, starch, or other inexpensive carbon sources would substantially lower the cost of organic acid production. Equally important would be the identification and cloning of fungal genes that play a role in citric acid overproduction and the construction of new production strains by plasmid transfer or cell fusion.

2. Yeasts

Citric acid is produced in significant amounts by yeasts, especially those of the genus *Candida* (Kapoor et al., 1982). *Candida lipolytica* has been studied most extensively because it has the capacity to produce citric acid from hydrocarbons and also to secrete enzymes. An isolate that exhibits a sexual phase (Wickerham et al., 1970) has been renamed *Saccharomycopsis lipolytica* (Yarrow, 1972) and more recently *Yarrowia lipolytica* (Barnett et al., 1984), and genetic techniques and a genetic map for this organism have been developed (Gaillardin et al., 1973; Esser and Stahl, 1976; Ogrydziak et al., 1978).

Most studies with *Candida* or *Saccharomycopsis* have employed various mutagens to derive strains that produce more citric acid but less isocitric acid, a common by-product of many yeast fermentations for citric acid (Kapoor et al., 1982). Fukuda et al. (1974) have also stressed the importance of deriving strains incapable of utilizing citric acid. Akiyama et al. (1972, 1973a,b) have obtained fluoroacetate-sensitive mutant strains with low aconitase activity that produce little isocitric acid and grow poorly on citric acid as a sole carbon source, while converting *n*-paraffins to citric acid at a yield as high as 145% (w/w). Such mutants may produce large amounts of 2-methylisocitric acid (Tabuchi and Hara, 1974). Takayama et al. (1982) have described a process with iron-requiring mutants of *Candida zeylanoides* that produce little isocitric acid while yielding significant amounts of purer citric acid that does not need extensive purification.

An abundant patent literature documents advances in the use of yeasts for organic acid production. Patents from Pfizer (1973), Kyowa Hakko Kogyo Co. (1982), and Takeda Chemical Industries (1983) have revealed that strains of *Candida, Torulopsis, Hansenula,* or *Pichia* produce citric acid under the proper fermentation conditions. Pfizer (1970) has claimed that *C. lipolytica* ATTC 8661 produces 56.8 g of citric acid per liter from $C_{12}-C_{18}$ alkanes and up to 110.4 g per liter from hexadecane, though the conversion is lowered from 94 to 11% in the latter case. Takeda Chemical Industries (1970) has claimed that mutants of *C. lipolytica* unable to utilize citric acid or L-lactic acid as the sole carbon source produce 112 g per liter after 3 days in medium containing $C_{12}-C_{18}$ alkanes, but only 1.3 g of isocitric acid per liter. A continuous fermentation process for the production of citric acid from hydrocarbons by *C. lipolytica* ATCC 20228 has been patented (Pfizer, 1974). Medium is continuously added and culture fluid continuously withdrawn, but the yeast cells and alkanes are not recycled. The increase in the cost of petroleum has influenced the economics of this novel process.

Very few studies on the genetics of citric acid-producing yeast strains have been published. Preliminary studies with *S. lipolytica* have attempted to cross auxotrophs of producing strains and have generated prototrophs of unknown ploidy (Shah et al., 1982). The results remain uncertain since the presence of neither diploids nor asci was demonstrated. The underlying difficulties in genetic experiments with this heterothallic, dimorphic yeast are low mating frequency, low ascospore viability, and the predominance of binucleate spores and two-spored asci (Esser and Stahl, 1976). A possible new approach involves protoplast fusion between *S. lipolytica* auxotrophs of like mating type (Stahl, 1978). After regeneration and selection of stable diploids, recombinant progeny have been obtained via the parasexual cycle or by sexual reproduction after transfer to conventional sporulation medium. Stahl (1978) reported that of 22 germinated spores from a cross of *lys ade* × *his*$^+$, 20 clones were recombinant for the auxotrophic markers. Apparently, the A and B mating type alleles of this yeast determine the initial phases of mating, but not karyogamy and meiosis. Thus, protoplast fusion followed by sporulation holds promise as a means of constructing and improving citric acid-producing *S. lipolytica* strains with industrial potential, in addition to facilitating genetic studies of enzyme secretion and alkane utilization by this yeast.

B. Itaconic Acid

Itaconic acid, a substituted methyl acrylic acid formed from *cis*-aconitic acid by various fungi, is marketed as a refined product or as esters, and has been used in the manufacture of plastics such as resin coatings and synthetic detergents. It can also be obtained by pyrolysis of citric acid. The microbial synthesis of itaconic acid was first described by Kinoshita (1929) in *Aspergillus itaconicus* and later in *Aspergillus terreus* (Calam *et al.*, 1939), *Ustilago zeae* (Haskins *et al.*, 1955), and *Helicobasidium mompa* (Araki *et al.*, 1957). Succinic acid and itatartaric acid are common by-products of itaconic acid production. Fermentation processes for itaconic acid have been reviewed elsewhere (Pfeiffer *et al.*, 1952; Lockwood, 1954; Lockwood and Schweiger, 1967; Kobayashi, 1967). High-yielding strains of *A. terreus*, the organism used in industrial fermentations, have been isolated from nature (Prescott and Dunn, 1959) and also improved after irradiation of conidia and screening (Hollaender *et al.*, 1945; Raper *et al.*, 1945; Pfeiffer *et al.*, 1953). More recent studies (Tabuchi *et al.*, 1981) report the formation of itaconic acid by wild-type and mutant strains of *Candida* and raise the possibility of industrial production by submerged cultures of yeast.

C. Gluconic Acid

About 45,000 metric tons of gluconic acid, an oxidation product of glucose, are manufactured annually worldwide, mainly as sodium gluconate, a versatile sequestering agent, or as the calcium salt, a pharmaceutical used in treating calcium deficiencies. It is also made as the free acid, a noncorrosive cleaning agent, or δ-gluconolactone, a slow-acting acidulant (Ward, 1967; Röhr *et al.*, 1983b). Fermentation processes are competitive with electrochemical synthesis.

Gluconic acid is produced by aspergilli and penicillia, in addition to other fungi (Foster, 1949). The first large-scale production of gluconic acid used *Penicillium purpurogenum* var. *rubri-sclerotium* and then *Penicillium chrysogenum* in surface culture (Smith, 1969). *Aspergillus niger*, the organism first found to produce this compound (Molliard, 1922), is the one utilized today in submerged fermentations with glucose and recycled mycelia. Advances in fermentation processes over the years have raised the yield of gluconic acid to about 95% (Blom et al., 1952; Prescott and Dunn, 1959; Ward, 1967; Röhr et al., 1983b).

Strain selection programs with A. niger have aimed to reduce the formation of citric and oxalic acids (Underkofler, 1954b). References to strain development by mutation and selection have appeared (Chopra et al., 1975; Kundu and Das, 1982), though available strains function at near-maximum theoretical efficiency.

Other efforts to improve A. niger strains relied on the derivation of boron-tolerant mutants that could ferment solutions containing up to 35% glucose in the presence of boron compounds. Since boron compounds increase the solubility of calcium gluconate and thus reduce the formation of a crust on recycled mycelia, the use of such strains allowed the addition of excess calcium carbonate to neutralize the increased amounts of gluconic acid produced (Underkofler, 1954b). However, the addition of boron was soon rejected owing to its undesirable side effects in animals (Lockwood, 1979).

Glucose oxidase, the enzyme that produces gluconic acid, is prepared commercially from A. niger, Penicillium notatum, or P. glaucum and is used to remove residual glucose or oxygen from foods (Ward, 1967; Marconi, 1974). Immobilized glucose oxidase may someday replace the fermentation process. Expression of the cloned fungal gene for glucose oxidase in a suitable host would facilitate production of this enzyme.

D. Fumaric Acid

Fumaric acid, a food acidulant and industrial chemical, has been manufactured by processes employing filamentous fungi, but is now produced mainly by chemical synthesis (Miall, 1978). Fumaric acid can be chemically converted to maleic acid, which is used to make polyester resins (Foster, 1954). Increasing costs of petrochemicals or demand for aspartic acid, which can be synthesized from fumaric acid, may again favor biotechnological production.

Several members of the order Mucorales have been found to produce significant amounts of fumaric acid since Ehrlich (1911) first reported the ability of *Rhizopus nigricans* to produce this Krebs cycle intermediate (Foster, 1954). Productivity has been associated with sexuality in this heterothallic organism with only the + type accumulating fumaric acid (Foster and Waksman, 1939). Genetic attempts to increase fumaric acid production have failed, owing to the inability to induce zygote germination and the multinucleate nature of the sporangiophores (Foster, 1954). Fermentation processes for fumaric acid production by *R. nigricans* and *Rhizopus japonicus* have been patented (Waksman, 1944) and more recent results with *Rhizopus arrhizus* indicate that a yield of 60% can be attained from glucose (Smith, 1969; Buchta, 1983).

Screening programs in Japan have demonstrated that the yeasts *Candida hydrocarbofumarica* (Yamada *et al.*, 1970) and *C. blankii* (Furukawa *et al.*, 1978) have the capacity to produce substantial amounts of fumaric acid from hydrocarbons.

E. Malic Acid

Fermentation processes for the production of L-malic acid, a flavor compound, are not competitive with cheaper enzymatic synthesis (Buchta, 1983) or the chemical synthesis of DL-malic acid (Irwin *et al.*, 1967). High yields of malic acid have been obtained with *Schizophyllum commune* from ethanol and CO_2 (Tachibana and Murakami, 1973); strains of *Paecilomyces varioti* from acetate, propionate, or ethanol (Takao *et al.*, 1977); *Candida brumptii* from *n*-paraffins (Sato *et al.*, 1977); and *R. arrhizus* when in mixed culture with a *Proteus vulgaris* having high fumarase activity (Takao and Hotta, 1977). Two-stage fermentations with *C. hydrocarbofumarica* and *Candida utilis* in media containing hydrocarbons have generated 72% weight yields of L-malic acid (Furukawa *et al.*, 1970).

Malic acid is produced by species of *Candida, Pullularia,* and *Pichia* (Miall, 1978), in addition to some *Aspergillus* and *Penicillium* species (Beuchat, 1978). The most productive strains have been identified by meticulous screening, as exemplified by a study of Takao (1965) with various basidiomycetes, but no genetic improvement programs have been reported.

L-Malic acid is produced in China from fumaric acid with *Candida rugosa* cells immobilized in polyacrylamide. The conversion results in an 82–85% yield during continuous operation (Zhang, 1982).

F. Lactic Acid

Lactic acid, which is used by the food, plastics, and leather industries, has been manufactured with homofermentative lactic acid bacteria, but can be made competitively by chemical means. Yields of 93–95% of the weight of glucose are common with species of *Lactobacillus* and comparable yields can be obtained with *Rhizopus oryzae*. Early processes used *R. oryzae* for D-lactic acid production in a rotary fermentor (Smith, 1969). The original studies with organisms of the genera *Rhizopus* and *Mucor* have been summarized by Ward *et al.* (1936). In the event that lactonitrile, the chemical precursor, becomes more expensive or in short supply, the *R. oryzae* fermentation may be preferred owing to simpler recovery, shorter fermentation times, and greater purity of the final product (Lockwood, 1979).

G. Other Organic Acids

Other organic acids can be made by fungal fermentations, but are obtained more cheaply by chemical synthesis. Succinic acid, a flavoring agent and general industrial chemical, is produced in large amounts by *C. brumptii* (Sato *et al.*, 1972) or by *Rhizopus* in mixed fermentations with bacteria (Sasaki *et al.*, 1970). Mutagenesis and screening of *C. rugosa* has produced strains capable of converting *n*-paraffins to succinic acid (Au Tian *et al.*, 1981). Tartaric acid, a food additive, is produced by strains of *A. niger, Aspergillus griseus*, and *P. notatum* (Beuchat, 1978). Pyruvic acid is produced by *Debaryomyces condertii* (Moriguchi, 1982) and *S. commune* (Takao and Tanida, 1982) in significant amounts, as is α-ketoglutaric acid by *C. lipolytica* (Tsugawa *et al.*, 1969).

A number of fungi have the capacity to form ascorbic acid-like substances. These include species of *Torula, Aspergillus, Penicillium*, and *Fusarium*. Erythorbic acid, an oxidation product of D-glucono-γ-lactone also known as isoascorbic acid, is the most important member of this class of compounds and is used as a food perservative. Of 4800 fungi that have been screened, all significant producers were found to be penicillia. A strain improvement program with *P. notatum* has employed monospore selection and mutagenesis with ultraviolet light, nitrous acid, and monoiodoacetic acid. Irradiation has successfully generated more potent strains with a 40% increased yield of erythorbic acid from glucose (Takahashi, 1969).

Lists of unusual organic acids produced by fungi (Shibata *et al.*, 1964) or by microorganisms in general (Miller, 1961) have been compiled.

IV. AMINO ACIDS

Almost all of the 20 standard amino acids are produced commercially and most are manufactured via bacterial fermentation, though chemical, enzymatic, and extraction processes are also used (Yoshinaga and Nakamori, 1983). The wide variety of applications for amino acids include food supplements, feedstuffs, flavoring agents, cosmetics, and starting materials for diverse pharmaceuticals and chemicals. Most of the total world output is used for nutrition: about 66% is used for food and 31% for feed (Soda *et al.*, 1983). Efforts to increase the efficiency of amino acid production have focused on lowering the cost of starting materials, selecting superior microbial strains, and improving fermentation processes.

Comprehensive reviews of amino acid fermentations have appeared in recent years (Kinoshita and Nakayama, 1978; Hirose and Okada, 1979; Soda *et al.*, 1983; Yoshinaga and Nakamori, 1983). Research on the production of amino acids by fungal fermentations has concentrated on tryptophan, lysine, and methionine. Fungal processes for other amino acids may become important in the future.

A. Tryptophan

Tryptophan, an essential amino acid, cannot be obtained in significant amounts by fermentation with sugars and is synthesized chemically. It has been produced with a strain of *Hansenula anomala* resistant to high levels of anthranilic acid (Terui, 1973) which produces up to 6 g of tryptophan per liter from the added precursor, anthranilic acid (Terui and Niizu, 1969), and as much as 14 g per liter from indole (Ebihara *et al.*, 1969). Indole-resistant producing strains have also been isolated. General control of amino acid biosynthesis appears to operate in this yeast since starvation of methionine or histidine mutants for the respective amino acid also elevates tryptophan excretion (Enatsu *et al.*, 1963). The tryptophan-hyperproducing strains all have lower anthranilic acid and tryptophan degrading activity and show altered repression and feedback inhibition by tryptophan. Attempts to improve available strains by crossing haploids have failed (Terui, 1973). In addition, a *Hansenula polymorpha* mutant resistant to 5-fluorotryptophan has been shown to excrete tryptophan into the medium (Denenu and Demain, 1975).

The production of tryptophan from indole or anthranilic acid has been reported in *Claviceps purpurea* (Malin and Westhead, 1959) and numerous strains of yeasts (Nickerson and Brown, 1965; Hütter, 1973). However, these processes are not commercially practical.

B. Lysine

Much interest has focused on the microbiological production of lysine because it is essential for the animal diet and because grain proteins are deficient in this amino acid (Champagnat *et al.*, 1963). To date, no processes for lysine production using fungi have been commercialized. Early experiments examined the growth of yeasts on α-ketoadipic acid, the immediate precursor of lysine, and lysine-rich cells of the yeasts *S. cerevisiae* and *C. utilis* were developed which were suitable for fortification of deficient foods (Broquist *et al.*, 1961). Feeding of other precursors has produced cells of various yeasts consisting of up to 20% lysine (Nickerson and Brown, 1965). Subsequent research has led to the discovery of various yeast mutants that overproduce lysine. Thialysine-resistant mutants of *S. cerevisiae* with pleiotropic properties obtained after mutagenesis excrete lysine (Haidaris and Bhattacharjee, 1978; Zwolshen and Bhattacharjee, 1981); thialysine strains of *Candida pelliculosa* produce up to 3.2 g of lysine per liter of medium (Takenouchi *et al.*, 1979); and *S. lipolytica lex$^-$* strains accumulate four times more intracellular lysine than wild-type strains and possess a desensitized first enzyme for lysine biosynthesis (Gaillardin et al., 1975).

Since Richards and Haskins (1957) screened 600 fungi for lysine excretion, a number of other species have been shown to overproduce lysine. Strains of *Gliocladium* sp., *Ustilago maydis* (Dulaney, 1957), and *Sphacelotheca sorghii* (Tauro et al., 1963) accumulate lysine in culture fluids, but in amounts generally less than 1 g per liter. However, mutants of *U. maydis* obtained with ultraviolet light and ethyleneimine produce 2.5 g of lysine per liter (Sanchez-Marroquin et al., 1971).

A novel enzymatic process for the synthesis of L-lysine from exogenous α-aminocaprolactam has been developed with *Aspergillus ustus* and then improved with acetone-dried cells of *Cryptococcus laurentii* (Fukumura, 1977). *Candida humicola* and *Trichosporon cutaneum* are also able to hydrolyze exogenous α-aminocaprolactam and may be useful for this process.

C. Methionine

Methionine, another amino acid that is low in cereal grains, is also a necessary supplement to the animal diet. Most methionine is synthesized chemically as the D,L form since this cheap process is sufficient for poultry feed. Microbial production has been found to be uneconomic (Sinskey, 1983), though research has been devoted to the development of methionine-rich microorganisms as food supplements. Mutants of *C. tropicalis* that have a 41% higher methionine content have been selected as small colonies on sulfur-deficient agar since overproducers have a greater sulfate requirement (Okanishi and Gregory, 1970). Mutants of a methanol-assimilating yeast with a high methionine content may also serve as a nutritional feed (Shay et al., 1983). Such strains were selected on the basis of their resistance to ethionine or norleucine and also their ability to crossfeed a methionine-requiring strain of *Bacillus subtilis*. The methionine content was increased to as much as 6.0% of the total protein compared to 1.2% for the parent. Ethionine- and selenomethionine-resistant strains of *S. lipolytica* (Morzycka et al., 1976) and an ethionine-resistant strain of *S. cerevisiae* (Sorsoli et al., 1964) that overproduce methionine also have been isolated. Finally, strains of *Hanseniospora valbyensis* resistant to ethionine, N-acetylnorleucine, polymyxin, or bacitracin have been shown to excrete methionine into the medium (Scherr and Rafelson, 1966).

D. Other Amino Acids

The production of glutamic acid, threonine, alanine, phenylalanine, and aspartic acid by fungal fermentation has also been reported.

More than 270,000 tons of glutamic acid are manufactured annually, mostly in

Japan by bacterial fermentation for use as a flavor potentiator (Kinoshita and Tanaka, 1972; Yoshinaga and Nakamori, 1983). A number of fungi, among them *Aspergillus oryzae, P. chrysogenum, Rhodotorula glutinis,* and species of *Cephalosporium* (Kinoshita *et al.*, 1957; Abe and Takayama, 1972) have been found to excrete this and other amino acids into culture broth.

Threonine is overproduced by amino acid auxotrophs of *Candida guilliermondii* obtained after mutagenesis and replica plating (Tsukada and Sugimori, 1971). Isoleucine, isoleucine–tryptophan, and isoleucine–tryptophan–methionine auxotrophs accumulated 1.8, 2.0, and about 4 g of threonine, respectively, per liter of medium. A variety of culture collection strains of *U. maydis* also excrete threonine, but in amounts less than 0.3 mg per liter (Ericson and Kurz, 1962).

Alanine, an amino acid that is manufactured primarily by enzymatic processes, is overproduced by a number of fungi (Krasil'nikov *et al.*, 1962; Chibata *et al.*, 1965; Pisano, 1966). Screening experiments have shown that 8 of 14 marine fungi tested accumulate alanine (Pisano *et al.*, 1964), as do species of *Rhizopus, Mucor, Rhodotorula, Fusarium, Torulopsis,* and *Willia* (Abe and Takayama, 1972). The most prolific producer among the fungi is a nicotinate$^-$, leu$^-$, met$^-$ auxotroph of *U. maydis* that accumulates 20 g of alanine per liter of fermentation broth (Dulaney *et al.*, 1964).

Phenylalanine can be produced by phenylalanine analog-resistant mutants of *Candida* (Okumura *et al.*, 1972) or by mutants of *Torula* sp. (Anonymous, 1983b). Genex produces 10–100 tons annually of phenylalanine with proprietary strains of *Torula* for use in the synthesis of the low-calorie sweetener aspartame (L-aspartyl-L-phenylalanyl methyl ester). A microbial process with suspended bacteria or *Sporobolomyces odorus* may someday replace the expensive chemical process for the synthesis of aspartame and recombinant DNA technology is expected to play a role (Anonymous, 1983a).

Aspartic acid is synthesized efficiently by a commercial process that enzymatically converts fumarate to aspartate (Yoshinaga and Nakamori, 1983). The same conversion, however, can be effected in mixed fermentations with *Rhizopus* and bacteria (Hotta and Takao, 1973). In addition, certain yeasts accumulate this amino acid in the medium (Abe and Takayama, 1972).

E. Single-Cell Protein

Microbial proteins are used to supplement human foods and animal feed. The often-used term single-cell protein (SCP) describes the dried cells of microorganisms which are used for this purpose. Although human consumption of baker's yeast dates to antiquity, economical large-scale processes for SCP, often using waste material as substrate, are quite recent. Present-day commercial SCP processes employ the fungi *C. utilis, P. varioti, Fusarium graminearum, K. fragilis,* and *Penicillium cyclopium,* while processes for the growth of

Cephalosporium eichhorniae, Scytalidium acidophilum, Chaetomium cellulyticum, and *A. niger* are in development (Litchfield, 1983). There are several excellent reviews on SCP and the use of microbial protein as food (Ratledge, 1975; Litchfield, 1979; Chen and Peppler, 1978; Oura, 1983; Solomons, 1983).

V. POLYSACCHARIDES

Polysaccharides of natural origin, also termed gums or biopolymers, are of great industrial importance, primarily for their thickening and gelling properties. Chemical synthesis of polysaccharides is presently not practical, though chemical modification is often performed. Polysaccharides of commercial significance are usually obtained directly from algae, plants, or microorganisms, or they are cellulosic derivatives. A number of review articles on many types of sugar polymers (Lawson and Sutherland, 1978; Kang and Cottrell, 1979) or microbial polysaccharides (Gorin and Spencer, 1968; Lawson, 1976; Margaritis and Zajic, 1978; Slodki and Cadmus, 1978; Sandford, 1979; Sutherland and Ellwood, 1979; Sutherland, 1982, 1983; Baird *et al.*, 1983) have considered the structure and processing of these biopolymers in more detail. The multitude of diverse applications that emphasize the potential of biopolymers for the future has been comprehensively reviewed by Sandford and Baird (1983).

Many of the microbial polysaccharides of commercial importance are bacterial. In fact, it was xanthan gum, an important surfactant derived from *Xanthomonas campestris,* that inaugurated the commercial production of microbial polymers about 20 years ago. Annual worldwide production of xanthan gum today exceeds 20,000 metric tons (Basta, 1984) and will increase as it replaces plant gums in numerous applications. Although fungi synthesize a wide variety of sugar polymers, many with unknown functions, this discussion will focus mainly on two types of neutral biopolymers, pullulan and scleroglucan, and then briefly consider other polysaccharides.

A. Pullulan

Pullulan, an extracellular, linear, high-molecular-weight homopolysaccharide composed of maltotriose and some maltotetraose units, is produced by the dimorphic "black yeast" *Aureobasidium pullulans,* formerly named *Pullularia pullulans* (Yuen, 1974). The molecular weight of pullulan, which is controlled by fermentation conditions and choice of strain, can range from 1×10^4 to 4×10^5. Pullulan is a nontoxic, nonnutritive, water-soluble α-D-glucan that forms strong films, affects the nature and dispersiveness of foods, and resembles styrene when molded into shapes, but is more elastic. This polymer holds promise as a food coating and packaging agent, an antioxidant, and perhaps even as a dietetic food ingredient. Pullulan has been produced on an industrial scale in Japan since 1974.

Aureobasidium pullulans and other members of the genus can efficiently convert more than 70% of the provided saccharide to pullulan during fermentations (Yuen, 1974). Pullulan elaboration appears to be correlated with blastospore formation and is greater in cultures with large numbers of spores (Catley, 1980). The physiology of pullulan synthesis has been discussed (Slodki and Cadmus, 1978; Catley, 1973) and early studies have been considered elsewhere (Gorin and Spencer, 1968). Other than *A. pullulans*, only one organism has been reported to synthesize pullulan. The haploid yeast stage of *Tremella mesenterica* accumulates pullulan in culture medium during late stages of growth (Fraser and Jennings, 1970).

A number of studies have shown that pullulan production can be increased by the genetic manipulation of *A. pullulans*. Ethidium bromide mutagenesis has generated mutants that produce more cells in the yeastlike state, the phase that favors pullulan production (Kelly and Catley, 1977). Increased pullulan production was observed. Exposure of *A. pullulans* to another mutagen, *N*-nitrosomethyl urea, has likewise generated superior pullulan-producing strains (Pronina *et al.*, 1980), as has exposure of a diploid strain to ultraviolet light (Imshenetsky *et al.*, 1982). Diploid and tetraploid strains of *A. pullulans* constructed with mitotic poisons have also been shown to produce more pullulan; the enhancement of biopolymer synthesis by the polyploid strains was dependent on carbon and nitrogen source (Imshenetsky *et al.*, 1981).

B. Scleroglucan

Scleroglucan is an extracellular, capsular, homopolysaccharide secreted by certain imperfect fungi, best exemplified by those of the genus *Sclerotium*. The molecular weight and extent of branching vary with strain. The structure, properties, and physiology of this β-linked glucan are reviewed by Rodgers (1973). The unique properties of pseudoplasticity over a broad range of temperature and pH, in addition to tolerance of salts (Rodgers, 1973), led to commercial production of scleroglucan in the United States under the trade name Polytran and, after 1976, under the trade name Actigum CS in France (Sandford and Baird, 1983). The organism used in production is *Sclerotium rolfsii*, while much of the basic research has been performed with the polysaccharide from *Sclerotium glucanicum*. Species of the genera *Sclerotinia, Corticum, Stromatinia,* and *Helotium* also produce scleroglucan (Rodgers, 1973; Halleck, 1967). No genetic manipulation of these organisms for biopolymer production has been reported.

The uses of scleroglucan in the food, cosmetics, and pharmaceutical industries have been discussed by Rodgers (1973). Industrial production of scleroglucan for tertiary oil recovery has also been proposed (Compere and Griffith, 1983). A simple growth medium and the tolerance of low pH by a number of *Sclerotium* strains, in the range of 1.5–3.0, reduce contamination, allow long-term continuous fermentation, and raise the possibility that on- or near-site biopolymer fermentation can provide purified culture broth for use in biopolymer-aided enhanced oil recovery, already a $100 million a year industry (Basta, 1984).

C. Other Fungal Polysaccharides

Baker's yeast glycan, also known by the trade name BYG, is composed mainly of glucose and mannose and has projected use in the food industry (Seeley, 1977; Sandford and Baird, 1983). This neutral polysaccharide has the mouthfeel of a fat or oil and thus may replace these substances in diet foods. Further development of this biopolymer from both baker's and brewer's yeasts will be needed, however. Since knowledge of the genetics of mannan biosynthesis in *S. cerevisiae* is at such an advanced state (Spencer *et al.*, 1971; Ballou, 1976, 1982), genetic manipulation should eventually have an impact on the improvement of production strains of yeast.

Zymosan, another yeast polysaccharide, has been used clinically for its ability to inactivate complement (Ferranto *et al.*, 1965). It is likely that other fungal polysaccharides with similar properties will be found (Nickerson and Brown, 1965).

Numerous fungal polysaccharides with commercial potential have been described and reviewed by Gorin and Spencer (1968), McNeely and Kang (1973), Slodki and Cadmus (1978), Margaritis and Zajic (1978), and Kang and Cottrell (1979). More specialized reviews on yeast phosphohexans (Slodki and Boundy, 1970) and yeast extracellular mannans (Slodki *et al.*, 1972) have also appeared.

VI. LIPIDS

Approximately 70% of the world's supply of fats and oils is obtained from plant sources. At present no lipids are produced by microbiological processes. Synthetic processes based on petrochemicals also contribute to the total amount, but are generally less competitive owing to increased petroleum prices and greater overhead expenses. Projected advances in agricultural genetics and enzyme engineering will have an impact on the competition between carbon source raw material and petroleum-based feedstock. Future economic considerations will dictate the feasibility of marketing microbial fats. A dwindling global food supply or periods of hardship may be an additional influence. For example, industrial production of fats by species of *Candida* and *Fusarium* was investigated and seriously considered in Germany during both world wars as a way to ease food shortages (Prescott and Dunn, 1959; Ratledge, 1978).

The wide variety of fat-producing yeasts and molds has been thoroughly reviewed by Ratledge (1978) and advances prior to 1959 have been discussed by Lundin (1950) and Woodbine (1959). The biochemistry of yeast lipids (Stodola *et al.*, 1967; Hunter and Rose, 1971; Rattray *et al.*, 1975) and the potential for commercialization of lipids from yeasts and molds have also been considered (Ratledge, 1970; Whitworth and Ratledge, 1974; Rattray, 1984).

Much research on microbial lipid production has concerned the screening of microorganisms for species with high fat content. A number of yeasts with potential for commercialization have been identified: *Rhodotorula gracilis*, *Lipomyces lipofer*, *Lipomyces starkeyi*, *Endomycopsis vernalis*, *Cryptococcus terricolus*, and *Candida* sp. no. 107. The molds *Mortierella vinacea*, *Mucor circinelloides*, *Aspergillus ochraceus*, *A. terreus*, and *Penicillium lilacinum* are also good producers of fat, as are certain fungi of the genera *Hansenula*, *Chaetomium*, *Cladosporium*, *Malbranchea*, *Rhizopus*, and *Pythium* (Whitworth and Ratledge, 1974; Ratledge, 1978). Ideal lipid-producing fungi should be rapidly growing organisms that efficiently produce concentrated, nontoxic lipids that are easily extracted and handled in bulk, in addition to being suitable alternatives to animal or plant fats. The lipid content of such oleaginous fungi can approach 60% of the cell dry weight, and as a rule about 80% of the total lipid content is represented by triglycerides. The remainder is composed of phospholipids, sterols, sterol esters, and conjugated lipids, while the saturated and unsaturated fatty acid composition is quite variable (Whitworth and Ratledge, 1974; Ratledge, 1978).

Few fungal mutants that show enhanced lipid production have been derived, though lipid biosynthesis has been analyzed by mutant methodology (Rattray *et al.*, 1975; Wakil *et al.*, 1983). Uchio and Shiio (1972b) have examined a mutant of *Candida cloacae* isolated after mutagenesis that does not grow on long-chain dicarboxylic acids but accumulates them when grown on *n*-alkanes as the sole carbon source. Mutagenic treatment of this strain and screening for inability to assimilate *n*-alkanes produced another strain with an even greater capacity for conversion of hydrocarbons to long-chain dicarboxylic acids (Uchio and Shiio, 1972a). A comparable strain of *Torulopsis candida* has also been isolated that converts *n*-decane to decanedioic acid (Ogata *et al.*, 1973). Similar fungal mutants with an increased ability to convert *n*-alkanes to fatty acids, or even grow on cheaper, more available carbon sources, would be interesting, but no such strains have been described to date. As suggested by Ratledge (1970), a major advance in microbial fat production would be the development of mutants of yeasts, molds, or bacteria that can efficiently convert hydrocarbons to fatty acids by an extracellular process. An extracellular process with resting cells of a strain of *Debaryomyces vanriji* incubated in buffer or growth medium has been described that converts normal paraffins in petroleum distillate to dicarboxylic acids or mixtures of dicarboxylic and monocarboxylic acids (Taoka and Uchida, 1981). No genetic manipulation of this strain has been reported. Several potentially useful strains of *Candida* have been isolated that show increased tolerance to *n*-alkanes (Otsuka *et al.*, 1966) or increased permeability to *n*-alkanes and constitutivity of the hydrocarbon-oxidizing enzymes (Ratledge, 1968). These strains may have application to more efficient lipid production from petrochemicals. More recently, a mutant of *C. lipolytica* that excretes long-chain fatty acids (Miyakawa *et al.*, 1984) and a mutant of *A. oryzae* that has a higher lipid content (31%) than the parent (15%) when grown on a synthetic glucose medium have been reported (Kaur and Worgan, 1982) and should stimulate the genetic investigation of lipid production by fungi.

Though environmental factors that control lipid accumulation during fermentation have been well studied (Whitworth and Ratledge, 1974), genetic improvement of lipid-producing strains could further increase the fat content, as mentioned above, or alter the nature of the fatty acid component. Since molds generally possess more polyenoic unsaturated fatty acids than yeasts (Ratledge, 1978), mutants of these organisms with an increased content of polyunsaturated fats would be especially useful. For example, *Mucor genevensis* (Gordon *et al.*, 1971) and *Mucor ramannianus* (Sumner *et al.*, 1969) accumulate γ-linolenic acid as one-third of their total fatty acids and linolenic and linoleic acids are the sole unsaturated fatty acids in some mucors. Fungi of the genus *Pellicularia* also have a high linoleic acid content and can make this fatty acid from carbohydrates or vegetable fiber as the carbon source (Suzuki *et al.*, 1981). Genetic studies with these fungi, and also aspergilli and penicillia that have a high polyunsaturated fatty acid content (Ratledge, 1978), would be valuable and perhaps could be applied to the production of the essential fatty acids, which are sometimes collectively termed vitamin F.

The new genetic technology could radically change the economics of lipid production by microbial fermentation. Cell fusion may allow the combination of desirable qualities of several microorganisms in a strain that efficiently converts cheap biomass into lipids. Genetic engineering may also substantially improve rates of lipid production, allow the utilization of diverse carbon sources, and perhaps allow conversion of yeasts with a low lipid content into oleaginous strains (Ratledge, 1982).

VII. NUCLEOTIDES AND NUCLEIC ACID-RELATED COMPOUNDS

A. Nucleotides

Industrial production of ribonucleotides for use as food seasonings has been investigated extensively in Japan. Most of the research has focused on three 5'-nucleotides, in order of decreasing flavor activity: guanylic acid (guanosine 5'-monophosphate or GMP), inosinic acid (inosine 5'-monophosphate or IMP), and xanthylic acid (xanthosine 5'-monophosphate or XMP) (Kuninaka, 1966). The 5'-nucleotides act alone or synergistically with monosodium L-glutamate and have been described as making some foods more "meaty" and "mouth-filling," in addition to suppressing undesirable flavors and imparting greater body and smoothness (Wagner et al., 1963; Titus and Klis, 1963). A great volume of literature on the flavor nucleotides has accumulated but unfortunately is not readily accessible to many scientists because much of it is written in Japanese. Articles by Kuninaka et al. (1964) on the history and development of flavor nucleotides and by Shimazono (1964) on their application to foods provide excellent background information. Other reviews also written in English discuss more recent advances in the microbial production of nucleic acid-related compounds (Demain, 1968; Ogata, 1971, 1975; Nakao, 1979).

More than 2000 and 1000 tons of the two most potent flavor nucleotides, 5'-IMP and 5'-GMP, respectively, are produced annually in Japan by direct fermentation with mutants of *Bacillus* spp. or *Brevibacterium* spp., or by the hydrolysis of yeast RNA. The fermentation process for inosinic acid produces 5'-IMP directly or inosine, which is then chemically phosphorylated. The process for 5'-GMP involves the production of 5-amino-4-imidazolecarboxamide ribotide following by its chemical conversion to 5'-GMP or the production of guanosine followed by its chemical or enzymatic phosphorylation (Ogata, 1975; Nakao, 1979).

In 1961, industrial production of 5'-IMP or 5'-GMP in Japan originally involved the production of RNA-rich yeast cells, extraction and hydrolysis of the RNA with specific enzymes, and then purification of the 5'-mononucleotides (Nakao, 1979). The RNA content of the yeast, usually *C. utilis,* can be elevated by optimizing culture conditions or by using mutants with increased RNA levels (Akiyama et al., 1975). The nucleolytic enzymes for digestion of the RNA are obtained from improved mutants of *Penicillium citrinum* or *Staphylococcus aureus.* Adenylic acid deaminase from *A. oryzae* is used to convert 5'-adenosine monophosphate to 5'-IMP. Finally, 5'-IMP and 5'-GMP are fractionated on ion exchange resins, treated with charcoal, and precipitated with ethanol. An alternative industrial procedure has involved the alkaline hydrolysis of yeast RNA followed by chemical phosphorylation of the nucleosides (Nakao, 1979).

Though fungal mutants with altered 5'-nucleotide metabolism have not been employed for the industrial production of flavor enhancers, several have been described. Tsukada and Sugimori (1964) have reported excretion of 5'-nucleotides by a mutant of *S. rouxii* apparently caused by polynucleotide phosphorylase. Horitsu et al. (1977) have characterized a strain of *C. lipolytica* and mutants derived from it that leak 5'-GMP into the medium. Wild-type yeasts also have been reported to excrete bases, nucleosides, nucleotides, or oligonucleotides during incubation in various buffers (Ogata, 1975).

B. Nucleic Acids

RNA is manufactured from yeasts and is used mainly for the production of flavor enhancers, though some unusual commercial uses for nucleic acids involve nutritional supplements and shampoos. Yeasts are the primary source for RNA, since their content ranges from 2.5 to 15%, while their DNA content is quite low. Though genetic selection of yeast strains for increased RNA content is difficult, potassium-sensitive strains of *Candida* and notably *C. utilis*, which is a common source of industrial RNA, have been reported to contain 10–15% RNA and lowered levels of DNA (Akiyama *et al.*, 1975). Other fungi, however, can have up to 28% RNA (Nakao, 1979).

VIII. VITAMINS

An abundant journal and patent literature documents the use of microorganisms for the production of vitamins, especially those of the B group, including biotin, folic acid, pantothenic acid, pyridoxine, riboflavin, thiamine, and vitamin B_{12}. However, of all microbial processes that have been described, only two have been successfully commercialized, the production of vitamin B_{12} by certain bacteria and the production of riboflavin by the ascomycetes *Eremothecium ashbyii* and *Ashbya gossypii* (Perlman, 1978).

A number of useful reviews on the microbiological production of vitamins should be consulted for a more comprehensive discussion (Hanson, 1967; Yamada *et al.*, 1971; Perlman, 1978).

A. Riboflavin

Present-day microbiological production of riboflavin, or vitamin B_2, with *A. gossypii* competes commercially with chemical synthesis. The annual production of more than 1.25 million kilograms is used to supplement the human diet and feedstuffs (Lago and Kaplan, 1981). A number of bacteria and fungi (Goodwin, 1959; Yamada *et al.*, 1971; Yoneda, 1984) can produce riboflavin, but *E. ashbyii* and *A. gossypii* are the most important since these fungi can make up to 2.5 and 7 g per liter, respectively, in industrial fermentations (Perlman, 1979). Details regarding the processes and pathways for microbial production of this vitamin have been discussed elsewhere (Pridham, 1952; Goodwin, 1959; Demain, 1972; Perlman, 1979).

Hyperproducing mutants of *A. gossypii* used in industrial fermentations have internal concentrations of the flavin nucleotides similar to those in wild-type strains but have almost 592 times more intracellular riboflavin. Early researchers generated such flavinogenic strains with various mutagens and reported a correlation between pigmentation and overproduction (Demain, 1972). Analysis of mutants has also indicated a correlation between sporulation and flavinogenesis. The best strains sporulated well and, in one study, all mutants incapable of overproducing riboflavin did not sporulate at all (Lago and Kaplan, 1981), suggesting that common factors may play a role in regulating both processes. In early studies Pridham and Raper (1952) employed sodium dithionate to eliminate low-yielding mutants in selection experiments, recommended frequent reisolation of flavinogenic strains, and advised immediate preservation of useful strains to cope with instability inherent in riboflavin-producing mutants.

Extensive studies on fermentation processes have complemented genetic manipulation of *A. gossypii*. Production by flavinogenic mutants is enhanced by a number of compounds, among them purines and surface-active agents (Demain, 1972).

B. Fungal Production of Vitamins

In addition to the flavinogenic strains discussed, other fungi excrete or accumulate vitamins; however, these processes have not been commercialized. *Saccharomyces cerevisiae* excretes nicotinic acid under favorable growth conditions (Tseng and Phillips, 1982) and during biotin starvation (Rose, 1960). Isonazid-resistant mutants of *Saccharomyces microsporus* excrete up to 6.8-fold more pyridoxine into the medium than does the parent strain (Scherr and Rafelson, 1963). *Candida albicans* accumulates vitamin B_6, especially pyridoxal, and *Candida, Brettanomyces,* and *Pichia* yeasts acquire a high content of pantothenic acid and coenzyme A when grown on hydrocarbons (Yamada *et al.,* 1971). In addition, *A. niger* has been reported to produce vitamin C as a metabolic product (Geiger-Huber and Galli, 1945), as have numerous yeasts (Heick *et al.,* 1972). Detailed reviews on the microbial production of flavin adenine dinucleotide, nicotinamide adenine dinucleotide, coenzyme A, and pyridoxal phosphate provide more examples of vitamin and coenzyme production (Ogata, 1975; Nakao, 1979; Sakai, 1980).

Dried baker's, brewer's, and torula yeasts, along with *Kluyveromyces fragilis* and *C. lipolytica,* are often used to enrich foods, vitamin supplements, or animal feed. The vitamin content of yeasts used for this purpose can be substantially elevated by growing desirable strains on thiamine, niacin, biotin, or other vitamins. Growth on thiazole and pyrimidine can also significantly raise the thiamine content of baker's yeast. Thus, whole cells of suitable fungi can serve as an important food source, providing vitamins, protein, and other diverse nutritional supplements (Harrison, 1968; Reed and Peppler, 1973; Peppler, 1967, 1979).

C. Vitamins A and D

1. Vitamin A

Carotenoids, a class of pigments considered secondary metabolites, are often precursors of vitamin A when ingested by animals and are added to a variety of foods and feed as a coloring agent. Though such additives are synthetic, much interest has focused on *Blakeslea trispora*, a heterothallic fungus of the Mucorales group which is extremely rich in mycelial β-carotene, as a potential source of carotenoids (Ciegler, 1965; Goodwin, 1972). The yield of β-carotene is as much as 1 g per liter of medium or 20 mg per gram of dry mycelia when both sexual forms are grown together (Ninet and Renaut, 1979). *Phycomyces blakesleeanus*, a fungus amenable to genetic manipulation, also produces substantial amounts of β-carotene. Araujo *et al.* (1982) have patented a process for the formation of intersexual heterokaryons of *P. blakesleeanus*, strains with nuclei of both the + and − sexual types, from mutants that overproduce β-carotene. The heterokaryon strains produce 0.5 mg of β-carotene per gram dry weight of cells, but are unstable, segregating the components homokaryotically. Recently, superproducing strains of *P. blakesleeanus* have been isolated which synthesize 25 mg of β-carotene per gram dry weight of cells (Davies, 1973; Murillo *et al.*, 1983). Future markets and prices will dictate the feasibility of microbiological production of carotenoids as a source of coloring agents and vitamin A.

2. Vitamin D

Sterols, which are secondary metabolites but essential for cellular functions, are widely distributed among the fungi (Weete, 1973). Ergosterol, a commercially important lipid, can be extracted from dehydrated yeast, irradiated with ultraviolet light, and thereby converted to vitamin D (Harrison, 1968). Efficient ergosterol-producing yeast strains have been isolated by screening natural isolates (Dulaney *et al.*, 1954), hybridization and polyploidization (Kosikov *et al.*, 1977), and mutagenesis followed by enrichment with echinocandin and nystatin (Parks *et al.*, 1982).

IX. POLYOLS

Polyols, or polyhydroxy alcohols, are synthesized by many microorganisms, but presently none are produced commercially by fermentation. The microbial production and biological role of polyols has been reviewed (Spencer *et al.*, 1957; Nickerson and Brown, 1965; Lewis and Smith, 1967; Spencer, 1968; Spencer and Spencer, 1978), as has the multitude of food uses (Griffin and Lynch, 1972).

A. Glycerol

Glycerol, a polyol with numerous applications (Kern, 1980), is presently made by saponification of fats or by chemical synthesis from propane or propylene. During World War I, however, it was produced in Germany by yeast fermentation for use in explosives. *Saccharomyces cerevisiae* was grown in the presence of highly refined beet sugar and the fermentation was "steered" with soluble sulfites, thus promoting glycerol formation. The presence of salts and difficulties in purifying the product made the process uneconomic (Underkofler, 1954a; Spencer *et al.*, 1957). Nevertheless, this and similar microbial processes provoked interest in the fermentative production of polyols, especially unsteered processes, and led to screening programs for microbes that overproduce polyhydroxy alcohols.

Screening programs for various species and strains of fungi that synthesize large quantities of glycerol with fewer by-products have identified the osmotolerant yeasts as prolific producers of glycerol (Hajny *et al.*, 1960; Vijaikishore and Karanth, 1984). A sizable percentage of metabolized sugar can be converted to glycerol by *Saccharomyces bailii*, 22%; *Torulopsis magnoliae*, 42.5%; and *Saccharomyces rouxii*, 40–50% (Onishi, 1959; Spencer and Spencer, 1978). Other salt- or sugar-tolerant yeasts that produce significant amounts of glycerol include species of *Pichia, Hansenula, Debaryomyces, Candida,* and *Trigomopsis* (Spencer, 1968; Spencer and Spencer, 1978).

The microbial production of glycerol is usually manipulated by adjusting environmental conditions; thus, very little study has been devoted to augmenting production by genetic means. Attempts have been made to enhance glycerol formation by blocking biochemical pathways. Wright *et al.* (1957) subjected *S. cerevisiae* and *C. utilis* to ultraviolet light, isolated nutritional and morphological mutants, and screened for increased glycerol production. None of the nutritional mutants of *C. utilis* produced more glycerol, while the yield for morphological mutants ranged from 12 to 28%, compared to 16% for the parent. The best producer utilized glucose poorly. Mutants of *S. cerevisiae* were also examined; three auxotrophic mutants and three ethanol-nonutilizers, apparently petites, did not make more glycerol. However, in fermentations containing sulfite, acid-producing strains synthesized 50 to 250% more glycerol than did the parent. In addition, one of the producing mutants appeared to tolerate sulfite better in steered fermentations. More recent studies by Johansson and Sjöström (1984) have shown that an allyl alcohol-resistant mutant of *S. cerevisiae* deficient in alcohol dehydrogenase I produces six to seven times more glycerol than the wild-type strain.

Future processes may employ immobilized cells of suitable strains of *S. cerevisiae* for glycerol production. In fact, carrageenan-immobilized yeast cells in column fermentors have been shown to synthesize glycerol in the presence of sodium sulfite (Bisping and Rehm, 1982).

B. Other Polyols

Fungi synthesize numerous other polyols, which accumulate in cells or in culture fluid. These include mannitol, arabitol, erythritol, xylitol, dulcitol, ribitol, threitol, iditol, heptitol, and sorbitol (Lewis and Smith, 1967; Nickerson and Brown, 1965; Spencer and Spencer, 1978). None of these compounds is produced by industrial fermentation, though the first four compounds have potential for the future. Fungi have been shown to accumulate significant amounts of these substances in culture fluids. *Aspergillus* spp. can excrete as much as 50% of the glucose consumed as mannitol, a humectant used in foods (Smiley *et al.*, 1967), and osmotolerant yeasts produce substantial quantities of arabitol and erythritol (Spencer and Spencer, 1978). Mutants of *Pichia haplophila* have been derived that produce D-arabitol in high yields from hydrocarbons or ethanol and under conditions where it can be easily separated from cuture fluid (Fujiwara and Masuda, 1981). In addition, *Candida polymorpha* (Onishi and Suzuki, 1966) and *Pichia quercibus* (Suzuki and Onishi, 1967) can convert up to 40% of the glucose consumed to xylitol, a sweet and pleasant-tasting polyol which, along with arabitol and iditol, may have commercial potential. Xylitol, a sugar substitute already used by some diabetics and as a sweetener in chewing gum with anticariogenic properties, can be produced chemically or microbiologically (Fratzke and Reilly, 1977). However, the cost of D-xylose makes the biological process expensive (Onishi and Suzuki, 1969).

The pathways involved in sugar alcohol production have been studied and biological mechanisms for polyol accumulation have been proposed. Polyol dehydrogenases, phosphatases, and pentose cycle enzymes involved in polyol synthesis have been examined in a number of osmotolerant yeasts (Spencer and Spencer, 1978), though much remains to be learned regarding the regulation of the pathways. Brown (1976) has proposed that polyols act as osmoregulators, protect enzymes against inactivation or inhibition at low levels of water activity, and may even serve as food reserves. A better understanding of the role of polyols and their synthesis should aid in the development of selection procedures for improved polyol-producing fungal strains. In addition, the identification of genes involved in polyol synthesis that could be manipulated *in vivo* and *in vitro* could be a first step to making polyol production by fermentation commercially feasible.

X. ETHANOL

Ethanol has a multitude of applications as an industrial solvent, extractant, antifreeze, component of gasohol, precursor for chemical syntheses, and beverage. Ethanol can be made from either petrochemical feedstocks or microbiologically by fermentation, but the cost of petroleum and the increased availability of corn and agricultural biomass has changed the economics of ethanol production markedly. Thus, interest in ethyl alcohol production by fermentation has grown substantially in the past 10 years and has been matched by a considerable literature devoted to topics ranging from biological aspects to chemical engineering. Consequently, recent advances in alcohol fermentation cannot be adequately summarized in a short review of fungal primary metabolites. Symposium volumes (Moo-Young and Robinson, 1981; Scott, 1984) and specialized books should be consulted by interested readers, along with recent review articles (Bu'lock, 1979; Kosaric et al., 1980, 1983; Jones et al., 1981; Detroy and St. Julian, 1983).

Notable advances have been made in strain improvement of ethanol-producing yeast using classical genetics and, recently, cell fusion, Using *S. cerevisiae* strains tolerant of high alcohol concentration, Rose and Beavan (1982) have achieved yields of ethanol greater than 90% of the consumed carbohydrate. A number of researchers have demonstrated the importance of protoplast fusion for the derivation of ethanol-tolerant strains (Seki et al., 1983), the construction of hybrids from polyploid strains (Panchal et al., 1982a), and the derivation of new osmotolerant yeast strains (Panchal et al., 1982b). Yeast strains with the capacity to convert pentoses (Gong et al., 1983) and disaccharides (Gondé et al., 1982) to ethanol have also been selected and respiratory-deficient mutants have been shown to be more productive (Moulin et al., 1982). Much effort is being devoted to introducing genes responsible for the breakdown of cellulose, xylan, and other readily available carbon sources into ethanol-producing fungi and bacteria. Progress in defining the regulatory sequences of cloned alcohol dehydrogenase genes and related loci (Young et al., 1982) and understanding signals that control ethanol metabolism will undoubtedly lead to superior ethanol-producing fungal strains, revolutionizing the field of alcohol fermentation.

The new yeast genetics (Struhl, 1983) has had an impact on the brewing industry, which has absorbed the recent advances in genetic technology. Application of these new methods has led to superior strains for the production of beer and wine, in addition to new strains for distilling, baking, and food purposes. A number of articles survey the progress in mutation and screening, sexual recombination, rare mating, the genetics of flocculence, the transfer of killer plasmids by cytoduction or liposome fusion, polyploidization, hybridization, spheroplast fusion, transformation, and, of course, the application of recombinant DNA methods to yeast genetics in industry (Molzahn, 1976; Tubb, 1979; Ouchi et al., 1979; Tubb et al., 1981; Stewart, 1978, 1981; Johnston and Oberman, 1982; Kielland-Brandt et al., 1983; Snow, 1983; Spencer and Spencer, 1983; Russell et al., 1984).

XI. THE PROMISE OF BIOTECHNOLOGY

This review has examined the application of genetics to industrial fungal fermentation. The volume and variety of primary metabolites produced by such fermentations clearly indicate that in specific cases traditional genetic approaches have been quite successful in generating fungal strains that can efficiently produce marketable substances. The disadvantages of industrial organisms have been overcome by persistence and ingenuity, and some industrial fermentations operate at near-maximal capability. Now, industrial strain improvement programs are embracing the new genetic technology and seeking even better strains and processes and also new products.

Can strain improvement programs for primary metabolites be expected to match the recent dramatic successes in producing valuable proteins in genetically engineered yeast? A number of such proteins, many with medical applications, have been synthesized in *S. cerevisiae* using cloned mammalian or viral genes. These scarce substances will eventually be marketed as high-value specialty chemicals or will be used to manufacture important new therapeutic products. Examples of some of these substances are: human leukocyte interferon D (Hitzeman *et al.*, 1981); human interferons $\alpha 1$, $\beta 1$, and γ (Tuite *et al.*, 1982; Hitzeman *et al.*, 1981; Dobson *et al.*, 1983) which can be secreted (Hitzeman *et al.*, 1983); human epidermal growth factor, urogastrone (Urdea *et al.*, 1983); human proinsulin (Stępien *et al.*, 1983); human preproparathyroid hormone (Born *et al.*, 1983); human interleukin-2 (Anonymous, 1984a); human blood factor VIII (Anonymous, 1984b) human alpha-1-antitrypsin (R. G. Woodbury, personal communication); human β-endorphin (Bitter *et al.*, 1984); calf chymosin (Mellor *et al.*, 1983); bovine growth hormone (Upjohn Co., 1984) and hepatitis B virus surface antigen (Valenzuela *et al.*, 1982). This list will certainly grow as more genes important in medicine, agriculture, and industry are cloned and expressed in suitable hosts.

Marketable primary metabolites differ from the specialty chemicals mentioned above in a number of ways. They are, in many cases, commodity chemicals already in production. Furthermore, they are mainly lower-value substances produced in large quantities, often on a massive scale compared to specialty products. Therefore, the introduction of improved genetically engineered strains would quite likely have an impact on existing facilities and processes and probably require substantial capital investment. These "design factors" (Hamilton, 1983) are relevant to the improvement of large-scale processes for commodity chemicals and are factors for consideration by molecular biologists, engineers, and business managers. Consequently, it may be unrealistic to expect processes for industrial primary metabolites to keep pace with recombinant DNA efforts directed toward the production of specialty proteins.

Other factors merit consideration in a comparison of industrial primary metabolites and the new specialty proteins. Cloning efforts for specialty proteins attempt to isolate and express individual genes, while efforts to manipulate pathways genetically involve an understanding of many genes and their integration into cell function. Thus, strain improvement programs must contend with a number of inherent complexities not encountered by projects for the cloning of individual genes encoding valuable proteins. Primary metabolism in fungi may involve the interlocking of many pathways, the concerted action of groups of genes, genes that may be scattered throughout the genome, and loci not yet identified that may play a critical enzymatic, regulatory, or organizational role. Genetic manipulation of primary metabolism will require traditional genetic approaches used in conjunction with recombinant DNA technology and probably the latest methods of protoplast fusion with yeasts (Morgan, 1983) and mycelial fungi (Anné, 1983; Croft and Dales, 1983).

Despite the complexity of primary metabolism, recombinant DNA research can be expected to contribute to the improvement of industrial fungi in the near future. Gene manipulation may (a) allow growth on cheaper nutrients, (b) amplify important genes via multicopy plasmids, especially genes for "bottleneck" enzymes, (c) introduce constitutive host or nonhost genes for important enzymes, (d) attach improved regulatory sequences to appropriate plasmid-encoded genes, (e) inactivate specific genes with deletions, thus creating desired biochemical blocks, (f) promote secretion of a valuable metabolite, (g) create novel pathways, or (h) introduce drug-resistance factors and thereby deter contamination. (Demain, 1981; Jackson, 1982; Hamilton, 1983). Recombinant DNA efforts may also allow the cloning of fungal genes that encode enzymes able to convert readily available biological or chemical intermediates to marketable substances *in vitro*. Other approaches will surely emerge with the development of fungal biotechnology.

The historical development of the fungal fermentation industry (Miall, 1975) and the emergence of fungal biotechnology (Kristiansen and Bu'lock, 1980) signal a bright future for industrial mycology. One cannot easily predict future directions in a field that is changing so rapidly. But, undoubtedly, fungal biotechnology will improve industrial fermentation, allow the economical production of new substances, and afford opportunities unimagined today.

ACKNOWLEDGMENTS

The author thanks Linda L. Lasure and John M. Baum for valuable discussions. Thanks also to Cheryl Kobold, Kim Wilson, and Shannon Clever of the Miles Library for their help in assembling the references and to Terri Collins for helping to prepare the manuscript for publication.

REFERENCES

Abe, S., and Takayama, K. (1972). Amino acid-producing microorganisms: variety and classification. *In* "The Microbial Production of Amino Acids" (K. Yamada, S. Kinoshita, T.Tsunoda, and K. Aida, eds.) pp. 3–37. Kodansha, Tokyo.

Abou-Zeid, A.-Z. A., and Ashy, M. A. (1984). Production of citric acid: a review. *Agric. Wastes* **9**, 51–76.
Akiyama, S., Suzuki, T., Sumino, Y., Nakao, Y., and Fukuda, H. (1972). Production of citric acid from n-paraffins by flouroacetate-sensitive mutant strains of *Candida lipolytica*. *Agric. Biol. Chem.* **36**, 339–341.
Akiyama, S., Suzuki, T., Sumino, Y., Nakao, Y., and Fukuda, H. (1973a). Induction and citric acid productivity of fluoroacetate-sensitive mutant strains of *Candida lipolytica*. *Agric. Biol. Chem.* **37**, 879–884.
Akiyama, S., Suzuki, T., Sumino, Y., Nakao, Y., and Fukuda, H. (1973b). Relationship between aconitate hydratase activity and citric acid productivity in fluoroacetate-sensitive mutant strain of *Candida lipolytica*. *Agric. Biol. Chem.* **37**, 885–888.
Akiyama, S., Doi, M., Arai, Y., Nakao, Y., and Fukuda, H. (1975). U.S. Pat. 3,909,352. classification. *IN*
Anné, J. (1983). Protoplasts of filamentous fungi in genetics and metabolite production. *In* "Protoplasts 1983" (I. Potrykus, C. T. Harms, A. Hinnen, R. Hütter, P. J. King, and R. D. Shillito, eds.) pp. 167–178. Birkhaeuser, Basel.
Anonymous (1983a). Japanese make aspartame sweetener microbially via two bacteria, one fungus. *McGraw-Hill Biotechnol. Newswatch* **3**(17), 3.
Anonymous (1983b). No letup in end-of-summer biotechnology business undertakings. *McGraw-Hill Biotechnol. Newswatch* **3**(18), 3.
Anonymous (1984a). Chiron announces production of interleukin-2. *Genet. Eng. News* **4**, 16.
Anonymous (1984b). Who's first with blood factor VIII? *Genet. Technol. News* **4**, 4.
Araki, T., Yamazaki, Y., and Suzuki, N. (1957). Production of itaconic acid by *Helicobasidium mompa*. *Nogyo Gijutsu Kenkyujo Hokoku C*. No. 8, 53–58.
Araujo, F. J. M., Calderon, I. L., Diaz, I. L., and Olmedo, E. C. (1982). U.S. Pat. 4,318,987.
Atticus (1975). Citric acid forges ahead as an industrial chemical. *Chem. Age India* **26**, 49–54.
Au Tian, J., Ku, M., Xu, S., Lin, Y., Wang, S., Chen, X., Xu, G., and Chen, L. (1981). Fermentative production of succinic acid from liquid n-paraffin by *Candida rugosa*. I. Screening and induced mutation of the microorganisms. *Wei Sheng Wu Hsueh Pao* **21**, 229–233. (In Chin.)
Azevedo, J. L., and Bonatelli, R., Jr. (1982). Genetics of the over-production of organic acids. *In* "Overproduction of Microbial Products" (V. Krumphanzl, B. Sikyta, and Z. Vaněk, eds.), pp. 437–450. Academic Press, New York.
Baird, J. K., Sandford, P. A., and Cottrell, I. W. (1983). Industrial applications of some new microbial polysaccharides. *Bio/Technology* **1**, 778–783.
Ball, C. (1967). Chromosome instability related to gene suppression in *Aspergillus nidulans*. *Genet. Res.* **10**, 173–183.
Ball, C. (1980). Genetic modification of filamentous fungi. *In* "Fungal Biotechnology" (J. E. Smith, D. R. Berry, and B. Kristiansen, eds.), pp. 43–54. Academic Press, New York.
Ball, C., and Azevedo, J. L. (1976). Genetic instability in parasexual fungi. *Proc.—Int. Symp. Genet. Ind. Microorg., 2nd, Sheffield, Engl., 1974* pp. 243–251.
Ballou, C. E. (1976). Structure and biosynthesis of the mannan component of the yeast cell envelope. *Adv. Microb. Physiol.* **14**, 93–158.
Ballou, C. E. (1982). Yeast cell wall and cell surface. *In* "The Molecular Biology of the Yeast Saccharomyces: Metabolism and Gene Expression" (J. N. Strathern, E. W. Jones, and J. R. Broach, eds.), pp. 335–360. Cold Spring Harbor Lab., Cold Spring Harbor, New York.
Barnett, J. A., Payne, R. W., and Yarrow, D. (1984). "Yeasts: Characteristics and Identification." Cambridge Univ. Press, London and New York.
Basta, N. (1984). Biopolymers challenge petrochemicals. *High Technol.* **4**, 66–71.
Bennett, J. W. (1983). Secondary metabolism and differentiation in fungi. *In* "Secondary Metabo-

lism and Differentiation in Fungi'' (J. W. Bennett and A. Ciegler, eds.), pp. 1–32. Dekker, New York.
Bernhauer, K. (1929). Über die Charakterisierung der Stämme von *Aspergillus niger*. II. Mitteilung: Die Bedeutung Säurer Substrate für die Charakterisierung und Züchtung der Pilzstämme. *Biochem. Z.* **205,** 240–244.
Beuchat, L. R. (1978). "Food and Beverage Mycology," pp. 368–396. Avi, Westport, Connecticut.
Bisping, B., and Rehm, H. J. (1982). Glycerol production by immobilized cells of *Saccharomyces cerevisiae*. *Eur. J. Appl. Microbiol. Biotechnol.* **14,** 136–139.
Bitter, G. A., Chen, K. K., Banks, A. R., and Lai, P.-H. (1984). Secretion of foreign proteins from *Saccharomyces cerevisiae* directed by α-factor gene fusions. *Proc. Natl. Acad. Sci. U.S.A.* **81,** 5330–5334.
Blom, R. H., Pfeifer, V. F., Moyer, A. J., Traufler, D. H., Conway, H. F., Crocker, C. K., Farison, R. E., and Hannibal, D. V. (1952). Sodium gluconate production. *Ind. Eng. Chem.* **44,** 435–440.
Bonatelli, R., Jr., Azevedo, J. L., and Valent, G. U. (1982). Citric acid production by *Aspergillus niger* mutants. *Rev. Bras. Genet.* **3,** 483–492.
Born, W., Freeman, M., Rapoport, A., Hendy, G., Klein, R., Khorana, H. G., Rich, A., Potts, J. T., Jr., and Kronenberg, H. M. (1983). Expression of human preproparathyroid hormone in *E. coli* and yeast. *Calcif. Tissue Int.* **35,** 679.
Broquist, H. P., Stiffey, A. V., and Albrecht, A. M. (1961). Biosynthesis of lysine from α-ketoadipic acid and α-aminoadipic acid in yeast. *Appl. Microbiol.* **9,** 1–5.
Brown, A. D. (1976). Microbial water stress. *Bacteriol. Rev.* **40,** 803–846.
Buchta, K. (1983). Organic acids of minor importance. *In* "Biotechnology. Microbial Products, Biomass, and Primary Products" (H. Dellweg, ed.), Vol. 3, pp. 467–478. Verlag Chemie, Weinheim.
Bu'lock, J. D. (1961). Intermediary metabolism and antibiotic synthesis. *Adv. Appl. Microbiol.* **3,** 293–343.
Bu'lock, J. D. (1979). Industrial alcohol. *In* "Microbial Technology. Current States and Future Prospects" (A. T. Bull, C. Ellwood, and C. Ratledge, eds.), pp. 309–325. Cambridge Univ. Press, London and New York.
Calam, C. T. (1964). The selection, improvement, and preservation of micro-organisms. *Prog. Ind. Microbiol.* **5,** 1–54.
Calam, C. T. (1970). Improvement of micro-organisms by mutation, hybridization, and selection. *In* "Methods in Microbiology" (J. R. Norris and D. W. Ribbons, eds.), Vol. 3A, pp. 435–459. Academic Press, New York.
Calam, C. T., Oxford, A. E., and Raistrick, H. (1939). Studies in the biochemistry of micro-organisms. 63. Itaconic acid, a metabolic product of a strain of *Aspergillus terreus* Thom. *Biochem. J.* **33,** 1488–1495.
Catley, B. J. (1973). The rate of elaboration of the extracellular polysaccharide, pullulan, during growth of *Pullularia pullulans*. *J. Gen. Microbiol.* **78,** 33–38.
Catley, B. J. (1980). The extracellular polysaccharide, pullulan, produced by *Aureobasidium pullulans:* a relationship between elaboration rate and morphology. *J. Gen. Microbiol.* **120,** 265–268.
Champagnat, A., Vernet, C., Laine, B., and Filosa, J. (1963). Biosynthesis of protein–vitamin concentrates from petroleum. *Nature (London)* **197,** 13–14.
Chang, L. T., and Terry, C. (1973). Intergenic complementation of glucoamylase and citric acid production in two species of *Aspergillus*. *J. Bacteriol.* **25,** 890–895.
Chen, S. L., and Peppler, H. J. (1978). Single-cell proteins in food applications. *Dev. Ind. Microbiol.* **19,** 79–94.

Chibata, I., Kakimoto, T., and Kato, J. (1965). Enzymatic production of L-alanine by *Pseudomonas dacunhae*. *Appl. Microbiol.* **13**, 638–645.
Chopra, C. L., Qazi, G. N., and Gaind, C. N. (1975). Calcium gluconate by submerged fermentation. *Res. Ind.* **20**, 1–3.
Ciegler, A. (1965). Microbial carotenogenesis. *Adv. Appl. Microbiol.* **7**, 1–33.
Ciegler, A., and Raper, K. B. (1957). Application of heterokaryons of *Aspergillus* to commercial-type fermentations. *Appl. Microbiol.* **5**, 106–110.
Cochrane, V. W. (1948). Commercial production of acids by fungi. *Econ. Bot.* **2**, 145–157.
Compere, A. L., and Griffith, W. L. (1983). Scleroglucan biopolymer production, properties, and economics. *In* "Advances in Biotechnology. Fermentation Products" (C. Vezina and K. Singh, eds.), pp. 441–446. Pergamon, New York.
Croft, J. H., and Dales, R. B. G. (1983). Interspecific somatic hybridization in Aspergillus. *In* "Protoplasts 1983" (I. Potrykus, C. T. Harms, A. Hinnen, R. Hütter, P. J. King, and R. D. Shillito, eds.), pp. 178–186. Birkhaeuser, Basel.
Das, A. (1972). Strain selection in citric acid fermentation—a review. *Curr. Sci.* **41**, 593–596.
Das, A., and Roy, P. (1978). Improved production of citric acid by a diploid strain of *Aspergillus niger*. *Can. J. Microbiol.* **24**, 622–625.
Das, A., and Roy, P. (1981). Rapid selection for citric acid production. *Adv. Biotechnol., [Proc. Int. Ferment. Symp.], 6th, London, Ont., 1980* **1**, pp. 51–55.
Davies, B. H. (1973). Carotene biosynthesis in fungi. *Pure Appl. Chem.* **35**, 1–28.
Demain, A. L. (1968). Production of purine nucleotides by fermentation. *Prog. Ind. Microbiol.* **8**, 35–72.
Demain, A. L. (1971). Overproduction of microbial metabolites and enzymes due to alteration of regulation. *Adv. Biochem. Eng.* **1**, 113–142.
Demain, A. L. (1972). Riboflavin oversynthesis. *Annu. Rev. Microbiol.* **26**, 369–388.
Demain, A. L. (1981). Industrial microbiology. *Science (Washington, D.C.)* **214**, 987–995.
Denenu, E., and Demain, A. (1975). Regulation of tryptophan by DL-fluorotryptophan resistant mutant strains of a methanol utilizing yeast, *Hansenula polymorpha*, DL-1. *Yeast* **24**, 37.
Detroy, R. W., and St. Julian, G. (1983). Biomass conversion: fermentation chemicals and fuels. *CRC Crit. Rev. Microbiol.* **10**, 203–228.
Dobson, M. J., Tuite, M. F., Mellor, J., Roberts, N. A., King, R. M., Burke, D. C., Kingsman, A. J., and Kingsman, S. M. (1983). Expression in *Saccharomyces cerevisiae* of human interferon-alpha directed by the *TRP1* 5' region. *Nucleic Acids Res.* **11**, 2287–2302.
Drew, S. W., and Demain, A. L. (1977). Effect of primary metabolites on secondary metabolism. *Annu. Rev. Biochem.* **31**, 343–356.
Dulaney, E. L. (1957). Formation of extracellular lysine by *Ustilago maydis* and *Gliocladium* sp. *Can. J. Microbiol.* **3**, 467–476.
Dulaney, E. L., Stapley, E. O., and Simpf, K. (1954). Studies on ergosterol production by yeasts. *Appl. Microbiol.* **2**, 371–378.
Dulaney, E. L., Jones, C. A., and Dulaney, D. L. (1964). Amino acid accumulation, principally alanine, by auxotrophs of *Ustilago maydis*. *Dev. Ind. Microbiol.* **5**, 242–249.
Ebihara, Y., Niitsu, H., and Terui, G. (1969). Fermentative production of tryptophan from indole by *Hansenula anomala*. *J. Ferment. Technol.* **47**, 733–738.
Ehrlich, F. (1911). Über die Bildung von Fumarsäure durch Schimmelpilze. *Ber. Dtsch. Chem. Ges.* **44**, 3737–3742.
Elander, R. P. (1982). Traditional *versus* current approaches to the genetic improvement of microbial strains. *In* "Overproduction of Microbial Products" (V. Krumphanzl, B. Sikyta, and Z. Vanêk, eds.), pp. 353–369. Academic Press, New York.
Elander, R. P., and Chang, L. T. (1979). Microbial culture selection. *In* "Microbial Technology"

(H. J. Peppler and D. Perlman, eds.), 2nd Ed., Vol. 2, pp. 243–302. Academic Press, New York.

Enatsu, T., Kiyoi, M., Matsushima, H., and Terui, G. (1963). Studies on the fermentative production of tryptophan from anthranilic acid by yeasts. V. Investigation of auxotrophic haploid mutants. *J. Ferment. Technol.* **41**, 500–508.

Ericson, L.-E., and Kurz, W. G. (1962). Microbial production of amino acids. I. Synthesis of lysine and threonine by *Ustilago* species. *Biotechnol. Bioeng.* **4**, 23–36.

Esser, K. (1974). Some aspects of basic genetic research on fungi and their practical implications. *Adv. Biochem. Eng.* **3**, 69–87.

Esser, K. (1978). Concerted breeding in fungi and its biotechnological application. *Endeavour,* **1**, 143–148.

Esser, K., and Stahl, U. (1976). Cytological and genetic studies of the life cycle of *Saccharomycopsis lipolytica*. *Mol. Gen. Genet.* **146**, 101–106.

Eveleigh, D. S. (1981). The microbiological production of industrial chemicals. *In* "Industrial Microbiology and the Advent of Genetic Engineering" (P. Morrison, ed.), pp. 70–79. Freeman, San Francisco, California.

Ferenczy, L. (1981). Microbial protoplast fusion. *In* "Genetics as a Tool in Microbiology" (S. W. Glover and D. A. Hopwood, eds.), pp. 1–34. Cambridge Univ. Press, London and New York.

Ferranto, A., Giuffrida, G., and Terranova, R. (1965). Fibrinolisi e fattori della immunita' aspecifica. *Boll.—Soc. Ital. Biol. Sper.* **41**, 1488–1491.

Fincham, J. R. S., Day, P. R., and Radford, A. (1979). "Fungal Genetics." Univ. of California Press, Berkeley.

Foster, J. W. (1949). "Chemical Activities of Fungi." Academic Press, New York.

Foster, J. W. (1954). Fumaric acid. *In* "Industrial Fermentations" (L. A. Underkofler and R. J. Hickey, eds.), Vol. 1, pp. 470–487. Chem. Publ. Co., New York.

Foster, J. W., and Waksman, S. A. (1939). Fumaric acid formation associated with sexuality in a strain of Rhizopus nigricans. *Science (Washington, D.C.)* **89**, 37.

Fraser, C. G., and Jennings, H. J. (1970). Glucan from *Tremella mesenterica* NRRL-Y6158. *Can. J. Chem.* **49**, 1804–1807.

Fratzke, A. R., and Reilly, P. J. (1977). Uses and metabolic effects of xylitol—Part I. *Proc. Biochem.* **12**, 27–30.

Fujiwara, A., and Masuda, S. (1981). U.S. Pat. 4,271,268.

Fukuda, H., Suzuki, T., Akiyama, S., and Sumino, Y. (1974). U.S. Pat. 3,799,840.

Fukumura, T. (1977). Conversion of D- and DL-α-amino-ε-caprolactam into L-lysine using both yeast cells and bacterial cells. *Agric. Biol. Chem.* **41**, 1327–1330.

Furukawa, T., Nakahara, T., and Yamada, K. (1970). Studies on utilization of hydrocarbons by microorganisms. Part XIX. Influence of several factors in fumaric acid production from *n*-paraffins by *Candida hydrocarbofumarica*. *Agric. Biol. Chem.* **34**, 1402–1406.

Furukawa, T., de Miranda, L. R., and Matsuyoshi, T. (1978). Fermentative production of fumaric acid from *n*-paraffins by *Candida blankii*. *J. Ferment. Technol.* **56**, 546–549.

Gaillardin, C. M., Charoy, V., and Heslot, H. (1973). A study of copulation, sporulation, and meiotic segregation in *Candida lipolytica*. *Arch. Microbiol.* **92**, 69–83.

Gaillardin, C. M., Sylvestre, G., and Heslot, H. (1975). Studies on an unstable phenotype induced by UV irradiation. The lysine excreting (lex$^-$) phenotype of the yeast *Saccharomycopsis lipolytica*. *Arch. Microbiol.* **104**, 89–94.

Gardner, J. F., James, L. V., and Rubbo, S. D. (1956). Production of citric acid by mutants of *Aspergillus niger*. *J. Gen. Microbiol.* **14**, 228–237.

Geiger-Huber, M., and Galli, H. (1945). Über den Nachweis der l'Ascorbinsäure als Stoffwechselprodukt von *Aspergillus niger*. *Helv. Chim. Acta* **28**, 248–250.

Golubtsova, V. M., Shcherbakova, E. Y., Runkovshaya, L. Y., and Ermakova, V. P. (1979). Change in the ratio of citric and oxalic acids in *Aspergillus niger* under the influence of mutagenic factors. *Mikrobiologiya* **48**, 1060–1065.

Gondé, P., Blondin, B., Ratomahenina, R., Arnaud, A., and Galzy, P. (1982). Selection of yeast strains for cellobiose alcoholic fermentations. *J. Ferment. Technol.* **60**, 579–584.

Gong, C.-S., Claypool, T. A., McCracken, L. D., Mann, C. M., Ueng, P. P., and Tsao, G. T. (1983). Conversion of pentoses by yeasts. *Biotechnol. Bioeng.* **25**, 85–102.

Goodwin, T. W. (1959). Production and biosynthesis of riboflavin in micro-organisms. *Prog. Ind. Microbiol.*, **1**, 137–177.

Goodwin, T. W. (1972). Carotenoids in fungi and non-photosynthetic bacteria. *Prog. Ind. Microbiol.* **11**, 29–88.

Gordon, P. A., Stewart, P. R., and Clark-Walker, G. D. (1971). Fatty acid and sterol composition of *Mucor genevensis* in relation to dimorphism and anaerobic growth. *J. Bacteriol.* **107**, 114–120.

Gorin, P. A. J., and Spencer, J. F. T. (1968). Structural chemistry of fungal polysaccharides. *Adv. Carbohydr. Chem.* **23**, 367–417.

Griffin, W. C., and Lynch, M. J. (1972). Polyhydric alcohols. In "CRC Handbook of Food Additives" (T. E. Furia, ed.), 2nd Ed., pp. 431–455. CRC Press, Cleveland, Ohio.

Haidaris, C. G., and Bhattacharjee, J. K. (1978). Lysine production by thialysine-resistant mutants of *Saccharomyces cerevisiae*. *J. Ferment. Technol.* **56**, 189–192.

Hajny, G. J., Hendershot, W. F., and Peterson, W. H. (1960). Factors affecting glycerol production by a newly isolated osmophilic yeast. *Appl. Microbiol.* **8**, 5–11.

Halleck, F. E. (1967). U.S. Pat. 3,301,848.

Hamilton, B. K. (1983). Design factors for construction of competitive production strains and manufacturing processes. In "Organic Chemicals from Biomass" (D. L. Wise, ed.), pp. 109–143. Benjamin/Cummings, Menlo Park, California.

Hanson, A. (1967). Microbial production of pigments and vitamins. In "Microbial Technology" (H. J. Peppler, ed.), pp. 222–250. Reinhold, New York.

Harrison, J. S. (1968). Yeast as a source of biochemicals. *Proc. Biochem.* **3**, 59–62.

Haskins, R. H., Thorn, J. A., and Boothroyd, B. (1955). Biochemistry of the Ustilaginales. XI. Metabolic products of *Ustilago zeae* in submerged culture. *Can. J. Microbiol.* **1**, 749–756.

Heick, H. M. C., Graff, G. L. A., and Humpers, J. E. C. (1972). The occurrence of ascorbic acid among the yeasts. *Can. J. Microbiol.* **18**, 597–600.

Hirose, Y., and Okada, H. (1979). Microbial production of amino acids. In "Microbial Technology, Vol. 1, Microbial Processes" (H. J. Peppler and D. Perlman, eds.), 2nd Ed., pp. 211–240. Academic Press, New York.

Hitzeman, R. A., Hagie, F. E., Levine, H. L., Goeddel, D. V., Ammerer, G., and Hall, B. D. (1981). Expression of a human gene for interferon in yeast. *Nature (London)* **293**, 717–722.

Hitzeman, R. A., Leung, D. W., Perry, L. J., Kohr, W. J., Levine, H. L., and Goeddel, D. V. (1983). Secretion of human interferons by yeast. *Science (Washington, D.C.)* **219**, 620–625.

Hollaender, A., Raper, K. B., and Coghill, R. D. (1945). The production and characterization of ultraviolet-induced mutations of *Aspergillus terreus*. I. Production of the mutations. *Am. J. Bot.* **32**, 110–165.

Hopwood, D. A. (1970). The isolation of mutants. In "Methods in Microbiology" (J. R. Norris and D. W. Ribbons, eds.), Vol. 3A, pp. 363–433. Academic Press, New York.

Horitsu, H., Satake, T., Ogawa, K., and Mikio, T. (1977). Leakage of 5′-guanosine monophosphate from *Candida lipolytica* (IFO, 0746) and its mutants. *Agric. Biol. Chem.* **41**, 1667–1672.

Hotta, K., and Takao, S. (1973). Conversion of fumaric acid fermentation to aspartic acid fermentation by the association of *Rhizopus* and bacteria. II. Production of aspartic acid by the combination of *Rhizopus* and *Proteus vulgaris*. *J. Ferment. Technol.* **51**, 12–18.

Hütter, R. (1973). Regulation of tryptophan biosynthetic enzymes in fungi. *In* "Genetics of Industrial Microorganisms. Actinomycetes and Fungi" (Z. Vaněk, Z. Hošťálek, and J. Cudlín, eds.), Vol. 2, pp. 109–124. Elsevier, Amsterdam.
Hunter, K., and Rose, A. H. (1971). Yeast lipids and membranes. *In* "The Yeasts" (A. H. Rose and J. S. Harrison, eds.), Vol. 2, pp. 211–270. Academic Press, New York.
Ikeda, Y. (1961). Potential application of parasexuality in breeding fungi. *Recent Adv. Bot.* pp. 383–386.
Ilczuk, Z. (1968). Genetics of citric acid producing strains of *Aspergillus niger*. I. Citric acid synthesis by morphological mutants induced by UV. *Acta Microbiol. Pol.* **17,** 331–336.
Ilczuk, Z. (1971a). Genetics of citric acid producing strains of *Aspergillus niger*. III. Citric acid synthesis by forced heterokaryons between auxotrophic mutants of *A. niger*. *Nahrung* **15,** 251–262.
Ilczuk, Z. (1971b). Genetics of citric acid producing strains of *Aspergillus niger*. IV. Citric acid synthesis by heterozygous diploids of *A. niger*. *Nahrung* **15,** 381–388.
Imshenetsky, A. A., Kondratyeva, T. F., and Smutko, A. N. (1981). Influence of carbon and nitrogen sources on pullulan biosynthesis by polyploid strains of *Pullularia pullulans*. *Mikrobiologiya* **50,** 102–105.
Imshenetsky, A. A., Kondratyeva, T. F., and Smutko, A. N. (1982). Spontaneous and UV-induced variability of the pullulan-synthesizing activity of *Pullularia (Aureobasidium) pullulans* strains having various ploidy. *Mikrobiologiya* **51,** 964–967.
Irwin, W. E., Kockwood, L. B., and Zienty, M. F. (1967). Malic acid. *In* "Kirk–Othmer Encyclopedia of Chemical Technology" (A. Standen, H. F. Mark, J. J. McKetta, Jr., and D. F. Othmer, eds.), Vol. 12, pp. 837–849. Wiley, New York.
Jackson, D. A. (1982). Molecular genetics and microbial fermentations. *In* "Trends in the Biology of Fermentations for Fuels and Chemicals" (A. Hollaender, P. Rabson, P. Rogers, A. San Pietro, R. Valentine, and R. Wolfe, eds.), pp. 187–200. Plenum, New York.
James, L. V., Rubbo, S. D., and Gardner, J. F. (1956). Isolation of high-yielding mutants of *Aspergillus niger* by a paper culture selection technique. *J. Gen. Microbiol.* **14,** 223–227.
Johansson, M., and Sjöström, J. E. (1984). Enhanced production of glycerol in an alcohol dehydrogenase (ADH I) deficient mutant of *Saccharomyces cerevisiae*. *Biotechnol. Lett.* **6,** 49–54.
Johnson, M. J. (1954). The citric acid fermentation. *In* "Industrial Fermentations" (L. A. Underkofler and R. J. Hickey, eds.), pp. 420–445. Chem. Publ. Co., New York.
Johnston, J. R. (1975). Strain improvement and strain stability in filamentous fungi. *In* "The Filamentous Fungi. Industrial Mycology" (J. E. Smith and D. R. Berry, eds.), Vol. 1, pp. 59–78. Arnold, London.
Johnston, J. R., and Oberman, H. (1982). Yeast genetics in industry. *Prog. Ind. Microbiol.* **15,** 151–205.
Jones, R. P., Pamment, N., and Greenfield, P. F. (1981). Alcohol fermentation by yeasts—the effect of environmental and other variables. *Proc. Biochem.* **16,** 42–49.
Kang, K. S., and Cottrell, I. W. (1979). Polysaccharides. *In* "Microbial Technology, Vol. 1, Microbial Processes" (H. J. Peppler and D. Perlman, eds.), 2nd Ed., pp. 417–481. Academic Press, New York.
Kapoor, K. K., Chaudhary, K., and Tauro, P. (1982). Citric acid. *In* "Prescott and Dunn's Industrial Microbiology" (G. Reed, ed.), 4th Ed. pp. 709–747. Avi, Westport, Connecticut.
Kaur, P., and Worgan, J. T. (1982). Lipid production by *Aspergillus oryzae* from starch substrates. *Eur. J. Appl. Microbiol. Biotechnol.* **16,** 126–130.
Kelly, P. J., and Catley, B. J. (1977). The effect of ethidium bromide mutagenesis on dimorphism, extracellular metabolism and cytochrome levels in *Aureobasidium pullulans*. *J. Gen. Microbiol.* **102,** 249–254.

Kern, J. C. (1980). Glycerol. In "Kirk–Othmer Encyclopedia of Chemical Technology" (M. Grayson and D. Ekroth, eds.), 3rd Ed., Vol. 11, pp. 921–932. Wiley, New York.
Kielland-Brandt, M. C., Nilsson-Tillgren, T., Petersen, J. G. L., Holmberg, S., and Gjermansen, C. (1983). Approaches to the genetic analysis and breeding of brewer's yeast. In "Yeast Genetics. Fundamental and Applied Aspects" (J. F. T. Spencer, D. M. Spencer, and A. R. W. Smith, eds.), pp. 421–437. Springer-Verlag, Berlin and New York.
Kinoshita, H. (1929). Study of the biosynthesis of teichoic acid and mannitol in a new filamentous fungus. Nippon Kagaku Kaishi (1921–47) **50**, 583–593.
Kinoshita, S., and Nakayama, K. (1978). Amino acids. In "Economic Microbiology. Primary Products of Metabolism" (A. H. Rose, ed.), Vol. 2, pp. 209–261. Academic Press, New York.
Kinoshita, S., and Tanaka, K. (1972). Glutamic acid. In "The Microbial Production of Amino Acids" (K. Yamada, S. Kinoshita, T. Tsunoda, and K. Aida, eds.), pp. 263–324. Kodansha, Tokyo.
Kinoshita, S., Udaka, S., and Shimono, M. (1957). Studies on amino acid fermentation. Part I. Production of L-glutamic acid by various microorganisms. J. Gen. Appl. Microbiol. **3**, 193–205.
Kobayashi, T. (1967). Itaconic acid fermentation. Proc. Biochem. **2**, 61–65.
Kosaric, N., Ng, D. C. M., Russell, I., and Stewart, G. S. (1980). Ethanol production by fermentation: an alternative liquid fuel. Adv. Appl. Microbiol. **26**, 147–227.
Kosaric, N., Wieczorek, A., Cosentino, G. P., Magee, R. J., and Prenosil, J. E. (1983). Ethanol fermentation. In "Biotechnology. Microbial Products, Biomass, and Primary Products" (H. Dellweg, ed.), Vol. 3, pp. 467–478. Verlag Chemie, Weinheim.
Kosikov, K. K., Lyapunova, T. S., Raevskaya, O. G., Semiknatova, N. M., Kochkina, I. B., and Meisel, M. N. (1977). Ergosterol synthesis in yeast hybrids and strains of the genus Saccharomyces of different ploidy. Mikrobiologiya **46**, 86–91.
Krasil'nikov, N. A., Aseeva, I. V., Bab'eva, I. P., Kaptereva, Y. V., Shirokov, O. G., and Korshunov, I. S. (1962). Biosynthesis of amino acids by soil microorganisms. Dokl. Biol. Sci. (English Transl.) **141**, 1046–1048.
Kresling, E. K., and Shtern, E. A. (1935). Effects of Ra and ultraviolet light on growth and citric acid formation in Aspergillus niger cultures. Proc. Inst. Sci. Res. Food Ind. **3**, 5–24.
Kristiansen, B., and Bu'lock, J. D. (1980). Developments in industrial fungal biotechnology. In "Fungal Biotechnology" (V. E. Smith, D. R. Berry, and B. Kristiansen, eds.), pp. 203–223. Academic Press, New York.
Kubicek, C. P., and Röhr, M. (1982). Novel trends in physiology and technology of citric acid production. In "Overproduction of Microbial Products" (V. Krumphanzl, B. Sikyta, and Z. Vaněk, eds.), pp. 253–262. Academic Press, New York.
Kundu, P. N., and Das, A. (1982). Calcium gluconate production by a nonconventional fermentation method. Biotechnol. Lett. **4**, 365–368.
Kuninaka, A. (1966). Recent studies of 5'-nucleotides as new flavor enhancers. In "Flavor Chemistry" (I. Hornstein, ed.), Advances in Chemistry Series, No. 56, pp. 261–274. Am. Chem. Soc., Washington, D.C.
Kuninaka, A., Kibi, M., and Sakaguchi, K. (1964). History and development of flavor nucleotides. Food Technol. **18**, 29–35.
Kyowa Hakko Kogyo Co., Ltd. (1982). U.S. Pat. 4,322,498.
Lago, B. D., and Kaplan, L. (1981). Vitamin fermentations: B2 and B12. In "Advances in Biotechnology. Fermentation Products" (C. Vezina and K. Singh, eds.), Vol. 3, pp. 241–257. Pergamon, New York.
Lawson, C. J. (1976). Microbial polysaccharides. Chem. Ind. (London) Mar., 258–261.
Lawson, C. J., and Sutherland, I. W. (1978). Polysaccharides. In "Economic Microbiology.

15. Fungal Primary Metabolism

Primary Products of Metabolism'' (A. H. Rose, ed.), Vol. 2, pp. 327–392. Academic Press, New York.

Lemke, P. A., Saksena, K. N., and Nash, C. H. (1976). Viruses of industrial fungi. *Proc.—Int. Symp. Genet. Ind. Microorg., 2nd Sheffield, Engl., 1974* pp. 323–337.

Lewis, D. H., and Smith, D. C. (1967). Sugar alcohols (polyols) in fungi and green plants. *New Phytol.* **66,** 143–184.

Lhoas, P. (1967). Genetic analysis by means of the parasexual cycle in *Aspergillus niger*. *Genet. Res.* **10,** 45–61.

Litchfield, J. H. (1979). Production of single cell protein for use in food or feed. *In* ''Microbial Technology, Vol. 1, Microbial Processes'' (H. J. Peppler and D. Perlman, eds.), 2nd Ed., pp. 93–155. Academic Press, New York.

Litchfield, J. H. (1983). Single-cell proteins. *Science (Washington, D.C.)* **219,** 740–746.

Lockwood, L. B. (1954). Itaconic acid. *In* ''Industrial Fermentations'' (L. A. Underkofler, ed.), Vol. 1, pp. 488–497. Chem. Publ. Co., New York.

Lockwood, L. B. (1979). Production of organic acids by fermentation. *In* ''Microbial Technology, Vol. 1, Microbial Processes'' (H. J. Peppler and D. Perlman, eds.), 2nd Ed., pp. 353–387. Academic Press, New York.

Lockwood, L. B., and Schweiger, L. B. (1967). Citric and itaconic acid fermentations. *In* ''Microbial Technology'' (H. J. Peppler, ed.), pp. 183–199. Reinhold, New York.

Lundin, H. (1950). Fat synthesis by micro-organisms and its possible applications in industry. *J. Inst. Brew.* **56,** 17–28.

McNeely, W. M., and Kang, K. S. (1973). Xanthan and some other biosynthetic gums. *In* ''Industrial Gums'' (R. L. Whistler and J. N. BeMiller, eds.), 2nd Ed., pp. 473–497. Academic Press, New York.

Malik, V. S. (1979). Genetics of applied microbiology. *Adv. Genet.* **20,** 37–126.

Malin, B., and Westhead, J. (1959). Production of L-tryptophan in submerged culture. *J. Biochem. Microbiol. Technol. Eng.* **1,** 49–57.

Marconi, W. (1974). Enzymes in the chemical and pharmaceutical industry. *In* ''Industrial Aspects of Biochemistry'' (B. Spencer, ed.), Vol. 30, Part I, pp. 139–181. Am. Elsevier, New York.

Margaritis, A., and Zajic, J. E. (1978). Mixing, mass transfer, and scale-up of polysaccharide fermentations. *Biotechnol. Bioeng.* **20,** 939–1001.

Martin, S. M. (1963). Production of organic acids by moulds. *In* ''Biochemistry of Industrial Microorganisms'' (C. Rainbow and A. H. Rose, eds.), pp. 415–451. Academic Press, New York.

Mellor, J., Dobson, M. J., Roberts, N. A., Tuite, M. F., Emtage, J. S., White, S., Lowe, P. A., Patel, T. Kingsman, A. J., and Kingsman, P. M. (1983). Efficient synthesis of enzymatically active calf chymosin in *Saccharomyces cerevisiae*. *Gene.* **24,** 1–14.

Metz, B., and Kossen, N. W. F. (1977). The growth of molds in the form of pellets—a literature review. *Biotechnol. Bioeng.* **19,** 781–799.

Miall, L. M. (1975). Historical development of the fungal fermentation industry. *In* ''The Filamentous Fungi. Industrial Mycology'' (J. E. Smith and D. R. Berry, eds.), Vol. 1, pp. 104–121. Arnold, London.

Miall, L. M. (1978). Organic acids. *In* ''Primary Products of Metabolism'' (A. H. Rose, ed.), Economic Microbiology, Vol. 2, pp. 47–119. Academic Press, New York.

Miles Laboratories, Inc. (1951). British Pat. 653,808.

Miller, M. W. (1961). ''The Pfizer Handbook of Microbial Metabolites.'' McGraw-Hill, New York.

Miyakawa, T., Nakajima, H., Hamada, K., Tsuchiya, E., Kamiryo, T., and Fukui, S. (1984). Isolation and characterization of a mutant of *Candida lipolytica* which excretes long-chain fatty acids. *Agric. Biol. Chem.* **48,** 499–504.

Molliard, M. (1922). Sur une nouvelle fermentation acide produite par le *Sterigmatocystis nigra*. *C. R. Hebd. Seances Acad. Sci.* **174,** 881–883.

Molzahn, S. W. (1976). A new approach to the application of genetics to brewing yeast. *J. Am. Soc. Brew. Chem.* **35,** 54–58.

Moo-Young, M., and Robinson, C. W., eds. (1981). "Advances in Biotechnology. Fuels, Chemicals, Foods and Waste Treatment," Vol. 2. Pergamon, New York.

Morgan, A. J. (1983). Yeast strain improvement by protoplast fusion and transformation. *In* "Protoplasts 1983" (I. Potrykus, C. T. Harmo, A. Hinnen, R. Hütter, P. J. King, and R. D. Shillito, eds.), pp. 155–166. Birkhaeuser, Basel.

Moriguchi, M. (1982). Fermentative production of pyruvic acid from citrus peel extract by *Debaryomyces condertii*. *Agric. Biol. Chem.* **46,** 955–961.

Morzycka, E., Sawnor-Korszynska, D., Paszewski, A., Grabski, J., and Raczynska-Bojanowska, K. (1976). Methionine overproduction in *Saccharomycopsis lipolytica*. *Appl. Environ. Microbiol.* **32,** 125–130.

Moulin, G., Boze, H., and Galzy, P. (1982). Utilization of a respiratory-deficient mutant for alcohol production. *J. Ferment. Technol.* **60,** 25–29.

Murillo, F. J., Calderon, I. L., Lopez-Diaz, I., and Cerda-Olmeda, E. (1983). Carotene superproducing strains of *Phycomyces*. *Appl. Environ. Microbiol.* **36,** 639–642.

Nakao, Y. (1979). Microbial production of nucleosides and nucleotides. *In* "Microbial Technology, Vol. 1, Microbial Processes" (H. J. Peppler and D. Perlman, eds.), 2nd Ed., pp. 311–354. Academic Press, New York.

Nga, B. H., Teo, S.-P., and Lim, G. (1975). The occurrence in nature of a diploid strain of *Aspergillus niger*. *J. Gen. Microbiol.* **88,** 364–366.

Nickerson, W. J., and Brown, R. G. (1965). Uses and products of yeasts and yeastlike fungi. *Adv. Appl. Microbiol.* **7,** 225–272.

Ninet, L., and Renaut, J. (1979). Carotenoids. *In* "Microbial Technology, Vol. 1, Microbial Processes" (H. J. Peppler and D. Perlman, eds.), 2nd Ed., pp. 529–544. Academic Press, New York.

Ogata, K. (1971). Industrial production of nucleotides, nucleosides, and related substances. *In* "Biochemical and Industrial Aspects of Fermentation" (K. Sakaguchi, T. Uemura, and S. Kinoshita, eds.), pp. 37–59. Kodansha, Tokyo.

Ogata, K. (1975). The microbial production of nucleic acid-related compounds. *Adv. Appl. Microbiol.* **19,** 209–247.

Ogata, K., Kaneyuki, H., Kato, N., Tani, Y., and Yamada, H. (1973). Accumulation of decanedioic acid from *n*-decane by *Torulopsis candida* no. 99. *Agric. Biol. Chem.* **51,** 227–235.

Ogrydziak, D., Bassel, J., Contopoulou, R., and Mortimer, R. (1978). Development of genetic techniques and the genetic map of the yeast *Saccharomycopsis lipolytica*. *Mol. Gen. Genet.* **163,** 229–239.

Okanishi, M., and Gregory, K. F. (1970). Isolation of mutants of *Candida tropicalis* with increased methionine content. *Can. J. Microbiol.* **16,** 1139–1143.

Okumura, S., Otsuka, S., Yamanoi, A., Yoshinaga, F., Honda, T., Kubota, K., and Tsuchida, T. (1972). U.S. Pat. 3,660,235.

Onishi, H. (1959). Studies on osmophilic yeasts. Part VI. Glycerol production by the salt-tolerant yeasts in medium with high concentration of sodium chloride. *Bull. Agric. Chem. Soc. Jpn.* **23,** 359–363.

Onishi, H., and Suzuki, T. (1966). The production of xylitol, L-arabitol, and ribitol by yeasts. *Agric. Biol. Chem.* **30,** 1139–1144.

Onishi, H., and Suzuki, T. (1969). Microbial production of xylitol from glucose. *Appl. Microbiol.* **18,** 1031–1035.

Otsuka, S.-I., Ishii, R., and Katsuya, N. (1966). Utilization of hydrocarbons as carbon sources in production of yeast cells. *J. Gen. Appl. Microbiol.* **12,** 1–11.

Ouchi, K., Wickner, R. B., Toh-e, A., and Akiyama, H. (1979). The breeding of killer yeasts for *sake* brewing by cytoduction. *J. Ferment. Technol.* **57,** 483–487.
Oura, E. (1983). Biomass from carbohydrates. In "Biotechnology. Microbial Products, Biomass, and Primary Products" (H. Dellweg, ed.), Vol. 3, pp. 3–41. Verlag Chemie, Weinheim.
Panchal, C. J., Harbison, A., Russell, I., and Stewart, G. G. (1982a). Ethanol production by modified strains of *Saccharomyces*. *Biotechnol. Lett.* **4,** 33–38.
Panchal, C. J., Peacock, L., and Stewart, G. G. (1982b). Increased osmotolerance of genetically modified ethanol producing strains of *Saccharomyces* sp. *Biotechnol. Lett.* **4,** 639–644.
Parks, L. W., Rodriguez, R. J., and McCammon, M. T. (1982). Sterols of yeast: a model for biotechnology in the production of fats and oils. *J. Am. Oil Chem. Soc.* **59,** 294A–295A.
Peberdy, J. F. (1979). Fungal protoplasts: isolation, reversion, and fusion. *Annu. Rev. Microbiol.* **33,** 21–39.
Peberdy, J. F. (1980). Protoplast fusion—a tool for genetic manipulation and breeding of industrial microorganisms. *Enzyme Microb. Technol.* **2,** 23–29.
Peppler, H. J. (1967). Yeast technology. In "Microbial Technology" (H. J. Peppler, ed.), pp. 145–171. Van Nostrand-Reinhold, Princeton, New Jersey.
Peppler, H. J. (1979). Production of yeasts and yeast products. In "Microbial Technology, Vol. 1, Microbial Processes" (H. J. Peppler and D. Perlman, eds.), 2nd Ed., pp. 157–185. Academic Press, New York.
Perlman, D. (1949). Mycological production of citric acid—the submerged culture method. *Econ. Bot.* **3,** 360–374.
Perlman, D. (1978). Vitamins. In "Primary Products of Metabolism" (A. H. Rose, ed.), Economic Microbiology, Vol. 2, pp. 303–326. Academic Press, New York.
Perlman, D. (1979). Microbial process for riboflavin production. In "Microbial Technology, Vol. 1, Microbial Processes" (H. J. Peppler and D. Perlman, eds.), 2nd Ed., pp. 521–527. Academic Press, New York.
Perlman, D., and Sih, C. J. (1960). Fungal synthesis of citric, fumaric, and itaconic acids. *Prog. Ind. Microbiol.* **2,** 169–194.
Pfeiffer, V. F., Vojnovich, C., and Heger, E. N. (1952). Itaconic acid by fermentation with *Aspergillus terreus*. *Ind. Eng. Chem.* **44,** 2975–2980.
Pfeiffer, V. F., Nelson, G. E. N., Vojnovich, C., and Lockwood, L. B., (1953). U.S. Pat. 2,657,173.
Pfizer, Inc. (1970). Br. Pat. 1,203,006.
Pfizer, Inc. (1973). U.S. Pat. 3,717,549.
Pfizer, Inc. (1974). Br. Pat. 1,369,295.
Pisano, M. A. (1966). The fermentative production of alanine. *Dev. Ind. Microbiol.* **7,** 35–40.
Pisano, M. A., Mihalik, J. A., and Catalano, G. R. (1964). Gelatinase activity by marine fungi. *Appl. Microbiol.* **12,** 470–474.
Pontecorvo, G., Roper, J. A., and Forbes, E. (1953a). Genetic recombination without sexual reproduction in *Aspergillus niger*. *J. Gen. Microbiol.* **8,** 198–210.
Pontecorvo, G., Roper, J. A., Hemmons, L. A., MacDonald, K. D., and Bufton, A. W. J. (1953b). The genetics of *Aspergillus nidulans*. *Adv. Genet.* **5,** 141–238.
Prescott, S. C., and Dunn, C. G. (1959). The itaconic acid fermentation. In "Industrial Microbiology," 3rd Ed., pp. 598–608. McGraw-Hill, New York.
Pridham, F. J., and Raper, K. B. (1952). Studies on variation and mutation in *Ashbya gossypii*. *Mycologia* **44,** 452–469.
Pridham, T. G. (1952). Microbial synthesis of riboflavin. *Econ. Bot.* **6,** 185–205.
Pronina, M. I., Elinov, N. P., and Dranishnikov, A. N. (1980). *N*-Nitrosomethyl urea-induced variability in *Aureobasidium pullulans* (De Bary) Arnaud. *Mikrobiologiya* **49,** 93–97.

Raper, K. B., Coghill, R. D., and Hollaender, A. (1945). The production and characterization of ultraviolet induced mutations in *Aspergillus terreus*. II. Cultural and morphological characteristics of the mutations. *Am. J. Bot.* **32,** 165–176.
Ratledge, C. (1968). Production of fatty acids and lipid by a *Candida* sp. growing on a fraction of *n*-alkanes predominantly in tridecane. *Biotechnol. Bioeng.* **10,** 511–533.
Ratledge, C. (1970). Microbial conversions of *n*-alkanes to fatty acids: a new attempt to obtain economical microbial fats and fatty acids. *Chem. Ind. (London)* June, 843–854.
Ratledge, C. (1975). The economics of single cell protein production. Substrates and processes. *Chem. Ind. (London)* Nov., 918–920.
Ratledge, C. (1978). Lipids and fatty acids. *In* "Primary Products of Metabolism" (A. H. Rose, ed.), Economic Microbiology, Vol. 2, pp. 263–302. Academic Press, New York.
Ratledge, C. (1982). Single cell oil. *Enzyme Microb. Technol.* **4,** 58–60.
Rattray, J. B. M. (1984). Biotechnology of the fats and oils industry—an overview. *J. Am. Oil Chem. Soc.* **61,** 1701–1712.
Rattray, J. B. M., Schibeci, A., and Kidby, D. K. (1975). Lipids of yeasts. *Bacteriol. Rev.* **39,** 197–231.
Reed, G., and Peppler, H. J. (1973). "Yeast Technology," pp. 328–354. Avi, Westport, Connecticut.
Richards, M., and Haskins, R. H. (1957). Extracellular lysine production by fungi. *Can. J. Microbiol.* **3,** 543–546.
Rodgers, N. E. (1973). Scleroglucan. *In* "Industrial Gums: Polysaccharides and Their Derivatives" (R. L. Whistler and J. N. BeMiller, eds.), pp. 499–511. Academic Press, New York.
Röhr, M., Stadler, P. J., Salzbrunn, W. O. J., and Kubichek, C. P. (1979). An improved method for characterization of citrate production by conidia of *Aspergillus niger*. *Biotechnol. Lett.* **7,** 281–286.
Röhr, M., Kubichek, C. P., and Kominek, J. (1983a). Citric Acid. *In* "Biotechnology. Microbial Products, Biomass and Primary Products" (H. Dellweg, ed.), Vol. 3, pp. 419–454. Verlag Chemie, Weinheim.
Röhr, M., Kubicek, C. P., and Kominek, J. (1983b). Gluconic acid. *In* "Biotechnology. Microbial Products, Biomass, and Primary Products" (H. Dellweg, ed.), Vol. 3, pp. 456–465. Verlag Chemie, Weinheim.
Roper, J. A. (1973). Mitotic recombination and mitotic nonconformity in fungi. *In* "Genetics of Industrial Microorganisms. Actinomycetes and Fungi" (Z. Vaněk, Z. Hoštálek, and J. Cudlín, eds.), Vol. 2, pp. 81–88. Elsevier, Amsterdam.
Rose, A. H. (1960). Excretion of nicotinic acid and nicotinic acid dinucleotide by biotin-deficient yeast. *Nature (London)* **186,** 139–140.
Rose, A. H., and Beavan, M. J. (1982). End-product tolerance and ethanol. *In* "Trends in the Biology of Fermentations for Fuel and Chemicals" (A. Hollaender, R. Rabson, P. Rogers, A. San Pietro, R. Valentine, and R. Wolfe, eds.), pp. 513–531. Plenum, New York and London.
Rowlands, R. T. (1984). Industrial strain improvement: mutagenesis and random screening procedures. *Enzyme Microb. Technol.* **6,** 3–11.
Russell, I., Jones, R. M., Panchal, C. J., Weston, B. J., and Stewart, G. G. (1984). Liposome-mediated DNA transfer in yeast. *Dev. Ind. Microbiol.* **25,** 475–484.
Sakai, T. (1980). Microbial production of coenzymes. *Biotechnol. Bioeng.* **22,** Suppl. 1, 143–162.
Sanchez-Marroquin, A., Ledezma, M., and Barreiro, J. (1971). Oxygen transfer and scale-up in lysine production by *Ustilago maydis* mutant. *Biotechnol. Bioeng.* **13,** 419–429.
Sandford, P. A. (1979). Exocellular, microbial polysaccharides. *Adv. Carbohydr. Chem.* **36,** 265–313.
Sandford, P. A., and Baird, J. (1983). Industrial utilization of polysaccharides. *In* "The Polysaccharides" (G. G. Aspinall, ed.), Vol. 2, pp. 411–490. Academic Press, New York.

15. Fungal Primary Metabolism

Sasaki, Y., Takao, S., and Hotta, K. (1970). Conversion of fumaric acid fermentation to succinic acid fermentation by the association of *Rhizopus* and bacteria. *Agric. Biol. Chem.* **48,** 776–781.

Sato, M., Nakahara, T., and Yamada, K. (1972). Fermentative production of succinic acid from *n*-paraffin by *Candida brumptii* IFO 0731. *Agric. Biol. Chem.* **36,** 1969–1974.

Sato, S., Nakahara, T., and Minoda, Y. (1977). Enzymatic studies on L-malic acid production from *n*-paraffins by *Candida brumptii*, IFO-0731. *Agric. Biol. Chem.* **41,** 1903 (1977).

Scherr, G. H., and Rafelson, M. E., Jr. (1963). The directed isolation of mutants having increased yields of fermentation products. *Dev. Ind. Microbiol.* **4,** 245–252.

Scherr, G. H., and Rafelson, M. E., Jr. (1966). Antimetabolites as selective agents for the isolation of biochemical mutants. *Dev. Ind. Microbiol.* **7,** 97–103.

Scott, C. D., ed. (1984). "Fifth Symposium on Biotechnology for Fuels and Chemicals." Wiley, New York.

Seeley, R. D. (1977). Fractionation and utilization of baker's yeast. *MBAA Tech. Q.* **14,** 35–39.

Seichertova, C., and Leopold, H. (1969). Activation of *Aspergillus niger* strains. II. Application of parasexual hybridization. *Zentralbl. Bakteriol., Parasitenkd. Infektionskr. Hyg., Abt. 2* **123,** 564–570.

Seki, T., Myoga, S., Limtong, S., Uedono, S., Kumnuanta, J., and Taguchi, H. (1983). Genetic construction of yeast strains for high ethanol production. *Biotechnol. Lett.* **5,** 351–356.

Shah, D. N., Purohit, A. P., and Sriprakash, K. S. (1982). Preliminary genetic studies on a citric acid producing strain of *Saccharomycopsis lipolytica*. *Enzyme Microb. Technol.* **4,** 116–117.

Shay, L. K., Wegner, E. H., and Reiter, S. E. (1983). Use of mutagenesis to enhance the methionine content of methanol-assimilating yeast. *Dev. Ind. Microbiol.* **14,** 305–311.

Shcherbakova, E. Y. (1964). A characterization of ultraviolet-induced biochemically active variants of *Aspergillus niger*. *Mikrobiologiya* **33,** 49–55.

Shcherbakova, E. Y., and Rezvaya, M. N. (1977). Formation of diploids in *Aspergillus niger* and their biosynthesis of citric acid. *Mikrobiologiya* **46,** 1064–1069.

Shibata, S., Natori, S., and Udagawa, S. (1964). "List of Fungal Products." Univ. of Tokyo Press, Tokyo.

Shimazono, H. (1964). Distribution of 5′-ribonucleotides in foods and their application to foods. *Food Technol.* **18,** 36–45.

Sinskey, A. J. (1983). Organic chemicals from biomass: an overview. *In* "Organic Chemicals from Biomass (D. Wise, ed.), pp. 1–67. Benjamin/Cummings, Menlo Park, California.

Slodki, M. E., and Boundy, J. A. (1970). Yeast phosphohexans. *Dev. Ind. Microb.* **11,** 86–91.

Slodki, M. E., and Cadmus, M. C. (1978). Production of microbial polysaccharides. *Adv. Appl. Microbiol.* **23,** 19–54.

Slodki, M. E., Ward, R. M., and Cadmus, M. C. (1972). Extracellular mannans from yeast. *Dev. Ind. Microbiol.* **13,** 428–435.

Smiley, K. L., Cadmus, M. C., and Liepins, P. (1967). Biosynthesis of D-mannitol from D-glucose by *Aspergillus candidatus*. *Biotechnol. Bioeng.* **19,** 365–374.

Smith, G. (1969). "An Introduction to Industrial Mycology," 6th Ed., pp. 310–326. Arnold, London.

Smith, J. E., Nowakowska-Waszczuk, A., and Anderson, J. G. (1974). Organic acid production by mycelial fungi. *In* "Industrial Aspects of Biochemistry" (B. Spencer, ed.), pp. 297–317. Elsevier, Amsterdam.

Snow, R. (1983). Genetic improvement of wine yeast. *In* "Yeast Genetics. Fundamental and Applied Aspects" (J. F. T. Spencer, D. M. Spencer, and A. R. W. Smith, eds.), pp. 439–459. Springer-Verlag, Berlin and New York.

Soda, K., Tanaka, H., and Esaki, N. (1983). Amino acids. *In* "Biotechnology. Microbial Products,

Biomass, and Primary Products'' (H. Dellweg, ed.), Vol. 3, pp. 467–478. Verlag Chemie, Weinheim.
Solomons, G. L. (1983). Single cell protein. *CRC Crit. Rev. Biotechnol.* **1,** 21–58.
Sorsoli, W. A., Spence, K. D., and Parks, L. W. (1964). Amino acid accumulation in ethionine-resistant *Saccharomyces cerevisiae. J. Bacteriol.* **88,** 20–24.
Spencer, J. F. T. (1968). Production of polyhydric alcohols by yeasts. *Prog. Ind. Microbiol.* **7,** 1–42.
Spencer, J. F. T., and Spencer, D. M. (1978). Production of polyhydric alcohols by osmotolerant yeasts. *In* "Primary Products of Metabolism" (A. H. Rose, ed.), Economic Microbiology, Vol. 2, pp. 392–425. Academic Press, New York.
Spencer, J. F. T., and Spencer, D. M. (1983). Genetic improvement of industrial yeasts. *Annu. Rev. Microbiol.* **37,** 121–142.
Spencer, J. F. T., Roxburgh, J. M., and Sallans, H. R. (1957). Factors influencing the production of polyhydric alcohols by osmophilic yeasts. *J. Agric. Food Chem.* **5,** 64–67.
Spencer, J. F. T., Gorin, P. A. J., and Rank, G. H. (1971). The genetic control of two types of mannan produced by *Saccharomyces cerevisiae. Can. J. Microbiol.* **17,** 1451–1454.
Stahl, U. (1978). Zygote formation and recombination between like mating types in the yeast *Saccharomycopsis lipolytica* by protoplast fusion. *Mol. Gen. Genet.* **160,** 111–113.
Stępien, P. P., Brousseau, R., Wu, R., Narang, S., and Thomas, D. Y. (1983). Synthesis of a human insulin gene. VI. Expression of the synthetic proinsulin gene in yeast. *Gene* **24,** 289–297.
Stewart, G. G. (1978). Application of yeast genetics within the brewing industry. *J. Am Soc. Brew. Chem.* **36,** 175–185.
Stewart, G. G. (1981). The genetic manipulation of industrial yeast strains. *Can. J. Microbiol.* **27,** 973–990.
Stodola, F. H., Deinema, M. H., and Spencer, J. F. T. (1967). Extracellular lipids of yeasts. *Bacteriol. Rev.* **31,** 194–213.
Strauss, E. S. (1981). Citric acid. *In* "Chemical Economics Handbook," pp. 5021–5023P. SRI Int. Menlo Park, California.
Struhl, K. (1983). The new yeast genetics. *Nature (London)* **305,** 391–397.
Sumner, J. L., Morgan, E. D., and Evans, H. C. (1969). The effect of growth temperature on the fatty acid composition of fungi in the order Mucorales. *Can. J. Microbiol.* **15,** 515–520.
Sutherland, I. W. (1982). Biosynthesis of microbial exopolysaccharides. *Adv. Microb. Physiol.* **23,** 79–150.
Sutherland, I. W. (1983). Extracellular polysaccharides. *In* "Biotechnology. Microbial Products, Biomass, and Primary Products" (H. Dellweg, ed.), Vol. 3, pp. 529–574. Verlag Chemie, Weinheim.
Sutherland, I. W., and Ellwood, D. C. (1979). Microbial exopolysaccharides—industrial polymers of current and future potential. *In* "Microbial Technology: Current State and Future Prospects" (A. T. Bull, D. C. Ellwood, and C. Ratledge, eds.), pp. 107–150. Cambridge Univ. Press, London and New York.
Suzuki, O., Jigami, Y., Nakasato, S., and Hashimoto, T. (1981). U.S. Pat. 4,281,064.
Suzuki, T., and Onishi, H. (1967). The production of xylitol and D-xylonic acid by yeast. *Agric. Biol. Chem.* **31,** 1233–1236.
Tabuchi, T., and Hara, S. (1974). Production of 2-methylisocitric acid from *n*-paraffins by mutants of *Candida lipolytica. Agric. Biol. Chem.* **38,** 1105–1106.
Tabuchi, T., Sugisawa, T., Ishidori, T., Nakahara, T., and Sugiyama, J. (1981). Itaconic acid fermentation by a yeast belonging to the genus *Candida. Agric. Biol. Chem.* **45,** 475–479.
Tachibana, S., and Murakami, T. (1973). L-Malate production from ethanol and calcium carbonate by *Schizophyllum commune. J. Ferment. Technol.* **51,** 858–864.

15. Fungal Primary Metabolism

Takahashi, T. (1969). Erythorbic acid fermentation. *Biotechnol. Bioeng.* **11,** 1157–1171.
Takao, S. (1965).Organic acid production by Basidiomycetes. I. Screening of acid-producing strains. *Appl. Microbiol.* **13,** 732–737.
Takao, S., and Hotta, K. (1977). L-Malic acid fermentation by mixed culture of *Rhizopus arrhizus* and *Proteus vulgaris. Agric. Biol. Chem.* **41,** 945–950.
Takao, S., and Tanida, M. (1982). Pyruvic acid production by *Schizophyllum commune. J. Ferment. Technol.* **60,** 277–280.
Takao, S., Tanida, M., and Kuwabara, H. (1977). L-Malic acid production frum non-sugar sources by *Paecilomyces varioti. J. Ferment. Technol.* **55,** 196–199.
Takayama, K., Adachi, T., Kohata, M., Hattori, K., and Tomiyama, T. (1982). U.S. Pat. 4,322,498.
Takeda Chemical Industries, Ltd. (1970). Belg. Pat. 744,416.
Takeda Chemical Industries, Ltd. (1983). U.S. Pat. 4,389,484.
Takenouchi, E., Yamamoto, T., Nikolova, D. T., Tanaka, H., and Soda, K. (1979). Lysine production by S-(β-aminoethyl)-L-cysteine resistant mutants of *Candida pelliculosa. Agric. Biol. Chem.* **43,** 727–734.
Taoka, A., and Uchida, S. (1981). U.S. Pat. 4,275,158.
Tauro, P., Rao, T. N. R., Johar, D. S., Sreenivasan, A., and Subrahmanyan, V. (1963). L-Lysine production by Ustilaginales fungi. *Agric. Biol. Chem.* **27,** 227–235.
Terui, G. (1973). Tryptophan. *In* "The Microbial Production of Amino Acids" (K. Yamada, S. Kinoshita, T. Tsunoda, and K. Aida, eds.), pp. 515–531. Kodansha, Tokyo.
Terui, G., and Niizu, H. (1969). Recent studies on the fermentative production of L-tryptophan. *Biotechnol. Bioeng. Symp.* No. 1, 33–52.
Thom, C., and Currie, J. N. (1916). Oxalic-acid production of species of Aspergillus. *J. Agric. Res.* **7,** 1–15.
Titus, D, S., and Klis, J. B. (1963). Product improvement with new flavor. *Food Process.* **24,** 128–129, 150–151.
Trumpy, B. H., and Millis, N. F. (1963). Nutritional requirements of an *Aspergillus niger* mutant for citric acid production. *J. Gen. Microbiol.* **30,** 381–393.
Tseng, M. M.-C., and Phillips, C. R. (1982). The kinetics of yeast growth and nicotinic acid production. *Biotechnol. Bioeng.* **24,** 1319–1325.
Tsugawa, R., Nakase, T., Kobayashi, T., Yamashita, K., and Okumura, S. (1969). Fermentation of n-paraffins by yeast. Part I. Fermentative production of α-ketoglutaric acid by *Candida lipolytica. Agric. Biol. Chem.* **33,** 158–167.
Tsukada, Y., and Sugimori, T. (1964). Excretion of ultraviolet-absorbing materials by *Zygosaccharomyces soja* mutant. Part I. Accumulation of 5′-nucleotides in the culture fluid of *Zygosaccharomyces soja* mutant during growth. *Agric. Biol. Chem.* **28,** 471–478.
Tsukada, Y., and Sugimori, T. (1971). Induction of auxotrophic mutants from *Candida* species and their application to L-threonine fermentation. *Agric. Biol. Chem.* **35,** 1–7.
Tubb, R. S. (1979). Applying yeast genetics in brewing. A current assessment. *J. Inst. Brew.* **85,** 286–289.
Tubb, R. S., Brown, A. J. P., Searle, B. D., and Goodey, A. R. (1981). Development of new techniques for the genetic manipulation of brewing yeasts. *Adv. Biotechnol.* **4,** 75–79.
Tuite. M. F., Dobson, M. J., Roberts, N. A., King, R. M., Burke, D. C., Kingsman, S. M., and Kingsman, A. J. (1982). Regulated high efficiency expression of human interferon-alpha in *Saccharomyces cerevisiae. EMBO J.* **5,** 603–608.
Turner, W. B. (1971). "Fungal Metabolites," pp. 2–10. Academic Press, New York.
Uchio, R., and Shiio, I. (1972a). Microbial production of long-chain dicarboxylic acids from n-alkanes. Part II. Production by *Candida cloacae* mutant unable to assimilate dicarboxylic acid. *Agric. Biol. Chem.* **36,** 426–433.

Uchio, R., and Shiio, I. (1972b). Production of dicarboxylic acids by *Candida cloacae* mutant unable to assimilate *n*-alkane. *Agric. Biol. Chem.* **36,** 1169–1175.
Underkofler, L. A. (1954a). Glycerol. *In* "Industrial Fermentations" (L. A. Underkofler and R. J. Hickey, eds.), pp. 252–270. Chem. Publ. Co., New York.
Underkofler, L. A. (1954b). Gluconic acid. *In* "Industrial Fermentations" (L. A. Underkofler and R. J. Hickey, eds.), pp. 446–469. Chem. Publ. Co., New York.
Upjohn Co. (1984). U.S. Pat. 4,443,539.
Urdea, M. S., Merryweather, J. P., Mullenbach, G. T., Coit, D., Heberlein, U., Valenzuela, P., and Barr, P. J. (1983). Chemical synthesis of a gene for human epidermal growth factor urogastrone and its expression in yeast. *Proc. Natl. Acad. Sci. U.S.A.* **80,** 7461–7465.
Valenzuela, P., Medina, A., Rutter, W. J., Ammerer, G., and Hall, B. D. (1982). Synthesis and assembly of hepatitis B virus surface antigen particles in yeast. *Nature (London)* **298,** 347–350.
Van Tieghem, P. (1867). Chimie vegetale; sur la fermentation gallique. *C. R. Hebd. Seances Acad. Sci.* **65,** 1091–1094.
Vijaikishore, P., and Karanth, N. G. (1984). Glycerol production by fermentation. *Appl. Biochem. Biotechnol.* **9,** 243–253.
Wagner, J. R., Titus, D. S., and Schade, J. E. (1963). New opportunities for flavor modification. *Food Technol.* **17,** 52–57.
Wakil, S. J., Stoops, J. K., and Joshi, V. C. (1983). Fatty acid synthesis and its regulation. *Annu. Rev. Biochem.* **52,** 537–579.
Waksman, S. (1944). U.S. Pat. 2,326,986.
Ward, G. E. (1967). Production of gluconic acid, glucose oxidase, fructose, and sorbose. *In* "Microbial Technology" (H. J. Peppler, ed.), pp. 200–221. Reinhold, New York.
Ward, G. E., Lockwood, L. B., May, O. E., and Herrich, H. T. (1936). Biochemical studies in the genus Rhizopus. I. The production of dextro-lactic acid. *J. Am. Chem. Soc.* **58,** 1286–1288.
Weete, J. D. (1973). Sterols of the fungi: distribution and biosynthesis. *Phytochemistry* **12,** 1843–1864.
Wehmer, C. (1903). "Beitrage zur Kenntnis einheimischer Pilze. Zwei neue Schimmelpilze als. Erreger einer Citronensäure-Gärung." Halm, Hannover/Leipzig.
Whitaker, A., and Long, P. A. (1973). Fungal pelleting. *Proc. Biochem.* **11,** 27–31.
Whitworth, D. A., and Ratledge, C. (1974). Microorganisms as a potential source of oils and fats. *Proc. Biochem.* **9,** 14–22.
Wickerham, L. J., Kurtzman, C. P., and Herman, A. I. (1970). Sexual reproduction in *Candida lipolytica. Science (Washington, D.C.)* **167,** 1141.
Woodbine, M. (1959). Microbial fat: micro-organisms as potential fat producers. *Prog. Ind. Microbiol.* **1,** 179–245.
Wright, R. E., Hendershot, W. F., and Peterson, W. H. (1957). Production and testing of yeast mutants for glycerol production. *Appl. Microbiol.* **5,** 272–279.
Yamada, K., Furukawa, T., and Nakahara, T. (1970). Studies on the utilization of hydrocarbons by microorganisms. Part XVIII. Fermentation production of fumaric acid by *Candida hydrocarbofumarica* n. sp. *Agric. Biol. Chem.* **34,** 670–675.
Yamada, K., Nakahara, T., and Fukui, S. (1971). Petroleum microbiology and vitamin production. *In* "Biochemical and Industrial Aspects of Fermentation" (K. Sakaguchi, T. Uemura, and S. Kinoshita, eds.), pp. 62–90. Kodansha, Tokyo.
Yarrow, D. (1972). Four new combinations in yeasts. *Antonie van Leeuwenhoek* **38,** 357–360.
Yoneda, F. (1984). Riboflavin (B2). *In* "Kirk–Othmer Encyclopedia of Chemical Technology" (M. Grayson and D. Ekroth, eds.), pp. 108–124. Wiley, New York.
Yoshinaga, F., and Nakamori, S. (1983). Production of amino acids. *In* "Amino Acids. Biosynthesis and Regulation" (K. M. Herrmann and R. L. Somerville, eds.), pp. 405–429. Addison-Wesley, Reading, Massachusetts.

Young, T., Williamson, V., Taguchi, A., Smith, M., Sledziewski, A., Russell, D., Osterman, J., Denis, C., Cox, D., and Beier, D. (1982). The alcohol dehydrogenase genes of the yeast, *Saccharomyces cerevisiae:* isolation, structure, and regulation. *In* "Genetic Engineering of Microorganisms for Chemicals" (A. Hollaender, R. D. Demoss, S. Kaplan, J. Konisky, D. Savage, and R. S. Wolfe, eds.), pp. 335–361. Plenum, New York.

Yuen, S. (1974). Pullulan and its applications. *Proc. Biochem.* **9,** 7–9, 22.

Zahorsky, B. (1913). U.S. Pat. 1,065,358.

Zhang, S.-Z. (1982). Industrial applications of immobilized biomaterials in China. *In* "Enzyme Engineering 6" (I. Chibata, S. Fukui, and L. B. Wingard, Jr., eds.), pp. 265–270. Plenum, New York.

Zwolshen, J. H., and Bhattacharjee, J. K. (1981). Genetic and biochemical properties of thialysine-resistant mutants of *Saccharomyces cerevisiae*. *J. Gen. Microbiol.* **122,** 281–287.

16

Mitochondrial DNA for Gene Cloning in Eukaryotes

PAUL TUDZYNSKI AND KARL ESSER
Lehrstuhl für Allgemeine Botanik
Ruhr-Universität
Bochum, Federal Republic of Germany

I. Introduction	403
II. Fungal Plasmids	404
A. Mitochondrial Plasmids	407
B. Nuclear Plasmids	408
C. Plasmids of Unknown Association	408
III. Cloning Vectors of Mitochondrial Origin	409
A. Mitochondrial Plasmids	409
B. mtDNA	410
IV. Biotechnological Implications	412
References	413

I. INTRODUCTION

Molecular cloning in eukaryotes is still in its beginnings, in spite of the enormous effects made in this field in the last 4 years. The motivation for developing eukaryotic cloning systems is the obvious difficulty of cloning and especially of expressing complex eukaryotic DNA in bacterial hosts. Problems arise from different ribosome- and RNA polymerase-binding sites and from different posttranscriptional and posttranslational processing systems in prokaryotes and eukaryotes (Esser and Lang-Hinrichs 1983).

In addition, the genetic information in complex metabolic pathways (e.g., the production of secondary metabolites) cannot be transferred as a whole to a prokaryotic host, thus necessitating a cloning system for the producing eukaryote itself.

An efficient eukaryotic cloning system must meet the following criteria:

1. A *transformation* system which ensures efficient uptake of DNA into the host cell.
2. A *selection* system which allows unambiguous identification of transformants (especially low mutation frequency and dominance of the selective marker) and a selective advantage for cells carrying the foreign DNA.
3. *Stable maintenance* of the introduced DNA, either by integration into the host genome, or by efficient autonomous replication of vector DNA.
4. A high level of *expression* of introduced genes, including transcription, correct splicing, and posttranslational processing of the gene product.

Vector–host systems of this kind are now at different stages of development in various eukaryotes, based on viruses like SV40 in animal cells (Razzaque *et al.*, 1983), gemini viruses in plants (Howarth and Goodman, 1982), transposons in *Drosophila* (Rubin and Spradling, 1982), pathogenic DNA-introducing mechanisms like the Ti plasmid of *Agrobacterium tumefaciens* for higher plants (Otten *et al.*, 1981), and endogenous plasmids like the 2-μm DNA in yeast (Guerineau, 1979).

The latter system is the only one so far which meets all the criteria mentioned above. Vector systems available for *Saccharomyces cerevisiae* include integrative vectors for introducing genes at specific sites of the genome as well as autonomously replicating vectors behaving as artificial chromosomes (Struhl, 1983). Unfortunately, the yeast system has turned out not to be useful for the expression of complex eukaryotic DNA from other species. Yeast is not able to process heterologous mosaic genes, such as the rabbit β-globin gene (Beggs *et al.*, 1980) and the alcohol dehydrogenase gene of *Drosophila* (Watts *et al.*, 1983). Moreover, attempts to use *S. cerevisiae* vectors for cloning in other eukaryotes have had only limited success, even in other yeasts (Gaillardin *et al.*, 1983). Obviously, a demand for vector systems suitable for other eukaryotes still remains. Since homologous systems (analogous to yeast) are likely to be most effective, a first prerequisite for the development of a cloning vector is to look for a potential natural vector DNA species in the organism of choice. In this chapter we will present the status of knowledge on potential vector DNA species in fungi. For a complete list, including plants and animal cells, see Esser *et al.* 1984).

Even in the absence of such DNA species the development of vectors is possible, because mitochondrial (mt) DNA (present in every eukaryotic cell) is well suited for this purpose (Esser *et al.*, 1983). Therefore, we will also give an outline of the potential for using mtDNA, and mitochondrial plasmids, for the construction of homologous vector systems in fungi.

II. FUNGAL PLASMIDS

There is accumulating evidence for the widespread occurrence of plasmids or plasmidlike DNA species in eukaryotes. Data presented in Table I give (to our

TABLE I.

Summary of Plasmid Screening Studies with Fungi

Species	Strain	Size of plDNA monomer, kb (μm)	Molecular structure	References
Mitochondria associated				
Independent of mitochondrial "chromosome"				
Claviceps purpurea	Wild strain K1	6.6 (2.1)	Linear	Tudzynski et al. (1983)
Neurospora crassa	"Mauriceville 1C"	5.3 (1.7)		Collins et al. (1981)
		3.6 (1.1)		
Neurospora intermedia	"Labelle"	4.1 (1.3)	ccc	Stohl et al. (1982)
	"Fiji"	5.2 (1.7)		
Homologous to parts of mitochondrial "chromosome"				
Aspergillus amstelodami	"ragged"	1.2 (0.4)	?(amplified)	Lazarus et al. (1980)
Neurospora crassa	"stopper"	Varying	ccc, partially amplified	Bertrand et al. (1980)
	"poky"			De Vries et al. (1981)
Podospora anserina	"senescent" pl	2.5 (0.8)	ccc amplified	Stahl et al. (1978)
	"senescent" pl β-sen	9.8 (3.1)		Wright et al. (1982)
	pl θ-sen	6.3 (2.0)		Jamet-Viermy et al. (1980)
Podospora curvicolla	"senescent"	10.9 (3.5)	(amplified)	B. Böckelmann (unpublished)
Nucleus-associated				
Dictyostelium discoideum		13.5 (4.3)	ccc	Metz et al. (1983)
Saccharomyces cerevisiae	Most strains (Ty 1)	6.4 (2.0)	ccc, chromatinlike	Guerineau (1979)
		6.0 (1.9)	ccc	Ballario et al. (1983)
Unknown association				
Ascobolus immersus	Unstable mutant (b10-B)	6.4 (2.0)	Linear	Francou (1981)
	Wild strain (India VII)	5.7 (1.8)		F. Meinhardt and Kempken (personal communication)
		7.9 (2.5)		
Cephalosporium acremonium		21.0 (6.7)	ccc	Minuth et al. (1982)

(*continued*)

TABLE I (*Continued*)

Species	Strain	Size of pIDNA monomer, kb (μm)	Molecular structure	References
Gäumannomyces graminis var. *tritici*		7.5 (2.4) 8.7 (2.8)	Linear?	Honeyman and Currier (1983)
Kluyveromyces lactis		8.9 (2.8) 13.4 (4.3)	Linear	Gunge et al. (1981)
Morchella conica		6.0 (1.9) 8.0 (2.5)	Linear	Meinhardt and Esser (1984)
Pichia membranaefaciens		6.4 (2.0)		Painting (1982)
Rhizoctonia solani		2.8 (0.9)	Linear	Hashiba et al. (1983)
Saccharomyces cerevisiae		9.4 (3.0)	ccc	Clark-Walker and Azad (1980)
Saccharomyces rouxii		6.3 (2.0)		
Saccharomyces bailii		5.5 (1.8) 6.3 (2.0)	?	Toh-e et al. (1983)
Saccharomyces bisporus		6.5 (2.1)	ccc	Tubb (1980)
Saccharomyces uvarum	Brewery strains	6.5 (2.1)		
Saccharomyces diastaticus		6.5 (2.1)		
Schizosaccharomyces pombe		6.4 (2.0) 9.6 (3.1)	ccc	Fournier et al. (1982)
Torulopsis glabrata		9.4 (3.0)	ccc	Clark-Walker and Azad (1980)

knowledge) the present status of plasmid screening studies in fungi. The plasmids are arranged according to their localization in the cell.

A. Mitochondrial Plasmids

Many plasmid species are localized within the mitochondrial membrane; two different types of these mitochondrial plasmids can be recognized:

1. *Plasmids that are independent of the high-molecular-weight mtDNA*, and show no homology to it, e.g., in two species of *Neurospora*. In this case, small ccc DNA molecules were found exclusively in wild strains, not in laboratory strains. It is tempting to speculate that laboratory strains of fungi will not be a promising source of plasmids, due to the absence of selective pressure for their maintenance. This idea is supported by investigations of the ergot fungus, *Claviceps purpurea*, in which different plasmids have been found in three wild strains, but none has been detected (so far) in any laboratory or commercial strain (P. Tudzynski, unpublished observations).

The two well-characterized plasmids of strain K1 of the ergot fungus (see Table I) have some interesting features: they are linear and partially homologous to each other, and they probably carry inverted repeats at their ends (A. Düvell, unpublished observations). They share these properties with mitochondrial plasmids of higher plants, especially of *Zea mays* and *Sorghum bicolor* (Pring *et al.*, 1977, 1982). Since *Claviceps* is an obligate parasite of cereals, these similarities deserve further analysis. In this context it is interesting that in another plant pathogenic fungus, *Gäumannomyces graminis*, comparable plasmids were detected, though their association with mitochondria is not yet proved (see Table I).

2. *Plasmids that are derived from the high-molecular-weight mtDNA*. These plasmid species are always correlated with impaired growth characteristics, such as the cytoplasmic mutants "ragged" in *Aspergillus*, "poky" and "stopper" in *Neurospora*, and "senescent" in *Podospora*. They result from a decay of high-molecular-weight mtDNA, comparable to the "petite" mtDNA molecules in yeast. Since these fungi are all obligatory aerobes, drastic alterations of the mtDNA lead to impaired growth. The best analyzed example is the *senescence syndrome* in *Podospora* (for reviews see Esser and Tudzynski, 1979; Esser, 1985; Esser and Böckelmann, 1985). Wild strains of *Podospora anserina* (and of *P. curvicolla*, Table I) show a limited vegetative growth capacity. After a defined period of time, growth stops irreversibly. The mycelial front shows the senescent phenotype: aberrant hyphal morphology and pigmentation. Senescent mycelia may contain several types of mitochondrial plasmids, each representing a defined part of the juvenile mitochondrial "chromosome," which decays in the course of senescence. The most prominent type, however, is the so-called plDNA (Stahl *et al.*, 1978), identical to the later described α-sen-DNA (Belcour *et al.*, 1981). By transformation experiments this small plasmid has been proved

to be causally correlated with the senescent phenotype (Tudzynski et al., 1980; Stahl et al., 1982). DNA sequence analyses have shown that plDNA is a "mobile intron" of the mitochondrial COI gene (Osiewacz and Esser, 1984). The molecular basis of senescence induction and plasmid liberation, however, is not yet understood in detail.

The data compiled in Table I show that mitochondrial plasmids are rather common in fungi. This is also the case in higher plants (Esser et al., 1983, 1984). It is probable that some of the plasmid species listed in Table I under "unknown association" actually are of mitochondrial origin; for example, the plasmids of *Morchella conica* copurify with mitochondrial preparations (F. Meinhardt, personal communication).

B. Nuclear Plasmids

So far only two plasmid species have been shown unequivocally to be localized within the nuclear membrane: the so-called 2-μm DNA in baker's yeast, *S. cerevisiae*, and, more recently, a large circular plasmid in the slime mould *Dictyostelium*. The first is the best characterized eukaryotic plasmid. It is present in almost all analyzed strains of *S. cerevisiae*. Probably, some of the 2-μm-sized plasmids found in other species of yeast (unknown association, Table I) also belong to the same class of plasmid. It is not possible to assign specific functions to this plasmid within the yeast cells. Cells that do not contain the plasmid, or cells from which the plasmid is removed by "curing," do not differ phenotypically from the cells carrying the plasmid (Erhart and Hollenberg, 1981).

The molecular structure of 2-μm DNA, however, has been characterized in detail. Its complete nucleotide sequence is known (Hartley and Donelson, 1980). So far, three prospective genes have been identified. Two of them seem to be responsible for stable maintenance of the plasmid. The third codes for a specific recombination system involving the plasmid's inverted repeats (Broach et al., 1982).

Interestingly, native 2-μm DNA possesses nucleosomic structure comparable to that of yeast chromosomal DNA. Therefore, this plasmid represents a true minichromosome (Seligy et al., 1980). As mentioned above, 2-μm-DNA was used to establish the first eukaryotic cloning system.

C. Plasmids of Unknown Association

Several members of this group have already been discussed, and as mentioned they probably belong to either the nuclear or the mitochondrial group. In most cases, localization has not been studied, or the available data are not convincing. The plasmids of at least three species deserve further discussion:

1. The "killer" plasmids in *Kluyveromyces lactis* are the only eukaryotic plasmids for which a defined function is known. They carry the information for

production of a killer toxin. Structurally, they are linear and possess inverted repeats, and the two types are dependent on each other (Gunge *et al.*, 1981). It is possible to transfer them into *S. cerevisiae,* where they are able to replicate and express their killer function (Gunge *et al.*, 1982). Thus, it should be possible to use them to establish a direct selection system.

2. The linear plasmids found in an unstable mutant of *Ascobolus immersus* (green ascospores) were initially thought to be identical to a genetically defined transposable element causing the mutant phenotype. Subsequently, comparable linear plasmids were also found in a wild strain of the same species and in several strains of the closely related *M. conica.* Since all these plasmids probably have inverted repeats, their possible role in transposition events is still open and deserves further investigation.

Since a detailed discussion of these various species of plasmids will be published elsewhere (Esser *et al.*, 1984), we have restricted this section to a brief summary. In principle, each of the plasmid species described could serve as a potential vehicle for a cloning system. In the following, however, we will focus on *mitochondrial plasmids* and *mtDNA* only, because we think that these replicons are the most promising for eukaryotic cloning in general.

III. CLONING VECTORS OF MITOCHONDRIAL ORIGIN

A. Mitochondrial Plasmids

The first mitochondrial hybrid vectors were constructed by integration of the plDNA of *P. anserina* in the bacterial plasmid pBR322 (Stahl *et al.*, 1980). The first transformation experiments were initiated using the idea that plDNA could induce senescence in "infected" or transformed mycelia; that is, the senescence phenotype was used as an indicative marker for transformed mycelia. Using a standard protoplast transformation method, it was indeed possible to obtain senescent mycelia by incubation with these hybrid vectors (Tudzynski *et al.*, 1982). These mycelia were shown to contain free plasmids by both Southern blotting and transformation of *Escherichia coli* with bulk DNA. Crude extracts of these mycelia showed β-lactamase activity due to expression of the bacterial *ampR* gene (Stahl *et al.*, 1982). These data proved unequivocally that these hybrid vectors (mtDNA + bacterial DNA) were replicated and expressed in *Podospora* (Fig. 1).

To overcome the problem of the induction of senescence in transformants, a nonaging mutant was used as recipient ("*i viv,*" Tudzynski *et al.*, 1982), and a hybrid vector based on pBR325, containing the bacterial chloramphenicol R gene, was used as the selective marker. Transformation of *Podospora* with this vector yielded chloramphenicol-resistant mycelia at the rate of about one per

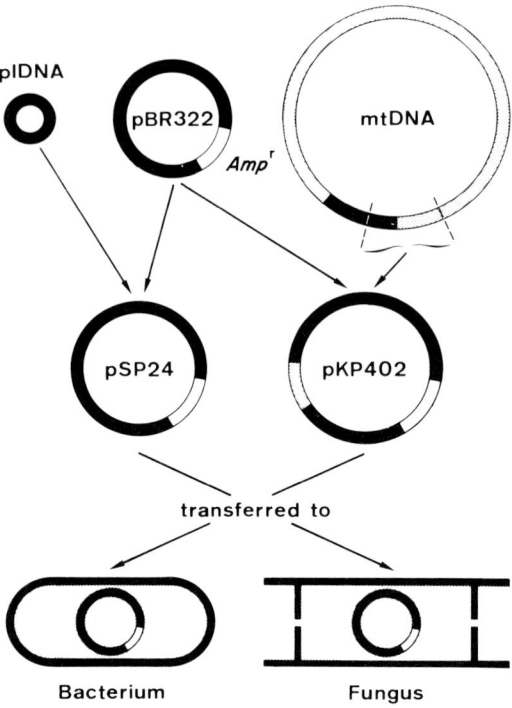

Fig. 1. Construction of shuttle vectors with mitochondrial DNA of *Podospora answerina* being integrated in bacterial plasmid pBR322. In using either the *Podospora* mitochondrial plasmid (above to the left) or a part of mitochondrial DNA containing an autonomously replicating sequence (above to the right), both hybrid vectors replicate and express equally in *E. coli* and *Podospora*.

microgram of DNA. Some of these were shown to contain free replicating plasmids (P. Tudzynski, unpublished observations).

To our knowledge the only other example of a hybrid vector based on a mitochondrial plasmid is *Neurospora intermedia* (the "Labelle" plasmid; see Table I). In this case the mitochondrial plasmid was integrated in pBR325, and the *Neurospora* gene *qa-2* was used as a selective marker. The resulting transformants were shown to contain free plasmids (Stohl and Lambowitz, 1983).

These data support our thesis that mitochondrial plasmids can be used for the construction of replicative vectors in fungi.

B. mtDNA

In *Podospora* plDNA is able to promote autonomous replication of hybrid vectors. In addition, a fragment of chromosomal mtDNA, carrying a small part of plDNA homologous sequences, is able to replicate autonomously (Stahl *et al.*,

1982; see also Fig. 1). Obviously, this mtDNA fragment contains a mitochondrial origin of replication. Therefore, a "normal" mitochondrial origin may be used as an essential part of an autonomously replicating vector. The consequence of this observation is that in the absence of endogenous plasmids a vector system may be established in any eukaryote on the basis of "chromosomal" mtDNA sequences.

A simple test is available to isolate such origin of replication sequences using the yeast transformation system (Beach *et al.*, 1980). A yeast integrative vector lacking a eukaryotic origin sequence (but being selectable in yeast) is combined with mtDNA restriction fragments, and the resulting hybrid vectors are tested for their transformation potential. If the integrated fragment contains an origin of replication (active in yeast), called an autonomously replicating sequence (ARS), the hybrid vector promotes a high transformation rate in yeast.

Using this system, several fungal mitchondrial ARS fragments have been identified: in *S. cerevisiae* (Blanc and Dujon, 1982; Lang-Hinrichs and Stahl, 1983), *Candida utilis* (Tikhomirova *et al.*, 1983), *P. anserina* (Lazdins and Cummings, 1982), *Penicillium chrysogenum* (U. Stahl, personal communication), *Cephalosporium acremonium* (Tudzynski and Esser, 1982), and *Claviceps purpurea* (P. Tudzynski, unpublished observations). The fungal mtDNA seems to contain several ARS; for instance, in yeast there are probably ARSs at intervals of 1700 bp (Hyman *et al.*, 1983). It should be pointed out, however, that these ARSs are defined only for yeast as putative replicons. They are not necessarily active in the donor organism itself.

Nevertheless, recent data from our group indicate that mitochondrial ARS regions do function in homologous and heterologous systems:

1. By the method outlined above, a hybrid vector was constructed (pCP2; see Fig. 2) which contains a mitochondrial ARS fragment of *C. acremonium* (Tudzynski and Esser, 1982). This vector replicates efficiently in yeast, is stable even under nonselective conditions, and is associated with the nucleus (Tudzynski and Esser, 1983). Preliminary experiments using fungicide resistance as a selective marker indicate that this mitochondrial hybrid vector is able to replicate in *Cephalosporium*. Furthermore, this vector may also be used to transform *P. anserina* (screening for *camR*, see above), It is both replicated and maintained in this heterologous system, and is even less prone to rearrangements than the homologous *Podospora* vectors (P. Tudzynski, unpublished observations). Thus, at least in this case, a mitochondrial ARS fragment of a filamentous fungus screened in yeast is active in the homologous (donor) system, as well as in other heterologous systems. The mitochondrial ARS may be well suited for the development of a general shuttle vector.

2. A hybrid vector containing a mitochondrial ARS of *S. cerevisiae* derived from mitochondrial DNA of a suppressive petite strain is also stably maintained in yeast. Preliminary data indicate that this vector also replicates in the hetero-

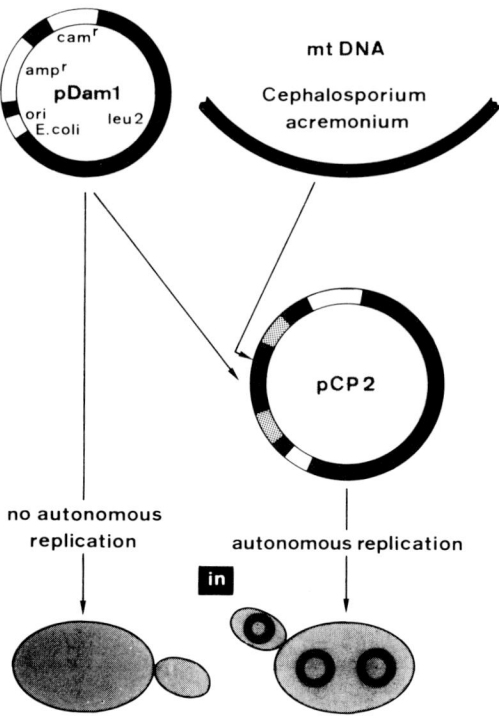

Fig. 2. Scheme of molecular cloning in yeast using an autonomously replicating sequence (ARS) of the mtDNA of the filamentous fungus *C. acremonium*. The vector pDAM1 is of bacterial origin. It carries the following marker genes: ampicillin resistance (amp^r), chloramphenicol resistance (cam^r). Its replication in *E. coli* is ascertained by its prokaryotic replication origin (*ori*). The leucine gene (*leu2*) originates from chromosomal DNA of *S. cerevisiae*. Only after integration of the eukaryotic ARS is the vector, now termed pCP2, able to replicate in the yeast cell.

logous yeast *Schizosaccharomyces pombe*, though it is largely rearranged (C. Lang-Hinrichs, unpublished observations).

Taken together, these results indicate that mtDNA may serve as replicons for eukaryotic replicative vectors. This is only a first, albeit essential, step in the development of a vector system. Several other problems, such as selection systems, rate of expression, and stability, remain to be solved.

IV. BIOTECHNOLOGICAL IMPLICATIONS

Filamentous fungi and yeasts are exploited in industry for the production of their secondary and primary metabolites (e.g., antibiotics, alkaloids, enzymes, amino acids). Their capacity to catalyze specific reactions, such as steroid hormone production, is also used industrially. Since most of these industrial fungi are imperfect, there are severe limits on the extent to which the techniques of classical genetics can be used for strain improvement. This explains the need for new methods, especially those which allow us to manipulate the producing eukaryote rather than to use its genetic information for the transformation of prokaryotes. The use of mitochondrial DNA to establish vectors for molecular cloning in eukaryotes, as presented in this chapter, may become an important tool for biotechnologists.

REFERENCES

Ballario, P., Filetici, P., Junakovic, N., and Pedone, F. (1983). Ty1 extrachromosomal circular copies in *Saccharomyces cerevisiae*. *FEBS Lett.* **155**, 225.
Beach, D., Piper, M., and Shall, S. (1980). Isolation of chromosomal origins of replication in yeast. *Nature (London)* **284**, 185–187.
Beggs, J. D., van den Berg, J., van Ooyen, A., and Weissmann, C. (1980). Abnormal expression of chromosomal rabbit β-globin gene in *Saccharomyces cerevisiae*. *Nature (London)* **283**, 835–840.
Belcour, L., Begel, O., Mosse, M. O., and Vierny, C. (1981). Mitochondrial DNA amplification in senescent cultures of *Podospora anserina*: variability between the retained, amplified sequences. *Curr. Genet.* **3**, 13–22.
Bertrand, H., Collins, R. A., Stohl, L. L., Goewert, R. R., and Lambowitz, A. M. (1980). Deletion mutants of *Neurospora crassa* mitochondrial DNA and their relationship to the "stop–start" growth phenotype. *Proc. Natl. Acad. Sci. U.S.A.* **77**, 6032–6036.
Blanc, H., and Dujon, B. (1982). Replicator regions of the yeast mtDNA active *in vivo* and in yeast transformants. *In* "Mitochondrial Genes" (P. Slonimski, P. Borst, and G. Atterdi, eds.), pp. 279–294. Cold Spring Harbor Lab., Cold Spring Harbor, New York.
Broach, J. R., Guarascio, V. R., and Jayaram, M. (1982). Recombination within the yeast plasmid 2 μ circle is site-specific. *Cell (Cambridge, Mass.)* **29**, 227–234.
Clark-Walker, G. D., and Azad, A. A. (1980). Hybridizable sequences between cytoplasmid ribosomal RNAs and 3 micron circular DNAs of *Saccharomyces cerevisiae* and *Torulopsis glabrata*. *Nucleic Acids Res.* **8**, 1009–1022.
Collins, R. A., Stohl, L. L., Cole, M. D., and Lambowitz, A. M. (1981). Characterization of a novel plasmid DNA found in mitochondria of *N. crassa*. *Cell (Cambridge, Mass.)* **24**, 443–452.
De Vries, H., de Jonge, J. C., van't Sant, P., Agsteribbe, E., and Arnberg, E. (1981). A "stopper" mutant of *Neurospora crassa* containing two populations of aberrant mitochondrial DNA. *Curr. Genet.* **3**, 205–211.
Erhart, E., and Hollenberg, C. P. (1981). Curing of *Saccharomyces cerevisiae* 2-μm DNA by transformation. *Curr. Genet.* **3**, 83–89.
Esser, K. (1985). "Genetic Control of Ageing: The Mobile Intron Model," 1984 Sandoz Lectures in Gerontology. Academic Press, New York. In press.
Esser, K., and Böckelmann, B. (1985). Senescence in fungi some years later. *In* "Models for Ageing Research" (F. A. Lints, ed.). In press.
Esser, K., and Lang-Hinrichs, C. (1983). Molecular cloning in heterologous systems. *Adv. Biochem. Eng./Biotechnol.* **26**, 143–173.

Esser, K., and Tudzynski, P. (1979). Genetic control and expression of senescence in *Podospora anserina*. In "Viruses and Plasmids in Fungi" (P. A. Lemke, ed.), pp. 595–615. Dekker, New York.

Esser, K., Kück, U., Stahl, U., and Tudzynski, P. (1983). Cloning vectors of mitochondrial origin for eukaryotes, a new concept in genetic engineering. *Curr. Genet.* **7,** 239–243.

Esser, K., Kück, U., Osiewacz, H. D., Stahl, U., and Tudzynski, P. (1984). "Plasmids in Eukaryotes." Springer-Verlag, Berlin and New York.

Fournier, P., Gaillardin, C., de Louvencourt, L., Heslot, H., Lang, B. F., and Kaudewitz, F. (1982). r-DNA plasmid from *Schizosaccharomyces pombe:* cloning and use in yeast transformation. *Curr. Genet.* **6,** 31–38.

Francou, F. (1981). Isolation and characterization of a linear DNA molecule in the fungus *Ascobolus immersus*. *Mol. Gen. Genet.* **184,** 440–444.

Gaillardin, C., Fournier, P, Budar, F., Kudla, B., Gerbaud, C., and Heslot, H. (1983). Replication and recombination of 2μm DNA in *Schizosaccharomyces pombe*. *Curr. Genet.* **7,** 245–253.

Guardiola, J., Grimaldi, G., Costantino, P., Micheli, G., and Cervone, F. (1982). Loss of nitrofuran resistance in *Fusarium oxysporum* is correlated with loss of a 46.7 kb circular DNA molecule. *J. Gen. Microbiol.* **128,** 2235–2242.

Guerineau, M. (1979). Plasmid DNA in yeast. In "Viruses and Plasmids in Fungi" (P. A. Lemke, ed.), pp. 540–593. Dekker, New York.

Gunge, N., Tamaru, A., Ozawa, F., and Sakaguchi, K. (1981). Isolation and characterization of linear deoxyribonucleic acid plasmids from *Kluyveromyces lactis* and the plasmid-associated killer character. *J. Bacteriol.* **145,** 382–390.

Gunge, N., Murata, K., and Sakaguchi, K. (1982). Transformation of *Saccharomyces cerevisiae* with linear DNA killer plasmids from *Kluyveromyces lactis*. *J. Bacteriol.* **151,** 462–464.

Hartley, J. L., and Donelson, J. E. (1980). Nucleotide sequence of the yeast plasmid. *Nature (London)* **286,** 860–864.

Hashiba, T., Hyakumachi, N., Honma, Y., and Matsuda, I. (1983). Isolation and characterization of a plasmid DNA in the fungus *Rhizoctonia solani*. *Abstr. IMC, 3rd, Tokyo* p. 92.

Honeyman, A. L., and Currier, T. C. (1983). The isolation and characterization of two linear DNA elements from *Gaeumannomyces graminis* var. *tritici*, the causative agent of "take-all-disease" of wheat. *Abstr. Annu. Meet. ASM, 83rd, New Orleans, La.* pp. 15.

Howarth, A. J., and Goodman, R. M. (1982). Plant viruses with genomes of single-stranded DNA. *Trends Biol. Sci.* May, 180–182.

Hyman, B. C., Cramer, J. H., and Rownd, R. H. (1983). The mitochondrial genome of *Saccharomyces cerevisiae* contains numerous, densely spaced autonomously replicating sequences. *Gene* **26,** 223–230.

Jamet-Vierny, C., Begel, O., and Belcour, L. (1980). Senescence in *Podospora anserina:* amplification of a mitochondrial DNA sequence. *Cell (Cambridge, Mass.)* **21,** 189–194.

Lang-Hinrichs, C., and Stahl, U. (1983). Mitochondrial DNA for yeast transformation. *FEBS Meet., 15th, Brussels* Abstra. p. 273.

Lazarus, C. M., Earl, A. J., Turner, G., and Küntzel, H. (1980). Amplification of a mitochondrial DNA sequence in the cytoplasmically inherited "ragged" mutant of *Aspergillus amstelodami*. *Eur. J. Biochem.* **106,** 633–641.

Lazdins, I. B., and Cummings, D. J. (1982). Autonomously replicating sequences in young and senescent mitochondrial DNA from *Podospora anserina*. *Curr. Genet.* **6,** 173–178.

Meinhardt, F., and Esser, K. (1984). Barrage formation in fungi. *Encycl. Plant Physiol., New Ser.* **17,** 350–361.

Metz, B. A., Ward, T. E., Welker, D. L., and Williams, K. L. (1983). Identification of an endogenous plasmid in *Dictyostelium discoideum*. *Embo J.* **2,** 515–519.

Minuth, W., Tudzynski, P., and Esser, K. (1982). Extrachromosomal genetics of *Cephalosporium acremonium.* I. Characterization and mapping of mitochondrial DNA. *Curr. Genet.* **5,** 227–231.

Osiewacz, H. D., and Esser, K. (1984). The mitochondrial plasmid of *Podospora anserina:* a mobile intron of a mitochondrial gene. *Curr. Genet.* **8,** 299–305.

Otten, L., De Greve, H., Hernalsteens, J. P., Van Montagu, M., Schieder, O., Straub, J., and Schell, J. (1981). Mendelian transmission of genes introduced into plants by the Ti plasmids of *Agrobacterium tumefaciens. Mol. Gen. Genet.* **183,** 209–213.

Painting, K. (1982). Plasmid screening in different yeast genera. *Int. Conf. Yeast Genet. Mol. Biol., 11th, Montpellier* p. 107.

Pring, D. R., Levings, C. S., III, Hu, W. W. L., and Timothy, O. H. (1977). Unique DNA associated with mitochondria of the S-type cytoplasm of male-sterile maize. *Proc. Natl. Acad. Sci. U.S.A.* **74,** 2904–2908.

Pring, D. R., Conde, M. F., Schertz, K. F., and Levings, C. S. (1982). Plasmid-like DNAs associated with mitochondria of cytoplasmic male-sterile sorghum. *Mol.Gen. Genet.* **186,** 180–184.

Razzaque, A., Mizusawa, H., and Seidman, M. M. (1983). Rearrangement and mutagenesis of a shuttle vector plasmid after passage in mammalian cells. *Proc. Natl. Acad. Sci. U.S.A.* **80,** 3010–3014.

Rubin, G. M., and Spradling, A. C. (1982). Genetic transformation of *Drosophila* with transposable element vectors. *Science (Washington, D.C.)* **218,** 348–353.

Seligy, V. L., Thomas, D. Y., and Miki, B. L. A. (1980). *Saccharomyces cerevisiae* plasmid, Scp or 2 μm: intracellular distribution, stability and nucleosomal-like packaging. *Nucleic Acids Res.* **8,** 3371

Stahl, U., Lemke, P. A., Tudzynski, P., Kück, U., and Esser, K. (1978). Evidence for plasmid like DNA in a filamentous fungus, the ascomycete *Podospora anserina. Mol. Gen. Genet.* **162,** 341–343.

Stahl, U., Kück, U., Tudzynski, P., and Esser, K. (1980). Characterization and cloning of plasmid like DNA of the ascomycete *Podospora anserina. Mol. Gen. Genet.* **178,** 639–646.

Stahl, U., Tudzynski, P., Kück, U., and Esser, K. (1982). Replication and expression of a bacterial–mitochondrial hybrid plasmid in the fungus *Podospora anserina. Proc. Natl. Acad. Sci. U.S.A.* **79,** 3641–3645.

Stohl, L. L., and Lambowitz, A. M. (1983). Construction of a shuttle vector for the filamentous fungus *Neurospora crassa. Proc. Natl. Acad. Sci. U.S.A.* **80,** 1058–1062.

Stohl, L. L., Collins, R. A., Cole, M. D., and Lambowitz, A. M. (1982). Characterization of two new plasmid DNAs found in mitochondria of wild-type *Neurospora intermedia* strains. *Nucleic Acids Res.* **10,** 1439–1458.

Struhl, K. (1983). The new yeast genetics. *Nature (London)* **305,** 391–396.

Tikhomirova, L. P., Kryukov, V. M., Strizhov, N. I., and Bayev, A. A. (1983). mtDNA sequences of *Candida utilis* capable of supporting autonomous replication of plasmids in *Saccharomyces cerevisiae. Mol. Gen. Genet.* **189,** 479–484.

Toh-e, A., Araki, H., Sakurai, T., Tatsumi, H., Jearnpipatkul, A., and Oshima, Y. (1983). New circular DNA plasmids in yeasts. *Abstra. IMC, 3rd, Tokyo* p. 311.

Tubb, R. S. (1980). 2μm DNA plasmid in brewery yeasts. *J. Inst. Brew.* **86,** 78–80.

Tudzynski, P., and Esser, K. (1982). Extrachromosomal genetics of *Cephalosporium acremonium.* II. Development of a mitochondrial DNA hybrid vector replicating in *Saccharomyces cerevisiae. Curr. Genet.* **6,** 153–158.

Tudzynski, P., and Esser, K. (1983). Nuclear association in yeast of a hybrid vector containing mitochondrial DNA. *Curr. Genet.* **7,** 165–166.

Tudzynski, P., Stahl, U., and Esser, K. (1980). Transformation to senescence with plasmid like DNA in the ascomycete *Podospora anserina. Curr. Genet.* **2,** 181–184.

Tudzynski, P., Stahl, U., and Esser, K. (1982). Development of a eukaryotic cloning system in *Podospora anserina*. I. Long-lived mutants as potential recipients. *Curr. Genet.* **6,** 219–222.

Tudzynski, P., Düvell, A., and Esser, K. (1983). Extrachromosomal genetics of *Claviceps purpurea*. I. Mitochondrial DNA and mitochondrial plasmids. *Curr. Genet.* **7,** 145–150.

Watts, F., Castle, C., and Beggs, J. (1983). Aberrant splicing of Drosophila alcohol dehydrogenase transcripts in *Saccharomyces cerevisiae*. *EMBO J.* **2,** 2085–2091.

Wright, R. M., Horrum, M. A., and Cummings, D. J. (1982). Are mitochondrial structural genes selectively amplified during senescence in *Podospora anserina? Cell (Cambridge, Mass.)* **29,** 505–515.

17

Molecular Bases of Fungal Pathogenicity to Plants

O. C. YODER AND B. GILLIAN TURGEON
Department of Plant Pathology
Cornell University
Ithaca, New York

I. Introduction	417
II. Infection Structures	418
III. Cutinase	421
IV. Pisatin Demethylase	422
V. Toxins	424
VI. Cloning and Analysis of Pathologically Important Genes from Fungi	427
A. Choosing a System	427
B. Vector Construction	428
C. Approaches to Isolation of Pathogenicity Genes	438
D. Uses for Cloned Pathogenicity Genes	438
References	441

I. INTRODUCTION

Plant pathogenic fungi confront two fundamental problems in their encounters with plants. First, the protective outer surface of the plant must be breached. Then, once within the plant, the fungus must determine whether or not the plant will support its growth and reproduction.

Fungi employ a number of strategies to deal with such problems (Table I). These include the elaboration of infection structures and enzymes needed for penetration of the plant epidermis, and the production of toxins, enzymes, and perhaps even hormones, which act as chemical messages to help determine whether or not the plant that has been attacked is an appropriate host.

Although many fungal structures and metabolites have been proposed as pathogenicity factors, few have been evaluated rigorously for roles in disease.

TABLE I.

Fungal Products Involved in Pathogenesis

Product	Role
Infection structures	Effect penetration of plant surface
Cutinase	Degrade cuticle on plant surface
Pisatin demethylase	Detoxify plant defense antibiotic
Toxins	Condition plant tissues for colonization

Certain host–parasite systems, however, have features that facilitate experimental analysis, and in these cases the roles in disease of particular fungal products have been described. The purpose of this chapter is first to discuss four different examples of fungal metabolites or structures (Table I) for which good experimental evidence indicates pathological significance, and second to outline methodologies needed for genetic dissection and molecular analysis of the disease process. These methodologies are currently being used on already characterized molecules and structures (Table I). An additional goal is to bring the power of molecular biological analysis to bear on products of fungal plant pathogens of unknown relevance to pathogenesis.

II. INFECTION STRUCTURES

Particular morphological changes appear necessary for fungal penetration into plants, although extracellular enzymatic action is also involved, at least in some cases (see Section III). Rust fungi provide favorable systems for studying prepenetration activities because their infection structures are distinct, elaborate, and inducible. The main features of rust infection structures are diagrammed in Fig. 1. Recent advances in understanding rust differentiation have been reviewed by Staples and Macko (1984). The emphasis in this section, following a brief description of the phenomenology, will be on the prospect for genetically dissecting the infection process and determining how the genes control pathogenicity.

Spores of the bean rust fungus *Uromyces appendiculatus* (also known as *U. phaseoli*) germinate on the leaf surface and produce a germ tube which quickly attaches itself snugly to the surface of the leaf (Fig. 1). This attachment is a critical step. If it fails, for example, if the spore germinates on the surface of a waxless leaf to which it cannot stick, none of the subsequent developmental changes occur (Wynn and Staples, 1981). As the germ tube grows across the surface of the leaf, it orients its direction of growth perpendicular to the surface ridges of the leaf. The result of this orientation is that germ tubes grow in straight

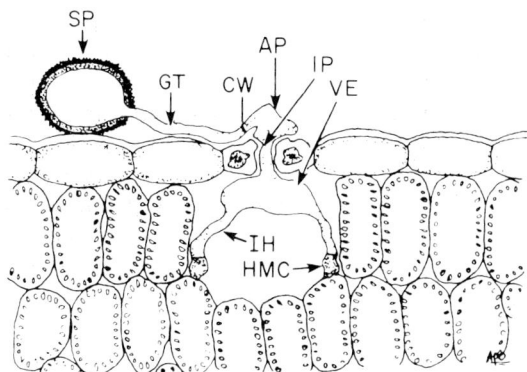

Fig. 1. Diagrammatic representation of *Uromyces appendiculatus* infection structures on a leaf surface. Arrows indicate spore (SP), germ tube (GT), cross wall (CW), appressorium (AP), infection peg (IP), vesicle (VE), infection hyphae (IH), and haustorial mother cell (HMC). Eventual penetration into host cells is achieved by HMCs. [From Staples and Macko (1984).]

lines, rather than randomly, so that the probability of finding a stomate, the site of penetration, is maximized (Wynn and Staples, 1981). Upon contacting a stomate, growth of the germ tube stops, the tip of the germ tube differentiates into a structure known as an appressorium, which covers the stomate, and growth resumes from the appressorium downward via a peg through the pore of the stomate into the substomatal cavity. There a substomatal vesicle forms. This gives rise to infection hyphae, which in turn produce haustorial mother cells (Fig. 1). Upon contact with a leaf mesophyll cell, haustoria form that are capable of breaching the wall. At this point, the fungus must determine whether or not the plant is a suitable host. If so, it will establish a nutritional relationship with the plant, colonize the tissues extensively, and eventually sporulate. If not, the fungus grows no further, fails to reproduce, and the plant recovers from the attack.

It is important to note that most if not all fungi that attack foliage perform on plant surfaces in a manner similar to that described above, although there are many variations on the theme. Some fungi, for example, penetrate directly through epidermal walls rather than through stomates, and in many cases the infection structures are not as distinct as they are in rust fungi. Conversely, saprophytic fungi undergo none of the differentiation phases described for parasitic ones. Saprophytes grow profusely on leaf surfaces but do not attach themselves to the epidermal layer, do not orient their growth with respect to the topography of the leaf surface, do not elaborate infection structures, and do not attempt to penetrate, even when they contact an opening such as a stomate by chance. This observation alone constitutes strong circumstantial evidence that production of infection structures is required for fungal pathogenicity to foliage.

Differentiation of infection structures is an inducible process. Differentiation does not occur on smooth surfaces. For *U. appendiculatus* the stimulus appears to be physical only; no chemicals are needed. If positive leaf replicas are made of chemically inert polystyrene and inoculated with spores, the fungus differentiates normally as though it were on a leaf surface (Wynn, 1976). However, on a smooth surface in the absence of physical stimuli certain chemicals or other treatments will induce differentiation. All or part of the infection structures will be produced on a noninducing surface in the presence of potassium ions (Staples *et al.*, 1983a) or cyclic AMP (Hoch and Staples, 1984), or after heat shock or treatment with ultrasound (Staples and Hoch, 1982). The most effective inducers are physical stimuli such as a stomate or a scratch on an artificial membrane (Staples *et al.*, 1983b).

Several biochemical changes occur in the fungal cell as it starts to differentiate. DNA synthesis begins, nuclei divide (Staples *et al.*, 1975), several new proteins appear (Huang and Staples, 1982), and the elements of the cytoskeleton take on a new orientation (Staples and Macko, 1984). It seems likely that at least some of the new proteins are components of the cytoskeleton (see below). The basic molecular question then is how the fungus recognizes a signal from the leaf surface and transmits that signal to the nucleus, where several genes are activated presumably to start the differentiation process.

There is indirect evidence that the cytoskeleton of *U. appendiculatus* (Hoch and Staples, 1983) plays an important role in differentiation. First, one of the proteins whose synthesis is greatly stimulated after induction is calmodulin, a component of the cytoskeleton (R. C. Staples, personal communication). Second, the greatest concentration of cytoskeletal filaments is near the interface between the fungus and the substrate. Third, chemicals such as colcemid, griseofulvin, and nocodazole, which depolymerize microtubules, or podophyllotoxin and the cytochalasins, which depolymerize microfilaments, induce the start of differentiation on a smooth surface, where differentiation would not normally occur (Staples and Hoch, 1982). Heat shock and ultrasound also depolymerize the cytoskeleton. Further, chemicals that stabilize microtubules (deuterium oxide) and microfilaments (phallotoxin or taxol) inhibit differentiation under conditions where it would normally occur.

It is possible to construct a model to explain differentiation of infection structures in terms of the cytoskeleton. In this model the cytoskeleton could be involved in any or all of three processes. First, the actual sensing of the shape of the substrate could be mediated by proteins associated with the cytoskeleton, since there is a high concentration of microfilaments at the growing tip of the germ tube (Hoch and Staples, 1983) and proteins in the extracellular matrix of the germ tube are necessary for sensing to occur (L. Epstein and R. C. Staples, personal communication). These extracellular proteins may be connected to the cytoskeleton. Second, cytoskeletal cables could transmit the signal from the germ tube tip to the nucleus, where differentiation-specific genes are activated. Third, the cytoskeleton could determine the particular shapes of the various infection structures (Fig. 1). To test this model it will be necessary to isolate genes that control production of the differentiation-specific proteins, as well as other genes that are activated during differentiation, and determine their metabolic functions as well as the mechanisms by which they are regulated in response to tactile stimuli.

III. CUTINASE

A long-standing issue regarding fungal penetration of plants is whether the process is strictly mechanical or involves enzymatic digestion of the plant cell surface. Although both mechanisms are probably involved, direct evidence in support of either of them has been difficult to obtain. Recently, Kolattukudy and his colleagues evaluated the role of cutinase in penetration and concluded that in certain cases at least this enzyme contributes to successful entry of the fungus into the plant.

The aerial surfaces of higher plants are covered with cutin, which is a hydroxy–epoxy fatty acid polymer impregnated with waxes (Kolattukudy, 1981). Cutinase is an esterase which has an active site containing a serine, a histidine, and a carboxyl residue. At least two isozymes of cutinase specific for cutin as well as a nonspecific esterase are known to degrade cutin (Purdy and Kolattukudy, 1975). Compounds such as diisopropyl fluorophosphate, which bind to active serines, inactivate cutinase.

Cutinase is produced by a variety of plant pathogenic fungi (Lin and Kolattukudy, 1980) but the most thoroughly investigated is from *Fusarium solani* f. sp. *pisi* (Koller and Kolattukudy, 1982). Both isozymes of cutinase from *F. solani* have been purified to homogeneity and antibodies have been prepared against cutinase I (Soliday and Kolattukudy, 1976). Production of cutinase is repressed by glucose and induced by cutin monomers to the extent that it constitutes >70% of the extracellular protein (Lin and Kolattukudy, 1978). The inducibility of the enzyme has permitted the isolation of an almost full-length cDNA clone of the gene that codes for cutinase (Soliday *et al.*, 1984). The primary structure of cutinase was determined from the nucleotide sequence of cDNA and the amino acid sequence of a portion of the cutinase polypeptide.

Several lines of evidence support the idea that cutinase plays an important role in penetration of fungi into plants (Table II). First, ferritin-conjugated antibody to cutinase was used to show that the enzyme is present at the site of penetration (Shaykh *et al.*, 1977). Second, antibodies to cutinase in the inoculum prevent infection of intact plants but not wounded plants (Maiti and Kolattukudy, 1979). Further, inhibitors of the enzyme such as diisopropyl fluorophosphate also pre-

TABLE II.

Evidence That Cutinase Is Required for Fungal Penetration

Cutinase is present at the site of penetration
Antibody against cutinase prevents infection
Chemical inhibitors of cutinase prevent infection
Mutants lacking cutinase are nonpathogenic

vent infection even though they have little or no effect on growth rate of the fungus. The mode of action of certain fungicides against *F. solani* may involve inhibition of cutinase rather than direct inhibition of the fungus (Koller *et al.*, 1982a,b). Finally, an isolate of *F. solani* highly virulent on intact tissue was found to produce large amounts of cutinase, whereas a different isolate that was weakly virulent on intact tissue (but fully virulent on wounded tissue) produced only a small amount of cutinase (Koller *et al.*, 1982c). However, the weakly virulent isolate may be deficient in other factors essential for penetration as well, since the addition of cutinase to its inoculum caused only partial restoration of virulence toward intact tissue.

The types of evidence supporting a role for *F. solani* cutinase in penetration also have been established for cutinase from *Colletotrichum gloeosporioides*, a pathogen of papaya fruit (Dickman *et al.*, 1982). In addition, it has recently been shown that mutants of *C. gloeosporioides* lacking the ability to secrete cutinase are not pathogenic on papaya (Dickman and Patil, 1984).

IV. PISATIN DEMETHYLASE

Nectria haematococca is a fungal pathogen of peas. When under attack, the pea plant responds by activating a biosynthetic pathway that gives rise to an antibiotic called pisatin. This metabolite is inhibitory to the growth in culture of most fungi, including some strains of *Nectria*.

Nectria isolates collected from nature exhibit a range of virulences toward pea. All of the strains with high virulence are tolerant of pisatin, whereas strains with low virulence range from tolerant to sensitive to pisatin in culture (VanEtten *et al.*, 1980).

Tolerance of pisatin is associated with the production by the fungus of an enzyme called pisatin demethylase (PDA), which is a cytochrome P-450 monooxygenase that detoxifies pisatin by removing a methyl group (Matthews and VanEtten, 1983). In pisatin-tolerant isolates PDA is inducible but the level of enzyme produced varies among isolates (VanEtten and Matthews, 1984). Chemicals that will induce PDA include only pisatin itself and certain closely related

17. Fungal Pathogenicity to Plants

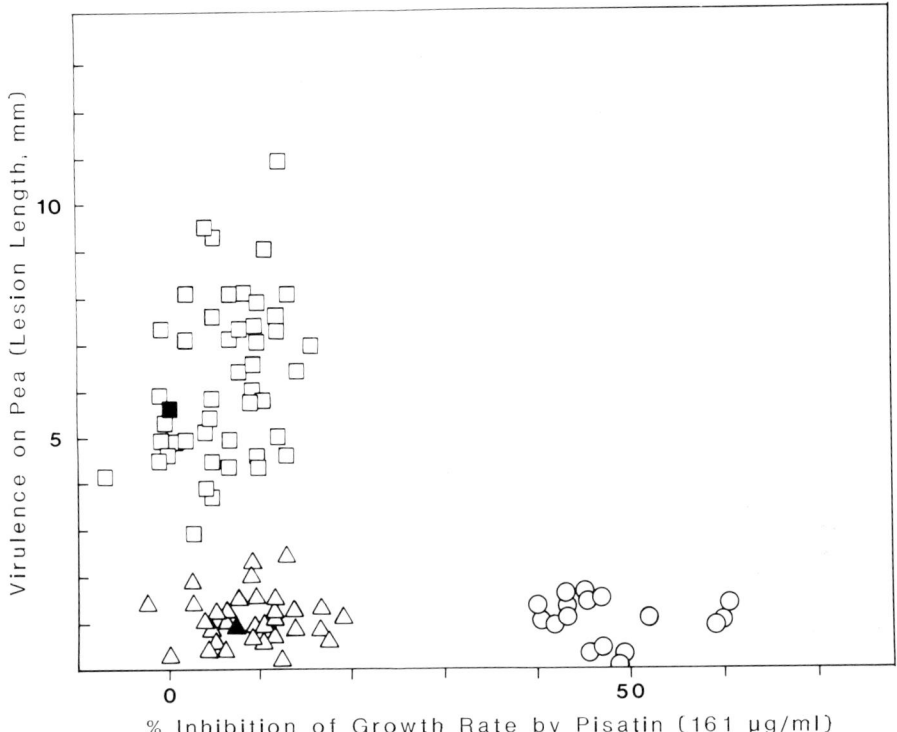

Fig. 2. Segregation of virulence toward pea, pisatin tolerance, and production of pisatin demethylase (PDA) in progeny of a *Nectria haematococca* cross. (■) Parental strain with high PDA; (▲) parental strain with low PDA; (□) progeny with high PDA; (△) progeny with low PDA; (○) progeny with no PDA. Only isolates with high PDA can be highly virulent on plants, although isolates with either low or high PDA are pisatin-tolerant in culture. The appearance of recombinants (the ○ symbols) that produce no PDA (and are sensitive to pisatin in culture) reflects the fact that two genes, one controlling high PDA (*Pda1*) and one controlling low PDA (*Pda2*), are segregating in this cross. Note that the number of □ symbols approximately equals the number of △ and ○ symbols combined, which is expected since *Pda1* is epistatic to *Pda2*. [From Kistler and VanEtten (1984a).]

compounds (VanEtten and Barz, 1981). Isolates that are sensitive to pisatin produce no PDA at all.

Genes controlling production of PDA and the involvement of PDA in the virulence of the fungus have been investigated by genetic analysis (Kistler and VanEtten, 1984a; Tegtmeier and VanEtten, 1982). The data in Fig. 2, representing progeny from a cross between a low and a high PDA producer, illustrate several points. First, several genes control PDA production. A single gene, *Pda1*, specifies high PDA production, and another gene, *Pda2*, determines low PDA production. Results of additional crosses (not shown) have demonstrated that a third gene, *Pda3*, also controls low PDA production (Kistler and VanEtten, 1984b). It is not yet known whether all three loci control production of the same enzyme. Second, high PDA production is required for high virulence of the fungus toward peas. In progeny of this cross (Fig. 2) and many others (Kistler and VanEtten, 1984a,b; Tegtmeier and VanEtten, 1982) it has been found that highly virulent isolates always produce high levels of PDA. Third, weakly virulent isolates may have either no detectable PDA, low PDA, or even high PDA; the latter class demonstrates that more than PDA alone is required for high virulence to peas. Fourth, only Pda^- strains, which lack PDA completely, are sensitive to pisatin in culture.

V. TOXINS

Toxic metabolites are commonly produced by plant pathogenic fungi. Many of these compounds are generally toxic to living cells (Rudolph, 1976; Stoessl, 1981), whereas some of them are specific to only certain genotypes of higher plants (Daly and Knoche, 1982; Nishimura and Kohmoto, 1983; Scheffer and Livingston, 1984; Yoder, 1980). In a number of cases, the toxin-sensitive plant is also the host of the toxin-producing fungus. Such fungal metabolites are called host-specific toxins because they have high activity only against fungal host plants.

None of the nonspecific toxins produced by pathogenic fungi have been critically evaluated for possible roles in plant disease, but several of the host-specific ones have undergone extensive analysis (Yoder, 1980). For the purpose of this discussion, only host-specific toxins will be considered because their unusual specificity for host plants alone implicates them as pathogenicity molecules. Further, as described below, some of them have been subjected to genetic analysis, which has demonstrated that they are causally involved in pathogenesis. A list of the host-specific toxins currently known is presented in Table III.

Chemically, the host-specific toxins are a diverse group of low-molecular-weight secondary metabolites (Macko, 1983). Two of them, HC-toxin and AM-toxin, are cyclic tetrapeptides of MW 436 and 445, respectively. HS-toxin is a sesquiterpene galactofuranoside (MW 884), T-toxin is a linear polyketol (MW 768), AK-toxin is an ester of epoxydecatrienoic acid (MW 413), and AAL-toxin is a dimethylheptadecapentol ester of propanetricarboxylic acid (MW 508). Several of these toxins are found in culture filtrates as families of isomers. Chemical structures of the remainder of the toxins listed in Table III are unknown. However, the chemistry and biosynthesis of many nonspecific toxins have been investigated (Stoessl, 1981).

The genetics of toxin production by many fungal pathogens has not been analyzed because most of them have no known sexual cycle. An exception to this

TABLE III.

Host-Specific Toxins Produced by Fungi[a]

Toxin	Fungus	Host
HV-toxin	*Cochliobolus*[b] *victoriae*	Oats
HC-toxin	*C. carbonum*	Corn
HS-toxin	*C. sacchari*	Sugarcane
T-toxin	*C. heterostrophus*	Corn
AK-toxin	*Alternaria alternata* f. sp. *kikuchiana*	Japanese pear
AM-toxin	*A. alternata* f. sp. *mali*	Apple
ACT-toxin	*A. alternata* f. sp. *citri* (tangerine race)	Dancy tangerine
ACL-toxin	*A. alternata* f. sp. *citri* (lemon race)	Rough lemon
AAL-toxin	*A. alternata* f. sp. *lycopersici*	Tomato
AF-toxin	*A. alternata* f. sp. *fragariae*	Strawberry
AT-toxin	*A. alternata* f. sp. *longipes*	Tobacco
PC-toxin	*Periconia circinata*	Sorghum
PM-toxin	*Phyllosticta maydis*	Corn
CC-toxin	*Corynespora cassiicola*	Tomato

[a] Each toxin affects only those genotypes of plant that are susceptible to the fungus that produces the toxin. This unusual specificity along with experiments showing cosegregation of toxin production and pathogenicity (in those cases where sexual genetic analysis is possible; see Yoder, 1980) demonstrate that these toxins are causally involved in pathogenesis. For reviews that discuss some or all of these toxins in more detail and cite original references see Daly (1982); Daly and Knoche (1982); Macko (1983); Nishimura and Kohmoto (1983); Scheffer (1976); Scheffer and Livingston (1984); and Yoder (1980).

[b] *Cochliobolus* designates the sexual stage of fungi having asexual stages known as *Helminthosporium, Bipolaris,* or *Drechslera.*

is the genus *Cochliobolus,* an ascomycete with a meiotic cycle much like that of *Neurospora* (Guzman *et al.,* 1982). Several members of the genus produce host-specific toxins, e.g., HV-toxin, HC-toxin, T-toxin, and HS-toxin (Table III). In each case, (except for HS-toxin, which has not been genetically analyzed), both Tox^+ and Tox^- strains are found to occur naturally. Tox^+ strains are always pathogenic (or highly virulent), whereas Tox^- strains are nonpathogenic (or weakly virulent). When a cross is made between a pathogenic Tox^+ strain and a nonpathogenic Tox^- strain there is 1 : 1 segregation for toxin production, indicating that a single gene is involved in each case. All toxin-producing progeny are pathogenic, whereas all toxinless progeny are nonpathogenic. Furthermore, only parental types and no recombinants are found among progeny. These correlations between pathogenicity and ability to produce toxin provide convincing evidence that each of the toxins examined in this way is causally involved in pathogenesis (Yoder, 1980).

The level of involvement in pathogenesis varies from toxin to toxin. HV-toxin and HC-toxin appear to be required by the respective toxin-producing fungi in order to cause any disease at all: these toxins are called pathogenicity factors (Yoder, 1980). T-toxin, on the other hand, is not required for pathogenicity because Tox$^-$ strains can cause low levels of disease. The toxin appears to regulate the level of virulence of the fungus since Tox$^+$ strains simply cause more disease than Tox$^-$; this toxin is known as a virulence factor.

The mechanisms by which toxins interact with plant cells to cause damage are generally unknown. Three cellular components have been proposed as sites of toxin action: the plasma membrane, the mitochondrion, and a specific enzyme.

The main evidence that the plasma membrane contains a toxin-sensitive site derives from studies of solute leakage from tissues after toxin treatment. The best example of this phenomenon is the effect of HV-toxin on susceptible oat tissues. Within 1–2 min after exposure to toxin there is more leakage from susceptible cells than from resistant cells or untreated controls (Scheffer, 1976). After an hour virtually all leachable materials are out of the tissue, indicating that the plasma membrane is no longer a permeability barrier. The speed of this effect alone suggests that the plasma membrane is the first organelle affected by the toxin and therefore contains a target site or receptor molecule for the toxin. Accordingly, a great deal of effort has been aimed at the isolation of a receptor for HV-toxin from susceptible plants (Scheffer, 1976; Scheffer and Livingston, 1984). To date, no such molecule has been found and no effects of HV-toxin on cell-free preparations of any organelle (plasmalemma, mitochondria, chloroplasts, nuclei, ribosomes) have been observed.

The only report of a binding protein for a pathogen-produced toxin is that of Strobel, working with HS-toxin, which specifically affects cultivars of sugarcane susceptible to *H. sacchari;* resistant sugarcane cultivars and other plant species are resistant to HS-toxin. It was claimed that a protein of 48 kilodaltons from susceptible plants bound HS-toxin. A slightly different protein found in resistant plants did not bind HS-toxin (Strobel, 1973). The significance of this reported differential binding was made questionable by a later observation that proteins from tobacco and mint, which are resistant to HS-toxin, also bind HS-toxin (Kenfield and Strobel, 1981). All of the work on binding proteins was done with a ^{14}C-labeled preparation of toxin called "helminthosporoside." The structure of helminthosporoside was reported to be cyclopropyl galactoside with a molecular weight of 236 (Steiner and Strobel, 1971). This structure was later found to be incorrect (Livingston and Scheffer, 1981). HS-toxin was shown by Macko *et al.* (1983) to be a sesquiterpene galactofuranoside with a molecular weight of 884. Moreover, the toxin-binding protein studies were repeated by Lesney *et al.* (1982), who showed (using a better preparation of toxin and extensive controls) that there is little or no toxin-binding activity in plant extracts or proteins prepared as described by Strobel. At present the mechanism of action of HS-toxin is unknown.

Other host-specific toxins, such as HC-toxin, PC-toxin, AK-toxin, and AM-toxin, also affect the properties of plasma membranes of susceptible cells, but none with the speed of HV-toxin. However, in no case has a plasma membrane toxin receptor molecule been identified. The site of action of PC-toxin may not be the plasma membrane, since toxin-induced cell death can occur without solute leakage (Dunkle and Wolpert, 1981). Recently, it has been found that PC-toxin causes four 16-kd proteins to be abundantly produced in susceptible but not resistant sorghum cells (Wolpert and Dunkle, 1983). It is possible that the damaging effect of PC-toxin is mediated by these proteins.

T-toxin, produced by *Cochliobolus heterostrophus,* clearly has a sensitive site in the mitochondrion of susceptible corn that is absent in mitochondria of resistant corn or other plant species (Gregory *et al.*, 1977). Both *in vivo* (Malone *et al.*, 1978) and *in vitro* (Gregory *et al.*, 1980), mitochondria are structurally and functionally destroyed by exposure to toxin. Further, cellular ATP levels in cells fall very rapidly immediately after exposure to toxin (Walton *et al.*, 1979). The toxin appears to exert its damaging effect on sensitive mitochondria by making the inner mitochondrial membrane leaky (Holden and Sze, 1984; Matthews *et al.*, 1979). The essential difference between sensitive and resistant mitochondria is not known, although some interesting observations have been made. The two types of mitochondria differ by restriction enzyme polymorphisms in their DNAs (Kemble and Pring, 1982), and sensitive mitochondria produce a 13-kd polypeptide that resistant mitochondria do not have (Forde *et al.*, 1978).

AAL-toxin from *Alternaria alternata* f. sp. *lycopersici* specifically affects only tomatoes susceptible to the fungus (Gilchrist and Grogan, 1976). It has no obvious effect on the plasma membrane but it does inhibit a particular enzyme, aspartate carbamoyltransferase, from susceptible plants, with a much smaller effect on the same enzyme from resistant plants (Gilchrist, 1983). It is possible that the differential effect of the toxin on this enzyme alone can explain susceptibility and resistance of tomatoes to the fungus.

VI. CLONING AND ANALYSIS OF PATHOLOGICALLY IMPORTANT GENES FROM FUNGI

A. Choosing a System

Before serious molecular biological work can be done with fungal plant pathogens, they must be sufficiently domesticated to permit efficient laboratory manipulation. Pathogens, as they are recovered from nature, are too variable and unpredictable for careful experimental work. It is therefore necessary to breed and select strains that perform well under laboratory conditions. But first an appropriate pathogen should be chosen. Criteria that are useful in deciding upon

TABLE IV.

Criteria for Choosing a Model Fungal Pathogen

Culturability
Easily manipulated sexual cycle
Rapid vegetative growth
Short life and disease cycles
Host plant that satisfies the above criteria

an experimental subject are listed in Table IV. In addition to the items listed, it is important that a fungus be amenable to mutation and selection of useful mutants such as auxotrophs and, more important, mutations that affect pathogenicity. Moreover, protocols must be developed for the production of large numbers of protoplasts that will regenerate at high frequency and methods for isolation of DNA that is of large size and digestible with restriction endonucleases must be available (Garber and Yoder, 1983).

Another important consideration is the ability of the fungus to withstand long-term storage. Many important pathogenic strains have been lost because of improper storage. In addition, it is clear that fungi can change rapidly after being isolated from nature and cultured (Griffiths and Bertrand, 1984). Thus, it is extremely important that all strains be metabolically immobilized in permanent storage immediately after acquisition and recovered directly from such storage for each experiment. In the past, convenient and reliable long-term storage methods have not been available. But with the development of inexpensive low-temperature (-80 or $-135°C$) freezers, it is now possible to store large numbers of strains in cryovials in glycerol (Maniatis et al., 1982). This method is advantageous because preparation of microbes for storage and their recovery from storage are simple rapid procedures and because the storage medium is aqueous, which obviates the need to maintain dry storage compartments. An alternative to low-temperature glycerol storage is the silica gel method (Perkins, 1977), which is known to preserve Neurospora for at least 20 years (Catcheside and Catcheside, 1979).

B. Vector Construction

Once the basic microbiological tools are in hand, it is necessary to construct a suitable transformation vector or set of vectors. Transformation systems for fungi are described in detail elsewhere in this volume (see Barnes and Thorner, Chapter 7; Turner and Ballance, Chapter 10; Rine and Carlson, Chapter 5; Boguslawski, Chapter 6; Arst and Scazzocchio, Chapter 13; Huiet and Case, Chapter 8;

TABLE V.
Fungi That Have Been Transformed[a]

Organism	Reference	Selectable gene[b]
Saccharomyces	Hinnen *et al.* (1978)	*LEU2* from yeast
Schizosaccharomyces	Beach and Nurse (1981)	*LEU2* from yeast
Kluyveromyces	Das and Hollenberg (1982)	*kan* from *E. coli*
Dictyostelium	Hirth *et al.* (1982)	*kan* from *E. coli*
Neurospora	Case *et al.* (1979)	*qa2* from *Neurospora*
Podospora	Stahl *et al.* (1982)	"Sen" from *Podospora*
Aspergillus	Yelton *et al.* (1984)	*TrpC* from *Aspergillus*

[a] The report by Banks (1983a) of *Ustilago* transformation awaits evidence that the "transformants" actually carried foreign DNA.

[b] LEU, leucine; kan, kanamycin resistance; qa, quinic acid; Sen, senescence; Trp, tryptophan.

and Weiss *et al.*, Chapter 11). The purpose of the discussion in this chapter is to emphasize the points that may be particularly significant in construction of vectors for undeveloped organisms.

At this time, transformation has not been satisfactorily achieved for plant pathogenic fungi. In fact, few eukaryotic microbes have been transformed at all, and only three of these are filamentous fungi (Table V); the modest number of successes so far reflects the unexpected difficulties that have been encountered by those who are developing transformation systems for fungi. From the information now available (Table V), it is apparent that the use of homologous sequences in vector construction is more likely to be successful than dependence on heterologous components. Thus, it is important to obtain selectable genes, promoters, and origins of DNA replication from the organism to be transformed.

Selectable genes are usually found in nuclear DNA and can sometimes be isolated by complementation of heterologous organisms (e.g., *E. coli* or yeast) with libraries of DNA constructed in a vector (see Weiss *et al.*, Chapter 11). Promoters and DNA origins of replication can be found on chromosomes or on native plasmids. Those on plasmids may be easier to isolate and more highly expressed than those from chromosomes. Indeed, one need only consider the achievements made possible in yeast by the resident 2-μm circle (see Chapter 6) to recognize the potential importance of native plasmids in vector construction. Native plasmids from filamentous fungi are discussed extensively both in this chapter and in those by Kinsey (Chapter 9) and by Tudzynski and Esser (Chapter 16) to emphasize the potential of plasmids in vector construction and to encourage more widespread investigation of plasmids in filamentous fungi.

1. Native Plasmids in Filamentous Fungi

Until recently, plasmids in filamentous fungi were unknown. Six years ago the first evidence for small circular DNA molecules was reported in *Podospora anserina* (Stahl *et al.* 1978). Now the existence of such sequences is well documented for at least three different fungal genera plus an additional one from our own recent work (Garber *et al.*, 1984; see also below), and there are indications that plasmids may exist in several more fungi (see Chapter 16). A summary is presented here for each of the four cases that have been extensively characterized.

a. *Podospora anserina*. This fungus undergoes a phenomenon known as senescence, which dictates a specific life span for each strain (Esser and Tudzynski, 1979). Germinated ascospores give rise to colonies that grow normally for a period that can be as short as 21 days (Tudzynski and Esser, 1979). During the so-called juvenile phase, growth of the mycelium may extend from 15 to 170 cm before senescence occurs, depending on the race of the fungus (Smith and Rubenstein, 1973a). At the end of the juvenile phase, growth stops and the cells at the edge of the colony begin to die. Mutations in nuclear genes can delay the onset of senescence, and certain recombinants appear to be immortal (Tudzynski and Esser, 1979). Senescence is inherited cytoplasmically (Smith and Rubenstein, 1973b), suggesting the involvement of mitochondria. Models to explain the molecular basis of senescence have invoked a causal role for alterations in mitochondrial DNA in the initiation of cellular death (Esser *et al.*, 1980; Belcour and Begel, 1978).

Efforts in several laboratories have indeed confirmed that changes in the mitochondrial chromosome occur as senescence progresses and that small circular DNAs appear concomitantly (Belcour *et al.*, 1981; Cummings *et al.*, 1979; Jamet-Vierny *et al.*, 1980; Kück *et al.*, 1981; Stahl *et al.*, 1980; Vierny *et al.*, 1982). These small circles, called senDNAs, amplify while the mitochondrial chromosome degenerates, until virtually all of the DNA in the mitochondrion is of the small circular type.

There are at least five different forms of senDNA (α, β, γ, δ, θ). They all occur as tandemly repeated multimeric circles, with monomer sizes of 2.0–9.8 kb. Each of them shares homology with the normal mitochondrial chromosome. In fact, each plasmid is derived from at least part of the coding region of a mitochondrial gene (Wright *et al.*, 1982). Furthermore, each one has a putative origin of DNA replication, called an autonomously replicating sequence (ARS), that was defined by complementation of yeast (Lazdins and Cummings, 1982).

Surprisingly, it has recently been observed that two forms of senDNA (α-plasmid and β-plasmid) transpose into the nucleus and are integrated into nuclear DNA during the senescence process (Wright and Cummings, 1983). This suggests that the sen plasmids have the properties of transposons, rendering them capable of nonhomologous recombination.

A partial sequence has been done for the α-plasmid (Osiewacz and Esser, 1983), which revealed an open reading frame and a noncoding region. But perhaps the most interesting features of the α-plasmid were observed upon sequencing the junctions between the plasmid and mitochondrial DNA (mtDNA) prior to excision and comparing them to the sequence of the splice site in the plasmid after excision (Cummings and Wright, 1983). The mtDNA sequences that flank the plasmid each contain a 10-bp palindrome, both of which are AT-rich. In the plasmid (after excision and circularization), there is a 5-bp sequence on one side of the splice site which is repeated 8 bp downstream. The splice itself produces a 7-bp sequence that is repeated 7 bp away. Cummings and Wright (1983) propose that the palindromes may provide sites that permit DNA binding proteins to initiate excision of plasmid DNA and that both the palindromes and the repeated sequences may be important in transposition to the nucleus as well as site-specific recombination.

b. *Neurospora* spp. Laboratory strains of *Neurospora* spp. are not known to have plasmids. However, when a worldwide group of strains recently isolated from nature was surveyed, four plasmids were found. One was in an *N. crassa* strain from Texas (Collins *et al.*, 1981); two were in *N. intermedia* strains isolated from Florida and from Fiji, respectively (Stohl *et al.*, 1982); an *N. intermedia* plasmid from an Indian strain (Collins and Lambowitz, 1983) was only recently discovered and has not yet been investigated.

The *Neurospora* plasmids are all multimeric circles composed of tandemly arranged head-to-tail arrays of monomer units. The monomer size is from 3.6 to 5.2 kb and the number of monomers per plasmid molecule ranges from one to at least six. In contrast to the *Podospora* plasmids, those from *Neurospora* have little or no homology with the mitochondrial chromosome, have no apparent effect on the maintenance of mtDNA, and confer no obvious phenotype on isolates that have them. Of particular interest is the Mauriceville plasmid, which is transcribed *in vivo* (Collins *et al.*, 1981) and which contains five repeats of an 18-bp palindromic sequence that is also found in the mitochondrial chromosome (Nargang *et al.*, 1983). Two of the repeats are identical with the mtDNA palindrome and three are very similar to it. Nargang *et al.* (1983) suggest that the plasmid may be a mobile genetic element and that the palindromes may be involved in a mechanism for transposition. An additional suggestion of mobility was reported recently, based on the observation that one of the *N. intermedia* plasmids is found in four different strains of *N. tetrasperma* (Natvig *et al.*, 1984). This could indicate interspecific exchange of plasmids, although other explanations are also possible. The practical significance of the *Neurospora* plasmids lies in their potential for the construction of shuttle vectors (Stohl and Lambowitz, 1983; Kinsey, Chapter 9, this volume).

c. *Aspergillus amstelodami*. Certain mutations in this fungus cause aberrant growth and are cytoplasmically inherited. One of them is called "ragged" (*rgd*) because of the uneven margin on colonies that carry the mutation. The effect of *rgd* is not lethal, as senDNA is in *Podospora*, and colonies can be maintained indefinitely in an actively growing state.

The mitochondrial chromosome in *rgd* mutants is indistinguishable from the wild type by restriction analysis (Lazarus *et al.*, 1980). However, *rgd* mitochondria have additional large circular DNA molecules that are not found in the wild type. These are composed of tandemly repeated reiterations of monomer units that can be of at least six different types, each type specific to a particular mutant. The monomers range in size from 0.9 to 2.7 kb, whereas the multimers can be as large as or larger than the mitochondrial chromosome itself. All monomer sequences are homologous with specific sites on the mitochondrial chromosome; *rgd1* occupies region 1, which is near the large rRNA gene, while *rgd3*–*rgd7* are overlapping sequences assigned to region 2 (Lazarus and Kuntzel, 1981). Several genes are found in region 2, including those for two tRNAs, subunit 6 of ATPase, and an unassigned reading frame which corresponds to URF4 of the human mitochondrial genome. There is a 215-bp sequence that is common to all region 2 plasmids. Within the 215 bp there is a 22-bp palindrome that is thought to be an origin of DNA replication because it can form a hairpin loop resembling origins in mtDNA of yeast, rat, mouse, and human (Lazarus and Kuntzel, 1981). This is consistent with the observations that the *rgd* DNAs apparently replicate independently of the mitochondrial chromosome. It is especially interesting that both *rgd1* and *rgd3* plasmids have homology with the mitochondrial chromosome of the related fungus *A. nidulans* and that *rgd3* is homologous with mtDNA of *N. crassa* as well.

d. *Cochliobolus heterostrophus*. A survey of 23 field isolates of *Cochliobolus* revealed one that contains plasmid DNA (Garber *et al.*, 1984). Characteristics of the plasmid are intermediate between those of the *Neurospora* and *Podospora* plasmids. The *Cochliobolus* plasmid (designated T40 because it was found in isolate T40, originally collected in Japan) is found as a series of head-to-tail multimeric circles composed of 1.9-kb monomer units. Multimer size ranges from 1 to at least 17 monomer units, as determined by electron microscopy. Presence of the T40 plasmid has no apparent effect on the structure of the mitochondrial chromosome. Mitochondrial DNAs from all 23 of the isolates initially surveyed, including T40, have homology to the T40 plasmid. There is one copy of the plasmid sequence integrated into each mitochondrial chromosome and in each isolate the 1.9-kb sequence maps at the same location. The T40

17. Fungal Pathogenicity to Plants

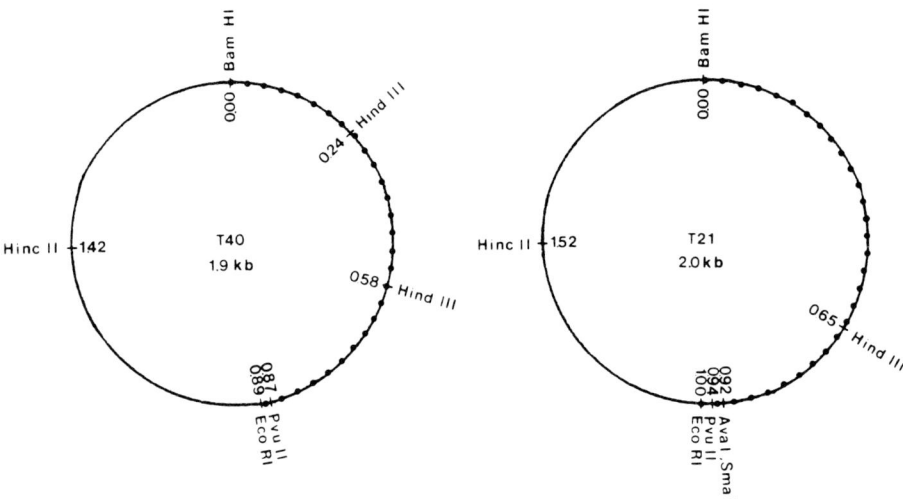

Fig. 3. Restriction enzyme maps of two *Cochliobolus* mitochondrial plasmids, T40 and T21. The plasmids are similar since they hybridize with each other and to a particular region of the mitochondrial chromosome. The restriction maps, however, are different: smooth line = region of apparent similarity, dotted line = region of apparent difference.

isolate has approximately 30 copies of the plasmid per mitochondrial chromosome.

The T40 plasmid has no homology with any of the mitochondrial plasmids from either *Neurospora* or *Podospora* (Garber *et al.*, 1984), but it does have homology, under conditions of high stringency, with mitochondrial DNA from normal (N) male-fertile maize. There is no hybridization to mtDNAs from maize carrying cytoplasmically determined male sterility of either the C, T, or S type (J. Bailey-Serres, personal communication). The significance of this differential homology is currently unknown.

Recently, a plasmid homologous to the T40 plasmid but slightly larger (2.0 kb) was isolated and cloned from a strain called T21 (J. Oard and O. C. Yoder, unpublished observations). Maps of the two plasmids are shown in Fig. 3. They are similar except that the sequence between 0 and ~0.89 kb in the T40 plasmid seems to have been replaced by a larger sequence in the T21 plasmid.

The T40 and T21 plasmids both carry promoters. Northern blots of RNA from strains T40 and T21 and from a plasmidless strain, probed with the cloned T40 plasmid, show that the integrated forms of the two plasmids each yield a transcript of about 900 bp. The excised, amplified version of the T40 plasmid produces a transcript of about 1000 bp, whereas that from the T21 plasmid yields a transcript of 1300 bp. Transcripts from the T40 strain that are homologous to the T40 plasmid probe are about 13 times more abundant than those from the plasmidless strain (J. J. Lin and O. C. Yoder, unpublished observations).

The T40 plasmid may contain an authentic *Cochliobolus* origin of DNA replication, although proof of this awaits a suitable transformation system for *Cochliobolus*. The evidence so far is that the plasmid occurs at high copy number free from the mitochondrial chromosome and that it acts as an ARS in yeast (see Section VI,B,3). It is possible that *Cochliobolus* plasmids will serve as sources of both DNA replication origins and promoters to be used in construction of vectors for *Cochliobolus*.

At this time we do not know if the T40 plasmid confers a phenotype on the fungus. Initially, we thought that the T40 plasmid had no effect on the fungus because the T40 isolate seemed to grow normally in culture. However, after more extensive observation of this isolate in 10-cm Petri dishes, it became clear that an occasional subculture would sector or stop growing. To examine this more carefully, the T40 and plasmidless strains (several subcultures of each) were compared by the "race tube" method (Ryan *et al.*, 1943), in which each isolate was placed at one end of a horizontal tube (24 in. long) containing agar medium. The cultures grew at different rates but the final result was that each plasmidless isolate arrived at the far end of its race tube and most of the T40 subcultures did not. Although not conclusive, this observation does leave open the possibility that the T40 plasmid affects growth. However, DNA preparations from both T40 sectors and normal-looking T40 colonies contain plasmid.

2. Selectable Genes

Availability of a variety of native plasmids from filamentous fungi will facilitate construction of vectors that are composed entirely of homologous sequences. An essential part of such a vector is a selectable gene. Most selectable genes that have been used for transformation of eukaryotic microbes are those that correct nutritional deficiencies (see Weiss *et al.*, Chapter 11; Turner and Ballance, Chapter 10). For example, in yeast the most frequently employed selectable genes are *URA3*, *LEU2*, *TRP1*, and *HIS3*; in *Neurospora* the most popular selectable genes are *qa2* and *am;* in *Aspergillus*, *trpC*, *argB*, and *amdS* are used. However, in order to use such genes the corresponding mutations must be available, yet particular mutants are often difficult to obtain in filamentous fungi.

Alternatives to nutritional genes are genes for drug-resistance. Such genes are preferable because they are selectable in all wild-type sensitive strains without the need for induced mutations. The most commonly used drug resistance gene for eukaryote transformation is the *kan* (kanamycin resistance) gene of *E. coli*. This gene codes for an enzyme (aminoglycoside phosphotransferase) that detoxifies not only kanamycin but the experimental drug G418 as well. G418 inhibits growth of both prokaryotic and eukaryotic cells. The *kan* gene has been shown to confer resistance to G418 in cells of yeast (Jimenez and Davies, 1980); it is also expressed in cells of higher plants (Fraley et al., 1983), and mammals (Gorman et al., 1983) if fused to an appropriate eukaryotic promoter. However, the *kan* gene has not been readily usable in filamentous fungi and only recently was reported to be functional in *Neurospora* (Bull and Wootton, 1984). One problem is that high concentrations (0.5–1.0 mg/ml) of G418 are needed to adequately inhibit growth of many fungi.

A more toxic drug, hygromycin, has been reported for use against both prokaryotes and eukaryotes. Hygromycin is effective against fungi at an approximately 10-fold lower concentration than is G418, and there is now a cloned gene, from *Streptomyces,* for resistance to it. This gene, *hyh,* confers hygromycin resistance in yeast provided any of several yeast promoters are fused to it (Gritz and Davies, 1983; Kaster et al., 1984). Thus, it is a likely prospect for selectability in filamentous fungi if coupled to a homologous promoter.

3. Origins of DNA Replication

One desirable feature of vectors is ability to replicate autonomously in the eukaryotic system. To do this the vector needs an appropriate origin of DNA replication. Sources of origins include native plasmids, the best example of which in eukaryotes is the yeast 2-μm circle, and chromosomes (see Chapters 5 and 10). There is a convenient method for testing ability of DNA sequences to replicate autonomously in yeast (Stinchcomb et al., 1980). The sequence to be tested is inserted into a vector such as YIp5, which carries a gene selectable in yeast (*URA3*) but no yeast origin of DNA replication, and transformed into yeast cells. Sequences that promote autonomous replication in yeast are called ARSs rather than origins, since they are not known to be authentic origins of DNA replication in the homologous systems.

Using the Stinchcomb et al. (1980) procedure, ARSs have been isolated from a variety of eukaryotes, including *Kluyveromyces lactis* (Sreekrishna et al., 1984), *Schizosaccharomyces pombe* (Fournier et al., 1982). *Candida utilis* (Hsu et al., 1983) and *C. utilis* mitochondrial DNA (Tikhomirova et al., 1983), *Candida maltosa* (Kawamura et al., 1983), *Ustilago maydis* (Banks, 1983b), *Physarum polycephalum* (Gorman et al., 1981), *Cephalosporium acremonium* (Skatrud and Queener, 1984), *Neurospora crassa* (Suzci and Radford, 1983), *Chlamydomonas reinhardii* (Loppes and Denis, 1983; Rochaix et al., 1984), tobacco chloroplasts (Ohtani et al., 1984), mouse (Roth et al., 1983), *Xenopus laevis* (Zakian, 1981), *Podospora anserina* (Lazdins and Cummings, 1982), *Oxytricha nova* (Colombo et al., 1984) and *Dictyostelium discoideum, Caenorhabditis elegans, Drosophila melanogaster,* and *Zea mays* (Stinchcomb et al., 1980).

We have isolated ARSs from *Cochliobolus* by using the Stinchcomb et al. (1980) procedure (Table VI). Approximately 750 yeast transformants were recovered from an *Eco*RI library of *Cochliobolus* DNA in YIp5. Forty transfor-

TABLE VI.

Cochliobolus heterostrophus Autonomously Replicating Sequences (ARSs)

Plasmid	Size of ARS fragment (kb)	Frequency of recovery (%)[a]	Location in genome	Copies per genome
pChARS6	1.05	20	mtDNA[d]	One
pChARS7	5.7	30	rDNA[d]	~130, tandemly repeated
pChARS8	3.6	10	nuDNA[d]	Undetermined
pChARS15	2.5	30	nuDNA	Undetermined
pChARS20	8.0	10	nuDNA	Dispersed multiple copies
pChARS19[b]	3.3		rDNA	~130, tandemly repeated
pYGT8[c]	1.9		T40 plasmid (mtDNA)	~30

[a] Percentages are based on the number of times each ARS was recovered among 10 plasmids that were arbitrarily chosen from a pool of 750.

[b] *ARS19* was recovered independently by subcloning the 3.3 kb *Eco*RI fragment of the cloned rDNA repeating unit.

[c] pYGT8 is YIp5 with the *Cochliobolus* T40 plasmid in the *Eco*RI site.

[d] mt, Mitochondrial; r, ribosomal; nu, nuclear.

mants were examined further and found to contain plasmids with *Cochliobolus* inserts; each plasmid promoted high-frequency transformation of yeast (compared to a low-frequency integrating plasmid) and was mitotically unstable (compared to a relatively stable plasmid carrying the yeast 2-μm circle replicon). Genomic locations and frequencies were established for ten of these ARSs by gel transfer analysis (Table VI). Two of the ten were a 1.05-kb fragment of the mitochondrial chromosome, three were the 5.7-kb *Eco*RI fragment of the 9-kb tandemly repeated unit that carries the 5.8 S, 17 S, and 25 S ribosomal RNA genes (R. C. Garber, B. G. Turgeon, E. U. Selker, and O. C. Yoder, unpublished observations), and the remaining five were from nuclear DNA that had no homology to the tandemly repeated rDNA. Of the latter five, *ARS15* was recovered three times from the group of ten, yet it hybridized to a single *Eco*RI fragment of genomic DNA, which hints that *ARS15* is on a tandemly repeated sequence. *ARS20* was found once among the ten and it hybridized to multiple *Eco*RI fragments in genomic DNA, suggesting a dispersed sequence.

Two additional ARSs were identified independently of those recovered from the *Eco*RI library. *ARS19* is on the 3.3-kb *Eco*RI fragment of the rDNA repeating unit. The T40 plasmid, which is homologous to a region of the mitochondrial chromosome (see Section VI,B,1), was subcloned into YIp5 and found to have ARS activity in yeast.

The value of ARSs from most eukaryotes, especially those that have been isolated by complementation of *S. cerevisiae,* has not yet been demonstrated. In general, ARSs returned to the organism of origin have not been found to function there as true replicons. This may be due in part to the lack of suitable transformation vectors for most of the organisms from which ARSs have been isolated. But in the case of *N. crassa ARS8*, which was isolated by complementation of *S. cerevisiae*, there was no detectable replicon function in *N. crassa* (Suzci and Radford, 1983). It is better to isolate ARSs by complementing the homologous system, which has been done for *K. lactis* (Sreekrishna *et al.*, 1984; Das and Hollenberg, 1982) and *S. pombe* (Beach and Nurse, 1981; Losson and Lacroute, 1983).

4. Centromeres

Chromosomal sequences from both *S. cerevisiae* (Carbon, 1984) and *S. pombe* (Carbon and Clarke, 1984) have been identified as functional centromeres. The effect of a chromosomal centromere in an ARS-containing plasmid is to maintain the copy number at about one per cell and to cause the plasmid to segregate stably in mitosis and meiosis as though it were a normal chromosome.

Stable vectors with low copy number are important for two reasons. First, some gene products at high concentration are toxic to cells. If such products are overproduced as the result of being encoded by a gene on a multicopy plasmid, cells containing the gene of interest can be killed. The low copy number of centromere-containing plasmids reduces the possibility of missing a gene because its product is toxic. Second, the stability provided by a centromere permits growth of cells containing an autonomously replicating plasmid on nonselective medium. This could be important for plant pathogenic fungi. For example, some toxins are produced in large amounts by fungi on a rich medium and little or not at all on a nutritionally exacting selective one. Thus, analysis of a gene controlling toxin production would be facilitated by a plasmid containing a centromere.

Centromeres have been isolated from yeasts either by "walking" from a cloned centromere-linked gene or by selecting for ability of DNA fragments to mitotically stabilize an otherwise unstable plasmid (Carbon, 1984). For plant pathogenic fungi, centromere-linked genes are generally unknown, so the alternative approach is more plausible. However, autonomously replicating vectors are so far not available, making heterologous systems necessary for centromere isolation. In choosing such a system it may be important to consider what is known about conservation of centromere sequences. *Saccharomyces cerevisiae* centromeres appear to be highly conserved among chromosomes within the strains examined and are very small (~200 bp). They are quite different from centromeres of *S. pombe*, which are several kilobases long (Carbon and Clarke, 1984) and similar in size to centromeres from higher eukaryotes. Thus, it may be found that isolation of centromeres by selecting for mitotic plasmid stability in a heterologous system will depend on choosing a system that recognizes the relevant centromere sequences.

C. Approaches to Isolation of Pathogenicity Genes

Isolation of genes controlling pathogenicity in fungi can be approached by a variety of methods. Functional complementation is most generally useful because it permits recovery of any gene for which there is a mutation that renders the gene nonfunctional. The main requirement for this approach is an efficient transformation system and the strategy used to isolate a pathogenicity gene (PAT^+) is the following. A library of DNA fragments from a PAT^+ strain is constructed in a vector carrying a gene (SEL^+) selectable in the pathogen of interest. The PAT^+ and SEL^+ alleles must be dominant. Protoplasts of a sel^-; pat^- strain are transformed with the library and transformants are selected for Sel$^+$. They are then screened for the presence of the PAT^+ allele. This may be done by an *in vitro* assay (Yoder, 1981) if the metabolic product of PAT^+ is known (e.g., if it is a toxin or an enzyme). If no product is known the transformants would be screened *in vivo* for ability to cause disease on a plant susceptible to the PAT^+ wild-type strain. When a transformant with the desired pathological phenotype is found, the transforming DNA must be recovered. If the vector is autonomously replicating, plasmid DNA can be isolated directly and amplified in *E. coli*. If an integrating vector is used, the transforming DNA can be recovered by packaging chromosomal sequences in a cosmid and selecting for the vector marker in *E. coli* (Lund *et al.*, 1982).

There are methods for isolating genes for which no mutant alleles are available. If the gene is inducible or developmentally regulated it can be isolated via an abundant mRNA. For example, a large number of developmentally regulated conidiation genes have been isolated from *A. nidulans* by virtue of their differential expression in conidiating versus mycelial cultures (Zimmermann *et al.*, 1980). This approach requires the removal by hybridization of all mRNAs common to both expressing and nonexpressing cultures. If the protein product of the gene is known and can be purified, antibodies to the purified protein can be prepared and used to screen a library of DNA fragments from a wild-type strain carried in an *E. coli* expression vector such as λgt11 (Young and Davis, 1983). Alternatively, the amino acid sequence of the protein can be determined and short synthetic oligonucleotides can be prepared which correspond to regions of the genetic code that are least degenerate. These synthetic probes can be used to screen a library constructed in *E. coli*. For example, the *Neurospora am* gene for glutamate dehydrogenase was isolated by this method (Kinnaird *et al.*, 1982).

D. Uses for Cloned Pathogenicity Genes

Possible ways in which cloned pathogenicity genes from filamentous fungi might be employed are listed in Table VII.

TABLE VII.

Exploitation of Cloned Pathogenicity Genes from Fungi

1. Analyze roles of metabolites in pathogenesis
2. Determine distribution of homologous sequences
3. Determine gene function and regulation
4. Construct agents for biological pest control

1. Analyze Roles of Metabolites in Pathogenesis

Physically defined genes provide the most powerful tools available to analyze whether or not a metabolite plays a role in pathogenesis and, if so, the nature and extent of the role. In the past, metabolites of fungal pathogens have been evaluated as factors in disease using experimental designs that do not exclude uncontrolled variables (Yoder, 1980). If a gene controlling production of a fungal metabolite is cloned, experiments can be performed in which there is a single variable, i.e., the cloned gene itself. To do such an experiment, protoplasts of a strain lacking the metabolite in question would be transformed with a plasmid containing the gene that determines production of the metabolite. Control protoplasts would receive the same plasmid but with an inactive gene for production of the metabolite, thus defining the single variable in the experiment. Colonies derived from the transformed protoplasts would be purified and tested both for the production of the metabolite and for any change in pathogenicity or virulence toward plants. If such a change is detectable, that change indicates the role of the metabolite in disease development. A complication in this experiment could arise if the gene had pleiotropic effects. This possibility could be reduced if the cloned sequence were known to be a structural gene, if there were no other apparent changes in the life cycle of the fungus, and if the metabolic effects of the gene and its product were fully understood.

2. Determine Distribution of Homologous Sequences

A cloned gene can be used as a probe for the isolation of other alleles at its locus, to determine the number of copies per genome, and to assess the phylogenetic boundaries of the gene's range. Isolation of homologous alleles from the same organism would lead to an understanding of whether the difference between mutant and wild-type strains is determined at the level of transcription or translation. It would also be of interest to probe genomic DNAs of a variety of organisms, both pathogens and nonpathogens, with a cloned pathogenicity gene to see which organisms have it. This information would bear on the issue of whether or not pathogens have unique genes for pathogenicity, and therefore unique metabolites for pathogenicity. Alternatively, both pathogens and nonpathogens may have the same sets of structural genes but different mechanisms of regulation. The latter view would be supported if it were found that a gene proved to be necessary for pathogenicity and first isolated from a pathogen also resided and functioned in the genomes of nonpathogens.

3. Determine Gene Function and Regulation

The nucleotide sequence of a cloned pathogenicity gene would reveal the coding and the regulatory regions. By comparison with sequences of other genes, it may be possible to find clues to the gene's function and to identify the elements involved in regulation. Suggestions about function may be found by scrutinizing the sequence (with computer assistance) and by analysis of its polypeptide product. Elements involved in regulation can be identified and studied by the construction of hybrid regulatory regions (Guarente et al., 1982) and assessment of their effects on expression of a gene in the homologous or heterologous genome. For example, when the regulatory region of the yeast *CYC1* gene (which is regulated by catabolite repression) is placed upstream of the *LEU2* gene (which is not normally regulated by catabolite repression), expression of the *LEU2* gene falls under control by catabolite repression (Guarante et al., 1984). It should be possible to identify the regulatory regions of pathogenicity genes by a similar approach.

4. Construct Agents for Biological Pest Control

Pathogenic fungi are potential biological control agents for pests such as weed plants and insects (Yoder, 1983), yet few fungi have been used successfully for biological pest control. There are a number of reasons for this, but one of the most important is that in general fungal pathogens, while being highly specific for their hosts, are not virulent enough to effectively reduce the host population under field conditions. It may be possible to remedy this deficiency by genetic engineering techniques once a supply of cloned pathogenicity genes is available.

Genes useful in construction of biocontrol agents can be classified according to their functions (Table VIII). One class of genes is involved in the production of specialized structures that are needed by the fungus to penetrate the host surface. These genes are thought to control general attacking mechanisms, since virtually all pathogenic fungi form some type of infection structure, whereas nonpathogens do not. Examples of attacking mechanisms include the enzyme cutinase and components of the cytoskeleton (described in Sections II and III).

A second class of genes is invoked after the fungus has penetrated into the plant and must decide whether the plant tissue is a congenial or hostile environment in which to complete its life cycle. Genes involved in this decision control molecular mechanisms that determine host range as well as levels of virulence. The types of metabolic gene products involved here include toxins that condition

TABLE VIII.

Classes of Pathogenicity Genes for Construction of Biocontrol Agents

Gene class	Examples of functional products
1. Penetration	Cutinase, cytoskeleton
2. Colonization	Toxins, enzymes
3. Fitness in nature	Conidia

the host tissue for colonization and enzymes that detoxify toxic compounds produced by the host (see Sections IV and V).

Yet another group of genes is needed by the pathogen for fitness in nature. These genes are probably not confined to pathogens and may confer qualities such as resistance to cold, heat, or desiccation, the quality and quantity of spores, and, for facultative organisms, the ability to compete saprophytically. Survivability in the field may be altered by genes that affect fitness. For example, a gene that changes the shape of the conidium may facilitate aerial dispersal. Conversely, if the life span of the fungus were to be limited, specific genes for senescence (see Section IV,B,1) may be added, or genes for cold-sensitive ribosomes (Russell *et al.*, 1980) may be used to prevent overwintering of inoculum.

There are currently projects under way in various laboratories which aim to isolate genes representing each of the three classes (Table VIII). When such genes are available, realistic attempts to design fungal biocontrol agents will be possible. The strategy will be to take advantage of the natural host-specificity of fungal pathogens to target the control to a particular pest or group of pests, and then to genetically engineer increased virulence by using genes of the various types listed in Table VIII.

REFERENCES

Banks, G. R. (1983a). Transformation of *Ustilago maydis* by a plasmid containing yeast 2-micron DNA *Curr. Genet.* **7,** 73–77.

Banks, G. R. (1983b). Chromosomal DNA sequences from *Ustilago maydis* promote autonomous replication of plasmids in *Saccharomyces cerevisiae. Curr. Genet.* **7,** 79–84.

Beach, D., and Nurse, P. (1981). High-frequency transformation of the fission yeast *Schizosaccharomyces pombe. Nature (London)* **290,** 140–142.

Belcour, L., and Begel, O. (1978). Lethal mitochondrial genotypes in *Podospora anserina:* a model for senescence. *Mol. Gen. Genet.* **163,** 113–123.

Belcour, L., Begel, O., Mosse, M., and Viering, C. (1981). Mitochondrial DNA amplification in

senescent cultures of *Podospora anserina:* variability between the retained, amplified sequences. *Curr. Genet.* **3,** 13–21.
Bull, J. H., and Wootton, J. C. (1984). Heavily methylated amplified DNA in transformants of *Neurospora crassa. Nature (London)* **310,** 701–704.
Carbon, J. (1984). Yeast centromeres: structure and function. *Cell (Cambridge, Mass.)* **37,** 351–353.
Carbon, J., and Clarke, L. (1984). Yeast centromeres and minichromosomes. *Mol. Basis Plant Dis. Conf., Univ. Calif., Davis* Abstr. p. 5.
Case, M. E., Schweizer, M., Kushner, S. R., and Giles, N. H. (1979). Efficient transformation of *Neurospora crassa* by utilizing hybrid plasmid DNA. *Proc. Natl. Acad. Sci. U.S.A.* **76,** 5259–5263.
Catcheside, D. E. A., and Catcheside, D. G. (1979). Survival of *Neurospora* conidia on silica gel. *Neurospora Newsl.* **26,** 24–25.
Collins, R. A., and Lambowitz, A. M. (1983). Structural variations and optional introns in the mitochondrial DNAs of *Neurospora* strains isolated from nature. *Plasmid* **9,** 53–70.
Collins, R. A., Stohl, L. L., Cole, M. D., and Lambowitz, A. M. (1981). Characterization of a novel plasmid DNA found in mitochondria of *N. crassa. Cell (Cambridge, Mass.)* **24,** 443–452.
Colombo, M. M., Swanton, M. T., Donini, P., and Prescott, D. M. (1984). Micronuclear DNA of *Oxytricha nova* contains sequences with autonomously replicating activity in *Saccharomyces cerevisiae. Mol. Cell. Biol.* **4,** 1725–1729.
Cummings, D. J., and Wright, R. M. (1983). DNA sequence of the excision sites of a mitochondrial plasmid from senescent *Podospora anserina. Nucleic Acids Res.* **11,** 2111–2119.
Cummings, D. J., Belcour, L., and Grandchamp, C. (1979). Mitochondrial DNA from *Podospora anserina.* II. Properties of mutant DNA and multimeric circular DNA from senescent cultures. *Mol. Gen. Genet.* **171,** 239–250.
Daly, J. M. (1982). The host-specific toxins of *Helminthosporia. In* "Plant Infection: The Physiological and Biochemical Basis" (Y. Asada *et al.*, eds.), pp. 215–234. Japan Sci. Soc. Press, Tokyo and Springer-Verlag, Berlin and New York.
Daly, J. M., and Knoche, H. W. (1982). The chemistry and biology of pathotoxins exhibiting host selectivity. *Adv. Plant Pathol.* **1,** 83–138.
Das, S., and Hollenberg, C. P. (1982). A high-frequency transformation system for the yeast *Kluyveromyces lactis. Curr. Genet.* **6,** 123.
Dickman, M. B., and Patil, S. S. (1984). Genetic and molecular analysis of a virulence determinant, cutinase from *Colletotrichum gloeosporioides. Mol. Basis Plant Dis. Conf., Univ. Calif., Davis* Abstr. No. 41.
Dickman, M. B., Patil, S. S., and Kolattukudy, P. E. (1982). Purification, characterization and role in infection of an extracellular cutinolytic enzyme from *Colletotrichum gloeosporioides* Penz. on *Carica papaya* L. *Physiol. Plant Pathol.* **20,** 333–347.
Dunkle, L. D., and Wolpert, T. J. (1982). Independence of milo disease symptoms and electrolyte leakage induced by the host-specific toxin from *Periconia circinata. Physiol. Plant Pathol.* **18,** 315–323.
Esser, K., and Tudzynski, P. (1979). Genetic control and expression of senescence in *Podospora anserina. In* "Viruses and Plasmids in Fungi" (P. A. Lemke, ed.), pp. 595–615. Dekker, New York.
Esser, K., Tudzynski, P., Stahl, U., and Kuck, U. (1980). A model to explain senescence in the filamentous fungus *Podospora anserina* by the action of plasmid like DNA. *Mol. Gen. Genet.* **178,** 213–216.
Forde, B. G., Oliver, R. J. C., and Leaver, C. J. (1978). Variation in mitochondrial translation

products associated with male-sterile cytoplasms in maize. *Proc. Natl. Acad. Sci. U.S.A.* **75**, 3841–3845.

Fournier, P., Gaillardin, C., de Louvencourt, L., Heslot, H., Lang, B. F., and Kaudewitz, F. (1982). r-DNA plasmid from *Schizosaccharomyces pombe:* cloning and use in yeast transformation. *Curr. Genet.* **6**, 31–38.

Fraley, R. T., et al. (1983). Expression of bacterial genes in plant cells. *Proc. Natl. Acad. Sci. U.S.A.* **80**, 4803–4807.

Garber, R. C., and Yoder, O. C. (1983). Isolation of DNA from filamentous fungi and separation into nuclear, mitochondrial, ribosomal, and plasmid components. *Anal. Biochem.* **135**, 416–422.

Garber, R. C., Turgeon, B. G., and Yoder, O. C. (1984). A mitochondrial plasmid from the plant pathogenic fungus *Cochliobolus heterostrophus. Mol. Gen. Genet.* **196**, 301–310.

Gilchrist, D. G. (1983). Molecular modes of action. *In* "Toxins and Plant Pathogenesis" (J. M. Daly and B. J. Deverall, eds.), pp. 81–136. Academic Press, New York.

Gilchrist, D. G., and Grogan, R. G. (1976). Production and nature of a host-specific toxin from *Alternaria alternata* f. sp. *lycopersici. Phytopathology* **66**, 165–171.

Gorman, C., Padmanabhan, R., and Howard, B. H. (1983). High efficiency DNA-mediated transformation of primate cells. *Science (Washington, D.C.)* **221**, 551–553.

Gorman, J. A., Dove, W. F., and Warren, N. (1981). Isolation of *Physarum* DNA segments that support autonomous relication in yeast. *Mol. Gen. Genet.* **183**, 306–313.

Gregory, P., Earle, E. D., and Gracen, V. E. (1977). Biochemical and ultrastructural aspects of southern corn leaf blight disease. *In* "Host Plant Resistance to Pests" (P. A. Hedin, ed.), ACS Symp. Ser., No. 62, pp. 90–114. Am. Chem. Soc., Washington, D.C.

Gregory, P., Earle, E. D., and Gracen, V. E. (1980). Effects of purified *Helminthosporium maydis* race T toxin on the structure and function of corn mitochondria and protoplasts. *Plant Physiol.* **66**, 477–481.

Griffiths, A. J. F., and Bertrand, H. (1984). Unstable cytoplasms in Hawaiian strains of *Neurospora intermedia. Curr. Genet.* **8**, 387–398.

Gritz, L., and Davies, J. (1983). Plasmid-encoded hygromycin B. resistance: the sequence of hygromycin B phosphotransferase gene and its expression in *Escherichia coli* and *Saccharomyces cerevisiae. Gene* **25**, 179–188.

Guarente, L., Yocum, R., and Gifford, P. (1982). A GAL10–CYC1 hybrid yeast promoter identifies the GAL4 regulatory region as an upstream site. *Proc. Natl. Acad. Sci. U.S.A.* **79**, 7410–7414.

Guarente, L., Lalonde, B., Gifford, P., and Alani, E. (1984). Distinctly regulated tandem upstream activation sites mediate catabolite repression of the *CYC1* gene of *S. cerevisiae. Cell (Cambridge, Mass.)* **36**, 503–511.

Guzman, D., Garber, R. C., and Yoder, O. C. (1982). Cytology of meiosis I and chromosome number of *Cochliobolus heterostrophus* (Ascomycetes). *Can. J. Bot.* **60**, 1138–1141.

Hinnen, A., Hicks, J. B., and Fink, G. R. (1978). Transformation of yeast. *Proc. Natl. Acad. Sci. U.S.A.* **75**, 1929–1933.

Hirth, K., Edwards, C. A., and Firtel, R. A. (1982). A DNA-mediated transformation system for *Dictyostelium discoideum. Proc. Natl. Acad. Sci. U.S.A.* **79**, 7356–7360.

Hoch, H. C., and Staples, R. C. (1983). Visualization of actin *in situ* by rhodamine-conjugated phalloin in the fungus *Uromyces phaseoli. Eur. J. Cell Biol.* **32**, 52.

Hoch, H. C., and Staples, R. C. (1984). Evidence that cAMP initiates nuclear division and infection structure formation in the bean rust fungus, *Uromyces phaseoli. Exp. Mycol.* **8**, 37–46.

Holden, M. J., and Sze, H. (1984). *Helminthosporium maydis* T toxin increased membrane permeability to Ca^{2+} in susceptible corn mitochondria. *Plant Physiol.* **75**, 235–237.

Hsu, W. H., Magee, P. T., Magee, B. B., and Reddy, C. A. (1983). Construction of a new yeast

cloning vector containing autonomous replication sequences from *Candida utilis*. *J. Bacteriol.* **154,** 1033–1039.
Huang, B.-F., and Staples, R. C. (1982). Synthesis of proteins during differentiation of the bean rust fungus. *Exp. Mycol.* **6,** 7–14.
Jamet-Vierny, C., Begel, O., and Belcour, L. (1980). Senescence in *Podospora anserina:* Amplification of a mitochondrial DNA sequence. *Cell (Cambridge, Mass.)* **21,** 189–194.
Jimenez, A., and Davies, J. (1980). Expression of a transposable antibiotic resistant element in *Saccharomyces*. *Nature (London)* **287,** 869–871.
Kaster, K. R., Burgett, S. G., and Ingolia, T. D. (1984). Hygromycin B resistance as dominant selectable marker in yeast. *Curr. Genet.* **8,** 353–358.
Kawamura, M., Takagi, M., and Yano, K. (1983). Cloning of a *LEU* gene and an *ARS* site of *Candida maltosa*. *Gene* **24,** 157–162.
Kemble, R. J., and Pring, D. R. (1982). Mitochondrial DNA associated with cytoplasmic male sterility and disease susceptibility in maize carrying Texas cytoplasm. *In* "Plant Infection: The physiological and Biochemical Basis" (Y. Asada *et al.*, eds.) pp. 187–197. Japan Sci. Soc. Press, Tokyo and Springer-Verlag, Berlin and New York.
Kenfield, D. S., and Strobel, G. A. (1981). α-Galactoside binding proteins from plant membranes: isolation, characterization, and relation to helminthosporoside binding proteins of sugarcane. *Plant Physiol.* **67,** 1174–1180.
Kinnaird, J. H., Keighren, M. A., Kinsey, J. A., Eaton, M., and Fincham, J. R. S. (1982). Cloning of the *am* (glutamate dehydrogenase) gene of *Neurospora crassa* through the use of a synthetic DNA probe. *Gene* **20,** 387–396.
Kistler, H. C., and VanEtten, H. D. (1984a). Regulation of pisatin demethylation in *Nectria haematococca* and its influence on pisatin tolerance and virulence. *J. Gen. Microbiol.* **130,** 2605–2613.
Kistler, H. C., and VanEtten, H. D. (1984b). Three non-allelic genes for pisatin demethylation in the fungus *Nectria haematococca*. *J. Gen. Microbiol.* **130,** 2595–2603.
Kolattukudy, P. E. (1981). Structure, biosynthesis, and biodegradation of cutin and suberin. *Annu. Rev. Plant Physiol.* **32,** 539–567.
Koller, W., and Kolattukudy, P. E. (1982). Mechanism of action of cutinase: chemical modification of the catalytic triad characteristic for serine hydrolases. *Biochemistry* **21,** 3083–3090.
Koller, W., Allan, C. R., and Kolattukudy, P. E. (1982a). Inhibition of cutinase and prevention of fungal penetration into plants by benomyl—a possible protective mode of action. *Pestic. Biochem. Physiol.* **18,** 15–25.
Koller, W., Allan, C. R., and Kolattukudy, P. E. (1982b). Protection of *Pisum sativum* from *Fusarium solani* f. sp. *pisi* by inhibition of cutinase with organophosphorus pesticides. *Phytopathology* **72,** 1425–1430.
Koller, W., Allan, C. R., and Kolattukudy, P. E. (1982c). Role of cutinase and cell wall degrading enzymes in infection of *Pisum sativum* by *Fusarium solani* f. sp. *pisi*. *Physiol. Plant Pathol.* **20,** 47–60.
Kück, U., Stahl, U., and Esser, K. (1981). Plasmid-like DNA is part of mitochondrial DNA in *Podospora anserina*. *Curr. Genet.* **3,** 151–156.
Lazarus, C. M., and Kuntzel, H. (1981). Anatomy of amplified mitochondrial DNA in "ragged" mutants of *Aspergillus amstelodami:* Excision points within protein genes and a common 215 bp segment containing a possible origin of replication. *Curr. Genet.* **4,** 99–107.
Lazarus, C. M., Earle, A. J., Turner, G., and Kuntzel, H. (1980). Amplification of a mitochondrial DNA sequence in the cytoplasmically inherited "ragged" mutant of *Aspergillus amstelodami*. *Eur. J. Biochem.* **106,** 633–641.
Lazdins, I. B., and Cummings, D. J. (1982). Autonomously replicating sequences in young and senescent mitochondrial DNA from *Podospora anserina*. *Curr. Genet.* **6,** 173–178.

Lesney, M. S., Livingston, R. S., and Scheffer, R. P. (1982). Effects of toxin from *Helminthosporium sacchari* on nongreen tissues and a reexamination of toxin binding. *Phytopathology* **22,** 844–849.

Lin, T. S., and Kolattukudy, P. E. (1978). Induction of a biopolyester hydrolase (cutinase) by low levels of cutin monomers in *Fusarium solani* f. sp. *pisi. J. Bacteriol.* **133,** 942–951.

Lin, T. S., and Kolattukudy, P. E. (1980). Isolation and characterization of a cuticular polyester (cutin) hydrolyzing enzyme from phytopathogenic fungi. *Physiol. Plant Pathol.* **17,** 1–15.

Livingston, R. S., and Scheffer, R. P. (1981). Isolation and characterization of host-selective toxin from *Helminthosporium sacchari*. *J. Biol. Chem.* **256,** 1705–1710.

Loppes, R., and Denis, C. (1983). Chloroplast and nuclear DNA fragments from *Chlamydomonas* promoting high frequency transformation of yeast. *Curr. Genet.* **7,** 473–480.

Losson, R., and Lacroute, F. (1983). Plasmids carrying the yeast OMP decarboxylase structural and regulatory genes: Transcription regulation in a foreign environment. *Cell (Cambridge, Mass.)* **32,** 371–377.

Lund, T., Grosveld, F. G., and Flavell, R. A. (1982). Isolation of Transforming DNA by cosmid rescue. *Proc. Natl. Acad. Sci. U.S.A.* **79,** 520–524.

Macko, V. (1983). Structural aspects of toxins. *In* "Toxins and Plant Pathogenesis" (J. M. Daly and B. J. Deverall, eds.), pp. 41–80. Academic Press, New York.

Macko, V., Acklin, W., Hildenbrand, C., Weibel, F., and Arigoni, D. (1983). Structure of three isomeric host-specific toxins from *Helminthosporium sacchari*. *Experientia* **39,** 343–347.

Maiti, I. B., and Kolattukudy, P. E. (1979). Prevention of fungal infection of plants by specific inhibition of cutinase. *Science (Washington, D.C.)* **205,** 507–508.

Malone, C. P., Miller, R. J., and Koeppe, D. E. (1978). The *in vivo* response of corn mitochondria to *Bipolaris (Helminthosporium) maydis* (race T) toxin. *Physiol. Plant.* **44,** 21–25.

Maniatis, T., Fritsch, E. F., and Sambrook, J. (1982). "Molecular Cloning: A Laboratory Manual." Cold Spring Harbor Lab., Cold Spring Harbor, New York.

Matthews, D. E., and Van Etten, H. D. (1983). Detoxification of the phytoalexin pisatin by a fungal cytochrome P-450. *Arch. Biochem. Biophys.* **224,** 494–505.

Matthews, D. E., Gregory, P., and Gracen, V. E. (1979). *Helminthosporium maydis* race T toxin induces leakage of NAD$^+$ from T cytoplasm corn mitochondria. *Plant Physiol.* **63,** 1149–1153.

Nargang, F. E., Bell, J. B., Stohl, L. L., and Lambowitz, A. M. (1983). A family of repetitive palindromic sequences found in *Neurospora* mitochondrial DNA is also found in a mitochondrial plasmid DNA. *J. Biol. Chem.* **258,** 4257–4260.

Natvig, D. O., May, G., and Taylor, J. W. (1984). Distribution and evolutionary significance of mitochondrial plasmids in *Neurospora* spp. *J. Bacteriol.* **159,** 288–293.

Nishimura, S., and Kohmoto, K. (1983). Host-specific toxins and chemical structures from *Alternaria* species. *Annu. Rev. Phytopathol.* **21,** 87–116.

Ohtani, T., Uchimiya, H., Kato, A., Harada, H., Sugita, M., and Sugiura, M. (1984). Location and nucleotide sequence of a tobacco chloroplast DNA segment capable of replication in yeast. *Mol. Gen. Genet.* **195,** 1–4.

Osiewacz, H. D., and Esser, K. (1983). DNA sequence analysis of the mitochondrial plasmid of *Podospora anserina*. *Curr. Genet.* **7,** 219–223.

Perkins, D. D. (1977). Details for preparing silica gel stocks. *Neurospora Newsl.* **24,** 16–17.

Purdy, R. E., and Kolattukudy, P. E. (1975). Hydrolysis of plant cuticle by plant pathogens. Properties of cutinase I, cutinase II, and a nonspecific esterase isolated from *Fusarium solani pisi*. *Biochemistry* **14,** 2832–2840.

Rochaix, J. D., van Dillewijn, J., and Rahire, M. (1984). Construction and characterization of autonomously replicating plasmids in the green unicellular alga *Chlamydomonas reinhardii*. *Cell (Cambridge, Mass.)* **36,** 925–931.

Roth, G. E., Blanton, H. M., Hager, L. J., and Zakian, V. A. (1983). Isolation and characterization of sequences from mouse chromosomal DNA with *ARS* function in yeast. *Mol. Cell. Biol.* **3,** 1898–1908.

Rudolph, K. (1976). Non-specific toxins. *Encycl. Plant Physiol., New ser.* **4,** 270–315.

Russell, P. J., Granville, R. R., and Tublitz, N. (1980). A cold-sensitive mutant of *Neurospora crassa* obtained using tritium-suicide enrichment that is conditionally defective in the biosynthesis of cytoplasmic ribosomes. *Exp. Mycol.* **4,** 23–32.

Ryan, F. J., Beadle, G. W., and Tatum, E. L. (1943). The tube method of measuring the growth rate of *Neurospora. Am. J. Bot.* **30,** 784–799.

Scheffer, R. P. (1976). Host-specific toxins in relation to pathogenesis and disease resistance. *Encycl. Plant Physiol., New Ser.* **4,** 247–269.

Scheffer, R. P., and Livingston, R. S. (1984). Host-selective toxins and their role in plant disease. *Science (Washington, D.C.)* **223,** 17–21.

Shaykh, M., Soliday, C., and Kolattukudy, P. E. (1977). Proof for the production of cutinase by *Fusarium solani* f. *pisi* during penetration into its host, *Pisum sativum. Plant Physiol.* **60,** 170–172.

Skatrud, P. L., and Queener, S. W. (1984). Cloning of a DNA fragment from *Cephalosporium* which functions as an autonomous replication sequence in yeast. *Curr. Genet.* **8,** 155–163.

Smith, J. R., and Rubenstein, I. (1973a). The development of "senescence" in *Podospora anserina. J. Gen. Microbiol.* **76,** 283–296.

Smith, J. R., and Rubenstein, I. (1973b). Cytoplasmic inheritance of the timing of "senescence" in *Podospora anserina. J. Gen. Microbiol.* **76,** 297–304.

Soliday, C. L., and Kolattukudy, P. E. (1976). Isolation and characterization of a cutinase from *Fusarium roseum culmorum* and its immunological comparison with cutinases from *F. solani pisi. Arch. Biochem. Biophys.* **176,** 334–343.

Soliday, C. L., Flurkey, W. H., Okita, T. W., and Kolattukudy, P. E. (1984). Cloning and structure determination of cDNA for cutinase, an enzyme involved for fungal penetration of plants. *Proc. Natl. Acad. Sci. U.S.A.* **81,** 3939–3943.

Sreekrishna, K., Webster, T. D., and Dickson, R. C. (1984). Transformation of *Kluyveromyces lactis* with the kanamycin (G418) resistance gene of Tn903. *Gene* **28,** 73–81.

Stahl, U., Lemke, P. A., Tudzynski, P., Kuck, U., and Esser, K. (1978). Evidence for plasmid like DNA in a filamentous fungus, ascomycete *Podospora anserina. Mol. Gen. Genet.* **162,** 341–343.

Stahl, U., Kuck, U., Tudzynski, P., and Esser, K. (1980). Characterization and cloning of plasmid like DNA of the ascomycete *Podospora anserina. Mol. Gen. Genet.* **178,** 639–646.

Stahl, U., Tudzynski, P., Kuck, U., and Esser, K. (1982). Replication and expression of a bacterial–mitochondrial hybrid plasmid in the fungus *Podospora anserina. Proc. Natl. Acad. Sci. U.S.A.* **79,** 3641–3645.

Staples, R. C., and Hoch, C. (1982). A possible role for microtubules and microfilaments in the induction of nuclear division in bean rust uredospore germlings. *Exp. Mycol.* **6,** 293–302.

Staples, R. C., and Macko, V. (1984). Germination of uredospores and differentiation of infection structures. *In* "The Cereal Rusts, Vol. 1, Origins, Specificity, Structure, and Physiology" (W. R. Bushnell and A. P. Roelphs, eds.), pp. 255–289. Academic Press, New York.

Staples, R. C., App, A. A., and Ricci, P. (1975). DNA synthesis and nuclear division during formation of infection structures by bean rust uredospore germlings. *Arch. Microbiol.* **104,** 123–127.

Staples, R. C., Grambow, H.-J., and Hoch, H. C. (1983a). Potassium ion induces rust fungi to develop infection structures. *Exp. Mycol.* **7,** 40–46.

Staples, R. C., Grambow, H.-J., Hoch, H. C., and Wynn, W. K. (1983b). Contact with membrane grooves induces wheat stem rust uredospore germlings to differentiate appressoria but not vesicles. *Phytopathology* **73,** 1436–1439.

Steiner, G. W., and Strobel, G. A. (1971). Helminthosporoside, a host-specific toxin from *Helminthosporium sacchari. J. Biol. Chem.* **246,** 4350–4357.
Stinchcomb, D. T., Thomas, M., Kelly, J., Selker, E., and Davis, R. W. (1980). Eukaryotic DNA segments capable of autonomous replication in yeast. *Proc. Natl. Acad. Sci. U.S.A.* **77,** 4559–4563.
Stoessl, A. (1981). Structure and biogenetic relations: fungal nonhost-specific. *In* "Toxins in Plant Disease" (R. D. Durbin, ed.), pp. 110–219. Academic Press, New York.
Stohl, L. L., and Lambowitz, A. M. (1983). Construction of a shuttle vector for the filamentous fungus *Neurospora crassa. Proc. Natl. Acad. Sci. U.S.A.* **80,** 1058–1062.
Stohl, L. L., Collins, R. A., Cole, M. D., and Lambowitz, A. M. (1982). Characterization of two new plasmid DNAs found in mitochondria of wild-type *Neurospora intermedia* strains. *Nucleic Acids Res.* **10,** 1439–1458.
Strobel, G. A. (1973). Biochemical basis of the resistance of sugarcane to eyespot disease. *Proc. Natl. Acad. Sci. U.S.A.* **70,** 1693–1696.
Suzci, A., and Radford, A. (1983). *ARS*8 sequences in the *Neurospora* genome. *Neurospora Newsl.* **30,** 13.
Tegtmeier, K. J., and VanEtten, H. D. (1982). The role of pisatin tolerance and degradation in the virulence of *Nectria haematococca* on peas: a genetic analysis. *Phytopathology* **72,** 608–612.
Tikhomirova, L. P., Kryukov, V. M., Strizhov, N. I., and Bayev, A. A. (1983). mtDNA sequences of *Candida utilis* capable of supporting autonomous relication of plasmids in *Saccharomyces cerevisiae. Mol. Gen. Genet.* **189,** 479–784.
Tudzynski, P., and Esser, K. (1979). Chromosomal and extrachromosomal control of senescence in the ascomycete *Podospora anserina. Mol. Gen. Genet.* **173,** 71–84.
VanEtten, H. D., and Barz, W. (1981). Expression of pisatin demethylating ability in *Nectria haematococca. Arch. Microbiol.* **129,** 56–60.
VanEtten, H. D., and Matthews, P. S. (1984). Naturally occurring variation in the inducibility of pisatin demethylating activity in *Nectria haematococca* mating population VI. *Physiol. Plant Pathol.* **25,** 149–160.
VanEtten, H. D., Matthews, P. S., Tegtmeier, K. J., Dietert, M. F., and Stein, J. I. (1980). The association of pisatin tolerance and demethylation with virulence on pea in *Nectria haematococca. Physiol. Plant Pathol.* **16,** 257–268.
Vierny, C., Keller, A.-M., Begel, O., and Belcour, L. (1982). A sequence of mitochondrial DNA is associated with the onset of senescence in a fungus. *Nature (London)* **297,** 157–159.
Walton, J. D., Earle, E. D., Yoder, O. C., and Spanswick, R. M. (1979). Reduction of adenosine triphosphate levels in susceptible maize mesophyll protoplasts by *Helminthosporium maydis* race T toxin. *Plant Physiol.* **63,** 806–810.
Wolpert, T. J., and Dunkle, L. D. (1983). Alterations in gene expression in sorghum induced by the host-specific toxin from *Periconia circinata. Proc. Natl. Acad. Sci. U.S.A.* **80,** 6576–6580.
Wright, R. M., and Cummings, D. J. (1983). Integration of mitochondrial gene sequences within the nuclear genome during senescence in a fungus. *Nature (London)* **30,** 86–88.
Wright, R. M., Horrum, M. A., and Cummings, D. J. (1982). Are mitochondrial structural genes selectively amplified during senescence in *Podospora anserina? Cell (Cambridge, Mass.)* **29,** 505–515.
Wynn, W. K. (1976). Appressorium formation over stomates by the bean rust fungus: response to a surface contact stimulus. *Phytopathology* **66,** 136–146.
Wynn, W. K., and Staples, R. C. (1981). Tropisms of fungi in host recognition. *In* "Plant Disease Control: Resistance and Susceptibility" (R. C. Staples and G. H. Toenniessen, eds.), pp. 45–69. Wiley, New York.
Yelton, M. M., Hamer, J. E., and Timberlake, W. E. (1984). Transformation of *Aspergillus nidulans* by using a *trp*C plasmid. *Proc. Natl. Acad. Sci. U.S.A.* **81,** 1470–1474.
Yoder, O. C. (1980). Toxins in pathogenesis. *Annu. Rev. Phytopathol.* **18,** 103–129.

Yoder, O. C. (1981). Assay. *In* "Toxins in Plant Disease" (R. D. Durbin, ed.), pp. 45–78. Academic Press, New York.

Yoder, O. C. (1983). Use of pathogen-produced toxins in genetic engineering of plants and pathogens. *In* "Basic Life Sciences, Vol. 26, Genetic Engineering of Plants: An Agricultural Perspective" (T. Kosuge, C. Meredith, and A. Hollaender, eds.), pp. 335–353. Plenum, New York.

Young, R. A., and Davis, R. W. (1983). Efficient isolation of genes by using antibody probes. *Proc. Natl. Acad. Sci. U.S.A.* **80,** 1194–1198.

Zakian, V. A. (1981). Origin of replication from *Xenopus laevis* mitochondrial DNA promotes high-frequency transformation of yeast. *Proc. Natl. Acad. Sci. U.S.A.* **78,** 3128–3132.

Zimmermann, C. R., Orr, W. C., Leclerc, R. F., Barnard, E. C., and Timberlake, W. E. (1980). Molecular cloning and selection of genes regulated in *Aspergillus* development. *Cell (Cambridge, Mass.)* **21,** 709–715.

18

Morphogenesis and Dimorphism of *Mucor*

RONALD L. CIHLAR
Department of Microbiology
Schools of Medicine and Dentistry
Georgetown University
Washington, D.C.

I.	Introduction	449
II.	Dimorphism of *Mucor*	450
	A. Enzyme Correlates of Morphogenesis	452
	B. Macromolecular Biosynthesis	454
III.	Molecular Analysis of *Mucor*	457
	A. Regulation of Protein Synthesis	457
	B. Genetic Analysis	459
	C. Molecular Genetics	460
IV.	Perspectives—Molecular Genetics	462
	References	464

I. INTRODUCTION

Dimorphic fungi are a group of organisms capable of either yeast or mycelial growth in response to environmental conditions. Experimentally, external factors such as growth medium composition, sparging gas, or temperature are most often manipulated in order to promote mycelial or yeast development and, more importantly for molecular investigations, to control a morphological transition from one actively growing cell type to another. Thus, not only can dimorphic fungi be exploited to examine questions pertaining to the biochemical basis and regulation of dimorphism itself, but investigation of the process may yield insights concerning cellular differentiation in general.

The zygomycetes *Mucor rouxii* and *Mucor racemosus* have been more rigorously studied in regard to the biochemical and molecular events that accompany morphogenesis than have most other dimorphic fungi. The *Mucor* system is particularly attractive since the yeast-to-hyphae transition can be either induced (Bartnicki-Garcia and Nickerson, 1962a; Mooney and Sypherd, 1976) or inhibited (Larsen and Sypherd, 1974; Ito *et al.*, 1982) in a variety of ways. This versatility allows for a distinction to be made between the physiological alterations that may be intimately related to morphogenesis (i.e., changes that occur under all conditions utilized to control the dimorphic response) and other secondary adjustments that arise solely in response to growth conditions (i.e., changes that occur under a limited number of the conditions available to control the dimorphic response).

Studies dealing with the yeast-to-hyphae transition of *Mucor* have concentrated largely on problems concerning cell wall biosynthesis (Bartnicki-Garcia, 1968a; Ruiz-Herrera and Bartnicki-Garcia, 1976; Domek and Borgia, 1981), the differential expression of key enzymes (Peters and Sypherd, 1979; Inderlied *et al.*, 1980), and the relationship between posttranslational modifications and protein function (Larsen and Sypherd, 1980; Hiatt *et al.*, 1982). In addition, the powerful tools provided by advances in molecular genetics have been applied to the *Mucor* system, as well as to studies dealing with other dimorphic fungi.

A primary aim of this review is to discuss the progress that has been made and approaches being employed in the latter type of investigation. The chapter will focus on *M. rouxii* and *M. racemosus;* however, observations concerning other dimorphic fungi will be mentioned where appropriate. The section concerning *Mucor* dimorphism is not intended to be comprehensive, but rather is designed to alert the reader to the more important physiological and biochemical correlates of morphogenesis that have been described to date. The reader is referred to recent reviews pertaining to dimorphism (Sypherd *et al.*, 1978; Stewart and Rogers, 1978; Farkas, 1979; Szanizlo *et al.*, 1983) for an in-depth treatment of the subject.

II. DIMORPHISM OF *MUCOR*

While a number of fungi exhibit yeastlike or mycelial growth, some fungi have the capacity to grow in either form in response to the environment. The property of dimorphism has potential clinical implications since, as a rule, the most severe systemic mycoses of man are caused by dimorphic fungi, although it is not clear that dimorphism in itself contributes to pathogenicity. Induction of either morphology is generally influenced by the environment. For example, *Candida albicans* exhibits yeastlike development in glucose-containing medium at 28°C (Barlow *et al.*, 1974). On the other hand, dilution of early stationary phase *C. albicans* cells into starvation buffer containing proline results in germ tube emergence (Dabrowa *et al.*, 1976). Increased CO_2 tension has been shown to favor the spherule phase of *Coccidioides immitis* (Sun and Huppert, 1976), while temperature and degree of agitation influence the uniformity of yeastlike growth of *Exophiala werneckii* (Hardcastle and Szaniszlo, 1974).

Clearly, the ability to mediate the yeast-to-hyphae transition provides an advantageous system in which to investigate the biochemical and molecular mechanisms that operate during differentiation. In this regard, investigations have shown that *M. rouxii* and *M. racemosus* are well suited for such studies. They are easily grown in a simple medium (Bartnicki-Garcia and Nickerson, 1962b), have a generation time of 2–4 hr depending upon the sparging gas, and the cellular morphology observed is a function of these growth conditions. For example, development as yeast requires a fermentable hexose and occurs under an atmosphere of CO_2 or CO_2/N_2 mixture (Bartnicki-Garcia and Nickerson, 1962b). Studies describing the effect of a 100% N_2 environment on *Mucor* growth (Haidle and Storck, 1966; Bartnicki-Garcia, 1968b) culminated in the work of Mooney and Sypherd (1976) which demonstrated that yeast growth under an N_2 atmosphere was dependent on the gas sparging rate rather than hexose concentration. It was also observed that if nitrogen anaerobiosis was maintained, hyphal development ensued upon elimination of gas flow (Mooney and Sypherd, 1976). The investigators, therefore, suggested that an unidentified volatile factor produced by *M. racemosus* might be responsible for triggering morphogenesis. However, subsequent investigations (P. T. Borgia, unpublished observations) demonstrated that more stringent scrubbing of N_2 to remove trace amounts of oxygen resulted in only yeast growth irrespective of flow rate, thereby discounting the importance of a volatile factor as a morphogen under these conditions.

While yeastlike development occurs when conditions favor fermentation, mycelial development occurs under conditions that allow for the emergence of respiratory capacity. In fact, early studies suggested that respiratory development was essential for mycelial growth. Experiments with *M. rouxii* have shown that morphogenesis can be inhibited by oxidative phosphorylation uncoupling agents (Terenzi and Storck, 1969). Similarly, inhibition of *M. rouxii* (Zorzopoulos *et al.*, 1973) and *M. genevensis* (Clark-Walker, 1973) mitochondrial protein biosynthesis with chloramphenicol also favored growth as yeast. Additional support for a link with respiratory capacity came from the characterization of respiratory-deficient mutants of *M. bacilliformis* that remained as yeast when grown in air (Storck and Morrill, 1971). In contrast, other experiments with *M. racemosus* (Paznokas and Sypherd, 1975) demonstrated that respiration could be divorced from hyphal growth, and it was concluded that morphogenesis was not a direct consequence of such capabilities. However, recent investigations of Ito *et al.* (1982) concerning lipid synthesis throughout the yeast-to-hyphae transition of *M. racemosus* led the authors to suggest that mitochondriogenesis may be important to the dimorphic response. Thus, in summary, although respiration per se may not be a prerequisite for hyphal development, a requirement for another mitochondrial function(s) in the process cannot be excluded.

While controversy exists concerning the role of mitochondria in *Mucor* morphogenesis, it has been demonstrated that the intracellular levels of cyclic 3′,5′-adenosine monophosphate (cAMP) are crucial in influencing cell morphology. Larsen and Sypherd (1974) and Paveto et al. (1975) found in *M. racemosus* and *M. rouxii*, respectively, that cAMP levels are approximately fourfold higher in CO_2-grown yeast than in aerobically grown hyphae. The lower intracellular cAMP level in *M. rouxii* hyphae was attributed to a corresponding increase in cAMP phosphodiesterase activity (Paveto et al., 1975). More importantly, both groups of investigators observed that supplementation of CO_2-grown yeast cultures with dibutyryl-cAMP (dbcAMP) concomitant with the introduction of air resulted in continued yeastlike growth rather than mycelial development. In this regard, the cAMP effect has been incorporated into several investigations in order to assess the importance of a number of biochemical correlates of morphogenesis (Orlowski and Sypherd, 1977; Peters and Sypherd, 1979; Domek and Borgia, 1981). It should also be noted that Orlowski (1980) examined cAMP metabolism during hyphal development from sporangiospores of *M. genevensis* and *M. mucedo*. It was observed that intracellular cAMP levels initially increased during the early stages of spore germination and then dropped dramatically prior to the appearance of germ tubes. In this case, changes in adenylate cyclase activity, rather than phosphodiesterase activity (Paveto et al., 1975), were judged to be responsible for regulating intracellular cAMP pool size. The contrasting results from the two investigations most probably reflects the different experimental conditions used by the individual groups. Finally, Orlowski (1980) demonstrated that hyphal development from sporangiospores could also be blocked by exogenous cAMP, thereby substantiating the correlation between cAMP levels and development.

Although the effect of cAMP on *Mucor* morphogenesis has been documented, the mechanism of its action remains obscure. One obvious explanation is that cAMP may be important in the regulation of cAMP-dependent kinases, which, in turn, may modulate the function of key proteins by affecting protein phosphorylation. Investigations designed to examine this possibility were conducted by Moreno et al. (1977), who showed the cochromatography of protein kinase activity and cAMP-binding protein activity of *M. rouxii*, and by Forte and Orlowski (1980), who measured cAMP-binding activity during hyphal development of *M. racemosus* and *M. genevensis*. The latter group found that such activity decreased during morphogenesis and that proteins of 51,000 and 65,000 daltons accounted for most of the observed cAMP-binding activity in both species. The relationship of this observation to dimorphism remains obscure.

A. Enzyme Correlates of Morphogenesis

With several defined conditions that allow for a rapid, highly reproducible morphological conversion (yeast to hyphae), studies have turned toward dissecting the biochemical changes that accompany the process. In particular, investigations concerning differential enzyme activity have focused on metabolic processes that might a priori be expected to directly influence cell morphology. These include enzymes required for sugar utilization (Paznokas and Sypherd, 1977; Inderlied and Sypherd, 1978; Borgia and Mehnert, 1982), nitrogen metabolism (Peters and Sypherd, 1979), and polyamine biosynthesis (Inderlied et al., 1980; Garcia et al., 1980).

Because of the strict requirement for the presence of a fermentable hexose during yeast growth, Inderlied and Sypherd (1978) reasoned that alterations in the flux of carbon through glucose catabolic pathways could point to key metabolites or enzymes important in influencing yeast or hyphal development. However, no significant differences in the catabolism of glucose during growth of either cell type were noted. The activity of pyruvate kinase, an enzyme that serves an important regulatory role in hexose utilization, has also been examined in M. rouxii and M. racemosus as a potential correlate of morphogenesis (Friedenthal et al., 1973; Paznokas and Sypherd, 1977). In both cases two differentially regulated isozymes predominated. Furthermore, it was ascertained that the appearance of activity of either isozyme in M. racemosus (Paznokas and Sypherd, 1977) was related solely to the nature of the carbon source, and consequently could not be directly linked to morphogenesis.

The nature of the nitrogen source available to Mucor also is important in promoting either yeast or mycelial growth. In general, growth as yeast requires an organic nitrogen source, while the requirement for hyphal growth can be satisfied by inorganic sources. Peters and Sypherd (1978) later showed that a minimal medium containing alanine, aspartate, glutamate, and ammonium chloride satisfied the nitrogen requirement for either yeast or hyphal cells of M. racemosus. On the other hand, yeast growth was inhibited in an identical medium lacking glutamate. This observation prompted experiments concerning the regulation of glutamate dehydrogenase (GDH) during morphogenesis (Peters and Sypherd, 1979). The investigators detected both NAD- and NADP-dependent GDH in both cell types; however, it was noted that only NAD-dependent GDH activity was enhanced after induction of hyphal development ($CO_2 \rightarrow$ air shift). In addition, the elevation in enzyme activity occurred prior to the appearance of germ tubes, and was prevented if yeast cultures were supplemented with dbcAMP concurrently with the introduction of air. These observations led to the conclusion that the differential expression of GDH during M. racemosus morphogenesis was a valid biochemical correlate of the yeast-to-hyphae transition, but evidence for a direct involvement of GDH in the process has yet to be provided.

Polyamines have been implicated in the mediation of a variety of alterations that accompany proliferation and differentiation in eukaryotic cells (Russell and Snyder, 1968; Cohen, 1978). Therefore, the possibility that either polyamines or the enzymes involved in their biosynthesis might play a role in inducing Mucor morphogenesis has been examined. Garcia et al. (1980) found that putrescine levels are approximately five times higher in hyphal cells than in yeasts, while spermidine levels remain constant in both cell types. As is the case in other filamentous fungi (Nickerson *et al.*, 1977), no spermine was detected. The activity of ornithine decarboxylase (ODC), the first enzyme leading to polyamine biosynthesis, was also studied throughout the yeast-to-hyphae transition of *M. racemosus* (Inderlied *et al.*, 1980). Enzyme activity increased 30- to 50-fold within 2–3 hr following a shift in incubation atmosphere from CO_2 to air and then rapidly declined. Supplementation of cultures with dbcAMP, putrescine, or cerulenin (Ito *et al.*, 1982) prevented the increase in ODC activity. Although enhanced ODC levels correlate well with morphogenesis, the relationship of the increase in ODC activity, as well as elevated putrescine levels, to morphogenesis remains unknown.

More recent experiments concerning the biochemical correlates of *Mucor* morphogenesis have been performed utilizing monomorphic mutants of *M. bacilliformis* that do not form hyphae (Ruiz-Herrera *et al.*, 1983). Fourteen mutants were isolated and then, subsequent to aerobic or anaerobic growth, were characterized with regard to respiratory capacity, activities of ODC and GDH, and intracellular cAMP levels. None of the mutants showed a normal pattern of oxidative energy metabolism. An aberrant cytochrome spectrum, as well as apparent differences in the sensitivity of respiration to cyanide or salicylhydroxamic acid, occurred with all but one of the morphological mutants. The remaining strain was postulated to be deficient in some aspect of energy transduction. Likewise, a strict correlation between low levels of ODC activity and the inability to form hyphae was observed. On the other hand, only two of the mutants examined had low levels of NAD-dependent GDH activity when grown in air. In addition, intracellular cAMP levels in aerobically grown mutant strains (yeast morphology) were generally comparable to levels observed for the wild type (hyphal morphology). Therefore, the suggestion was put forth that ODC activity and mitochondrial function might be important in morphogenesis of *M. bacilliformis*. The direct association of cAMP and GDH with morphogenesis was not ruled out, but it was concluded that their participation must be an early event in the cascade leading to hyphal development.

B. Macromolecular Biosynthesis

Other investigations concerning biochemical correlates of *Mucor* morphogenesis have focused on determining the pattern of the biosynthesis of macromolecules as the yeast-to-hyphae transition proceeds. Detailed analysis of adjustments in the rate of protein synthesis during *M. racemosus* hyphal differentiation induced by a CO_2-to-air shift has been provided by Orlowski and Sypherd (1977, 1978). It was first determined that the rate of accumulation of protein throughout morphogenesis, as measured by the incorporation of

18. Morphogenesis and Dimorphism of *Mucor*

[^{14}C]leucine, corresponded to the increased growth rate brought about by the change in incubation atmosphere. However, measurement of the instantaneous rate of protein synthesis revealed a rapid burst in the acceleration of protein production, followed by a gradual decrease as hyphal growth continued. These observations were later attributed to fluctuations in the speed with which the ribosome traverses mRNA (Orlowski and Sypherd, 1978). Although providing insight concerning possible regulatory mechanisms of protein synthesis in *Mucor*, the alterations observed do not directly influence morphogenesis, but rather are more likely a reflection of changes in growth rate (Ross and Orlowski, 1982a,b).

A variety of quantitative differences have been reported in the chemical composition of yeast and hyphal cell walls. The most dramatic change is the marked decrease of cell wall mannans as mycelial growth proceeds (Bartnicki-Garcia and Nickerson, 1962c; Bartnicki-Garcia, 1968b; Sypherd *et al.,* 1978). A closer examination of cell wall polysaccharides by Dow and Rubery (1977) revealed that high-molecular-weight polymers of hyphal cells were comprised of a higher percentage of galactose and fucose than their yeast counterparts. The functional significance, if any, of these alterations is not known.

By far the most intensively investigated aspect of cell wall biosynthesis of *Mucor* concerns chitin deposition and the enzyme responsible for its polymerization, chitin synthase. This stems from the role of chitin as the skeletal structure of the *Mucor* cell wall. Since the morphological transition primarily reflects a change in cell shape, knowledge of the mechanisms that regulate the sites of chitin deposition, as well as its synthesis, should lead to an understanding of the factors that control the dimorphic response (Bartnicki-Garcia, 1973). The amounts of chitin in yeast and hyphal cells of *M. rouxii* are essentially identical (Bartnicki-Garcia, 1968b). However, significant differences were observed in both the activity and stability of chitin synthase found in the two cell types (Ruiz-Herrera and Bartnicki-Garcia, 1976). In crude yeast cell extracts the enzyme existed primarily as a zymogen, while in mycelial extracts the enzyme was largely in an active state. Conversely, the mycelial enzyme appeared to be more readily inactivated than the yeast enzyme. Since the absolute amount of enzyme (active enzyme plus proteolytically activated zymogen as measured in crude extracts) was approximately the same in both cell types, the regulation of chitin synthase may occur at least in part at the posttranslational level. As pointed out by Ruiz-Herrera and Bartnicki-Garcia (1976), the relative instability of the mycelial enzyme compared to the situation in yeast might help explain the change in chitin synthesis from the uniform pattern of deposition observed throughout the yeast cell wall to the polarized pattern of elongation that occurs at the hyphal tip of growing mycelial forms of *Mucor* (Bartnicki-Garcia and Lippman, 1969; McMurrough *et al.,* 1971). Whole-cell studies by Domek and Borgia (1981) have shown that the rate of chitin and chitosan synthesis is also accelerated in mycelial cells of *M. racemosus*. In addition, it was suggested that cAMP may be the signal responsible for regulating both the site and rate of chitin synthesis. The validity of this hypothesis awaits testing through additional investigations.

Only a few studies have been reported concerning lipid biosynthesis and plasma membrane function in *Mucor* morphogenesis. However, the association of a number of cell wall biosynthetic enzymes with the membrane and the participation of membrane components in transport and secretory processes point to involvement of the membrane in morphogenesis and cell wall biosynthesis. Early studies examined the fatty acid and sterol composition of whole-cell membranes extracted from *M. genevensis* (Gordon *et al.*, 1971) and *M. rouxii* (Safe and Caldwell, 1975). Both groups demonstrated that hyphal cells contained a higher proportion of fatty acids and sterols than did yeast cells. In the case of *M. genevensis* it was also shown that aerobically grown yeast cells maintained by the respiration inhibitor phenethyl alcohol had a fatty acid and sterol composition similar to aerobically grown mycelium. Therefore, lipid composition was thought not to influence cell morphology directly. Studies with *M. racemosus* by Ito *et al.* (1982) made use of the antilipogenic agent cerulenin to prevent morphogenesis under aerobic conditions while allowing yeast growth to continue. It was found that supplementation of cultures with cerulenin during the morphological transition prevented a characteristic increase in lipid synthesis that occurs as hyphal development begins. In addition, the turnover rates of phosphatidyl choline and phosphatidyl ethanolamine were significantly reduced in the presence of the drug. Whether these alterations influence the change from a symmetric to a polarized mechanism of cell wall growth remains under study. The relationship between plasma membrane function and cell wall biosynthesis has also been examined by Dow *et al.* (1981) in regard to secretion of cell wall matrix polymers in hyphal cells of *M. rouxii*. The investigations established several precursor–product relationships between membrane-associated polyuronides and glycoproteins and their eventual extracellular location. Similar studies with yeast cells have yet to be reported.

Thus, it is clear that the rates of biosynthesis of RNA and protein, as well as constituents of the cell wall and membrane-containing structures, all increase during the yeast-to-hyphae transition as mediated by shifting the incubation atmosphere from CO_2 to air. Unfortunately, the relationship of these changes to morphogenesis remains unknown. It now seems probable that the elevation in the rate of protein synthesis is a result of the increased rate of growth brought about by the experimental conditions. It is also likely that at least some of the observed increase in the rate of chitin and phospholipid synthesis is a consequence of the same factors. However, the obvious importance of cell wall biosynthesis to dimorphism and the intimate relationship of the plasma membrane to cell wall biosynthetic processes suggest that further work in these areas will provide clues to the regulation of the yeast-to-hyphae transition.

III. MOLECULAR ANALYSIS OF *MUCOR*

Although the experiments discussed in the preceding sections have provided a wealth of information concerning the dimorphic response of *Mucor*, it remains a difficult task to identify those events that directly affect morphogenesis. It is, therefore, necessary to complement our present view of *Mucor* differentiation by applying new experimental approaches to the problem. Of course, foremost in this respect are the methodologies of molecular biology, and these have now been employed in a number of studies involving *Mucor*. These experiments will be described by first discussing those directed toward the study of differential protein synthesis, and then mentioning the progress that has been made in both gene isolation and classical genetic investigations.

A. Regulation of Protein Synthesis

The extent of differential protein synthesis during morphogenesis has been assessed by Hiatt *et al.* (1980). In these experiments, [^{35}S]methionine was used to label proteins of yeast and hyphal cells, as well as of yeast cells induced to undergo hyphal development. Proteins were subsequently resolved by two-dimensional polyacrylamide gel electrophoresis. In all cases, approximately 500 proteins were observed, and the overall patterns of proteins synthesized were very similar irrespective of cell type; however, a few proteins were specific to the morphological form exhibited. More significant alterations were noted in the rates of synthesis of individual polypeptides. In this investigation morphogenesis was promoted under both aerobic and anaerobic conditions, leading to the conclusion that most of the alterations in the pattern of protein synthesis were not made in response to an aerobic environment. The possibility that some of the changes occurred as a result of an increased rate of growth during the switch to filamentous development remains to be investigated.

The regulation of the activity of individual proteins need not occur solely at the transcriptional level, but might also result from posttranslational modifications of existing proteins. The significance of posttranslational modification to dimorphism has been addressed in a number of studies that have characterized protein components of the translational apparatus of *M. racemosus*. The rationale for these investigations has been that modifications of such proteins might directly influence gene expression by altering message selection, peptide initiation and elongation rates, etc. In particular, ribosomal proteins have been analyzed by two-dimensional gel electrophoresis (Larsen and Sypherd, 1979). The protein compositions of ribosomes isolated from yeast, young germlings, and hyphae were identical. However, tailing in the migration pattern of the 40 S ribosome subunit protein, S6, was observed. Further investigations showed that three distinct derivatives of S6 existed and that these reflected differing levels of phosphorylation. Experiments aimed at defining the relationship of adjustments in the level of phosphorylation to morphogenesis showed that cell morphology was not dependent on the degree of phosphorylation of S6. Larsen and Sypherd (1980) later showed that S6 phosphorylation was influenced by changes in the metabolic state and intracellular ATP concentrations of the cell. Interestingly, no phosphorylation of S6 was observed in sporangiospores.

Other studies have examined the extent of protein methylation. Sypherd et al. (1979) described a conditional yeast mutant, designated *coy*, that required supplementation with exogenous methionine in order to carry out hyphal development, but did not require methionine for growth. This observation led to experiments that measured the intracellular levels of S-adenosylmethionine (SAM) (Garcia et al., 1980), which functions as the methyl-group donor for most proteins that are modified by methylation. The investigators found that SAM concentrations increased threefold during germ tube emergence and, furthermore, that total protein methylation increased by a similar amount during this period. Proteins labeled by the addition of L-[*methyl*-^{3}H]methionine were then resolved by two-dimensional gel electrophoresis with the aim of determining whether quantitative or qualitative changes in the methylation of specific proteins could be detected (Garcia et al., 1980). The data suggested that a number of proteins may exhibit increased methylation as germ tube emergence proceeds, and that qualitative changes occur in the spectrum of proteins that become methylated.

The most abundant protein in hyphal cells also was one of the more highly methylated proteins detected by Garcia et al. (1980). Because of its potential importance it was chosen for a more detailed analysis (Hiatt et al., 1982). The protein was basic with a pH of approximately 9.5, and had a molecular weight of about 53,000 daltons. Smaller amounts of the protein (2.5-fold less) were present in yeast cells then were found in hyphal cells. The extent of methylation of the protein was assessed by two-dimensional thin-layer chromatography and high-pressure liquid chromatography. Surprisingly, the protein was found to be highly methylated with approximately 20% of the lysine moieties of the protein having undergone methylation. Because of the physical characteristics of the protein, the investigators reasoned that the protein might be the α subunit of the protein synthesis elongation factor, EF-1. Subsequent purification and assay of the protein for phe-tRNA binding activity confirmed that the protein was indeed EF-1α. The identification of the protein as EF-1α opens the intriguing possibility that its state of methylation could influence gene expression during *Mucor* morphogenesis.

Methylation of EF-1α during the yeast-to-hyphae transition remains under investigation; however, changes in the methylation of the factor have been examined during the time course of aerobic germination of sporangiospores to young hyphal cells (Hiatt et al., 1982, Sypherd et al., 1984). EF-1α from sporangiospores was essentially unmethylated. Methylation of EF-1α was first observed 3 hr after the onset of germination and increased to a maximum of about 16% methylation of the total number of lysine residues after 9 hr of incubation in air. In order to determine whether activity was, in part, controlled by the extent of methylation, EF-1α activity was assayed with purified protein isolated from sporangiospores (unmethylated) and hyphal germlings (methylated). No significant difference in EF-1α activity from the two sources was noted when using a polyuridylic acid-directed *in vitro* translation assay (Sypherd *et al.*, 1984). Thus, the function of methylation of EF-1α remains unclear. It should also be mentioned that the pattern of methylated tryptic peptides of EF-1α, isolated from the *M. racemosus* morphological mutant, *coy*, was altered in comparison to the pattern of EF-1α from wild-type cells. This suggested that aberrant methylation of EF-1α may adversely affect *Mucor* development. Further investigations should lead to a definitive answer to the role, if any, of EF-1α during morphogenesis.

B. Genetic Analysis

An understanding of *Mucor* morphogenesis has been hampered by the lack of a system that allows for formal genetic analysis. Certainly, the organisms are heterothallic and have two morphologically identical mating types termed (+) and (−) (Bergman *et al.*, 1969). The mating of *M. mucedo* strains has received the most attention, and it has been found that the first steps in the reaction require the presence of diffusible "hormones" that have been identified as trisporic acids (TA) (Austin *et al.*, 1969; Bu'Lock *et al.*, 1974). Wurtz and Jockusch (1975, 1978) have described a variety of *M. mucedo* mutants that differ in their ability to respond to and undergo differentiation in the presence of TA and their precursors. These studies (a) confirmed that TA were the inducers of sexual differentiation, (b) supported other investigations (Mesland *et al.*, 1974) that suggested a requirement for the compounds, not only for the induction of sexual differentiation but also for the continuation of the process, (c) indicated that a precursor leading to TA synthesis serves as an antagonist to TA in regulating zygospore growth, and (d) described the sequence of events and morphological structures that culminate in the formation of macrosporangia.

Although the process leading to sexual differentiation of the mucorales is at least partially understood, standard genetic analysis of the progeny of matings has proved prohibitive to date. As pointed out by Lasker and Borgia (1980), this stems from several factors, including a dearth of suitable mutant strains, difficulty in obtaining meaningful data due to low germination frequency of zygospores, and to a dormancy of zygospores that can extend for several months (Bergman *et al.*, 1969). The isolation of mutant strains does not seem to pose an unmanageable problem. For example, Peters and Sypherd (1978) described a freeze–thaw enrichment technique used with *M. racemosus* that, when coupled with N-methyl-N'-nitro-N-nitrosoguanidine (NTG) mutagenesis, increased the frequency of auxotrophs in the surviving population to 2–3%. Likewise, NTG mutagenesis alone resulted in the isolation of a variety of *M. mucedo* monomorphic mutants locked in the yeast phase (Ruiz-Herrera *et al.*, 1983). Similar results have been obtained with *M. racemosus* (P. Sypherd, personal communication). Despite the problems inherent in analysis of zygospores, alternative approaches for subjecting *Mucor* to conventional genetic analysis are promising. In particular, Genther and Borgia (1978) have described a protocol for promoting spheroplast fusion and heterokaryon formation with *M. racemosus*. Spheroplasts of auxotrophic strains were formed by using a combination of commercially available chitinase and chitosanase isolated from Myxobacter Al-1. Spheroplasts were then fused in a solution of $Ca(NO_3)_2$ and, after wall regeneration, heterokaryons were observed at a frequency of 1.45×10^{-4}. In a subsequent study (Lasker and Borgia, 1980), the efficiency of prototrophic heterokaryon formation was increased to approximately 5% when polyethylene glycol and $CaCl_2$ were used to mediate cell fusions. With the availability of developmental mutants, this system should prove useful in grouping strains into complementation groups as a first step in their genetic characterization.

C. Molecular Genetics

The use of recombinant DNA methodologies will facilitate genetic studies that would be impossible in the mucorales if only classical genetic techniques could be relied upon. The problems facing the investigator desiring to isolate a particular gene from this group of organisms are largely the same as those facing individuals working in other experimental systems, that is, the ability to devise a suitable strategy or obtain an appropriate probe to allow detection of the gene of interest.

Mucor racemosus gene libraries constructed with either the plasmid vector pBR322 (Cihlar and Sypherd, 1980) or λ Charon 4A as the cloning vehicle (T. Leathers and C. Katayama, personal communication), are now available for use. The first *M. racemosus* genes to be isolated were those encoding the rRNA genes (Cihlar and Sypherd, 1980). In this case, the various species of rRNA were obtained for use as probes by *in vivo* labeling of rRNAs with ^{32}P, followed by their isolation either by velocity gradient centrifugation (25 S and 18 S RNAs) or by polyacrylamide gel electrophoresis (5.8 S and 5 S RNAs). Bacterial clones harboring the homologous rDNA sequences were detected by colony hybridization, and the recombinant plasmids obtained from them were later analyzed in order to localize the rRNA genes. The rDNA was found to be arranged as a tandem 10.2-kb repeat unit, and the 18 S–5.8 S–25 S arrangement of the rRNA genes is consistent with that of the rDNA region of other eukaryotes examined to date. A number of clones have also been isolated on the basis of harboring *Mucor*

18. Morphogenesis and Dimorphism of *Mucor*

tRNA genes (R. L. Cihlar and P. S. Sypherd, unpublished observations). Again, *in vivo* ^{32}P-labeled tRNA served as the probe in colony hybridization experiments. Preliminary experiments to characterize the recombinant plasmids carrying tRNA genes have shown the genes to be scattered at diverse genomic locations, but precedent suggests that localized clustering of some tRNA genes will be found (Abelson, 1980).

An approach that has proved fruitful for obtaining certain genes of *Saccharomyces cerevisiae* (Struhl *et al.*, 1976; Clarke and Carbon, 1978), *Neurospora crassa* (Vapnek *et al.*, 1977), and *Aspergillus nidulans* (Yelton *et al.*, 1983) has been through the complementation of the analogous genes of *E. coli*. This method of gene isolation has been attempted with recombinant plasmids containing *M. racemosus* DNA (Cihlar and Sypherd, 1982). Groups of 250 bacterial clones from the *M. racemosus* pBR322 gene library were pooled and recombinant plasmids were extracted. These were subsequently utilized to transform appropriate auxotrophic strains of *E. coli*. Transformants were obtained using plasmids from only one pool and only for *E. coli* strains with the *leuB* marker (isopropylmalate dehydrogenase). Characterization of the complementing activity suggested that the structural gene for isopropylmalate dehydrogenase had not been cloned, but rather that transformants arose through an unknown mechanism of suppression of the *leuB* mutation. More comprehensive investigations are required to determine whether the isolation of *Mucor* genes by complementation of mutations in *E. coli,* or in other suitable hosts, is a feasible method for obtaining genes of interest. In this respect, it should be noted that isolation of a LEU gene of *Candida maltosa* by complementation of the *leu-2* lesion of *S. cerevisiae* and the *leuB* lesion of *E. coli* has been reported (Kawamura *et al.*, 1983).

Actin is a structural protein that can influence cellular architecture and therefore may be important in *Mucor* morphogenesis. Since the sequence of actin is highly conserved, it has been possible to utilize an actin gene, cloned from *Dictyostelium discoideum* genomic DNA (Kindle and Firtel, 1978), as a probe to isolate actin-encoding sequences from other organisms (Ng and Abelson, 1980; Tobin *et al.* 1980). An identical approach has proved successful in identifying a recombinant plasmid containing an actin gene from *Candida albicans* (Lasker and Riggsby, 1984) and also in detecting the actin-encoding sequences in an *M. racemosus* λ Charon 4A genomic library (R. L. Cihlar, unpublished observations). In both cases the genes were isolated by utilizing the *S. cerevisiae* actin gene (Ng and Abelson, 1980) as a probe. *Candida albicans* was found to harbor one actin gene, while the number of actin genes in *M. racemosus* has not yet been determined. The availability of these genes should yield information concerning their structure, as well as the regulation of their expression during morphogenesis.

Advances have also been made recently in obtaining and characterizing an

EF-1α gene of *M. racemosus* (Sypherd *et al.*, 1984). In these studies a partial cDNA library was constructed in pBR322. Plasmid DNA was obtained from individual clones and pooled in groups of six. Message selection by hybridization, coupled with *in vitro* translation of selected mRNA in a rabbit reticulocyte system and subsequent two-dimensional polyacrylamide gel analysis, identified a pool that putatively contained EF-1α cDNA. This was confirmed, and the EF-1α cDNA clone was identified by screening the individual cDNA clones that comprised the pool used in the preliminary characterization. In addition, the EF-1α cDNA clone has been used to examine the transcription of the EF-1α gene during hyphal germination from spores (Fonzi *et al.*, 1985). EF-1α cDNA was used to probe Northern blots of mRNA obtained at 3-hr intervals during germination. This analysis showed that EF-1α is encoded by a single transcript of about 1.45 kb, and that a constant amount of EF-1α mRNA can be detected throughout germination. It was, therefore, concluded that the *in vivo* regulation of EF-1α must occur posttranscriptionally, and is probably mediated in some fashion by protein methylation. Finally, EF-1α cDNA has been used to isolate the homologous genomic DNA sequences from an *M. racemosus* λ Charon 4A gene library (P. S. Sypherd, personal communication). Analysis of an EF-1α gene is currently in progress.

A dispersed repetitive DNA sequence has also been isolated from cloned *M. racemosus* genomic DNA (Cihlar *et al.*, 1984). The sequence is approximately 3.0 kb and is reiterated 25 to 35 times in genomic DNA. Sequences homologous to the repeated unit were also found in *M. genevensis*, *M. hiemalis*, and *M. mucedo*, indicating that at least portions of the element have been conserved across species lines. *In vivo* evidence has not shown the element to be transposable, but the possibility that the repeated DNA sequence may be a transposon has not been excluded.

In summary, a variety of *Mucor* genes have been isolated and partially characterized. The approaches used in gene isolation have largely been extensions of the methodologies developed in other systems. With the exception of the cloning of EF-1α encoding sequences, it is unlikely that analysis of the genes isolated to date will provide much insight concerning *Mucor* morphogenesis; however, valuable information concerning gene structure and regulation of *Mucor* gene expression during growth apart from morphogenesis should result from these ongoing investigations.

IV. PERSPECTIVES—MOLECULAR GENETICS

The increasing number of investigations concerning dimorphism that employ the methodologies of molecular genetics can only lead to a greater understanding of the factors that regulate morphogenesis. However, as these studies pertain to

Mucor, a number of avenues of research must be expanded before the true potential of analysis in the system can be tapped. Primary in this respect is the development of vectors suitable for cloning directly in the dimorphic fungi. Plasmid cloning vectors have been designed with certain features in mind. In particular (a) vectors have been constructed so as to contain selectable markers that allow easy detection of clones harboring the vector, and (b) as a general rule, with the exception of YIp vectors of *Saccharomyces* (Botstein and Davis, 1982), plasmid vectors are self-replicating in the host organism. As has been shown in the above discussion, available vectors are useful in isolating genes of interest from the filamentous fungi, provided a suitable probe is available. However, in order to study the contribution of a particular gene to morphogenetic processes in the filamentous fungi, it is necessary to be able to introduce, maintain, and manipulate such genes in the homologous genetic background. Efforts must, therefore, be directed toward the construction of a vector that will fulfill these requirements. Initial basic work in this area might be aimed at isolating genes that can be used as selective markers, e.g., amino acid biosynthetic genes, and obtaining the companion auxotrophic host strains that are also necessary. Other work might be performed to develop an integrative type vector comprised, in part, of the repeated DNA sequence of *Mucor* that has been described (Cihlar *et al.,* 1984). Many other possibilities can be envisioned as well. Of course, an effective transformation protocol must be developed in order to introduce DNA into the cell. A prerequisite for transformation is the development of successful spheroplasting and wall regeneration protocols, and these already are in hand (Genther and Borgia, 1978; Lasker and Borgia, 1980). Suitable conditions for transformation must be developed empirically, and will also rely on the availability of genetically marked strains. Finally, screening protocols must be devised that will allow the isolation of genes that may be more intimately involved in morphogenesis. In the short term, differential screening techniques based on conditions that promote maximal expression of the gene product of interest might be devised. For example, the level of expression of NAD-dependent GDH can be manipulated by the nature of the nitrogen source and the incubation atmosphere (Peters and Sypherd, 1978). Thus, if the enzyme is regulated primarily at the transcriptional level, mRNAs isolated from cells grown under conditions that allow high- and low-level expression of the GDH gene might be used to construct cDNA probes. Subsequent screening of duplicates of *Mucor* gene libraries might allow detection of clones harboring genes that are differentially expressed under the designated conditions. In the longer term, the development of host–vector systems coupled with the availability of developmental mutants should allow for the direct selection of morphogenetically important genes. No insurmountable problems exist in addressing these questions, and approaches such as these, or of other design, will certainly lead to a greater understanding of dimorphism in the near future.

NOTE ADDED IN PROOF

Heeswijck and Roncero (1984, *Carlsberg Res. Commun.* **49,** 691–702) have recently reported the successful transformation of *Mucor circinelloides (racemosus)* and complementation of auxotrophs with recombinant plasmids consisting of the *S. cerevisiae* vector, YRp17, and *M. circinelloides* DNA.

ACKNOWLEDGMENTS

I would like to thank my colleagues who provided information prior to its publication. Thanks also to Dr. Kathryn Hoberg for critical reading of this manuscript, and to Stephanie Coleman for manuscript preparation.

REFERENCES

Abelson, J. (1980). The organization of tRNA genes. *In* "Transfer RNA: Biological Aspects" (D. Soll, J. N. Abelson, and P. R. Schimmel, eds.), pp. 211–220. Cold Spring Harbor Lab., Cold Spring Harbor, New York.

Austin, D. J., Bu'Lock, J. D., and Gooday, G. W. (1969). Trisporic acids: sexual hormones from *Mucor mucedo* and *Blakeslea trispora*. *Nature (London)* **223,** 1178–1179.

Barlow, A. J. E., Aldersley, T. A., and Chattaway, F. W. (1974). Factors present in serum and seminal plasma which promote germ-tube formation and mycelial growth in *Candida albicans*. *J. Gen. Microbiol.* **82,** 261–272.

Bartnicki-Garcia, S. (1968a). Cell wall chemistry, morphogenesis and taxonomy of fungi. *Annu. Rev. Microbiol.* **22,** 87–108.

Bartnicki-Garcia, S. (1968b). Control of dimorphism in *Mucor* by hexoses: inhibition of hyphal morphogenesis. *J. Bacteriol.* **96,** 1586–1594.

Bartnicki-Garcia, S. (1973). Fundamental aspects of hyphal morphogenesis. *Symp. Soc. Gen. Microbiol.* **23,** 245–267.

Bartnicki-Garcia, S., and Lippman, E. (1969). Fungal morphogenesis: cell wall construction in *Mucor rouxii*. *Science (Washington, D.C.)* **165,** 302–304.

Bartnicki-Garcia, S., and Nickerson, W. J. (1962a). Induction of yeast-like development in *Mucor* by carbon dioxide. *J. Bacteriol.* **84,** 820–840.

Bartnicki-Garcia, S., and Nickerson, W. (1962b). Nutrition, growth and morphogenesis of *Mucor rouxii*. *J. Bacteriol.* **84,** 841–858.

Bartnicki-Garcia, S., and Nickerson, W. J. (1962c). Isolation, composition, and structure of cell walls of filamentous and yeast-like forms of *Mucor rouxii*. *Biochim. Biophys. Acta* **58,** 102–119.

Bergman, K., Burke, P. V., Cerda-Olmedo, E., David, C. N., Delbruck, M., Foster, K. W., Goodell, E. W., Heisenberg, M., Meissner, G., Zalokar, M., Dennison, D. S., and Shropshire, W., Jr. (1969). Phycomyces. *Bacteriol. Rev.* **33,** 99–157.

Borgia, P. I., and Mehnert, D. W. (1982). Purification of a soluble and wall bound form of β-glucosidase from *Mucor racemosus*. *J. Bacteriol.* **149,** 515–522.

Botstein, D., and Davis, R. W. (1982). "The Molecular Biology of the Yeast *Saccharomyces*: Metabolism and Gene Expression," pp. 607–636. Cold Spring Harbor Lab., Cold Spring Harbor, New York.

Bu'Lock, J. D., Jones, B. E., Taylor, D., Winskill, N., and Quarrie, S. A. (1974). Sex hormones in mucorales: the incorporation of C_{20} and C_{18} precursors into trisporic acids. *J. Gen. Microbiol.* **80,** 301–306.

Cihlar, R. L., and Sypherd, P. S. (1980). The organization of the ribosomal RNA genes in the fungus *Mucor racemosus. Nucleic Acids Res.* **8,** 793–804.

Cihlar, R. L., and Sypherd, P. S. (1982). Complementation of the *leu*B6 mutation of *Escherichia coli* by cloned DNA from *Mucor racemosus. J. Bacteriol.* **151,** 521–523.

Cihlar, R. L., Katayama, C., Dewar, R., and Sypherd, P. S. (1984). Organization of the *Mucor racemosus* genome. *In* "Microbiology 1984" (L. Leive and D. Schlessinger, eds.), pp. 126–128. Am. Soc. Microbiol., Washington, D.C.

Clarke, L., and Carbon, J. (1978). Functional expression of cloned yeast DNA in *Escherichia coli*: specific complementation of arginosuccinate lyase (*argH*) mutations. *J. Mol. Biol.* **120,** 517–532.

Clark-Walker, G. D. (1973). Relationship between dimorphology and respiration in *Mucor genevensis* studies with chloramphenicol. *J. Bacteriol.* **116,** 972–980.

Cohen, S. S. (1978). The functions of the polyamines. *In* "Advances in Polyamine Research" (R. A. Campbell, D. R. Morris, D. Bartos, G. D. Daves, and F. Bartos, eds.), pp. 1–10. Raven, New York

Dabrowa, N., Taxer, S. S. S., and Howard, D. H. (1976). Germination of *Candida albicans* induced by proline. *Infect. Immun.* **13,** 830–835.

Domek, D. B., and Borgia, P. T. (1981). Changes in the rate of chitin-plus-chitosan synthesis accompany morphogenesis of *Mucor racemosus. J. Bacteriol.* **146,** 945–951.

Dow, J. M., and Rubery, P. H. (1977). Chemical fractionation of the cell walls of mycelial and yeast-like forms of *Mucor rouxii:* a comparative study of the polysaccharide and glycoprotein components. *J. Gen. Microbiol.* **99,** 29–41.

Dow, J. M., Carreon, R. R., and Villa, V. D. (1981). Role of membranes of mycelial *Mucor rouxii* in synthesis and secretion of cell wall matrix polymers. *J. Bacteriol.* **145,** 272–279.

Farkas, V. (1979). Biosynthesis of cell walls of fungi. *Microbiol. Rev.* **43,** 117–144.

Fonzi, W. A., Katayama, C., Leathers, T., and Sypherd, P. S. (1985). Regulation of elongation factor 1α in *Mucor racemosus. Mol. Cell. Biol.* **4,** 1100–1103.

Forte, J. W., and Orlowski, M. (1980). Profile of cyclic adenosine 3′,5′-monophosphate-binding proteins during the conversion of yeasts to hyphae in the fungus *Mucor. Exp. Mycol.* **4,** 78–86.

Friedenthal, M., Roselino, E., and Passerone, S. (1973). Multiple molecular forms of pyruvate kinase from *Mucor rouxii:* immunological relationship among the three isozymes and nutritional factors affecting the enzymatic pattern. *Eur. J. Biochem.* **35,** 145–158.

Garcia, J. R., Hiatt, W. R., Peters, J., and Sypherd, P. S. (1980). *S*-Adenosylmethionine levels and protein methylation during morphogenesis of *Mucor racemosus. J. Bacteriol.* **142,** 196–201.

Genther, F. J., and Borgia, P. T. (1978). Spheroplast fusion and heterokaryon formation in *Mucor racemosus. J. Bacteriol.* **134,** 349–352.

Gordon, P. A., Stewart, P. R., and Clark-Walker, G. D. (1971). Fatty acid and sterol composition of *Mucor genevensis* in relation to dimorphism and anaerobic growth. *J. Bacteriol.* **107,** 114–120.

Haidle, C. W., and Storck, R. (1966). Control of dimorphism in *Mucor rouxii. J. Bacteriol.* **92,** 1236–1244.

Hardcastle, R. V., and Szaniszlo, P. J. (1974). Characterization of dimorphism in *Cladosporium werneckii. J. Bacteriol.* **119,** 294–302.

Hiatt, W. R., Inderlied, C. B., and Sypherd, P. S. (1980). Differential synthesis of polypeptides during morphogenesis of *Mucor. J. Bacteriol.* **141,** 1350–1359.

Hiatt, W. R., Garcia, R., Merrick, W. C., and Sypherd, P. S. (1982). Methylation of elongation factor 1α from the fungus *Mucor. Proc. Natl. Acad. Sci. U.S.A.* **79,** 3433–3437.

Inderlied, C. B., and Sypherd, P. S. (1978). Glucose metabolism and dimorphism in *Mucor. J. Bacteriol.* **133,** 1282–1286.

Inderlied, C. B., Cihlar, R. L., and Sypherd, P. S. (1980). Regulation of ornithine decarboxylase during morphogenesis of *Mucor racemosus*. *J. Bacteriol.* **141,** 699–706.

Ito, E., Cihlar, R. L., and Inderlied, C. B. (1982). Lipid synthesis during morphogenesis in *Mucor racemosus*. *J. Bacteriol.* **152,** 880–887.

Kawamura, M., Takagi, M., and Yano, K. (1983). Cloning of a *leu* gene and an *ars* site of *Candida maltosa*. *Gene* **24,** 157–162.

Kindle, K. L., and Firtel, R. A. (1978). Identification and analysis of *Dictyostelium* actin genes, a family of moderately repeated genes. *Cell (Cambridge, Mass.)* **15,** 763–778.

Larsen, A. D., and Sypherd, P. S. (1974). Cyclic adenosine 3′,5′-monophosphate and morphogenesis in *Mucor racemosus*. *J. Bacteriol.* **117,** 432–438.

Larsen, A. D., and Sypherd, P. (1979). Ribosomal proteins of the dimorphic fungus, *Mucor racemosus*. *Mol. Gen. Genet.* **175,** 99–109.

Larsen, A. D., and Sypherd, P. S. (1980). Physiological control of phosphorylation of ribosomal protein S6 in *Mucor racemosus*. *J. Bacteriol.* **141,** 20–25.

Lasker, B. A., and Borgia, P. T. (1980). High frequency heterokaryon formation by *Mucor racemosus*. *J. Bacteriol.* **141,** 565–569.

Lasker, B. A., and Riggsby, W. S. (1984). Cloning and characterization of the actin gene of *Candida albicans*. *Annu. Meet. Am. Soc. Microbiol.* Abstr. H117.

McMurrough, I., Flores-Carreon, A., and Bartnicki-Garcia, S. (1971). Pathway of chitin synthesis and cellular localization of chitin synthetase in *Mucor rouxii*. *J. Biol. Chem.* **246,** 3999–4007.

Mesland, D. A. M., Huisman, J. G., and van den Ende, H. (1974). Volatile sexual hormones in *Mucor mucedo*. *J. Gen. Microbiol.* **80,** 111–117.

Mooney, D. T., and Sypherd, P. S. (1976). Volatile factor involved in the dimorphism of *Mucor racemosus*. *J. Bacteriol.* **126,** 1266–1270.

Moreno, S., Paveto, C., and Passerone, S. (1977). Multiple protein kinase activities in the dimorphic fungus *Mucor rouxii*. *Arch. Biochem. Biophys.* **180,** 225–231.

Ng, R., and Abelson, J. (1980). Isolation and sequence of the gene for actin in *Saccharomyces cerevisiae*. *Proc. Natl. Acad. Sci. U.S.A.* **77,** 3912–2916.

Nickerson, J. W., Dunkle, L. D., and Van Etten, J. L. (1977). Absence of spermine in filamentous fungi. *J. Bacteriol.* **129,** 173–176.

Orlowski, M. (1980). Cyclic adenosine 3′,5′-monophosphate and germination of sporangiospores from the fungus *Mucor*. *Arch. Microbiol.* **126,** 133–140.

Orlowski, M., and Sypherd, P. S. (1977). Protein synthesis during morphogenesis of *Mucor racemosus*. *J. Bacteriol.* **132,** 209–218.

Orlowski, M., and Sypherd, P. S. (1978). Regulation of translation rate during morphogenesis in the fungus *Mucor*. *Biochemistry* **17,** 569–575.

Paveto, C., Epstein, A., and Passeron, S. (1975). Studies on cyclic adenosine 3′,5′-monophosphate levels, adenylate cyclase, and phosphodiesterase activities in the dimorphic fungus *Mucor rouxii*. *Arch. Biochem. Biophys.* **169,** 449–457.

Paznokas, J. L., and Sypherd, P. S. (1975). Respiratory capacity, cyclic adenosine 3′,5′-monophosphate and morphogenesis of *Mucor racemosus*. *J. Bacteriol.* **124,** 134–139.

Paznokas, J. L., and Sypherd, P. S. (1977). Pyruvate kinase isozymes of *Mucor racemosus:* control of synthesis by glucose. *J. Bacteriol.* **130,** 661–666.

Peters, J., and Sypherd, P. S. (1978). Enrichment of mutants of *Mucor racemosus* by differential freeze-killing. *J. Gen. Microbiol.* **105,** 77–81.

Peters, J., and Sypherd, P. S. (1979). Morphology associated expression of nicotinamide adenine dinucleotide-dependent glutamate dehydrogenase in *Mucor racemosus*. *J. Bacteriol.* **137,** 1134–1139.

Ross, J. F., and Orlowski, M. (1982a). Growth-rate-dependent adjustment of ribosome function in chemostat-grown cells of the fungus *Mucor racemosus*. *J. Bacteriol.* **149,** 650–653.

Ross, J. F., and Orlowski, M. (1982b). Regulation of ribosome function in the fungus *Mucor:* growth rate vis-à-vis dimorphism. *FEMS Microbiol. Lett.* **13,** 325–328.

Ruiz-Herrera, J., and Bartnicki-Garcia, S. (1976). Proteolytic activation and inactivation of chitin synthetase from *Mucor rouxii*. *J. Gen. Microbiol.* **97,** 241–249.

Ruiz-Herrera, J., Ruiz, A., and Lopez-Romero, E. (1983). Isolation and biochemical analysis of *Mucor bacilliformis* monomorphic mutants. *J. Bacteriol.* **156,** 264–272.

Russell, D. H., and Snyder, S. H. (1968). Amine synthesis in rapidly growing tissues:ornithine decarboxylase activity in regenerating rat liver, chick embryos and various tumors. *Proc. Natl. Acad. Sci. U.S.A.* **66,** 1420–1427.

Safe, S., and Caldwell, J. (1975). The effect of growth environment on the chloroform–methanol and alkali-extractable cell wall and cytoplasmic lipid levels of *Mucor rouxii*. *Can. J. Microbiol.* **21,** 79–84.

Stewart, P. R., and Rogers, P. J. (1978). Fungal dimorphism: a particular expression of cell wall morphogenesis. *In* "The Filamentous Fungi" (J. E. Smith and D. R. Berry, eds.), pp. 164–196. Wiley, New York.

Storck, R., and Morrill, R. C. (1971). Respiratory-deficient yeast like mutant of *Mucor. Biochem. Genet.* **5,** 467–479.

Struhl, K., Cameron, J. R., and Davis, R. W. (1976). Functional genetic expression of eukaryotic DNA in *Escherichia coli*. *Proc. Natl. Acad. Sci. U.S.A.* **73,** 1471–1475.

Sun, S. H., and Huppert, M. (1976). A cytological study of morphogenesis in *Coccidioides immitis. Saboruaudia* **14,** 185–198.

Syph erd, P. S., Borgia, P. T., and Paznokas, J. L. (1978). Biochemistry of dimorphism in the fungus *Mucor. Adv. Microb. Physiol.* **18,** 67–104.

Sypherd, P. S., Orlowski, M., and Peters, J. (1979). Models of fungal dimorphism: control of dimorphism in *Mucor racemosus*. *In* "Microbiology 1979" (D. Schlessinger, ed.), pp. 224–227. Am. Soc. Microbiol., Washington, D.C.

Sypherd, P. S., Fonzi, W. A., Katayama, C., and Leathers, T. D. (1984). EF-1α of *Mucor racemosus:* regulation and molecular gene cloning. *In* "Microbiology 1984" (L. Levie and D. Schlessinger, eds.), pp. 152–154. Am. Soc. Microbiol., Washington, D.C.

Szanizlo, P. J., Jacobs, C. W., and Geis, P. A. (1983). Dimorphism: morphological and biochemical aspects. *In* "Fungi Pathogenic for Humans and Animals" (D. H. Howard, ed.), pp. 323–436. Dekker, New York.

Terenzi, H. F., and Storck, R. (1969). Stimulation of fermentation and yeast-like morphogenesis in *Mucor rouxii* by phenethyl alcohol. *J. Bacteriol.* **97,** 1248–1261.

Tobin, S. L., Zulauf, E., Sanchez, F., Craig, E. A., and McCarthy, B. J. (1980). Multiple actin-related sequences in the *Drosophila melanogaster* genome. *Cell (Cambridge, Mass.)* **19,** 121–131.

Vapnek, D., Hautala, J. A., Jacobson, J. W., Giles, N. H., and Kushner, S. R. (1977). Expression in *Escherichia coli* K-12 of the structural gene for catabolic dehydroquinase of *Neurospora crassa*. *Proc. Natl. Acad. Sci. U.S.A.* **74,** 3508–3512.

Wurtz, T., and Jockusch, H. (1975). Sexual differentiation in *Mucor:* trisporic acid response mutants and mutants blocked in zygospore development. *Dev. Biol.* **43,** 213–220.

Wurtz, T., and Jockusch, H. (1978). Morphogenesis in *Mucor mucedo:* mutation affecting gamone response and organ differentiation. *Mol. Gen. Genet.* **159,** 249–257.

Yelton, M. M., Hamer, J. E., de Souza, E. R., Mullaney, E. J., and Timberlake, W. E. (1983). Developmental regulation of the *Aspergillus nidulans* trpC gene. *Proc. Natl. Acad. Sci. U.S.A.* **80,** 7576–7580.

Zorzopoulos, J., Jobbagy, A. J., and Terenzi, H. F. (1973). Effects of ethylenediaminetetraacetate and chloramphenicol on mitochondrial activity and morphogenesis in *Mucor rouxii*. *J. Bacteriol.* **115,** 1198–1204.

19

Toward Gene Manipulations with Selected Human Fungal Pathogens

W. LAJEAN CHAFFIN

Department of Microbiology
Texas Tech University Health Sciences Center
Lubbock, Texas

I. Introduction	469
II. *Cryptococcus neoformans*	471
A. Introduction	471
B. Spheroplast Fusion	472
C. Recombinant DNA Techniques	473
III. *Histoplasma capsulatum*	473
A. Introduction	473
B. Recombinant DNA Techniques	474
IV. *Wangiella dermatitidis*	474
A. Introduction	474
B. Spheroplast Fusion	475
C. Recombinant DNA Techniques	475
V. *Candida albicans*	476
A. Introduction	476
B. Spheroplast Fusion	477
C. Recombinant DNA Techniques	484
VI. Concluding Remarks	486
References	487

I. INTRODUCTION

Fungi are medically important in two general ways. The first way results from the use of products produced by fungi as pharmacological agents. Incalculable benefits have resulted from the introduction of penicillin and other antibiotics and drugs of fungal origin into medical practice. Fungi are also important as agents of disease. Mycosis is a term used to indicate a disease of fungal etiology.

Fungal diseases may be separated into three general categories. Some fungi produce toxic products, which may be secreted or retained in the fungus. Disease results from ingesting contaminated material or the fungi themselves. A second way in which fungi are implicated in disease is as a source of antigen eliciting an allergic reaction. There are many other sources of environmental allergens such as plant pollens, animal hair, and food, which may cause allergic responses. It is not the allergen but the allergic response that characterizes the disease and it is this response that is treated. Fungi are also important as agents of infectious disease in which the fungus grows on the host.

Fewer than 200 species of fungi are recognized as human pathogens and fewer than half of that number are frequently isolated from infections. Most of these fungi are saprobes growing on dead organic substrates and normal inhabitants in soil and plant environments. A few have evolved a parasitic relationship with man (anthropophilic dermatophytes), and two are components of the normal human microbial flora. Sites of infection range from superficial surfaces, such as some infections of hair, to internal tissues such as brain. There is also a range in potential harm to the human host from the cosmetically undesirable to generally life-threatening. Fungi have a relatively low virulence for human hosts and the normal host has a high resistance to infection. Even those fungi which are associated with systemic infections of a normal host usually elicit an asymptomatic or subclinical infection. With opportunistic infections, the natural defense of the host must be abridged before the fungus finds a sufficiently hospitable environment to establish an infection. Contemporary medical practice with its improved methods of diagnosis and use of drugs such as those with cytotoxic and immunosuppressive effects has contributed to increasing both the number of compromised patients and the severity of the breach in normal host defense. Under these conditions the potential exists for many organisms, on occasion, to be agents of opportunistic infection. Some of the fungi infectious to man are also agents of animal disease and may cause significant veterinary problems. In this chapter, the perspective of infection used is that of human disease. Human mycoses are discussed more extensively in many medically related texts as well as specific texts of medical mycology (Emmons *et al.*, 1977; Rippon, 1982).

The application of genetic approaches to biological questions has proved a very powerful tool. This should also prove true in the application of such techniques to pathogenic fungi. The fact that, to date, genetics has been somewhat neglected in the study of pathogenic fungi is due to a combination of circumstances. Some of the pathogenic fungi have not yet been shown to have sexual reproduction. Included in this group are *Coccidioides immitis* and *Candida albicans*. For others, e.g., *Cryptococcus neoformans,* the teleomorph form has been described only in the past decade. The recruitment of more investigators, some with specific interest in genetics, and the increased development and application of genetically related nonsexual techniques with other organisms have contributed to the increased interest and usage of such techniques in recent years with fungal pathogens. The following discussion has been limited to fungi in which the application of nonsexual genetic techniques has been initiated, although some studies using sexual genetics are included. The observations from studies employing nonsexual techniques are divided into two categories: spheroplast fusion and recombinant DNA techniques. The latter terminology is used broadly to include associated procedures such as restriction mapping.

II. *CRYPTOCOCCUS NEOFORMANS*

A. Introduction

Cryptococcus neoformans is the etiologic agent of cryptococcosis. Infection is initiated in the lungs. Most cases are probably asymptomatic, although the absence of an immunological screen has inhibited the ascertainment of the prevalence of infection. The organism and the infection are found worldwide. The organism is often present in the debri associated with pigeon roosts and thus is common in urban environments. Most infections are diagnosed following dissemination from the lung to secondary sites, among which the central nervous system is the most common. Disseminated infections occur in both normal individuals and those with impaired defenses.

Cryptococcus neoformans is unusual among pathogenic fungi in that it is encapsulated. The presence of an extracellular capsule has been identified as one of the factors that promote infection. Additional factors that have been identified are the ability to grow at 37°C and phenoloxidase activity, the latter being involved in the production of melanin. *Cryptococcus neoformans* shares many physiological and morphological characteristics with other species of *Cryptococcus*. Most isolates of other species do not grow at 37°C and none produces melanin. The perfect phase of *C. neoformans* produces basidiospores and is called *Filobasidiella neoformans*. Consequently, conventional genetic analysis is possible in this organism. Crosses between isolates and between auxotrophic or marked parental strains have been used to examine questions of taxonomy and cytogenetics of basidiospore formation. Conventional crosses have also been used in the study of virulence factors. To obtain a number of strains carrying the same defect in melanin production, as well as melanin producers of the same background, Rhodes *et al.* (1982) selected basidiospores from one cross. These mutant strains and their revertants were studied for their ability to produce infection in mice. Cosegregation and reversion of the melanin phenotype and virulence supported a role for melanin production in pathogenesis. Melanin production and growth at 37°C were tested together (Kwon-Chung *et al.*, 1982). Progeny of a cross between a melanin-negative and temperature-sensitive (no growth at 37°C) strain and a wild-type strain of opposite mating type were obtained. Mating type, melanin production, and temperature sensitivity behaved as nuclear genes, with melanin production and temperature sensitivity showing a weak linkage. Only progeny growing at 37°C and producing melanin were virulent for mice. An analysis of random basidiospores has been used to map defects in seven capsule-negative mutants (Still and Jacobson, 1983). The production of wild-type recombinants in crosses of the capsular mutants was determined. Mutations represented by the seven mutant strains mapped as a single linkage group with several loci. In addition to the usual haploid basidiospore products of meiosis, there are some unusual products that appear to be diploid (Kwon-Chung, 1978). Putative diploids isolated from a cross between two auxotrophs were stable and uninucleate and sporulated to give parental strains (Jacobson *et al.*, 1983). Such diploids may be useful in genetic studies, e.g., in complementaion tests for alleles (Jacobson *et al.*, 1983; Still and Jacobson, 1983).

B. Spheroplast Fusion

Spheroplasts have been produced from cells of *C. neoformans* using two different lytic enzyme preparations. Cells treated with β-mercaptoethanol and snail enzyme preparations were rendered osmotically sensitive in about 3 hr (Polacheck *et al.*, 1982). The protoplasts were stablized with salt (KCl). A more rapid procedure was reported in which cells exposed to a reducing agent were treated with mutanase (Novozyme 234, Novo Industri A/S, Copenhagen, Denmark) for 30 min. Osmotic fragility was observed by 10 min but additional incubation increased the percentage of cell wall removed. Cells were stabilized with sorbitol and showed >20% regeneration when plated on regeneration medium (Rhodes and Kwon-Chung, 1982). The procedure was reported effective with *Cryptococcus laurentii*.

Spheroplast fusion has not been reported for *C. neoformans*. However, this procedure has been successfully applied to the closely related species *C. laurentii* (Samad and Brady, 1984). A methionine- and adenine-requiring strain and a lysine- and adenine-requiring strain were each fused with an arginine- and uracil-requiring strain. The effect of several different osmotic stabilizers was examined and sorbitol was chosen for subsequent experiments as the stabilizer least affecting both parental auxotrophic strains. Prototrophic fusion products were obtained at frequencies ranging from 0.0004 to 0.1%. The fusion products had the same size dimensions and colonial morphology as wild-type cells. Estimation of DNA content of products from the two fusions described above showed only a slight increase (1.05- to 1.1-fold) above that of the wild type. Initial experiments suggested that the fusion products were stable and no segregants were observed.

Species of *Rhodotorula* are very rarely implicated in infections. The teleomorph state described for some of the species is in the genus *Rhodosporidium,* which reproduces sexually with formation of basidiospores. Spheroplasts of *Rhodotorula glutinis* (which has been isolated from some animal infections) were prepared with an enzyme preparation from *Penicillium lilacum* and stabilized with citrate. Doubly auxotrophic parental strains (requiring arginine and uracil or methionine and adenine) were fused with polyethylene glycol and $CaCl_2$. Prototrophic fusion products recovered from stabilized minimal medium possessed a cell and colonial phenotype and a DNA content indistinguishable from those of parental cells. Prototrophic products appeared stable without production of segregants (Storts and Brady, 1983; Samad and Brady, 1984). Efforts to produce intergeneric fusion products between *C. laurentii* and *R. glutinis* were not successful.

C. Recombinant DNA Techniques

The observations that phenoloxidase is a virulence factor and that deficient strains can be obtained which show loss of activity apparently as a result of a nuclear mutation (Rhodes *et al.,* 1982) make isolation of the phenoloxidase gene particularly attractive.

A transformation system in *C. neoformans* would be very useful not only to isolate genes such as phenoloxidase but for the study of the interaction of such a gene with other virulence factors. In the absence of a homologous transformation system, it may be possible to detect the activity of this enzyme in transformed cells of other organisms. With the objective of isolating the phenoloxidase gene, a genomic DNA library of *C. neoformans* was constructed in plasmid YEp13 (J. C. Rhodes, personal communication). The isolation of the phenoloxidase gene or the development of a transformation system has not been reported.

III. *HISTOPLASMA CAPSULATUM*

A. Introduction

Histoplasma capsulatum is a thermally dimorphic fungus which is capable of causing disease in a normal, healthy host. However, most infections resolve as asymptomatic or subclinical pulmonary infections. For some individuals infection is more than a symptomatic respiratory infection. Disseminated infections may have a variety of clinical manifestations. Among adults, the most common disseminated form is chronic disease, which is more frequent among males. The organism and infections are found worldwide, although, there are endemic areas. For example, in the United States this area is, in broad terms, the eastern half of the country. In areas of highest endemicity, skin test epidemiologic surveys show a history of infection of 80–90%. The organism grows as a mold in soil with a high nitrogen content, such as that mixed with the guano of birds and bats. In infection the organism reproduces as a yeast.

B. Recombinant DNA Techniques

Treatment of repetitive DNA (mitochondrial DNA and ribosomal DNA) of various strains of *H. capsulatum* and *H. capsulatum* var. *duboisii* with restriction enzymes suggested three classes (Vincent *et al.*, 1983). One class included only an avirulent strain (class 1), one class included virulent strains of *H. capsulatum* and the variant (class 2), and one class had only virulent strains of *H. capsulatum* (class 3). A circular mitochondrial DNA map has been constructed for strains of each class (Vincent *et al.*, 1984). The sizes are 33, 37, and 53 kb for class 3, class 2, and class 1, respectively. There are insertions and deletions when maps are compared to each other, with maps of class 2 and class 3 being more similar to each other than to class 1.

IV. WANGIELLA DERMATITIDIS

A. Introduction

Saprobic dematiaceous fungi have been implicated as etiologic agents in a number of cutaneous, subcutaneous, and systemic infections. There is no consensus on the appropriate nomenclature for subcutaneous infections in which pigmented fungal elements are observed. To further complicate the issue, the taxonomy of some of these organisms has also been in flux. For convenience, the terms used in this discussion follow those used in a standard text (Rippon, 1982). *Wangiella dermatitidis* is most frequently isolated from subcutaneous nodular lesions (phaeomycotic cysts) and other subcutaneous infections (phaeohyphomycosis). Like agents associated with classical chromoblastomycosis, lesions may contain planate dividing yeastlike structures known as sclerotic cells, although this is infrequent (McGinnis, 1983; Nishimuira and Miyaji, 1983; Matsumoto *et al.*, 1984). The multicellular form of the organism strongly resembles the sclerotic body, whose formation is thought to be a factor promoting pathogenesis. The formation of the multicellular form has been studied in the laboratory of Szaniszlo and co-workers (Roberts and Szaniszlo, 1978, 1980; Jacobs, 1983; Jacobs and Szaniszlo, 1982). Three mutants that did not continue yeast growth at 37°C but formed the multicellular form at the restrictive temperature were isolated and designated Mc1, Mc2, and Mc3 (Roberts and Szaniszlo, 1978). In the absence of a sexual stage, these mutants are being studied by using nonsexual genetic techniques.

B. Spheroplast Fusion

Spheroplast fusion techniques were examined as a basis for establishing a parasexual system for genetic analysis (Schafer et al., 1984). To facilitate the use and interpretation of results of fusions, additional lesions were introduced in the Mc2 and Mc3 strains. The melanin pigment (Geis et al., 1984) that gives these fungi their characteristic dark color renders cells relatively resistant to lytic enzyme preparations such as Glusulase and Zymolyase. However, melanin-deficient albino mutants are susceptible (Jacobs, 1983). Albino derivatives were obtained for each morphological mutant. Subsequent experiments demonstrated that some pairs of albino derivatives were able to crossfeed and produce pigmented cells when in physical contact. In addition, in order to facilitate selection of fusion products and reduce the probability of obtaining revertants, auxotrophic strains were obtained. Two auxotrophic albino derivatives of Mc2 (requiring either arginine or biotin) were each fused with an albino auxotrophic derivative of Mc3 (requiring uracil). Spheroplasts were produced by treating cells with Zymolyase following treatment with β-mercaptoethanol. Spheropolasts were stabilized with salt ($MgSO_4$ or $CaCl_2$) and fused during centrifugation in the presence of polyethylene glycol and subsequently plated on minimal medium. Fusion products that grew on minimal medium and were pigmented were examined for complementation of the morphological defect. Such isolates reproduced as yeasts at 37°C, suggesting that the defects resulting in a multicellular phenotype at 37°C were complementary and in different loci. This genetic complementation confirmed other observations suggesting that Mc2 and Mc3 mutant strains contained defects in different loci (Roberts and Szaniszlo, 1978; Jacobs and Szaniszlo, 1982). Further analysis of fusion products showed that they were 1.9–3.1 times larger than parental cells. Nuclear staining with mithramycin showed that virtually all cells of both parental and fusion strains contained a single nucleus. Cells with multiple nuclei appeared to be in the process of nuclear segregation. At the nonrestrictive temperature, 25°C, the fusion products had a slightly slower growth rate than parental strains, while at 37°C, where parental strains did not form budding yeasts, the fusion products did. Several fusion products were examined for segregants. For convenience, albino colonies showing segregation for the melanin defect were examined. For each clone, segregants of only one parental phenotype were observed, although the other phenotype was observed from another clone. Among the nonparental types observed for each clone were segregants with genotypes derived from both parents.

C. Recombinant DNA Techniques

As with other organisms, the ability to transform mutants with recombinant DNA and select desired transformants offers a potential method for obtaining and studying genes implicated in pathogenesis and morphogenesis. However, a transformation system is not yet available for this organism. An alternative approach is to isolate genes of *W. dermatitidis* by complementation in another organism.

A variety of auxotrophs have been derived in *W. dermatitidis* which could be used as the basis of a nutritional complementation method for selecting transformants. With this objective, experiments were undertaken to isolate genes of *W. dermatitidis* complementing defects in *Escherichia coli* resulting in leucine and uracil growth requirements. Genomic DNA was cloned into the plasmid YRp7 (Struhl *et al.*, 1979) and the resulting recombinant DNA used to transform *E. coli*. The ampicillin-resistant transformants were screened for the complementation of either the leucine or uracil requirement. Plasmid preparations obtained from ampicillin-resistant colonies growing in the absence of leucine or uracil contained inserts in the YRp7 plasmid and were able to produce ampicillin-resistant transformants of *E. coli*. However, depending on the colony from which the preparation was made, only 10–50% of these transformants were complemented for the specific growth defect W. L. Chaffin and P. J. Szaniszlo, unpublished observations). Additional experiments will be required to confirm that genes of *W. dermatitidis* complementing the *leuB* and *pyrF* defects of *E. coli* have been isolated.

Interspecies complementation may also be used to isolate genes of morphological interest. The yeast cell cycles of *Saccharomyces cerevisiae* and *W. dermatitidis* are very similar (Roberts and Szaniszlo, 1980; Jacobs and Szaniszlo, 1982). The phenotypic expressions of the defect of *cdc24* strains of *S. cerevisiae* and Mc3 strains of *W. dermatitidis* are also very similar. These similarities as well as success in intergeneric cell cycle complementation between *S. cerevisiae* and *Schizosaccharomyces pombe* (Beach *et al.*, 1982; Barker and Johnston, 1983) suggested that a gene complementing the *cdc24* defect of *S. cerevisiae* might be isolated from *W. dermatitidis*. This possibility is being investigated (P. J. Szaniszlo and C. R. Cooper, personal communication). Whether a gene complementing *cdc24* will also complement defects in Mc3 strains of *W. dermatitidis* will require additional confirmation.

V. CANDIDA ALBICANS

A. Introduction

Candida albicans is a dimorphic fungus which may be part of normal human flora and which is a common cause of mycotic infections. These infections are extremely diverse in their manifestation, varying from irritation and inflammation of a cutaneous surface to life-threatening systemic infection. These infections are generally considered to be opportunistic since some aspect of normal host defense is usually impaired. This impairment may range from minor irritations such as exposure to warmth and moisture to more serious insults such as immunosuppression. The prognosis is affected by the severity of the host impairment. Treatment or control of the underlying condition as well as of the infection itself may influence the outcome. Several species of the imperfect genus *Candida* cause infections. Because of the frequency with which *C. albicans* is the etiologic agent, the disease is most often associated with that species and it is the most studied. Other species encountered with some frequency are *C. guilliermondii, C. krusei, C. parapsilosis, C. pseudotropicalis, C. stellatoidea*, and *C. tropicalis*. Although any species may be isolated from any form of infection, some species are isolated more frequently from certain forms. For example, *C. parapsilosis* is frequently associated with candidal endocarditis. For some of these species mating strains have been observed and the teleomorph stage identified.

Some factors have been suggested to promote the pathogenesis of *C. albicans*. As part of the normal flora, the organism is generally present in the yeast form, whereas the presence of the hyphal form is associated with infection. There is an apparent relationship between morphology and pathogenesis. The loss of an extracellular protease reduced the virulence of mutant strains (MacDonald and Odds, 1983). Adherence of the organism to tissue is also considered to be a factor in colonization and tissue invasion. For example, species that adhere more poorly than *C. albicans* are encountered less frequently as agents of infections (King *et al.*, 1980).

B. Spheroplast Fusion

Protoplast fusion in polyethylene glycol has been used to develop a parasexual genetic system in *C. albicans* and *C. tropicalis*. Fusion products were detected by complementation of auxotrophic defects in the parental strains and were selected under conditions that permit fusion products but not parental strains to grow. With *C. albicans* the fusion system has been used to examine the characteristics of the fusion product as well as to establish complementation among mutants with the same auxotrophic phenotype.

For production of protoplasts of *C. albicans*, whole cells are generally treated with a reducing agent, β-mercaptoethanol (Sarachek *et al.*, 1981; Poulter *et al.*, 1981; Kakar and Magee, 1982) or dithiothreitol (Evans *et al.*, 1982). Snail enzyme preparations (Sarachek *et al.*, 1981; Pesti and Ferenczy, 1982), Zymolase (Poulter *et al.*, 1981; Kakar and Magee, 1982), and β-glucuronidase (Evans *et al.*, 1982) have been used to produce osmotically sensitive cells. The osmotically sensitive cells were stabilized with sugar, sorbitol (Kakar and Magee, 1982) or mannitol (Pesti and Ferenczy, 1982), or salts such as KCl

(Sarachek *et al.*, 1981; Evans *et al.*, 1982) or $MgSO_4$ (Poulter *et al.*, 1981; Pesti and Ferenczy, 1982). Strains with differing auxotrophic requirements were fused in polyethylene glycol and $CaCl_2$ (Sarachek *et al.*, 1981; Poulter *et al.*, 1981; Kakar and Magee, 1982; Evans *et al.*, 1982; Pesti and Ferenczy, 1982). Following fusion, cells were plated and regenerated using variants of different protocols: Cells were suspended in osmotically stabilized salt buffer and plated on plates containing osmotically stabilized medium (Sarachek *et al.*, 1981) or 0.5% agar was added to the cell suspension and the mixture poured as a soft overlay onto selective medium agar plates (Poulter *et al.*, 1981). In a second general protocol, cells were mixed with sorbitol-stabilized medium in a test tube and poured into a plate containing stabilized medium (Kakar and Magee, 1982) or the cells were mixed with the stabilized medium directly in the Petri plate (Evans *et al.*, 1982). The procedure used in regeneration may affect the sensitivity of cells to some selective conditions (unpublished observations). Protoplast regeneration frequencies varied from 3 to >90% (Sarachek *et al.*, 1981; Kakar and Magee, 1982) with differences in regeneration varying with serotype and phenotype of mutant derivatives (Sarachek *et al.*, 1981).

Protoplasts of *C. tropicalis* were produced by using general procedures (Fournier *et al.*, 1977; Vallin and Ferenczy, 1978). Protoplasts were obtained in buffer containing β-mercaptoethanol or dithiothreitol with a snail enzyme preparation and with salt for osmotic stability of the protoplasts. Following treatment with polyethylene glycol and $CaCl_2$, cells were mixed with mannitol-containing agar medium and poured onto the surface of plates containing the same solidified medium (Vallin and Ferenczy, 1978). Alternatively, after fusion, cells were incubated with medium and mannitol and subsequently washed in high-salt buffer and spread on solid medium (Fournier *et al.*, 1977).

The development of spheroplast fusion as a means of establishing a parasexual genetic system for *C. albicans* has been pursued in several laboratories. Characteristics of fusion products and their formation have been examined as part of complementation studies as well as specific studies. Nutritional complementation between parental auxotrophs has been used to select fusion products on unsupplemented media. Most investigators report that fusion mixtures yield rapidly growing prototrophic colonies as well as smaller colonies with multiple cell phenotypes (Sarachek *et al.*, 1981; Poulter *et al.*, 1981; Kakar and Magee, 1982). The first class is thought to represent prototrophic, uninucleate cells which arose as a result of chromosome mixing as an early event and which multiply rapidly to produce a stable prototrophic colony. The smaller, slower growing colonies appeared to be heterokaryons from which may segregate parental types (Poulter *et al.*, 1981) as well as stable uninucleate fusion products (Sarachek *et al.*, 1981). Kakar and Magee (1982) also reported small colonies that were fully prototrophic or fully auxotrophic. Among colonies isolated by Evans *et al.* (1982), several auxotrophic colonies were detected. Magee and co-workers (Kakar and Magee, 1982; Kakar et al., 1983) have analyzed the nutritional requirements of cells in the small colonies containing both prototrophic and auxotrophic phenotypes. These colonies were obtained from fusion of doubly auxotrophic parents and were selected for complementation of one requirement of each parent. The auxotrophic component(s) of these colonies was most frequently one or the other parental type or a mixture of both. Another class were auxotrophs requiring the unselected nutrient. Occasionally, auxotrophs were recovered which were doubly auxotrophic with a nutritional requirement derived from each parent and were prototrophic for the selected requirements.

In a similar type of analysis using random stable products, Sarachek et al. (1981) reported finding an auxotroph with a nutritional requirement derived from each parent and thus apparently recombinant. No parental auxotrophs were observed. In a subsequent analysis Sarachek and Weber (1984) did not observe any double auxotrophs, either parental or recombinant for the two unselected requirements. This analysis was performed with several different parental pairs. The fusion products were selected on medium requiring complementation of one requirement of each parent and the isolation carried out at different temperatures between 25 and 41°C. The frequency of prototrophs and the frequency of prototrophs showing complementation for the unselected markers increased with increasing temperature of selection. More than 95% of isolates tested at 41°C were prototrophic for all parental pairs. At 25°C, the frequency of prototrophs was 41–84% depending on the parental pair. This temperature effect was independent of the pair of parental requirements complemented in the selection. Sarachek and Weber (1984) suggested that complementation resulted from transfer of genetic material between nuclei rather than nuclear fusion and that transfer was increased at higher temperatures. Such transfer had been previously noted in certain mutants of *S. cerevisiae* (Dutcher, 1981). Sarachek and Weber (1984) argued that lysine auxotrophs, isolated from a fusion requiring complementation of the threonine and cysteine–methionine parental defects of parents requiring lysine and cysteine–methionine or adenine and threonine, were unlikely to have obtained the unselected adenine-requiring alleles. In other words, such auxotrophic isolates should be heterozygous for selected markers and homozygous for unselected markers. When lysine auxotrophs described above were examined for the ability to segregate adenine-requiring strains, none was observed. This observation was consistent with the postulated homozygosity of the wild-type allele at the adenine locus. In previous studies focused on spheroplast fusion as a system of genetic analysis, investigators had considered the chromosome interactions in the fusion products (Sarachek et al., 1981; Kakar and Magee, 1982; Kakar et al., 1983; Poulter et al., 1981, 1982; Poulter and Hanrahan, 1983). The analysis of products suggested that chromosomes from both parents were mixed in a nucleus and that stable products were uninucleate. While some of these products were consistent with complete nuclear fusion or karyogamy, other observations suggested that aneuploidy in some form was a likely explanation for at least some products. Suggested mechanisms included chromosome loss from a nucleus of elevated ploidy or from either type of nucleus in a heterokaryon or by transfer of material between nuclei (Kakar *et al.*, 1983).

Two unexpected observations were noted. In crosses in which fusion products were selected for complementation of one requirement of each parent, the auxotrophic derivatives were asymmetrically distributed. Magee and co-workers observed that among double auxotrophs there was a predominance of one parental phenotype (Kakar *et al.*, 1983). Sarachek and Weber (1984) observed only single auxotrophs among the auxotrophic segregants of fusion products selected for partial complementation of the parental requirements. However, in most crosses, the auxotrophic characteristic of one parent predominated by as much as 10 to 1. A particularly unusual observation was noted by Kakar *et al.* (1983) in crosses with one strain with a requirement for tryptophan. The parent strain required histidine and tryptophan. Many of the tryptophan-requiring products from the fusion of this parent with other strains also required lysine. Although the parental strain was subsequently demonstrated to be heterozygous for the lysine requirement, the mechanism by which fusion rendered a strain homozygous for this requirement is unclear.

Saracheck *et al.* (1981) reported that the regeneration frequency of strains varied with serotype and phenotype. Prototrophic fusion products were obtained between serotypes with complementing nutritional requirements and the frequency of fusion product isolation was not related to serotype. Some strains that regenerated at low efficiency were highly productive in generating fusion products. The authors concluded that the nature of the complementing auxotrophies was the major factor in the production of hybrid fusion products. The heterokaryon formed by fusion is sensitive to low temperatures (Sarachek and Rhoads, 1982). The viability of both heterokaryons and monokaryotic segregants produced by them was reduced by holding at low temperature, although the heterokaryons were more sensitive than the segregants. Both the temperature at which heterokaryons were grown following fusion and the temperature of growth after replating affected survival (Sarachek and Rhoads, 1983). Heterokaryons replated more efficiently at temperatures below that at which they initially grew and with declining efficiency at increasing temperatures. Formation of heterokaryons was further examined by varying the cell ratio of parental strains with good or poor regenerative abilities (Sarachek and Rhoads, 1981). Productive fusions required interaction of multiple protoplasts of each parent and were unaffected by regenerative ability in the range of 25–41°C. These temperature effects were further investigated using different carbon sources to support growth, metabolic inhibitors, and antibiotics. Those additions which most affected the observations were the ones which could be associated with mitochondrial function but their identities were not the same for cold and growth temperature effects. The authors

19. Gene Manipulation with Human Fungal Pathogens 481

noted that cold-sensitive defects in eukaryotes are often associated with ribosome assembly in the cytoplasm or mitochondria. Both cold and growth temperature effects could be attributed to temperature-dependent but different disturbances of normal mitochondrial functions in fused cells (Sarachek and Rhoads, 1982, 1983).

The size of fusion products has also been examined but there is little consensus among these reports. Poulter *et al.* (1981) reported that fusion prototrophs were larger than wild-type or auxotrophic parental strains. The long axis of one prototrophic product was 6.5 μm, compared to 4 μm for the wild type. Pesti and Ferenczy (1982) also reported that fusion products were larger than parental cells, although cell size was also affected by nystatin resistance or sensitivity. Sarachek *et al.* (1981) reported that the volume of fusion products varied from less than parental size to double parental size. Evans *et al.* (1982) observed a variable size for both parental cells and fusion products. Microscopic examination of a fusion product and the parental isolate showed no obvious difference (Gibbons, 1983). Since parental cell size is affected by growth rate, differences in size between parental cells and fusion products could be affected by growth rate differences (Chaffin, 1984). The DNA content of prototrophic fusion products was reported to be approximately twice that of parental strains (Pesti and Ferenczy, 1982) or 1.1–1.8 times greater than parental levels (Sarachek *et al.*, 1981). There was no correlation between size and DNA content in fusion products (Sarachek *et al.*, 1981). Stable fusion products, prototrophs, and auxotrophs were uninucleate (Sarachek *et al.*, 1981; Pesti and Ferenczy, 1982). Fusion products were also able to produce germ tubes and chlamydospores normally (Sarachek *et al.*, 1981; Poulter *et al.*, 1981).

Hybridization of auxotrophic strains with the same phenotype has been used to define complementation groups. Magee and colleagues (Kakar and Magee, 1982; Kakar *et al.*, 1983) have identified isoleucine–valine, methionine, arginine, and histidine alleles. In these experiments each parental strain with an auxotrophic requirement to be tested for allelism also required a second nutrient. Fusion products were selected on medium lacking the second requirement of each parental strain and supplemented with the common nutrient requirement to reduce the selection of revertants. Colonies growing on the selective medium were then tested for complementation of the auxotrophic requirement in question by replating on appropriate media. Of four independent isoleucine–valine auxotrophs tested, at least three complementation groups were observed. Three methionine auxotrophs yielded two complementation groups and two arginine auxotrophs showed complementation. Four histidine auxotrophs also yielded four complementation groups. In other cases, fusion between two adenine auxotrophs (Kakar and Magee, 1982), between two arginine auxotrophs, and between two histidine auxotrophs (Gibbons, 1983) failed to produce prototrophic products, suggesting that the defects were in the same allele. Spheroplast fusion, along with analysis of products of reversion and mitotic crossing-over, demonstrated that red, adenine-requiring mutants can be resolved into two classes (Poulter and Rikkerink, 1983). Defects in the *ade2* but not the *ade1* locus showed intragenic complementation. Similar phenotypic and genotypic observations with red, adenine-requiring mutants have been shown in other fungi. The *ade1* and *ade2* genes probably correspond to phosphoribosylaminoimidazole-succinocarboxamide synthetase and phosphoribosylaminoimidazole carboxylase, respectively (Poulter and Rikkerink, 1983).

Spheroplast fusion has also been used to investigate interactions between strains sensitive and resistant to the polyene antibiotic nystatin (Pesti and Ferenczy, 1982). The resistant strains were classified into groups based on their sterol content. Nystatin sensitivity and sterol production of fusion products suggested that drug resistance was a recessive characteristic. Complementation between resistant mutants showed three patterns. Hybrids between two different sterol production classes were sensitive. Within one class of resistance mutants, crosses showed no complementation between certain isolates, suggesting allelism. Within that class, however, one isolate produced a semiresistant fusion product, suggesting either modifier genes (Ahmed and Woods, 1967) or perhaps an intragenic complementation. Fusion of strains resistant and sensitive to 5-fluorocytosine produced sensitive heterokaryons and a majority of sensitive prototrophic products (Sarachek and Weber, 1984). This observation confirmed the nuclear determination of 5-fluorocytosine sensitivity and the recessive characteristic of resistance. A previous study suggested that resistance to another drug, nalidixic acid, resulted from mitochondrial mutations (Sarachek, 1979). Several independent, nalidixic acid-resistant derivatives of two doubly auxotrophic strains were crossed in combinations of resistant with resistant mutants of each strain and sensitive parents with resistant mutants (Haught *et al.*, 1984). Analysis of parental phenotype auxotrophs arising from fusion heterokaryons suggested that the resistance to nalidixic acid was associated with a nuclear locus, and analysis of prototrophic products suggested that resistance was a recessive characteristic. An unexpected finding was that resistance to nalidixic acid was at the same locus for the independent mutations in each strain but that the locus in the two strains was different.

The production of auxotrophic strains and analysis of fusion products along with other techniques have contributed to establishing linkage groups among loci and the ploidy of natural isolates. The five linkage groups that have been described are the following: (1) linkage of requirements for isoleucine–valine, cytosine, and serine (Kakar *et al.*, 1983); (2) linkage of an isoleucine and adenine requirement (Kakar *et al.*, 1983); (3) linkage of tryptophan, lysine, and methionine requirements (Kakar *et al.*, 1983); (4) linkage between a methionine requirement and canavanine sensitivity with the locus for canavanine sensitivity suggested to be distal to the centromere (Crandall, 1983); and (5) linkage of a requirement for arginine, adenine (producing red colonies), and methionine. This last linkage group has been studied most frequently. Sarachek *et al.* (1981) originally reported that requirements for arginine and adenine were located on the same chromosome with arginine located between the centromere and the adenine locus. Poulter *et al.* (1982) observed linkage between adenine and methionine requirements with the methionine locus between the centromere and the adenine locus. Subsequently, an arginine locus was added to this linkage group with a suggested location between the methionine and adenine loci (Poulter and Hanarahan, 1983). Poulter and Hanrahan (1983) used spheroplast fusion to demonstrate that the adenine requirement in their strain failed to complement the adenine requirement described by Sarachek *et al.* (1981) and by that test the two were allelic. This adenine requirement was shown to result from a defect in the *ade1* locus (Poulter and Rikkerink, 1983). Poulter and Hanrahan (1983) used fusions between strains that were isogenic to facilitate demonstration of linkage. After fusion, linked genes present in one strain were rendered heterozygous by complementation with a second isolate derived from the same parent. Mitotic crossing-over was induced in the fusion product and the distribution of phenotypes analyzed. The frequent coincident appearance of these traits is suggestive of linkage and the relative frequency of the various combinations can be used to order the linked genes with the centromere.

Having established the linkage group of arginine, methionine, and adenine, Poulter and Hanrahan (1983) examined the conservation of linkage among different isolates. Seven different isolates contained the linkage group with the same order. Among the isolates tested were strains used in other laboratories. In addition, an interspecies fusion with *C. stellatoidea* suggested that the linkage of centromere–methionine–adenine was conserved in that organism. The conservation of one linkage group, if extended to others, will facilitate the use of spheroplast fusion as a parasexual genetic system. The ability to form interspecies hybrids with *C. stellatoidea* reflects on the taxonomy and relatedness of these two species. In addition, interspecies hybridization may assist in studying the factors that effect the virulence of the two species.

Analyses of auxotrophs and products of spheroplast fusion have contributed to elucidating the ploidy of this imperfect yeast. Currently there is general agreement that the organism is diploid (Whelan and Magee, 1981; Poulter *et al.*, 1982; Sarachek and Weber, 1984).

Candida tropicalis has been less extensively studied. Singly auxotrophic parental strains requiring adenine or cysteine were fused (Vallin and Ferenczy, 1978). Pink, slow-growing colonies appeared on minimal medium only under conditions favoring fusion. A smaller number of white, fast-growing prototrophs was also observed. Cells of the pink colonies most frequently contained three or four nuclei and could be propagated indefinitely on minimal medium. Such colonies regularly gave rise to white stable prototrophs and on rich medium parental phenotypes appeared. The stable prototrophs that segregated from pink colonies were larger than parental cells, contained approximately 1.3–1.6 times as much DNA, and were uninucleate. The percentage of auxotrophic cells in prototrophic colonies decreased with time after initial plating (Fournier et al., 1977). From the unstable heterokaryons resulting from a fusion of doubly auxotrophic parents, stable auxotrophic isolates with parental phenotypes were recovered. Auxotrophic isolates were also recovered from stable prototrophic fusion products as a result of spontaneous events or following treatment with p-fluorophenylalanine or γ-rays. In addition to isolates with parental phenotypes, other singly and doubled auxotrophic strains were obtained. This second class of double auxotrophs contained isolates with a requirement from each parent and thus were recombinant for these requirements (Fournier et al., 1977). Stable prototrophic fusion products from a fusion between double auxotrophs, one sensitive and one resistant to 5-fluorocytosine, were examined for drug sensitivity. Both resistant and sensitive isolates were observed. The sensitive isolates were replated and resistant colonies appeared at a frequency similar to that of the appearance of auxotrophic requirements. The investigators suggested that distribution of sensitivity and resistance, an unselected marker, among the selected fusion products might represent a random assortment of chromosomes and that resistance to 5-fluorocytosine was recessive (Fournier et al., 1977).

C. Recombinant DNA Techniques

Techniques associated with recombinant DNA technology have been used to probe the organization of *C. albicans* DNA, to examine sequence heterogeneity among isolates and species, and to isolate specific genes. As part of an examination of the organization of the mitochondrial DNA (mtDNA) of *C. albicans*, Wills et al. (1985) cloned five of six restriction fragments produced by an *Eco*RI digest of mtDNA into the plasmid pBR322. Restriction sites within the cloned sequences assisted in constructing a restriction map of mtDNA and in observing repetitive sequences. The restriction map of mtDNA was circular, which confirmed previous observations of circular mtDNA visualized by transmission electron microscopy (Wills et al., 1984). The mtDNA contained an inverted duplication with the repetitive regions separated by nonduplicated regions. The nonduplicated regions occurred in both orientations with respect to each other, so that mtDNA exists as a mixed population with respect to sequence organization (W. S. Riggsby, personal communication).

The actin gene of *C. albicans* has been cloned and partially characterized. The *C. albicans* gene was isolated by utilizing homology with the actin gene from *S. cerevisiae* (Lasker and Riggsby, 1984). Genomic DNA enriched for the actin sequence was obtained by sucrose gradient fractionation of DNA digested with *Sal*I. The isolated fragments were cloned into pBR322. Colonies of *E. coli* transformed with the recombinant DNA were then screened with the *S. cerevisiae* gene probe and presumptive colonies confirmed by hybridization to the *Drosophila melanogaster* actin gene. Subsequent hybridization experiments with the cloned *C. albicans* gene suggested that it was present at one copy per haploid genome. The absence of common restriction sites between the *C. albicans* actin gene and the *S. cerevisiae* gene suggested that there is a considerable evolutionary distance between the two genes.

Complementation of nutritional defects of *E. coli* has also been used to isolate genes of *C. albicans*. A genomic DNA library was constructed in YRp7 and used to transform an *E. coli* strain defective in *trpC*, *pyrF*, and *leuB* genes (unpublished observations). The ampicillin-resistant transformants were examined in the absence of the various nutritional supplements. Several colonies that grew in the absence of uracil or leucine were further characterized. Plasmid DNA isolated from these colonies contained DNA inserts and was able to transform *E. coli* to ampicillin resistance. These resistant colonies were then examined for cotransformation of either uracil or leucine. Depending on the colony used for the plasmid isolation, 33–50% of the ampicillin-resistant transformants were also able to grow in the absence of leucine. With preparations isolated from colonies growing in the absence of uracil, 30–60% of the ampicillin transformants were able to grow in the absence of the supplement. All of the transformants could grow in the absence of tryptophan, the required activity being supplied by a gene in the YRp7 vector (Struhl *et al.*, 1979). Additional characterization of these recombinant plasmids will be required to determine whether they contain complementing sequences of *C. albicans*. These experiments were initiated to obtain a gene from *C. albicans* which might be used as a selectable marker in constructing a vector for transformation in this organism. Several attempts have been made to develop a transformation system in *C. albicans*. Various plasmids, most developed for *S. cerevisiae*, with potential for selection in *C. albicans* have been used. To date, these efforts to develop and verify transformation have been unsuccessful (P. T. Magee, personal communication; Y. Koltin, personal communication).

The information obtained about the organization of mtDNA and the isolated actin gene have been used to examine questions of relatedness between isolates and species (Mason *et al.*, 1984). Whole-cell DNA preparations from several isolates of *C. albicans* were digested with *Eco*RI and the separated fragments probed with radiolabeled mtDNA. All isolates showed the same pattern, with a little size heterogeneity in the smaller fragments. Whole-cell DNA preparations from single isolates of *C. tropicalis*, *C. parapsilosis*, *C. pseudotropicalis*, *C. stellatoidea*, and *Torulopsis glabrata* were probed with radiolabeled *C. albicans* mtDNA under conditions of reduced stringency. DNA from some species *(T. glabrata, C. pseudotropicalis*) showed no homology, while that of two other species *(C. tropicalis, C. parapsilosis)* showed some hybridization but not identity of fragment sizes. On the other hand, DNA from *C. stellatoidea* had essentially the same pattern as DNA from *C. albicans*. A similar experiment was performed with the radiolabeled actin gene from *C. albicans*. Several isolates of *C. albicans* and an isolate of *C. stellatoidea* showed hybridization to the same size fragments in two restriction enzyme digests. Hybridization was observed with restriction fragments of DNA from the other species but none of the bands of these species were common with the bands hybridizing in digests of DNA from *C. albicans* and *C. stellatoidea*. The authors suggested that these observations supported the view that *C. stellatoidea* is phylogenetically indistinguishable from *C. albicans*.

VI. CONCLUDING REMARKS

This brief survey demonstrates that nonsexual genetic techniques can be used to address a range of interests and problems, including mechanisms of drug resistance, metabolic pathways, control of morphology, and genome organization in terms of gene and sequence organization and taxonomy. As with any pathogenic organism, observations have dual implications, one for the organism itself and one for the associated disease. Consequently, studies of genomic organization may both reflect on the relationship between species and strains and on gene expression and at the same time have epidemiologic implications such as correlation with course or source of infection.

For imperfect fungi, spheroplast fusion provides a means for genetic analysis. The advantages of this system need not be elaborated. Even for fungi with conventional sexual genetics, these techniques offer an alternative method for producing diploid or polyploid cells. Such fusions also offer opportunities to test gene interactions.

Currently, perhaps the major problem with fully exploiting the potential of recombinant DNA technology is the development of the transformation systems in medically important fungi. Established techniques and transformation systems can be utilized with various problems of interest in pathogenic fungi. This is exemplified by the isolation of the actin gene from *C. albicans*. This approach should continue to be useful for the study of certain problems. Genes may be isolated with this approach, if their expression can be detected in a heterologous system or if specific probes can be obtained. Other questions may require the homologous transformation system. For example, when only a phenotype is known, e.g., the loss of a morphological or virulence trait, homologous transformation may lead to isolation of gene(s) restoring function. Another area in which transformation may be particularly useful is the study of the interaction of virulence factors. The generation of mutants may show the involvement of a given factor in promoting pathogenesis. However, when the trait is in a different genetic background, a correlation between the level of expression of the trait and virulence may not be demonstrable. Frequently, virulence of an organism is the end result of several interacting factors. The ability to separate or isolate genes associated with virulence factors and to place them in different backgrounds should assist in the study of the phenomenon of virulence. The requirements for homologous transformation systems in pathogenic fungi are the same as for established systems. Vector DNA must enter the cell, be replicated, be transmitted to daughter cells, and contain a gene whose expression can be detected. The first and the last of these may present fewest problems. For spheroplast-based transformation systems, techniques for spheroplast formation have already been described for several fungi and similarities in cell wall structure suggest that these techniques could be extended to other species. The genes for expression may be from the homologous or a heterologous source. In the latter case, reports of intergeneric expression suggest that at least some genes should be expressed. The difficulty in developing such systems is that the inability to obtain transformants could result from failure of any one of the requirements. In the absence of successful transformation it may be difficult to determine which steps are defective.

These nonsexual genetic techniques should assist medical mycologists in studying pathogenic fungi both as unique organisms and as agents of infection. Such studies may contribute to understanding fundamental biological processes as well as provide insights that may be utilized to improve diagnostic and prognostic procedures and treatment of the disease.

ACKNOWLEDGMENTS

I thank many individuals for stimulating discussion and communicating unpublished data. I am particularly indebted to R. J. Brady, D. H. Howard, Y. Koltin, P. T. Magee, J. C. Rhodes, S. W. Riggsby, A. Sarachek, and P. J. Szaniszlo.

REFERENCES

Ahmed, K. A., and Woods, R. A. (1967). A genetic analysis of resistance to nystatin in *Saccharomyces cerevisiae*. *Genet. Res.* **9,** 179–193.

Barker, D. G., and Johnston, L. H. (1983). *Saccharomyces cerevisiae cdc*9, a structural gene for yeast DNA ligase which complements *Schizosaccharomyces pombe cdc*17. *Eur. J. Biochem.* **134,** 315–319.

Beach, D., Durkazc, B., and Nurse, P. (1982). Functionally homologous cell cycle control genes in budding and fission yeast. *Nature (London)* **300,** 706–709.

Chaffin, W. L. (1984). The relationship between cell size and cell division in *Candida albicans*. *Can. J. Microbiol.* **30,** 192–203.

Crandall, M. (1983). UV-induced mitotic co-segregation of genetic markers in *Candida albicans*: evidence for linkage. *Curr. Gene.* **7,** 167–173.

Dutcher, S. K. (1981). Internuclear transfer of genetic information in kar1-1/KAR1 heterokayrons in *Saccharomyces cerevisiae. Mol. Cell. Biol.* **1**, 245–253.
Emmons, C. W., Binford, C. H., Utz, J. P., and Kwon-Chung, K. J. (1977). "Medical Mycology," 3rd Ed. Lea & Febiger, Philadelphia, Pennsylvannia.
Evans, K. O., Adeniji, A., and McClary, D. O. (1982). Selection and fusion of auxotrophic protoplasts of *Candida albicans. Antonie van Leeuwenhoek* **48**, 169–182.
Fournier, P., Provost, A., Bouguignon, C., and Heslot, H. (1977). Recombination after protoplast fusion in the yeast *Candida tropicalis. Arch. Microbiol.* **115**, 143–149.
Geis, P. A., Wheeler, M. H., and Szaniszlo, P. J. (1984). Pentaketide metabolites of melanin synthesis in the dematiaceous fungus *Wangiella dermatidis. Arch. Microbiol.* **137**, 324–328.
Gibbons, G. R. (1983). Complementation studies in arginine auxotrophs of *Candida albicans*. M.S. Thesis, California State Univ., Long Beach.
Haught, M. A., Bish, J. T., and Sarachek, A. (1984). Transmission of nalidixic acid resistance through protoplast-fusion crosses of *Candida albicans. Abstr., Annu. Meet. Am. Soc. Microbiol.* p. 295.
Jacobs, C. W., and Szaniszlo, P. J. (1982). Microtubule function and its relation to cellular development and the yeast cell cycle in *Wangiella dermatitidis. Arch. Microbiol.* **133**, 155–161.
Jacobson, E. S., White, C. W., and Nicholas, C. C. (1983). Construction of stable diploid strains of *Cryptococcus neoformans. Abstr., Annu. Meet. Am. Soc. Microbiol.* p. 384.
Kakar, S. N., and Magee, P. T. (1982). Genetic analysis of *Candida albicans:* Identification of different isoleucine–valine, methionine and arginine alleles by complementation. *J. Bacteriol.* **151**, 1247–1252.
Kakar, S.N., Partridge, R.M., and Magee, P.T. (1983). A genetic analysis of *Candida albicans:* isolation of a wide variety of auxotrophs and demonstrations of linkage and complementation. *Genetics* **104**, 241–255.
King, R. D., Lee, J. C., and Morris, A. L. (1980). Adherence of *Candida albicans* and other *Candida* species to mucosal epithelial cells. *Infect. Immun.* **27**, 667–674.
Kwon-Chung, K. J. (1978). Heterothallism vs. self-fertile isolates of *Filobasidiella neoformans (Cryptococcus neoformans). Sci. Publ.—Pan Am. Health Organ.* No. 356, 204–213.
Kwon-Chung, K. J., Polachek, I., and Popkin, T. J. (1982). Melanin-lacking mutants of *Cryptococcus neoformans* and their virulence for mice. *J. Bacteriol.* **150**, 1414–1421.
Lasker, B. A., and Riggsby, W. S. (1984). Cloning and characterization of the actin gene of *Candida albicans. Abstr., Annu. Meet. Am. Soc. Microbiol.* p. 111.
MacDonald, F., and Odds, F. C. (1983). Virulence for mice of a proteinase-secreting strain of *Candida albicans* and a proteinase-deficient mutant. *J. Gen. Microbiol.* **129**, 431–438.
McGinnis, M. R. (1983). Chromoblastomycosis and phaeohyphomycosis: New concepts, diagnosis and mycology. *J. Am. Acad. Dermatol.* **8**, 1–16.
Mason, M. M., Lasker, B. A., and Riggsby, W. S. (1984). Specific DNA homologies between *Candida albicans* and *Candida stellatoidea. Abstr., Annu. Meet. Am. Soc. Microbiol.* p. 295.
Matsumoto, T., Padhye, A. A., Ajello, L., Standard, P. G., and McGinnis, M. R. (1984). Critical review of human isolates of *Wangiella dermatitidis. Mycologia* **76**, 232–249.
Nishimuira, K., and Miyaji, M. (1983). Defense mechanisms of mice against *Exophiala dermatitidis* infection. *Mycopathologia* **81**, 9–21.
Pesti, M., and Ferenczy, L. (1982). Protoplast fusion hybrids of *Candida albicans* sterol mutants differing in nystatin resistance. *J. Gen. Microbiol.* **128**, 123–128.
Polacheck, I., Hearing, V., and Kwon-Chung, K. J. (1982). Biochemical studies of phenoloxidase and utilization of catecholamines in *Cryptococcus neoformans. J. Bacteriol.* **150**, 1212–1220.
Poulter, R. T. M., and Hanrahan, V. (1983). Conservation of genetic linkage in nonisogenic isolates of *Candida albicans. J. Bacteriol.* **156**, 498–506.

Poulter, R. T. M., and Rikkerink, E. H. A. (1983). Genetic analysis of red, adenine-requiring mutants of *Candida albicans*. *J. Bacteriol.* **156**, 1066–1077.

Poulter, R., Jeffery, K., Hubbard, M. J. K., Shepherd, M. G., and Sullivan, P. A. (1981). Parasexual genetic analysis of *Candida albicans* by spheroplast fusion. *J. Bacteriol.* **146**, 833–840.

Poulter, R., Hanrahan, V., Jeffery, K., Markie, D., Shepherd, M. G., and Sullivan, P. A. (1982). Recombination analysis of naturally diploid *Candida albicans*. *J. Bacteriol.* **152**, 969–975.

Rhodes, J. C., and Kwon-Chung, K. J. (1982). A new efficient method for protoplast formation in *Cryptococcus neoformans*. *Abstr., Annu. Meet. Am. Soc. Microbiol.* p. 336.

Rhodes, J. C., Polacheck, I., and Kwon-Chung, K. J. (1982). Phenoloxidase activity and virulence in isogenic strains of *Cryptococcus neoformans*. *Infect. Immun.* **36**, 1175–1184.

Rippon, J. W. (1982). "Medical Mycology: The Pathogenic Fungi and the Pathogenic Actinomycetes," 2nd Ed. Saunders, Philadelphia, Pennsylvania.

Roberts, R., and Szaniszlo, P. J. (1978). Temperature-sensitive multicellular mutants of *Wangiella dermatitidis*. *J. Bacteriol.* **135**, 622–632.

Roberts, R. L., and Szaniszlo, P. J. (1978). Temperature-sensitive multicellular mutants of *Wangiella dermatitidis*. *J. Bacteriol.* **135**, 622–632.

Samad, S. A., and Brady, R. J. (1984). Protoplast fusion in *Cryptococcus laurentii*. *Abstr., Annu. Meet. Am. Soc. Microbiol.* p. 118.

Sarachek, A. (1979). Population changes induced in *Candida albicans* by nalidixic acid. *Mycopathologia* **68**, 105–120.

Sarachek, A., and Rhoads, D. D. (1981). Production of heterokaryons of *Candida albicans* by protoplast fusions: Effects of differences in proportions and regenerative abilities of fusion partner. *Curr. Genet.* **4**, 221–222.

Sarachek, A., and Rhoads, D. D. (1982). Cold-sensitivity of heterokaryons of *Candida albicans*. *Sabouraudia* **20**, 251–260.

Sarachek, A., and Rhoads, D. D. (1983). Effects of growth temperatures on plating efficiencies and stabilities of heterokaryons of *Candida albicans*. *Mycopathologia* **83**, 87–95.

Sarachek, A., and Weber, D. A. (1984). Temperature-dependent internuclear transfer of genetic material in heterokaryons of *Candida albicans*. *Curr. Genet.* **8**, 181–189.

Sarachek, A., Rhoads, D. D., and Schwarzhoff, R. H. (1981). Hybridization of *Candida albicans* through fusion of protoplasts. *Arch. Microbiol.* **129**, 1–8.

Schafer, R. C., Cooper, C. R., and Szaniszlo, P. J. (1984). Complementation of two multicellular genes from *Wangiella dermatitidis* by spheroplast fusion. *Abstr., Annu. Meet. Am. Soc. Microbiol.* p. 293.

Still, C. N., and Jacobson, E. S. (1983). Recombinational mapping of capsule mutations in *Cryptococcus neoformans*. *J. Bacteriol.* **156**, 460–462.

Storts, D. R., and Brady, R. J. (1983). Fusion-dependent exchange of genetic material in auxotrophic protoplasts of *Rhodotorula glutinis*. *Abstr., Annu. Meet. Am. Soc. Microbiol.* p. 119.

Struhl, K., Stinchcomb, D. T., Schere, S., and Davis, R. W. (1979). High-frequency transformation of yeast: autonomous replication of hybrid DNA molecules. *Proc. Natl. Acad. Sci. U.S.A.* **76**, 1035–1039.

Vallin, C., and Ferenczy, L. (1978). Diploid formation of *Candida tropicalis* via protoplast fusion. *Acta Microbiol. Acad. Sci. Hung.* **25**, 209–212.

Vincent, R. D., Goldman, W. E., Goewert, R., Schlessinger, D., Lambowitz, A., Kobayashi, G. S., and Medoff, G. (1983). Differences in the mitochondrial and ribosomal DNAs among strains of *Histoplasma capsulatum*. *Abstr., Annu. Meet. Am. Soc. Microbiol.* p. 384.

Vincent, R. D., Lambowitz, A., Kobayashi, G., and Medoff, G. (1984). Comparison of mtDNAs from strains of three classes of *H. capsulatum*. *Abstr., Annu. Meet. Am. Soc. Microbiol.* p. 295.

Whelan, W. L., and Magee, P. T. (1981). Natural heterozygosity in *Candida albicans*. *J. Bacteriol.* **145**, 896–903.

Wills, J. W., Lasker, B. A., Sirotkin, L., and Riggsby, W. S. (1984). Repetitive DNA of *Candida albicans:* Nuclear and mitochondrial DNA components. *J. Bacteriol.* **157**, 918–924.

Wills, J. W., Troutman, W. B., and Riggsby, W. S. (1985). The circular mitochondrial genome of *Candida albicans* contains a large inverted duplication. *J. Bacteriol.* In press.

20

Fungal Carbohydrases: Amylases and Cellulases

BLAND S. MONTENECOURT
Department of Biology and the Biotechnology Research Center
Lehigh University
Bethlehem, Pennsylvania

DOUGLAS E. EVELEIGH
Department of Biochemistry and Microbiology
Cook College, Rutgers University
New Brunswick, New Jersey

I. Introduction	491
II. Fungal Amylases	494
A. Enzymes and Substrates	494
B. Molecular Biology of Fungal Amylases	495
III. Fungal Cellulases	501
A. Enzymes and Substrates	501
B. Molecular Biology of Fungal Cellulases	502
IV. Summary and Outlook	505
References	508

I. INTRODUCTION

Industrial strain development is predominantly concerned with the understanding of the subtle complexities and the advantageous manipulation of the genetic organization of hyperproducing microorganisms used in the biosynthesis of fermentation products. The basic aim of strain development programs is to reduce the cost of the fermentation products, which in this review are fungal carbohydrases. Commercial enzymes include amylases (α-amylase and glucoamylase), widely employed for the saccharification of starch in the food processing

TABLE I.

Commercial Fungal Carbohydrases[a]

Enzyme	Source	Application
Glucoamylase	*Aspergillus awamori* *Aspergillus oryzae* *Aspergillus niger* *Endomyces* sp. *Rhizopus niveus* *Saccharomyces distaticus* *Schwanniomyces castelli*	Saccharification in the bread making, soft drink, and brewing industries
α-Amylase	*Aspergillus oryzae* *Aspergillus niger*	Saccharification in the bread making, soft drink, and brewing industries
Cellulase	*Aspergillus niger* *Trichoderma reesei* *Trichoderma viride*	To facilitate extraction in the food industry (green tea, soy protein) and as a digestive aid
β-Glucanase	*Aspergillus niger*	To enhance filtration in brewing
Dextranase	*Penicillium* sp. *Trichoderma* sp.	Hydrolysis of dental plaque; removal of *Leuconostoc* slimes in sugarcane/beet processing.
α-Galactosidase	*Aspergillus* sp. *Mortierella vinca* *Saccharomyces uvarum*	In sugar beet processing to convert raffinose to sucrose to enhance crystallization
Invertase	*Saccharomyces cerevisiae* *Saccharomyces uvarum*	Sucrose conversion in the confectionery industry (soft-centered candy)
Lactase (β-galactosidase)	*Aspergillus niger* *Kluyveromyces fragilis* *Kluyveromyces lactis*	Utilization of whey and improved quality of milk products in the dairy industry
Naringinase	*Aspergillus niger*	Debittering of citrus products
Pectinase	*Aspergillus* spp.	To aid in the filtration of fruit pulp and to clarify fruit juices and wine
Pentosanase	*Aspergillus niger* *Aspergillus* sp.	Digestive aid; haze removal in wines
Yeast lytic enzyme	*Basidiomyces* sp. *Trichoderma hartzianum*	Research studies

[a] From Aunstrup (1978, 1983); Aunstrup *et al.* (1979); Barbesgaard (1977); Boing (1982); Lambert (1983); Rose (1980).

industry, and cellulases (endoglucanase and exoglucanase), which are of limited application in food processing (digestive aids in geriatric and baby foods) but of major potential use in the hydrolysis of renewable cellulosic biomass. Other applications of fungal carbohydrases are outlined in Table I. Our focus is on amylases and cellulases. These are enzyme complexes which include exo- and endo-splitting polysaccharidases plus a terminally acting enzyme responsible for

20. Fungal Carbohydrates: Amylases and Cellulases

POLYSACCHARIDE HYDROLASES

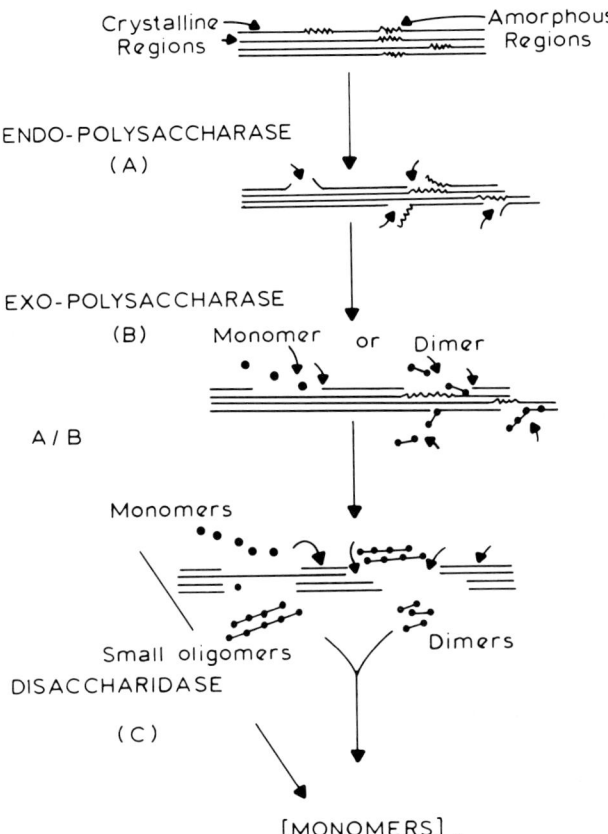

Fig. 1. Generalized scheme for polysaccharide hydrolysis. The polymer substrate is initially attacked by endo-polysaccharases (A), thus generating multiple sites for attack by exo-polysaccharases (B). With cellulase this sequence appears absolute, with initial scission of the amorphous zones. The degradation of noncrystalline polysaccharides (e.g., starch) is not subject to such strict sequential activities. In both instances, continued cooperative action between exo- and endo-splitting polysaccharides continues, combined with the terminal action of disaccharidases (C) to yield monomeric sugars. Exo- and endo-polysaccharases can each occur in multiple forms, e.g., glucoamylase 1, which attacks raw starch and maltodextrins, and glucoamylase II, which only degrades the latter substrate; similarly, endo-cellulases exhibit different specificities in relation to the degree of polymerization of the substrate. In all instances, the combined activities of the multiple forms result in enhanced degradation.

splitting the disaccharide products (maltose and cellobiose) into glucose, and are necessary for the complete solubilization of the substrate (Fig. 1). Strain development for these latter enzymes, respectively α- and β-glucosidase, will be considered only briefly.

Since there are estimates of annual existing biomass production in the United States of 1×10^9 dry tons (Bungay, 1981), fungal carbohydrases are of major significance in potential industrial application in addition to their role in the ecological recycling of cellulosic, hemicellulosic, and starchy biomass materials. Amylases and cellulases represent the largest potential industrial use of these extracellular enzymes, and therefore fungal producers have been the subject of extensive gene manipulation to increase the enzyme yield and activity. Several genetic approaches have been employed in the development of improved strains. By far the most successful approach to date has been the isolation of hyperproductive mutant strains. However, sexual, parasexual, and protoplast recombination systems are gradually being developed. These systems will become essential in the breeding of healthy production strains, including the means for correcting unwanted mutations. Methods of fungal strain improvement have been discussed in considerable detail (Cuskey *et al.*, 1983; Johnston, 1975, 1981; Rowlands, 1983, 1984; Stewart *et al.*, 1984). Recently, fungal carbohydrase genes have been cloned and sequenced, and efforts have been made to obtain their expression in both prokaryotic and eukaryotic hosts. Substantial insights into the mechanism controlling carbohydrase expression and secretion have been gained and yield considerable understanding of this facet of organization of the fungal genome. This chapter focuses on these recombinant DNA advances in the molecular biology of amylases and cellulases, which open a cornucopia of opportunities for further insights and industrial development.

II. FUNGAL AMYLASES

A. Enzymes and Substrates

Starch is a major agricultural commodity of both industrial and developing nations. This substrate for amylases is a complex polysaccharide consisting of amylose and amylopectin, two polymers that differ greatly in their physical properties. Amylose is a linear molecule of α-1,4-linked D-glucose units with a degree of polymerization of approximately 10^3 units. This unbranched polymer, following aqueous solution, can retrograde to become highly insoluble. Amylopectin, on the other hand, is a highly branched molecule consisting of α-1,4-D-glucose chains with α-1,6 branchpoints. It has a degree of polymerization of 10^4–10^5, an average chain length of the branches of 25–30 glucose moieties, and remains solubilized following dissolution.

Amylases are ubiquitous among the fungi. The principal fungal amylolytic activities found are α-amylase (endo-α-1,4-glucan glucanohydrolase, EC 3.2.1.1) and glucoamylase, sometimes called amyloglucosidase (exo-1,4-α-glucanohydrolase, EC 3.2.1.3). No fungal β-amylases (exo-α-1,4-glucan maltohydrolase) have been described to date. Most studies have focused on the imperfect fungi, especially the genus *Aspergillus*, since species such as *A. awamori, A. niger,* and *A. japonicus* have long been recognized as safe in the food industry. α-Amylases attack the internal α-1,4 bonds of amylose, amylopectin, and glycogen to yield maltodextrin products. Glucoamylases sequentially attack these substrates and also the branched maltodextrins, from the nonreducing termini with inversion of the configuration to yield β-D-glucose. The result of the two amylolytic activities is the efficient solubilization of starch.

The mechanism of action of amylases and of their synergism has been reviewed (Fogarty and Kelly, 1980; French, 1981) and a model for the subsite structure and ligand-binding mechanism of glucoamylase has been put forward (Hiromi *et al.,* 1983). Thus, the hydrolysis of starch is reasonably well understood from a biochemical point of view. One unique enzymological aspect is the synergistic action between the amylases in relation to the ever-changing array of substrates. Fungal glucoamylases, for example, have been found to exist extracellularly in two forms. One form is capable of attacking raw starch, while the other degrades only solubilized maltodextrins. Their cooperative action aids the efficiency of hydrolysis. The dynamic status of the substrate, the required synergism, and the different forms of glucoamylase may be interrelated, since in *Aspergillus* the genome contains only a single gene for the enzyme and the two forms are generated by posttranscriptional and posttranslational modification.

B. Molecular Biology of Fungal Amylases

α-Amylase and glucoamylases are extracellular glycoprotein enzymes of considerable industrial interest, capable of releasing glucose from the nonreducing ends of starch and other related polysaccharides, including branched-chain dextrins. The major current commercial sources of α-amylases are of prokaryotic origin and thus the molecular biology of these enzymes will not be considered here. Although glucoamylases are produced by a number of filamentous fungi and certain yeasts, only the enzymes of *Aspergillus* and *Rhizopus* have been extensively investigated at the molecular level. Most fungal amylases exist in multiple forms (Ueda, 1981). The glucoamylase of *Aspergillus* (*A. awamori* and *A. niger*) has two forms, a large form (apparent molecular mass of about 70,000 daltons), which degrades raw starch, and a smaller form (apparent molecular mass of 60,000 daltons), which is restricted to attacking maltodextrins. Both large and small forms are produced in roughly equal amounts and appear equal in their degradative ability toward soluble starch and oligosaccharides (Svensson *et al.*, 1982). The close taxonomic relationship between *A. awamori* and *A. niger* is well recognized, and is further evident from the inability to distinguish the glucoamylases from these two species by immunological criteria. The two glucoamylases from *A. niger*, designated G1 and G2, have been isolated and shown to share a common NH_2-terminal polypeptide chain but differ in their COOH-terminal region (Svensson *et al.*, 1982). The complete amino acid sequence of the G1 isozyme has been determined (Svensson *et al.*, 1983a,b). The majority of the O-linked carbohydrate was associated with 70 amino acid residues approximately 100 residues from the COOH-terminal end of the molecule. Two N-glycosylation sites were found in the central portion of the enzyme.

Initial observations indicated that the *A. niger* glucoamylases, G1 and G2, were closely related based upon their partial amino acid sequences. It was suggested that G1 contains an extension into the COOH-terminal region which is absent in the smaller G2 form (Svensson *et al.*, 1982). The two forms were shown previously to be immunologically cross-reactive (Lineback *et al.*, 1969). Similar structural relationships have been inferred for the glucoamylases M1 and M2 from *A. saitoi* (Inokuchi *et al.*, 1981, 1982; Takahashi *et al.*, 1978, 1981). In contrast, the multiple forms of *Rhizopus* glucoamylase have been shown to possess different NH_2-terminal regions (Takahashi *et al.*, 1982), suggesting that they are products of different genes. In addition to the two basic G1 and G2 forms of glucoamylase, smaller forms are often found in the culture filtrates and are probably formed by limited proteolysis. Treatment of *A. awamori* glucoamylases G1 and G2 *in vitro* with protease generated species that were apparently identical by electrophoretic characterization to the forms isolated from culture filtrates (Hayashida and Yoshino, 1978; Yoshino and Hayashida, 1978). On this basis, it was suggested that both posttranscriptional and posttranslational processing may occur.

In an effort to elucidate the molecular relationship between G1 and G2 of *A. niger*, glucoamylase-specific cDNA was synthesized from poly(A)$^+$ mRNA and the primary structure of the G1 mRNA was determined (Boel *et al.*, 1984a). These workers showed that precursor glucoamylase G1 (640 amino acids) contained an 18-amino-acid putative signal peptide and a short 6-amino-acid propeptide, which are presumably cleaved by a trypsin-like protease, to yield a final 616-amino-acid mature primary amino acid sequence. At the 3' end of the G1 mRNA there is an untranslated region 124 nucleotides upstream from the poly(A) tail. This region lacks the AATAAA poly(A) recognition signal found in the mRNA of higher eukaryotes. A second shorter form with an apparent molecular mass of 61,000 daltons was immunoprecipitable, as was G1, following *in vitro* translation of the poly(A)$^+$ mRNA. By sequencing the overlapping cDNA clones, it was shown that G2 mRNA lacked a 169-bp intervening sequence that was present at the 3' end of the G1 mRNA (Boel *et al.*, 1984a). The deletion of this 169-bp intervening sequence causes a shift in the open reading frame and, overall, results in a smaller protein having an additional octapeptide at the COOH-terminal end of the G2 glucoamylase due to the more distant termination site.

With the availability of the cloned fragments of the G1 and G2 genes, Boel and co-workers (1984b) were able to show by Southern blot hybridization analysis that a single glucoamylase gene existed in the chromosomal DNA of *A. niger*. A similar observation was subsequently made by Nunberg and co-workers (1984) with respect to *A. awamori* glucoamylase. The *A. niger* gene was isolated from a genomic library of *A. niger* and found to contain five intervening sequences (Fig. 2). Four short introns ranging in length from 55 to 75 bp have not been found to persist in any translatable mRNA. Four intervening sequences of *A. awamori* were similarly identified (Nunberg *et al.*, 1984). Both *A. niger* and *A. awamori* contain a fifth intervening sequence. This long 169-bp sequence was found to be involved in the differential mRNA processing of the G1 and G2 forms of glucoamylase (Boel *et al.*, 1984b). The exon/intron junctions of all the intervening sequences conform to the consensus sequences found at the splice junction of other eukaryotic genes (Mount, 1982) (Fig. 3). The small intron (IVS D) closest to the 3' end of the gene carries an additional conserved sequence TACTAAC (Boel *et al.*, 1984b), which has been identified in yeast chromosomal introns (Langford and Gallwitz, 1983; Pikielny *et al.*, 1983). In yeast, it has been proposed that this sequence may function in a self-splicing mechanism by substituting for the U1 snRNAs, allowing proper alignment for appropriate mRNA processing. It should be noted that only one (IVS D) of the five introns contains this consensus sequence. Presumably the other introns, including the 169-bp intron, are spliced by a mechanism similar to the trans-acting U1 snRNAs described in higher eukaryotes and recently detected in yeast (Tollervey *et al.*, 1983; Wise *et al.*, 1983) or by an as yet unidentified mechanism. Short intervening sequences have been found in the genes of other Ascomycetes, e.g., the glutamate dehydrogenase (Kinnaird *et al.*, 1982), histone 3, and histone 4 genes of *Neurospora crassa* (Woudt *et al.*, 1983), and in certain nuclear genes

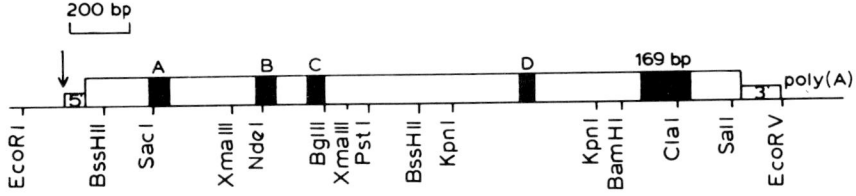

Fig. 2. The glucoamylase gene of *Aspergillus niger*. (↓) Initiation of transcription; (■) intervening sequences; (□) half-boxes at 5' and 3' ends represent untranslated regions. Splicing out of introns A–D yields G1 mRNA, while the further removal of the 169-bp intron yields G2 mRNA. [From Boel *et al.* (1984b); with permission.]

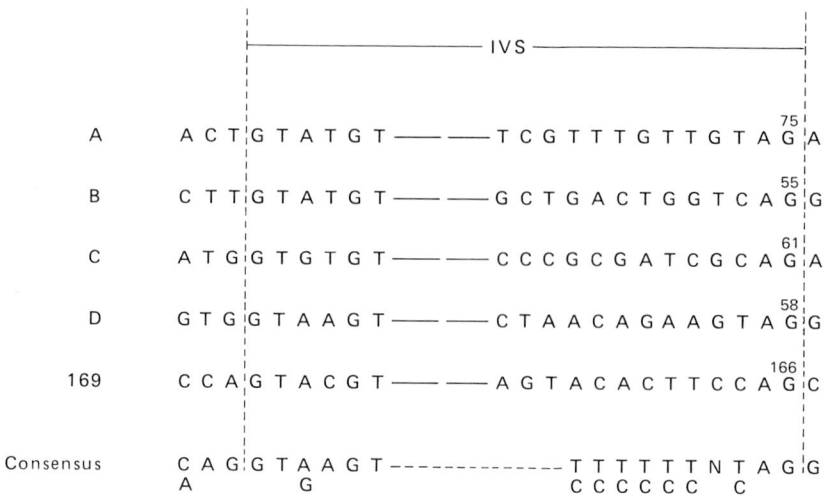

Fig. 3. Intron/exon junctions from the five intervening sequences in the *A. niger* glucoamylase gene. The junctions are aligned with the corresponding consensus sequences for eukaryotic splice sites (Mount, 1982). The length of the intervening sequences is indicated. [From Boel *et al.* (1984a); with permission.]

of yeast (Gallwitz and Sures, 1980; Miller, 1984). However, other genes of Ascomycetes, e.g., the *trp-I* gene of *N. crassa,* lack of intervening sequences (Schechtman and Yanofsky, 1983). It would appear from these restricted examples that in lower eukaryotic organisms, unlike higher eukaryotes, the existence of intervening sequences may be limited to a smaller number of specific genes.

A similar cloning approach has been undertaken to study the glucoamylase of *A. awamori* (Nunberg *et al.,* 1984). Under inducing conditions (starch-grown cells) a several hundredfold increase in a 2.3-kb mRNA encoding the larger G1 glucoamylase was demonstrated. This mRNA species was absent under noninducing conditions (xylose-grown cells). A cDNA probe was prepared from the 2.3-kb glucoamylase specific mRNA and used to locate a 3.4-kb *Eco*RI fragment from a genomic library of *A. awamori*. This 3.4-kb DNA fragment was subcloned into pACYC184 to identify the flanking sequences from other phage libraries. The nucleotide sequence of the structural gene located on the 3.4-kb *Eco*RI fragment was determined and found to be identical to the known amino acid sequence of *A. awamori* and also identical to the G1 sequence of *A. niger* (Svensson *et al.,* 1983a,b).

With respect to the promoter region, the glucoamylase genes of both *A. niger* and *A. awamori* have been shown to contain a typical TATA box in the -35 region and a CAAT run related to the canonical CCAAT in the -100 region (Boel *et al.,* 1984a; Innis *et al.,* 1985). This promoter resembles those described for other eukaryotes (Breathnach and Chambon, 1981) and that of *Trichoderma*

TABLE II.

Codon Utilization in the Glucoamylase Gene *Aspergillus awamori*[a]

Phe	UUU	4	Ser	UCU	16	Tyr	UAU	6	Cys	UGU	3
	UUC	18		UCC	19		UAC	21		UGC	7
Leu	UUA	0		UCA	4	Ter	UAA	0	Ter	UGA	0(1)
	UUG	6(7)		UCG	14		UAG	1(0)	Trp	UGG	19
Leu	CUU	3	Pro	CCU	4	His	CAU	0	Arg	CGU	4
	CUC	17		CCC	10		CAC	4		CGC	7(8)
	CUA	2		CCA	0	Gln	CAA	4(3)		CGA	4
	CUG	20(19)		CCG	8		CAG	13		CGG	3(2)
Ile	AUU	12	Thr	ACU	20	Asn	AAU	6	Ser	AGU	12(11)
	AUC	11		ACC	39		AAC	19		AGC	23(24)
	AUA	1		ACA	5	Lys	AAA	0	Arg	AGA	1
Met	AUG	3		ACG	10		AAG	13		AGG	1
Val	GUU	6	Ala	GCU	25	Asp	GAU	21	Gly	GGU	14(15)
	GUC	15		GCC	19		GAC	23		GGC	22
	GUA	2		GCA	10	Glu	GAA	9		GGA	7
	GUG	19		GCG	11		GAG	17		GGG	4(3)

[a] Data from Nunberg et al. (1984). Data in parentheses from Boel et al. (1984a) for *A. niger*.

(see below). In *A. niger* two different transcription initiation sites have been identified in the untranslated 5' region which are separated from each other by 20 nucleotide base pairs (Boel et al., 1984b). Multiple transcription initiation sites have been identified in the genes of several other eukaryotic organisms, for example, the alcohol dehydrogenase I gene of yeast (Bennetzen and Hall, 1982a), the tryptophan synthase gene of yeast (Zalkin and Yanofsky, 1982), and cytochrome c of *Schizosaccharomyces pombe* (Russell, 1983). Multiple transcription initiation sites seem to be a common property of highly transcribed genes in lower eukaryotes.

Codon bias will be an important factor in expression of cloned heterologous genes (Table II). Analysis of the sequenced gene shows that 57 of the 61 possible coding triplets are utilized (Boel et al., 1984a,b), and in this respect the extreme codon bias noted in certain yeast genes was not found (Bennetzen and Hall, 1982b; Russell and Hall, 1983).

One of the goals of genetic manipulation is to construct useful industrial microorganisms that exhibit new metabolic capabilities. Thus "industrial" hosts (e.g., yeast) have been sought which efficiently express cloned heterologous genes. Initial attempts to gain expression of the subcloned *A. awamori* glucoamylase gene in an autonomously replicating *Escherichia coli*/yeast shuttle vector YEp13 were unsuccessful (Innis et al., 1985). This was surprising in that the putative regulatory sequences within the 5' and 3' flanking regions of the *A. awamori* glucoamylase gene show strong similarities with the consensus sequences identified for structural genes of *Saccharomyces cerevisiae*. However, neither enzymatic activity, immunologically precipitable glucoamylase specific protein, nor glucoamylase specific mRNA could be detected in the transformants. Other *A. niger* genes that have been cloned into yeast cosmid vectors and subsequently transformed into yeast have yielded similar results (Penttila *et al.*, 1984). Of the five genes tested, β-galactosidase, glucoamylase, *ura3*, *leu2*, and β-glucosidase, only the β-glucosidase was minimally expressed in yeast and at

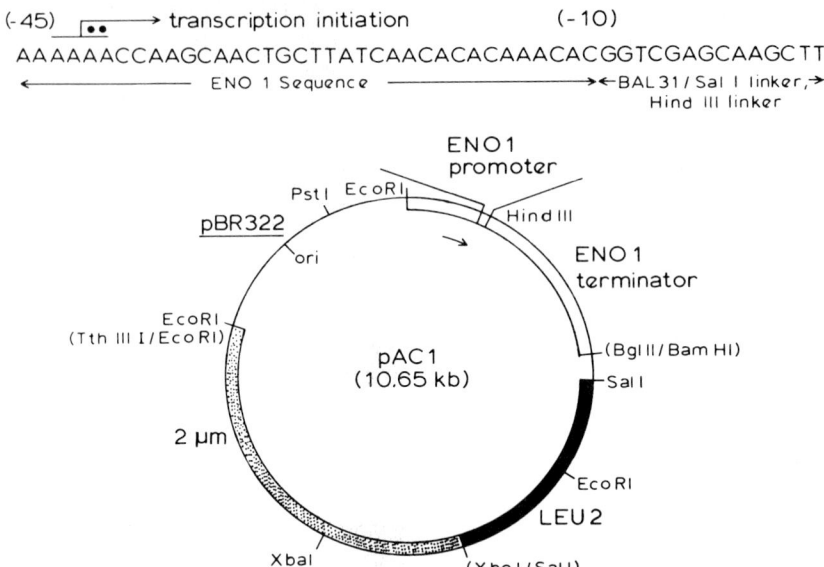

Fig. 4. Construction of yeast expression plasmid pAC1 (Innis *et al.*, 1985). The yeast *ENO1* structural gene (14.6-kb fragment) was obtained from plasmid *peno46* (Holland *et al.*, 1981) following digestion with *Hin*dIII. Following a partial digestion with *BAL*31 the DNA was ligated to phosphorylated *Sal*I linkers, cleaved with *Sal*I, religated in the presence of T4 DNA ligase, and used to transform *E. coli* RR101. Plasmid DNA was isolated from individual transformants and analyzed by *Sal*I/*Xba*I digestion to determine the approximate size of each deletion. A deletion extending from the *Hin*dIII site within the coding region of *ENO1* to a site 10 bp upstream from the translational initiation codon was identified by DNA sequence analysis and used for the construction of the *ENO1* promotor region of pAC1. The *Sal*I site at −10 was adapted with *Hin*dIII linkers regenerating a unique *Hin*dIII site between the *ENO1* promoter/terminator of pAC1. An expanded view of the DNA sequence surrounding the transcription initiation region of *ENO1* is shown. The other components of pAC1 include: (———) pBR322 sequences; (▦) yeast 2-μm sequences from the 3.7-kb *Eco*RI/*Sal*I fragment of pDB248 (Beach and Nurse, 1981); (■) 2.2-kb *Xho*I/*Sal*I fragment of yeast *LEU2* (Broach *et al.*, 1979); and (□) *ENO1* promoter/terminator sequences; the terminator is the natural *Hin*dIII/*Bgl*II fragment of *peno46* (Holland *et al.*, 1981). [From Innis *et al.* (1985), *Science* **228**, 21–26. Copyright 1985 by the AAAS.]

an extraordinarily low level. On the other hand, Yamashita and Fukui (1983) successfully cloned the glucoamylase gene from *Saccharomyces diastaticus* into *S. cerevisiae* and obtained a strain capable of simultaneous saccharification and fermentation of starch. It would seem, therefore, that there is nothing specific about the glucoamylase gene itself, such as codon bias, which is preventing its expression in *S. cerevisiae*.

Subsequent to their initial attempts to gain expression of the glucoamylase gene under the control of the natural *A. awamori* promoter, Innis and co-workers (1985) constructed a hybrid shuttle vector that contained the essential additions of the promoter and termination regions of the yeast enolase gene, *ENO1*. The vector also contained the origin of replication from *E. coli*, the *bla* gene from pBR322, the origin of replication from the yeast 2-μm plasmid, and a yeast *leu2* structural gene, thus permitting its replication and selection in both *E. coli* and yeast. Furthermore, a unique *Hin*dIII site was positioned between the promoter and the termination sites (Fig. 4). The previously isolated glucoamylase gene fragment was cloned into this site both with and without removal of the intervening sequences. Yeast transformants were isolated which could grow on starch as the sole source of carbon. However, only transformants containing the gene in which the intervening sequences had been removed expressed the glucoamylase activity. This suggests that, in spite of the similarities in consensus sequences between the *A. awamori* and yeast intron/exon junctions, yeast cannot efficiently process the intervening sequences of this filamentous fungus. Furthermore, hybridization experiments indicated that in hosts carrying the chimeric plasmid, termination occurred at the distal *ENO1* terminator rather than at the natural glucoamylase termination site. These results imply that neither the promoter nor the termination sites of *Aspergillus* are recognized by *Saccharomyces* in spite of their reasonably close phylogenetic relationship.

III. FUNGAL CELLULASES

A. Enzymes and Substrates

Although a wide variety of fungi actively degrade crystalline cellulose in nature, relatively few have been shown to efficiently secrete large amounts of these complexes (Reese and Mandels, 1984). The latter cellulolytic fungi include representatives of the Deuteromycetes/Ascomycetes (*Fusarium*, *Penicillium*, *Trichoderma* species) and Basidiomycetes (*Phanerochaete* and *Schizophyllum*). The genera most extensively studied are *Trichoderma* and *Phanerochaete*. The current concept of the activities involved in the degradation of crystalline cellulose assumes an initial attack by endoglucanases (endo-β-1,4-glucan glucanohydrolase, EC 3.2.1.4), which randomly cleave the β-1,4-glucosidic linkages in regions of the cellulose where stringent hydrogen bonding between the adjacent cellulose chains is lacking. This initial attack is followed by attachment of cellobiohydrolase (exo-β-1,4-glucan cellobiohydrolase, EC 3.2.1.91) and hydrolysis of cellobiosyl moieties from the nonreducing ends of the nicked chains. Continued synergistic interaction of these two enzymatic activities yields cellodextrins and cellobiose (see Fig. 1). In *Trichoderma* there is a multiplicity of both endoglucanases (EGI, EGII, EGIII, EGIV) (Shoemaker and Brown, 1978a,b) and cellobiohydrolases (CBHI, CBHII) (Fägerstam and Pettersson, 1979, 1980; Gritzali and Brown, 1979). Each of these enzymes appears to have a preferred substrate specificity and mode of action. For example, endoglucanases can be classified as saccharifying or liquefying based on their mode of action. In reality, optimal hydrolysis rates will depend on the ratio of the exo- and endosplitting enzymes and on the proportions of long and short oligosaccharides, both of which change throughout a total saccharification. There is obviously an optimal practical ratio of exo- and endocellulases to gain efficient hydrolysis. This restraint of the enzyme ratios does not apply to the commercial use of amylases, since α-amylase and glucoamylase from two different microbial sources are used sequentially and in proportions defined by the operator.

In the basidiomycete *Schizophyllum* the two endoglucanases (EGI and EGII) show similar physical and chemical properties and differ only in a 16-residue alanine-rich amino acid sequence at the amino-terminal end (Paice *et al.*, 1984). These authors suggest that EGII may arise following posttranslational proteolytic cleavage of EGI. A similar observation has been made with respect to *Phanerochaete*, where proteolytic cleavage has been implicated in the activation of the cellulases (Eriksson, 1983), but in this instance no sequence data are available to clarify a precursor–product relationship.

As a result of the multiplicity of the cellulase components, the synergism required for hydrolysis of the native substrate and the implication that posttranslational modification may be necessary for optimal or differential activity, direct genetic cloning techniques generally envisioned for a prokaryotic host may not be appropriate when considering the cloning of fungal genes. An additional complication that may hinder genetic manipulation is that the vast majority of fungal extracellular enzymes are glycoproteins. Whether the carbohydrate moieties of the enzyme are required for secretion, correct intermolecular folding, binding and recognition of the substrate, or simply to protect from extensive proteolysis or thermal denaturation is unknown at this time, though hyperglycosylated *Trichoderma reesei* cellobiohydrolase produced by yeast, following cloning and expression of this gene, has a specific activity essentially equivalent to that of the native enzyme (Shoemaker *et al.*, 1984).

B. Molecular Biology of Fungal Cellulases

Recently, there have been several exciting biochemical advances related to fungal cellulases. The group at the University of Uppsala (Fägerstam *et al.*,

1984) has determined the complete amino acid sequence of CBHI. Of the five asparagine residues which usually serve as recognition sites for glycosylating enzymes, three were found to possess carbohydrate moities. The carbohydrate contained glucose, glucosamine, and mannose. Previous workers (Gum and Brown, 1976) had reported that galactose was also found in CBHI, but this was not confirmed by Pettersson and co-workers. O-Linked carbohydrate was estimated to be attached at 22 serine and threonine sites, 10 of which are very near the carboxy-terminal end of the enzyme. The carbohydrate-rich carboxy terminus was shown to contain several repetitive sequences that were also proline-rich. Although the location of the carbohydrate moieties within the primary structure of the enzyme has been identified, structures of the O-linked and N-linked carbohydrate have not yet been determined.

The NH_2-terminal sequences of CBHII and EGII have also been determined (Pettersson *et al.*, 1981) and homologous regions were found between CBHI and EGII (Bhikhabhai and Pettersson, 1984). In addition, the NH_2-terminal sequences of two endoglucanases of the basidiomycete *Schizophyllum commune* have been elucidated (Paice *et al.*, 1984) The two endoglucanases from *S. commune* display similar chemical and physical properties but different molecular masses and isoelectric points. Their amino-terminal amino acid "sequences" are identical except that EGI contains an additional 16 amino acid residues (alanine-rich) at the amino-terminal end. The authors suggest that one explanation for the molecular heterogeneity of endoglucanases is extracellular proteolytic cleavage, a hypothesis previously put forward by other workers (Nakayama *et al.* 1976; Gong *et al.*, 1979). The amino acid sequence of the catalytic site of hen egg white lysozyme was compared with that of the endoglucanase I of *S. commune* (Yaguchi *et al.*, 1983) and the CBHI of *T. reesei* (Paice *et al.*, 1984). The results indicated a conserved region within the active sites, suggesting that carbohydrases which act as general acid catalysts share common amino acid sequences at their active sites.

In addition to the advances made in elucidating the biochemistry of fungal cellulases, two laboratories have cloned the CBHI gene of *T. reesei* (Shoemaker *et al.*, 1983a,b; Teeri *et al.*, 1983). The complete nucleotide sequence has been determined (Shoemaker *et al.*, 1983b) and extremely close agreement was found with the amino acid sequence data of Fägerstam *et al.* (1984). The CBHI gene was found to contain two introns within the coding region. Within each intron, 21 bp upstream from the 3' junction are found sequences homologous to a yeast sequence known to be necessary for accurate splicing. The intron/exon junctions were found to be similar to consensus sequences of other higher eukaryotic organisms. A putative promoter region, followed by a pyrimidine-rich region, was found 131 bp upstream from the translation initiation site. A 17-amino-acid signal peptide characterized the amino-terminal end. This is apparently cleaved during processing since the mature enzyme contains a pyroglutamyl residue at the amino terminus (Fägerstam *et al.*, 1984). The *T. reesei* gene bears some similarities to yeast genes and can be stably maintained in yeast (Shoemaker *et* al., 1984). A summary of the CBHI map is given (Fig. 5). Rearrangements of heterologous genes can also occur in yeast; thus *T. reesei* inserts, 30 kb in the pBTI-1 cosmid, were modified to leave residual 6-kb insertions, but the mechanism of this rearrangement is unclear (Picataggio, 1983).

Following the initial success that Innis and co-workers (1985) achieved in expressing the glucoamylase gene from *A. awamori* in *S. cerevisiae,* the CBHI and EGII genes of *T. reesei* lacking introns were cloned into the yeast expression vector carrying the enolase promoter and terminator regions. With the appropriate yeast transcription recognition signals, these cellulase genes were expressed in *S. cerevisiae* and approximately 70–80% of the activity was free in the extracellular medium. However, unlike the *A. awamori* glucoamylase, the *T. reesei* cellulases were significantly overglycosylated in yeast, yielding a CBHI with an "approximate" molecular mass of 100,000 (Shoemaker *et al.,* 1984), which is nearly twice the molecular mass of the native enzyme. This provides additional evidence that posttranscriptional and posttranslational processing in yeast and filamentous fungi are different.

Cellobiose, the last product in the sequential biodegradation of cellulase, is converted into two molecules of glucose by the enzyme β-glucosidase (cellobiase). One potential application of genetically manipulated fungi is the construction of cellulolytic yeast that can simultaneously saccharify and ferment cellulose to ethanol. Although the *S. cerevisiae* genome contains the structural genes for β-glucosidase (Duerksen and Halvorson, 1958), it is poorly expressed and this yeast is unable to grow on cellobiose. There are several yeasts which are able to grow on cellobiose, for example, *Kluyveromyces fragilis.* Raynal and Guerineau (1984) have prepared a genomic bank of *K. fragilis* DNA in a yeast

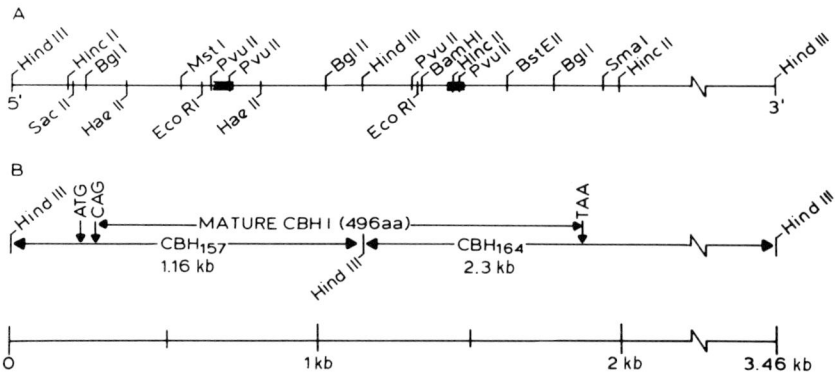

Fig. 5. Genomic map of CBHI. (A) Restriction sizes and introns (■); (B) the two adjacent *Hin*dIII genomic fragments that encode the entire CBHI protein. The relative positions of the initiating methionine (ATG), the glutamine codon (CAG) of the processed protein, and the TAA stop codon are marked. [From Shoemaker *et al.* (1983b); with permission.]

cosmid in *E. coli* and identified a recombinant plasmid carrying the β-glucosidase gene. The initial plasmid carried a 35-kb insert, but following subcloning, the gene was localized to a 3.5-kb fragment. The plasmid was subsequently used to transform *S. cerevisiae*. The transformants synthesized β-glucosidase (as measured by activity of cell-free extracts) at much higher levels than the wild *K. fragilis*. However, the transformants were still unable to grow on cellobiose. In *K. fragilis*, β-glucosidase is an intracellular enzyme and therefore must have some mechanism for transporting the cellobiose across the cell membrane. *Saccharomyces cerevisiae* appears to lack this transport facility. Only when this recombinant plasmid was transferred into a yeast strain that had been rendered leaky by mutation could growth on cellobiose be detected. Consequently, in order to construct yeast strains able to utilize cellobiose, it will be necessary to manipulate both the amount of β-glucosidase produced and either the permeability of the yeast strain to cellobiose or the secretion of cellobiase. The *Kluyveromyces* β-glucosidase gene is efficiently expressed in *Saccharomyces* although it is not secreted. In contrast, the genes encoding *Aspergillus* β-glucosidase are only marginally expressed in *Saccharomyces* (Penttila *et al.*, 1984).

IV. SUMMARY AND OUTLOOK

The recent advances in the cloning of carbohydrases have resulted in several new insights and some surprises. To date, these fungal genes have been shown to contain small introns, but no statement can yet be made regarding the generality of this feature because relatively few genes have been analyzed. The demonstration that *A. niger* has a single glucoamylase gene that results in two products, glucoamylases G1 and G2, is a novel finding for lower eukaryotes (Boel *et al.*, 1984b), but it will probably not be too surprising to find this result extended to other fungal systems. The fungal nucleotide sequences of the putative regulatory sites in both 5' and 3' flanking regions of the genes show considerable similarity to the yeast sequences. However, they are either incorrectly, poorly, or not processed in yeast, leading to the need for both a yeast promoter and terminator (ENO1) to gain expression of the heterologous fungal genes (Innis *et al.*, 1985; Shoemaker *et al.*, 1984). Furthermore, in spite of the similarity of fungal and yeast intron consensus sequences, yeast is unable to correctly splice the fungal genes, and it was necessary to delete the fungal introns in order to gain expression of glucoamylase and cellulase (Innis *et al.*, 1985; Shoemaker *et al.*, 1984). Penttila *et al.* (1984) also noted that of five *A. niger* genes studied only one, β-glucosidase, was expressed, and this at extremely low levels. A major difference between the filamentous fungi and yeast was the codon utilization pattern. The codon usage of a *A. awamori* and *A. niger* for glucoamylase closely resembles that for *Trichoderma* cellobiohydrolase I (Shoemaker *et al.*, 1983b) and also the *Neurospora trp-1* gene (Schechtman and Yanofsky, 1983). For filamentous fungi the codon bias may reflect the generalization that extremely abundant proteins such as glucoamylase and cellobiohydrolase exhibit a different codon bias from proteins synthesized more conservatively. Nunberg *et al.* (1984) speculate that the differences in codon bias may also indicate evolutionary distance, and thus *Saccharomyces* is quite distinct from *Aspergillus, Neurospora,* and *Trichoderma*. Although yeast and the two molds *Aspergillus* and *Trichoderma* are all Ascomycetes, they may not be as closely related as one might have expected.

In contrast to these differences, further surprises were found which indicated a close relationship between the genera. With regard to secretion, yeast recognized the three different heterologous leader peptide sequences, processed them correctly (even including two processing steps), and secreted glucoamylase, cellobiohydrolase, and endoglucanase. The efficiency of the system is seen in that 70–80% of the cellulases (Shoemaker *et al.*, 1984) and 90% of the glucoamylase (Meade *et al.*, 1984) were secreted. The proportion of these heterologous proteins was 20–30% of the total excreted proteins and, in the case of glucoamylase, resulted in a general increase in protein secretion compared to that of the native yeast cell. The overall synthesis yielded about 1% total cell protein. These leader sequences can therefore be considered as tools with which to gain secretion of other heterologous proteins. Perhaps their use should be considered to gain export of the routinely intracellular cellobiase of *K. fragilis* (Raynal and Guerineau, 1984).

As fungal glycosidases are glycoproteins, the question of whether the expression of their genes will yield glycosylated products has been raised. The results to date are mixed. With O-serine or O-threonine glycosylation, little change was detected with the recombinant forms of either the cellulases or glucoamylase. N-Glycosylation of the recombinant form of glucoamylase was decreased perhaps 4–5%. In contrast, the cellulases had highly decorated asparagine residues yielding glycoproteins within the molecular mass range of 100,000 daltons. The question of whether yeast will glycosylate a foreign gene product has been answered affirmatively, though in one instance (glucoamylase) there is normal glycosylation while in the other (cellobiohydrolase) hyperglycosylation occurs. Perhaps more important, these enzymes, whether normally or hyperglycosylated, are active and indeed have specific activities in the same range as the native enzyme.

A major interest in fungal carbohydrases stems from their commercial use. One research facet has focused on cloning them into yeast in order to extend its substrate range. The results have been most satisfying with the demonstration of an amylolytic yeast gained through recombinant DNA protocols (Fig. 6; Innis *et al.*, 1985). The growth rate of this yeast on maltodextrins is relatively slow, due

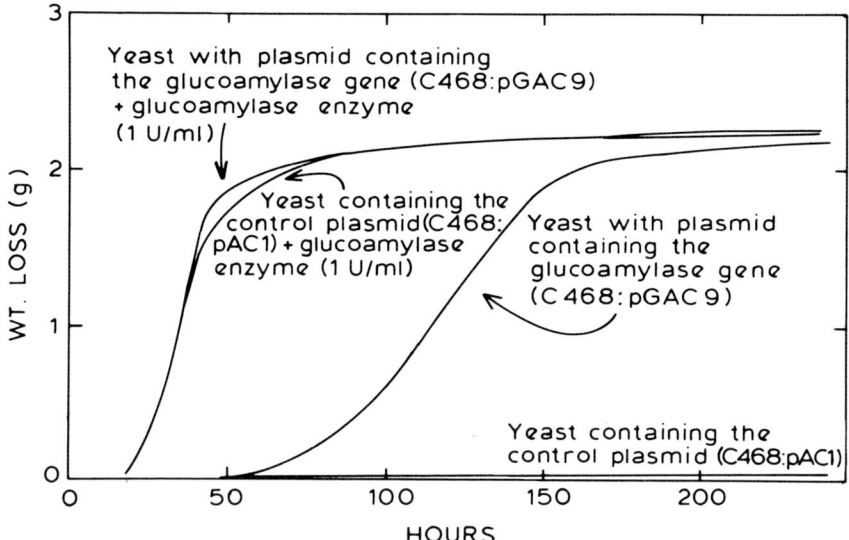

Fig. 6. Fermentation of starch by recombinant yeast strains. Strains with control plasmid (C468:pAC1) and with the plasmid containing the glucoamylase gene (C468:pGAC9) were grown in yeast minimal medium (40 μg/ml histidine, minus leucine) with 10% washed Maltrin M150 as the sole carbon source (Maltose oligosaccharides, av. degree of polymerization 13–17; Maltrin M150, Grain Processing Corporation, Muscatine, Iowa). Cultures were incubated at 30°C and agitated at 200 RPM, with and without the addition of glucoamylase (Sigma Co., St. Louis, Missouri) at 1 unit/ml. Fermentation was measured by weight loss. [From Innis et al. (1985), Science **228**, 21–26. Copyright 1985 by the AAAS.]

in part to glucoamylase acting extremely slowly toward small oligomers (the pentamer and smaller oligomers). However, overall this is an excellent beginning. Furthermore, the glucoamylase displays no change in K_m and V_{max} when assayed at concentrations of alcohol up to 15% (Meade et al., 1984). These results open the way for effective saccharification of starch combined with high ethanol yield.

These advances in the understanding of cellulases are significant in relation to the potential utilization of biomass. In this regard, it is important to note that the major bottleneck to development is due to the recalcitrant nature of lignocellulose. An optimal approach is to devise an economically attractive scheme in which the three components of lignocellulose are dissociated, e.g., by steam explosion. No such process has yet been commercialized. An enzymatic approach is also possible. True ligninases have been demonstrated (Kutsuki and Gold, 1982; Tien and Kirk, 1983) and cloning systems for the white rot fungi *Schizophyllum commune* (Seligy et al., 1983) and *Phanerochaete chrysosporium* (Paterson et al., 1984; Rao and Reddy, 1984) have recently been developed. These advances, taken in conjunction with the ability to clone cellulases, should permit consideration of efficient lignocellulose utilization, either by conversion to degradative products or by partial degradation to yield lignin and/or cellulosic polymeric components.

In summary, differences between regulatory signals of *Aspergillus* and *Trichoderma* and those of yeast have indicated the apparent necessity of using the host organism regulatory functions in gaining expression of foreign proteins. However, the way is now clear for further development, be it academic or applied, of carbohydrase synthesis in either the native organism or a foreign host.

ACKNOWLEDGMENTS

This chapter was supported by the New Jersey Agricultural Experiment Station (Publication F-01111-03-84), supported by state funds, the U.S. Hatch Act, and U.S. Department of Energy contracts (S.E.R.I.) EG-78-S-4591-A000 (D.E.E.) and (Office of Basic Energy Sciences) DE-AC02-83ER13106 (B.S.M.). The authors also gratefully acknowledge the contributions of the Cetus Corp. (M. Innis, J. Meade, and S. P. Shoemaker) in making available to us manuscripts in advance of publication. Their generosity has allowed this chapter to reflect the state of the art.

REFERENCES

Aunstrup, K. (1978). Enzymes of industrial interest: traditional products. *Annu. Rep. Ferment. Processes* **2**, 125–154.

Aunstrup, K. (1983). Enzymes of industrial interest: traditional products. *Annu. Rep. Ferment. Processes* **6**, 175–201.

Aunstrup, K., Andresen, O., Falch, E. A., and Neilsen, T. K. (1979). Production of microbial enzymes. *In* "Microbial Technology" (H. J. Peppler and D. Perlman, eds.), 2nd Ed., Vol. 1, pp. 282–309. Academic Press, New York.

Barbesgaard, P. (1977). Industrial enzymes produced by the genus *Aspergillus*. *In* "Genetics and Physiology of *Aspergillus*" (J. E. Smith and J. A. Pateman, eds.), pp. 391–404. Academic Press, New York.

Beach, D., and Nurse, P. (1981). High frequency transformation of the fission yeast *Schizosaccharomyces pombe*. *Nature (London)* **290**, 140–142.

Bennetzen, J. L., and Hall, B. D. (1982a). Codon selection in yeast. *J. Biol. Chem.* **257**, 3018–3025.

Bennetzen, J. L., and Hall, B. D. (1982b). Codon selection in yeast. *J. Biol. Chem.* **257**, 3026–3031.

Bhikhabhai, R., and Pettersson, L. G. (1984). The cellulolytic enzymes of *Trichoderma reesei* as a system of homologous proteins. *FEBS Lett.* **167**, 301–308.

Boel, E., Hijort, I., Svensson, B., Norris, F., and Fiil, N. P. (1984a). Glucoamylase G1 and G2 from *Aspergillus niger* are synthesized from two different but closely related mRNAs. *EMBO J.* **3**, 1097–1102.

Boel, E., Hansen, M. T., Hoegh, I., and Fiil, N. P. (1984b). Two different types of intervening sequences in the glucoamylase of *Aspergillus niger*. *EMBO J.* **3**, 1581–1585.

Boing, J. T. P. (1982). Enzyme production. *In* "Prestcott and Dunn's Industrial Microbiology" (G. Reed, ed.), 4th Ed., pp. 634–708. Avi, Westport, Connecticut.

Breathnach, R., and Chambon, P. (1981). Organization and expression of eucaryotic split genes coding for proteins. *Annu. Rev. Biochem.* **50,** 349–383.
Broach, J. R., Strathern, J. N., and Hicks, J. B. (1979). Transformation in yeast: Development of a hybrid cloning vector and isolation of the *CAN1* gene. *Gene* **8,** 121–133.
Bungay, H. R. (1981). "Energy, The Biomass Options." Wiley (Interscience). New York.
Cuskey, S. M., Montenecourt, B. S., and Eveleigh, D. E. (1983). Screening for cellulolytic mutants. *In* "Liquid Fuel Developments" (D. L. Wise, ed.), pp. 31–48. CRC Press, Boca Raton, Florida.
Duerksen, J. D., and Halvorson, H. (1958). Purification and properties of an inductible β-glucosidase of yeast. *J. Biol. Chem.* **233,** 1113–1120.
Eriksson, K. E. (1983). Advances in enzymatic degradation of lignocellulosics. *Proc. Int. Symp. Ethanol Biomass, Winnipeg, Manit., 1982* pp. 345–369.
Fägerstam, L. G., and Pettersson, L. G., (1979). The cellulolytic complex of *Trichoderma reesei* QM 9414. An immunological approach. *FEBS Lett.* **98,** 363–367.
Fägerstam, L. G., and Pettersson, L. G. (1980). The 1,4-β-glucan cellobiohydrolase of *Trichoderma reesei*. A new type of synergism. *FEBS Lett.* **119,** 97–110.
Fägerstam, L. G., Pettersson, L. G., and Engstrom, J. A. (1984). The primary structure of a 1,4-β-glucan cellobiohydrolase from the fungus *Trichoderma reesei*. *FEBS Lett.* **167,** 309–315.
Fogarty, W. M., and Kelly, C. T. (1980). Amylases, amyloglucosidases and related glucanases. *In* "Economic Microbiology" (A. H. Rose, ed.), Vol. 5, pp. 115–170. Academic Press, New York.
French, D. (1981). Amylases: enzymatic mechanism. *In* "Trends in the Biology of Fermentation for Fuels and Chemicals" (A. Hollaender, ed.), pp. 151–182. Plenum, New York.
Gallwitz, D., and Sures, I. (1980). Structure of a split yeast gene: Complete nucleotide sequence of the actin gene in *Saccharomyces cerevisiae*. *Proc. Natl. Acad. Sci. U.S.A.* **77,** 2546–2550.
Gong, C. S., Ladisch, M. R., and Tsao, G. T. (1979). Biosynthesis, purification and mode of action of cellulases of *Trichoderma reesei*. *Adv. Chem. Ser.* No. 191, 261–287.
Gritzali, M., and Brown, R. D., Jr. (1979). The cellulase system of *Trichoderma*. *Adv. Chem. Ser.* No. 181, 237–260.
Gum, E. K., and Brown, R. D., Jr. (1976). Structural characterization of a glycoprotein cellulase, 1,4-β-D glucan cellobiohydrolase C from *Trichoderma viride*. *Biochim. Biophys. Acta* **446,** 371–386.
Hayashida, S., and Yoshino, E. (1978). Formation of active derivatives of glucoamylase I during the digestion with fungal acid protease and α-mannosidase. *Agric. Biol. Chem.* **42,** 927–933.
Hiromi, K., Ohnishi, M., and Tanaka, A. (1983). Subsite structure and ligand binding mechanism of glucoamylase. *Mol. Cell. Biochem.* **51,** 79–95.
Holland, M. J., Holland, J. P., Thill, G. P., and Jackson, K. A. (1981). The primary structure of two yeast enolase genes. Homology between the 5' non-coding flanking regions of yeast enolase and glyceraldehyde phosphate dehydrogenase. *J. Biol. Chem.* **256,** 1385–1395.
Innis, M. A., Holland, M. J., McCabe, P. C., Cole, G. E., Wittman, V. P., Tal, R., Watt, K. W. K., Gelfand, D. H., Holland, J. P., and Meade, J. H. (1985). Expression, glycosylation and secretion of an *Aspergillus* glucoamylase from *Saccharomyces cerevisiae*. *Science (Washington, D.C.)* **228,** 21–26.
Inokuchi, K., Takahashi, T., and Irie, M. (1981). Purification and characterization of a minor glucoamylase from *Aspergillus saitoi*. *J. Biochem. (Tokyo)* **90,** 1055–1067.
Inokuchi, M., Iwama, K. M., and Takahashi, T. (1982). Modification of a glucoamylase from *Aspergillus saitoi* with 1-cyclohexyl-3-(2-morpholinyl-(4)-ethyl) carbodiimide. *J. Biochem. (Tokyo)* **89,** 125–134.
Johnston, J. R. (1975). Strain improvement and strain stability in filamentous fungi. *In* "The Filamentous Fungi, Vol. 1, Industrial Mycology" (J. E. Smith and D. R. Berry, eds.), pp. 340–364. Wiley, New York.

Johnston, J. R. (1981). Genetic manipulation of industrial yeasts. *Genet. Eng. Biotechnol.: Proc. Int. Symp. Genet. Eng., Piracicaba, Braz.* pp. 13–18.

Kinnaird, J. H., Keighren, M. A., Kinsey, J. A., Eaton, M., and Fincham, J. R. S. (1982). Cloning of the *am* (glutamate dehydrogenase) gene of *Neurospora crassa* through the use of a synthetic DNA probe. *Gene* **20**, 387–396.

Kutsuki, H., and Gold, M. H. (1982). Generation of hydroxyl radical and its involvement in lignin degradation by *Phanerochaete crysosporium*. *Biochem. Biophys. Res. Commun.* **109**, 320–322.

Lambert, P. W. (1983). Industrial enzyme production and recovery from filamentous fungi. *In* "The Filamentous Fungi, Vol. 4, Fungal Technology" (J. E. Smith, D. R. Berry, and B. Kristiansen, eds.), pp. 401–437. Arnold, London.

Langford, C. J., and Gallwitz, D. (1983). Evidence for an intron-contained sequence required for the splicing of yeast RNA polymerase II transcripts. *Cell (Cambridge, Mass.)* **33**, 519–527.

Lineback, D. R., Russel, I. J., and Rasmussen, C. (1969). Two forms of the glucoamylase of *Aspergillus niger*. *Arch. Biochem. Biophys.* **134**, 539–553.

Meade, J. H., McCabe, P. C., Cole, G. E., Wittman, V. P., Tal, R., Gelfand, D. H., Holland, J. P., and Innis, M. A. (1984). Cloning and expression of the *Aspergillus awamori* glucoamylase gene in yeast. *Am. Soc. Microbiol. Conf. Genet. Mol. Biol. Ind. Microorg.*, 3rd, Bloomington, Indiana. (Abstr.)

Miller, A. M. (1984). The yeast *MATa1* gene contains two introns. *EMBO J.* **3**, 1061–1065.

Mount, S. M. (1982). A catalogue of splice junction sequences. *Nucleic Acids Res.* **10**, 459–472.

Nakayama, M. Y., Tomita, H., Suzuki, H., and Nisizawa, K. (1976). Partial proteolysis of some cellulase components from *Trichoderma viride* and the substrate specificity of the modified products. *J. Biochem. (Tokyo)* **79**, 955–966.

Nunberg, J. H., Meade, J. H., Cole, G., Lawyer, F. C., McCabe, P., Schweickart, V., Tal, R., Wittman, V. P., Flatgaard, J. E., and Innis, M. A. (1984). Molecular cloning and characterization of the glucoamylase gene of *Aspergillus awamori*. *Cell. Mol. Biol.* **4**, 2306–2314.

Paice, M. G., Desrochers, M., Rho, D., Jurasek, L., Roy, M. C., Rollin, C. F., De Miguel, E., and Yaguchi, Y. (1984). Two forms of endoglucanase from the basidiomycete *Schizophyllum commune* and their relationship to other β-1,4-glycoside hydrolases. *Biotechnology* **1**, 535–539.

Paterson, A., McCartny, A. J., and P. Broda. (1984). The application of molecular biology to lignin degradation. *In* "Microbiological Methods for Environmental Biotechnology" (J. M. Grainger and L. M. Lynch, eds.), pp. 33–68. Academic Press, New York.

Penttila, M. E., Nevalainen, K. M. H., Raynal, A., and Knowles, J. C. K. (1984). Cloning of *Aspergillus niger* genes in yeast. Expression of the gene coding *Aspergillus* β-glucosidase. *Mol. Gen. Genet.* **194**, 494–499.

Pettersson, L. G., Fagerstam, L. G., Bhikhabhai, R., and Leandoer, K. (1981). The cellulase complex of *Trichoderma reesei*. *Symp. Wood Pulping Chem. (Stockholm)* **3**, 39–42.

Picataggio, S. (1983). Maintenance and expression of cloned *Trichoderma reesei* DNA in the yeast *Saccharomyces cerevisiae*. Ph.D. Thesis, Rutgers Univ., New Brunswick, New Jersey.

Pikielny, C. W., Teem, J. L., and Rosbash, M. (1983). Evidence for the biochemical role of an internal sequence in yeast nuclear mRNA introns: Implication for U1 RNA and metazoan mRNA splicing. *Cell (Cambridge, Mass.)* **34**, 395–403.

Rao, T. R., and Reddy, C. A. (1984). DNA sequences from a ligninolytic filamentous fungus, *Phanerochaete chrysosporium*, capable of automomous replication in yeast. *Biochem. Biophys. Res. Commun.* **118**, 821–827.

Raynal, A., and Guerineau, M. (1984). Cloning and expression of the structural gene for β-glucosidase of *Kluyveromyces fragilis* in *Escherichia coli* and *Saccharomyces cerevisiae*. *Mol. Gen. Genet.* **195**, 108–115.

Reese, E. T., and Mandels, M. (1984). Rolling with the times: Production and applications of *Trichoderma reesei* cellulase. *Annu. Rep. Ferment. Process* **7**, 1–20.

Rose, A. H. (1980). "Economic Microbiology, Vol. 5, Microbial Enzymes and Bioconversions." Academic Press, New York.

Rowlands, R. T. (1983). Industrial fungal genetics and strain selection. *In* "The Filamentous Fungi, Vol. 6, Fungal Technology" (J. E. Smith, D. R. Berry, and B. Kristiansen, eds.). Arnold, London.

Rowlands, R. T. (1984). Industrial strain improvement: rational screens and genetic recombination techniques. *Enzyme Microb. Technol.* **6**, 290–300.

Russell, P. R. (1983). Evolutionary divergence of the mRNA transcription initiation mechanism in yeast. *Nature (London)* **301**, 167–169.

Russell, P. R., and Hall, B. D. (1983). The primary structure of the alcohol dehydrogenase gene from the fission yeast *Schizosaccharomyces pombe*. *J. Biol. Chem.* **258**, 143–149.

Schechtman, M. G., and Yanofsky, C. (1983). Structure of the tri-functional *trp*1 gene from *Neurospora crassa* and its aberrant expression in *Escherichia coli*. *J. Mol. Appl. Genet.* **2**, 83–89.

Seligy, V. L., Barbier, J. R., Dimock, K. D., Dove, M. J., Moranelli, F., Morosoli, R., Willick, G. E., and Yaguchi, M. (1983). Current status in the application of recombinant DNA technology in the construction of cellulolytic yeast strains. *In* "Gene Expression in Yeast" (M. Korhola and E. Väisanen, eds.), pp. 167–185. Found. Biotech. Ind. Ferment. Res., Helsinki.

Shoemaker, S. P., and Brown, R. G., Jr. (1978a). Enzymatic activities of endo-1,4-β-D-glucanases purified from *Trichoderma viride*. *Biochim. Biophys. Acta* **523**, 133–146.

Shoemaker, S. P., and Brown, R. G., Jr. (1978b). Characterization of endo,1,4 β-D glucanases purified from *Trichoderma viride*. *Biochim. Biophys. Acta* **523**, 147–161.

Shoemaker, S., Watt, K., Tsitovsky, G., and Cox, R. (1983a). Characterization of cellulases purified from *Trichoderma reesei* strain L 27. *Biotechnology* **1**, 687–690.

Shoemaker, S., Schweickart, V., Ladner, M., Gelfand, D., Kwok, S., Myambo, K., and Innis, M. A. (1983b). Molecular cloning of exo-cellobiohydrolase I derived from *Trichoderma reesei* strain L 27. *Biotechnology* **1**, 691–696.

Shoemaker, S. P., Gelfand, D. H., Kwok, S., Schweickart, V., Watt, K., van Arsdell, and Innis, M. A. (1984). Expression of *Trichoderma reesei* L27 cellulases in *Saccharomyces cerevisiae*. *Am. Chem. Soc., Natl. Meet., 188th.* (abstr.)

Stewart, G. G., Panchal, C. J., Russell, I., and Sills, A. M. (1984). Biology of ethanol producing microorganisms. *CRC Crit. Rev. Biotechnol.* **1**, 161–188.

Svensson, B. T., Pedersen, G., Svendsen, I., Sakai, T., and Ottesen, M. (1982). Characterization of two forms of glucoamylase from *Aspergillus niger*. *Carlsberg Res. Commun.* **47**, 55–69.

Svensson, B., Larsen, K., and Svendsen, I. (1983a). Amino acid sequence of tryptic fragments of glucoamylase G1 from *Aspergillus niger*. *Carlsberg Res. Commun.* **48**, 517–527.

Svensson, B., Larsen, K., Svendsen, I., and Boel, E. (1983b). The complete amino acid sequence of the glycoprotein, glucoamylase G1 from *Aspergillus niger*. *Carlsberg Res. Commun.* **48**, 529–544.

Takahashi, T., Tsuchida, Y., and Irie, M. (1978). Purification and properties of three forms of glucoamylase from *Rhizopus* species. *J. Biochem. (Tokyo)* **84**, 1183–1194.

Takahashi, T., Inokuchi, N., and Irie, M. (1981). Purification and characterization of a glucoamylase form *Aspergillus saitoi*. *J. Biochem. (Tokyo)* **89**, 125–134.

Takahashi, T., Tsuchida, Y., and Irie, M. (1982). Isolation of two inactive fragments of a *Rhizopus* sp. glucoamylase: Relationship among three forms of the enzyme. *J. Biochem. (Tokyo)* **92**, 1623–1633.

Teeri, T., Salovuori, I., and Knowles, J. (1983). The molecular cloning of the major cellulase gene from *Trichoderma reesei*. *Biotechnology* **1**, 696–699.

Tien, M., and Kirk, T. K. (1983). Lignin-degrading enzyme from the hymenomycete *Phanerochaete chrysosporium* Burds. *Science (Washington, D.C.)* **221,** 661–663.

Tollervey, D. J., Wise, A., and Guthrie, C. (1983). A U4-like small nuclear RNA is dispensible in yeast. *Cell (Cambridge, Mass.)* **35,** 753–762.

Ueda, S. (1981). Fungal glucoamylases and raw starch digestion. *TIBS* **6,** 89–90.

Wise, J. A., Tollervey, D., Maloney, D., Swerdlow, H., Dunn, E. J., and Guthrie, C. (1983). Yeast contains small nuclear RNAs encoded by single copy genes. *Cell (Cambridge, Mass.)* **35,** 743–751.

Woudt, L. P., Pastink, A., Kempers-Veenstra, A. E., Jansen, A. E. M., Mager, W. H., and Planta, R. J. (1983). The genes coding for histone H3 and H4 in *Neuropsora crassa* are unique and contain intervening sequences. *Nucleic Acids Res.* **11,** 5347–5360.

Yaguchi, Y., Roy, M. C., Rollin, C. F., Paice, M. G., and Jurasek, L. (1983). A fungal cellulase shows sequence homology with the active site of hen egg-white lysozyme. *Biochem. Biophys. Res. Commun.* **116,** 408–411.

Yamashita, I., and Fukui, S. (1983). Molecular cloning of a glucoamylase-producing gene in the yeast *Saccharomyces. Agric. Biol. Chem.* **47,** 2689–2692.

Yoshino, E., and Hayashida, S. (1978). Enzymatic modification of glucoamylase of *Aspergillus awamori* var. *kawachi. J. Ferment. Technol.* **56,** 289–295.

Zalkin, H., and Yanofsky, C. (1982). Yeast TRP 5: Structure, function, regulation. *J. Biol. Chem.* **257,** 1491–1500.

V
Postscript

21

Prospects for a Molecular Mycology

J. W. BENNETT
Department of Biology
Tulane University
New Orleans, Louisiana

I.	Introduction	515
II.	A Primer in Mycology	516
	A. Circumscription	516
	B. Classification	517
	C. Fungi as Experimental Organisms for Molecular Genetics	518
III.	Molecular Mycology	519
	A. Molecular Biology for Mycologists	519
	B. Opportunities	520
	C. Conclusions	524
	References	526

I. INTRODUCTION

Mycology is the study of fungi. Respectively, "mycology" and "fungus" are derived from Greek and Latin words for "mushroom." Mushrooms are, indeed, among the largest and most conspicuous fungi, but microscopy reveals many other minute organisms with mushroom-like attributes. These organisms are at the interface of microscopic and macroscopic life, too small to be seen with the naked eye unless they are growing as a colony. Nevertheless, these microscopic fungi frequently make themselves known by what they do: ferment, putrefy, and cause disease. Many of these organisms are recognized by laymen and have labels in the vernacular: "mold," "mildew," "blight," "smut," "rust," and so on.

Fungi have always troubled taxonomists, because they do not fit easily into the traditional schemes for classifying organisms, nor do they constitute an easily defined group. Taxonomists at different times in history have classified fungal species in various ways. Some of this confusion can be traced to Linnaeus, who was notoriously vague and basically wrong in his interpretation of the group. Linnaeus originally classified the fungi as members of the botanical class Cryptogamia, plants with "concealed" reproductive structures. Ever since, mycology as a formal study has remained a subdivision of botany. Fungal taxonomy, unlike bacterial taxonomy, is governed by the rules of botanical nomenclature. During this time, the fungi as a group have been reappraised by each generation of taxonomists, all the while gradually moving their way up the hierarchy of formal ranks. During the 19th century, fungi gained class rank within the plant kingdom, and were usually grouped alongside the algae as members of the division Thallophyta. During the early 20th century, fungi were accorded divisional status within the plant kingdom. Then, during the 1950s and 1960s the traditional two kingdom classification of plants and animals began to fall apart, and all but the most conservative systematists were ready for a new scheme. The important dichotomous division became the prokaryotes and eukaryotes, replacing the old dichotomy of plants and animals.

The most popular of the modern hierarchical schemes for eukaryotes was the five-kingdom scheme proposed by Whittaker (1969). In this system, fungi became a separate kingdom, on equal footing with the plants. This system accepted and legitimized the numerous unique and unplantlike attributes of fungi and simultaneously caused a professional identity crisis among some mycologists.

If fungi are not plants, why link their study to the rest of botanical research? The new taxonomic status validly reflects our increased understanding of the singular—some would say bizarre—attributes of the group. What are these singular attributes? What kind of organisms does a mycologist study? What role will fungi play in the new molecular biology and, conversely, how will the new molecular biology influence the study of fungi? In short, what are the prospects for a molecular mycology?

II. A PRIMER IN MYCOLOGY

A. Circumscription

Fungi are eukaryotic. They contain membrane-bound nuclei, mitochondria, 80 S ribosomes, and the other features of eukaryotic cells. Fungi are like plants in that they are usually not motile and they usually possess a rigid cell wall; they are like animals in that they are heterotrophic. Heterotrophic organisms cannot fix their own carbon from inorganic sources, but must obtain carbon from organic sources.

The circumspection of fungi is difficult. They encompass "a bewildering array of forms and kinds of organisms: uninucleate, multinucleate, acellular, coenocytic, cellular, spherical, filamentous, simple, highly complex, motile, nonmotile, parasitic, saprobic, microscopic, macroscopic, through almost any combination of extremes'' (Ross, 1979, p. 6).

Recognizing the impossibility of a simple delineation of the kingdom Fungi, we can, nevertheless, make some generalizations.

The vegetative phase of most fungi is dominated by a threadlike, filamentous cell called a hypha. Many hyphae together are called a mycelium. Sometimes, harking back to the botanical era, the mycelium is referred to as a "thallus." Certain primitive fungi lack hyphae, while the yeasts, considered an advanced group, have reverted to a unicellular state. Most fungi are both multicellular and multinucleate. Most fungi have rigid, chitinous cell walls.

The fungal mode of nutrition is absorptive, based on extracellular digestion. Enzymes are secreted across the cell walls into the environment; then nutrients are absorbed back into the cell. This absorptive mode of nutrition was the primary criterion used by Whittaker (1969) to justify placing the fungi in a separate kingdom.

B. Classification

Mycologists do not agree on a common scheme for the classification of fungi; moreover, new discoveries lead to revisions and rearrangements of existing taxonomic schemes. Criteria for fungal classification are largely morphological, with particular emphasis on the morphology of reproductive structures. Unfortunately, the fossil record is scant, so evolutionary relationships must be inferred from living representatives. In the absence of hard data, acrimonious disputes about the "correct" evolutionary relationships, and the most "natural" classification of fungi, are an occupational hazard among the mycologists who specialize in systematics.

Most classifications distinguish two major groups within the kingdom Fungi: the Myxomycota ("slime molds," organisms with plasmodia or pseudoplasmodia) and the Eumycota ("true fungi," organisms with walled, usually hyphal cells). The slime molds are an ambiguous group, retained in the kingdom Fungi more by tradition than by logic. Unlike the "true" fungi, slime molds ingest particulate food. Furthermore, their vegetative phase consists of naked cells or masses of protoplasm (the plasmodium) that move about in disconcertingly animal-like fashion. When slime molds reproduce, however, they form fungal-like spores with thick cell walls. Whittaker included the slime molds in his version of the kingdom Fungi with this admonition: "The slime molds cross the distinctions of the kingdoms in both nutrition and organization, and offer a free choice of treatment as aberrant fungi, eccentric protists, or very peculiar animals" (Whittaker, 1969). Those of us who hope for a more molecular mycology may wish, possessively, to retain slime molds under the rubric of mycology, since this group contains *Physarium* and *Dictyostelium*. Species from both genera have figured prominently in modern developmental biology.

The Eumycota are usually divided into "lower" and "higher" fungi. The lower fungi generally have hyphae without cross walls (they are nonseptate) and include several groups that possess flagellated zoospores. With the exception of the yeasts, the higher fungi have septate hyphae and many of them produce complex reproductive structures.

For the benefit of the novice to the field, an elementary outline of fungal taxonomy is presented in Appendix I. It follows Ainsworth (1971) and Ainsworth *et al.* (1973a,b).

C. Fungi as Experimental Organisms for Molecular Genetics

A good contender for *the* birthday of molecular biology is the publication of Beadle and Tatum's (1941) "one gene–one enzyme" hypothesis. This research utilized *Neurospora crassa* and introduced many of the basic strategies of microbial genetics, such as the use of conditional lethal mutants, later expanded in bacterial systems.

Today, three species of fungi—all ascomycetes—qualify as genuine "model systems" in modern genetics: *Neurospora crassa, Aspergillus nidulans,* and last, but certainly not least, *Sarracharomyces cerevisiae*. What is so special about these organisms?

In many ways, bacteria, yeasts, and molds share common advantages as experimental organisms. Because they are all heterotrophic and absorptive, they are easily grown in the laboratory on simple defined media. They are fast-growing, so experiments do not take long. They are hardy (as evidenced by their frequent isolation as laboratory contaminants when they are not wanted). They neither take up much room nor cost much to feed. Mutants of all sorts, especially conditional lethals, are easily isolated.

Yeasts and molds are genuine eukaryotes. As such, they possess eukaryotic attributes that simply cannot be studied in *Escherichia coli* and other prokaryotic model systems. With fungi, you do not have to give up the advantages of a microbial system to do eukaryotic genetics. Moreover, fungi have a number of specific features that make them especially useful as eukaryotic models. They are "simple" eukaryotes. The genome is small. Most of the life cycle is spent in the haploid state, and this facilitates both mutant isolation and gene mapping. Morphological differentiation responds to triggers from the environment, so that cell cycles can be controlled experimentally by appropriate manipulation of culture conditions. As ascomycetes, *A. nidulans, N. crassa,* and *S. cerevisiae* produce their meiotic spores in an ascus. All the nuclei derived from a single meiotic event are packaged into separate spores and retained in the ascus; in *Neurospora* the products of meiosis are even arranged in linear order. Detailed genetic analysis of crossing-over and direct mapping of centromeres become uncommonly easy. Transformation systems are available for all three species, and it is relatively straightforward to isolate their DNA. Most of the preceding chapters in this book serve as a testimonial to the force and flexibility of these three fungi in modern genetic analysis. The more traditional role of fungi in genetics is discussed in texts by Esser and Kuenen (1967), Burnett (1975), and Fincham et al. (1979).

III. MOLECULAR MYCOLOGY

A. Molecular Biology for Mycologists

Erwin Chargaff once defined molecular biology as "the practice of biochemistry without a license." This cynical epigram reflects the way in which many biological disciplines have converged by examining problems using the common language of chemistry. At the center of molecular biology is one magnificent molecule, DNA. New breakthroughs continue to increase our understanding of the structure and function of DNA. The recent and newest revolution is the result of our ability to isolate, amplify, move about, sequence, and otherwise manipulate and characterize individual segments of DNA molecules.

Unfortunately, it is still possible to receive an advanced degree in biology and avoid exposure to molecular biology. Yet molecular biology permits powerful new ways of approaching many questions that have been considered difficult or even intractable. Fortunately, the techniques of working with recombinant DNA, unlike some others, have become progressively easier to execute as the field has advanced. With the help of books of well-tested recipes and commercial kits of reagents for most of the important techniques, a person who aspires to bring the molecular dimension to a classical question can do so in a surprisingly short time. Just as it is unnecessary to be able to program in order to benefit from a word processor, it is unnecessary to obtain a degree in biochemistry in order to benefit from recombinant DNA techniques. For traditional mycologists who want to "raise their consciousness," a brief reading list is outlined below.

For a review of classical molecular genetics, I recommend both Watson (1976) and Stent and Calendar (1978). For easy introductions to recombinant DNA and modern genetic engineering, try Watson et al. (1983) and/or Drlica (1984).

Specific methodologies and detailed protocols are outlined in several laboratory manuals. It is worth investing in a copy of the one of Maniatis et al. (1982). Also recommended are Davis et al. (1980), Rodriguez and Tait (1983), Hackett et al. (1984), and Pühler and Timmis (1984).

B. Opportunities

Fungi can be used to solve certain problems in biology. It is predictable that those problems with direct economic impact will receive more attention than others, although, unfortunately, they are not necessarily the most interesting or heuristic ones. The following list highlights some aspects of the fungal life-style that offer exceptional opportunities for combining recombinant DNA techniques with fungal models. The list is extremely general, far from exhaustive, and reflects my own biases.

1. Nuclear–Cytoplasmic Relations

The nuclear–cytoplasmic relationships encountered among the majority of fungi are uncommon among eukaryotes. Multinuclear cells are the rule, not the exception. Among the lower (aseptate) hyphal forms, the entire mycelium functions as a large coenocyte. In higher (septate) hyphal forms, pores are present in the septa, allowing nuclear migration from cell to cell. In addition, anastomoses between adjacent hyphae are often found. Heterokaryons, composed of dissimilar nuclei in a common cytoplasm, are encountered in nature and can be synthesized in the laboratory. Among the basidiomycetes, a special binucleate heterokaryon, the dikaryon, is found. The parasexual cycle also allows the study of somatic interactions between genetically dissimilar nuclei. Overall, the fungi are outstanding systems for studying extrachromosomal inheritance and for examining somatic interactions between dissimilar nuclei.

2. Pathogenicity

Mycoses are diseases caused by fungi actively growing in animal tissues. Fungi also cause disease indirectly by producing mycotoxins and by acting as allergens. Finally, drugs that reduce resistance to infection, such as immunosuppressants and broad-spectrum antibiotics, which kill harmless commensal bacteria, may lead the way for opportunistic fungal infections. A startling number of fungal pathogens affecting humans show a yeast–mycelial dimorphism (see Chapter 18). There are two reasons to hope for progress in medical mycology: (1) the growing clinical importance of mycoses in transplant and cancer patients, which has generated increased interest in and support for medical mycology, and (2) the scientific attractiveness of the developmental model offered by dimorphism. Although progress to date has been minimal (see Chapter 19), it is likely that researchers working with medically important fungi will have the incentive, the background, and the funding to benefit from new techniques in molecular biology.

In contrast to animal disease, where viruses and bacteria tend to be the most important etiological agents, fungi are the major cause of plant disease. Roughly 70–80% of cross losses are caused by fungi (Deacon, 1980). The hyphal morphology and pattern of assimilation is naturally suited to growing through the vascular system of higher plants. In addition, fungi have a number of adaptations for plant parasitism, often involving special recognition phenomena. In response, many plants produce phytoalexins, special fungitoxic or fungistatic chemicals. There is an enormous descriptive literature in plant pathology about these phenomena, yet the molecular bases of fungal–plant recognition mechanisms remain elusive (see Chapter 17). As with medically important fungi, the economic incentives for doing research in this field are high, and there is a large critical mass of scientific talent trained in plant pathology.

3. Morphogenesis

As simple eukaryotes with simple developmental cycles, fungi offer many advantages for the study of cellular differentiation. Since spores are both a beginning and an end of fungal development, most studies with fungi involve some aspect of spore formation or spore germination. Among the fungi as a whole, spores come in diverse forms and shapes, each type specifically adapted for a different means of dispersal (water, wind, arthropod, etc). Any given spore type can provide a useful developmental model system. In addition, numerous fungi exhibit pleomorphism, the formation of more than one type of spore by the same organism, usually at different parts of the life cycle. Sexual and asexual spores form in response to entirely different environmental stimuli.

One theory concerning development that has grown out of animal studies proposes that significant evolutionary change in multicellular organisms may arise from mutations that control spatial and temporal patterns of differentiation. Even radical morphological differences between species may be due to small gene changes affecting controls of developmental timing (Gould, 1977). The nematode *Caenorhabditis elegans* is being touted as *the* model system for testing this theory (Ambros and Horvitz, 1984). Fungi, with their richness of existing spore diversity, advanced genetics, ease of experimental manipulation, and unfolding molecular insights into developmentally controlled genes (Yelton *et al.*, 1983, 1984), may be even better suited for testing this and other theories in developmental biology.

4. Fungal Associations

In addition to their activities as saprophytes and parasites, many fungi form associations with other organisms that can be categorized, loosely, as mutualistic. The classical example of a mutualistic interaction is lichens. They are a distinct life form, composed of a symbiotic association of a fungus and an alga. An intact lichen is quite different from its component parts. Relatively few species of algae have been identified from lichens but many species of fungi are involved. Lichens provide living laboratories for studying a close interaction between the genetics and physiology of two widely differing species. Perhaps because lichens have few modern economic uses, they are not receiving much attention. Another fungal association, the mycorrhiza, is more likely to benefit from molecular biology. A mycorrhiza ("fungus root") is a symbiotic, non-pathogenic or feebly pathogenic association between a fungus and the roots of a plant; for instance, with some orchids seed germination occurs only in the presence of the right fungus. Forest trees have extensive mycorrhizal associations with basidiomycetes, and many crops display optimal growth only in the presence of appropriate mycorrhizal associations. In some soils, mycorrhizal symbionts are important in making phosphate and other nutrients into soluble forms that are accessible to the plant.

Microbiologists are so well trained in the use of pure culture methodologies that they often underestimate the importance of interspecific dependencies in nature. Lichens and mycorrhiza are but two examples of fungal associations that offer unique experimental opportunities for establishing, to coin another label, a "molecular ecology."

5. *Fungal Metabolism and Biotechnology*

The fungi possess remarkable catabolic and anabolic capabilities. Both aspects of their metabolism make them singularly attractive in biotechnological applications.

Fungi are good at degrading refractory substrates: cellulose, lignin, chitin, keratin. Mycelia can penetrate a solid substrate by apical growth. Together with this mechanical infiltration, molds secrete hydrolytic enzymes that digest the substrate. The resultant heterogeneous mass of mold and substrate is a major stage in the decomposition of organic matter in all terrestrial habitats, and is particularly important in soil formation. The hydrolytic enzymes themselves have direct use in industry, and the genes encoding several celluloses and amylases have already been cloned and manipulated so that they are expressed in bacteria and yeast (see Chapter 20).

Filamentous fungi secrete enzymes more efficiently than do most bacteria or yeasts, making them very attractive as potential hosts for gene cloning. If molds could be developed as host systems for mammalian and other eukaryotic genes, the gene products might be secreted directly into the culture medium, reducing expensive purification steps. Another practical consideration is that species such as *Aspergillus niger* and *A. oryzae* are "generally regarded as safe" (GRAS) by the Food and Drug Administration, hence presenting fewer regulatory problems than are posed by many prokaryotic hosts.

A third reason for wanting to develop filamentous fungi as hosts for genetic manipulations is their range of anabolic capability. Fungi, along with green plants and bacteria, are the major producers of secondary metabolites. Antibiotics such as penicillin and cephalosporin, as well as mycotoxins such as aflatoxins and trichothecenes, are of major economic significance. Thousands of other secondary metabolites have been characterized from fungi; many of them have pharmacological activity (Turner, 1971; Bennett and Ciegler, 1983; Turner and Aldridge, 1983).

All secondary metabolites are formed via multistep biochemical pathways. Since it would be difficult to clone all the enzymes for a pathway into a bacterium or yeast, the strategy for increasing yields of economically important secondary metabolities would be to modify existing pathways in the host in which they naturally occur, probably by effecting changes in regulatory genes. The pharmaceutical industry has already developed efficient technologies for the large-scale production of antibiotics from molds. The financial investment in these technologies and facilities is enormous and, therefore, it would make sense to find genetically engineered, high-yielding production strains.

A number of companies are actively engaged in research on economically important filamentous species. A transformation system has been reported from *Cephalosporium acremonium* (Queener *et al.*, 1984). Because so much of this research is proprietary, it is difficult to assess the state of the art. General overviews of past, present, and projected uses of filamentous fungi in industrial fermentations are addressed in several of the recent monographs on biotechnology (Crueger and Crueger, 1982; Whelan and Black, 1982; Vining, 1983; Ball, 1984).

6. Reproduction and Sexuality

Nonbiologists, and even some biologists, must be continually reminded that sex and reproduction are not necessarily the same thing. Moreover, the evolutionary advantages of sexual reproduction are not obvious. In sexual reproduction, an organism transmits only half its genes to each progeny. It "costs" something for an organism to evolve a sexual reproductive mode over an asexual one in which all of its genes are passed on.

The origin and evolution of sexual reproduction and genetic recombination are fundamental issues in evolutionary theory. We do not know why organisms reproduce sexually. The matter has received extensive theoretical attention and much of the empirical work has been conducted with animals (Williams, 1975; Smith, 1971, 1978).

Theoreticians interested in studying the evolution of sex and recombination would do well to pay more attention to fungal models. Fungi have the widest range of reproductive modes of any kingdom. In some species, reproduction occurs by asexual methods only (so far as is known). In others, there is always a sexual process. Probably one-third of all fungi display more than one type of reproduction, frequently in two well-marked phases, variously called imperfect/anamorphic and perfect/teleomorphic (Ainsworth, 1971). A variety of genetic devices that favor outbreeding or inbreeding have evolved. Hormones are frequently involved and complex recognition systems are known in a number of water molds. The mating patterns of basidiomycetes may involve thousands of distinct mating types, interacting via bipolar and tetrapolar incompatibility systems.

According to the orthodox interpretation, sex and recombination favor the fixation of favorable mutations and thus facilitate natural selection in a changing environment. The opportunities for testing this theory with haploid pleomorphic fungi are enormous. I repeat, theoreticians interested in studying the evolutionary advantages of sex would do well to study fungi. In addition to the diversity of reproductive modes and the advanced state of the formal genetics available for certain systems, fungi have the advantage of freeing researchers from the emotional and anthropomorphic preconceptions that taint much of the existing literature derived from animal models. The molecular basis of fungal mating-type loci and incompatibility factors may help us understand the evolution of sexuality in other forms of life.

C. Conclusions

There are two popular ways to divide biological sciences into subdivisions. In one framework, the organism is the defining unit, and we get groupings such as botany, zoology, mycology, protozoology, entomology, and the like. In the second system, the division is based on approach or technique rather than the kinds of organisms. In this sytem, categories such as genetics, ecology, physiology, morphology, and molecular biology are used. Molecular mycology, then, combines the two systems and is the application of a certain set of experimental approaches to the study of fungi.

Mycology has never been a "glamor" field, and most people who work with fungi do not even call themselves mycologists. They are plant pathologists, food technologists, natural products chemists, industrial microbiologists, geneticists, and so on. Often they are less interested in the biology of their system than in some practical application. Yet it is this same group of nonmycologists working with fungi who offer the greatest hope for bringing molecular biology to the study of fungi. This is because research is stimulated by the profit motive (and its academic variant, the availability of funding) as well as by intellectual curiosity. Sometimes both stimuli reinforce one another. Economics will be the driving force behind much of the new mycological research, and many of the recombinant DNA techniques tailored for use on fungi will come out of industrial and agricultural laboratories.

What will the new techniques do for mycology? Recombinant DNA technology makes many new experiments possible. Rarely before has the introduction of a new methodology had such an immediate impact. There are those who believe we now have the magic elixir that will resolve all biological engimas. This stance has been labeled "molecular biology macho" (Lewin, 1984). Here

21. Prospects for a Molecular Mycology

is an example. Sydney Brenner, in reflecting on the publication of the complete developmental lineages of *C. elegans,* was quoted during the summer of 1984:

Peter Medawar has written that science is the art of the soluble. My reply is that molecular biology is the art of the inevitable. If you do it, it's inevitable you will find out how it works—in the end. Maybe it won't be until you are able to put the last period on the page, when you know everything, but you will be able to say, aha, now I understand! But you will get there in the end. [Lewin, 1984, p. 1329.]

Where is "there?" If "there" is the attainment of a "molecular mycology," how will we get there? Can we get there?

Someone once said that experiments are the questions we put to nature. Recombinant DNA technology allows us to perform clever new experiments, and thereby ask certain questions. It remains to be seen whether recombinant DNA technology will inevitably reveal the answers.

I confess that I am temperamentally unable to share Brenner's "molecular biology macho." On a philosophical level, I distrust the Faustian notion that we can "know everything." On a practical level, we already have considerable evidence that cellular and organismal development involves an order of complexity that does not reduce readily to the genetic blueprint. Topographic and organizational parameters still elude most of our experimental approaches.

Moreover, talking about "a" molecular mycology—singular—is simply a linguistic device. There is no archetypical fungus out there in nature, waiting to have every bit of its DNA sequenced and analyzed. The kingdom Fungi is comprised of over 100,000 species (Ainsworth, 1971). Among these many life forms we find extraordinary diversity. We will not be able to observe every aspect of every species in detail, so we must discriminate and try to select the significant. But even if we cannot "know everything," we can learn many things, especially, as I said earlier, if we ask the right questions. As one who has spent a professional lifetime being beguiled by fungi, here are a few of the questions I would like to see posed:

What kinds of DNA specify a mycelium and direct apical growth?
How do polynucleotides interact to direct the production of a spore?
What is the molecular nature of an incompatibility factor?
What biochemical signals turn on secondary metabolism?

Literally hundreds of other important mycological questions can be addressed with the help of recombinant DNA technology. As already stated in the preface of this book, the prospects for a long and happy marriage between molecular biology and mycology are very, very good.

ACKNOWLEDGMENTS

Many colleagues were helpful as I pondered about molecular mycology. I especially thank Linda Lasure, Bob Metzenberg, Ian Ross, Claudio Scazzocchio, and Bill Timberlake for their inputs and insights. Ron Cape, Linda Lasure, and Bob Metzenberg reviewed and improved the final manuscript. Since I did not accept all of their suggestions, I take sole responsibility for the material presented.

REFERENCES

Ainsworth, G. C. (1971). "Ainsworth & Bisby's Dictionary of the Fungi," 6th Ed. Commonw. Mycol. Inst., Kew, Surrey, England.
Ainsworth, G. C., Sparrow, F. K., and Sussman, A. S. (1973a). "The Fungi: An Advanced Treatise," Vol. 4A, A Taxonomic Review with Keys: Ascomycetes and Fungi Imperfecti." Academic Press, New York.
Ainsworth, G. C., Sparrow, F. K., and Sussman, A. S. (1973b). "The Fungi: An Advanced Treatise, Vol. 4B, A Taxonomic Review with Keys: Basidiomycetes and Lower Fungi." Academic Press, New York.
Ambros, V., and Horvitz, H. R. (1984). Heterochronic mutants of the nematode *Caenorhabditis elegans. Science (Washington, D.C.)* **226,** 409–416.
Ball, C., ed. (1984). "Genetics and Breeding of Industrial Microorganisms." CRC Press, Boca Raton, Florida.
Beadle, G. W., and Tatum, E. L. (1941). Genetic control of biochemical reactions in *Neurospora. Proc. Natl. Acad. Sci. U.S.A.* **27,** 499–506.
Bennett, J. W., and Ciegler, A., eds. (1983). "Secondary Metabolism and Differentiation in Fungi." Dekker, New York.
Burnett, J. H. (1975). "Mycogenetics: An Introduction to the General Genetics of Fungi." Wiley, New York.
Crueger, W., and Crueger, A. (1982). "Biotechnology: A Textbook of Industrial Microbiology" (C. Haessly, transl. from German). Sinauer, Sunderland, Massachusetts.
Davis, R. W., Botstein, D., and Roth, J. R. (1980). "A Manual for Genetic Engineering: Advanced Bacterial Genetics." Cold Spring Harbor Lab., Cold Spring Harbor, New York.
Deacon, J. W. (1980). "Introduction to Modern Mycology." Wiley, New York.
Drlica, K. (1984). "Understanding DNA and Gene Cloning: A Guide for the Curious." Wiley, New York.
Esser, K., and Kuenen, R. (1967). "Genetics of Fungi" (E. Steiner, transl.). Springer-Verlag, Berlin and New York.
Fincham, J. R. S., Day, P. R., and Radford, A. (1979). "Fungal Genetics," 4th Ed. Blackwell, Oxford.
Gould, S. J. (1977). "Ontogeny and Phylogeny." Harvard Univ. Press, Cambridge, Massachusetts.
Hackett, P. B., Fuchs, J., and Messing, J. W. (1984). "An Introduction to Recombinant DNA Techniques." Benjamin/Cummings, Menlo Park, California.
Lewin, R. (1984). Why is development so illogical? *Science (Washington, D.C.)* **224,** 1327–1329.
Maniatis, T., Fritsch, E. F., and Sambrook, J. (1982). "Molecular Cloning: A Laboratory Manual." Cold Spring Harbor Lab., Cold Spring Harbor, New York.
Pühler, A., and Timmis, K. N., eds. (1984). "Advanced Molecular Genetics." Springer-Verlag, Berlin and New York.
Queener, S. W., Ingolia, T. D., Skatrud, P. L., Chapman, J. L., and Kaster, K. R. (1984). Recombinant DNA studies in *Cephalosporium acremonium. ASM Conf. Genet. Mol. Biol. Ind. Microorg.* Abstr. p. 29.
Rodriguez, R. L., and Tait, R. C. (1983). "Recombinant DNA Techniques: An Introduction." Addison-Wesley, Reading, Massachusetts.
Ross, I. K. (1979). "Biology of the Fungi." McGraw-Hill, New York.

Smith, J. M. (1971). What use is sex? *J. Theor. Biol.* **30,** 319–345.
Smith, J. M. (1978). "The Evolution of Sex." Cambridge Univ. Press, London and New York.
Stent, G. S., and Calendar, R. (1978). "Molecular Genetics: An Introductory Narrative." Freeman, San Francisco, California.
Turner, W. B. (1971). "Fungal Metabolites." Academic Press, New York.
Turner, W. B., and Aldridge, D. C. (1983). "Fungal Metabolites II." Academic Press, New York.
Vining, L. C., ed. (1983). "Biochemistry and Genetic Regulation of Commercially Important Antibiotics." Addison-Wesley, Reading, Massachusetts.
Watson, J. D. (1976). "Molecular Biology of the Gene," 3rd Ed. Benjamin, New York.
Watson, J. D., Tooze, J., and Kurtz, T. (1983). "Recombinant DNA: A Short Course. Freeman, San Francisco, California.
Whelan, W. J., and Black, S., eds. (1982). "From Genetic Experimentation to Biotechnology—the Critical Transition." Wiley, New York.
Whittaker, R. H. (1969). New concepts of kingdoms of organisms. *Science (Washington, D.C.)* **163,** 150–161.
Williams, G. (1975). "Sex and Evolution." Princeton Univ. Press, Princeton, New Jersey.
Yelton, M. M., Hamer, J. E., De Souza, E. R., Mullaney, E. J., and Timberlake, W. E. (1983). Developmental regulation of the *Aspergillus nidulans trp* C gene *Proc. Natl. Acad. Sci. U.S.A.* **80,** 757–758.
Yelton, M. M., Timberlake, W. E., and van den Hondel, C.A.M.J.J. (1985). A cosmid for selecting genes by complementation in *Aspergillus nidulans:* selection of the developmentally regulated *yA* locus. *Proc. Natl. Acad. Sci. U.S.A.* **82,** 834–838.

Appendixes

I

Fungal Taxonomy

LINDA L. LASURE
Bioproduct Research
Miles Laboratories, Inc.
Elkhart, Indiana

J. W. BENNETT
Department of Biology
Tulane University
New Orleans, Louisiana

I. Introduction	531
II. Outline of Fungal Taxonomy	534
References	534

I. INTRODUCTION

The purpose of this section is to provide a brief explanation of fungal taxonomy and a reference table for identifying major taxa. Remember that taxonomic schemes are neither static nor universally accepted. The one presented below follows Ainsworth (1971) and Ainsworth *et al.* (1973a,b). Other authorities may present quite different hierarchies and headings. Nomenclatural convention for fungi demands that subdivisions end in "-mycotina," classes in "-mycetes," orders in "-ales," and families in "-aceae." Depending on the authority and the scheme adopted, you may find the same group accorded differential rank. For example, the ascus-producing fungi may be viewed as a class, Ascomycetes, or as a subdivision, Ascomycotina.

If you are interested in exposure to other taxonomic arrangements and in learning more about mycology in general, consult one of the comprehensive, recent mycology texts such as Burnett (1968), Alexopoulos and Mims (1979), Ross (1979), or Moore-Landecker (1982).

The kingdom Fungi is divided here into two divisions. The Myxomycota, commonly called the "slime molds," are a varied group of organisms having a plasmodium at some point in their life cycle. One contemporary mycologist pointed out that "the very words *slime mold* reflect the confusion that has surrounded this group of organisms, because they are certainly not molds and they are not particularly slimy" (Ross, 1979, p. 178). A number of taxonomic questions remain unanswered as to whether the members of the Myxomycota really belong with the fungi.

Members of the division Eumycota, commonly called the "true fungi," usually have a filamentous or yeastlike form, and no plasmodium. Our scheme divides the group into five subdivisions. The Mastigomycotina and Zygomycotina constitute the "lower fungi"; the Ascomycotina, Basidiomycotina, and Deuteromycotina constitute the "higher fungi."

The lower fungi are distinguished by hyphae without cross-walls (nonseptate), the formation of asexual spores by cleavage of cytoplasm with sporangia, and include several groups that possess flagellated zoospores. For many years, the lower fungi were grouped together in a single class, the Phycomycetes. Phycomete means "algal fungus" and the name stems from the theory that these fungi were degenerate algae that had lost their chlorophyll. The term phycomycete no longer has official taxonomic status, but is still encountered in older texts and in works by authors who have not kept up with trends in fungal systematics. The classification we present here puts the lower fungi into two subdivisions, both of which encompass a diverse and composite group of organisms.

The subdivision Mastigomycotina includes species often identified with animals because of the defining characteristic of the group, motile spores. Many of these organisms are called water molds because of the prevalent aquatic growth habit.

Zygomycotina contains nonseptate fungi which lack a motile stage and are only rarely aquatic. Members of this subdivision exhibit gametangial fusion and zygospore formation.

Taxonomically, the higher fungi are easier to delineate. With the exception of the yeasts, they have septate hyphae and often produce elaborate fruiting bodies. They are divided here into three subdivisions: the Ascomycotina, the Basidiomycotina, and the Deuteromycotina. The Ascomycotina and Basidiomycotina are distinguished by their sexual spores; the Deuteromycotina reproduce entirely by asexual means.

The Ascomycotina form ascospores inside a specialized reproductive structure called an ascus. Two haploid nuclei fuse within the immature ascus and then the diploid fusion nucleus immediately undergoes meiosis, resulting in four haploid spores. One mitotic division usually ensues so that most members of the Ascomycotina have eight-spored asci. The retention of the products of meiosis within a single morphological structure has facilitated many elegant studies on chromosomal mechanisms of crossing-over. The three premier species for fungal genetics, *Aspergillus nidulans, Neurospora crassa,* and *Saccharomyces cerevisiae,* are all members of this group. Special features of fungal genetic analysis are discussed in detail by Esser and Kuenen, (1967), Burnett (1975), and Fincham *et al.* (1979).

The Basidiomycotina form sexual basidiospores on a basidium. Basidiospore formation closely resembles ascospore development, except that the spores are borne externally. Fusion of haploid nuclei results in a transient diploid that immediately undergoes meiosis to form four haploid basidiospores. An unusual cytological feature of the basidiomycete life cycle is the formation of a special binucleate cell called the dikaryon. This subdivision contains the majority of conspicuous, macroscopic fungi such as mushrooms, puffballs, and shelf fungi. It also contains the important plant pathogens known collectively as rusts and smuts.

The Deuteromycotina, or Fungi Imperfecti, are distinguished by the absence of any known sexual form. They reproduce largely by asexual conidiospores. Taxonomists consider this an "artificial" group, and often highlight this artificiality by using the prefix "form" with reference to the taxa within this subdivision (e.g., form-class, form-family, form-genus, form-species). Many species originally classified as imperfects are eventually shown to possess a sexual stage, usually within the Ascomycotina or, more rarely, within the Basidiomycotina. The sexual phase, also called the perfect stage or teleomorph, is given a separate name. The rules of botanical nomenclature specify that sexual names should have precedence over the asexual (also called imperfect or anamorphic) names. This creates both practical and philosophical problems. The genus names *Aspergillus, Penicillium,* and *Fusarium* are all imperfect epithets. According to the internationally adopted rules of nomenclature, any time a sexual stage is found for a member of one of these genera, the name of that species should be changed to that of the sexual form. For example, according to these rules, *Aspergillus nidulans* should be called *Emericella nidulans.* In practice, despite the fact that this species regularly forms ascospores, virtually everyone still calls it *Aspergillus nidulans.*

Many economically important fungi are classified in the Deuteromycotina. For more details about the taxonomy of *Aspergillus* see Raper and Fennell (1965); for *Fusarium* see Nelson *et al.* (1981); and for *Pencillium* see Pitt (1979) and Ramirez (1982). The majority of important human pathogens also belong to this group; see Rippon (1982). Finally, for a discussion of the issues and problems surrounding nomenclatural conventions in the Fungi Imperfecti, see Bennett (1985).

II. OUTLINE OF FUNGAL TAXONOMY

Kingdom: Fungi
  Division: Myxomycota (plasmodium or pseudoplasmodium present)
    Class: Acrasiomycetes ("cellular slime molds")
      Example: *Dictyostelium*
    Class: Myxomycetes ("acellular slime molds")
      Example: *Physarum*
  Division: Eumycota (assimilative phase typically filamentous or yeastlike)
    Subdivision: Mastigomycotina (nonseptate mycelium, motile spores)
      Examples: *Achlya, Allomyces, Blastocladiella, Phythium, Phytophthora, Saprolegnia*
    Subdivision: Zygomycotina (nonseptate mycelium, zygospores)
      Examples: *Absidia, Blakeslea, Mortierella, Mucor, Pilobolus, Rhizopus*
    Subdivision: Ascomycotina ("sac fungi"; septate mycelium or yeast: sexual spores borne in an ascus)
      Examples: *Saccharomyces, Saccharomycopsis (Yarrowia), Schizosaccharomyces; Neurospora, Podospora, Sordaria;* the sexual stages of both *Aspergillus* and *Penicillium; Ascobolus;* truffles and morels
    Subdivision: Basidiomycotina ("club fungi"; septate mycelium or yeast; sexual spores borne exogenously on a basidium)
      Examples: *Puccinia, Ustilago,* jelly fungi, rusts, smuts; *Agaricus, Coprinus, Schizophyllum,* mushrooms, puffballs, shelf fungi
    Subdivision: Deuteromycotina (the Fungi Imperfecti; septate mycelium or yeast; no known sexual phase)
      Examples: *Aspergillus, Fusarium, Penicillium; Candida, Histoplasma, Wangiella*

REFERENCES

Ainsworth, G. C. (1971). "Ainsworth & Bisby's Dictionary of the Fungi," 6th Ed. Commonw. Mycol. Inst., Kew, Surrey, England.

Ainsworth, G. C., Sparrow, F. K., and Sussman, A. S. (1973a). "The Fungi: An Advanced Treatise, Vol. 4A, A Taxonomic Review with Keys: Ascomycetes and Fungi Imperfecti." Academic Press, New York.

Ainsworth, G. C., Sparrow, F. K., and Sussman, A. S. (1973b). "The Fungi: An Advanced Treatise, Vol. 4B, A Taxonomic Review with Keys: Basidiomycetes and Lower Fungi." Academic Press, New York.

Appendix I. Fungal Taxonomy

Alexopoulos, C. J., and Mims, C. W. (1979). "Introductory Mycology," 3rd Ed. Wiley, New York.
Bennett, J. W. (1985). Taxonomy of fungi and biology of the aspergilli. In "Biology of Industrial Microorganisms" (A. L. Demain and N. Soloman, eds.), pp. 359–406. Cummings, Menlo Park, California.
Burnett, J. H. (1968). "Fundamentals of Mycology." Arnold, London.
Burnett, J. H. (1975). "Mycogenetics: An Introduction to the General Genetics of Fungi." Wiley, New York.
Esser, K., and Kuenen, R. (1967). "Genetics of Fungi" (E. Steiner, transl.). Springer-Verlag, Berlin and New York.
Fincham, J. R. S., Day, P. R., and Radford, A. (1979). "Fungal Genetics," 4th Ed. Blackwell, Oxford.
Moore-Landecker, E. (1982). "Fundamentals of the Fungi," 2nd Ed. Prentice-Hall, Englewood Cliff, New Jersey.
Nelson, P. E., Toussoun, T. A., and Cook, R. J., eds. (1981). "Fusarium; Diseases, Biology, and Taxonomy." Pennsylvania State Univ. Press, University Park.
Pitt, J. I. (1979). "The Genus Penicillium and Its Teleomorphic States Eupenicillium and Talaromyces." Academic Press, New York.
Ramirez, C. (1982). "The Manual and Atlas of the Penicillia." Elsevier, Amsterdam.
Raper, K. B., and Fennell, D. I. (1965). "The Genus Aspergillus." Williams & Wilkins, Baltimore, Maryland.
Rippon, J. W. (1982). "Medical Mycology," 2nd Ed. Saunders, Philadelphia, Pennsylvania.
Ross, I. I. (1979). "Biology of the Fungi." McGraw-Hill, New York.

II

Conventions for Gene Symbols

J. W. BENNETT
Department of Biology
Tulane University
New Orleans, Louisiana

LINDA L. LASURE
Bioproduct Research
Miles Laboratories, Inc.
Elkhart, Indiana

I. *Aspergillus nidulans*	538
II. *Neurospora crassa*	539
III. *Saccharomyces cerevisiae*	540
IV. Other Fungi	542
References	542

Recommendations for uniform conventions of genetic nomenclature have been published for bacteria (Demerec *et al.*, 1966), *Aspergillus nidulans* (Clutterbuck, 1973), *Saccharomyces cerevisiae* (Sherman, 1981), and *Neurospora crassa* (Perkins *et al.*, 1982). In this volume, we make no attempt to impose a uniform standard of genetic symbols, but rather allow our authors to utilize the conventions of their particular organism and laboratory.

Although the designations for gene symbols and phenotypes are not the same for bacteria, yeasts, and molds, enough similarity exists to mislead the unwary reader. Since the publication of the proposals for bacterial genetics by Demerec *et al.* (1966), most primary gene symbols have been designated by three-letter, italicized symbols (e.g., *arg* for a locus affecting arginine biosynthesis). Some *Neurospora* and *Aspergillus* symbols predate the proposals for standardization of genetic nomenclature in bacteria and have fewer or more than three letters.

The conventions for distinguishing different loci that produce the same phenotypic change show minor, but confusing, variation from system to system. In bacteria and *A. nidulans* an italicized capital letter immediately follows the three-letter symbol (*argA, argB,* etc.), while in yeast nonhyphenated numbers are used (*arg1, arg2,* etc.). In *N. crassa,* hyphenated numbers are used to distinguish loci (*arg-1, arg-2,* etc.). In yeast, hyphenated numbers designate alleles (*arg1-37*); in *A. nidulans,* unhyphenated numbers designate alleles (*argA2*) and hyphenated numbers designate unmapped mutants (*arg-51*).

The conventions for phenotype, dominance, mating-type loci, designation of wild type, and other genetic symbols also show subtle differences between the systems. The most important recommendations from Clutterbuck (1973), Sherman (1981), and Perkins *et al.* (1982) are summarized in Sections I–III. Some representative examples are given to illustrate each system. Section IV cites a few additional systems of fungal genetic nomenclature. See the references for more complete explanations of all these nomenclatural conventions.

I. *ASPERGILLUS NIDULANS*

The recommendations for the nomenclature and conventions used for *A. nidulans* follow those of bacterial genetics and are published in Clutterbuck (1973). All genetic loci and mutants introduced subsequent to this publication are designated by three-letter symbols in italics (e.g., *arg*). Older symbols, previously adopted in the literature, are retained and consist of one to five italic letters (e.g., *y* = yellow; *panto* = pantothenic acid requirement). Nonallelic loci that have the same primary symbols are distinguished by an italic capital letter following the symbol, e.g., *argA, argB*. Alleles are distinguished by italic serial mutant numbers after the symbol and locus letter, e.g., *argA1, argA2*. Where the allelic relationships of a mutant have not yet been determined, the capital letter is replaced by a hyphenated number (e.g., *arg-51*). Mitochondrial gene symbols are enclosed in square brackets, e.g., [*oliA1*].

Wild-type alleles are indicated by a superscript "plus," e.g., *argA*$^+$. Occasionally, dominant mutants are designated by capitalizing the first of the three letters in a symbol (*Acr* for acriflavine resistance). In general, dominance is not indicated in the primary gene symbol. Symbols for phenotypes are distinguished from symbols for genes. Often the phenotype is simply written out in unabbreviated fashion (e.g., "arginine requirement"); alternatively, a nonitalic version of the gene symbol with the first letter capitalized is used, e.g., Arg$^-$.

Suppressors used to be designated by complex symbols including the locus and/or allele suppressed, e.g., *suA1adE20,* but now simple symbols are encouraged, e.g., *suaA1* allele-specific, locus-nonspecific suppressor. It is important to note that the wild-type, nonsuppressing allele is designated with a symbol "plus," as in *suaA*$^+$, opposite to the usage for bacteria.

Superscripts are used to indicate mutants with specific properties; for instance, *areA*^{d}*18* is an *areA* allele giving derepressed phenotypes for ammonium-repressed genes, while *areA*^{r}*1* gives correspondingly repressed phenotypes.

The following examples illustrate the conventions used in the genetic nomenclature for *A. nidulans*:

argA	A specific locus or mutation that produces a requirement for arginine as the phenotype
argA$^+$	The wild-type allele
argA2	A specific allele or mutation in the *argA* gene
arg-51	An arginine-requiring mutant not yet tested for allelism, whose locus is unknown
Arg$^+$	A strain not requiring arginine
Arg$^-$	A strain requiring arginine

A list of *A. nidulans* loci is given in Clutterbuck (1974), genetic maps are given in Clutterbuck (1984), and the mitochondrial genome is summarized by Spooner and Turner (1984).

II. *NEUROSPORA CRASSA*

A summary of conventions, gene symbols, and map locations of *N. crassa* genes is presented in Perkins *et al.* (1982), following Barratt and Perkins (1965). These conventions antedate bacterial genetic nomenclature and more closely follow those of *Drosophila*. Three-letter gene symbols are used most frequently, but symbols of one to four letters are also found. Two-letter symbols are quite common (e.g., *ad*, adenine requirement; *qa*, quinate utilization). Recessive gene symbols are written entirely in lowercase italics. When the mutant allele is known to be dominant, the first letter is capitalized (e.g., *Sk*, Spore killer).

Symbols without superscripts are used to represent mutant alleles. The same symbol with a superscript "plus" designates the wild-type allele, e.g., *ad*$^+$. Alleles differing in resistance or sensitivity, or allelic series having no definitive wild type, may be distinguished by other superscripts (e.g., *cyh-1*^{R}, cycloheximide resistance; *cyh-1*^{S}, cycloheximide sensitivity).

Nonallelic loci are distinguished from one another by numbers, separated from the symbol for the locus by a hyphen, e.g., *ad-1, ad-2*. The use of hyphens to distinguish nonallelic gene symbols differs sharply from the conventions for bacteria, *Aspergillus,* and yeast. In *Neurospora,* the allele number is "not usually displayed with the gene symbols, except when necessitated by the use of several alleles, when it is included in parentheses after the full locus symbol, e.g. *pyr-3* (KS43), or when a new mutant gene has not yet been assigned a locus number pending tests for allelism with similar genes at previously established loci. In the latter situation, a mutant gene is temporarily designated by an appropriate letter symbol followed immediately by the allele number in parentheses, e.g. *ilv(STL6)"* (Perkins *et al.*, 1982, p. 427).

Mating-type alleles are called *A* and *a*. Suppressors are designated *su*, followed immediately by the symbol of the suppressed gene in parentheses; nonallelic suppressors of the same gene are distinguished by hyphenated numbers following the parentheses, e.g., *su(met-7)-1*, *su(met-7)-2*. Following the *Drosophila* convention, su^+ designates the wild type and *su* designates the mutant suppressor allele.

The following examples illustrate the major conventions used in the genetic nomenclature for *N. crassa:*

arg	Any locus or mutation that produces a requirement for arginine as the phenotype
arg-1	A specific locus that produces a requirement for arginine
arg-1$^+$	The wild-type allele of the *arg-1* gene
arg-1 (JWB7)	A specific allele of the *arg-1* gene
arg (JWB22)	An arginine-requiring mutant not yet tested for allelism, whose locus is unknown
Arg$^+$	A strain not requiring arginine
Arg$^-$	A strain requiring arginine

The Perkins *et al.* (1982) reference includes a detailed compendium of *N. crassa* loci and linkage maps. The maps are updated in Perkins (1984) and the mitochondrial genome is summarized by Collins and Lambowitz (1984).

III. *SACCHAROMYCES CEREVISIAE*

The recommendations for the nomenclature and conventions used in yeast genetics are summarized by Sherman (1981) and Sherman and Lawrence (1974). Gene symbols are consistent with the proposals of Demerec *et al.* (1966), whenever possible, and are designated by three italicized letters, e.g., *arg*. Contrary to the proposals of Demerec *et al.* (1966), the genetic locus is identified by a number (not a letter) following the gene symbol, e.g., *arg2*. Dominant alleles are denoted by using uppercase italics for all three letters of the gene symbol, e.g., *ARG2*. Lowercase letters symbolize the recessive allele, e.g., the auxotroph *arg2*. Wild-type genes are designated with a superscript "plus," ($sup6^+$ or $ARG2^+$). Alleles are designated by a number separated from the locus number by a hyphen, e.g., *arg2-14*. Locus numbers are consistent with the original assignments; however, allele numbers may be specific to a particular laboratory.

Phenotypic designations are written out or denoted by cognate symbols, with-

Appendix II. Conventions for Gene Symbols

out italics, and by the superscripts "plus" and "minus." For example, independence of and requirement for arginine can be symbolized, respectively, as Arg⁺ and Arg⁻.

Gene clusters, complementation groups within a gene, or domains within a gene having different properties are designated by capital letters following the locus number, e.g., *his4A, his4B*. (Note that in the conventions of Demerec *et al.*, 1966, capital letters following the gene symbol designate different loci.)

Wild-type and mutant alleles of the mating-type and related loci do not follow the standard rules. The two wild-type alleles at the mating-type locus are designated *MAT*a and *MAT*α. The two complementation groups of the *MAT*α locus are denoted *MAT*α*1* and *MAT*α*2*. Mutations of the *MAT* genes are denoted, e.g., *mat*a*-1, mat*α*1-1*. The wild-type homothallic alleles at the *HMR* and *HML* loci are denoted *HMR*a, *HMR*α, *HML*a, and *HML*α. Mutations at these loci are denoted, e.g., *hmr*a*-1, hml*α*-1*.

The following examples illustrate the conventions used in the genetic nomenclature for *S. cerevisiae*:

ARG2	A locus or dominant allele
arg2	A locus or recessive allele that produces a requirement for arginine as the phenotype
ARG2⁺	The wild-type allele of this gene
arg2-9	A specific allele or mutation at the *ARG2* locus
Arg⁺	A strain not requiring arginine
Arg⁻	A strain requiring arginine

For information on yeast mitochondrial genomes, see Grivell (1984).

For most structural genes that code for proteins, the nonmutant ("wild-type") allele is usually dominant to the mutant form of a gene. In yeast, the convention for dominant, "normal" genes utilizes capitalized italic symbols such as *HIS4* and *LEU2*. In traditional genetics, we learn about genes through their mutations, and linkage maps are created by following mutant alleles in crosses. Published linkage data, therefore, consist of gene symbols for the mutant, usually recessive, alleles [e.g., on linkage group III, *his4* and *leu2*. Those mutant alleles that are dominant to their nonmutant, "normal" alleles will appear on linkage maps in capital letters (*SUP22* and *FLD1* on chromosome IX)]. In addition, capital letters are used to represent dominant wild-type genes that control the same character and that are used for mapping (*SUC2, SUC1*, etc.), as well as DNA segments whose locations have been determined by a combination of recombinant DNA techniques and classical mapping procedures, e.g., *RDN1*, the segment encoding ribosomal RNA.

Detailed yeast linkage maps have been published by Mortimer and Schild (1980, 1984).

IV. OTHER FUNGI

Genetic conventions in other fungi sometimes follow one of the systems outlined above. In the past, workers with "less popular" species tended to follow some version of the bacterial–*A. nidulans* conventions; more recently, the yeast system has been gaining in popularity. For example, yeast conventions are used for the plant pathogen *Cochliobolus heterostrophus* (O. Yoder, personal communication). Regrettably, many workers adopt idiosyncratic symbols.

Both the "Handbook of Genetics, Vol. 1, Bacteria, Bacteriophages, and Fungi" (King, 1974) and "Genetic Maps 1984" (O'Brien, 1984) contain information about some of the better studied of the less popular fungi. Specific references, in alphabetic order by genus, follows:

Ascobolus immersus (Decaris *et al.*, 1974)
Dictyostelium discoideum (Newell, 1984)
Phycomyces (Cerdá-Olmedo, 1974)
Podospora anserina (Esser, 1974; Marcou *et al.*, 1984)
Schizosaccharomyces pombe (Gutz *et al.*, 1974)
Sordaria (Olive, 1974)
Ustilago maydis (Holliday, 1974)

Two species of Basidiomycetes, *Coprinus radiatus* and *Schizophyllum commune*, have been studied intensively, especially with respect to their incompatibility factors. Consult the following references for more information about these systems: Raper (1966), Guerdoux (1974), Raper and Hoffman (1974), and Schwalb and Miles (1978).

ACKNOWLEDGMENTS

We thank A. J. Clutterbuck, R. Metzenberg, D. Perkins, and F. Sherman for reviewing sections of this appendix.

REFERENCES

Barratt, R. W., and Perkins, D. D. (1965). Neurospora genetic Nomenclature. *Neurospora Newsl.* **8**, 23–24.
Cerdá-Olmedo, E. (1974). Phycomyces. *In* "Handbook of Genetics, Vol. 1, Bacteria, Bacteriophages, and Fungi" (R. C. King, ed.), pp. 343–357. Plenum, New York.
Clutterbuck, A. J. (1973). Gene symbols in *Aspergillus nidulans*. *Genet. Res.* **21**, 291–296.
Clutterbuck, A. J. (1974). *Aspergillus nidulans*. *In* "Handbook of Genetics, Vol. 1, Bacteria, Bacteriophages, and Fungi" (R. C. King, ed.), pp. 447–510. Plenum, New York.
Clutterbuck, A. J. (1984). Loci and linkage map of the filamentous fungus *Aspergillus nidulans*

Appendix II. Conventions for Gene Symbols

(Eidam) Winder ($n = 8$). In "Genetic Maps 1984" (S. J. O'Brien, ed.), pp. 265–273. Cold Spring Harbor Lab., Cold Spring Harbor, New York.

Collins, R. A., and Lambowitz, A. M. (1984). The physical and genetic map of mitochrondrial DNA from *Neurospora crassa* strain 74-OR23-1A. In "Genetic Maps 1984" (S. J. O'Brien, ed.), pp. 274–276. Cold Spring Harbor Lab., Cold Spring Harbor, New York.

Decaris, B., Girard, J., and Leblon, G. (1974). Ascobolus. In "Handbook of Genetics, Vol. 1, Bacteria, Bacteriophages, and Fungi" (R. C. King, ed.), pp. 563–573. Plenum, New York.

Demerec, M., Adelberg, E. A., Clark, A. J., and Hartman, P. E. (1966). A proposal for a uniform nomenclature in bacterial genetics. *Genetics* **54**, 61–76.

Esser, K. (1974). *Podospora anserina*. In "Handbook of Genetics, Vol. 1, Bacteria, Bacteriophages, and Fungi" (R. C. King, ed.), pp. 531–551. Plenum, New York.

Grivell, L. A. (1984). Restriction and genetic maps of yeast mitochondrial DNA. In "Genetic Maps 1984" (S. J. O'Brien, ed.), pp. 234–247. Cold Spring Harbor Lab., Cold Spring Harbor, New York.

Guerdoux, J. L. (1974). Coprinus. In "Handbook of Genetics, Vol. 1, Bacteria, Bacteriophages, and Fungi" (R. C. King, ed.), pp. 627–636. Plenum, New York.

Gutz, H., Heslot, H., Leupold, U., and Oprieno, N. (1974). *Schizosaccharomyces pombe*. In "Handbook of Genetics, Vol. 1, Bacteria, Bacteriophages, and Fungi" (R. C. King, ed.), pp. 395–446. Plenum, New York.

Holliday, R. (1974). *Usilago maydis*. In "Handbook of Genetics, Vol. 1, Bacteria, Bacteriophages, and Fungi" (R. C. King, ed.), pp. 575–595. Plenum, New York.

King, R. C., ed. (1974). "Handbook of Genetics, Vol. 1, Bacteria, Bacteriophages and Fungi." Plenum, New York.

Marcou, D., Picard-Bennoun, and Simonet, J.-M. (1984). Genetic map of *Podospora anserina*. In "Genetic Maps 1984" (S. J. O'Brien, ed.), pp. 252–261. Cold Spring Harbor Lab., Cold Spring Harbor, New York.

Mortimer, R. K., and Schild, D. (1980). Genetic map of *Saccharomyces cerevisiae*. *Microbiol. Rev.* **44**, 519–571.

Mortimer, R. K., and Schild, D. (1984). Genetic map of *Saccharomyces cerevisiae*. In "Genetic Maps 1984" (S. J. O'Brien, ed.), pp. 224–233. Cold Spring Harbor Lab., Cold Spring Harbor, New York.

Newell, P. C. (1984). Genetic loci of the cellular slime mold *Dictyostelium discoideum*. In "Genetic Maps 1984" (S. J. O'Brien,ed.), pp. 248–251. Cold Spring Harbor Lab., Cold Spring Harbor, New York.

O'Brien, S. J., ed. (1984). "Genetic Maps 1984. A Compilation of Linkage and Restriction Maps of Genetically Studied Organisms," Vol. 3. Cold Spring Harbor Lab., Cold Spring Harbor, New York.

Olive, L. S. (1974). Sordaria. In "Handbook of Genetics, Vol. 1, Bacteria, Bacteriophages, and Fungi" (R. C. King, ed.), pp. 553–562. Plenum, New York.

Perkins, D. D. (1984). *Neurospora crassa* genetic maps. In "Genetic Maps 1984" (S. J. O'Brien, ed.), pp. 277–285. Cold Spring Harbor Lab., Cold Spring Harbor, New York.

Perkins, D. D., Radford, A., Newmeyer, D., and Björkman, M. (1982). Chromosomal loci of *Neurospora crassa*. *Microbiol. Rev.* **46**, 426–570.

Raper, J. R. (1966). "Genetics of Sexuality in Higher Fungi." Ronald Press, New York.

Raper, J. R., and Hoffman, R. M. (1974). *Schizophyllum commune*. In "Handbook of Genetics, Vol. 1, Bacteria, Bacteriophages, and Fungi" (R. C. King, ed.), pp. 597–626. Plenum, New York.

Schwalb, M. N., and Miles, P. G., eds. (1978). "Genetics and Morphogenesis in the Basidiomycetes." Academic Press, New York.

Sherman, F. (1981). Genetic nomenclature. In "Molecular Biology of the Yeast, Saccharomyces cerevisiae'' (J. N. Strathern, E. W. Jones, and J. R. Broach, eds.), pp. 639–640. Cold Spring Harbor Lab., Cold Spring Harbor, New York.

Sherman, F., and Lawrence, C. W. (1974). Saccharomyces. *In* "Handbook of Genetics, Vol. 1, Bacteria, Bacteriophages, and Fungi" (R. C. King, ed.), pp. 359–393. Plenum, New York.

Spooner, R. A., and Turner, G. (1984). Mitochondrial genetic loci of *Aspergillus nidulans*. *In* "Genetic Maps 1984" (S. J. O'Brien, ed.), pp. 262–264. Cold Spring Harbor Lab., Cold Spring Harbor, New York.

Index

A

Abortive transformants, 283
Absidia, 534
Accumulation of protein, 454
Acetamidase, 346
Acetate requirement, 323
Achlya, 534
Achlya ambisexualis, 70
Achlya bisexualis, 43
Actin, 484
  actin gene, 461
acuD, 273, 347
Additive integration, 169
Adenine nucleotide translocator, 88
Adenylosuccinase, 8
Aflatoxin, 49
Agaricus, 534
Alanine, 370
alcA alcR cluster, 350
Alcohol dehydrogenase I, 346
Alcohol utilization gene cluster, 346
alcR, 313, 330, 331, 336
Aldehyde dehydrogenase, 346
Allele-specific suppression, 321, 330
Allelic complementation, 8, 11
Allomyces arbuscula, 42, 49, 534
Allosteric regulation, 22
Alpha-aminoadipate, 129
Alpha-aminoadipate aminotransferase, 201
Alpha-factor, 215
Alpha-ketoglutaric acid, 368
am^+ 248
Ambrosiozyma 54
amdS, 264, 266, 267, 539, 547, 555
Amino acid
  control, general, 179
  control, of transcription, 210
  sequence homology, 290
  uptake, 9
DL-α-Aminoadipic acid, 201
p-Aminobenzoic acid biosynthesis, 347
Aminoglycoside phosphotransferase, 199
Amphidiploids, 46

cAMP-binding proteins, 452
cAMP-dependent kinases, 452
Amplification, 198, 305
Amylases, 49, 496, 497
Amyloglucosidase, 497
Amylopectin, 496, 498
Amylose, 496, 497
Aneuploid, 46, 144, 298
Ans1, 271
Antibiotic yield, 294
Antibiotic-resistant transformants, 200
Antibiotics, 198, 293, 294, 305
Anticodon, 7, 8
aplA, 313, 316, 318, 319, 330, 331, 349
Apocytochrome *b,* 70, 80
L-Arabinose regulon, 315
D-Arabitol, 382
araC, 315
areA, 313, 316, 319, 320, 321, 322, 330, 334
areB, 322, 328
arg-12, 284, 285, 286, 290
Arginine metabolism, 282
Arginine permease gene, 217
aro, 11
ARS (autonomously replicating sequence), 136, 171, 172, 173, 177, 178, 198, 210, 211, 245, 271, 272, 435, 436
  Cochliobolus, 435
  mitochondrial, 411
Artificial chromosomes, 177
Ascobolus, 532
Ascobolus immersus, 540
Ascomycotina (Ascomycetes), 36, 53, 499, 531, 532, 534
Ascospores, 54
Ashbya gossypii, 378
Aspartic acid, 370, 371
Aspergillus, 49, 72, 82, 83, 87, 264, 265, 382, 497, 507, 510, 534
Aspergillus amstellodami, 84, 86, 272, 432
Aspergillus awamori, 498, 506, 507
Aspergillus flavus, 49
Aspergillus fonsecaens, 362
Aspergillus genes, 297

545

Aspergillus griseus, 368
Aspergillus itaconicus, 365
Aspergillus leporis, 49
Aspergillus microgynus, 49
Aspergillus nidulans, 43, 67, 68, 71, 72, 73, 74, 75, 281, 282, 283, 284, 286, 288, 290, 291, 295, 298, 299, 309, 310, 312, 313, 314, 322, 334, 350, 362, 518, 533, 537, 538
Aspergillus niger, 67, 72, 268, 350, 361, 362, 363, 365, 366, 368, 372, 497, 507, 522
Aspergillus orchraceus, 375
Aspergillus oryzae, 49, 51, 371, 376, 522
Aspergillus parasiticus, 49
Aspergillus sojae, 49
Aspergillus tamarii, 49, 51
Aspergillus terreus, 365, 375
Aspergillus ustus, 370
Aspergillus wentii, 51
ATP synthetase, 347
ATPase 6, 70, 72, 73, 75, 76, 78, 79, 86
ATPase 8, 71, 72, 75, 78
ATPase 9, 71, 75, 79, 82, 83, 88
Aureobasidium pullulans, 372, 373
Autogenous regulation, 315
Autolysis, 37
Autonomous plasmid, 167, 177, 183, 280
Autonomous replication, 198
Autonomously replicating plasmid (YRp31), 208
Auxotrophic mutants, 3, 4, 6, 127, 207, 478

B

Balanced interactions, 304
Balanced lethal genotypes, 298
Base pairing, 41
Basidiomycotina (Basidiomycetes), 36, 52, 53, 55, 532, 534
Benomyl, 10, 131
Beta-carotene, 380
Beta-galactosidase, 274
Beta-galactosidase gene, 179, 180, 219
Beta-glucosidase, 502, 506, 507
Biomass, 496
Biopolymers, 372
bla gene, of *E. coli,* 503
Blakeslea trispora, 380, 534
Blastocladiella emersonii, 49, 534

Brettanomyces abstines/B. custersii, 51
Brettanomyces anomalus/B. clausenii, 51
Brettanomyces galactosidase, 151
Brettanomyces tubulin, 10
Britten and Davidson, 325
Brute force cloning method, 350

C

cAMP-binding proteins, 452
cAMP-dependent kinases, 452
Candida, 47, 48, 534, 364, 381
Candida albicans, 379, 476
Candida blankii, 367
Candida brumptii, 367, 368
Candida cloacae, 375
Candida guilliermondii, 371
Candida humicola, 370
Candida hydrocarbofumarica, 367
Candida lipolytica (Yarrowia), 363, 364, 368, 376, 377, 379
Candida maltosa, 461
Candida pelliculosa, 369
Candida polymorpha, 382
Candida rugosa, 367, 368
Candida sp. no. 107, 375
Candida stellatoidea, 477, 483, 485, 486
Candida tropicalis, 370, 477, 478, 483, 485
Candida utilis, 367, 369, 371, 377, 378, 381
Candida zeylanoides, 364
CAN1, 128, 199, 218
Capsule, 471
Carbamoyl phosphate, 281
Carbamoyl-phosphate synthetase, 288
Carbohydrases, 494, 505, 508
Carbon catabolite repression, 22, 322, 323, 328, 329
Carbon sources, 323
Carotenoids, 379, 380
Cascade experiments, 349
Cascade, of regulatory genes, 22
CAT, 219
CAT cartridges, 219, 220
cat gene, 219
catabolic dehydroquinase, 248
CBHI, 505, 506
CBPI locus, 89
cdc6, 131
CDC8, 180

Index

cDNA, 349
cDNA clone, 330
Cell
  cycle, 476
  morphology, 452
  size, 481
  surface antigens, 54
  wall, 455
  wall, mannans, 53, 54
  wall, matrix polymers, 456
Cell-cycle mutants, 10
Cellobiohydrolase (exo-β-1,4-glucan cellobiohydrolase, E.C. 3.2.191), 504
Cellodextrins, 504
Cellular localization, 151
Cellular senescence, 83
Cellulase, 494, 507, 509
Cellulose, 503
CEN, 177, 178
Centromere, 168, 174, 177, 437
Centromere plasmids, 136
Cephalosporium acremonium, 68, 523
Cephalosporium eichhorniae, 372
Ceratocystis, 48
Cerulinin, 456
Cesium chloride gradient, 38
Chaetomium cellulyticum, 372, 375
Chain termination mutations, 11, 336
Charon 4a clone, 333
Chimeric plasmid DNA, 162
Chitin synthase, 455
Chloramphenicol acetylase, 265
Chloramphenicol acetyltransferase, 135, 219
Chromatin, 27, 337
Chromosome
  chromosomal aberration screening method, 347, 350
  chromosomal aberrations, 350
  chromosomal DNA, 207
  chromosomal rearrangements, 152, 322, 328
  duplication, 295
  ends (telomeres), 177
  exchanges, 185
  loss, 147
  loss mapping, 130
  mini-chromosome, 175
  replicators, 172
  sequences, 165
  stability, 177
Chytridiomycetes, 52, 53

cis-acting
  receptor site, 325
  regulatory elements, 309, 315, 321, 322, 327, 328, 329, 330, 333
cis effects of mutations, 12, 22, 23, 24, 179
Citrate synthase, 90
Citric acid, 361, 365
Citrulline, 281
Cladosporium, 375
Claviceps purpurea, 68, 83, 84, 369
Clone, 330, 333, 336
Cloning
  by complementation, 198
  gene, 141
  methodologies, 310, 345, 352
  S. cerevisiae mating type, 26
  strategy, 345
  vectors, 409
Cluster genes, 10, 12, 14, 16, 332, 346
Cochliobolus heterostrophus, 69, 83, 84, 432, 542
Codominant, 317
Codon usage, 27, 234, 501, 507
COI, COII and COIII, 72, 73, 74, 75, 78, 80, 87, 89, 90
Coinducer, 327
Cold-sensitive mutants, 10
Colinearity, 6
Colony hybridization, 165
Complementation, 9, 25, 34, 143, 152, 179, 198, 208, 281, 311, 315, 319, 333, 346, 461, 478, 481, 485
Consensus, 503
Conspecific, 51
Constitutive, 316, 317
Constitutive mutations, 13, 313
Construction
  of genetic map, 129
  of library, 142
Control of gene expression, 309
Copper-binding protein, 200
Coprinus, 534
Coprinus cinereus, 67, 69
Coprinus radiatus, 542
Coprinus stercorarius, 69
Copy number, 297, 305
  control, 137
  effects, 144
Correction-deficient mutants (cor), 183
Cosmid, 274

Cosmid clones, 350
Cot, 44
Cotransformation, 267
cpc, 288
creA, 322
Cross-pathway regulation, 288
Cryptococcosis, 471
Cryptococcus laurentii, 370, 472
Cryptococcus laurentii var. *flavescens*, 55
Cryptococcus laurentii var. *magnus*, 55
Cryptococcus laurentii var. *laurentii*, 55
Cryptococcus neoformans, 471
Cryptococcus terricolus, 375
CUP1 (copper resistance), 185, 200
Cutinase, 421
Cyclic 3′,5′-adenosine monophosphate (cAMP), 452
Cycloheximide, 10
CYC1, 219
CYC1-*lacZ* fusions, 179
CYH2, 199, 218
Cytochrome
  cytochrome *b*, 72, 73, 75, 76, 78, 79, 80, 81, 89, 90
  cytochrome *bc*, 88
  cytochrome *c*, 6, 7, 25, 88
  cytochrome *c* oxidase, 83, 88
  cytochrome *c* oxidase subunits, 70, 72, 73, 74, 75, 78, 80, 87, 89, 90
  cytochrome *c* peroxidase, 88
  cytochrome oxidase, 289
Cytoduction, 137
Cytoplasmic gene expression, 281
Cytoplasmic mutants, 407
Cytosolic marker enzyme, 289

D

D-Arabitol, 382
DAPI, 43
Debaryomyces, 381
Debaryomyces condertii, 368
Debaryomyces marama, 39
Debaryomyces vanriji, 375
Dekkera bruxellensis/*Brettanomyces lambicus*, 51
Deletion, 317, 329, 332, 333, 334
Deletion mapping, 311

Delta–delta recombination, 186
Dematiaceous fungi, 474
2-Deoxygalactose, 199
Deoxyribose nucleic acid, *see* DNA
Derepression, 288
Detection, of gene, 460
Deuteromycotina (Deuteromycetes), 36, 503, 532, 534
Developmentally regulated gene clusters, 347
Dicistronic mRNA, 333, 335
Dictyostelium, 518, 534
Dictyostelium discoideum, 542
Diethyl pyrocarbonate, 38
Differential hybridization, 346
Differential protein synthesis, 457
Dimorphism, 450
p-Diphenoloxidase, 347
Diploids, 198, 207, 311, 319
Diploid, partial, 295, 296
Direct selection, 127
Direction, of transcription, 332
Disease, 469
Disomic, 298, 299, 300, 302, 303, 304, 305
Disomic strain stability, 295, 298
DL-α-Aminoadipic acid, 201
DNA
  base composition, 35, 38, 39
  complementarity, 45
  conformation, 28
  dispersed repetetive, 462
  divergence, 46
  fungal, 37
  genomic library, 205
  hetero-duplex, 182, 183, 184
  homology, 46, 54
  isolation, 70
  ligase, 10
  linear, 42, 176
  mitochondrial, 39, 176
  preparation for cloning, 142
  purity, 38
  reassociation, 44
  relatedness, 41, 44, 149, 183, 184
  renatured, 41, 44
  repair, 41
  replication origins, 171
  sequence analysis, 210
  telomeric, 177
  transforming, 165

Index

uptake, 163
uptake by yeast cells, 162
DNA–DNA hybridization, 290
DNA–DNA reassociation, 46
DNases, 38
Dolipores, 52, 54
Domains, 314, 316
Dominance, 311, 317
Dominant markers, 134
Dose effects, 317, 319
Double-strand gap, 148, 183, 184
Down-promoter alleles, 329
Drug resistance, 9, 264
Duplication, 149
 of genes, 129
 of *penB2* locus, 299

E

Echinocandin B, 128
EF-1α, 458, 462
Effector, 317, 323
Elongation factor Tu, 88
Emericella nidulans, 533
Endo I, 505
Endo II, 506
Endoglucanases (endo-β-1,4-glucan glucanohydrolase E.C. 3.2.1.4), 503, 505, 508
Endomycopsis vernalis, 375
Endosymbiont hypothesis, 81
Enhancer-like element, 333
Enolase promoter, 506
ENO1, 200
Enrichment, 127, 128
Enzyme correlates of morphogenesis, 452
Epidermophyton, 41
Episomal, 168
Eremothecium ashbyii, 378
Ergosterol, 380
Erythorbic acid, 368
Escherichia coli, 198, 296, 315
Ethanol, 382, 383
 supplementation, 323
 utilization, 331
Ethidium bromide, 43
Eukaryotic cloning system, 404
Eumycota, 517, 532, 534

Evolutionary clock, 52
Evolutionary relationships, 55
Excision, 186
Exo-β-glucanases, 55
Exon, 182
Exon/intron junctions, 499
Expression, 1, 151, 180

F

Fatty acid, 375, 376, 456
Fatty acid-less death, 128
Fiji plasmid, 246
Filamentous fungi, 280, 281, 291, 345
Filobasidiella bacillispora, 46
Filobasidiella neoformans, 46, 471
Fine-structure mapping, 311, 316
Flanking markers, 311
Flavor nucleotides, 377
Flow cytometry, 352
Fluoroacetate, 152
5-Fluorocytosine, 482, 484
5-Fluoro-orotic acid, 199
Flux, 106
Fragment patterns, 49
Frameshift mutations, 25
Fumaric acid, 366, 367
Fungi
 associations of, 521
 carbohydrases of, 493
 Fungi Imperfecti, 533
 of industrial importance, 352
 plasmids of, 297
 taxonomy of, 531
Fusarium, 371, 503, 534
Fusarium graminearum, 371
Fusion protein, 151
F1 progeny, 46
F1-ATPase, 88, 90

G

G + C content, 38
Galactose permease-negative (*gal2*), 199
Galactose-negative mutants, 199
gal mutants
 gal1, 199
 gal4, 199
 GAL4c, 199
 gal80, 199

Gapped
  molecules, 183
  plasmid, 210
GDH, 454
Gelasinospora austosteira, 48
Gene
  amplification, 185, 296, 297
  bank, 273
  cluster, 10, 12, 18, 330, 332, 334, 349
  conversion, 169, 182, 183, 185, 213
  disruption, 145, 179, 213, 217
  dosage, 296, 297
  dosage effects, 145, 179
  dosage on penicillin production, 298
  duplication, 296, 305
  essential, 146
  eviction, 148, 179
  expression, 151, 180
  fine-structure mapping, 6
  function, 53
  fusion, 151, 179
  heterologous, 182, 281, 291
  homologous, 290
  isolation, 178
  isolation by self-cloning, 273
  order, 74, 75
  regulation, 13
  replacement, 210, 212, 213
  transplacement, 179, 210, 211
  tRNA, 71, 74
Genetic
  backgrounds, 301
  code, 71, 74
  divergence, 36
  hybridization, 35, 44
  map, 127, 147, 207
  recombination, 182, 184
  relatedness, 45
Genome size, 43, 66
Genomic "libraries," 25
Geotrichum, 41
Glucoamylase, 495, 497, 507, 508, 509
Gluconic acid, 365, 366
Glucose catabolic pathways, 453
Glucose-6-phosphate dehydrogenase, 289
Glusulase, 37, 163, 237
Glutamate dehydrogenase, 6, 7, 8, 25, 250, 321, 453
Glutamic acid, 370
Glutamine synthetase, 55, 321

L-Glutamine, 321
Glycerol, 380, 381
Glycogen, 497
Glycoprotein, 495, 504, 508
Glycosidases, 508
5'-GMP, 377
Gratuitous inducers, 315
Growth habits, 312
Growth testing, 312
Guanylic acid, 376
G418, 135, 199, 200, 265
G418-resistance gene, 200, 219

H

Hanseniospora valbyensis, 370
Hansenula, 53, 364, 375, 381
Hansenula alni, 46
Hansenula americana, 46
Hansenula anomala, 369
Hansenula beckii, 54
Hansenula bimundalis, 46, 53, 54
Hansenula canadensis, 48, 53, 54
Hansenula jadinii, 47
Hansenula minuta, 47, 48
Hansenula mrakii, 66, 69
Hansenula petersonii, 69
Hansenula polymorpha, 369
Hansenula wingei, 42, 46, 48, 53, 54
Haploid genome, 127
Helicobasidium mompa, 365
Hetero-duplex, 182, 183, 184
Heterokaryon, 138, 152, 311, 319, 362, 478, 484
Heterokaryon formation, 460
Heterologous expression, 345
Heterologous genes, 281, 291
Heterothallic strains, 45
Hierachy of structural genes, 321
High-copy-number vectors, 208
his genes
  his-3, 11
  HIS4, 179, 186
  his7, 207
Histoplasma, 534
Histoplasma capsulatum, 473
Holliday junction, 183, 184
Homologous genes, 208, 290
Homologous recombination, 147, 205, 211, 266
Homologous transformation, 286

Index

Promoter, 76, 78, 79, 235, 274, 321, 335, 498, 503, 507
   fusion, 152, 218
   qa, 235
   region, 505
Protein
   methylation, 458
   polymorphism, 55
   synthesis, 455
Proton magnetic resonance (PMR), 53
Protoplast, 36, 260, 472, 477, 478, *see also* Spheroplast
   fusion, 364, 365
   regeneration, 163
Pseudoconstitutive mutations, 316, 318, 331
Pseudogene, 81, 322
*Pst*I palindromes, 247
Puccinia, 534
Pullulan, 372, 373
Pulsed field gradient electrophoresis, 352
Purine degradation, 312, 315, 331
pVK55, 248
pyr gene
   pyr-3, 11
   pyr-44, 264, 267, 269, 271
   pyrF, 476, 485
   pyrG, 264, 267
Pyridoxine, 379
Pyroglutamyl residue, 505
Pyruvate dehydrogenase, 323
Pyruvic acid, 368
Pythium, 375
Pythium ultimum, 68

Q

qa gene cluster, 229, 233, 234
qa-2$^+$, 248
Quinate utilization, 347
Quinic acid, 230
qutBCE, 347

R

rad52, 131, 183, 184, 185
Ragged mutants, 84, 86
Rate-limiting steps, 105
Rational approaches for strain development, 305
recA, 270

Receptor sites, 309, 317, 321, 327, 328, 330
Recombinant DNA (rDNA), 36, 360, 384, 385, 460, 519, 541
Recombinant DNA methodology, 223
Regeneration, 480
Regulation of gene expression, 13, 309, 310, 321, 327
Regulatory circuit, 325
Regulatory gene, 309, 312, 313, 318, 322, 325
Regulatory gene *alcR*, 317
Repair, 149, 183, 184
Repeat sequences, 43, 67
Repetitive DNA, 325, 474
Replicating vectors, 270
Replication origins, 77, 78, 271
Reproduction and sexuality, 523
Restriction endonucleases, 49, 50, 210
Restriction pattern analysis, 51
Reverse transcription, 25
Rhizopus, 368, 371, 375, 497, 534
Rhizopus arrhizus, 366, 367
Rhizopus japonicus, 366
Rhizopus nigricans, 366
Rhizopus oryzae, 367
Rhodotorula, 371, 472
Rhodotorula glutinis, 371
Rhodotorula gracilis 375
Riboflavin, 378, 379
Ribosomal DNA, 474
Ribosomal proteins, 10, 52, 457
Ribosomal repeat, 267
Ribosomal RNA, 71, 72, 73, 75, 80, 87, 89
Ribosomal RNA relatedness, 51
Ribosome
   subunit protein, S6, 457
Ribosome-binding site, 321
RNA, 38, 85
   5 S, 52
   9 S, 78, 79
   16 S, 51
   18 S, 52
   blot analysis, 333
   polymerase, 77
   ribosomal, 76, 77, 78, 79, 86, 89, *see also* Ribosomal RNA
   L-rRNA, 71, 72, 73, 75, 80, 86, 87, 89
   S-rRNA, 71, 74, 75, 85
   transfer, 7, 71, 74, 77, 78, 86
Rusts, 534

S

S nuclease, 42
S-adenosylmethionine, 458
Saccharomyces, 52, 54, 508, 534
Saccharomyces, industrial strains, 222
Saccharomyces bailii, 381
Saccharomyces carlsbergensis, 72
Saccharomyces cerevisiae, 6, 10, 42, 43, 67, 69, 71, 72, 80, 84, 87, 126, 197, 198, 199, 205, 206, 208, 223, 281, 282, 297, 310, 369, 370, 374, 379, 381, 383, 384, 506, 518, 533, 537, 540, 555
Saccaromyces diastaticus, 503
Saccharomyces exiguus, 67, 69, 75
Saccharomyces microsporus, 379
Saccharomyces pombe, 72, 73, 75
Saccharomyces rouxii, 381
Saccharomyces telluris, 47
Saccharomycopsis, 534
Saccharomycopsis lipolytica (Yarrowia), 54, 69, 363, 364, 365, 369, 370
Saprolegnia, 41, 68, 534
Saprolegnia ferax, 49, 68
Scanning electron microscopy 54
Schizophyllum, 534
Schizophyllum commune, 69, 367, 368, 505, 509, 542
Schizosaccharomyces, 534
Schizosaccharomyces pombe, 7, 66, 69, 143, 542
Schwanniomyces, 47, 54
Schwanniomyces castellii, 48
Schwanniomyces occidentalis, 48
Scleroglucan, 373
Sclerotinia, 373
Sclerotium glucanicum, 373
Sclerotium rolfsii, 373
Scytalidium acidophilum, 372
Secondary metabolism, 358
Secondary metabolites, 293, 294, 296, 305, 522
Secreted enzymes, 324
Selectable markers (genes), 134, 198, 200, 211, 213, 264, 434
Selection of specific mutations, 128
Self-splicing, 80
Selfing, 310, 334
Semidominance, 317, 318
senDNAs, 87

Senescence, 84, 87, 88, 430
Septum ultrastructure, 52, 54
Sequence divergence, 51
Sequences, 503
Sexual cycle, 269
Sexual stages, 44
Shotgun 178, 200
Shuttle vectors, 10, 26, 142, 209, 253
  construction of, 410, *see also* Two-micron (2-μm) circle
Signal peptide, 505
Single-cell protein, 371
Single-copy nuclear DNA, 42
Single-stranded circular (ssc) DNA, 176
Site-specific recombination gene (FLP), 171
Slime mold, 350, 532
Smuts, 534
sod (stabilization of disomy) mutants, 298
$sod^{III}A$, 298, 299, 300, 302, 304
$sod^{III}A^+$, 304
$sod^{III}A1$, 304
Sordaria, 534
Southern blot hybridization, 166, 207, 211, 221, 284, 286, 329, 350
Species, 35, 36, 46
Sphacelotheca sorghii, 370
Spheroplast, 37, 133, 163, 280, 460, 472, 475
  fusion, 460, 472, 475, 477, 478
  transformation, 163
Splicing system, 182
spm (SPT), 186
Sporangiospores, 458
Sporobolomyces odorus, 371
Sporothrix schenckii, 48
Starch, 496
Steady-state model, 105, 106
Sterol, 456
Stopper mutant, 67, 84, 85, 86
Strain G96, 300
Strain improvement, 293, 294
Stromatinia 373
Structural gene, 312, 315, 325, 327, 333
suAmeth, 324, 329
Subunit 9, 347
Succinic acid, 368
SUC2, 217, 219, 220
Sugar metabolism, 154
Suillus grisellus, 66, 69
Sulfur repression, 324

Superoxide dismutase, 55
Superstrains, 130
  fungal species, 429
Suppression, 7
SUP11, 217
SUP4, 217
Surface structure, 54
Surrogate genetics, 153
Systems analysis, 106

T

tamA, 322
Tandem duplications, 211, 290
Tandem repeat, 268
Tartaric acid, 368
Taxonomy, 516
tcm1 (trichodermin resistance), 199
Telomeres, 137, 177
Telomeric DNA, 177, 178
Temperature-conditional mutants, 5, 10
Termination regions, 503, 506
Terminator, 503, 507
Tetrads, 183, 198, 207
  analysis, 129, 165, 216, 286
  unordered, 286
Tetrahymena, 80
Tetrahymena rDNA, 137
Tetrahymena thermophila, 52
Threonine, 370, 371
Thymidine kinase gene, 180
Tn601, 135, 199, 219
Tn9, 219
Tn903 (Tn55), 199
Torula, 371
Torulaspora, 52, 54
Torulopsis, 47, 66, 74, 364, 371
Torulopsis bovina, 47
Torulopsis candida, 375
Torulopsis glabrata, 42, 69, 75
Torulopsis magnoliae, 381
Torulopsis pintolopessi, 47, 48
Toxins, host-specific, 425
trans-Acting genes, 13, 19, 20, 313
Transcript processing, 78
Transcription, 77, 335
Transcriptional regulation, 208, 216
Transformation, 133, 201, 236, 260, 262, 280, 310, 317, 319, 330, 333, 334, 345, 351, 463
  abortive, 266, 271
  Aspergillus nidulans, 347
  efficiency, 162, 266, 271
  factors in *Neurospora*, 239
  high frequency, 135, 352
  integrative, 183
  intergeneric, 241
  lithium acetate, 238
  of nutritional mutants, 264
  Podospora, 409
  procedures 263
  qa and *Neurospora* 236
  types of *Neurospora* transformants, 240
  whole cell, 166
  yeast, 161, 162, 167, 178, 197, 198
Transition model, 105
Translational suppressors, 332
Translocation, 299, 311, 317, 320, 322, 335, 350
Transmission electron microscopy, 54
Transposable element, 81, 186, 328
Transposition, 83
Transposons, 199
Tremella mesenterica, 373
Trichoderma, 503, 504, 506, 510
Trichoderma cellobiohydrolase I, 508
Trichoderma reesei, 506
Trichophyton, 41
Trichosporon cutaneum, 370
Trigomopsis, 381
Trisomic, 298
Trisporic acids, 459
tRNA, 7, 71, 74, 77, 78, 84
tRNA genes, 461
tRNA processing locus, 79
trpC, 264, 266, 269, 274
TRP1, 207, 211, 312
TRP1 RI circle, 174
trp1-289, 211
Truffles, 534
Tryptic peptides of EF-1α, 459
Tryptophan, 369
Tryptophan synthetase, 6, 7
Tubulin, 10
Tubulin genes, 347
Two-micron (2-μm) circle, 26, 136, 169, 202, 245
Two-micron (2-μm) DNA, 207
Two-micron (2-μm) DNA plasmid, 198

Two-micron (2-μm) vector, 200
Ty, 169, 186, 222, 248

U

uaY, 315, 316, 317, 318, 319, 330, 331
uaZ, 349
Unequal sister chromatid exchanges, 185
Unique sequences, 42
Up-promoter effect, 330
Up-promoter mutation, 316, 329
Ura^+Lys^+ 205
ura3, 198, 199, 205, 264
URA3, 198, 199, 207, 213, 222
ura3 lys2, 207
URA5, 199
URF, 71, 72, 74, 86, 87
Uromyces appendiculatus, 418, 419
Ustilago, 534
Ustilago cynodontis, 69
Ustilago maydis, 370, 371, 542
Ustilago zeae, 365
U1 snRNAs, 499

V

var1, 67, 70, 72, 75, 76, 77, 79, 81, 89
Vectors, 135, 208, 463
  integrative, 135, 208, 211
  manipulation, 137
  recovery, 137
Vitamin A, 379
Vitamin B6, 379
Vitamin C, 379
Vitamin D, 380
Vitamins, 378, 379

W

ω-Amino acids, 325
Wangiella, 534
Wangiella dermatitidis, 474
Whittaker, 515, 517
Whole-genome comparisons, 44
Wide domain regulation, 309, 313, 315, 319, 324, 325

X

X-gal, 151, 220, 274
Xanthylic acid, 376

Y

ya, 347
Yarrowia lipolytica, 363, 534, *see also* *Saccharomycopsis; Candida*
YCp19, 168
YCp630, 208
Yeast, 506, 507, 508, 510, *see also* *Saccharomyces*
  cloning vectors, 209
  cosmid vectors, 502
  DNA, 222
  enolase gene ENO1, 503
  glycan, 374
  hyphal cell walls, 455
  leu2 structural gene, 503
  plasmids, 167, 168
  telomeres, 173
  transformation, 197, 198
  2-μm plasmid, 503
  YARp (yeast acentric replicating plasmids), 173
  YCp (yeast centromeric plasmid), 135, 174
  YEp (yeast episomal plasmid), 135, 169, 176, 177
  YIp (yeast integrating plasmids), 135, 171, 176, 177, 186, 211
  YRp (yeast replicating plasmids), 135, 177
YEp13, 168, 501
YEp24, pFL1, 205
YEp620, 208, 209, 222
YIp333, 208
YIp5, 168, 208, 436
YIp600, 209
YLP, 177
YLp21, 168
YRMp, 177
YRp610, 208, 209, 211
YRp7, 168, 171
YRp7, 476, 485

Z

Zoosporic fungi, 48
Zygomycotina (zygomycetes), 36, 39, 53, 532, 534
Zygosaccharomyces, 52
Zymolyase, 37, 163
Zymosan, 374

SETON HALL UNIVERSITY
McLAUGHLIN LIBRARY
SO. ORANGE, N. J.